Basic Radar Analysis

Second Edition

For a listing of recent titles in the *Artech House Radar Library*, turn to the back of this book.

Basic Radar Analysis

Second Edition

Mervin C. Budge, Jr.
Shawn R. German

ARTECH HOUSE

BOSTON | LONDON
artechhouse.com

Library of Congress Cataloging-in-Publication Data
A catalog record for this book is available from the U.S. Library of Congress.

British Library Cataloguing in Publication Data
A catalog record for this book is available from the British Library.

ISBN: 978-1-60831-555-4

Cover design by John Gomes

© 2020 Artech House
685 Canton Street
Norwood, MA 02062

All rights reserved. Printed and bound in the United States of America. No part of this book may be reproduced or utilized in any form or by any means, electronic or mechanical, including photocopying, recording, or by any information storage and retrieval system, without permission in writing from the publisher.

All terms mentioned in this book that are known to be trademarks or service marks have been appropriately capitalized. Artech House cannot attest to the accuracy of this information. Use of a term in this book should not be regarded as affecting the validity of any trademark or service mark.

10 9 8 7 6 5 4 3 2 1

Contents

Preface to the Second Edition xv
Preface to the First Edition xvii
Acknowledgments xix

Chapter 1   Radar Basics 1
   1.1 Introduction 1
   1.2 Radar Types 1
   1.3 Range Measurement 3
   1.4 Ambiguous Range 4
   1.5 processing Window and Instrumented Range 7
   1.6 Range-Rate Measurement: Doppler 7
   1.7 Decibels 11
   1.8 dB Arithmetic 11
   1.9 Complex Signal Notation 12
   1.10 Radar Block Diagram 14
   1.11 Exercises 15

Chapter 2 Radar Range Equation 19
   2.1 Introduction 19
   2.2 Basic Radar Range Equation 19
         2.2.1 Derivation of ES 21
         2.2.2 Derivation of E_N 28
   2.3 A Power Approach to SNR 31
   2.4 Radar Range Equation Example 32
   2.5 Detection Range 34
   2.6 Search Radar Range Equation 34
   2.7 Search Radar Range Equation Example 37
   2.8 Radar Range Equation Summary 39
   2.9 Exercises 40
   Appendix 2A:Derivation of Search Solid Angle Equation 43

Chaper 3   Radar Cross Section 45
   3.1 Introduction 45
   3.2 RCS of Simple Shapes 46

3.3 Swerling RCS Models 51
    3.3.1 Swerling Statistics 52
    3.3.2 Swerling Fluctuation Models... 53
    3.3.3 Math Behind the Fluctuation Model 55
3.4 Relation of Swerling Models to Actual Targets 57
3.5 Simulating Swerling Targets 58
3.6 Frequency Agility and SW2 or SW4 Targets... 63
3.7 Exercises 66

Chapter 4   Noise 69

4.1 Introduction 69
4.2 Noise in Resistive Networks 70
    4.2.1 Thevenin Equivalent Circuit of a Noisy Resistor 70
    4.2.2 Multiple Noisy Resistors 71
4.3 Equivalent/Effective Noise Temperature for Active Devices 72
4.4 Noise Figure 73
    4.4.1 Derivation of Noise Figure 73
    4.4.2 Attenuators... 74
4.5 Noise Figure of Cascaded Devices... 76
4.6 An Interesting Example.. 79
4.7 Output Noise Energy When the Source Temperature Is Not T_0 80
4.8 A Note About Cascaded Devices and the Radar Range Equation 80
4.9 Cascaded Attenuators 81
4.10 Exercises 81

Chapter 5   Radar Losses 85

5.1 Introduction 85
5.2 Transmit Losses 85
5.3 Antenna Losses 90
5.4 Propagation Losses... 95
5.5 Receive Antenna and RF Losses 98
5.6 Processor and Detection Losses 100
5.7 Exercises 107
5B.1 Function tropatten.m 118
5B.2 Function troprefract.m... 123
5B.3 Function troploss.m... 124
5B.4 Function rainAttn2way.m.. 124

Chapter 6   Detection Theory... 127

6.1 Introduction 127
6.2 Noise in Receivers.. 129
    6.2.1 IF Configuration 129
    6.2.2 Baseband Configuration 132
6.3 Signal in Receivers... 133
    6.3.1 Introduction and Background 133
    6.3.2 Signal Model for SW0/SW5 Targets 135

6.3.3 Signal Model for SW1/SW2 Targets 136
6.3.4 Signal Model for SW3/SW4 Targets 138
6.4 Signal-Plus-Noise in Receivers 139
6.4.1 General Formulation 139
6.4.2 Signal-Plus-Noise Model for SW1/SW2 Targets......................... 139
6.4.3 Signal-Plus-Noise Model for SW0/SW5 Targets......................... 140
6.4.4 Signal-Plus-Noise Model for SW3/SW4 Targets......................... 142
6.5 Detection Probability 144
6.5.1 Introduction 144
6.5.2 Amplitude Detector Types 145
6.5.3 Detection Logic 146
6.5.4 Calculation of Pd and Pfa... 147
6.5.5 Behavior versus Target Type 152
6.6 Determination of False Alarm Probability.. 154
6.6.1 *Pfa* Computation Example 156
6.6.2 Detection Contour Example 156
6.7 Summary... 160
6.8 Exercises... 160

Chapter 7   CFAR Processing... 163
7.1 Introduction 163
7.2 Cell-Averaging CFAR 167
7.2.1 Estimation of Interference Power... 169
7.2.2 CA-CFAR Analysis 170
7.2.3 CA-CFAR Example 174
7.2.4 CA-CFAR FIR Implementation.. 179
7.2.5 CFAR Processing at the Edges of Instrumentation 181
7.3 CA-CFAR with Greatest-of Selection 181
7.3.1 GO-CFAR Example 183
7.4 CA-CFAR with Smallest of Selection 184
7.4.1 SO-CFAR Example... 186
7.5 Ordered Statistic CFAR.. 188
7.5.1 OS-CFAR Example... 189
7.6 Minimum Selected CA-CFAR 192
7.6.1 MSCA-CFAR Algorithm 193
7.6.2 MSCA-CFAR Analysis... 194
7.6.3 MSCA-CFAR Example 197
7.7 Summary... 198
7.7.1 CFAR Problems and Remedies... 198
7.7.2 CFAR Scale Factors.. 199
7.8 Exercises... 200
Appendix 7A: Maximum Likelihood Estimation 204
Appendix 7B: Toeplitz Matrix and CFAR 205

Chapter 8   Matched Filter 209
    8.1 Introduction 209
    8.2 Problem Definition.. 209
    8.3 Problem Solution.. 210
    8.4 Matched Filter Examples 215
        8.4.1 General Formulation... 216
        8.4.2 Response for an Unmodulated Pulse 216
        8.4.3 Response for an LFM Pulse.. 219
    8.5 Summary 221
    8.6 Closing Comments 222
    8.7 Exercises 224

Chapter 9   Detection Probability Improvement Techniques......................... 227
    9.1 Introduction 227
    9.2 Coherent Integration.. 228
        9.2.1 SNR Analysis 228
        9.2.2 Detection Analysis.. 232
    9.3 Noncoherent Integration.. 235
        9.3.1 Coherent and Noncoherent Integration Comparison 240
        9.3.2 Detection Example with Coherent and Noncoherent Integration . 245
    9.4 Cumulative Detection Probability... 248
        9.4.1 Cumulative Detection Probability Example 249
    9.5 m-of-n Detection 250
        9.5.1 m-of-n Detection Example for SW0/SW5, SW2 and SW4
             Targets... 255
        9.5.2 m-of-n and Noncoherent Comparison for SW1 and SW2 Targets 256
    9.6 Exercises 258
    Appendix 9A: Noise Autocorrelation at the Output of a Matched Filter 261
    Appendix 9B: Probability of Detecting SW1 and SW3 Targets on m Closely
        Spaced Pulses 262
        9B.1 Marcum Q Function 267
    Appendix 9C: Cumulative Detection Probability 267

Chapter 10   Ambiguity Function 271
    10.1 Introduction 271
    10.2 Ambiguity Function Development.. 272
    10.3 Example 1: Unmodulated Pulse 274
    10.4 Example 2: LFM Pulse.. 278
    10.5 Numerical Techniques 281
    10.6 Ambiguity Function Generation Using the FFT 282
    10.7 Exercises 283

Chapter 11   Waveform Coding.. 285
    11.1 Introduction 285
    11.2 FM Waveforms 287

  11.2.1 LFM with Amplitude Weighting 287
  11.2.2 Nonlinear FM 289
 11.3 Phase-coded Pulses 294
  11.3.1 Frank Polyphase Coding 295
  11.3.2 Barker-coded Waveforms 297
  11.3.3 PRN-coded Pulses 300
 11.4 Step Frequency Waveforms 307
  11.4.1 Doppler Effects 311
 11.5 Costas Waveforms 313
  11.5.1 Costas Waveform Example 316
 11.6 Closing Comments 319
 11.7 Exercises 320
 Appendix 11A: LFM and the $sinc^2(x)$ Function 326

Chapter 12 Stretch Processing 329
 12.1 Introduction 329
 12.2 Stretch Processor Configuration 332
 12.3 Stretch Processor Operation 334
 12.4 Stretch Processor SNR 336
  12.4.1 Matched Filter 336
  12.4.2 Stretch Processor 337
 12.5 Practical Implementation Issues 338
  12.5.1 Stretch Processor Example 339
 12.6 Range-rate Effects 340
  12.6.1 Expanded Transmit and Receive Signal Models 340
  12.6.2 Stretch Processor Modification 342
  12.6.3 Slope Mismatch Effects 344
  12.6.4 Range-rate Effects on Range Bias 348
  12.6.5 Doppler Frequency Measurement Effects 351
  12.6.6 A Matched Filter Perspective 352
 12.7 Exercises 355

Chapter 13 Phased Array Antenna Basics 357
 13.1 Introduction 357
 13.2 Two-Element Array Antenna 358
  13.2.1 Transmit Perspective 358
  13.2.2 Receive Perspective 361
 13.3 N-Element Linear Array 363
 13.4 Directive Gain Pattern (Antenna Pattern) 367
 13.5 Beamwidth, Sidelobes, and Amplitude Weighting 370
 13.6 Steering 372
  13.6.1 Time-delay Steering 372
  13.6.2 Phase Steering 375
  13.6.3 Phase Shifters 376

13.7 Element Pattern 376
13.8 Array Factor Relation to the Discrete-time Fourier Transform 377
13.9 Planar Arrays... 380
    13.9.1 Weights for Beam Steering.. 383
    13.9.2 Array Shapes and Element Locations (Element Packing) 383
    13.9.3 Feeds... 384
    13.9.4 Amplitude Weighting 387
    13.9.5 Computing Antenna Patterns for Planar Arrays 387
    13.9.6 Directive Gain Pattern 389
    13.9.7 Grating Lobes 391
13.10 Polarization 397
13.11 Reflector Antennas.. 398
13.12 Other Antenna Parameters 401
13.13 Exercises 403
Appendix 13A: An Equation for Taylor Weights 406
Appendix 13B: computation of antenna patterns 407
    13B.1 Linear Arrays 407
    13B.2 Planar Arrays 408
        13B.2.1 Rectangular Packing.. 408
        13B.2.2 Triangular Packing 409

Chapter 14 AESA Basics and Related Topics 411

14.1 Introduction... 411
14.2 T/R Module 413
14.3 Time-delay Steering and Wideband Waveforms 416
    14.3.1 Subarray Size, Scan Angle, and Waveform Bandwidth 418
    14.3.2 Subarray Pattern Distortion Examples.................................... 422
    14.3.3 Array Beam Forming with TDUs... 423
14.4 Simultaneous Multiple Beams 427
    14.4.1 Overlapped Subarrays... 433
    14.4.2 Nonuniform Subarray Sizes 437
    14.4.3 Transmit Array Considerations.. 438
14.5 AESA Noise figure 440
    14.5.1 T/R Module Noise Figure... 440
    14.5.2 Subarray Gain and Noise Figure.. 441
    14.5.3 Array Gain and Noise Figure.. 443
    14.5.4 AESA Noise Figure Example 444
14.6 Exercises 446

Chapter 15 Signal Processors... 455

15.1 Introduction... 455
15.2 Signal Processor Structure 457

Chapter 16 Signal Processor Analysis 463

16.1 Introduction... 463
16.2 Signal Model Generation 463

16.2.1 Signal Model: Time Domain Analysis 466
16.2.2 Signal Model: Frequency Domain Analysis 468
16.2.3 Relation of PC and PS to the Radar Range Equation 472
16.3 Signal Processor Analyses 472
    16.3.1 Background 472
16.4 Exercises 474
Appendix 16A: Derivation Signal Processor Input Spectrum 475
Appendix 16B: Proof that $r(t)$ is Wide-Sense Cyclostationary 481

Chapter 17 Clutter Model 483

17.1 Introduction 483
17.2 Ground Clutter model 483
    17.2.1 Ground Clutter RCS Model 483
    17.2.2 Ground Clutter Spectrum Model 489
17.3 Rain Clutter Model 491
    17.3.1 Rain Clutter RCS Model 491
    17.3.2 Rain Clutter Spectral Model 494
17.4 Exercises 495

Chapter 18 Moving Target Indicator (MTI) 497

18.1 Introduction 497
    18.1.1 MTI Response Normalization 498
18.2 MTI Clutter Performance 500
    18.2.1 Clutter Attenuation 500
    18.2.2 SCR Improvement 503
18.3 Ground Clutter Example 504
18.4 Rain Clutter Example 507
18.5 Phase Noise 508
    18.5.1 Higher Order MTI Processors 510
    18.5.2 Staggered PRIs 513
    18.5.3 MTI Transients 518
18.6 Exercises 518

Chapter 19 Digital Pulsed Doppler Processors 521

19.1 Introduction 521
19.2 Pulsed Doppler Clutter 523
19.3 Signal Processor Configuration 528
19.4 Digital Signal Processor Analysis Techniques 530
    19.4.1 Phase Noise and Range Correlation Effects 533
    19.4.2 ADC Considerations 538
19.5 Summary and Rules of Thumb 539
19.6 HPRF Pulsed Doppler Processor Example 541
19.7 MPRF Pulsed Doppler Processor Example 546
19.8 LPRF Pulsed Doppler Signal Processor Example 551
19.9 Exercises 556

Chapter 20   Analog Pulsed Doppler Processors 565
   20.1 Introduction 565
   20.2 Analog Pulsed Doppler Signal Processor Example 571
   20.3 Exercises 573

Chapter 21   Chaff Analysis 575
   21.1 Introduction 575
   21.2 Chaff Analysis Example 578
   21.3 Exercises 581

Chapter 22   Radar Receiver Basics 583
   22.1 Introduction 583
   22.2 Single-Conversion Superheterodyne Receiver 584
   22.3 Dual-Conversion Superheterodyne Receiver 592
   22.4 Receiver Noise 594
   22.5 The 1-dB Gain Compression Point 597
   22.6 Dynamic Range 599
         22.6.1 Sensitivity 600
         22.6.2 Minimum Detectable and Minimum Discernable Signal 602
         22.6.3 Intermodulation Distortion 602
         22.6.4 Required Dynamic Range 605
   22.7 Cascade Analysis 606
         22.7.1 Cascade Analysis Conventions 607
         22.7.2 Procedure 608
         22.7.3 Power Gain 609
         22.7.4 Noise Figure and Noise Temperature 610
         22.7.5 1-dB Compression Point 611
         22.7.6 Second-Order Intercept 612
         22.7.7 Third-Order Intercept 613
   22.8 Digital Receiver 617
         22.8.1 Bandpass Sampling 618
         22.8.2 Digital Down conversion 621
         22.8.3 Practical DDC 626
         22.8.4 CIC Filter Structure 628
         22.8.5 Analog-to-Digital Converter 634
   22.9 Receiver Configurations 645
   22.10 Exercises 652
   Appendix 22: A Digital Down conversion Using Band-pass Sampling 660

Chapter 23   Introduction to Synthetic Aperture Radar Signal Processing 667
   23.1 Introduction 667
   23.2 Background 668
         23.2.1 Linear Array Theory 668
         23.2.2 Transition to SAR Theory 672
   23.3 Development of SAR-Specific Equations 673
   23.4 Types of SAR 675

23.4.1 Theoretical Limits for Strip Map SAR 676
23.4.2 Effects of Imaged Area Width on Strip Map SAR Resolution ... 677
23.5 SAR Signal Characterization 678
23.5.1 Derivation of the SAR Signal 678
23.5.2 Examination of the Phase of the SAR Signal 680
23.5.3 Extracting the Cross-Range Information 681
23.6 Practical Implementation 683
23.6.1 A Discrete-Time Model 683
23.6.2 Other Considerations 685
23.7 An Algorithm for Creating a Cross-Range Image 686
23.8 Example 1 - Generation of a Cross-range SAR Image 687
23.9 Down-Range and Cross-Range Imaging 691
23.9.1 Signal Definition 691
23.9.2 Preliminary Processing Considerations 696
23.9.3 Quadratic Phase Removal and Image Formation 701
23.10 Algorithm for Creating a Cross- and Down-Range Image 703
23.11 Example 2: Cross- and Down-range SAR Image 704
23.12 An Image-Sharpening Refinement 706
23.13 Closing Remarks 710
23.14 Exercises 711

Chapter 24 Introduction to Space-Time Adaptive Processing 717
24.1 Introduction 717
24.2 Spatial Processing 718
24.2.1 Signal Plus Noise 719
24.2.2 Signal Plus Noise and Interference 722
24.2.3 Example 1: Spatial Processing 724
24.3 Temporal Processing 726
24.3.1 Signal 726
24.3.2 Noise 727
24.3.3 Interference 728
24.3.4 Doppler Processor 728
24.3.5 Example 2: Temporal Processing 730
24.4 Adaptivity Issues 732
24.5 Space-Time Processing 732
24.5.1 Example 3: Space-Time Processing 734
24.5.2 Example 4: Airborne Radar Clutter Example 739
24.6 Adaptivity Again 743
24.7 Practical Considerations 744
24.8 Exercises 745

Chapter 25 Sidelobe Cancellation 747
25.1 Introduction 747
25.2 Interference Canceller 748
25.3 Interference Cancellation Algorithm 750

   25.3.1 Single Interference Signal.. 750
   25.3.2 Simple Canceler Example... 751
   25.3.3 Multiple Interference Sources.. 754
  25.4 SLC Implementation Considerations 755
   25.4.1 Form of $v_m(t)$ and $v_a(t)$.. 755
   25.4.2 Properties of $v_s(t)$, $v_I(t)$, $n_m(t)$, and $n_{an}(t)$................................. 757
   25.4.3 Scaling of Powers 758
   25.4.4 Two Auxiliary Channel Open-Loop SLC Example 759
   25.4.5 Performance Measures... 761
   25.4.6 Practical Implementation Considerations 762
   25.4.7 Two Auxiliary Channel Open-Loop SLC Example with SMI ... 765
  25.5 Howells-Applebaum Sidelobe Canceller 766
   25.5.1 Howells-Applebaum Implementation.. 766
   25.5.2 IF Implementation 768
   25.5.3 Single-Loop Howells-Applebaum SLC Example....................... 769
   25.5.4 Two-Loop Howells-Applebaum SLC Example 770
  25.6 Sidelobe Blanker 772
  25.7 Exercises 773
  Appendix 25A: Derivation of ϕ (25.40).. 775

Chapter 26 Advances in Radar 779
  26.1 Introduction.. 779
  26.2 MIMO Radar.. 779
  26.3 Cognitive Radar 780
  26.4 Other Advancements in Radar Theory.. 781
  26.5 Hardware Advancements 782
  26.6 Conclusion.. 784

Appendix A Data Windowing Functions... 787

About the Authors 797

Index 799

Preface to the Second Edition

At the end of the preface of our first book, we asked for feedback on our book. We are very thankful that, indeed, many readers provided such feedback. Much of it was to point out errors, and suggestions on how we might better organize the book. Others were requests to add material that was not present in the first edition. We have tried to address the comments in this edition. For one, we corrected errors readers, and we, found. We most likely did not find all errors, so we ask readers to continue pointing out errors they find. To that end, our email addresses are merv@thebudges.com and shawnrg@att.net. We will publish errata on the Artech website as we make the recommended corrections.

The major reorganization was to divide Chapter 13, Signal Processing, which was very long, into seven, shorter chapters (new Chapters 15 through 21). We also added new information and examples to the new chapters, and retitled some of the sections to be more descriptive of their contents.

Several of our readers and coworkers asked that we add chapters on constant false alarm rate (CFAR) processing and active electronically scanned arrays (AESAs). We did that by adding two new chapters. One is Chapter 7 on CFAR processing. It includes discussions about the basics of generic CFAR processing in general, but primarily focuses on cell averaging (CA) CFAR, including basic modifications such as smallest of, greatest of, and ordered statistic. Additionally, Chapter 7 touches on topics related to CFAR processing such as clutter models, detector law, implementation, and processing at the edges of instrumentation. The second new chapter, Chapter 14, is on AESAs. It contains a basic description of AESA structure and transmit/receive (T/R) modules. It also contains fairly detailed discussions of time delay steering, simultaneous multiple beams, overlapped subarrays, non-uniform subarrays, and AESA noise figure.

In Chapter 13, Phased Array Antenna Basics, we added a discussion of the relation between the array aperture function and the discrete Fourier transform, and discussed how the fast Fourier transformer (FFT) could be used to generate aperture functions of linear and planar phased arrays. We moved the aperture function generation method discussed in the first edition to an appendix. We still feel it is useful, but we received several comments about it being difficult to implement.

In Chapter 12, Stretch Processing, we addressed the impact of Doppler frequency from the stretch processing perspective rather than the matched filter perspective as in the first edition. That approach seemed more logical since the chapter contained a discussion of stretch processing, not matched filters.

Some of the other new material includes

- Costas frequency coding in the Waveform Coding chapter (new Chapter 11);

- An expanded discussion of digital down conversion, CIC filters, and decimation in the Radar Receiver Basics chapter (new Chapter 22);
- An example of a two-loop Howells-Applebaum SLC in the Sidelobe Cancellation and Sidelobe Blanking chapter (new Chapter 25).

As we stated at the end of the first edition preface, we hope you find this book useful, and we welcome your feedback.

Preface to the First Edition

This book is based on lecture notes for a three-course sequence in radar taught by Dr. Budge at the University of Alabama in Huntsville. To create this book, we filled in some details that are normally covered in lectures and added information in the areas of losses, waveforms, and signal processing. We also added a chapter on receiver basics.

The first of the three courses, which focuses on the radar range equation and its various progressions, provides an introduction to basic radar analysis covered in Chapters 1 through 9 of this book. Chapter 1 begins with definitions of radar-related terms and terminology, which is followed in Chapter 2 by a detailed derivation of the radar range equation and discussions of its various parameters. Following that, in Chapter 3, we discuss radar cross section (RCS) with emphasis on the Swerling RCS models. We next discuss noise, noise temperature, and noise figure in Chapter 4 and losses in Chapter 5 to round out our treatment of the radar range equation. Following this, in Chapter 6, we discuss one of the main uses of radar, which is the detection of target signals embedded in noise. We address detection theory for several radar receiver configurations and Swerling RCS models. This leads naturally to matched filter theory, discussed in Chapter 7, and its extension to the ambiguity function of Chapter 9. We complete discussions of the radar range equation and detection theory with discussions of methods of increasing detection probability and decreasing false alarm probability in Chapter 8. This includes coherent integration, noncoherent integration, m-of-n detectors, and cumulative detection probability.

The second course covers the material in Chapters 12 and 13. Chapters 12 and 13 include analysis of phased array antennas and signal processing. The phased array discussions include linear and planar phased arrays and provide explanations of efficient methods for generating antenna radiation patterns and computing directivity. The phased array discussions also include discussions of time delay steering, phase steering, phase shifters, element patterns, grating lobes, feeds, and polarization. The signal processor studies of Chapter 13 include ground and rain clutter modeling, and the analysis of the clutter rejection and signal-to-noise improvement of moving target indicator (MTI) and pulsed-Doppler signal processors. Also included are detailed discussions of phase noise and range correlation, plus chaff modeling and analysis. Chapter 14 contains a discussion of basic receiver analysis, which we plan to include in future courses.

Finally, the third course covers advanced topics that include stretch processing, covered in Chapter 11; phase-coded waveforms, discussed in Chapter 10; synthetic aperture radar (SAR) processing, discussed in Chapter 15; space-time adaptive processing (STAP), covered in Chapter 16; and sidelobe cancellation (SLC), covered in Chapter 17. In all of these areas we focus on implementation. For example, we discuss how to implement a SAR processor and process actual SAR data from the RADARSAT I SAR platform. We show how to implement a stretch processor, STAP processors, and both open- and closed-loop SLCs.

The main audience for the courses, and the intended audience for this book, consists of practicing radar engineers who are pursuing an advanced degree with a radar specialization, or have a need for a detailed understanding of radar analysis. As such, the courses and this book focus on providing the theory and tools radar engineers need to perform their day-to-day work in the fields of radar analysis, radar modeling and simulation, and radar design. The homework exercises and the examples in this book are derived from real-world analysis problems. In fact, one of the common phrases of radar engineers working at Dynetics, Inc., the authors' company, is that the project they are working on is "Homework 16."

This book focuses on analysis of radars and developing a firm understanding of how radars and their various components work. As such, it does not avoid the sometimes complicated mathematics needed to fully understand some of the concepts associated with the analysis and design of radars. However, we try to summarize the results of mathematical derivations into easily usable equations and, in some instances, convenient rules of thumb. We hope you find this book useful, and we welcome your feedback.

Acknowledgments

We would like to express thanks to several people who contributed to this book. First, our Artech reviewer for reviewing the chapters and providing valuable suggestions on how to improve the material in the book. We would also like to thank others who reviewed portions of the book and/or offered technical ideas, including Dr. B. K. Bhagavan, Dr. Steve Gilbert, Dr. Jeff Skinner, Dr. Greg Coxson, Dr. Colin McClelland, Alan Volz, David Hardaker, John A. Cribbs, Joshua Robbins, Bill Myles, Vernon Handley, Alexandria Carr, Zach Hubbard, Cooper D. Barry, Sara Meadows, and Belinda Biron. We also thank Stacy Thompson, Shawn's sister, for reviewing several chapters for grammatical errors.

On the publishing side, we thank the Dynetics Corporate Development department, especially Joyce Walters, Virginia Elmer, Janet Pickens, Melanie Kelley, and Julie Wypyszynski for preparing the manuscript. We also thank Todd German for the photo used in the SAR homework of Chapter 23.

We would also like to thank Professor Emeritus Ian G. Cumming of the University of British Columbia, who authored *Digital Processing of Synthetic Aperture Radar Data: Algorithms and Implementation,* for allowing us use of his SAR data and programs in authoring this book. For providing the RADARSAT data of a scene of Vancouver used in the SAR homework, we would like to thank Mr. Gordon Staples of MDA and the Canadian Space Agency/Agence Spatiela Canadienne.

Finally, we thank Carmie, Merv's wife, and Karen German, Shawn's mother, for their unwavering encouragement and support during the preparation of this book.

Chapter 1

Radar Basics

1.1 INTRODUCTION

According to Skolnik and other sources, the first attempt to detect targets using electromagnetic radiation took place in 1904 (patent date for the telemobiloscope), when Düsseldorf engineer Christian Hülsmeyer bounced waves off a ship [1–5]. During the 1920s, several researchers, including R. C. Newhouse, G. Breit, M. A. Tuve, G. Marconi, L. S. Alder, and probably many others in the United States and other countries, were obtaining patents on, and conducting experiments with, radar. Although these appear to be the first instances of radar usage, the term "radar" was not applied then. The name for radar was coined in 1940 by two U.S. Navy officers (Lieutenant Commanders Samuel M. Tucker and F. R. Furth) as a contraction of RAdio Detection And Ranging [6–8]. As with many other technological advancements, significant early achievements in radar occurred during World War II. Since then, radar technology has grown rapidly and continues to advance at a quick pace. We now see wide application of radars in both commercial (airport radars, police radars, weather radars) and military (search radars, track radars) arenas.

1.2 RADAR TYPES

Radars can use two types of signals:

- Pulsed, where the radar transmits a sequence of pulses of radio frequency (RF) energy;
- Continuous wave (CW), where the radar transmits a continuous signal.

When Hülsmeyer bounced electromagnetic waves off a ship, he used a CW radar; Breit and Tuve used a pulsed radar [9].

CW radars typically use separate transmit and receive antennas because it is not usually possible to receive with full sensitivity through an antenna while it is transmitting a high-power signal. Pulsed radars avoid this problem by using what we might think of as time multiplexing. Specifically, the antenna connects to the transmitter while the pulse transmits and connects to the receiver after the transmit phase. A transmit/receive (T/R) switch in the radar performs this switching function. Such pulsed-signal radars constitute the most common type because they require only one antenna.

The two basic types of radars are *monostatic* and *bistatic* radars:

- In a monostatic radar, the transmitter and receiver (as well as their associated antennas) are collocated. This is the most common type of radar because it is the most compact. A pulsed monostatic radar usually employs the same antenna for transmit and receive. A CW monostatic radar usually employs separate transmit and receive antennas, with a shield between them.
- In a bistatic radar, the transmitter and receiver are separated, often by very large distances (>1 km). Such a radar might be used in a missile seeker, with the transmitter located on the ground or in an aircraft and the receiver located in the missile.

As indicated previously, the word radar is a contraction of RAdio Detection And Ranging. This contraction implies that radars both detect the presence of a target and determine its location. The contraction also implies that the quantity measured is range. While these suppositions are correct, modern radars can also measure range rate and angle. Measuring such parameters permits reasonably accurate calculations of the x-y-z location and velocity of a target, and in some instances, reasonable estimates of the higher derivatives of x, y, and z.

Radars operate in the RF band of the electromagnetic spectrum between about 5 MHz (high frequency, HF) and 300 GHz (millimeter wave, mm). Table 1.1 lists frequency bands [Institute of Electrical and Electronics Engineers (IEEE) waveband specifications] and associated frequencies. Another set of waveband specifications, the European and U.S. ECM (electronic countermeasure), experienced some popularity in recent times, but has waned in the past 10 or so years.

Typically, but not always:

- Search radars operate at very high frequency (VHF) to C-band;
- Track radars operate in S-, C-, X-, and Ku-bands, and sometimes in Ka-band;
- Active missile seekers operate in X-, Ku-, K-, and Ka-bands; and
- Instrumentation radars and short-range radars sometimes operate in the Ka-band and above.

Some notes on operating frequency considerations [10]:

- Low-frequency radars require large antennas or have broader beams (broader distribution of energy in angle space—think of the beam of a flashlight). They are not usually associated with accurate angle measurement.
- Low-frequency radars have limitations on range measurement accuracy because fine range measurement implies large instantaneous bandwidth of the transmit signal. This causes problems with the transmitter and receiver design because the bandwidth could represent a significant percentage of transmit frequency.

Table 1.1
Radar Frequency Bands

Band	Frequency Range	Origin of Name
HF	3–30 MHz	High frequency
VHF	30–300 MHz	Very high frequency
UHF	300–1,000 MHz	Ultrahigh frequency
L	1–2 GHz	Long wave
S	2–4 GHz	Short wave
C	4–8 GHz	Compromise between S- and X-bands
X	8–12 GHz	Described fire control radars in World War II. X stands for "cross," as in "crosshairs"
Ku	12–18 GHz	Kurz—under
K	18–27 GHz	Kurz—German for "short wave"
Ka	27–40 GHz	Kurz—above
V	40–75 GHz	Very short
W	75–110 GHz	W follows V in the alphabet
mm	110–300 GHz	millimeter

Source: [11–13].

- Range-rate measurement is not accurate in low-frequency radars because Doppler frequency is directly related to transmit frequency.
- High power is easier to generate at low frequencies because the devices can be larger, thus accommodating higher voltages and currents.
- Search calls for high power but not necessarily fine range or angle measurement. Thus, search radars tend to use lower frequencies.
- Track calls for fine range and angle measurement but not necessarily high power. Thus, track radars tend to use higher frequencies.

The above notes often lead to the assignment of search and track functions to different radars. However, modern radars tend to be multifunction, incorporating both purposes in one. This usage often leads to tradeoffs in operating frequency and in search-and-track functions.

1.3 RANGE MEASUREMENT

The common way to measure range with a radar is to measure the time delay from transmission to reception of a pulse. Figure 1.1 illustrates this notion. Since RF energy travels at the speed of light, $c \approx 3 \times 10^8$ m/s,[1] the time required for the transmit pulse to travel to a target at a range of R is

$$\tau_{out} = R/c \tag{1.1}$$

---

[1] The exact value for the speed of light in a vacuum is 299,792,458 m/s; $c \approx 3 \times 10^8$ m/s provides a useful rule of thumb.

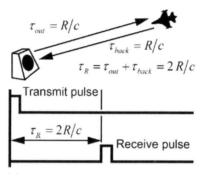

Figure 1.1 Illustration of range delay.

The time required for the pulse to return from the target back to the radar is

$$\tau_{back} = R/c \tag{1.2}$$

Thus, the total round-trip delay between transmission and reception of the pulse is

$$\tau_{total} = \tau_R = \tau_{out} + \tau_{back} = 2R/c \tag{1.3}$$

Since we can measure τ_R in a radar, we can compute range by solving (1.3) for R; that is,

$$R = \frac{c\tau_R}{2} \tag{1.4}$$

As a note, the term *slant range* suggests measurement along a line (often slanted) from the radar to the target; as such, the term applies here to R. The term *ground range*, the distance from the radar to the vertical projection of the target onto the ground, will be discussed in a later chapter.

A rule of thumb for range measurement can be derived as follows. Suppose $\tau_R = \tau$ μs. In other words, suppose we express the time delay, often called *range delay*, in units of microseconds. We can then write

$$R = \frac{c}{2}\tau_R = \frac{3\times 10^8 \text{ m/s}}{2}\left(\tau \text{ μs} \times 10^{-6} \text{ s/μs}\right) = 150\tau \text{ m} \tag{1.5}$$

Thus, we can compute the range by multiplying the range delay, in microseconds, by 150 m; the range computation scaling factor is 150 m/μs.

1.4 AMBIGUOUS RANGE

Since pulsed radars transmit a sequence of pulses, the determination of range to the target poses a problem. The issue is where we choose $t = 0$ to compute range delay. The common method is to choose $t = 0$ at the beginning of a transmit pulse; thus, $t = 0$ resets on each transmit pulse.

To define the problem, consider Figure 1.2, which shows transmit pulses spaced 400 μs apart. Given a target range of 90 km, the range delay to the target is

$$\tau_R = \frac{2R}{c} = \frac{2 \times 90 \times 10^3}{3 \times 10^8} = 60 \times 10^{-5} \text{ s} = 600 \text{ μs} \tag{1.6}$$

This means the return from pulse 1 is not received until after pulse 2 is transmitted; the return from pulse 2 is not received until after pulse 3 is transmitted; and so on. Since all transmit pulses are the same and all received pulses are the same, we have no way of associating received pulse 1 with transmit pulse 1. In fact, since the radar resets $t = 0$ on each transmit pulse, it will associate received pulse k with transmit pulse $k + 1$. Further, it would measure the range delay as 200 μs and conclude, in error, that the target range is

$$R_A = 150\tau = 150 \times 200 = 30{,}000 \text{ m or } 30 \text{ km} \tag{1.7}$$

Because of this, we say that we have an *ambiguity*, or uncertainty, in measuring range.

If the spacing between pulses is τ_{PRI}, we say the radar has an *unambiguous range* of

$$R_{amb} = \frac{c\tau_{PRI}}{2} \tag{1.8}$$

If the target range is less than R_{amb}, the radar can measure its range unambiguously. For a target range greater than R_{amb}, the radar cannot measure its range unambiguously.

In the notation above, the term PRI stands for *pulse repetition interval*, or the spacing between transmit pulses. A related term, *pulse repetition frequency*, or PRF, is the reciprocal of the PRI.

To avoid range ambiguities, radar designers typically choose the PRI of a pulse train (a group of two or more pulses) to exceed the range delay of the longest range targets of interest. They also select transmit power to reduce the possibility for long-range target detection by the radar.

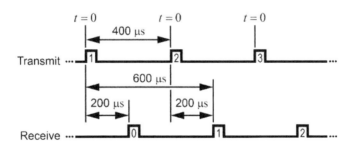

Figure 1.2 Illustration of ambiguous range.

Ambiguous range sometimes presents a problem in search, but generally not in track. In track, the radar tracking filters or algorithms provide an estimate of target range, which enables the radar to "look" in the proper place, even given ambiguous returns.

Using waveforms with multiple PRIs provides another method for circumventing ambiguous range problems. Figure 1.3 depicts an example of a multiple PRI transmit waveform and appropriate received signal. In this example, the spacings between pulses are 400 μs, 300 μs, and 350 μs. As in the previous example, a target range delay of 600 μs is posited. It will be noted that the time delay between the number 1 received pulse and the number 2 transmit pulse is 200 μs, and the time delay between the number 2 received pulse and the number 3 transmit pulse is 300 μs. The fact that the time delay between the most recently transmitted and received pulses is changing can be used to indicate ambiguous range operation. The radar can use this property to ignore the ambiguous returns.

Alternatively, the radar could use the measured range delays in a *range resolve* algorithm to compute the true target range. Such an approach is used in pulsed Doppler radars because the PRIs used in these radars almost always result in ambiguous range operation.

Changing the operating frequency, f_c, on each pulse provides yet another method of circumventing the ambiguous range problem. In this case, if the return from pulse k arrives after the transmission of pulse $k + 1$, the receiver will be tuned to the frequency of pulse $k + 1$ and will not "see" the return from pulse k.

Phased array radars, which steer the antenna beam electronically rather than mechanically, often transmit a single pulse and then re-steer the beam to a different angular position. In this situation, the concept of a PRI, and thus unambiguous range, is not strictly defined since there is only one pulse. The term is used in such cases, nevertheless. The unambiguous range is taken to be the range delay during which the beam stays in one position before moving to another position. The time the beam stays in one position is termed a *beam dwell*.

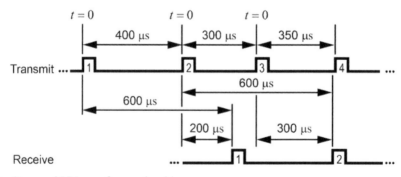

Figure 1.3 Staggered PRI waveform and ambiguous range.

1.5 PROCESSING WINDOW AND INSTRUMENTED RANGE

The preceding discussion of ambiguous range might lead one to conclude that a radar can detect (and track) targets at all ranges between 0 and R_{amb}. However, in practice, this is not the case. The pulse received from a target at a range of zero would arrive at the radar simultaneously with the transmission of the sounding pulse. Since the receiver is off during this time, it cannot process the pulse. The minimum usable range is therefore equal to

$$R_{min} = \frac{c\tau_p}{2} \tag{1.9}$$

where τ_p is the radar pulsewidth.[2]

Similarly, a received pulse arriving at the receiver too near the next transmit pulse prevents the entire pulse from entering the receiver before the receiver turns off for the next transmit pulse. Thus, the receiver cannot completely process the received pulse. This leads to the conclusion that the maximum usable range is

$$R_{max} = \frac{c(\tau_{PRI} - \tau_p)}{2} \tag{1.10}$$

That is, the maximum usable range extends to one pulsewidth before the next transmit pulse. We define the time interval between R_{min} and R_{max} as the *processing window*. Occasionally, these bounds can be exceeded somewhat; however, this does not often occur because it can lead to processing difficulties.

Although R_{max} defines the maximum usable range, most radars operate over a shorter range interval, termed the *instrumented range*. The instrumented range is set by system requirements and allows for factors such as display limits, circuit transients, radar calibration, radar mode changes, and the like.

1.6 RANGE-RATE MEASUREMENT: DOPPLER

In addition to measuring range, radars can also measure the rate of change of range, or *range rate*. The radar accomplishes this by measuring the *Doppler frequency*; that is, the frequency difference between the transmitted and received signals. To examine this further, consider the geometry of Figure 1.4. The aircraft in this figure moves in a straight line at a velocity of v. As a result, the range to the target changes continually. Indeed, over a differential time of dt, the range changes by an amount dR, from R to $R + dR$. The resulting range rate is

$$\dot{R} = \frac{dR}{dt} \tag{1.11}$$

---

[2] We refer to the range between 0 and R_{min} as the *blind range* of the radar.

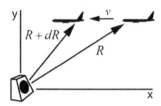

Figure 1.4 Geometry for Doppler calculation.

We note that the range rate of Figure 1.4 is negative because the range decreases from time t to time $t + dt$. Also, in general, $\dot{R} \neq v$. Equality holds only if the target flies along a radial path relative to the radar.

We will briefly digress to consider the relationship of range rate to target position and velocity in a Cartesian coordinate system centered at the radar. Suppose the target position and velocity state vector is given by

$$X = \begin{bmatrix} x & y & z & \dot{x} & \dot{y} & \dot{z} \end{bmatrix}^T \tag{1.12}$$

where the superscript T denotes the transpose. We can write the target range as

$$R = \sqrt{x^2 + y^2 + z^2} \tag{1.13}$$

and the range rate as

$$\dot{R} = \frac{dR}{dt} = \frac{x\dot{x} + y\dot{y} + z\dot{z}}{R} \tag{1.14}$$

Now, let's return to the problem of measuring range rate with a radar. To start, think about the nature of the transmit pulse. In its simplest form, the transmit pulse constitutes a snippet of a sinusoid, whose frequency is equal to the operating frequency of the radar (e.g., 100×10^6 Hz or 100 MHz for a radar operating at VHF, 10^9 Hz or 1 GHz for a radar operating at L-band, or 10×10^9 Hz or 10 GHz for a radar operating at X-band). We normally term the operating frequency the *carrier frequency* of the radar and denote it as f_c or f_o. Figure 1.5 depicts a (amplitude normalized) transmit pulse example. This figure is not to scale, as it shows only 10 cycles of the carrier over the duration of the pulse. For an X-band radar and a pulse duration, or pulsewidth, of $\tau_p = 1$ µs, there will be 10,000 cycles of the carrier over the duration of the pulse.

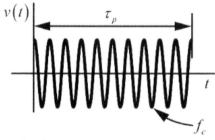

Figure 1.5 Depiction of a transmit pulse.

We can mathematically represent the transmit pulse as

$$v_T(t) = \text{rect}\left[\frac{t - \tau_p/2}{\tau_p}\right]\cos(2\pi f_c t) \tag{1.15}$$

where

$$\text{rect}[t] = \begin{cases} 1 & |t| \leq 1/2 \\ 0 & \text{otherwise} \end{cases} \tag{1.16}$$

is termed the *rectangle function*.

Ideally, the radar receives an attenuated, delayed version of the transmit signal, that is,

$$v_R(t) = A v_T(t - \tau_R) \tag{1.17}$$

where the amplitude scaling factor, A, comes from the radar range equation (see Chapter 2), and the delay, τ_R, is the range delay discussed earlier. To compute Doppler frequency, we acknowledge that the range delay in (1.17) is a function of time, t, and write

$$\tau_R(t) = \frac{2R(t)}{c} \tag{1.18}$$

Substituting (1.18) into (1.17) gives

$$\begin{aligned}v_R(t) &= A\text{rect}\left[\frac{t - \tau_R(t) - \tau_p/2}{\tau_p}\right]\cos\{2\pi f_c[t - \tau_R(t)]\} \\ &= A\text{rect}\left[\frac{t - \tau_R(t) - \tau_p/2}{\tau_p}\right]\cos[\phi(t)]\end{aligned} \tag{1.19}$$

The argument of the cosine term contains the necessary information.[3] We wish, then, to examine

$$\phi(t) = 2\pi f_c[t - \tau_R(t)] = 2\pi f_c[t - 2R(t)/c] \tag{1.20}$$

Expanding $R(t)$ into its Taylor series about $t = 0$ gives

$$\begin{aligned}R(t) &= R(0) + \dot{R}(0)t + \ddot{R}(0)t^2/2! + \dddot{R}(0)t^3/3! + \cdots \\ &= R + \dot{R}t + g(t^2, t^3, \cdots)\end{aligned} \tag{1.21}$$

Substituting (1.21) into (1.20) yields

---

[3] The impact of the time-varying range delay on the rect[•] function is that the returned pulse will be slightly shorter or longer than the transmit pulse. This difference in pulsewidths is usually very small and can be neglected.

$$\begin{aligned}\phi(t) &= 2\pi f_c t - 2\pi f_c \left[2R + 2\dot{R}t + 2g\left(t^2, t^3, \cdots\right)\right]/c \\ &= 2\pi f_c t - 2\pi f_c (2R/c) - 2\pi f_c (2\dot{R}/c)t - 4\pi f_c g\left(t^2, t^3, \cdots\right)/c\end{aligned} \quad (1.22)$$

or

$$\phi(t) = 2\pi f_c t + \phi_R + 2\pi f_d t + g'\left(t^2, t^3, \cdots\right) \quad (1.23)$$

In (1.23),

- $\phi_R = -2\pi f_c(2R/c)$ is a phase shift due to range delay and is of little use in practical radars.
- $g'(t^2, t^3, \cdots) = -4\pi f_c g(t^2, t^3, \cdots)/c$ is a nonlinear phase term usually ignored (until it creates problems in advanced signal processors).
- $f_d = -f_c(2\dot{R}/c)$ is the *Doppler frequency*, or *Doppler shift,* of the target return.

Recalling the radar *wavelength*, λ, is given by

$$\lambda = c/f_c \quad (1.24)$$

allows us to rewrite f_d in its more standard form

$$f_d = -2\dot{R}/\lambda \quad (1.25)$$

Dropping the $g'(t^2, t^3, \cdots)$ term and substituting (1.23) into (1.19), we get

$$v_R(t) = A\,\mathrm{rect}\left[\frac{t - \tau_R - \tau_p/2}{\tau_p}\right]\cos\left[2\pi(f_c + f_d)t + \phi_R\right] \quad (1.26)$$

where we have used $\tau_R = 2R(0)/c$. In (1.26), we note that the frequency of the returned signal is $f_c + f_d$, instead of simply f_c. Thus, comparing the frequency of the transmit signal to the frequency of the received signal permits us to determine the Doppler frequency, f_d. Obtaining f_d allows computation of the range rate from (1.25).

In practice, it is not as easy to measure Doppler frequency as the calculations above imply. The problem lies in the relative magnitudes of f_d and f_c. Consider the following example, whereby a target travels at approximately Mach 0.5, or about 150 m/s. For now, assume the target flies directly toward the radar, so $\dot{R} = -v = -150$ m/s. Assume, furthermore, that the radar operates at X-band with a specific carrier frequency of $f_c = 10$ GHz. With these assumptions, we get

$$\lambda = \frac{c}{f_c} = \frac{3 \times 10^8}{10 \times 10^9} = 0.03 \text{ m} \quad (1.27)$$

and

$$f_d = -\frac{2\dot{R}}{\lambda} = -\frac{2(-150)}{0.03} = 10^4 \text{ Hz} = 10 \text{ kHz} \quad (1.28)$$

Comparing f_c to f_d yields the observation that f_d is a million times smaller than f_c.

While measuring Doppler frequency is not easy, it is achievable. Such Doppler frequency measurement requires a very long transmit pulse (on the order of milliseconds rather than microseconds) or the processing of several pulses.

1.7 DECIBELS

A measurement convention commonly used in radar analyses is the *decibel*. Engineers at Bell Telephone Laboratories (now Bell Labs) originally formulated the concept of the decibel to measure losses over given distances of telephone cable [14]. By definition, a decibel representation of a quantity equals 10 times the logarithm to the base 10 of that quantity. As implied by its name, a decibel is 1/10 of a bel, a logarithm of quantity that was coined by the Bell Labs engineers to honor Alexander Graham Bell.

The decibel, abbreviated dB, is useful in radar analyses because of the large range of numbers encountered in such analyses. The abbreviation dB, by itself, means 10log(power ratio). Thus, the signal-to-noise ratio in decibels is computed using

$$SNR\big|_{dB} = 10\log\left(\frac{P_S}{P_N}\right) \qquad (1.29)$$

where P_S is the signal power [in watts (W), milliwatts (mW), kilowatts (kW), and so forth] and P_N is the noise power in the same units as P_S.

Equation (1.29) provides the "standard" use for the decibel, as originally conceived by Bell Labs. Since then, analysts in the fields of radar, electronics, and communications have expanded the definition of the decibel to include many other forms:

- The abbreviation dBW denotes power level relative to 1 watt, or 10log(P), where P is power in watts.
- The abbreviation dBm" denotes power level relative to 1 milliwatt, or 10log(P/0.001) = 30 + 10log (P).
- The abbreviation dBV denotes voltage level relative to 1 volt root mean square (rms), or 20log($|V|$).[4]
- The abbreviation dBsm denotes area in square meters relative to 1 square meter, or 10log(A), where A is area in square meters. We use this to represent the radar cross section of a target.
- The abbreviation dBi denotes antenna directivity (gain) relative to the directivity of an isotropic antenna, or 10log(G), where G is the antenna directivity in watts per watt (W/W). The gain of an isotropic antenna is taken to be 1 W/W.

1.8 DB ARITHMETIC

In radar analyses, it is often convenient or necessary to perform conversions from ratios to dB values without a calculator. To aid in this, some common relations between ratios

---
[4] Differing applications for dBV use 1 volt root mean square, peak-to-peak, or peak as reference.

and dB are contained in Table 1.2. These relations can be used to find other conversions by remembering that multiplication of ratios translates to addition of dB values. For example, to compute the dB value for a ratio of 4, we recognize that 4 = 2 × 2. Thus, the dB value corresponding to a ratio of 4 is 3 + 3 = 6 dB. As another example, the dB value for a ratio of 50 can be found by recognizing that 50 = 10 × 10/2, and thus the dB value is 10 + 10 − 3 = 17 dB.

Table 1.2
Relation Between Ratios and Decibels

dB	Power Ratio
−10	0.1000
−9	0.1259
−6	0.2512
−3	0.5012
0	1.0000
3	1.9953
6	3.9811
10	10
20	100
30	1,000

1.9 COMPLEX SIGNAL NOTATION

When we wrote the equations for $v_T(t)$ and $v_R(t)$ in (1.15) and (1.19), we used what we term *real signal notation*. With this notation, the signal equations are real functions of time that include sines and cosines. We find that when we need to work with such real functions, we are faced with the manipulation of these sines and cosines, which can be cumbersome because of the need to use trigonometric identities.

To circumvent the problems associated with the manipulation of sines and cosines, radar analysts commonly use an alternate signal notation termed *complex signal notation*. With this notation, signals are represented as complex functions through the use of exponentials with complex arguments. For example, we would write the transmit signal of (1.15) in complex signal notation as

$$v_T^c(t) = \text{rect}\left[\frac{t - \tau_p/2}{\tau_p}\right] e^{j2\pi f_c t} \quad (1.30)$$

and the receive signal as

$$v_R^c(t) = A\,\text{rect}\left[\frac{t - \tau_R(t) - \tau_p/2}{\tau_p}\right] e^{j2\pi f_c [t - \tau_R(t)]} \quad (1.31)$$

where the superscript c is used to distinguish these signals from $v_T(t)$ and $v_R(t)$. In these equations, $j = \sqrt{-1}$.

Through the use of the Euler identity

$$e^{j\theta} = \cos\theta + j\sin\theta \tag{1.32}$$

we can relate $v_T(t)$ and $v_R(t)$ to their complex signal counterparts as

$$v_T(t) = \text{real}\left[v_T^c(t)\right] = \text{real}\left[\text{rect}\left(\frac{t-\tau_p/2}{\tau_p}\right)e^{j2\pi f_c t}\right]$$
$$= \text{rect}\left[\frac{t-\tau_p/2}{\tau_p}\right]\cos(2\pi f_c t) \tag{1.33}$$

and

$$v_R(t) = \text{real}\left[v_R^c(t)\right] = \text{real}\left[A\,\text{rect}\left(\frac{t-\tau_R(t)-\tau_p/2}{\tau_p}\right)e^{j2\pi f_c[t-\tau_R(t)]}\right]$$
$$= A\,\text{rect}\left[\frac{t-\tau_R(t)-\tau_p/2}{\tau_p}\right]\cos\{2\pi f_c[t-\tau_R(t)]\} \tag{1.34}$$

In these equations, real[x] denotes the real part of the complex number x. The imaginary part would be denoted as imag[x].

The primary reason for using complex signal notation is ease of mathematical manipulation. Specifically, multiplying exponentials is easier than multiplying sines and cosines. However, we also find that complex signal notation often provides a convenient and clear means of describing signal properties. For example, a complex signal notation for (1.26) might be

$$v_R^c(t) = A\,\text{rect}\left[\frac{t-\tau_R-\tau_p/2}{\tau_p}\right]e^{j\phi_R}e^{j2\pi f_d t}e^{j2\pi f_c t} \tag{1.35}$$

In this equation, we recognize $A\,\text{rect}[(t-\tau_R-\tau_p/2)/\tau_p]$ as the magnitude of the complex (and real) signal. We recognize ϕ_R as a constant phase part of the signal (since it is the argument of the complex exponential, $e^{j\phi_R}$). The term $2\pi f_d t$ is a phase that depends on the Doppler frequency. We recognize the last term as the carrier frequency part of the signal. Thus, with complex signal notation, we can conveniently characterize the various properties of the signal by separating them into separate complex exponential and magnitude terms.

An extension of complex signal notation is *baseband signal notation*. With this notation, we drop the carrier exponential. This is the approach commonly used in alternating current (AC) circuit analysis (steady-state analysis of resistor-inductor-capacitor circuits excited by a sinusoid). In those analyses, the sinusoid is not explicitly used. Instead, the voltages and currents in the circuit are represented by their amplitude and phase, and the circuit analyses are performed using complex mathematics.

A means of dropping the carrier term is to set f_c to zero. This is the basis of the word "baseband." That is, baseband signals are assumed to have a carrier frequency of zero.

We will use all three signal notations in this book, depending on need. Generally, we will use real signals when we need to specifically address the real properties of signals. We will use complex signal notation when we need to explicitly discuss the operating frequency but do not want to have to manipulate real signals. We will use baseband signal notation when we are not specifically concerned with the operating frequency of the signal. We will also use baseband signal notation when we discuss signal processing, since many digital signal processors explicitly operate on baseband signals.

1.10 RADAR BLOCK DIAGRAM

Figure 1.6 contains a generic radar block diagram that includes the major areas we will discuss in this book. We start in Chapter 2 by tracing a signal from the transmitter through the antenna to the target and back to the matched filter (through the antenna and receiver) to derive one of the key equations of radar theory: the radar range equation. We follow that in Chapters 3, 4, and 5 with discussions of the radar cross section, noise, and loss terms of the radar range equation. In Chapter 6, we present detailed discussions of false alarm probability and detection probability for the Swerling target types and targets with a constant radar cross section. In Chapter 7 we discuss constant false alarm rate (CFAR), processing, which is the mechanism modern radars use to determine the detection threshold. The discussions of detection theory naturally lead to matched filter theory, which we cover in Chapter 8. In Chapter 9, we discuss signal processing from the perspective of improving detection probability. Later, in Chapters 15 through 21, we provide detailed discussions of signal processing from the perspective of clutter mitigation. In Chapter 22, we discuss receivers, including the modern field of digital receivers.

In Chapters 10 and 11, we address the waveform generator portion of the radar by discussing the ambiguity function and an assortment of waveform codings. In Chapter 13, we move to the antenna and present a discussion of phased array antennas. We follow this in Chapter 14 with an introductory discussion of active electronically scanned arrays (AESAs). The remainder of the text is devoted to a discussion of advanced topics, including stretch processing (Chapter 12), synthetic aperture radar (Chapter 23), space-time adaptive processing (Chapter 24), and sidelobe cancellation (Chapter 25).

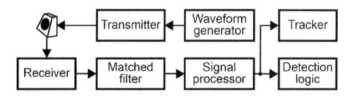

Figure 1.6 Generic radar block diagram.

1.11 EXERCISES

1. Find the round-trip time delay for a target at the following ranges:

    a) 15 km
    b) 37 mi
    c) 350 kft
    d) 673 nmi (nautical miles)

2. What minimum PRIs does a radar require in order to operate unambiguously in range for the target cases of Exercise 1? What PRFs correspond to these minimum PRIs? Ignore pulsewidth in formulating your answer. What would your answers be if you include a pulsewidth of 200 μs?

3. Find the Doppler frequencies for an 8-GHz (low X-band) radar at the following target range rates:

    a) −100 m/s
    b) 150 mph
    c) −30 m/s

4. A target has a state vector given by

$$\begin{bmatrix} x \\ y \\ z \\ \dot{x} \\ \dot{y} \\ \dot{z} \end{bmatrix} = \begin{bmatrix} 20\,\text{km} \\ 15{,}000\,\text{ft} \\ 22\,\text{kft} \\ 200\,\text{m/s} \\ 50\,\text{m/s} \\ -15\,\text{m/s} \end{bmatrix} \quad (1.36)$$

    where the state vector is referenced to the radar. Find the range (R) and range rate ($\dot{R}$) of the target. Find the round-trip time delay (τ_R) to the target. Find the Doppler frequency for a radar operating at 8.5 GHz.

5. Skolnik [1] poses an interesting problem: if the moon is located approximately 384,400 km from the Earth, what is the range (time) delay (τ_R) to the moon? What PRF should we use to operate unambiguously in range?

6. In Section 1.6, we chose to ignore the term

$$g'(t^2, t^3, \cdots) = -4\pi f_c \left(\ddot{R} t^2/2! + \dddot{R} t^3/3! + \cdots \right)/c \quad (1.37)$$

We want to verify this as a valid assumption, at least for the second derivative term. In particular, we want to show, for a specific, realistic example, that the variation across the pulse of the phase

$$\phi_g(t) = g(t^2) = 4\pi \frac{f_c}{c} \ddot{R} t^2/2! \quad (1.38)$$

is small compared to the variation of the phase across the pulse,

$$\phi_d(t) = 2\pi f_d t \tag{1.39}$$

To do this, compare

$$\Delta\phi_g = \phi_g(t_0 + \tau_p) - \phi_g(t_0) \tag{1.40}$$

to

$$\Delta\phi_d = \phi_d(t_0 + \tau_p) - \phi_d(t_0) \tag{1.41}$$

We intend for you to show $\Delta\phi_g \ll \Delta\phi_d$. For the equation above, $t_0 = \tau_R = 2R/c$, the time it takes for the pulse to return from the target. For this exercise, use

$$\begin{bmatrix} x \\ y \\ z \\ \dot{x} \\ \dot{y} \\ \dot{z} \end{bmatrix} = \begin{bmatrix} 90\,\text{km} \\ 0 \\ 10\,\text{km} \\ -500\,\text{mph} \\ 0 \\ 0 \end{bmatrix} \tag{1.42}$$

a pulsewidth of $\tau_p = 1$ µs, and a carrier frequency of $f_c = 10$ GHz (X-band). Assume the higher derivatives of x, y, and z equal zero. Verify the assumption by computing the following:

a) $\dot{R}$
b) $\ddot{R}$
c) f_d
d) $\Delta\phi_g$
e) $\Delta\phi_d$

Is the assumption valid? Explain briefly.

7. If a radar generates a power of 100 kW (100,000 W), what is the power in dBW? What is the power in dBm?

8. If the radar in Exercise 7 receives a return target power of −84 dBm, what is the received power in decibels relative to 1 W? What is the received power in watts? What is the received power in milliwatts (1 milliwatt = 10^{-3} W)?

References

[1] Skolnik, M. I., *Introduction to Radar Systems*, 3rd ed., New York: McGraw-Hill, 2001, pp. 14–16.

[2] Fink, D. G., *Radar Engineering*, New York: McGraw-Hill, 1947, pp. 3–8.

[3] Ridenour, L. N., *Radar System Engineering*, vol. 1 of MIT Radiation Lab. Series, New York: McGraw-Hill, 1947; Norwood, MA: Artech House (CD-ROM edition), 1999, pp. 13–17.

[4] Guarnieri, M., "The Early History of Radar," *IEEE Ind. Electron. Mag.*, vol. 4, no. 3, Sept. 2010, pp. 36–42.

[5] Pritchard, D., *Radar War: Germany's Pioneering Achievement,* 1904–45, Wellingborough, Northamptonshire, England: Patrick Stephens, 1989, Chapter 1.

[6] Page, R. M., *The Origin of Radar,* Garden City, NY: Anchor Books, 1962.

[7] Buderi, R., *The Invention that Changed the World: How a Small Group of Radar Pioneers Won the Second World War and Launched a Technical Revolutio*n, New York: Simon & Schuster, 1996.

[8] Gebhard, L. A., "Evolution of Naval Radio-Electronics and Contributions of the Naval Research Laboratory," Naval Research Laboratory, Washington, D.C., Rep. No. 8300, 1979. Available from DTIC as ADA084225.

[9] Sarkar, T. K. and Salazar Palma, M. "A History of the Evolution of RADAR," 2014 44th European Microwave Conference, Rome, 2014, pp. 734–737.

[10] Budge, M. C., Jr., "EE 619: Intro to Radar Systems," www.ece.uah.edu/courses/material/EE619/index.htm.

[11] *IEEE Standard Dictionary of Electrical and Electronic Terms*, 6th ed., New York: IEEE, 1996.

[12] Barton, D. K., *Radar System Analysis and Modeling,* Norwood, MA: Artech House, 2005.

[13] P-N Designs, Inc., "Frequency Letter Bands." www.microwaves101.com/encyclopedias/frequency-letter-bands.

[14] Martin, W.H., "Decibel–The Name for the Transmission Unit," *Bell Syst. Tech. J.,* vol. 8, no.1, Jan. 1929, pp.1–2.

Chapter 2

Radar Range Equation

2.1 INTRODUCTION

One of the simpler equations of radar theory is the radar range equation. Although it is one of the simpler equations, ironically, it is an equation that is easily misunderstood and misused. The problem lies not with the equation itself, but with the various terms that it is composed of. It is our belief that an understanding of the radar range equation leads to a solid foundation in the fundamentals of radar theory. Because of the difficulties associated with using and understanding the radar range equation, a considerable portion of this book is devoted to its terms and the items it impacts, such as antennas, receivers, matched filters, signal processors, and detection theory.

According to David K. Barton, the radar range equation was developed during World War II, with the earliest associated literature subject to military security restrictions [1]. Kenneth A. Norton and Arthur C. Omberg of the U.S. Naval Research Laboratory authored the first published paper on the radar range equation in 1947, titled "The Maximum Range of a Radar Set" [1–3].

2.2 BASIC RADAR RANGE EQUATION

One form of the basic radar range equation is (it is also sometimes termed the radar equation and the single-pulse signal-to-noise equation, among other names)

$$SNR = \frac{E_S}{E_N} = \frac{P_T G_T G_R \lambda^2 \sigma \tau_p}{(4\pi)^3 R^4 k T_s L} \text{ W-s/W-s or joules/joule (J/J)} \quad (2.1)$$

where [4]

- SNR denotes the signal-to-noise ratio in units of joules per joule, or J/J. The equivalent units are watt-seconds per watt-seconds, or W-s/W-s.

- E_S denotes the signal energy, in joules (J) or watt-seconds (W-s), at some point in the radar receiver—usually at the output of the matched filter or the signal processor.
- E_N denotes the noise energy, in joules, at the same point that E_S is specified.
- P_T, termed the *peak transmit power*, denotes the average power, in watts, during radar signal transmission. Although P_T can be specified at the output of the transmitter or at some other point, such as the output of the antenna feed, we specify it here as the power at the output of the transmitter.
- G_T denotes the directivity, or directive gain, of the transmit antenna in units of watts per watt.
- G_R denotes the directivity, or directive gain, of the receive antenna in units of watts per watt. In many cases, $G_R = G_T$.
- λ denotes the radar wavelength in units of meters.
- σ denotes the average target *radar cross section (RCS)* in units of square meters.
- τ_p is the transmit pulsewidth, in seconds. In this book, we assume the transmit pulse has a rect[x] function envelope.[1] τ_p is the width of that rect[x] function.
- R denotes the slant range from the radar to the target in units of meters.
- k denotes Boltzmann's constant and is equal to $1.3806503 \times 10^{-23}$ W/(Hz K) or (W-s)/K, although it is often truncated to 1.38×10^{-23} (W-s)/K.
- $T_s = T_a + (F_n - 1) T_0$ Kelvin (K) is the system noise temperature, where T_a is an antenna noise temperature used to characterize environment noise.
- T_0 denotes a reference temperature. The IEEE defines noise figure in terms of a noise temperature of $T_0 = 290$ K [6, p. 2-5], which results in the rule of thumb approximation of $kT_0 = 4 \times 10^{-21}$ W/Hz.
- F_n denotes the overall radar *noise figure*[2] and is dimensionless or has units of watts per watt.
- L denotes all the losses one takes into consideration when using the radar range equation. This term, which accounts for losses that apply to the signal and not the noise, has the units of watts per watt. L accounts for a multitude of factors that degrade radar performance, including those related to the radar itself, the environment in which the radar operates, the operator of the radar, and, often, the inexperience of the radar analyst. Loss factors are covered in more detail in Chapter 5.

The radar range equation of (2.1) is termed the *single-pulse* radar range equation because the *SNR* calculation is based on a single transmit pulse. If the radar transmits and processes several pulses, the equation can be modified by adding a multiplicative term, G_{SP}, that accounts for the transmission and processing of multiple pulses. This is addressed in Chapter 9.

The form of the radar range equation of (2.1) is nontraditional [6–8] in that *SNR* is specified as an energy ratio rather than a power ratio. The latter form is briefly discussed

---

[1] An assumption of this form of the radar range equation is that the radar is pulsed, not CW. For CW radars, it would be more appropriate to use the form of Section 2.3.
[2] Noise figure and noise factor are often treated as synonyms, although some authors make a distinction [5, pg. 46]. Specifically, the term "noise figure" is used when in logarithmic form, while noise factor is used when in linear form. We will use noise figure for both the W/W and dB version in this book.

in Section 2.3. The energy ratio form is used in this book because the power ratio form requires specification of a noise bandwidth. It has been the authors' experience that this sometimes causes confusion for both students and practicing radar analysts. The energy ratio formulation circumvents this problem by using τ_p instead of a noise bandwidth. As a note, τ_p is the *uncompressed pulsewidth*; it is the width of the envelope of the transmit pulse. Any phase (or frequency) modulation on the pulse is not a factor in the radar range equation, except possibly as a loss due to differences between the received pulse and the matched filter impulse response (see Chapter 5).

At this point we might ask: "Why compute SNR, or even be concerned with noise?" We answer the second question first by noting that by the time the signal returned from a target gets to the point of interest in a radar, it will be embedded in noise. Part of that noise comes from the environment (e.g., galactic noise and earth noise) and part of it comes from the components of the receiver. As to the first question, the reason we are interested in SNR is the probability of detecting the signal returned from a target is directly dependent on SNR. We discuss this in Chapter 6. By the way, the reason we talk about the probability of detecting the signal is because noise and the signal are random processes, which means we need to use probability theory to characterize them.

We will derive the radar range equation and attempt to carefully explain its various terms and their origins. We start by deriving E_S, the signal energy component, and follow this by deriving E_N, the noise energy component.

2.2.1 Derivation of ES

2.2.1.1 The Transmitter

We begin the derivation at the transmitter output and go through the waveguide and antenna out into space (see Figure 2.1). For now, assume the radar is in free space. We can account for the effects of the atmosphere in the loss term, L. We assume the transmitter generates a single pulse with a rectangular envelope that has a width of τ_p. Figure 2.2 contains a simplified representation of this pulse. In this example, the pulse is modulated with a constant frequency of f_c, the *carrier* frequency.[3] The power indicated in Figure 2.2 is normalized power, the power delivered to a 1-ohm resistor.

The average transmit power in the signal *over the duration of the pulse* is termed the *peak* transmit power and is denoted as P_T. We term this power the peak transmit power because later we will consider the transmit power averaged over many pulses.

The waveguide in Figure 2.1 carries the signal from the transmitter to the antenna feed input. The waveguide's only feature of interest in the radar range equation is that it is a lossy device that attenuates the signal. Although we only refer to the "waveguide" here, there are several devices included between the transmitter and antenna feed of a practical radar (see Chapter 5).

---

[3] We assume nothing in the transmit, propagation, or receive path of the radar, up to the matched filter, distorts the rectangular pulse envelope. Clearly, this will not be the case, since a rectangular pulse has infinite bandwidth and the transmitter, environment, and receiver have finite bandwidth. As discussed in Chapters 5 and 7, we accommodate envelope distortion by including a loss factor. As a note, in practical radars, the loss due to pulse envelope distortion is usually small (< 1 dB).

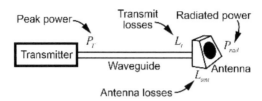

Figure 2.1 Transmit section of a radar.

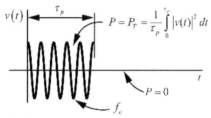

Figure 2.2 Depiction of a transmit pulse.

Because it is a lossy device, we characterize the waveguide in terms of its loss, which we denote as L_t and term *transmit loss*. Since L_t is a loss, it is greater than unity. With this the power at the input to the antenna feed takes the form

$$P_{feed} = \frac{P_T}{L_t} \text{ W} \qquad (2.2)$$

Generally, the feed and other components of the antenna attenuate the signal further. If we consolidate all these losses into an antenna loss term, L_{ant}, the radar finally radiates the power

$$P_{rad} = \frac{P_{feed}}{L_{ant}} = \frac{P_T}{L_t L_{ant}} \text{ W} \qquad (2.3)$$

Since the pulse envelope width is τ_p, the *energy* radiated by the antenna is

$$E_{rad} = P_{rad} \tau_p \text{ W-s or J} \qquad (2.4)$$

2.2.1.2 The Antenna

The purpose of the radar antenna is to concentrate, or focus, or collimate, the radiated energy in a small angular sector of space. As an analogy, the radar antenna works much like the reflector in a flashlight. Like a flashlight, a radar antenna does not perfectly focus the beam. For now, however, we will assume it does. Later, we will account for imperfect focusing by using a scaling term.

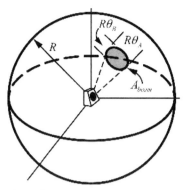

Figure 2.3 Radiation sphere with antenna beam.

Given the purpose above, we assume all the radiated energy is concentrated in the area, A_{beam}, indicated in Figure 2.3. With this the energy density over A_{beam} is

$$S_R = \frac{E_{rad}}{A_{beam}} = \frac{P_T \tau_p / L_t L_{ant}}{A_{beam}} \text{ W-s/m}^2 \quad (2.5)$$

To extend (2.5) to the next step, we need an equation for A_{beam}. Given lengths for the major and minor axes of the ellipse in Figure 2.3 of $R\theta_A$ and $R\theta_B$, we can write the area of the ellipse:

$$A_{ellipse} = \frac{\pi}{4} R^2 \theta_A \theta_B \text{ m}^2 \quad (2.6)$$

We recognize that the energy is not uniformly distributed across $A_{ellipse}$ and that some of the energy will "spill" out of the area $A_{ellipse}$ (i.e., the antenna does not focus the energy perfectly, as indicated earlier). We account for this by replacing $\pi/4$ with a scale factor K_A. Further discussion of K_A will follow shortly. We can write A_{beam}, then, as follows:

$$A_{beam} = K_A R^2 \theta_A \theta_B \text{ m}^2 \quad (2.7)$$

Substituting (2.7) into (2.5) produces

$$S_R = \frac{\tau_p P_T / L_t L_{ant}}{K_A R^2 \theta_A \theta_B} \text{ W-s/m}^2 \quad (2.8)$$

We now define a term, G_T, the transmit *antenna directivity*, or directive gain, as

$$G_T = \frac{4\pi}{K_A \theta_A \theta_B} \text{ W/W} \quad (2.9)$$

Using (2.9) to rewrite (2.8), we get

$$S_R = \frac{G_T P_T \tau_p}{4\pi R^2 L_t L_{ant}} \text{ W-s/m}^2 \quad (2.10)$$

We reiterate: the form of antenna directivity given in (2.9) depends on the assumption that L_{ant} captures the losses associated with the feed and other components of the antenna.

Some analysts combine the feed and antenna losses with the transmit antenna directivity and term the result the *power gain*, or simply *gain*, of the antenna [9]. We will avoid doing so here, owing to the confusion it produces when using (2.9) and the difficulties associated with another form of directivity that is presented shortly.

The form of G_T in (2.10) and the radar range equation tacitly assume an antenna pointed directly at the target. If the antenna is not pointed at the target, we must modify G_T to account for this. We do this by means of an *antenna pattern*, which is a function that gives the value of G_T at the target, relative to the antenna's pointing direction.

2.2.1.3 Effective Isotropic Radiated Power

We temporarily interrupt our derivation to define the quantity termed *effective isotropic radiated power*. To do so, we ask the question: What power would we need at the output of an isotropic radiator to produce an energy density of S_R at all points on a sphere of radius R? An isotropic radiator (ideal point source) is a hypothetical antenna that does not focus energy but instead distributes it uniformly over the surface of a sphere centered on the antenna. Though it cannot exist in the real world, the isotropic radiator serves a mathematical and conceptual function in radar theory, not unlike that of the impulse function in mathematical theory.

By denoting the effective isotropic radiated power as P_{eff} and recalling the surface area of a sphere of radius R is $4\pi R^2$, we can write the energy density on the surface of the sphere (assuming lossless propagation) as

$$S_R = \frac{P_{eff} \tau_p}{4\pi R^2} \text{ W-s/m}^2 \qquad (2.11)$$

If we equate (2.10) and (2.11) and solve for P_{eff}, we obtain

$$P_{eff} = \frac{P_T G_T}{L_t L_{ant}} \text{ W} = EIRP \qquad (2.12)$$

as the effective isotropic radiated power (EIRP).

We emphasize that P_{eff} is not the power at the output of the antenna. The power at the output of the antenna is $P_T/L_t L_{ant}$. The antenna's purpose is to focus this power over a relatively small angular sector.

2.2.1.4 Antenna Directivity

We turn next to the factor K_A in (2.9). As we indicated, K_A accounts for the properties of the antenna. Specifically, it accounts for two facts:

- The energy is not uniformly distributed over the ellipse.
- Not all of the energy is concentrated in the antenna beam (the ellipse of Figure 2.3). Some energy "spills" out the ellipse into what we term the *antenna sidelobes*.

The value 1.65 is a somewhat common value for K_A [10, p. 143]. Using this value, we can write the antenna directivity as

$$G_T = \frac{4\pi}{1.65\theta_A\theta_B} \text{ W/W} \qquad (2.13)$$

We term the quantities θ_A and θ_B the *antenna beamwidths*, which have the units of radians. In many applications, θ_A and θ_B are specified in degrees. In this case, we can write the directivity as

$$G_T = \frac{25,000}{\theta_A^\circ \theta_B^\circ} \text{ W/W} \qquad (2.14)$$

where the two beamwidths in the denominator are in degrees. The derivation of (2.14) is straightforward and left as an exercise.

While (2.14) uses a numerator of 25,000, various authors provide alternative approximations to account for factors such as antenna type, beamshape, sidelobe characteristics, and so on [11, p. 125]. For example, some authors use 41,253, which would apply to a rectangular beam pattern with no sidelobes and would be indicative of an ideal antenna with maximum directivity [9]. Similarly, some authors use 32,383 for a rectangular aperture with uniform illumination and 33,709 for circular apertures with uniform illumination [12, p. 366]. As still another variant, some authors prefer 26,000 over 25,000 [9]. It has been the authors' experience that 25,000 or 26,000 apply well to antennas that use some type of weighting to reduce sidelobes. As a note, the different approximations correspond to different values of K_A.

To visualize the concept of beamwidth, consider Figure 2.4, which is a plot of $G_T(\alpha,\varepsilon)$ versus α for $\varepsilon = 0$. As discussed in Chapter 13, $G_T(\alpha,\varepsilon)$ is a means of representing antenna directivity as a function of target location relative to antenna pointing angles. If $\alpha = 0$ and $\varepsilon = 0$, the beam is pointed directly at the target and the directivity is maximum. As illustrated in Chapter 13, α and ε are orthogonal angles roughly related to azimuth and elevation, respectively.

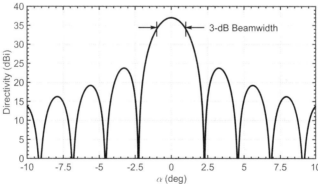

Figure 2.4 Sample antenna pattern.

The unit of measurement on the vertical axis of Figure 2.4 is dBi, or decibels relative to isotropic (see Chapter 1), the common unit of measurement for G_T in radar applications. We define the antenna beamwidth as the distance between the *3-dB points*[4] of Figure 2.4. These 3-dB points are the angles where $G_T(\alpha,\varepsilon)$ is 3 dB below its maximum value. With this we find the antenna represented in Figure 2.4 has a beamwidth of 2°. With this we say the beamwidth, θ_B, of Figure 2.3 and (2.14) is 2°. If we were to plot $G_T(\alpha,\varepsilon)$ versus ε for $\alpha = 0$, and find a distance between the 3-dB points of 2.5°, we would then say the beamwidth, θ_A, was 2.5° The antenna directivity would be computed as

$$G_T = G_T(0,0) = \frac{25,000}{2 \times 2.5} = 5,000 \text{ W/W or 37 dBi} \qquad (2.15)$$

In subsequent sections, we drop the notation dBi and use dB.

The humps on either side of the central antenna beam depicted in Figure 2.4 are the antenna sidelobes discussed above.

2.2.1.5 The Target and Radar Cross Section

To return to our derivation, we have an equation for S_R, the energy density at the location of the target. As the electromagnetic wave passes the target, the target captures some of its energy and reradiates it toward the radar. More accurately, the electromagnetic wave induces currents on the target, and the currents generate another electromagnetic wave that propagates away from the target. Analysts occasionally designate this as energy reflection, a technically incorrect term. The process of capturing and reradiating energy is very complicated and the subject of much research. For now, we take a simplified approach to the process by using the concept of *radar cross section* (RCS).

We note that S_R has the units of W-s/m². Therefore, if we were to multiply S_R by an area, we would convert it to an energy. This is what we do with RCS, which we denote by σ and ascribe the units of m², or dBsm if represented in dB units. Hence, we represent the energy captured and reradiated by the target as

$$E_{tgt} = \sigma S_R \text{ W-s} \qquad (2.16)$$

To continue our idealized assumption, we posit the target acts as an isotropic radiator and radiates E_{tgt} uniformly in all directions. The target, in fact, behaves much like an actual antenna and radiates energy with different amplitudes in different directions.

Given the assumption that E_{tgt} is the energy radiated by a target and the target acts as an isotropic antenna, we can represent the energy density at the radar as

$$S_{rec} = \frac{E_{tgt}}{4\pi R^2} \text{ W-s/m}^2 \qquad (2.17)$$

or, by substituting (2.10) into (2.16) and the result into (2.17),

---

[4] The concept of 3-dB points should be familiar from control and signal processing theory as the standard measure used to characterize bandwidth.

$$S_{rec} = \frac{P_T G_T \sigma \tau_p}{(4\pi)^2 R^4 L_t L_{ant}} \text{ W-s/m}^2 \qquad (2.18)$$

2.2.1.6 Antenna Again

As the electromagnetic wave from the target passes the radar, the radar antenna captures a part of the energy in this wave and sends it to the radar receiver. If we extend the logic we applied to the target, we can formulate the energy at the output of the antenna feed as

$$E_{ant} = S_{rec} A_e \text{ W-s} \qquad (2.19)$$

where A_e denotes the *effective* area of the antenna and is a measure describing the antenna's ability to capture the returned electromagnetic energy and convert it into usable power. A more common term for A_e is *effective aperture* of the antenna.

The effective aperture is related to the physical area of the antenna by

$$A_e = \rho_{ant} A_{ant} \text{ m}^2 \qquad (2.20)$$

where A_{ant} is the area of the antenna projected onto a plane placed directly in front of the antenna and ρ_{ant} denotes the antenna efficiency. We make this clarification of area because we do not want to confuse it with the actual surface area of the antenna. For example, if the antenna is a parabola of revolution (a paraboloid), a common type of antenna, the actual area of the antenna would be the area of the paraboloidal surface of the antenna, whereas A_{ant} is the area of the disc defined by the front rim of the antenna. In most phased array antennas (flat-face phased array antennas), A_{ant} is the area of the part of the antenna containing the array elements.

While the antenna efficiency can take on any value between 0 and 1, it is seldom below 0.4 or above 0.7 [11, p. 124]. A rule of thumb for the antenna efficiency value is $\rho_{ant} = 0.6$.

Substituting (2.18) into (2.19) yields

$$E_{ant} = \frac{P_T G_T \sigma A_e \tau_p}{(4\pi)^2 R^4 L_t L_{ant}} \text{ W-s} \qquad (2.21)$$

2.2.1.7 Antenna Directivity Again

Equation (2.21) is not very easy to use because of the A_e term. We can characterize the antenna more conveniently by using directivity, much as we did on transmit. According to antenna theory, we can relate antenna directivity to effective aperture by the equation [12, p. 108; 8, p. 4]

$$G_R = \frac{4\pi A_e}{\lambda^2} \text{ W/W} \qquad (2.22)$$

Substituting (2.22) into (2.21) produces the following:

$$E_{ant} = \frac{P_T G_T G_R \lambda^2 \sigma \tau_p}{(4\pi)^3 R^4 L_t L_{ant}} \text{ W-s} \tag{2.23}$$

We next need to propagate the signal through the receiver. We do this by including a gain term, G, which accounts for all of the receiver components up to the point where we measure SNR. With this we get

$$E_{rec} = \frac{P_T G_T G_R \lambda^2 \sigma \tau_p}{(4\pi)^3 R^4 L_t L_{ant}} G \text{ W-s} \tag{2.24}$$

2.2.1.8 Losses

As a final step in this part of the development, we need to account for losses we have ignored thus far. There are many losses that we will need to account for (see Chapter 5). For now, we will consolidate all these losses with $L_t L_{ant}$ and denote them by L. Using this approach, we say the signal energy in the radar is given by

$$E_S = \frac{P_T G_T G_R \lambda^2 \sigma \tau_p}{(4\pi)^3 R^4 L} G \text{ W-s} \tag{2.25}$$

which is E_{rec}, with the additional losses included.

We said E_S denotes the signal energy in the radar, although we did not say where in the radar. We will defer this discussion for now and turn our attention to the noise energy term, E_N.

2.2.2 Derivation of E_N

The two main contributors to noise in radars are the environment and the electronic components of the receiver. Environment noise includes radiation from the earth, galactic and intergalactic noise, atmospheric noise, and, in some instances, man-made noise such as noise jammers. Galactic and intergalactic noise includes cosmic background radiation and solar or other star noise. The environment noise we consider is earth, galactic, and intergalactic noise.

Electronic equipment noise is termed *thermal noise* (also known as Johnson noise) and arises from agitation of electrons caused by heat [13, p. 752; 15]. This form of noise was discovered by Johnson [14] and first analyzed by Nyquist [15]. One of the equations in Nyquist's paper leads to a definition of noise power spectral density, or energy, for resistive devices as

$$N_0 = kT \text{ W-s} \tag{2.26}$$

where $k = 1.38 \times 10^{-23}$ W-s/K denotes Boltzmann's constant and T denotes the *noise temperature* of the resistor in kelvin (K).

Equation (2.26) is actually a limiting case of one form of Planck's law. This is discussed further in Chapter 4. An implication of (2.26) is that the noise energy (in resistive devices) is independent of frequency.

Device manufacturers and communication analysts [16, 17] have adopted a modified form of (2.26) for electronic devices given by

$$N_0 = kT_0 F \text{ W-s} \tag{2.27}$$

where F is termed the *noise figure* of the device and T_0 is the previously discussed reference temperature normally referred to as "room temperature." It is interesting to note that $kT_0 = 4\times10^{-21}$ W/Hz, which makes one think the (somewhat arbitrary) value of $T_0 = 290$ K was chosen to make kT_0 a "nice" number, and not because it is room temperature. While $T_0 = 290$ K is now the standard (the IEEE defines noise figure in terms of a noise temperature of 290 K [6]), other reference temperatures have been used in the past (e.g., 291, 292, 293, and 300 K [2, 18–22]).

The N_0 terms of (2.26) and (2.27) were developed for electronic components and not environment noise. However, radar analysts have adopted (2.26) as a way of characterizing the energy in a radar due to noise in the environment as well as in the electronics. We will do the same here. Thus, we define the noise energy at the input to the matched filter as

$$E_N = GkT_s \text{ W-s} \tag{2.28}$$

where

$$T_s = T_a + (F_n - 1)T_0 . \tag{2.29}$$

In (2.28), G is the same overall receiver gain that appeared in (2.25). T_s is termed the *system noise temperature* and T_a [in (2.29)] is termed the *antenna temperature*. F_n is the overall noise figure of the radar from the "antenna face" to the input to the matched filter. It includes the noise figures of all active and passive devices in the radar, including any antenna components (e.g., radome, phase shifters, waveguides, feeds) that exhibit an ohmic, or dissipative, loss. Equations (2.28) and (2.29) are derived in Chapter 4.

The antenna temperature, T_a, provides a means of characterizing the environment noise in the radar. Blake [23] provides an equation for T_a for the case where the radar beam is pointing into the sky but not directly at the sun or a star (an example of the latter is given in Chapter 4). His equation is

$$T_a \approx 0.876 T'_a + 36 \text{ K} \tag{2.30}$$

This equation also takes into consideration that earth noise is entering through the antenna sidelobes and backlobes. It assumes an antenna without ohmic losses, which would be the case here since the ohmic losses of the antenna are included in F_n. The temperature, T'_a, is determined from Figure 2.5, which comes from [23]. In the figure, θ is the elevation angle of the radar beam relative to the horizon. The assumptions on which the figure is based are provided as a quote below. It is a quote from Blake's Naval Research Laboratory (NRL) report [23].

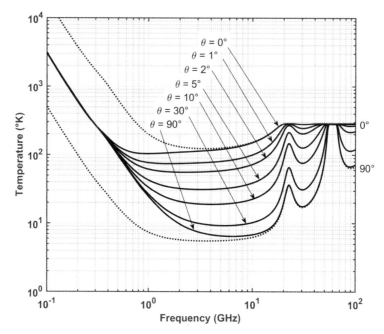

Figure 2.5 Noise temperature of an idealized antenna – surface-based radar. [*Source:* L. V. Blake, "A Guide to Basic Pulse-Radar Maximum-Range Calculation," NRL Report 6930, Naval Research Report Laboratory (1969).]

"noise temperature of an idealized antenna (lossless, no earth-directed side lobes) at the earth's surface as a function of frequency for a number of beam elevation angles. The solid curves are for the geometric-mean galactic temperature, sun noise 10 times the quiet level, the sun in a unity-gain side lobe, a cool temperate-zone troposphere, 30°K cosmic blackbody radiation, and zero ground noise. The upper dashed curve is for maximum galactic noise (center of galaxy, narrow-beam antenna), sun noise 100 times the quiet level, zero elevation angle, and other factors the same as for the solid curves. The lower dashed curve is for minimum galactic noise, zero sun noise, and a 90° elevation angle. The slight bump in the curves at about 500 MHz is due to the sun noise characteristic. The curves for low elevation angles lie below those for high angles at frequencies below 400 MHz because of the reduction of galactic noise by atmospheric absorption. The maxima at 22.2 GHz and 60 GHz are due to water-vapor and oxygen absorption resonances."

An alternative to the system noise temperature, $T_s = T_a + (F_n - 1)T_0$, used in (2.28) is

$$T_s = F_n T_0 \qquad (2.31)$$

This form uses the assumption $T_a = T_0$ and would be a reasonable approximation for the case where F_n was large (greater than about 7 dB) or where one was performing preliminary radar range equation calculations. Otherwise, the T_s of (2.29) should be used.

As a note, [10, p. 11] points out that T_a would equal T_0 if the radar beam was pointing directly at the ground, an unlikely event in ground-based radars.

An important reminder is that the overall radar noise figure, F_n, contains all of the ohmic loss terms of the receive path of the radar, including antenna ohmic losses. As a result, those losses should not be included in the loss term, L, of (2.25). This is a common mistake that is easily made by both novice and experienced radar analysts.

Combining (2.28) and (2.25) with the relation $SNR = E_S/E_N$ results in (2.1), or

$$SNR = \frac{E_S}{E_N} = \frac{P_T G_T G_R \lambda^2 \sigma \tau_p}{(4\pi)^3 R^4 k T_s L} \quad \text{W-s/W-s} \tag{2.32}$$

We note that the G terms of (2.25) and (2.28) canceled. In other words, the overall gain on the radar receiver does not affect SNR. This is a convenient circumstance that allows us to ignore the details of the receiver structure when computing SNR.

Neither (2.1) nor (2.32) states where the radar characterization of the SNR takes place. Such characterization occurs at the matched filter output, as discussed in Chapter 8.

2.3 A POWER APPROACH TO SNR

The power approach defines the SNR as the ratio of signal power to noise power. Recall that (2.25) denotes the signal energy in the radar while (2.28) denotes the noise energy. We use these to write (2.32) in a different form as

$$SNR = \frac{E_S}{E_N} = \frac{\left(P_T G_T G_R \lambda^2 \sigma \tau_p\right) / \left[(4\pi)^3 R^4 L\right]}{k T_s} \quad \text{W-s/W-s} \tag{2.33}$$

If we move τ_p from the numerator to denominator and *define*

$$B_{eff} = 1/\tau_p \tag{2.34}$$

we get

$$SNR = \frac{P_T G_T G_R \lambda^2 \sigma}{(4\pi)^3 R^4 k T_s L B_{eff}} = \frac{P_S}{P_N} \quad \text{W/W} \tag{2.35}$$

which is SNR expressed as a power ratio.

Note that we *defined* B_{eff} as $1/\tau_p$. It must be emphasized that B_{eff} may not be an actual bandwidth anywhere in the radar. Because of the possibility of misinterpreting B_{eff}, readers are advised to avoid using (2.35) and use only (2.1). An exception to this recommendation would be for the case of CW radars. In these types of radars, it would be appropriate to use (2.35) with B_{eff} equal to the bandwidth of the Doppler filter of the signal processor. In that case, SNR would be the SNR at the output of the Doppler filter.

2.4 RADAR RANGE EQUATION EXAMPLE

To illustrate the use of the radar range equation, we consider an example of a monostatic radar with the parameters given in Table 2.1.

We wish to compute the SNR on a target with an RCS of 6-dBsm and at a range of 60 km. To perform the computation, we need to find the parameters of the radar range equation [(2.1) and (2.32)] and ensure that they are in consistent units. Most of the parameters are in Table 2.1 or can be derived from the parameters of Table 2.1. The two parameters not in Table 2.1 are the target range and the target RCS, which are given above.

We will need to compute the wavelength, λ, the total losses, and the system noise temperature, T_s. Table 2.2 gives the appropriate parameters in dB units and MKS units and show the calculation of λ and L.

Table 2.1
Radar Parameters

Radar Parameter	Value
Peak transmit power at power tube, P_T	1 MW
Transmit losses, including feed and antenna, $L_t L_{ant}$	2 dB
Pulsewidth, τ_p	0.4 µs
Antenna directivity, G_T, G_R	38 dB
Operating frequency, f_c	8 GHz
System noise figure F_n	8 dB
Other losses, L_{other}	2 dB

Table 2.2
Radar Range Equation Parameters

Radar Range Equation Parameter	Value (MKS)	Value (dB)
P_T	10^6 W	60 dBW
G_T	6,309.6 W/W	38 dB
G_R	6,309.6 W/W	38 dB
$\lambda = c/f_c$	0.0375 m	−14.26 dB(m)
σ	3.98 m²	6 dBsm
R	60×10³ m	47.78 dB(m)
k	1.38×10⁻²³ W-s/K	−228.6 dB(W-s/K)
τ_p	0.4×10⁻⁶ s	−64 dB(s)
$L = L_t L_{ant} L_{other}$	2.51 W/W	4 dB
T_s	1,602 K	32 dB(K)

T_s was computed from [see (2.29)]

$$T_s = T_a + (F_n - 1)T_0 \tag{2.36}$$

and [see (2.30)]

$$T_a \approx 0.876 T_a' + 36 = 0.876 \times 30 + 36 = 62.3 \tag{2.37}$$

$T_a' \approx 30$ K was obtained from Figure 2.5 using $\theta = 5°$.

Substituting the MKS values from Table 2.2 into (2.32) yields

$$\begin{aligned} SNR &= \frac{P_T G_T G_R \lambda^2 \sigma \tau_p}{(4\pi)^3 R^4 k T_s L} \\ &= \frac{(10^6)(6309.6)(6309.6)(0.0375)^2 (3.98)(0.4 \times 10^{-6})}{(4\pi)^3 (60 \times 10^3)^4 (1.38 \times 10^{-23})(1602)(2.51)} \\ &\approx 62.5 \text{ W-s/W-s or } 17.96 \text{ dB} \end{aligned} \tag{2.38}$$

To double check, we compute (2.32) using dB values, as

$$\begin{aligned} SNR &= (P_T + G_T + G_R + 2\lambda + \sigma + \tau_p) \\ &\quad - [30\log(4\pi) + 4R + k + T_s + L] \end{aligned} \tag{2.39}$$

where all quantities are the dB units from Table 2.2. Substituting yields

$$\begin{aligned} SNR &= [60 + 38 + 38 + 2(-14.26) + 6 - 64] \\ &\quad - [32.98 + 4(47.78) + (-228.6) + 32.0 + 4] \\ &= 17.98 \text{ dB or } 62.8 \text{ W-s/W-s} \end{aligned} \tag{2.40}$$

which agrees with (2.38).

Figure 2.6 contains a plot of *SNR* vs. *R* for the problem of this example.

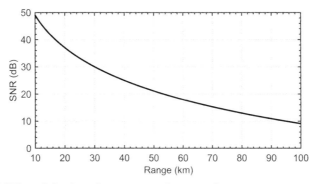

Figure 2.6 Plot of *SNR* vs. *R* for the radar range equation example.

2.5 DETECTION RANGE

An important use of the radar range equation is the determination of *detection range*, or the maximum range at which a target has a given probability of being detected by a radar. The criterion for detecting a target is that the SNR be above some threshold value. If we consider the above radar range equation, we note that SNR varies inversely with the fourth power of range. This means that if the SNR is a certain value at a given range, it will increase as range decreases. We therefore define the detection range as the range at which we achieve a certain SNR. To find the detection range, we can use Figure 2.6 or solve the radar range equation for R. The latter yields

$$R = \left[\frac{P_T G_T G_R \lambda^2 \sigma \tau_p}{(4\pi)^3 (SNR) k T_s L} \right]^{1/4} \text{ m} \qquad (2.41)$$

Suppose, for example, we want the range at which the SNR, given a 6-dBsm target, to be 13 dB. Using the Table 2.2 values in (2.41) yields

$$R = \left[\frac{P_T G_T G_R \lambda^2 \sigma \tau_p}{(4\pi)^3 (SNR) k T_s L} \right]^{1/4} \qquad (2.42)$$

$$R = \left[\frac{(10^6)(6309.57)^2 (0.0375)^2 (3.98)(0.4 \times 10^{-6})}{(4\pi)^3 (19.95)(1.38 \times 10^{-23})(1602)(2.51)} \right]^{1/4} \approx 80 \text{ km} \qquad (2.43)$$

This means target detection occurs at a maximum range of 80 km, or at all ranges of 80 km or less.

The value of 13 dB used in this example is a somewhat standard detection threshold. In Chapter 6, we show that an SNR threshold of 13 dB yields a single-pulse detection probability of 0.5 on an aircraft-type target (a Swerling 1 target).

2.6 SEARCH RADAR RANGE EQUATION

We now want to discuss an extension to the radar range equation used to analyze and design search radars. Its most common use is in the initial sizing of search radars in terms of power and physical size. In fact, the measure of performance usually used to characterize these types of radars is *average power-aperture product*, $P_A A_e$ [7, p. 89], which is the product of the *average power* times the effective aperture of the radar.

We begin by assuming the radar searches an angular region, or sector, denoted Ω. The term Ω takes units of steradians. One of the more common search sectors is a section of the surface of a sphere bounded by some elevation and azimuth extents; Figure 2.7 shows an example of such a surface. The figure indicates an azimuth extent of $\Delta \alpha$ and an elevation extent from ε_1 to ε_2. As shown in the appendices, the angular area of this search sector is

$$\Omega = \Delta\alpha\left(\sin\varepsilon_2 - \sin\varepsilon_1\right) \text{ steradians} \tag{2.44}$$

where all angles are in radians.

In Section 2.2.1.2, it was shown that the area of the beam on the surface of a sphere of radius R could be written as

$$A_{beam} = K_A R^2 \theta_A \theta_B \text{ m}^2 \tag{2.45}$$

Dividing by R^2 results in an angular beam area of

$$\Omega_{beam} = K_A \theta_A \theta_B \text{ steradians} \tag{2.46}$$

This gives the number of beams required to cover the search sector as

$$n = \Omega/\Omega_{beam} = \Omega/K_A \theta_A \theta_B \tag{2.47}$$

Equation (2.47) is ideal in that it essentially assumes a rectangular search sector and rectangular beams. In practice, the number of beams required to fill a search sector is given by

$$n = K_{pack} \Omega/\Omega_{beam} \tag{2.48}$$

where K_{pack} denotes the *packing factor* and accounts for how the beams are arranged within the search sector [24]. For the simple case of rectangular beams, or elliptical beams that touch at their 3-dB points, $K_{pack} = 1$. If the radar uses anything other than rectangular packing or if the beams touch somewhere other than their 3-dB points, K_{pack} will deviate from unity.

Recall that one of the parameters of $P_A A_e$ is the average power, P_A. If the radar has a pulsewidth of τ_p and a PRI of T, the average power is

$$P_A = P_T\left(\tau_p/T\right) = P_T d \text{ W} \tag{2.49}$$

where d represents the *duty cycle* of the radar.

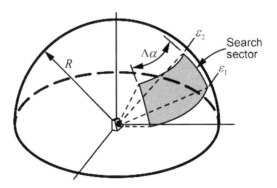

Figure 2.7 Search sector illustration.

One of the requirements imposed on a search radar is that it must cover the search sector in T_{scan} seconds. This means that the radar must process signals from n beams in T_{scan} seconds. Given this requirement, the time allotted to each beam is

$$T_{beam} = T_{scan}/n \text{ s} \tag{2.50}$$

Allowing one PRI per beam gives

$$T = T_{beam} = T_{scan}/n \text{ s} \tag{2.51}$$

Equation (2.48) suggests $n = K_{pack}\Omega/K_A\theta_A\theta_B$, which we can combine with (2.51) to get

$$T = T_{scan}K_A\theta_A\theta_B/K_{pack}\Omega \text{ s} \tag{2.52}$$

We use $d = \tau_p/T$ to obtain

$$d = \frac{\tau_p K_{pack}\Omega}{T_{scan}K_A\theta_A\theta_B} \tag{2.53}$$

Substituting (2.53) and (2.49) into the radar range equation (2.1) produces

$$\begin{aligned} SNR &= \frac{P_T G_T G_R \lambda^2 \sigma \tau_p}{(4\pi)^3 R^4 k T_s L} = \frac{P_A}{d}\frac{G_T G_R \lambda^2 \sigma \tau_p}{(4\pi)^3 R^4 k T_s L} \\ &= \frac{P_A}{\tau_p K_{pack}\Omega \Big/ T_{scan}K_A\theta_A\theta_B}\frac{G_T G_R \lambda^2 \sigma \tau_p}{(4\pi)^3 R^4 k T_s L} \\ &= \frac{T_{scan}K_A\theta_A\theta_B}{K_{pack}\Omega}\frac{P_A G_T G_R \lambda^2 \sigma}{(4\pi)^3 R^4 k T_s L} \text{ W-s/W-s} \end{aligned} \tag{2.54}$$

Finally, we arrive at the search radar range equation by using (2.9) and (2.22) in (2.54):

$$SNR = \frac{P_A A_e \sigma}{4\pi R^4 k T_s L}\frac{T_{scan}}{K_{pack}\Omega} \text{ W-s/W-s} \tag{2.55}$$

We leave the details of deriving (2.55) as an exercise.

Note (2.55) does not explicitly depend on operating frequency (via λ), antenna directivity, or pulse width—as does the standard radar range equation. This can be of value in performing preliminary search radar designs because we need not specify a lot of parameters.

It must be emphasized that the search radar range equation leads to a preliminary radar design. At best, it provides a starting point for a more detailed design in which the specific parameters not in (2.54) are defined. This will be discussed further in the following example.

2.7 SEARCH RADAR RANGE EQUATION EXAMPLE

As an interesting example[5], we consider a requirement placed on search radars used for ballistic missile defense. Specifically, the Strategic Arms Limitation Talks I (SALT I) treaty specifies that the power aperture product be limited to 3×10^6 W-m² [25]. Given this limitation, we wish to perform a first-cut design of a radar to be used for ballistic missile search.

We begin by assuming the search will cover a region of space that extends from 0° to 45° in elevation and 30° in azimuth. Further, we wish to traverse the search sector in 10 s. The targets of interest have an RCS of −10 dBsm and we must achieve an SNR of 13 dB to declare a detection. Current technology for this hypothetical radar supports a noise figure of 4 dB and total losses of 6 dB. We assume an average beam (elevation) angle of 10°, which, from Figure 2.5, gives $T_a' = 15$ K. This, with (2.36) and (2.37), leads to $T_s = 487$ K. We assume $K_{pack} = 1$ for this preliminary design. Table 2.3 summarizes these parameters.

To determine the detection range of the radar, we first solve (2.55) for R:

$$R = \left[\frac{P_A A_e \sigma}{4\pi (SNR) k T_s L\, K_{pack} \Omega} T_{scan}\right]^{1/4} \text{ m} \quad (2.56)$$

We compute Ω from (2.44):

$$\Omega = \Delta\alpha \left(\sin\varepsilon_2 - \sin\varepsilon_1\right)$$
$$= \left(\frac{\pi}{6}\right)\left(\sin\frac{\pi}{4} - \sin 0\right) = 0.118\pi \text{ steradians} \quad (2.57)$$

When we combine this with the values in Table 2.3, we get

$$R = \left[\frac{3\times10^6 \times 0.1}{4\pi (20)\times 1.38\times 10^{-23}\times 487\times 10^{0.6}} \frac{10}{0.118\pi}\right]^{1/4} = 1{,}047 \text{ km} \quad (2.58)$$

which we hope will prove sufficient.

Table 2.3
Search Radar Range Equation Parameters

Parameter	Value
Azimuth search extent	30°
Elevation search extent	0–45°
Power aperture product	3×10^6 W-m²
Search scan time, T_{scan}	10 s
Target RCS, σ	−10 dBsm
Detection SNR	13 dB
Total losses, L	6 dB
Packing factor, K_{pack}	1
System noise temperature, T_s	487 K

---

[5] This example is adapted from lecture notes by Dr. Stephen Gilbert.

We extend the example and establish some additional characteristics for this radar. We start by requiring the radar operate unambiguously in range. The means that we need to choose the PRI, T, to satisfy the following equation:

$$T > \frac{2R}{c} = \frac{2 \times 1.047 \times 10^6}{3 \times 10^8} = 0.007 \text{ s} = 7 \text{ ms} \quad (2.59)$$

We choose $T = 7.5$ ms.

If we devote one PRI per beam, over the course of 10 s we would need to transmit and receive

$$n = \frac{T_{scan}}{T} = \frac{10}{0.0075} = 1,333 \text{ beams} \quad (2.60)$$

By assuming a circular beam, we can use (2.47) to calculate the beamwidth:

$$\theta_A = \theta_B = \sqrt{\frac{\Omega}{K_A n}} = \sqrt{\frac{0.118\pi}{1.65 \times 1,333}} = 0.013 \text{ rad} = 0.74° \quad (2.61)$$

If we operate the radar at a frequency of 1 GHz (L-band), we get a wavelength of

$$\lambda = \frac{c}{f_c} = \frac{3 \times 10^8}{10^9} = 0.3 \text{ m} \quad (2.62)$$

From (2.14) we have

$$G_T = G_R = \frac{25,000}{\theta_B^2} = \frac{25,000}{(0.74)^2} = 45,654 \text{ W/W} = 46.6 \text{ dB} \quad (2.63)$$

Using (2.22), we can write

$$A_e = \frac{G_R \lambda^2}{4\pi} = \frac{45,654(0.3)^2}{4\pi} = 327 \text{ m}^2 \quad (2.64)$$

If we assume an antenna efficiency of 60%, we compute a physical area:

$$A_{physical} = A_e / \rho_{ant} = A_e / 0.6 = 327 / 0.6 = 545 \text{ m}^2 \quad (2.65)$$

Finally, by assuming a circular aperture, we obtain an antenna diameter of

$$D = \sqrt{\frac{4 A_{physical}}{\pi}} = 26.3 \text{ m} \quad (2.66)$$

which is approximately the height of a seven-story building.

Our final calculation yields the peak power of the radar. We assume we want a range resolution of 150 m, which translates to a 1-μs pulsewidth if we use an unmodulated pulse. With the computed PRI of 7.5 ms, we can calculate the duty cycle as

$$d = \tau_p / T = 10^{-6} / 7.5 \times 10^{-3} = 0.0133\% \quad (2.67)$$

From (2.64) and the given average power aperture of 3×10^6 W-m^2, we compute an average power of

$$P_A = \frac{P_A A_e}{A_e} = \frac{3\times10^6}{327} = 9{,}174 \text{ W} \tag{2.68}$$

Combining this result with (2.67) leads us to compute a peak power of

$$P_T = P_A/d = 9{,}174/0.000133 = 69 \text{ MW} \tag{2.69}$$

which is larger than desired.

We can reduce the peak power by using a longer pulse and pulse compression (see Chapter 11). A 100-μs pulse, with pulse compression, would reduce peak power to the more reasonable value of 690 kW.

This completes our preliminary design for a search radar. In practice, this would serve as a starting point for a much more detailed design where we would specifically revisit all of the terms of the radar range equation (not the search radar range equation) with actual hardware constraints.

2.8 RADAR RANGE EQUATION SUMMARY

Table 2.4 and Table 2.5 summarize various equations related to the radar range equation and the search radar range equation.

Table 2.4
Radar Range Equation Summary

Equation Name	Equation
Radar range equation	$SNR = \dfrac{P_T G_T G_R \lambda^2 \sigma \tau_p}{(4\pi)^3 R^4 k T_s L}$
Antenna directivity (G_T, G_R)	$\dfrac{25{,}000}{\theta_A^\circ \theta_B^\circ}$ or $\dfrac{4\pi A_e}{\lambda^2}$ where $A_e = \rho A_{ant}$ and $\rho = 0.6$
Effective isotropic radiated power	$EIRP = \dfrac{P_T G_T}{L_t L_{ant}}$
System noise temperature	$T_s = T_a + (F_n - 1)T_0$
Antenna temperature	$T_a = 0.876 T_a' + 36$ where T_a' is from Figure 2.5

Table 2.5
Search Radar Range Equation Summary

Equation Name	Equation
Search radar range equation	$SNR = \dfrac{P_A A_e \sigma}{4\pi R^4 k T_s L} \dfrac{T_{scan}}{K_{pack} \Omega}$
Average power	$P_A = P_T \tau_p / T, \quad T = \text{PRI}$
Effective aperture	$A_e = \rho_{ant} A_{ant}$
Scan period	T_{scan}: time to cover search volume
Search solid angle	$\Omega = 2\Delta a(\sin\varepsilon_2 - \sin\varepsilon_1)$
	Δa: azimuth extent of search sector
	ε_1: lower elevation limit of search sector
	ε_2: upper elevation limit of search sector

2.9 EXERCISES

1. Derive the equation

$$G = \frac{25{,}000}{\theta_A^\circ \theta_B^\circ} \tag{2.70}$$

from

$$G = \frac{4\pi}{K_A \theta_A^{rad} \theta_B^{rad}} \tag{2.71}$$

In these equations, θ_A° and θ_B° denote beamwidths in degrees and θ_A^{rad} and θ_A^{rad} denote beamwidths in radians. $K_A = 1.65$.

2. A radar has a peak power of 1 MW, combined transmit and antenna losses of 1 dB, and a transmit antenna directivity of 41 dB. The radar is operating in free space so there is nothing to absorb the radiated energy. It uses a pulse with an envelope width of 1 μs.

a) Calculate the total energy on the surface of a (hypothetical) sphere with a radius of 100 km centered on the radar.
b) Repeat part a) for a sphere with a radius of 200 km centered on the radar.
c) Do your answers make sense? Explain.

3. Consider a monostatic radar with the following parameters:

- Peak transmit power at the power amp output—10 kW
- Transmit losses—1 dB
- Antenna losses—1 dB (transmit)
- Antenna losses—1 dB (receive)
- Operating frequency—6 GHz
- PRF—1,000 Hz
- Pulsewidth—100 μs

- Transmit antenna effective aperture—0.58 m²
- Receive antenna beamwidth—1.2° Az × 2.5° El (the radar has separate transmit and receive antennas positioned next to each other)
- Other losses—8 dB
- System noise temperature, T_s—1,155 K

a) Calculate the transmit antenna directivity, in dB.
b) Calculate the effective aperture, in square meters, for the receive antenna, given an antenna efficiency of 60%.
c) Calculate the EIRP for the radar, in dBW.
d) Given a detection threshold of 20 dB, what is the detection range, in km, for a target with a radar cross section of 10 dBsm?
e) Plot *SNR* in dB, versus *R* in km for a target with an RCS of 10 m². Let *R* vary from the minimum usable range to the maximum usable range (see Chapter 1).

4. Consider a monostatic radar that has the following parameters:

- Peak transmit power at power amp output—100 kW
- Transmit and antenna losses—2 dB
- Operating frequency—10 GHz
- PRF—2,000 Hz
- Antenna diameter—1.5 m (circular aperture)
- Antenna efficiency—60%
- Other losses—12 dB
- Noise figure—4 dB
- The radar transmits a 10-µs rectangular pulse.
- The beam elevation angle is in the range of 1° to 5°.

a) Create a table containing all parameters necessary for the radar range equation. Derive those parameters missing explicit values above. List as TBD those parameters with insufficient information for entering a value.
b) Calculate the unambiguous range of the radar.
c) Plot SNR, in dB, versus target range, in km, for a 6-dBsm target. Vary the range from 5 km to the radar's unambiguous range.
d) Given a 13-dB SNR requirement for detection, calculate the detection range, in km, for a 6-dBsm RCS target.
e) What is the maximum detection range, in km, if the minimum SNR required for detection is raised to 20 dB?
f) Calculate the antenna beamwidth, in degrees.

5. A radar generates 200 kW of peak power at the power tube and has 2 dB of loss between the power tube and the antenna. The radar is monostatic with a single antenna that has a directivity of 36 dB and a loss of 1 dB. The radar operates at a frequency of 5 GHz. Determine the EIRP, in dBW, for the radar. Determine the EIRP in watts. Determine the power at the receive antenna output, in dBm, for the following conditions:

a) A 1.5-m² RCS target at a range of 20 km
b) A 20-dBsm target at a range of 100 km

6. How does doubling the range change the powers in Exercise 5? Give your answer in dB. This problem illustrates an important rule of thumb for the radar range equation.

7. A radar with losses of 13 dB and a noise figure of 8 dB must detect targets within a search sector 360° in azimuth and from 0° to 20° in elevation. The radar must cover the search sector in 6 s. The targets of interest have an RCS of 6 dBsm, and the radar requires 20 dB of SNR to declare a detection. The radar must have a detection range of 75 km. Calculate the average power aperture ($P_{avg}A_e$), in W-m^2, required by the radar to satisfy the search requirements above.

8. The radar of Exercise 7 uses an antenna with a fan beam that has beamwidths of 1° in azimuth and 5° in elevation. The radar operates at a frequency of 4 GHz. What average power, in kW, must the radar have? Given an antenna efficiency of 60%, calculate the approximate antenna dimensions, in m. Hint: The relative height and width of the antenna are inversely proportional to the relative beamwidths.

9. Assuming the radar of Exercise 7 uses one PRI per beam, determine the PRI for the radar. Can the radar operate unambiguously in range? Explain.

10. We typically describe the range resolution of a radar as the width of its pulses, if the radar uses unmodulated pulses. What pulse width does the radar of Exercise 7 require for a range resolution of 150 m? What is the peak power of the radar, in MKS units?

11. Derive (2.55).

References

[1]   Barton, D. K., *Radar Equations for Modern Radar*, Norwood, MA: Artech House, 2013.

[2]   Norton, K. A., and A. C. Omberg, "The Maximum Range of a Radar Set," Proc. IRE, vol. 35, no. 1, Jan. 1947, pp. 4–24. First published Feb. 1943 by U.S. Army, Office of Chief Signal Officer in the War Department, in Operational Research Group Report, ORG-P-9-1.

[3]   Barton, D. K., ed., *Radars*, Vol. 2: The Radar Range Equation (Artech Radar Library), Dedham, MA: Artech House, 1974.

[4]   Budge, M. C., Jr., "EE 619: Intro to Radar Systems." www.ece.uah.edu/courses/material/EE619/index.htm.

[5]   Erst, S. J., *Receiving Systems Design*, Dedham, MA: Artech House, 1984.

[6]   Skolnik, M. I., ed., *Radar Handbook*, New York: McGraw-Hill, 1970.

[7]   Skolnik, M. I., *Introduction to Radar Systems*, 3rd ed., New York: McGraw-Hill, 2001.

[8]   Hovanessian, S. A., *Radar System Design and Analysis*, Norwood, MA: Artech House, 1984.

[9]   Stutzman, W. L., "Estimating Directivity and Gain of Antennas," IEEE Antennas Propagat. Mag., vol. 40, no. 4, Aug. 1998, pp. 7–11.

[10] Barton, D. K., *Radar System Analysis and Modeling*, Norwood, MA: Artech House, 2005.

[11] Barton, D. K., *Radar System Analysis*, Englewood Cliffs, NJ: Prentice-Hall, 1964.

[12] Stutzman, W. L., and G. A. Thiele, *Antenna Theory and Design*, 3rd ed., New York: Wiley & Sons, 2013.

[13] Ziemer, R. E., and W. H. Tranter, *Principles of Communications*, 3rd ed., Boston, MA: Houghton Mifflin, 1990.

[14] Johnson, J. B., "Thermal Agitation of Electricity in Conductors," *Phys. Rev.*, vol. 32, Jul. 1928, pp. 97–109.

[15] Nyquist, H., "Thermal Agitation of Electric Charge in Conductors," *Phys. Rev.*, vol. 32, Jul. 1928, pp. 110–113.

[16] Rohde, U. L., J. Whitaker, and T. T. N. Bucher, *Communications Receivers*, 2nd ed., New York: McGraw-Hill, 1997.

[17] Losee, F. A., *RF Systems, Components, and Circuits Handbook*, 2nd ed., Norwood, MA: Artech House, 2005.

[18] IEEE 100, The Authoritative Dictionary of IEEE Standards Terms, 7th ed., New York: IEEE, 2000.

[19] Ridenour, L. N., *Radar System Engineering*, vol. 1 of MIT Radiation Lab. Series, New York: McGraw-Hill, 1947. Reprinted: Norwood, MA: Artech House (CD-ROM edition), 1999, p. 33.

[20] Uhlenbeck, G. E., and J. L. Lawson, *Threshold Signals*, vol. 24 of MIT Radiation Lab. Series, New York: McGraw-Hill, 1950. Reprinted: Norwood, MA: Artech House (CD-ROM edition), 1999, p. 99.

[21] Lathi, B. P., Signals, *Systems and Communication*, New York: Wiley & Sons, 1965, pp. 548–551.

[22] Van Voorhis, S. N., *Microwave Receivers*, vol. 23 of MIT Radiation Lab. Series, New York: McGraw-Hill, 1948. Reprinted: Norwood, MA: Artech House (CD-ROM edition), 1999, p. 4.

[23] Blake, L. V., "A Guide to Basic Pulse-Radar Maximum-Range Calculation, Part 1—Equations, Definitions, and Aids to Calculation," Naval Research Laboratory, Washington, D.C., Rep. No. 6930, Dec. 23, 1969, p. 49. Available from DTIC as 701321.

[24] Curry, G. R., *Radar System Performance Modeling*, Norwood, MA: Artech House, 2001, pp. 117–118.

[25] Mahan, E.R., and E. C. Keefer, eds., *Foreign Relations of the United States*, 1969–1976, vol. 32, SALT I, 1969–1976, Washington, D.C.: U.S. Government Printing Office, 2010, p. 814.

APPENDIX 2A: DERIVATION OF SEARCH SOLID ANGLE EQUATION

We can write the area of the small square in Figure 2A.1 as

$$dA = ds\,dr = \left[R\cos(\varepsilon)\,d\alpha\right](R\,d\varepsilon) \tag{2A.1}$$

or

$$dA = R^2 \cos(\varepsilon)\,d\varepsilon\,d\alpha \tag{2A.2}$$

To get the total area over the angles $[\varepsilon_1, \varepsilon_2]$ and $[\alpha_0 - \Delta\alpha/2,\ \alpha_0 + \Delta\alpha/2]$, we integrate (2A.1) and (2A.2) over these angle ranges. This yields

$$A = \int_{\varepsilon_1}^{\varepsilon_2} \int_{\alpha_0 - \Delta\alpha/2}^{\alpha_0 + \Delta\alpha/2} R^2 \cos(\varepsilon)\,d\alpha\,d\varepsilon \tag{2A.3}$$

Evaluating the integral results in

$$A = R^2 \left[\sin(\varepsilon_2) - \sin(\varepsilon_1) \right] \Delta\alpha \qquad (2A.4)$$

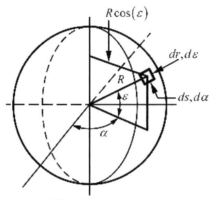

Figure 2A.1 Geometry for computing solid angle.

Dividing by R^2 yields the solid angle as

$$\Omega = A/R^2 = \Delta\alpha \left[\sin(\varepsilon_2) - \sin(\varepsilon_1) \right] \text{ steradians} \qquad (2A.5)$$

Chapter 3

Radar Cross Section

3.1 INTRODUCTION

In this chapter, we discuss radar cross section, or RCS. The concept of scattering of electromagnetic waves by objects (what the concept of RCS attempts to quantify) dates back to 1861, when Alfred Clebsch discussed the topic in a memoir [1]. In 1871, Lord Rayleigh,[1] whose name is often associated with electromagnetic scattering and RCS, published a paper titled "On the Incidence of Aerial and Electric Waves On Small Obstacles" [2].

Although many authors wrote about electromagnetic scattering in the late 1800s and early 1900s, the first mention of RCS did not occur until 1947, when Ridenour introduced it in an article in the MIT Radiation Laboratory Series [3, p. 21]. In that reference, Ridenour provided a definition of RCS as

$$\sigma = 4\pi \frac{\text{power reradiated toward the source per unit solid angle}}{\text{incident power density}} \quad (3.1)$$

The units of the numerator of (3.1) are watts while the units of the denominator are watts/m^2. Thus, the units of RCS are m^2.

Two of the key phrases in the definition of (3.1) are "reradiated" and "toward the source." This says the RCS parameter attempts to capture, in a single number, the ability of the target to capture energy from the radar and reradiate it back toward the radar.

In general, computation of RCS is very complicated. In fact, except for some very simple surfaces, RCS can be only approximately computed. This may explain why there is a large amount of current research in methods to more reliably predict the RCS characteristics of practical targets [4–10].

---
[1] John William Strutt, 3rd Baron Rayleigh.

3.2 RCS OF SIMPLE SHAPES

In general, the RCS of a target depends on its physical size. However, this is not always the case. An example of the case where RCS depends on physical size is a sphere. Specifically, the RCS of a perfectly conducting sphere of radius r is

$$\sigma = \pi r^2 \qquad (3.2)$$

provided $r \gg \lambda$ [1, p. 65].

A case where RCS does not depend on physical size is a cone where the nose of the cone is facing toward the radar, as shown in Figure 3.1. For the case of Figure 3.1, the RCS is given by [11, p. 89]

$$\sigma = \frac{\lambda^2}{16\pi} (\tan \theta)^4 \qquad (3.3)$$

It will be noted that the RCS is proportional to wavelength but is not dependent on the overall size of the cone. If the cone had any other orientation relative to the line of sight (LOS) to the radar (see Figure 3.1 for a definition of LOS), its RCS would depend on the length of the cone and the diameter of the base [12]. Also, if the point of the cone is not perfectly sharp, the RCS will depend on the radius of the nose (see Figure 3.2) [13–15]. In this case, the RCS [14] is the sum of the expressions for a cone [16] and the rounded tip [17]. That is, $\sigma = \sigma_{cone} + \sigma_{tip}$ where

Figure 3.1 Cone geometry.

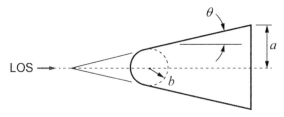

Figure 3.2 Ideal reentry vehicle RCS (blunted cone)—nose-on aspect.

$$\sigma_{cone} = \frac{1}{\pi}\left[\frac{ka\sin(\pi/n)}{n}\right]^2 \left|\frac{1}{\cos(\pi/n)-\cos(3\pi/n)} + \frac{\sin(\pi/n)e^{j(2ka-\pi/4)}}{n(\pi ka)^{1/2}\left[\cos(\pi/n)-\cos(3\pi/2n)\right]^2}\right|^2 \quad (3.4)$$

$$\sigma_{tip} = \pi b^2 \left[1 - \frac{\sin\left[2kb(1-\sin\theta)\right]}{kb\cos^2\theta} + \frac{1+\cos^4\theta}{4(kb)^2\cos^4\theta} - \frac{\cos\left[2kb(1-\sin\theta)\right]}{2(kb)^2\cos^4\theta}\right] \quad (3.5)$$

$k = 2\pi/\lambda$, θ is the cone half angle, $n = 3/2 + a/\pi$, a is the cone base radius, and b is the sphere radius.

In most cases, the RCS is dependent on both the size of the object and the radar wavelength. It also depends on what the object is made of, as metal objects generally have a larger RCS than nonmetallic objects of the same size.

Examples of other simple shapes and their RCSs are contained in Figure 3.3 [11, 12, 18].[2] For the case of the chaff dipole, the given equation for RCS as a function of only wavelength results from the assumption that the length of the chaff dipole is equal to the wavelength and that the dipole is oriented normal to the LOS. If one were to consider all orientations of a chaff dipole, the average RCS would be $\sigma_{avg} = 0.153\lambda^2$, $\sigma_{avg} = 0.166\lambda^2$, and $\sigma_{avg} = 0.184\lambda^2$ for half-wave, full-wave, and 1.5-wave dipoles, respectively [19].

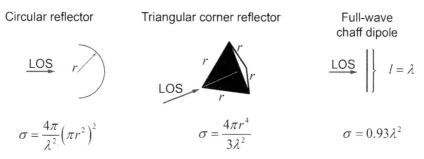

Figure 3.3 RCSs of some simple shapes.

---

[2] Figure 3.3 shows a full wavelength dipole. For a half-wavelength dipole, $\sigma = 0.86\lambda^2$.

A classical plot in RCS theory is shown in Figure 3.4 [3, p. 65]. This figure contains a plot of normalized RCS versus normalized radius for a perfectly conducting sphere. It illustrates that the RCS of an object is generally a complicated function of both the size of the object and the wavelength of the electromagnetic wave that impinges on the object.

The equation for the curve of Figure 3.4 is [20, pp. 35–36]

$$\sigma = \frac{\lambda^2}{\pi}\left|\sum_{n=1}^{\infty}(-1)^n\left(n+\frac{1}{2}\right)\left[\frac{krJ_{n-1}(kr)-nJ_n(kr)}{krH^{(1)}_{n-1}(kr)-nH^{(1)}_n(kr)}-\frac{J_n(kr)}{H^{(1)}_n(kr)}\right]\right|^2$$

$$= \frac{\lambda^2}{\pi}\left|\sum_{n=1}^{\infty}(-1)^n\left(n+\frac{1}{2}\right)(b_n-a_n)\right|^2 \qquad (3.6)$$

where
- r is the radius of the sphere.
- $k = 2\pi/\lambda$ is the wave number.
- $J_n(kr)$ is the spherical Bessel function of the first kind of order n and argument kr.
- $Y_n(kr)$ is the spherical Bessel function of the second kind of order n and argument kr (also called Weber's function).
- $H_n^{(1)}(kr) = J_n(kr) + jY_n(kr)$ is the spherical Bessel function of the third kind of order n and argument kr (also called a Hankel function).

Equation (3.6) is usually referred to as the Mie[3] series and is one of the few tractable RCS equation [21].

If the object size (e.g., radius for a sphere) is less than a wavelength, we say that the object is in the Rayleigh region of the incident electromagnetic wave. In this region, the RCS of the object is a function of the size of the object relative to a wavelength. As an example, the normalized RCS of a perfectly conducting sphere whose radius places it in the Rayleigh region (see Figure 3.4) is given by [22, p. 101]

$$\frac{\sigma_{ray}}{\pi r^2} \approx 9\left(\frac{2\pi r}{\lambda}\right)^4 \quad \lambda > 10r \qquad (3.7)$$

The most common example objects that are in the Rayleigh region for many radars are rain and clouds [23, p. 149; 24, p. 41]. Another example would be insects.

---

[3] After Gustav Adolf Feodor Wilhelm Ludwig Mie [20].

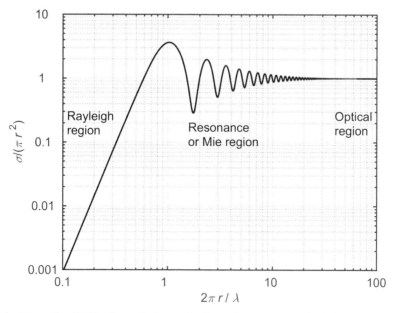

Figure 3.4 Normalized RCS of a perfectly conducting sphere vs. normalized size.

The center region of Figure 3.4 is termed the resonance, or Mie, region. The Mie designation is in honor of Gustav Mie, who first gave the exact equation for the curve of Figure 3.4 [20]. (This equation was later detailed by Stratton [25].) In this region, the object size is on the order of a wavelength and the RCS is transitioning from being dependent on both object size and wavelength to being dependent mainly on object size. As indicated in Figure 3.4, the RCS of a sphere can appear to be larger than dictated by its size. Typical objects that could be in the resonance region would be birds, bullets, artillery shells, some missiles, and very small aircraft, depending on frequency.

The third RCS region is termed the "optical region" and is where most large objects fall. In this region, the object is much larger than a wavelength. Further, the RCS is (or can be) a strong function of the size of the object.

In general, the RCS of an object depends on the orientation of the object relative to the LOS. As an example, the RCS of the flat plate illustrated in Figure 3.5 is given by [22, p. 105; 26, p. 457]

$$\sigma = \frac{(kdw)^2}{\pi} \text{sinc}^2\left[kd\sin(\theta)\cos(\phi)\right]\text{sinc}^2\left[kw\sin(\theta)\sin(\phi)\right]\cos^2(\theta) \quad (3.8)$$
$$k = 2\pi/\lambda, \quad kw \gg 1, \quad kd \gg 1$$

where $\text{sinc}(x) = \sin(\pi x)/(\pi x)$.

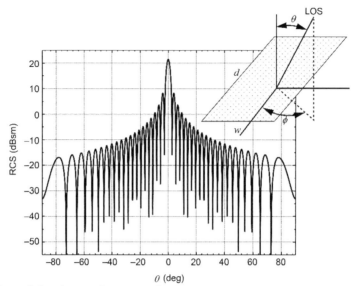

Figure 3.5 RCS of a 1-m² flat plate at a frequency of 1 GHz.

The plot of Figure 3.5 was created for a flat plate with $d = w = 1$ m and $\phi = 0$ and $\lambda = 0.3$ m (L-band). As can be seen, the RCS varies significantly as the angle of the LOS changes. Also note that the peak RCS is significantly larger than the 1-m² area of the plate.

Most targets of interest are not the simple shapes indicated thus far. In fact, targets such as airplanes consist of many different shapes that are in different orientations. Further, as the targets move relative to the radar LOS, the relative orientations of the various shapes change significantly. As a result, a typical plot of target RCS versus orientation relative to the LOS has a very complex appearance.

A classical plot that illustrates this variation of RCS is in Figure 3.6 [3, p. 77]. This figure shows the measured variation in RCS of an AT-11 Kansan [a twin-engine aircraft used during World War II for bombing and gunnery training by the United States Army Air Forces (USAAF)] as a function of azimuth orientation relative to the LOS. As can be seen, the RCS varies by quite a large amount and in a random-looking fashion. If one considers that the orientation of the aircraft will change continually as the aircraft flies toward the radar, the angular variation in RCS will translate to a time variation of RCS that would look random.

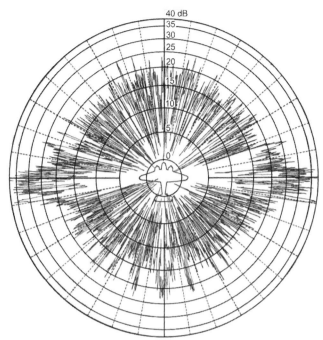

Figure 3.6 Experimental RCS of an AT-11 Kansan. (*Source*: L. N. Ridenour, *Radar System Engineering*, Vol. 1 of MIT Radiation Laboratory Series, 1947. Reprinted with permission.)

3.3 SWERLING RCS MODELS

In an attempt to capture target RCS fluctuation effects in a mathematical model that could be easily used in detection studies, Peter Swerling [27] developed statistical representations of RCS, which are commonly referred to as the *Swerling RCS models*. There are four Swerling models, termed Swerling 1, Swerling 2, Swerling 3, and Swerling 4. Many radar analysts refer to a fifth Swerling model that is termed Swerling 0 or Swerling 5. The fifth Swerling model is defined as a target that has a constant RCS.[4] This Swerling model would be representative of a sphere since the ideal RCS of a sphere is a constant function of orientation angle.

The four Swerling models attempt to represent both statistical and temporal variations of RCS. The statistical properties of Swerling 1 and Swerling 2 RCS variations (which we will refer to Swerling 1 or Swerling 2 *targets*, or SW1 and SW2 targets) are the same and are governed by the density function

$$f(\sigma) = \frac{1}{\sigma_{AV}} e^{-\sigma/\sigma_{AV}} U(\sigma) \tag{3.9}$$

---

[4] If we were to give credit where credit is due, the "Swerling 0/5" model should be termed the Rice model after the analyst [28] who discussed it in terms of single-pulse detection, or the Marcum model after the analyst [29] who first considered it in terms of multiple pulse detection.

where $U(\sigma)$ is the unit step function. Equation (3.9) is the equation for an exponential density function. σ_{AV} is the average RCS of the target and is the value that would be used in the radar range equation.

The statistical properties of SW3 and SW4 targets are also the same and are governed by the density function

$$f(\sigma) = \frac{4\sigma}{\sigma_{AV}^2} e^{-2\sigma/\sigma_{AV}} U(\sigma) \tag{3.10}$$

Again, σ_{AV} is the average RCS of the target.

Equations (3.9) and (3.10) are special cases of the chi-squared density function [3]. Equation (3.9) is a chi-squared density function with two degrees of freedom, and (3.10) is a chi-squared density function with four degrees of freedom. The general, k-degree-of-freedom, chi-squared density function is the density function of the sum of the squares of k, independent, zero-mean, equal variance, Gaussian random variables.

The difference between a SW1 and a SW2 target lies in the difference in the time variation of RCS. The same is true for a SW3 and a SW4 target. With a SW1 or SW3 target, the RCS fluctuates slowly over time and with a SW2 or SW4 target, the RCS fluctuates rapidly over time. In the classical definitions given by Swerling [27], SW1 and SW3 targets maintain a constant RCS during the time the radar illuminates it on a particular scan, but its RCS changes independently (in a random fashion) on a scan-to-scan basis. For SW2 and SW4 targets, the RCS changes independently (and randomly) on a pulse-to-pulse basis.

Scan-to-scan means that the radar "looks" at the target infrequently—on the order of once every several seconds. Pulse-to-pulse means that the radar "looks" at the target every PRI. The phrase scan-to-scan derives from search radar terminology where the radar constantly rotates and "scans" by the target only every few seconds.[5]

3.3.1 Swerling Statistics

Plots of the density functions of (3.9) and (3.10) are shown in Figure 3.7. These plots indicate that the RCS values for SW1 and SW2 targets vary about a value below σ_{AV}, whereas the RCS values for SW3 and SW4 targets are concentrated at values fairly close to the average RCS. This is further illustrated in Figure 3.8, which contains sample plots of RCS versus dimensionless time for SW1/SW2 and SW3/SW4 targets with an average RCS of 1 m² or 0 dBsm. As will be noted, the RCS values for the SW1/SW2 case tend to vary significantly about a value below the average RCS of 0 dBsm, whereas the RCS values for the SW3/SW4 case tend to cluster more tightly around a value slightly below the average RCS of 0 dBsm.

---

[5] As a note, Swerling was considering pulsed, presumably rotating, search radars in his analyses [27].

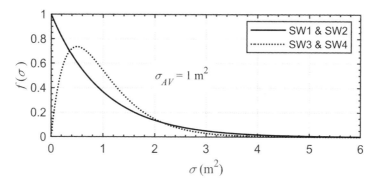

Figure 3.7 Density functions for Swerling RCS models.

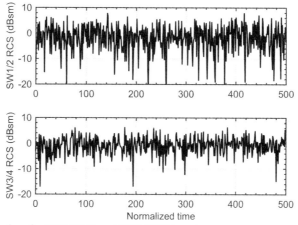

Figure 3.8 RCS vs. time for SW1/SW2 and SW3/SW4 targets.

3.3.2 Swerling Fluctuation Models

As was indicated earlier, the difference between SWodd (SW1, SW3) targets and SWeven (SW2, SW4) targets lies in the rate at which the RCS is assumed to vary. It was stated that the SWodd model assumes that the RCS changes on a scan-to-scan basis. In the search radar range equation discussions of Chapter 2, we referred to a search volume and indicated that the search radar covers the search volume within a certain time we termed T_{scan}. This process of covering the search volume is termed a scan, and T_{scan} is termed the scan time. If we were using a SWodd target model in the search radar analysis, we would assume the RCS changed from scan to scan but stayed constant during the scan. Thus, we would assume that the RCS changed every T_{scan} seconds, but stayed constant over any specific T_{scan} interval. If we were using a SWeven target model, we would assume the target RCS changed every PRI, or every T seconds. Thus, the SWeven RCS models imply rapid RCS fluctuation, whereas the SWodd models imply slow RCS fluctuation. This difference in RCS fluctuation is illustrated notionally in Figure 3.9, which is a plot of RCS versus pulse number, or PRI, for the two cases. For the SW1 model, the RCS changes every 500 pulses and the RCS changes every pulse for the SW2 model.

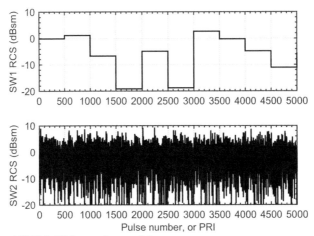

Figure 3.9 SW1 and SW2 RCS fluctuation models.

The concept of SWodd and SWeven represents an idealization that is not achieved in practice. Actual targets exhibit RCS variations that lie somewhere between SWodd and SWeven. How close the fluctuation lies to either model depends on the complexity of the target, the operating frequency of the radar, and the time between RCS observations. As an example, we consider a target that we can model by five spheres, or point sources. In this example, we "fly" the target model toward the radar with a constant x velocity of 75 m/s and a y and z velocity of zero. The center starts at $x = 20,100$ m, $y = 5,000$ m, and $z = 0$ m. We assume all of the scatterers have the same RCS (1 m^2) and compute the composite RCS (the total RCS of all five scatterers) as a function of time.

Figure 3.10 contains plots of composite RCS over a 3-second interval for cases where the carrier frequency is 8.136 GHz (low X-band) and 97.632 GHz (W-band).

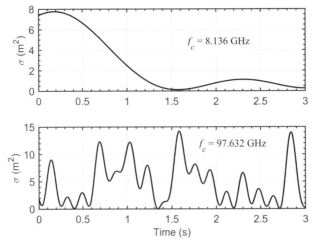

Figure 3.10 Sample RCS variation.

For the X-band case, the RCS remains fairly constant for time periods of tens of milliseconds. However, over periods of seconds, the RCS variation becomes unpredictable (i.e., random). Thus, at X-band, this target exhibits an RCS behavior that is consistent with a SWodd target.

For the W-band case, the RCS variation (with time) is much more rapid so that the RCS varies significantly over time intervals of tens of milliseconds. In this case, it might be appropriate to represent the target with a SWeven model.

In general, RCS variation also depends on target size and speed. The RCS fluctuation of large targets is more rapid than the RCS fluctuation of small targets. Also, the RCS fluctuation of fast targets is more rapid than the fluctuation of slow targets. Demonstration of this is considered in the exercises.

3.3.3 Math Behind the Fluctuation Model

To understand the above relation between RCS variation rate and operating frequency, we need to consider how the signals from the scatterers combine to form the composite signal in the radar. We start by considering the unmodulated pulse we discussed earlier. For this case, we can write the voltage pulse at the transmitter output as

$$v_T(t) = V_T e^{j2\pi f_c t} \text{rect}\left[\frac{t}{\tau_p}\right] \tag{3.11}$$

This voltage is converted to an electric field by the antenna and propagates to the target, which creates another electric field. The electric field created by the target propagates back to the radar, where the antenna converts it to a voltage. If the target is a sphere (a point scatterer), the voltage at the antenna output or some point in the receiver (before the matched filter) can be written as[6]

$$v_R(t) = V_R e^{j2\pi f_c(t-2R/c)} \text{rect}\left[\frac{t-2R/c}{\tau_p}\right] \tag{3.12}$$

In (3.11) and (3.12)

$$V_T \propto \sqrt{P_T} \tag{3.13}$$

and

$$V_R \propto \sqrt{P_S} = \sqrt{\frac{P_T G_T G_R \lambda^2}{(4\pi)^3 R^4 L}\sigma} \tag{3.14}$$

where P_T is the transmit power and P_S is the received signal power. The notation $\propto$ means "proportional to."

If we have N point targets clustered close together, relative to the pulse width, their electric fields, at the radar, will add. Because of this, the total voltage in the radar receiver is

---

[6] We are treating the transmitter, antenna, receiver, and environment as ideal so that the pulse is rectangular when it leaves the antenna and rectangular when it reaches the input to the matched filter.

$$v_R(t) = \sum_{k=1}^{N} V_{Rk} e^{j2\pi f_c(t-2R_k/c)} \text{rect}\left[\frac{t-2R_k/c}{\tau_p}\right] \quad (3.15)$$

where

$$V_{Rk} \propto \sqrt{\frac{P_T G_T G_R \lambda^2}{(4\pi)^3 R_k^4 L} \sigma_k} \quad (3.16)$$

In (3.16), σ_k is the RCS of the kth scatterer (sphere, target) and R_k is the range to that scatterer.

Since we assumed the scatterers are close together, the various R_k are close to the average range to the cluster, R, and we can write

$$v_R(t) = K\left(\sum_{k=1}^{N} \sqrt{\sigma_k} e^{j2\pi f_c(t-2R_k/c)}\right) \text{rect}\left[\frac{t-2R/c}{\tau_p}\right] \quad (3.17)$$

where

$$K \propto \sqrt{\frac{P_T G_T G_R \lambda^2}{(4\pi)^3 R^4 L}} \quad (3.18)$$

With some manipulation, (3.17) becomes [30]

$$v_R(t) = K\left(\sum_{k=1}^{N} \sqrt{\sigma_k} e^{-j4\pi R_k/\lambda}\right) e^{j2\pi f_c t} \text{rect}\left[\frac{t-2R/c}{\tau_p}\right] \quad (3.19)$$

Finally, if we define σ as the net RCS of the N scatterers, we can write

$$v_R(t) = K\sqrt{\sigma} e^{j\phi} e^{j2\pi f_c t} \text{rect}\left[\frac{t-2R/c}{\tau_p}\right] \quad (3.20)$$

where

$$\sigma = \left|\sum_{k=1}^{N} \sqrt{\sigma_k} e^{-j4\pi R_k/\lambda}\right|^2 \quad (3.21)$$

and

$$\phi = \arg\left(\sum_{k=1}^{N} \sqrt{\sigma_k} e^{-j4\pi R_k/\lambda}\right) \quad (3.22)$$

With some thought, it will be observed that σ is a strong function of R_k. Indeed, variations in R_k of $\lambda/2$ can cause the phase of the voltage from the kth scatterer to vary by 2π. Thus, it does not take much relative movement of the scatterers to dramatically affect the value of the sum in (3.21). Also, as the carrier frequency increases, λ decreases and smaller changes in the relative positions of the scatterers can have larger effects on the variations of σ, the total RCS.

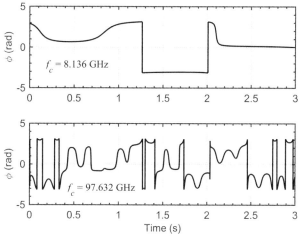

Figure 3.11 Sample signal phase variation.

The above is what led to the difference in RCS variation demonstrated in Figure 3.10. In both cases (top plot and bottom plot), the changes in the relative positions of the scatterers is the same over the 3-second period considered. However, at the lower carrier frequency (top plot), the relative positions change by less than a wavelength over the 3-second period. On the other hand, for the higher frequency, the relative positions change by several wavelengths.

In addition to the changes in net RCS, the variation in the phase of the return signal, as given by (3.22), will exhibit similar differences in temporal behavior. This is illustrated in Figure 3.11. As can be seen, the phase variations are more rapid for the higher carrier frequency than for the lower carrier frequency. We will make use of this property when we discuss how to simulate the various types of Swerling targets.

3.4 RELATION OF SWERLING MODELS TO ACTUAL TARGETS

Our discussions of the Swerling RCS models have thus far been theoretical. To be of use in practical radar problems, we need to attempt to relate the various models to actual targets. One of the standard assumptions is that the SW1/SW2 RCS fluctuation model is associated with complex targets such as aircraft, tanks, ships, and cruise missiles. These would be targets that have a large number of surfaces and joints, all with different orientations. In practice, detection measurements indicate that, indeed, the SW1/SW2 model provides a reasonably good representation of complex targets [31]. Interestingly, in his paper, Swerling has an underlined statement that states, "Most available observational data on aircraft targets indicates agreement with the exponential density..." [27]. His phrase "exponential density" is referring to an equation of the same form as (3.9), ; that is, the SW1/SW2 model.

The standard assumption concerning the SW3/SW4 fluctuation model is that it applies to somewhat simple targets such as bullets, artillery shells, reentry vehicles and the like. According to Swerling, the SW3/SW4 model is consistent with a target that consists of a predominant scatterer and several smaller scatterers, or one large scatterer with small changes in orientation [27]. In terms of application to practical targets, Swerling goes on to say, "More definite statements as to actual targets for which [the SW3/SW4] or the nonfluctuating [SW0/SW5] model apply must await further experimental data."

3.5 SIMULATING SWERLING TARGETS

Analysts and radar testers often have a need to simulate the returns from fluctuating targets. This might occur in simulation when attempting to reconcile the detection performance of radar simulations with predictions based on theory. It can also occur when evaluating the impact of target RCS fluctuations on target acquisition and tracking. In tower testing of actual radars (testing with signals generated from a test tower on a test range or through RF or IF injection in a laboratory environment), the use of fluctuating target returns provides more realistic estimates of detection performance than does the use of constant amplitude target returns.

Because of this perceived need, we present methods of simulating target returns with Swerling-like fluctuation characteristics. The methods make use of the fact that Swerling fluctuation statistics are governed by chi-squared probability density functions. As indicated earlier, the RCS (probability) density functions for SW1 and SW2 targets is a chi-squared density with two degrees of freedom. This means the density results from summing the square of two independent, zero-mean, equal variance, Gaussian random variables. In equation form, if x_1 and x_2 are random variables with the properties just described, then the random variable

$$RCS = \sigma = \frac{1}{2}\left(x_1^2 + x_2^2\right) \qquad (3.23)$$

will be governed by a chi-squared, two-degree-of-freedom density function. This further tells us that if we want to generate random numbers that have statistics consistent with the SW1/SW2 RCS model, we can obtain them by generating two independent, zero-mean, equal variance, Gaussian random numbers, squaring them and taking the average of the squares. The variance of the random numbers should be equal to the average RCS of the target, σ_{AV}. The resulting random variable will be governed by the density function of (3.9).

To simulate a SW2 target, we would create a new random number on every return pulse. This stems from the fact that SW2 RCS values are, by definition, independent from pulse to pulse.

To simulate a SW1 target, we would generate a random number once every group of N pulses and maintain that as the RCS over the N pulses. Here N would be the number of pulses processed by the coherent and/or noncoherent processor (see Chapter 9). The idea of maintaining the RCS constant over the N pulses stems from the definition of SW1 RCS fluctuations, which states that the RCS remains constant during the time the radar beam scans by the target on a particular scan, but changes randomly from scan to scan.

As a note, the phase of the SW2 target also varies randomly from pulse to pulse and the phase of the SW1 target remains constant over the N pulses, but varies randomly from one group of N pulses to the next. We can achieve this phase behavior by defining the phase as

$$\Phi = \tan^{-1}(x_2, x_1) \qquad (3.24)$$

where the $\tan^{-1}$ is the four-quadrant arctangent. An alternate way of thinking about the above is to treat x_1 and x_2 as the real and imaginary parts of a complex number, x, and defining the RCS magnitude as $|x|^2/2$ and the RCS phase as $\arg(x)$ where $\arg(x)$ denotes the angle of the complex number, x.

While the above method of generating SW1 RCS fluctuations is accurate in terms of the SW1 fluctuation model, it can be cumbersome from an implementation perspective and is not representative of the fluctuation of RCS for actual targets. As illustrated in Figures 3.10 and 3.11, RCS tends to fluctuate continuously over time at rates that depend on carrier frequency.

A method of achieving such a temporal characteristic and maintaining the SW1 statistics is to filter the Gaussian random numbers before squaring and adding them. Filtering the random numbers correlates them but does not change their Gaussian statistics.[7] Thus, when the random numbers at the output of the filter are squared and added, the result will be a set of correlated, chi-squared, two degree-of-freedom, random variables that change fairly slowly over time.

A block diagram of the proposed method for generating SW1-like RCS values is shown in Figure 3.12. Sequences of independent, zero-mean, unit variance, Gaussian random numbers are generated and combined into a sequence of complex random numbers. The complex sequence is then filtered by a lowpass filter (LPF). The output of the LPF is then scaled so that the variance of the real and imaginary parts is equal to σ_{AV}. After scaling, the square of the magnitude is computed and divided by two [in compliance with (3.23)] to obtain the RCS. The angle of the complex number is formed to obtain the phase of the voltage that would result when the RCS is used to generate the complex return signal from the target.

In computer simulations, we sometimes implement the filter as an ideal "brick wall" LPF using the Fast Fourier transformer (FFT).[8] We prefer the FFT approach over a recursive filter approach because of the need to consider filter transients in the latter. We use the brick wall LPF because it is easy to implement. The length of the FFT is determined by the number of RCS samples needed in one execution of the simulation.

In field test applications, it would be better to use recursive digital filters to generate the RCS values because the signals must persist over long time periods.

---

[7] This is an interesting property of Gaussian random processes that does not apply to random processes governed by other density functions.
[8] A brick wall response is essentially a rect[x] function. It is unity over a given frequency range and zero elsewhere.

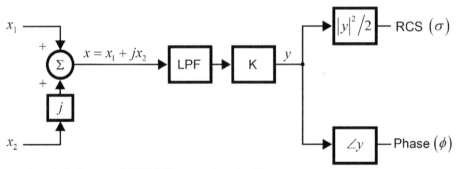

Figure 3.12 Block diagram of SW1 RCS generation algorithm.

To set the filter bandwidth, we need the time between RCS samples. We normally choose this as the radar PRI for testing detection. For tracking studies, we use the track update period or the PRI, depending on whether or not we are modeling the signal processor.

As indicated by Figures 3.10 and 3.11, the bandwidth of the filter should be based on the operating frequency of the radar. If we assume the behavior in Figures 3.10 and 3.11 is representative, we would choose a bandwidth between about 0.5 and 2 Hz for radars operating in the S- to X-band and scale the bandwidth according to frequency and size from there. We would use 0.5 Hz for small, slow targets and 1 to 2 Hz for large, fast targets.

Figures 3.13 and 3.14 contain plots that were generated by the technique of Figure 3.12, where we used an FFT with a brickwall LPF. The filter bandwidth was set to 0.5 Hz for the top plot of the figures and 5 Hz for the bottom plot. As can be seen, the behavior is similar to the five-scatterer example of Figures 3.10 and 3.11.

A suggested RCS generation technique for SW3 and SW4 targets is similar to the method used for SW1 and SW2 targets, except that the RCS is based on the sum of four Gaussian terms instead of two. This is because SW3 and SW4 RCS fluctuations are governed by a chi-squared density with four degrees of freedom. In equation form,

$$RCS = \sigma = \frac{1}{4}\left(x_1^2 + x_2^2 + x_3^2 + x_4^2\right) \qquad (3.25)$$

To simulate a SW4 target, we would create a new random number on every return pulse. To simulate a SW3 target, we would generate a random number once every group of N pulses and maintain that as the RCS over the N pulses.

It is not clear how the phase should be modeled for this case. One approach would be to use (3.24). An alternative might be to use

$$\Phi = \frac{1}{2}\left[\tan^{-1}(x_2, x_1) + \tan^{-1}(x_4, x_3)\right] \qquad (3.26)$$

That is, average the phase from two complex numbers represented by $x_a = x_1 + jx_2$ and $x_b = x_3 + jx_4$.

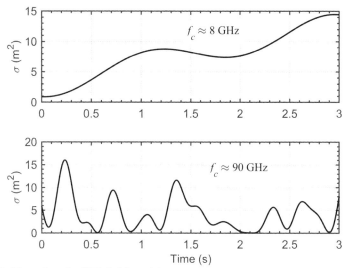

Figure 3.13 RCS vs. time for SW1 RCS model.

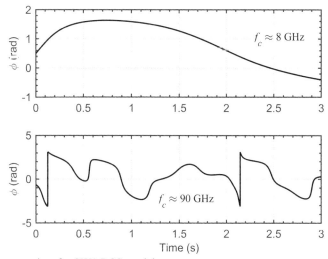

Figure 3.14 Phase vs. time for SW1 RCS model.

An extension for the SW3 case would be to use an extension of the filter method suggested for SW1 targets. A block diagram of this method is shown in Figure 3.15. As can be seen, the method uses two of the SW1 generators and then averages the outputs of the magnitude square and angle computation blocks.

Figures 3.16 and 3.17 contain plots of RCS and phase generated by the model of Figure 3.15. It is interesting that the RCS variations of Figure 3.16 appear to be smaller than those of Figure 3.13 and tend to be closer to the average RCS of 5 m². This is consistent with the expected difference in RCS behavior between SW1 and SW3 targets.

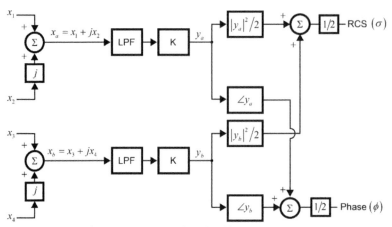

Figure 3.15 Block diagram of SW3 RCS generation algorithm.

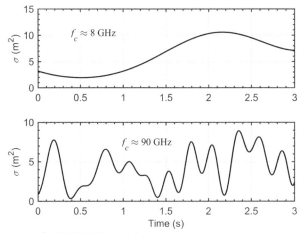

Figure 3.16 RCS vs. time for SW3 RCS model.

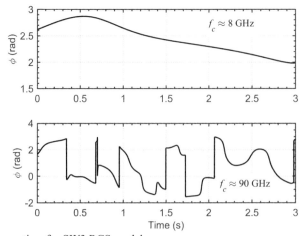

Figure 3.17 Phase vs. time for SW3 RCS model.

3.6 FREQUENCY AGILITY AND SW2 OR SW4 TARGETS

In Chapter 9, we show that if targets exhibit SW2 or SW4 fluctuation statistics, noncoherent integration can provide a significant increase in detection probability relative to that which can be obtained with a single pulse. As an example, the single-pulse Signal-to-Noise ratio (SNR) required to provide a detection probability of 0.9 with a false alarm probability of 10^{-6} on a SW2 target is 21 dB. If the radar noncoherently integrates 10 pulses, the single-pulse SNR required to achieve the same detection and false alarm probabilities is reduced by 14 dB, to 7 dB. This leads us to consider whether there is anything that can be done in the radar to change the fluctuation statistics from SW1/SW3 to SW2/SW4. One way is to change the operating frequency on a pulse-to-pulse basis. The question is: How large of a frequency change is needed?

To be rigorously applicable to target detection theory, a statistical approach is used here to address this problem. The results derived here indicate that for a target with a length L, the frequency separation required for target returns at the two frequencies to be statistically uncorrelated is given by

$$\Delta f \geq \frac{c}{2L} \tag{3.27}$$

where c is the speed of light.

As a note, the fact that the target returns are statistically uncorrelated does not imply that they are statistically independent. However, this is the standard assumption.

We assume the target consists of N scatterers distributed across range in some fashion. In particular, we assume the ranges to the scatterers are random and that all of the ranges are governed by the same density function. We assume each scatterer has a different RCS and that the RCSs are random, mutually independent, and independent of the ranges.

Let the target be illuminated by a pulse whose compressed pulsewidth is larger than the target extent, L, in range. With this we can write the peak of the complex voltage at the output of the matched filter as

$$v(f_c) = \sum_{k=1}^{N} V_k e^{-j4\pi f_c R_k / c} \tag{3.28}$$

where f_c is the carrier frequency, V_k is the complex voltage associated with the kth scatterer, and R_k is the (slant) range to the kth scatterer. The magnitude of V_k is related to the RCS through the radar range equation, and it is assumed that the phase of V_k is a random variable uniformly distributed over 2π. The phases are assumed to be mutually independent and independent of the ranges and RCSs. With this the V_k are independent, complex, random variables. The V_k are also independent of the R_k.

We will be interested in deriving the correlation coefficient between $v(f_c)$ and $v(f_c + \Delta f)$. We assume that Δf is small relative to f_c (tens of MHz versus GHz). We use this assumption so we can further assume V_k is the same at both frequencies. We write $v(f_c + \Delta f)$ as

$$v(f_c + \Delta f) = \sum_{k=1}^{N} V_k e^{-j4\pi(f_c+\Delta f)R_k/c} \tag{3.29}$$

The correlation coefficient between $v(f_c)$ and $v(f_c + \Delta f)$ can be written as

$$r(\Delta f) = \frac{C(\Delta f)}{\sqrt{\sigma^2(f_c)\sigma^2(f_c + \Delta f)}} \tag{3.30}$$

where $C(\Delta f)$ is the covariance between $v(f_c)$ and $v(f_c + \Delta f)$, $\sigma^2(f_c)$ is the variance on $v(f_c)$, and $\sigma^2(f_c + \Delta f)$ is the variance on $v(f_c + \Delta f)$.

To ease the computation of the three elements of $r(\Delta f)$, we show that the means of $v(f_c)$ and $v(f_c + \Delta f)$ are zero. We write

$$E\{v(f)\} = E\left\{\sum_{k=1}^{N} V_k e^{-j4\pi f_c R_k/c}\right\} = \sum_{k=1}^{N} E\{V_k\} E\{e^{-j4\pi f_c R_k/c}\} \tag{3.31}$$

by virtue of the fact that the phase of V_k is uniform on $[0, 2\pi]$, $E\{V_k\} = 0$, and thus $E\{v(f)\} = 0$.

Because of (3.31) we can write

$$\sigma^2(f) = E\{|v(f)|^2\} = \sum_{k=1}^{N}\sum_{l=1}^{N} E\{V_k V_l^*\} E\{e^{j4\pi f(R_l - R_k)/c}\} \tag{3.32}$$

Recognizing that $e^{j4\pi f(R_l - R_k)/c} = 1$ when $l = k$, (3.32) can be rewritten as

$$\begin{aligned}\sigma^2(f) &= \sum_{k=1}^{N} E\{|V_k|^2\} + \sum_{k=1}^{N}\sum_{l=1, l\neq k}^{N} E\{V_k V_l^*\} E\{e^{j4\pi f(R_l - R_k)/c}\} \\ &= \sum_{k=1}^{N} P_k\end{aligned} \tag{3.33}$$

where P_k is the average power due to scatterer k, which is derived from the radar range equation using the average RCS of scatterer k. The double sum of (3.33) is zero by virtue of the fact that

$$E\{V_k V_l^*\} = E\{V_k\} E\{V_l^*\} = 0, \ l \neq k \tag{3.34}$$

since V_k and V_l are independent for $l \neq k$, and $E\{V_k\} = E\{V^*_l\} = 0$.

We can write the covariance as

$$\begin{aligned}C(\Delta f) &= E\{v(f_c) v^*(f_c + \Delta f)\} \\ &= \sum_{k=1}^{N}\sum_{l=1}^{N} E\{V_k V_l^*\} E\{e^{j4\pi f_c(R_l - R_k)/c}\} E\{e^{j4\pi \Delta f R_l/c}\}\end{aligned} \tag{3.35}$$

From above, we recognize that $E\{V_k V^*_l\} = 0$ for $l \neq k$, $E\{V_k V^*_l\} = P_k$ for $l = k$ and $E\{e^{j4\pi f_c(R_l - R_k)/c}\} = 1$ for $l = k$. With this (3.35) reduces to

$$C(\Delta f) = \sum_{k=1}^{N} P_k E\{e^{j4\pi \Delta f R_k/c}\} \tag{3.36}$$

Or, recognizing that $E\{e^{j4\pi \Delta f R_k/c}\}$ is the same for all k, because the R_k have the same density function,

$$C(\Delta f) = r(\Delta f) \sum_{k=1}^{N} P_k = r(\Delta f) \sigma^2(f) \tag{3.37}$$

Substituting (3.37) and (3.33) into (3.30) yields

$$r(\Delta f) = E\{e^{j4\pi \Delta f R_k/c}\} \tag{3.38}$$

which we can use to discuss the frequency separation, Δf, required to convert a SWodd RCS to a SWeven RCS.

For the case where the target scatterers are uniformly distributed over some $R \pm L/2$, we can compute a specific function for $r(\Delta f)$. Specifically, if the R_k are governed by the density

$$f(R_k) = \frac{1}{L} \text{rect}\left[\frac{R_k - R}{L}\right] \tag{3.39}$$

where

$$\text{rect}[x] = \begin{cases} 1 & |x| \leq 1/2 \\ 0 & |x| > 1/2 \end{cases} \tag{3.40}$$

we get

$$r(\Delta f) = E\{e^{j4\pi \Delta f R_k/c}\} = \int_{-\infty}^{\infty} e^{j4\pi \Delta f R_k/c} f(R_k) dR_k = \int_{R-L/2}^{R-L/2} \frac{e^{j4\pi \Delta f R_k/c}}{L} dR_k \tag{3.41}$$

$$= \text{sinc}(2\Delta f L/c) e^{j4\pi \Delta f R/c}$$

If we say that $v(f_c)$ and $v(f_c + \Delta f)$ become uncorrelated for all Δf greater than the Δf where the sinc function first goes to zero, then $v(f_c)$ and $v(f_c + \Delta f)$ become uncorrelated for

$$|\Delta f| \geq \frac{c}{2L} \tag{3.42}$$

This is the same as (3.27). Thus, if the frequency separation on two pulses is greater than the speed of light divided by twice the target length, the target will transition from having an approximate SWodd RCS fluctuation to an approximate SWeven RCS fluctuation.

As another example, we consider the case where the R_k obey a Gaussian distribution. That is,

$$f(R_k) = \frac{1}{\sigma_L \sqrt{2\pi}} e^{(R_k - R)^2 / \sigma_L^2} \qquad (3.43)$$

In this case, $r(\Delta f)$ becomes

$$r(\Delta f) = e^{-(4\pi \Delta f \sigma_L)^2 / 2c^2} e^{j 4\pi R \Delta f / c} \qquad (3.44)$$

If we let $\sigma_L = L/2$, 95.5 % of the scatterers will lie between $\pm L/2$. At $\Delta f = c/2L$, $r(\Delta f) = 0.007$. Thus, we can say that the returns derived from carrier frequencies separated by this Δf are uncorrelated, and the RCS fluctuation will transition from an approximate SWodd characteristic to an approximate SWeven characteristic.

The results presented herein indicate that Δf does not need to be large to cause target RCS fluctuation to transition from an approximate SWodd characteristic to an approximate SWeven characteristic. For example, a target with a range extent of 15 m requires a Δf of only 10 MHz from pulse to pulse. For larger aircraft such as a Boeing 747, which is about 71 m long, only a 2.1-MHz frequency change from pulse to pulse is needed. These examples used the assumption that the scatterers were distributed across the length of the target. In practice, it is likely that this will not be the case. Instead, it is likely that the scatterers will be grouped along different parts of the target (e.g., near the nose, near the wings, and near the tail for aircraft). Because of this, the lengths of the groups of scatterers will be smaller than the length of the aircraft. This means that the frequency changes indicated above are most likely low. More reasonable values may be in the range of tens of megahertz.

3.7 EXERCISES

1. A classic example in RCS discussions is termed the two-scatterer problem. In this exercise, we seek to find the composite RCS of two equal-size scatterers separated by a distance of $2d$. The geometry for this exercise is shown in Figure 3.18.

   Show that the composite RCS is given by

   $$\sigma = 4\sigma_0 \cos^2 \left[\frac{4\pi d \cos(\theta)}{\lambda} \right] \qquad (3.45)$$

   where σ_0 is the RCS of each scatterer. Generate plots of σ versus θ for $d/\lambda = 1$ and $d/\lambda = 3$ with $\sigma_0 = 1$ m^2. These plots will demonstrate that the degree to which σ varies as a function of θ depends on the separation of the scatterers relative to the radar wavelength.

2. Repeat the example of Section 3.3.2 for the case of 10 scatterers randomly located in a square with x and y dimensions of 10 m. Assume all of the scatterers have an equal RCS of 1 m^2.

3. Repeat Exercise 2 with an x velocity of vx = –150 m/s. Use the same 10 scatterers and initial conditions as in Exercise 2. How do the results of Exercises 2 and 3 compare?

4. Repeat Exercise 2 with all scatterer positons multiplied by two. This will produce a target with the same scatterer arrangement as in Exercise 2, but the target will be twice as large as the target of Exercise 2. Keep all other parameters the same as in Exercise 2. How do the results of Exercises 2 and 4 compare?

5. Implement a SW1-like RCS model as discussed in Section 3.4 and generate curves like Figures 3.13 and 3.14.

6. Repeat Exercise 5 for a SW3-like RCS model and generate curves like Figures 3.16 and 3.17.

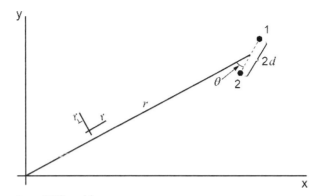

Figure 3.18   Two-scatterer RCS problem.

References

[1] Logan, N. A., "Survey of Some Early Studies of the Scattering of Plane Waves by a Sphere," *Proc. IEEE*, vol. 53, no. 8, 1965, pp. 773–785.

[2] Strutt, J. W. (3rd Baron Rayleigh), "On the Incidence of Aerial and Electric Waves Upon Small Obstacles in the Form of Ellipsoids or Elliptic Cylinders and on the Passage of Electric Waves Through a Circular Aperture in a Conducting Screen," *Philosophical Mag.*, vol. 44, Jul. 1897, pp. 28–52.

[3] Ridenour, L. N., *Radar System Engineering*, vol. 1 of MIT Radiation Lab. Series, New York: McGraw-Hill, 1947; Norwood, MA: Artech House (CD-ROM edition), 1999.

[4] Jernejcic, R. O., A. J. Terzuoli, Jr., and R. F. Schindel, "Electromagnetic Backscatter Predictions Using XPATCH," in *IEEE/APS Int. Symp. Dig.*, vol. 1, Seattle, WA, Jun. 20–24, 1994, pp. 602–605.

[5] Song, J. M., et al., "Fast Illinois Solver Code (FISC)," *IEEE Antennas Propagat. Mag.*, vol. 40, no. 3, Jun. 1998, pp. 27–34.

[6] Hastriter, M. L., and W. C. Chew, "Comparing Xpatch, FISC, and ScaleME Using a Cone-Cylinder," *IEEE Antennas Propag. Soc. Int. Symp.*, vol. 2, Monterey, CA, Jun. 20–25, 2004, pp. 2007–2010.

[7] Emerson, W. H., and H. B. Sefton, Jr., "An Improved Design for Indoor Ranges," *Proc. IEEE*, vol. 53, no. 8, Aug. 1965, pp. 1079–1081.

[8] Shields, M., "The Compact RCS/Antenna Range at MIT Lincoln Laboratory," *3rd European Conf. Antennas Propag., 2009 (EuCAP 2009)*, Berlin, Mar. 23–27, 2009, pp. 939–943.

[9]   Zhang, L. et al., "High-Resolution RCS Measurement Inside an Anechoic Chamber," *Int. Forum Inf. Technol. Applicat. (IFITA) 2010,* vol. 3, Washington, D.C., Jul. 16–18, 2010, pp. 252–255.

[10] Sevgi, L., Z. Rafiq, and I. Majid, "Testing Ourselves: Radar Cross Section (RCS) Measurements," *IEEE Antennas Propagat. Mag.,* vol. 55, no. 6, Dec. 2013, pp. 277–291.

[11] Barton, D. K., *Radar System Analysis and Modeling,* Norwood, MA: Artech House, 2005.

[12] Ruck, G. T., et al., eds., *Radar Cross Section Handbook,* New York: Plenum Press, 1970.

[13] Weiner, S., and S. Borison, "Radar Scattering from Blunted Cone Tips," *IEEE Trans. Antennas Propag.,* vol. 14, no. 6, Nov. 1966, pp. 774–781.

[14] Blore, W. E., "The Radar Cross Section of Spherically Blunted 8° Right-Circular Cones," *IEEE Trans. Antennas Propag.,* vol. 21, no. 2, Mar. 1973, pp. 252–253.

[15] Roscoe, B. J., and Banas, J. F., "Cross Section of Blunted Cones," *Proc. IEEE,* vol. 61, no. 11, Nov. 1973, p. 1646.

[16] Bechtel, M. E., "Application of Geometric Diffraction Theory to Scattering from Cones and Disks," *Proc. IEEE,* vol. 53, no. 8, Aug. 1965, pp. 877–882.

[17] Stuart, W. D., "Cones, Rings, and Wedges," in *Radar Cross Section Handbook,* vol. 1 (G. Ruck et al., eds.), New York: Plenum Press, 1970.

[18] Mack, C. L., Jr., and B. Reiffen, "RF Characteristics of Thin Dipoles," *Proc. IEEE,* vol. 52, no. 5, May 1964, pp. 533–542.

[19] Peebles, P. Z., Jr., "Bistatic Radar Cross Sections of Chaff," *IEEE Trans. Aerosp. Electron. Syst.,* vol. 20, no. 2, Mar. 1984, pp. 128–140.

[20] Mie, G., "Beiträge zur Optik trüber Medien, spezeill kolloidaler Metallösungen," *Annalen der Physik, Seris IV,* vol. 25, no. 3, 1908, pp. 377–445.

[21] Currie, N. C., R. D. Hayes, and R. N. Trebits, *Millimeter-Wave Radar Clutter,* Norwood, MA: Artech House, 1992.

[22] Blake, L. V., *Radar Range-Performance Analysis,* Norwood, MA: Artech House, 1986.

[23] Nathanson, F. E., J. P. Reilly, and M. N. Cohen, eds, *Radar Design Principles,* 2nd ed., New York: McGraw-Hill, 1991.

[24] Skolnik, M. I., *Introduction to Radar Systems,* 3rd ed., New York: McGraw-Hill, 1962.

[25] Stratton, J. A., *Electromagnetic Theory,* New York: McGraw-Hill, 1941.

[26] Kerr, D. E., *Propagation of Short Radio Waves,* vol. 13 of MIT Radiation Lab. Series, New York: McGraw-Hill, 1951; Norwood, MA: Artech House (CD-ROM edition), 1999.

[27] Swerling, P., "Probability of Detection for Fluctuating Targets," RAND Corp., Santa Monica, CA, Res. Memo. RM-1217, Mar. 17, 1954. Reprinted: *IRE Trans. Inf. Theory,* vol. 6, no. 2, Apr. 1960, pp. 269–308.

[28] Rice, S. O., "Mathematical Analysis of Random Noise," *Bell Syst. Tech. J.,* vol. 23, no. 3, Jul. 1944, pp. 282–332; vol. 24, no. 1, Jan. 1945, pp. 461–556.

[29] Marcum, J. I., "A Statistical Theory of Target Detection by Pulsed Radar," RAND Corp., Santa Monica, CA, Res. Memo. RM-754-PR, Dec. 1, 1947. Reprinted: *IRE Trans. Inf. Theory,* vol. 6, no. 2, Apr. 1960, pp. 59–144. Reprinted: *Detection and Estimation* (S. S. Haykin, ed.), Halstad Press, 1976, pp. 57–121.

[30] Budge, M. C., Jr., "EE 619: Intro to Radar Systems," www.ece.uah.edu/courses/material/EE619/index.htm.

[31] Wilson, J. D., "Probability of Detecting Aircraft Targets," *IEEE Trans. Aerosp. Electron. Syst.,* vol. 8, no. 6, Nov. 1972, pp. 757–76.

Chapter 4

Noise

4.1 INTRODUCTION

In this chapter, we discuss the noise, noise temperature, and noise figure terms of the radar range equation. We start with the basic definition of noise as it applies to radar theory and then progress to the topics of noise temperature and noise figure.

The type of noise of interest in radar theory is termed thermal noise or Johnson noise and is generated by the random motion of charges in conductors. John Bertrand Johnson and Harry Theodor Nyquist discovered this type of noise in 1927 [1, 2]. Johnson observed the noise in experiments and Nyquist developed a theoretical basis for Johnson's measurements. Their papers do not make clear whether Nyquist developed the theory to support Johnson's observations or Johnson performed the experiments to verify Nyquist's theory. We suspect a somewhat collaborative effort, given the dates of the papers.

One of the equations in Nyquist's paper defines the mean-square voltage appearing across the terminals of a resistor of R ohms at a temperature T kelvin, in a (differential) frequency band dv hertz wide, as

$$E^2 dv = 4RkTdv \ \text{V}^2 \tag{4.1}$$

where $k = 1.38 \times 10^{-23}$ W-s/K is Boltzmann's constant. Johnson had a similar equation, but for mean-square current.

We retained Nyquist's notation in (4.1); however, from here on, we will adopt a more common notation. To that end, we denote the noise voltage generated by a resistor in a differential frequency interval, df, as $v(t)$. We stipulate $v(t)$ is a zero-mean, wide-sense stationary, real random process with a mean-square value and variance of [1]

$$\sigma_v^2 = E\left\{v^2(t)\right\} = 4RkTdf \ \text{V}^2 \tag{4.2}$$

---

[1] As a note, $E\{x\}$ denotes the expected, or mean, value of x. The E in (4.1) is volts-squared per Hertz. As indicated, we kept the E in (4.1) to be consistent with Nyquist's original paper. Also, in the rest of this chapter we also use E to denote energy, or power spectral density. E used alone is energy (or power spectral density), the notation $E\{\bullet\}$ denotes expected value, or mean.

Since $v(t)$ is zero-mean, its mean-square value equals its variance.

The stipulation of zero-mean says the noise voltage does not have a direct current (DC) component, which is reasonable since such a component would have been noted by Johnson. The stipulation of wide-sense stationary implies the mean and variance are constant. This is reasonable since we already stipulated a mean of zero and we expect the noise power (i.e., mean-square value, variance) to be constant over any time period of interest to us.

Nyquist showed that the noise energy term of (4.2), kT, is a limiting case of one form of the more general Planck's law, which is

$$E = \frac{hf}{\exp(hf/kT) - 1} \quad \text{W-s} \tag{4.3}$$

where E denotes energy in W-s, or power spectral density in W/Hz (the two associations are equal)[2]. $h = 6.6254 \times 10^{-34}$ W-s^2 is the Planck constant and f is frequency, in Hz.

As $f \to 0$, (4.3) degenerates to

$$E = kT \quad \text{W-s} \tag{4.4}$$

An implication of (4.4) is that E is constant over frequencies applicable to most radars (see Exercise 10). This further implies that σ_v^2 is independent of frequency, unless R is a function of frequency (see footnote 2).

This background prepares us to consider noise energy in a radar receiver. However, we will first discuss how σ_v^2 translates to power and energy delivered to a load.

4.2 NOISE IN RESISTIVE NETWORKS

4.2.1 Thevenin Equivalent Circuit of a Noisy Resistor

Figure 4.1 shows the Thevenin equivalent circuit of a noisy resistor. It consists of a noise source with a voltage characterized by (4.2) and a noiseless resistor with a value of R.

If we connect the noisy resistor to a noiseless resistor, R_L, we can find the power delivered to R_L by the noisy resistor using the equivalent circuit of Figure 4.1 to compute the voltage across R_L. We then use this voltage to find the power delivered to R_L. Figure 4.2 shows the resulting circuit.

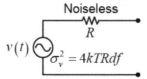

Figure 4.1 Thevenin equivalent circuit of a noisy resistor.

---

[2] A caveat is necessary here. Energy and power spectral density are the same, *if the signal (noise in this case) spectrum is flat*. If the power spectrum varies as a function of frequency, it is more appropriate to use $S(f)$ for the power spectral density.

Noise 71

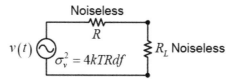

Figure 4.2 Diagram for computing the power delivered to a load.

The voltage across R_L is given by

$$v_{R_L}(t) = v(t) \frac{R_L}{R_L + R} \text{ V} \tag{4.5}$$

Using (4.2), the power delivered to R_L in a differential bandwidth, df, is

$$P_L = \frac{E\{v_{R_L}^2(t)\}}{R_L} = \frac{E\{v^2(t)\} R_L}{(R_L + R)^2} = \frac{4kTRR_L df}{(R_L + R)^2} \text{ W} \tag{4.6}$$

If the load is matched to the source resistance (i.e., if $R_L = R$), we have

$$P_L = \frac{4kTR^2 df}{(2R)^2} = kTdf \text{ W} \tag{4.7}$$

If we divide P_L by df, we obtain the energy delivered to the load as

$$E_L = kT \tag{4.8}$$

which is the form given by (4.4), and the form used in the radar range equation.

4.2.2 Multiple Noisy Resistors

If we have a network consisting of multiple noisy resistors, we can find its Thevenin equivalent circuit by using superposition. To see this, consider the example of Figure 4.3. The left schematic of the figure shows two parallel noisy resistors, and the center schematic shows their equivalent circuits based on Figure 4.1. The right schematic shows the overall Thevenin equivalent circuit for the pair of resistors.

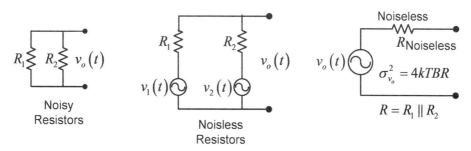

Figure 4.3 Schematic diagrams for the two-resistor problem.

To find $v_o(t)$, we first consider one voltage source at a time and short all other sources. Thus, with only source $v_1(t)$, we get

$$v_{o1}(t) = v_1(t) \frac{R_2}{R_1 + R_2} \text{ V} \tag{4.9}$$

and with only source $v_2(t)$, we get

$$v_{o2}(t) = v_2(t) \frac{R_1}{R_1 + R_2} \text{ V} \tag{4.10}$$

By superposition, we have

$$v_o(t) = v_{o1}(t) + v_{o2}(t) = v_1(t) \frac{R_2}{R_1 + R_2} + v_2(t) \frac{R_1}{R_1 + R_2} \text{ V} \tag{4.11}$$

To get the equivalent resistance, we short both voltage sources of the center figure and find the equivalent resistance across the terminals. When we short the sources, we note that R_1 and R_2 are in parallel, which allows us to compute the equivalent resistance as

$$R = R_1 \parallel R_2 = \frac{R_1 R_2}{R_1 + R_2} \text{ } \Omega \tag{4.12}$$

We next need to compute the mean-square value of $v_o(t)$. To facilitate this, we must further stipulate that the noise voltages generated by the noisy resistors are independent. We justify this restriction by rationalizing that the random motion of charges in one resistor should be independent of the random motion of charges in any other resistor. With this restriction, we are able to say

$$E\{v_1(t) v_2(t)\} = E\{v_1(t)\} E\{v_2(t)\} = 0 \tag{4.13}$$

where the last equality is because $v_1(t)$ and $v_2(t)$ are zero-mean. With this and some algebraic manipulation, we have

$$\sigma_{v_o}^2 = \sigma_{v_1}^2 \frac{R_2^2}{(R_1 + R_2)^2} + \sigma_{v_2}^2 \frac{R_1^2}{(R_1 + R_2)^2} = 4kTRdf \text{ V}^2 \tag{4.14}$$

where we have made use of (4.12), $\sigma_{v_1}^2 = 4kTR_1 df$, and $\sigma_{v_2}^2 = 4kTR_2 df$. The details of (4.14) are left as an exercise.

4.3 EQUIVALENT/EFFECTIVE NOISE TEMPERATURE FOR ACTIVE DEVICES

For passive devices, such as resistive attenuators, it is possible to find the noise energy delivered to a load by extending the technique used in the above example. For active devices, this is not possible. Measurement provides the only method for determining the noise energy an active device delivers to a load.

In general, the noise energy delivered to the load depends on the input noise energy to the device and the internally generated noise. The standard method of representing this is to write the noise energy delivered to the load as the sum of the amplified input noise and the noise generated internally by the active device [3, 4]:[3]

$$E_{n\_out} = GE_{n\_in} + E_{n\_int} = GkT_a + GkT_e \qquad (4.15)$$

where

- G denotes the gain of the device.
- kT_a denotes the input, or source, noise energy.
- T_a denotes the noise temperature of the source.
- GkT_e denotes the noise energy generated by the device.
- T_e denotes the *equivalent/effective noise temperature* of the device, referred to the input of the device.[4]

In (4.15), the term GE_{n_in} represents the portion of the output noise energy due only to the noise into the device. This component of the output noise is the input noise amplified by the gain of the device. The term GkT_e represents the energy of the noise generated by the device. Its form is chosen to be consistent with the standard kT representation discussed above. Including G in this term is a convenience that allows us to write a consistent expression for the noise energy as

$$E_{n\_out} = GE_{n\_in} + E_{n\_int} = GkT_a + GkT_e = Gk(T_a + T_e) = GkT_s \qquad (4.16)$$

In (4.16), T_s denotes the noise temperature, or combined noise temperature, of the device. It is the combined temperature of the noise source and the equivalent/effective noise temperature of the device. We termed this the *system noise temperature* in Chapter 2. For a radar, T_a represents the temperature of the noise entering the antenna from the environment. The value of T_a ranges from tens of kelvin when the antenna beam points at clear sky, to thousands of kelvin when the beam points at the sun [5, p. 208].

For resistors, T_e is the actual temperature of the resistors. For active devices, it is not an actual temperature, but the temperature necessary for a resistor to produce the same noise energy as the active device—thus the origin of the words *equivalent* or *effective*.

4.4 NOISE FIGURE

4.4.1 Derivation of Noise Figure

An alternative to using equivalent/effective noise temperature is to use *noise figure*. Harald Trap Friis formalized the early research on noise figure in a 1944 paper [6] that defined noise figure as the ratio of the SNR at the input of the device to the SNR at the output of the device. In equation form,

---

[3] Until now, we have been diligent in providing units for the various quantities we use. Henceforth, we will assume the reader is familiar with the appropriate units and no longer include them in equations, unless necessary to avoid confusion.
[4] We use the terminology "equivalent/effective noise temperature" because equivalent noise temperature and effective noise temperature are used interchangeably in the literature.

$$F_{n\_Friss} = \frac{SNR_{in}}{SNR_{out}} = \frac{P_{s\_in}/P_{n\_in}}{P_{s\_out}/P_{n\_out}} \qquad (4.17)$$

where P_{s_in} denotes the signal power into the device; P_{n_in} denotes the noise power into the device; P_{s_out} denotes the signal power out of the device; and P_{n_out} denotes the noise power out of the device.

In this book, we use the IEEE definition [7]. An interpretation of that definition is: noise figure is the noise energy delivered to a load by the actual device divided by the noise energy delivered to the load by an ideal device with the same gain. In equation form

$$F_n = \frac{E_{n\_out\_actual}}{E_{n\_out\_ideal}} \qquad (4.18)$$

where

$$E_{n\_out\_actual} = GE_{n\_in} + E_{n\_int} \qquad (4.19)$$

and

$$E_{n\_out\_ideal} = GE_{n\_in} \qquad (4.20)$$

The IEEE definition goes further to say that the noise figure equation is defined for the case where the noise temperature of the input to the device, (i.e., T_a) is the reference value of $T_0 = 290$ K. Using this and (4.16) (with $T_a = T_0$) gives

$$E_{n\_out\_actual} = GkT_0 + GkT_e \qquad (4.21)$$

and

$$E_{n\_out\_ideal} = GkT_0 \qquad (4.22)$$

which leads to

$$F_n = \frac{GkT_0 + GkT_e}{GkT_0} = 1 + \frac{T_e}{T_0} \qquad (4.23)$$

Alternately, we can solve for T_e in terms of F_n as

$$T_e = T_0(F_n - 1) \qquad (4.24)$$

An important point from (4.23) is that the minimum noise figure of a device is $F_n = 1$.

4.4.2 Attenuators

For most devices, noise figure is determined by measurement. Attenuators represent the exception to this rule. For attenuators, the noise figure is normally taken to be the attenuation. Thus, for an attenuator with an attenuation of L (a number greater than 1), the noise figure is assumed to be

$$F_n = L \qquad (4.25)$$

The rationale behind this is that an attenuator matched to the source and the load impedances (which are assumed identical) produces a noise energy out of the attenuator equal to the noise energy into the attenuator [8, 9]. To show this we use (4.15) with $E_{n\_out} = E_{n\_in} = kT_0$ and $G = 1/L$ to write

$$E_{n\_out} = kT_0 = (1/L)E_{n\_in} + (1/L)kT_e \qquad (4.26)$$

If we further assume a source temperature of T_0 (recall the necessity for using this temperature when computing noise figure), we get

$$kT_0 = kT_0/L + kT_e/L \qquad (4.27)$$

or

$$T_e = T_0(L-1) \qquad (4.28)$$

and, by association with (4.24), $F_n = L$.

The authors have always been concerned with the assumption that the noise energy out of an attenuator is identical to the noise energy into the attenuator. To investigate this further, we analyzed a T-type attenuator that consisted of noisy resistors. Figure 4.4 contains a schematic of the circuit we analyzed. The values of R_1, R_2, and R_3 were computed so that the input and output resistance of the attenuator was R and the attenuation was L W/W. When the temperature of the source and the three resistors of the attenuator was the same, we found the energy out of the attenuator was the same as the energy into the attenuator. However, when the temperature of the source differed from the temperature of the resistors, the energy into the attenuator did not equal the energy out of the attenuator. Thus, for this simple example, we verified that the noise energy into and out of the attenuator are equal, if the source and resistors are at the same temperature. We assume this is also the case for a general attenuator.

We carried the T-type attenuator example a step further and considered some cases where the source temperature was T_a but the temperature of the attenuator resistors was some other temperature, T_R. We found, at least for the example cases we considered, an output noise energy given by

$$E_{n\_out} = kT_a/L + kT_R(L-1)/L \qquad (4.29)$$

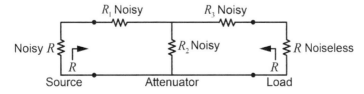

Figure 4.4 Schematic of a T-type attenuator.

We derived (4.29) from (4.26) and (4.28) with the temperature T_0 replaced by T_R and $E_{n\_in} = kT_a$. This handy equation lets us analyze attenuators with different source noise temperatures and circumstances where the attenuator is not at a temperature of T_0. We caution that we have not proved (4.29) valid for a general attenuator, only for our T-type resistive attenuator. However, it agrees with a similar equation in Blake's NRL report [10].

For those (ambitious) readers who are interested, we included the above problem as Exercise 7.

4.5 NOISE FIGURE OF CASCADED DEVICES

Since a typical radar has several devices that contribute to the overall equivalent/effective noise temperature or noise figure, we need a method of computing the equivalent/effective noise temperature and noise figure of a cascade of components. To this end, we consider the block diagram of Figure 4.5. In this figure, the circle to the left denotes a noise source, represented in a radar by the antenna or other radar components. For the purpose of computing noise figure, we assume the temperature of the noise source is T_0 (consistent with the definition of noise figure). The blocks following the noise source represent various radar components, such as amplifiers, mixers, attenuators, and so on. The various blocks are characterized by their gain, G_k, noise figure, F_k, and equivalent/effective noise temperature, T_k (the T_k are the T_e of the devices).

To derive the equation for the overall noise figure and equivalent/effective noise temperature of the N devices, we will consider only Device 1, then Devices 1 and 2, then Devices 1, 2, and 3, and so forth. This will allow us to develop a pattern we can extend to N devices.

Recalling that we always assume a source temperature of T_0 when computing noise figure, we posit an input noise energy for Device 1 of

$$E_{n\_in\_1} = kT_0 \qquad (4.30)$$

The noise energy out of Device 1 is [see (4.15)]

$$E_{n\_out\_1} = G_1 E_{n\_in\_1} + E_{n\_int\_1} = kT_0 G_1 + kT_1 G_1 \qquad (4.31)$$

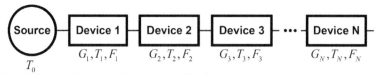

Figure 4.5 Block diagram for computing system noise figure.

From (4.23), the equivalent/effective noise temperature of Device 1 is $T_{e1} = T_1$ and the noise figure from the source through Device 1 is

$$F_{n1} = 1 + \frac{T_1}{T_0} = F_1 \tag{4.32}$$

For Device 2, the input noise energy is

$$E_{n\_in\_2} = E_{n\_out\_1} = k(T_0 + T_1)G_1 \tag{4.33}$$

and the noise energy out of Device 2 is

$$\begin{aligned} E_{n\_out\_2} &= G_2 E_{n\_in\_2} + E_{n\_int\_2} = k(T_0 + T_1)G_1 G_2 + kT_2 G_2 \\ &= kT_0 G_1 G_2 + kT_1 G_1 G_2 + kT_2 G_2 \\ &= kT_0 G_1 G_2 + k\left(T_1 + \frac{T_2}{G_1}\right) G_1 G_2 \end{aligned} \tag{4.34}$$

From (4.34), we see the equivalent/effective noise temperature of the cascade of Devices 1 and 2 is

$$T_{e2} = T_1 + \frac{T_2}{G_1} \tag{4.35}$$

and $G_1 G_2$ is the total gain of Devices 1 and 2.

The noise figure from the source through Device 2 is

$$F_{n2} = 1 + \frac{T_{e2}}{T_0} = 1 + \frac{T_1}{T_0} + \frac{1}{G_1}\frac{T_2}{T_0} \tag{4.36}$$

or, with the definition $T_2 = T_0(F_2 - 1)$,

$$F_{n2} = F_1 + \frac{F_2 - 1}{G_1} \tag{4.37}$$

Notice how the gain of the first device reduces the contribution of the equivalent/effective noise temperature and noise figure of the second device. We will examine this concept again in an example. For now, we proceed with determining the noise figure from the source through the third device.

The noise energy at the output of a cascade of Devices 1, 2, and 3 is

$$\begin{aligned} E_{n\_out\_3} &= G_3 E_{n\_in\_3} + E_{n\_int\_3} = G_3 E_{n\_out\_2} + E_{n\_int\_3} \\ &= k\left(T_0 + T_1 + \frac{T_2}{G_1}\right) G_1 G_2 G_3 + kT_3 G_3 \end{aligned} \tag{4.38}$$

Rearranging the terms yields

$$E_{n\_out\_3} = kT_0 G_1 G_2 G_3 + k\left(T_1 + \frac{T_2}{G_1} + \frac{T_3}{G_1 G_2}\right) G_1 G_2 G_3 \qquad (4.39)$$

With this result, the equivalent/effective noise temperature of the cascade of Devices 1, 2, and 3 is

$$T_{e3} = T_1 + \frac{T_2}{G_1} + \frac{T_3}{G_1 G_2} \qquad (4.40)$$

and the total gain of Devices 1, 2,` and 3 is $G_1 G_2 G_3$.

The system noise figure from the source through Device 3 is

$$F_{n3} = 1 + \frac{T_{e3}}{T_0} = 1 + \frac{T_1}{T_0} + \frac{1}{G_1}\frac{T_2}{T_0} + \frac{1}{G_1 G_2}\frac{T_3}{T_0} \qquad (4.41)$$

or, using (4.24),

$$F_{n3} = F_1 + \frac{F_2 - 1}{G_1} + \frac{F_3 - 1}{G_1 G_2} \qquad (4.42)$$

Here, we note the product of the gains of the preceding two devices reduces the contribution of the noise figure and temperature of Device 3.

With some thought, we can extend (4.40) and (4.42) to write the equivalent/effective noise temperature of the system, from the source through Device N, as [11]

$$T_{eN} = T_1 + \frac{T_2}{G_1} + \frac{T_3}{G_1 G_2} + \cdots \frac{T_N}{G_1 G_2 \cdots G_{N-1}} \qquad (4.43)$$

The noise figure of the system, from the source through Device N, is

$$F_{nN} = F_1 + \frac{F_2 - 1}{G_1} + \frac{F_3 - 1}{G_1 G_2} + \frac{F_4 - 1}{G_1 G_2 G_3} + \cdots + \frac{F_N - 1}{G_1 G_2 G_3 \cdots G_{N-1}} \qquad (4.44)$$

and the overall system gain is $G_1 G_2 G_3 \ldots G_N$.

In the equations above, we found the noise figure between the input of Device 1 through the output of Device N. If we wanted the equivalent/effective noise temperature and noise figure between the input of any other device (say, Device k) and the output of some other succeeding device (say, Device m), we would assume the source of Figure 4.5 (at a temperature of T_0) is connected to the input of Device k and we would include terms like (4.43) and (4.44) that would carry to the output of Device m. Thus, for example, the equivalent/effective noise temperature of only Devices 2, 3, and 4 is

$$T_{e4} = T_2 + \frac{T_3}{G_2} + \frac{T_4}{G_2 G_3} \qquad (4.45)$$

and the noise figure from the input of Device 2 to the output of Device 4 is

$$_2F_{n4} = F_2 + \frac{F_3-1}{G_2} + \frac{F_4-1}{G_2G_3} \qquad (4.46)$$

and the total gain of the three devices is $G_2G_3G_4$. We leave the derivation of (4.45) and (4.46) as an exercise.

4.6 AN INTERESTING EXAMPLE

We now consider an example of why, as a general rule of thumb, radar designers normally include an RF amplifier as an early element in a receiver. In this example, we consider the two options of Figure 4.6. In the first option, we have an amplifier followed by an attenuator, and in the second option we reverse the order of the two components. The gains and noise figures of the two devices are the same in both configurations.

For Option 1, the noise figure from the input of the first device to the output of the second device is

$$F_{n2\_option2} = F_1 + \frac{F_2-1}{G_1} = 4 + \frac{100-1}{100} \approx 5 \text{ W/W or 7 dB} \qquad (4.47)$$

For the second option, the noise figure from the input of the first device to the output of the second device is

$$F_{n2\_option2} = F_1 + \frac{F_2-1}{G_1} = 100 + \frac{4-1}{0.01} = 400 \text{ W/W or 26 dB!} \qquad (4.48)$$

This is a dramatic difference in noise figure of the combined devices. In general, the contribution of a device to overall noise figure will be reduced by the total gain of the preceding devices. In Option 1, the contribution of the 20-dB noise figure of the attenuator was reduced by the 20 dB gain of the amplifier. As a result, it added only one dB to the noise figure of the amplifier to result in an overall noise figure of 7 dB.

In Option 2, the 6 dB noise figure of the amplifier was increased by the attenuator attenuation (negatively reduced by the negative gain—in dB—of the attenuator). The result was that the overall noise figure was the sum of the noise figures of the attenuator and amplifier.

This example demonstrates why radar designers like to include an amplifier early in the receiver chain: it essentially sets the noise figure of the receiver. As a general rule of thumb, a nominal gain of 20 to 25 dB in the RF amplifier usually ensures the noise figure of the receiver is primarily due to the noise figure of the RF amplifier.

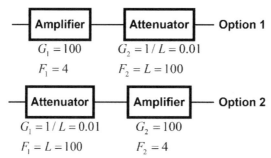

Figure 4.6 Two configurations options.

4.7 OUTPUT NOISE ENERGY WHEN THE SOURCE TEMPERATURE IS NOT T_0

In the discussion above, we considered a source temperature of T_0. We now want to examine how to compute the noise energy out of a device for a source temperature other than T_0. This is the situation we encountered when computing SNR in Chapter 2, where we used a source temperature of T_a.

From (4.15), we have

$$E_{n\_out} = GE_{n\_in} + E_{n\_int} = GkT_a + GkT_e \qquad (4.49)$$

where T_a is the noise temperature of the source. If we were to rewrite (4.49) using noise figure, we would have

$$E_{n\_out} = GE_{n\_in} + E_{n\_int} = GkT_a + GkT_e = GkT_a + GkT_0(F_n - 1) = GkT_s \qquad (4.50)$$

where

$$T_s = T_a + (F_n - 1)T_0 \qquad (4.51)$$

Equations (4.50) and (4.51) are the same equations we encountered when we computed the noise energy term of the radar range equation in Section 2.2.2.

If we have a cascade of N devices, G denotes the combined gain of the N devices; T_e denotes the equivalent/effective noise temperature of the N devices; and F_n denotes the noise figure of the N devices. In the exercises, we consider specific examples of how different source temperatures can affect E_{n_out}. Relative to Chapter 2, G and F_n are the gain and noise figure of the entire radar, up to the point we are interested in SNR.

4.8 A NOTE ABOUT CASCADED DEVICES AND THE RADAR RANGE EQUATION

Sometimes radar analysts are uncertain about whether to include the loss of lossy, resistive, passive components between the antenna face and the first active device in the loss term of the radar range equation, or in the equivalent/effective noise temperature, T_e, and noise figure of the radar. The simple answer is: if $T_a = T_0$, it does not matter, as long as it is not included in both places. If $T_a \neq T_0$, the loss should be included in T_e and the noise figure. If it is not, (2.29)—which is used to compute T_s in Chapter 2—would be invalid. As defined in Chapter 2, the passive components between the face of the antenna and the first active device include the radome and all other passive components up to the RF amplifier (which is usually the first active device).

4.9 CASCADED ATTENUATORS

As an extension of the general cascaded noise figure equation [(4.44)], we consider the case where we have N, cascaded, resistive type of lossy devices. Recall that the gain of a resistive type of lossy device is $G = 1/L$, where L is the loss of the device. Its noise figure is $F = L$ (see Section 4.4.2). If we use this in (4.44), we get

$$\begin{aligned} F &= F_1 + \frac{F_2 - 1}{G_1} + \frac{F_3 - 1}{G_1 G_2} + \cdots + \frac{F_N - 1}{G_1 G_2 \cdots G_{N-1}} \\ &= L_1 + \frac{L_2 - 1}{1/L_1} + \frac{L_3 - 1}{1/(L_1 L_2)} + \cdots + \frac{L_N - 1}{1/(L_1 L_2 \cdots L_{N-1})} \\ &= L_1 + L_1(L_2 - 1) + L_1 L_2 (L_3 - 1) + \cdots + L_1 L_2 \cdots L_{N-1}(L_N - 1) \\ &= L_1 L_2 \cdots L_N = L_{TOT} \end{aligned} \quad (4.52)$$

That is, the noise figure of the N, cascaded lossy devices is the total loss of the N devices. The gain of the cascaded devices is $1/L_{TOT}$.

4.10 EXERCISES

1. A radar receiver has the components and parameters indicated in Table 4.1. Their relative locations are as in the table.

Table 4.1
Receiver Components

Device	Gain (dB)	Noise Figure (dB)
Waveguide (attenuator)	−2	2 ($F = L = 1/G$)
RF amplifier	20	6
First mixer	−3	10
IF amplifier	100	20

a) Compute the noise figure in dB through each of the four devices, referenced to the waveguide input. Note the increase in noise figure each device causes.

b) Repeat part a) for the case where the RF amplifier has a low noise figure of 2 dB. Again, note the increase in noise figure created by each device. Note how the devices following the low-noise RF amplifier seem to have more effect on noise figure than does the RF amplifier with the higher noise figure.

c) What is the equivalent/effective noise temperature, in kelvin, of the RF amplifier of part a)?

d) Based on the values used for part a), what is the equivalent/effective noise temperature, in kelvin, of the receiver, through the IF amplifier, referenced to the waveguide input?

e) Compute the noise power, in dBm, at the output of the IF amplifier, assuming a noise temperature for the antenna (input to the waveguide) of 290 K. Use the values from part a). Assume a bandwidth of 1 MHz.

f) Repeat part e) using an antenna noise temperature of 100 K. Note that this is almost the same as for 290 K. This indicates that the internal noise of the receiver is the major contributor the total system noise energy for this particular case.

g) Repeat part e) using an antenna noise temperature of 6,000 K. This result indicates that the antenna noise propagated through the receiver is the major contributor to the total system noise energy.

2. Assume a radar with noise figure $F_n = 6$ dB referenced to the antenna face.

a) Compute the equivalent/effective noise temperature of the receiver.

b) Assume a noise bandwidth of 1 MHz and a receiver gain of 100 dB. Compute the noise power in dBW at the receiver output.

3. The radar of Exercise 2 has an SNR of 20 dB for a particular scenario. Suppose it operates at night and looks into a clear sky with a T_a' of 10 K (see Chapter 2).

a) Assume a noise bandwidth of 1 MHz and a receiver gain of 100 dB. Compute the noise power, in dBW at the receiver output.

b) What is the change in SNR, in dB, relative to the case of Exercise 2? In Exercise 2, we assumed a $T_a = T_0$.

4. Repeat Exercise 3 with $T_a = 20{,}000$ K.

5. Derive (4.14).

6. Derive equations for the three resistors of the attenuator from Figure 4.4, in terms of the input and output resistance, R, and the loss, L. The loss has the units of W/W.

7. Derive an equation for the noise energy delivered to the attenuator of Figure 4.4 by the source and the noise energy delivered to the load by the attenuator. Assume a source noise temperature of T_a and a noise temperature for all attenuator resistors of T_R.

8. Derive (4.43).

9. Derive (4.45) and (4.46).

10. Plot E in (4.3) versus f. Let f vary logarithmically from 10 MHz to 1,000 GHz. Plot E with the units of dB relative to a milli-joule. Generate curves for temperatures of 2.9, 29, and 290 K. Does your plot support the statement that E is insensitive to frequency in the range of frequencies used in radars?

11. Prove (4.43) and (4.44) by induction.

References

[1] Johnson, J. B., "Thermal Agitation of Electricity in Conductors," *Phys. Rev.*, vol. 32, July 1928, pp. 97–109.

[2] Nyquist, H., "Thermal Agitation of Electric Charge in Conductors," *Phys. Rev.*, vol. 32, Jul. 1928, pp. 110–113.

[3] Erst, S. J., *Receiving Systems Design*, Dedham, MA: Artech House, 1984, p. 49.

[4] Gao, J., *RF and Microwave Modeling and Measurement Techniques for Field Effect Transistors*, Raleigh, NC: SciTech, 2010, p. 104.

[5] Barton, D. K., *Radar Equations for Modern Radar*, Norwood, MA: Artech House, 2013.

[6] Friis, H.T., "Noise Figures of Radio Receivers," *Proc. IRE*, vol. 32, no. 7, Jul. 1944, pp. 419–422.

[7] *IEEE Standard Dictionary of Electrical and Electronic Terms*, 6th ed., New York: IEEE, 1996.

[8] Egan, W. F., *Practical RF System Design*, New York: Wiley & Sons, 2003, pp. 55–56.

[9] Pozar, D. M., *Microwave Engineering*, New York: Addison-Wesley, 1990, pp. 590–591.

[10] Blake, L. V., "A Guide to Basic Pulse-Radar Maximum-Range Calculation, Part 1—Equations, Definitions, and Aids to Calculation," Naval Research Laboratory, Washington, D.C., Rep. No. 6930, Dec. 23, 1969, p. 49. Available from DTIC as 701321.

[11] Budge, M. C., Jr., "EE 619: Intro to Radar Systems," www.ece.uah.edu/courses/material/EE619/index.htm.

Chapter 5

Radar Losses

5.1 INTRODUCTION

For our last radar range equation-related topic, we address the loss term, L. Losses have been included in the radar range equation since it first appeared in Norton and Omberg's 1947 paper [1]. Losses have been continuously studied and new loss factors have been introduced as radar technology has advanced [2–11].

In an attempt to organize our discussion of losses, we will trace the losses through the steps we used to derive the radar range equation in Chapter 2. That is, we start with the transmitter and antenna and progress to propagation losses. On receive, we will address losses between the antenna and RF amplifier, which we term *RF losses*, and then proceed to losses associated with the matched filter, signal processor, and detection process.

5.2 TRANSMIT LOSSES

Transmit losses are losses in components between the final RF power source and the antenna feed. Radars that use reflector antennas, space-fed phased arrays, or constrained feed-phased arrays could have the components shown in Table 5.1 [12–17]; however, not all radars have all of the devices listed. As examples, Figure 5.1 contains two different transmitter block diagrams illustrating various devices that might be included in each. The two example transmitters shown in Figure 5.1 are merely intended to be representative in general.

Waveguide run (sometimes referred to as "plumbing") refers to all of the pieces of waveguide used between components. Waveguide is typically used because of its extremely low loss and high power-handling capability.

Waveguide switches (which can be manual or automatic) are used for routing signals. For the example shown in Figure 5.1.a, the transmitter can be switched to a dummy load (high power terminator) for test purposes. Similarly, a waveguide switch (usually manual) can be used to provide a means of test signal injection.

Table 5.1
Representative RF Losses

Component	Loss (dB)
Waveguide run	0.1–0.3
Waveguide switch	0.7
Power divider	1.6
Duplexer	0.3–1.5
TR switch	0.5–1.5
Circulator/isolator	0.3–0.5
Receiver protection	0.2–1.0
Preselector (receive only)	0.5–2.5 (0.5 typical)
Directional coupler	0.3–0.4
Rotary joint	0.2–0.5
Mode adapter	0.1
Waveguide step attenuator	0.8
Limiter	0.2–1.0
Feed (monopulse or simple)	0.2–0.5

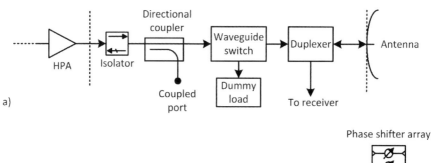

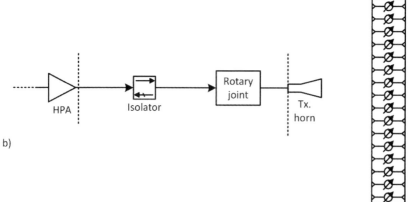

Figure 5.1 Representative transmitter block diagrams.

Power dividers are passive devices used for power distribution [18, pp. 328–333] [19, pp. 146–153]. A power divider usually splits a signal into n equal-phase equal-amplitude parts, where n can be odd or even [20, 21]. Power dividers are usually reciprocal devices, allowing a power divider to serve as power combiner.[1] One application for power dividers is a corporate antenna feed.

Power dividers incur two losses, a *coupling loss* and an *insertion loss*. Coupling loss, stated in dB, is the loss from input to output as the result of splitting the input n ways. For example, consider a 1:4 power divider. The ideal coupling loss, from input to output, is equal to $-10 \cdot \log(n) = -6$ dB. If the input to a 1:4 power divider is -20 dBm, the output will be -26 dBm. Coupling loss is recoverable.

A power divider's insertion loss is power lost within the device, dissipated as heat, due primarily to reflections, dielectric absorption, radiation effects, and conductor losses [21]. Insertion loss is not recoverable. Depending on the class of power divider, insertion loss can theoretically be zero, but is usually not more than 3 dB in practical devices [20, 21]. Unlike coupling loss, insertion loss is not necessarily a strong function of n. For example, given a 1:4 power divider, with an insertion loss of 0.5 dB, the output signals will be -6.5 dB relative to the input.

One common class of power divider is the Wilkinson power divider, invented in 1960 by Ernest J. Wilkinson [22] [18, p. 328] [19, pp. 146–147]. The Wilkinson power divider offers good isolation (usually ~20 dB) between outputs and equal output impedances. Wilkinson power dividers can be designed to operate at frequencies ranging from hundreds of MHz to tens of GHz [21].

The duplexer shown in Figure 5.1.a is a fast-acting, nonreciprocal device that allows for a common antenna to be used for transmit (Tx) and receive (Rx) in a radar using pulsed waveforms. The duplexer protects the receiver from the high-power transmit signal, including power reflected from impedance mismatches at the antenna.[2] Often a high-power circulator serves as duplexer, but the duplexer can also be implemented using a balanced network of transmit-receive (TR) switches and a receiver protector [23]. In transmit-receive (T/R) modules, the duplexer is usually a circulator. In high-power radars, the duplexer can be a TR switch of the gas discharge type (TR tube, see below) [23]. A TR switch is an automatic device employed in a radar for preventing the transmitted energy from reaching the receiver but allowing the receive energy to reach the receiver without appreciable loss [23].

Circulators are three-port devices where the signal into one port can only leave the next port (like a roundabout where you must exit on the street following the street you entered). Terminating one port of a circulator results in an isolator, which is a two-port device where the signal can travel in only one direction. Any reflected energy returned to the isolator is directed to a terminating load. Isolators are often used prior to poorly matched components (e.g., a filter or switch).

---

[1] In general, it is difficult to design a nonreciprocal power divider.
[2] A separate device, such as a limiter, is often used to protect the receiver from high-power returns occurring after the end of the transmit pulse.

While the duplexer provides a certain amount of protection for the receiver against high-power returns, it does not always provide enough receiver protection. Receiver protection in Table 5.1 refers to devices specifically used to protect the receiver, such as diode or ferrite limiters. A TR tube is a gas-filled RF switching tube. When high power from the transmitter enters the TR tube, the tube arcs, shorting out. It reflects the incoming power, thus protecting the receiver. TR tubes are very fast acting.

A preselector is a filter (often implemented in waveguide) used in the receiver to limit bandwidth. For frequency-agile radars, the agility bandwidth is passed; for single-frequency radars, the preselector is matched to the channel bandwidth.

Directional couplers, shown in Figure 5.1.a and b, are used to sample, or couple, signals out of the transmitter for test purposes. The power ratio between the input signal and the sampled signal is a calibrated amount (e.g., 10, 20, 40 dB). A low coupling ratio (e.g., 40 dB) allows transmitter power measurements to be made using low-power test equipment.

A rotary joint, shown in Figure 5.1.b, is a device used to couple RF energy from a fixed transmission line to a device that is rotating, such as an antenna. A rotary joint can also be used with antennas that stow or pack themselves.

A mode adapter is generally any device that changes the mode of propagation (e.g., from coaxial line transmission to rectangular waveguide transmission). Waveguide attenuators are sometimes used in front of the receiver RF low noise amplifier (LNA) for automatic gain control/sensitivity time control (AGC/STC) (see Chapter 14).

Figure 5.2 contains plots of theoretical waveguide loss versus frequency for several standard waveguides [18, 24]. From this we see that the waveguide losses indicated in Table 5.1 are representative of radars that contain a total 1 to 2 m of waveguide connecting the various components of the transmitter.

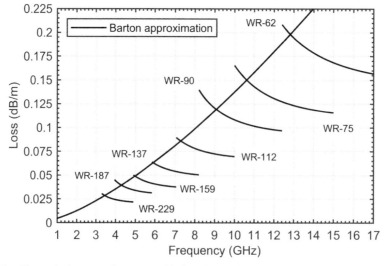

Figure 5.2 Theoretical rectangular waveguide loss (copper).

The calculations associated with the theoretical loss in a rectangular waveguide, which are dependent on the broad and short wall dimensions, the permeability and permittivity of the dielectric filling the waveguide, and the waveguide material, can be cumbersome (see Appendix 5A). Barton presents a convenient approximation for waveguide loss in dB/m of [9, p. 359] [3]:

$$K_{\alpha W} = 0.0045 \cdot f^{1.5} - 0.00003 \cdot f^{2.2} \quad \text{(dB/m)} \qquad (5.1)$$

where f is the frequency in GHz ($f < 200$ GHz). This approximation is plotted in Figure 5.2 for comparison as the dashed curve.

For active phased arrays that use T/R modules [25], the losses are primarily due to a switch or circulator used to route signals from the power amplifier to the antenna and from the antenna back to the receiver LNA. This is illustrated in Figure 5.3. Because of the collocation of the T/R module with associated array elements, the RF losses associated with active phased arrays are generally much lower than those associated with radars that use passive antennas such as reflectors, space-fed phased arrays, and constrained feed-phased arrays.

Table 5.2 contains a summary of transmit RF losses for the three transmitter configurations of Figures 5.1.a & b and 5.3. In computing the waveguide losses, we assume the radar with the reflector (Figure 5.1.a) is an L-band search radar. The space-fed phased array in Figure 5.1.b is an S-band multifunction radar and the T/R module in Figure 5.3 is used in an X-band multifunction radar. The difference in operating frequencies is the reason for the different waveguide losses. We also assume the transmitter and receiver of the L-band search radar is located in the rotating assembly that houses the antenna. Thus, we did not assign a rotary joint to the L-band loss budget.

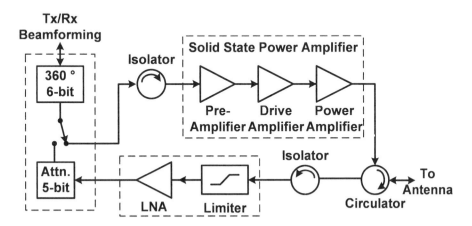

Figure 5.3 Example T/R module block diagram.

---

[3] Barton notes that (5.1) matches the line losses plotted in Saad and Hansen, *Microwave Engineers' Handbook*, Artech House, 1971 [15, 16]. Equation (5.1) appears in the file titled "10-1 Loss Factors" included on the accompanying DVD of [9].

Table 5.2
Example Transmit RF Losses

Loss Term	L-Band Search Radar	S-Band Multifunction Radar	X-Band Active Array Radar
Circulator	—	—	0.4
Isolator	0.4	0.4	—
Rotary joint	—	0.3	—
Directional coupler	0.3	—	—
Waveguide switch	0.7	—	—
Duplexer	0.7	—	—
Waveguide	0.1	0.2	—
Total	2.2 dB	0.9 dB	0.4 dB

5.3 ANTENNA LOSSES

The next element of the transmit chain is the antenna and its associated feed. A representative list of losses associated with the various feed and antenna components is contained in Table 5.3. The entries for waveguide and stripline feed apply to antennas that use constrained feeds, and the difference between parallel and series feed networks is illustrated in Figure 5.4 [26, p. 291]. In the series feed, the energy enters on one end of an RF transmission line (such as a rectangular waveguide or stripline) and is extracted at different points along the line. In a parallel feed, the energy enters an RF transmission line and is subsequently split several times before being delivered to the radiating elements. As a note, it is possible for an antenna to use both series and parallel feed networks [27, pp. 5–8]. As an example, the rows of an array could be fed by a series feed, while the elements in each row would be fed by a parallel feed network.

A feed loss of 0 dB is sometimes assigned to active arrays. This is because the radiating element driven by a T/R module is very close to the T/R module's power amplifier. However, T/R modules have losses associated with the radiating element (feed, mismatch, and ohmic losses). Although these losses can be relatively small (<1 dB) they are not zero.

The phase shifter losses apply to passive and constrained feed phased arrays. As a note, the losses apply to the entire array and not to each phase shifter of the array. The losses are shown as 0 dB for active phased arrays because the phase shifter is not in the path between the antenna and the power amplifier or LNA, where loss is important (see Figure 5.3).

Table 5.3
Antenna Losses

Location	Component	Typical Loss (dB)
Feed system	Feed horn for reflector or lens	0.1
	Waveguide series feed	0.7
	Waveguide parallel feed	0.4
	Stripline series feed	1.0
	Stripline parallel feed	0.6
	Active module at each element	0.0
Phase shifter	Nonreciprocal ferrite, or Faraday rotator	0.7
	Reciprocal ferrite	1.0
	Diode (3- or 4-bit)	1.5
	Diode (5- or 6-bit)	2.0
	Diode (per bit)	0.4
	Active module at each element	0.0
Array	Mismatch[4]	0.2
	Scan	1.3–1.7
Exterior	Radome	0.5–1.0

Source: [9, 28].

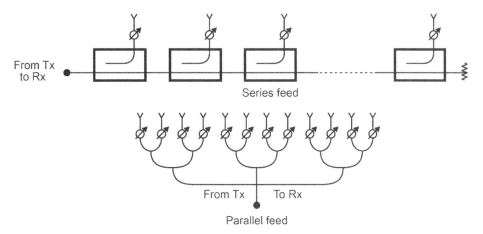

Figure 5.4 Series and parallel feeds. (*After*: [26].)

---

[4] In a passive electronically scanned array the mismatch loss can depend on the scan angle. This variability in the mismatch loss could be lumped in with the scan loss.

Mismatch loss is a loss due to impedance mismatch between free space and the feed and/or radiating elements of the antenna. Mismatch loss is given by

$$L_\Gamma = \frac{1}{1-\Gamma^2} \qquad (5.2)$$

where

$$\Gamma = \frac{VSWR-1}{VSWR+1} \qquad (5.3)$$

Γ is the reflection coefficient and VSWR is the voltage standing wave ratio [18, 24, 19]. The value of 0.2 dB in Table 5.3 corresponds to a VSWR of approximately 1.5.

When the beam of a phased array antenna is scanned off of broadside (off of array normal), the antenna directivity decreases. If this is not explicitly included when generating the antenna pattern at the scanned angle, it should be included as a loss. Barton suggests a factor of

$$L_{scan} = \cos^{-\beta}\theta \qquad (5.4)$$

where θ is the scan angle [9, p. 369]. Equation (5.4) is usually used as a model for the one-way scan loss. Figure 5.5 contains a plot of (5.4) for β = 1.25, 1.5, 1.75 and 2.0[5]. The average scan loss is 1.3, 1.6, 1.9, and 2.2 dB for β = 1.25, 1.5, 1.75, and 2, respectively (over 60-deg scan).

While (5.4) is a commonly used model for the element factor roll-off, using the actual embedded element pattern is preferable if available. However, (5.4) and the actual embedded element factor roll-off function should not be used together since this would amount to double-counting the scan loss. It is possible for the element factor, and the exponent in (5.4), to be different in the E and H planes.

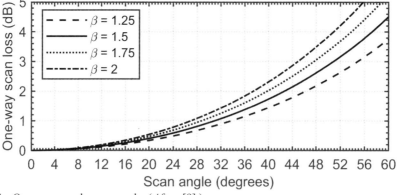

Figure 5.5 One-way scan loss vs. angle. (*After:* [9].)

---

[5] Some elements, such as dipoles, can have an exponent (β) of less than 1 in the H plane.

The next loss we discuss is beamshape loss.[6] This loss is associated with the situation where the antenna beam is not pointed directly at the target or where the beam is scanning across the target during the time the radar is coherently or noncoherently integrating a sequence of pulses (coherent and noncoherent integration is discussed in Chapter 8). In both cases, the full effect of the antenna directivity (G_T and G_R) terms of the radar range equation will not be realized. This most often happens during search. It is not applicable during track because it is assumed the target is very close to beam center during track.

Unlike the other antenna losses, beamshape loss is a two-way loss. As such, it should be included in the loss budget only once.

Historically, radar analysts have used average beam shape loss values of 1.6 or 3.2 dB for 1-D or 2-D scanning, respectively, as derived by Blake in his 1953 paper [29]. However, Hall and Barton [5, 9, 30] revisited this problem in the 1960s and derived revised loss numbers of 1.24 and 2.48 dB. It should be noted that Barton and Hall indicate that there are many factors that affect beam shape loss, such as beam step size in phased array radars, beam packing scheme (e.g., rectangular or triangular), number of pulses noncoherently integrated, whether or not the radar is continuously scanning, and detection probability. As such, the values of 1.24 and 2.48 dB should be considered rule-of-thumb numbers that would be suitable for preliminary radar analysis or design. In a more detailed analysis, these numbers should be revised based on the factors discussed by Barton and Hall.

The value of 1.24 dB is related to what is termed 1-D scanning, and the value of 2.48 dB relates to 2-D scanning. 1-D scanning would be associated with search radars that use a fan beam (a beam with a large beamwidth in one dimension (usually elevation) and a narrow beam in the other dimension). The radar would then rotate (or nod) in the narrow beam dimension but remain fixed in the wide beam direction. An example of such a radar is considered in Example 2 of Chapter 6, where we analyze a search radar with a cosecant squared elevation beam. In these fan beam types of radars, we assume the antenna directivity does not change much in the wide direction and thus use a 1-D loss. If this is not the case, we would want to use the 2-D beamshape loss.

An example of where the use of the 2-D beamshape loss would be appropriate is in phased arrays radars (such as the second and third examples of Table 5.2) that scan a sector by stepping the beam in orthogonal directions (azimuth and elevation or u and v—see Chapter 13). In this situation, the radar would move to a beam position and transmit a pulse, or burst of pulses, and then move to another beam position. Because of this action, it is likely that the target could be off of beam center in two dimensions, thus the need for the 2-D beamshape loss. In this situation, it may also be appropriate to include the scanning loss of (5.4) if the angular extent of the search sector is large.

A situation where we might want to use only a 1-D beamshape loss with a phased array radar is where we are generating a detection contour (see Example 2 of Chapter 6). In such a case, we would use the antenna directivity plot in, for example, elevation, and have the directivity as a function of elevation. However, we would need to account for

---

[6] Synonym: antenna-pattern loss.

the fact that the target is not on beam center in azimuth. Thus, we would include a 1-D beamshape loss in the radar range equation.

In discussing the phased array examples, we made the tacit assumption that the beams of the search sector were spaced close together as illustrated by Figure 5.6. This is similar to what Barton terms *dense packing* [9] and is characterized by the fact that there is no angular region that is not covered by the 3-dB beam contour of the antenna (the 3-dB beam contours are the circles in Figure 5.6). Barton discusses another type of packing he terms *sparse packing*, wherein there may be parts of the angle space that are not covered by beams on any one scan (but hopefully will be covered on successive scans). In this situation, he points out that the beamshape loss now becomes a function of detection probability. This is something that should be considered in detailed studies of the impact of search methodology on detection performance of the radar.

We continue our previous example by adding antenna losses to Table 5.2 to generate Table 5.4. We assumed the L-band search radar is a rotating search radar with a cosecant squared antenna beam. Therefore, we included only the feed, mismatch and 1-D beamshape loss. We assume that the S- and X-band radars are conducting a wide sector search (±60°) and include mismatch and scan losses. For the scan loss we used the average value indicated earlier for the case of $\beta = 1.5$. We assumed the beams in the S-band radar were tightly packed and used Barton's 2-D scan loss of 2.48 dB. For the X-band radar, we assumed the beams were not as tightly packed and thus used the historical 2-D scan loss of 3.2 dB. We assumed the radome on the S-band radar was cloth and use a fairly low value of radome loss. We assumed a hard radome on the X-band array and used a larger value of radome loss.

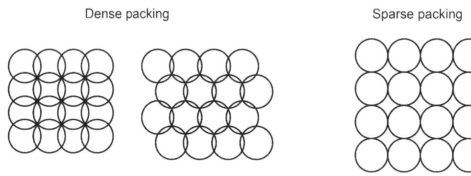

Figure 5.6 Examples of dense and sparse beam packing.

Table 5.4
Example Transmit RF and Antenna Losses

Loss Term	L-Band Search Radar	S-Band Multifunction Radar	X-Band Active Array Radar
XMIT RF	2.2	0.9	0.4
Feed horn	0.1	0.1	—
Phase shifter	—	0.7	—
Mismatch	0.2	0.2	0.2
Scan	—	1.6	1.6
Radome	—	0.5	0.8
1-D beamshape	1.24	—	—
2-D beamshape	—	2.48	3.2
Total	3.74 dB	6.48 dB	6.2 dB

5.4 PROPAGATION LOSSES

The next losses we consider are propagation losses. The two main sources of propagation loss are those due to oxygen and water vapor absorption and those due to rain. Absorption losses depend on operating frequency, elevation angle of the target, and range to the target. They also depend on temperature, humidity, atmospheric pressure, and other such atmospheric conditions. However, the atmospheric conditions are usually ignored and a *standard atmosphere* is used for performance prediction purposes [9, 31, 32].

Atmospheric losses were historically determined from graphs [3, 4, 10, 11]. However, with today's computers, they are easily calculated using the equations given in Appendix 5B. For illustration purposes, Figures 5.7, 5.8, and 5.9 contain plots (applicable to surface-based radars) of two-way atmospheric loss versus target range for different elevation angles, and frequencies of 1, 3, and 10 GHz (L-, S-, and X-band). The plots were generated using the equations in Appendix 5B.

Figure 5.10 contains plots of two-way absorption loss, due to rain only, in dB/km, versus frequency for different rainfall rates. This plot was also generated from equations presented in Appendix 5B. As a note, a somewhat standard rainfall rate for modeling purposes appears to be 4 mm/hr. According to Blake, this corresponds to moderate rain. For comparison, rainfall rates of 0.25 mm/hr, 1 mm/hr, and 16 mm/hr are considered a drizzle, a light rain, and a heavy rain, respectively [32, p. 219] [33].

96　　　　　　　　　　　　　　　　Basic Radar Analysis

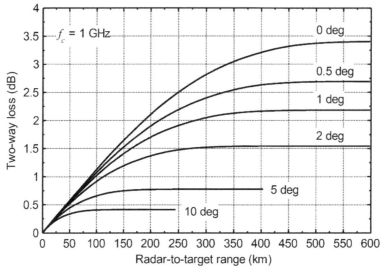

Figure 5.7 Atmospheric propagation loss for various beam elevation angles—standard atmosphere, 1 GHz.

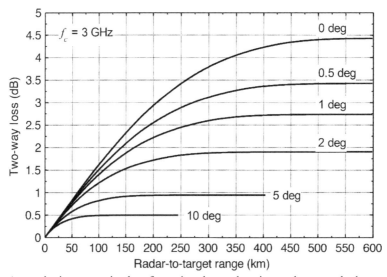

Figure 5.8 Atmospheric propagation loss for various beam elevation angles—standard atmosphere, 3 GHz.

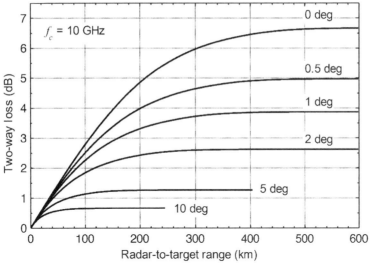

Figure 5.9 Atmospheric propagation loss for various beam elevation angles—standard atmosphere, 10 GHz.

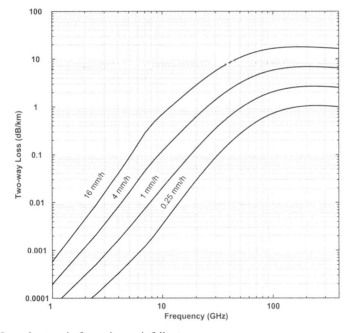

Figure 5.10 Loss due to rain for various rainfall rates.

We will assume the two phased array radars operate at ranges out to about 100 km and at elevation angles of 0° to 60°. We will ignore rain in this particular analysis. For this case, the atmospheric propagation loss for the S-band radar can vary from 0 dB to about 1.7 dB (see Figure 5.8). For the X-band radar, the atmospheric propagation loss can vary from 0 dB to about 2.8 dB. We will use a compromise value of 1 dB for the S-band radar and 2 dB for the X-band radar. With this, our loss table is now as shown in Table 5.5.

5.5 RECEIVE ANTENNA AND RF LOSSES

In general, the receive antenna losses will be the same as the transmit antenna losses. The possible exception to this is the case where the radar uses separate transmit and receive antennas or separate transmit and receive feeds. In that case, it may be necessary to derive a separate set of losses for the receive antenna. Also, active phased array antennas often use different aperture weightings on transmit and receive (for example, uniform weighting on transmit and Taylor weighting on receive). The aperture weighting loss can either be included in the loss budget or in the antenna gain, but not both.

Like the antenna, the RF components in the receive path will generally be the same as in the transmit path. Thus, the losses in Table 5.1 apply to receive, with the addition of the preselector losses. We will assume that the RF portions of the L- and S-band radars are as shown in Figure 5.11. The block diagram of the X-band T/R module used in the active phased array is as shown in Figure 5.3.

Because of the requirements for computing noise figure (see Chapter 4) and system noise temperature (see Chapter 2), receive antenna and RF losses must be apportioned between the loss term of the radar range equation (L) and noise figure (F_n). Losses not due to resistive-type devices or active devices must be included in the loss term. Losses due to resistive or active devices must be included in the noise figure. For the examples we have considered, the only RF and antenna losses not due to resistive or active devices are mismatch loss, scan loss, and beamshape loss. Mismatch loss on receive is caused by energy being reflected back into the environment because of impedance mismatch and scan loss is due to changes in directivity as the antenna beam is scanned off of array normal.

Continuing with our example, the loss table now becomes that shown in Table 5.6. We have assumed that the antennas and feeds are the same in the three radars so that the receive antenna losses will be the same as the transmit antenna, except for the beamshape loss. Recall that beamshape loss is not included in Table 5.6 because it is a two-way loss and needs only be applied on one or the other. In this example it was already applied on transmit.

Table 5.5
Example Transmit RF, Antenna, and Propagation Losses

Loss Term	L-Band Search Radar	S-Band Multifunction Radar	X-Band Active Array Radar
Prior losses	3.74	6.48	6.2
Propagation	2.0	1.0	2.0
Total	5.74 dB	7.48 dB	8.2 dB

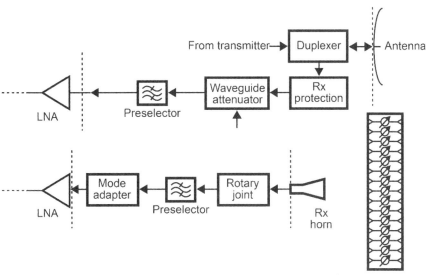

Figure 5.11 Representative receiver RF block diagrams.

For the L- and S-band radars, we will need to account for the losses of the various components between the antenna face and the RF amplifier (the LNA in Figure 5.11). The waveguide attenuator loss (the WG attenuator entry) shown in Table 5.6 for the L-band radar applies to the case where the attenuation is set to 0 dB. The loss will increase as the attenuation increases, on a dB for dB basis.

Table 5.6
Example Transmit RF, Antenna, Propagation, and Receive Losses

Loss Term	L-Band Search Radar	S-Band Multifunction Radar	X-Band Active Array Radar
Prior losses	5.74	7.48	8.2
Antenna (resistive)	0.1	1.3	0.8
Antenna (nonresistive)	0.2	1.8	1.8
Duplexer	0.7	—	—
Rx protection	0.2	—	—
WG attenuator	0.8	—	—
Rotary joint	—	0.2	—
Mode adapter	—	0.1	—
Circulator	—	—	0.4
Limiter	—	—	0.2
Preselector	0.5	0.5	—
Total in L	5.94 dB	9.28 dB	10.0 dB
Total in F_n	2.3 dB	2.1 dB	1.4 dB

In both the L- and S-band radars, we used the typical loss of 0.5 dB for the preselector. Also, we used 0.2 dB for the rotary joint loss. We used the transmit values for the components that are common to the transmitter and receiver.

As indicated earlier, the receive RF and antenna losses must be apportioned between losses that go into the loss term of the radar range equation and loss terms that go into the noise figure calculation. These are indicated as Antenna (resistive) and Antenna (nonresistive) in Table 5.6. For the L-band radar, the Antenna (nonresistive) losses included only the scan loss of 0.2 dB. When this value is added to the prior losses of 5.74, the total losses included in the loss term of the radar range equation (L) is 5.94 dB. The sum of the remaining losses in the L-band radar column (2.3 dB) go into the noise figure. For the S- and X- band radar, the nonresistive losses include the mismatch loss of 0.2 dB and the scan loss of 1.6 dB. When these are added to the prior losses for these two radars, the total losses included in the radar range equation become $7.48 + 0.2 + 1.6 = 9.28$ dB and $8.2 + 0.2 + 1.6 = 10.0$ dB. Loss included in the noise figure is the sum of the remaining losses.

5.6 PROCESSOR AND DETECTION LOSSES

The final set of losses we discuss are those associated with the matched filter, the signal processor, and the constant false alarm rate (CFAR) processing (Table 5.7).

The loss associated with the matched filter is a mismatch loss that mainly applies to matched filters for unmodulated pulses or for the chips of phase coded pulses (see Chapter 11). This loss occurs because the ideal rectangular pulse generated by the transmitter becomes distorted because of the bandwidth limiting that takes place as the pulse travels through the transmitter and antenna, to and from the target and back through the antenna and receiver to the matched filter. The estimate provided in Table 5.7 was derived by considering rectangular pulse that has been passed through different types of bandlimiting devices. A summary of the results of the analysis is shown in Table 5.8. In that table, the N-stage tuned filters are filters of different orders that have a bandwidth equal to the reciprocal of the pulsewidth. As can be seen, the nominal loss is about 0.5 dB.

The sidelobe reduction weighting loss applies to waveforms that use linear frequency modulation (LFM) for pulse compression (see Chapter 11). It is an amplitude taper used to reduce the range sidelobes of the compressed pulse. Since it is an amplitude taper, it also reduces the peak of the matched filter output. The amount of reduction generally depends on the type of weighting and the desired sidelobe levels. A list of various types of amplitude tapers and the associated loss is shown in Table 5.9 [34–36]. (For a summary of some common weighting functions, see Appendix A.)

Table 5.7
Processor and Detection Losses

Source	Typical Values (dB)
Matched filter loss	
Mismatch loss	0.5
Sidelobe reduction weighting loss	1.5
MTI loss with staggered waveforms	
Without noncoherent integration	0–1
With noncoherent integration	1.5–2.5
Doppler filter sidelobe reduction loss	1–3
Range straddle loss	0.3–1.0
Doppler straddle loss	0.3–1.0
CFAR loss	1–2.5

Table 5.8
Matched Filter Mismatch Loss

Input Signal	Filter	Mismatch Loss (dB)
Rectangular pulse	Gaussian	0.51
Rectangular pulse	1-stage single-tuned[7]	0.89
Rectangular pulse	2-stage single-tuned	0.56
Rectangular pulse	3-stage single-tuned	0.53
Rectangular pulse	5-stage single-tuned	0.50
Rectangular pulse	Matched	0.00

The table also contains the peak sidelobe level associated with the weighting, along with the associated range straddle loss. Range straddle loss will be discussed later in this chapter. Some common weightings used with LFM are Hamming, Hann, and Gaussian. The other amplitude tapers are often used for sidelobe reduction in antennas and in Doppler processors. Amplitude tapers are not used with phase coded waveforms because the phase coding sets the sidelobe levels. In fact, if an amplitude taper was used with a phase coded waveforms (see Chapter 11), it is likely that the compression properties of the waveform would be destroyed.

With the increasing processing power of digital signal processors, renewed attention is being given to the use of phase weighting with LFM waveforms to produce nonlinear LFM waveforms [37–40]. These waveforms have a desirable property of reduced sidelobes without the attendant weighting loss. They have the disadvantages of being difficult to generate and process. Nonlinear LFM is discussed further in Chapter 11.

---

[7] Multistage tuned filters have largely been replaced by modern filter designs (e.g., Bessel filters for short-pulse and phase-coded waveforms).

As is discussed in Chapter 18, for radars that use moving target indicator (MTI) processors, it is common practice to use waveforms with staggered PRIs [41, 42]. That is, waveforms with PRIs that change from pulse to pulse. The reason is that radars that use MTI processors and constant PRIs have frequency responses that have nulls in the range of expected target Doppler frequencies. The range rates corresponding to these nulls are termed *blind velocities.*

With a staggered PRI waveform, the nulls are "filled in" by the stagger so that the nulls move out of the range of expected target Doppler frequencies. With staggered PRIs, the MTI frequency response will vary quite a bit (5–10 dB) over the range of velocities. However, for reasonable sets of PRIs, the average response will be close to 0 dB across the frequency range of interest. Thus, the average SNR loss across the frequency range is between 0 and 1 dB, and most of the time is closer to 0 dB than to 1 dB. If the output of an MTI is noncoherently integrated, the noise correlation effect of the MTI will cause an additional loss [6, 7]. Barton indicates that this loss is approximately 1.5 and 2.5 dB for two- and three-pulse MTIs, respectively [9, p. 384].

As with LFM waveforms, amplitude weighting is also used to reduce the sidelobes of Doppler signal processors. In this case, the sidelobe reduction is needed in order to increase the clutter rejection capability of the Doppler processor. This topic is discussed further in Chapters 19 and 20. As with LFM weighting, use of amplitude weighting in Doppler processors causes a loss in SNR (and spectral broadening) relative to the case of no weighting.

A common amplitude weighting in modern radars that use digital signal processing and FFTs is Chebyshev with a sidelobe level determined by the clutter rejection requirements. However, Blackman and Blackman-Harris are also used. These amplitude weightings are attractive because of the low sidelobe levels that can be obtained.

For illustration, the Doppler response of a 45-dB Chebyshev-weighted FFT processor is presented in Figure 5.12. We note Figure 5.12 indicates a 1.4 dB-Doppler weighting loss. The Doppler filter responses are dotted, and the straddle loss, which we discuss below, is represented by the heavy black line. The scalloped shape of the straddle loss is why the term *scalloping loss* is sometimes used. A single Doppler filter centered at 10 kHz is shown by a solid line. For radar range equations purposes, we use the average of the straddling loss (see Table 5.9 [34–36]).

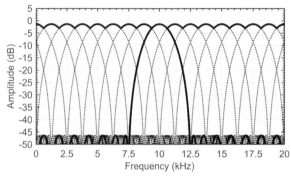

Figure 5.12 Straddle (scalloping) loss—45 dB Chebyshev weighting—$N = 16$, $F_s = 20$ kHz.

Table 5.9

Amplitude Weighting and Associated Properties[d]

Weighting	Max SLL (dB)	SNR Loss (dB)	Max. Straddle Loss (dB)	Mean Straddle Loss (dB)
Rectangular	−13.26	0.0	3.92	1.11
Bartlett	−26.52	1.3	1.82	0.57
Hamming	−42.67	1.3	1.75	0.55
$\cos^\kappa(x)$				
$\kappa = 1$	−23.00	0.9	2.09	0.64
$\kappa = 2$ (Hann)	−31.47	1.8	1.42	0.45
$\kappa = 3$	−39.30	2.4	1.07	0.34
$\kappa = 4$	−46.74	2.9	0.86	0.23
$\cos^\kappa(x)$ on pedestal[a]				
$\kappa = 1.0, \alpha = 0.04$	−23.02	0.8	2.20	0.68
$\kappa = 1.8, \alpha = 0.07$	−35.27	1.3	1.79	0.56
$\kappa = 2.0, \alpha = 0.16$	−34.31	1.0	2.04	0.63
$\kappa = 2.2, \alpha = 0.09$	−39.93	1.4	1.72	0.54
$\kappa = 3.0, \alpha = 0.02$	−40.78	2.2	1.19	0.38
Gaussian[b]				
$\alpha = 2.5$	−43.30	1.6	1.58	0.50
$\alpha = 3.0$	−56.64	2.3	1.16	0.37
Blackman	−58.11	2.4	1.10	0.35
Blackman-Harris				
Exact	−68.23	2.3	1.15	0.34
3-term	−70.52	2.3	1.13	0.36
4-term	−92.01	3.0	0.82	0.35
Nuttall	−98.16	3.0	0.85	0.28
Chebyshev				
60 dB	−60.00	1.8	1.42	0.45
80 dB	−80.00	2.4	1.09	0.35
100 dB	−100.00	2.9	0.88	0.23
Taylor[c]				
30 dB, $\bar{n} = 6$	−30.22	0.66	2.39	0.72
35 dB, $\bar{n} = 4$	−35.17	0.91	2.12	0.65
40 dB, $\bar{n} = 3$	−39.20	1.00	2.04	0.63

[a] The parameter κ controls pedestal height.

[b] The parameter α is inversely proportional to sidelobe level.

[c] The parameter $\bar{n}$ controls the extent of constant level sidelobes, specified in dB, nearest the main lobe.

[d] See Appendix A for windowing functions.

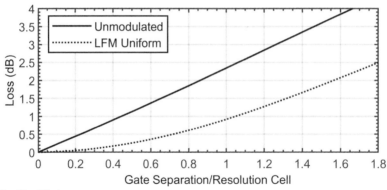

Figure 5.13 Straddle loss.

Detection decisions in radars are made by sampling the output of the matched filter or signal processor in range, and sometimes, in Doppler. Generally, the range samples are spaced between ½ and 1 range resolution cell width apart and the Doppler samples are spaced ½ to 1 Doppler resolution cell apart. Because of this finite spacing, it is likely that the samples will not occur at the peak of the target's range or Doppler response. The result is a loss in SNR.

Representative curves for this loss, which is called straddle loss, are indicated in Figure 5.13. The dashed curve applies to Doppler straddle loss and to range straddle loss when the radar uses LFM pulses. Nominal values of loss for these cases vary from about 0.3 to 1 dB for typical sample spacings of 0.5 to 1 resolution cell. For unmodulated pulses, or pulses with phase modulation, the loss is somewhat more severe and ranges from about 1 to 2.4 dB.

The final loss in Table 5.9 is CFAR loss. In modern radars, the detection threshold is computed by a CFAR circuit or algorithm because this circuit or algorithm can easily adapt to different noise (and jammer) environments. The CFAR attempts to determine the desired threshold-to-noise ratio (TNR—see Chapter 6) based on a limited number of samples of the noise at the output of the signal processor. Because of the limited number of samples used, the threshold will not be precisely set relative to theory. This impreciseness is accommodated by adding a CFAR loss to L.

The precise CFAR loss value depends on the type of CFAR and the number of noise samples (number of reference cells)[8] used to determine the threshold. It also depends upon the desired false alarm probability (P_{fa}), the detection probability (P_d) (though minimally), and the type of target (Swerling model—0 through 5—see Chapter 3) [43] [44, p. 639]. CFAR is discussed in Chapter 7.

For preliminary designs, we choose an expression that is applicable in general. One such expression is provided by Hansen and Sawyers for the CFAR loss of a greatest of (GO) cell averaging (CA) CFAR, given a square law detector and a Swerling 1 target [8] is

---

[8] While the exact number of reference cells depends on the application and dimensionality of the CFAR, from ~10 to ~40 cells is fairly typical. There is a trade-off between more reference cells lowering CFAR loss but resulting in slower threshold transitions at clutter boundaries. One rule of thumb is to use enough reference cells for ~ 1-dB CFAR loss.

$$L_{cfar} = \frac{P_{fa}^{-1/M} - P_{d}^{-1/M}}{P_{d}^{-1/M} - 1} \cdot \frac{\ln(P_{d})}{\ln(P_{fa}) - \ln(P_{d})} \tag{5.5}$$

where P_{fa} is the desired probability of false alarm, P_d is the desired probability of detection, and M is the number of reference cells used to form the noise estimate.

It turns out that (5.5) is also a reasonably good approximation when considering linear and log detectors, SO CFAR (smallest of CFAR) and CA CFAR (cell-averaging CFAR), as well as the other Swerling targets. An example of the dependency of CFAR loss on various parameters is shown in Figure 5.14 for a CA-CFAR.

To complete our example loss table, we will add processor and detection losses. For the L-band search radar we assume the radar is using LFM pulses with Hamming weighting in the matched filter to reduce the range sidelobes.[9] Since it is a search radar, we assume that the range samples are spaced one range resolution cell apart. The radar has the ability to use MTI processing, but for the long-range search uses only the LFM pulses (because the targets are expected to be beyond the horizon and we are not considering rain). The radar uses a CA-CFAR with a reference window of 18 range cells. The desired P_{fa} is 10^{-6}.

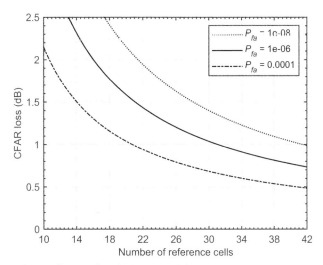

Figure 5.14 Loss for a cell averaging CFAR, $P_d = 0.9$.

---

[9] Weighting makes it a mismatched filter.

Table 5.10
Total Losses for the Example

Loss Term	L-Band Search Radar	S-Band Multifunction Radar	X-Band Active Array Radar
Prior losses	5.94	9.28	10.0
Filter mismatch	0.5	0.5	0.5
MTI	0	0.2	—
LFM weighting	1.3	1.3	—
Doppler weighting	—	—	2.9
Average range straddle	1	1	1
Average Doppler straddle	—	—	0.3
CFAR	1.8	1.9	0.64
Total	10.54 dB	14.18 dB	15.34 dB

The S-band radar also uses an LFM pulse with Hamming weighting. Since this radar may need to operate in ground clutter, it uses an MTI processor with a staggered PRI waveform. Analyses of the frequency response of the MTI indicates that the average SNR loss across the range rates of interest is about 0.2 dB. During search, the radar spaces the range samples one range resolution cell apart. It uses a GO-CFAR designed to provide a Pfa of 10^{-8}. The CFAR processor uses a reference window of 22 cells.

The X-band radar uses phase coded waveforms and a pulsed-Doppler signal processor. Since the radar has a stringent clutter rejection requirement, the pulsed-Doppler processor uses 100-dB Chebyshev weighting. The radar samples in range at one range resolution cell and in Doppler at ½ Doppler resolution cell. The radar uses a GO-CFAR with 32 reference cells and a Pfa of 10^{-4}. Even though the X-band radar uses Doppler processing, it performs CFAR and detection in only the range direction. Specifically, it performs CFAR and detection on each Doppler cell.

Table 5.10 contains the total losses with the processor and detection losses included.

The losses introduced in this chapter are what we consider representative of those one would use in a preliminary radar design or analysis. We did not attempt to present an exhaustive list of losses, as that would require hundreds of pages instead of the few devoted to this chapter. For more detailed expositions of the many loss terms that would need to be considered in a final radar design, the reader is directed to [1–8, 28, 30, 41, 45–53]. A very good reference is Barton's 2013 text [9], which contains approximately 200 pages dedicated to the discussion of losses. Another notable reference is Blake [32].

5.7 EXERCISES

1. The figures of merit for window functions (with respect to a rectangular window) are given by

Table 5.11
Window Figures of Merit—Equations

Description	Equation	Description	Equation
Peak signal coherent gain	$\dfrac{1}{N}\sum_{k=0}^{N-1} w(n)$	Processing gain	$\dfrac{1}{N}\dfrac{\left|\sum_{k=0}^{N-1} w(n)\right|^2}{\sum_{n=0}^{N-1} w^2(n)} = \dfrac{1}{\text{ENBW}}$
Peak signal power gain	$\dfrac{1}{N^2}\left[\sum_{k=0}^{N-1} w(n)\right]^2$	Equivalent noise bandwidth	$N\dfrac{\sum_{n=0}^{N-1} w^2(n)}{\left|\sum_{k=0}^{N-1} w(n)\right|^2}$
Peak noise power gain	$\dfrac{1}{N}\sum_{n=0}^{N-1} w^2(n)$	Scalloping loss	$\dfrac{\left|\sum_{n=0}^{N-1} w(n)\, e^{-\left(j\frac{\pi n}{N}\right)}\right|}{\sum_{n=0}^{N-1} w(n)}$

Source: [34]

Calculate the parameters in Table 5.11 for Hamming, Hann, and Gaussian windows for $N = 32$. Relate these parameters to those in Table 5.9.

2. There are two forms for a window function, referred to as symmetric and periodic.[10] Appendix B lists some window functions in causal symmetric forms (identical endpoints) is generally used for finite impulse response (FIR) filter design. Periodic forms, characterized by a missing (implied) endpoint to accommodate periodic extension, are generally used for spectral estimation (divide by N versus $N - 1$).

Pick one window function from Appendix A (at the end of the book) and plot the symmetric and periodic forms on the same chart for $N = 16$. Select parameters using Table 5.9 as necessary. Generate a separate chart showing the FFT of the periodic and symmetric forms. Zero pad as necessary for a clear plot. What differences are evident in the frequency domain?

3. Generate Figure 5.12 for a 32-point Gaussian weighing.

---

[10] For some windows (but not all), MATLAB® can generate either symmetric (default) or periodic windows, via a flag in the particular window function, with symmetric being the default. The spectral difference is minimal, decreasing as N increases.

4. The parameters in Table 5.9 are a weak function of N. For a Hamming weighting, generate plots of SLL, SNR loss, and peak straddle loss versus N. Let N vary from 0 to 500.

5. Several antenna pattern models often used for analysis are listed in Table 5.12. Plot all of the power patterns on the same figure. Let $\theta_3 = 2.3°$. How do they compare? What is the beamshape loss for each?

Table 5.12
Antenna Pattern Models

Pattern	Voltage Pattern	Sidelobe Level
Gaussian	$f_g(\theta) = e^{-2(\ln 2)\frac{\theta^2}{\theta_3^2}}$	N/A
Uniform rectangular	$f_u(\theta) = \dfrac{\sin(\pi K_\theta \theta/\theta_3)}{\pi K_\theta \theta/\theta_3} = \dfrac{\sin(0.8859\pi \theta/\theta_3)}{0.8859\pi \theta/\theta_3}$	-13.6 dB
Uniform rectangular	$f_{\cos}(\theta) = \dfrac{\cos(\pi K_\theta \theta/\theta_3)}{1-(2K_\theta \theta/\theta_3)^2} = \dfrac{\cos(3.7353\theta/\theta_3)}{1-(2.3779\theta/\theta_3)^2}$	-23 dB
Uniform circular	$f_{\text{cir}}(\theta) = \dfrac{2J_1(\pi K_\theta \theta/\theta_3)}{\pi K_\theta \theta/\theta_3} = \dfrac{2J_1(1.0290\pi \theta/\theta_3)}{1.0290\pi \theta/\theta_3}$	-17.6 dB

$J_1(\bullet)$ is the first-order Bessel function of the first kind.

Source: [9].

References

[1] Norton, K. A., and A. C. Omberg, "The Maximum Range of a Radar Set," *Proc. IRE,* vol. 35, no. 1, Jan. 1947, pp. 4–24. First published Feb. 1943 by U.S. Army, Office of Chief Signal Officer in the War Department, in Operational Research Group Report, ORG-P-9-1.

[2] Hall, W. M., "Prediction of Pulse Radar Performance," *Proc. IRE,* vol. 44, no. 2, Feb. 1956, pp. 224–231. Reprinted: Barton, D. K., *Radars, Vol. 2: The Radar Range Equation*, Norwood, MA: Artech House, 1974, pp. 31–38.

[3] Blake, L. V., "Recent Advancements in Basic Radar Range Calculation Technique," *IRE Trans. Mil. Electron.,* vol. 5, no. 2, Apr. 1961, pp. 154–164.

[4] Blake, L. V., "Curves of Atmospheric-Absorption Loss for Use in Radar Range Calculation," Naval Research Laboratory, Washington, D.C., Rep. No. 5601, Mar. 23, 1961. Available from DTIC as AD255135.

[5] Hall, W. M., and D. K. Barton, "Antenna Pattern Loss Factor for Scanning Radars," *Proc. IEEE,* vol. 53, no. 9, Sept. 1965, pp. 1257–1258.

[6] Hall, W. M., and H. R. Ward, "Signal-to-Noise Loss in Moving Target Indicator," *Proc. IEEE,* vol. 56, no. 2, Feb. 1968, pp. 233–234.

[7] Trunk, G. V., "MTI Noise Integration Loss," *Proc. IEEE,* vol. 65, no. 11, Nov. 1977, pp. 1620–1621.

[8] Hansen, V. G., and J. H. Sawyers, "Detectability Loss Due to 'Greatest Of' Selection in a Cell-Averaging CFAR," *IEEE Trans. Aerosp. Electron. Syst.,* vol. 16, no. 1, Jan. 1980, pp. 115–118.

[9] Barton, D. K., *Radar Equations for Modern Radar,* Norwood, MA: Artech House, 2013.

[10] Blake, L. V., "A Guide to Basic Pulse-Radar Maximum-Range Calculation, Part 1—Equations, Definitions, and Aids to Calculation," Naval Research Laboratory, Washington, D.C., Rep. No. 6930, Dec. 23, 1969, p. 49. Available from DTIC as 701321.

[11] Blake, L. V., "A Guide to Basic Pulse-Radar Maximum-Range Calculation, Part 2—Derivations of Equations, Bases of Graphs, and Additional Explanations," Naval Research Laboratory, Washington, D.C., Rep. No. 7010, Dec. 31, 1969. Available from DTIC as 703211.

[12] Mini-Circuits, *RF/IF Designer's Handbook,* Brooklyn, NY: Mini-Circuits, 1992.

[13] Watkins-Johnson Co., *RF and Microwave Component Designer's Handbook,* Palo Alto, CA: Watkins-Johnson Co., 1988.

[14] Losee, F. A., *RF Systems, Components, and Circuits Handbook,* 2nd ed., Norwood, MA: Artech House, 2005.

[15] Saad, T. S., R. C. Hansen, and G. J. Wheeler, eds., *Microwave Engineers' Handbook,* vol. 1, Dedham, MA: Artech House, 1971.

[16] Saad, T. S., R. C. Hansen, and G. J. Wheeler, eds., *Microwave Engineers' Handbook,* vol. 2, Dedham, MA: Artech House, 1971.

[17] Kahrilas, P. J., *Electronic Scanning Radar Systems (ESRS) Design Handbook,* Norwood, MA: Artech House, 1976.

[18] *IEEE Standard Dictionary of Electrical and Electronic Terms,* 6th ed., New York: IEEE, 1996.

[19] Gardiol, F. E., *Introduction to Microwaves,* Dedham, MA: Artech House, 1984.

[20] Pozar, D. M., *Microwave Engineering,* New York: Addison Wesley, 1990.

[21] Agrawal, A. K., and E. L. Holzman, "Beamformer Architectures for Active Phased-Array Radar Antennas," *IEEE Trans. Antennas Propag.,* vol. 47, no. 3, Mar. 1999, pp. 432–442.

[22] Mailloux, R. J., *Phased Array Antenna Handbook,* 2nd ed., Norwood, MA: Artech House, 2005.

[23] Hansen, R. C., ed., *Microwave Scanning Antennas, Vol. III: Array Systems,* New York: Academic Press, 1966.

[24] Brookner, E., "Right Way to Calculate Reflector and Active-Phased-Array Antenna System Noise Temperature Taking into Account Antenna Mismatch," *IEEE Int. Symp. Phased Array Syst. and Technology 2003,* Boston, MA, Oct. 14–17, 2003, pp. 130–135.

[25] Blake, L. V., "The Effective Number of Pulses Per Beamwidth for a Scanning Radar," *Proc. IRE,* vol. 41, no. 6, Jun. 1953, pp. 770–774.

[26] Hall, W. M., "Antenna Beam-Shape Factor in Scanning Radars," *IEEE Trans. Aerosp. Electron. Syst.,* vol. 4, no. 3, May 1968, pp. 402–409.

[27] United States Committee on Extension to the Standard Atmosphere (COESA), *U.S. Standard Atmosphere, 1976,* Washington, D.C.: U.S. Government Printing Office, Oct. 1976.

[28] Blake, L. V., *Radar Range-Performance Analysis,* Norwood, MA: Artech House, 1986.

[29] Humphreys, W. J., *Physics of the Air,* 3rd ed., New York: McGraw-Hill, 1940.

[30] Vizmuller, P., *RF Design Guide: Systems, Circuits and Equations,* Norwood, MA: Artech House, 1995.

[31] Cook, C. E., and M. Bernfeld, *Radar Signals: An Introduction to Theory and Application,* New York: Academic Press, 1967. Reprinted: Norwood, MA: Artech House, 1993.

[32] Lewis, B. L., et al., *Aspects of Radar Signal Processing,* Norwood, MA: Artech House, 1986.

[33] Boukeffa, S., Y. Jiang, and T. Jiang, "Sidelobe Reduction with Nonlinear Frequency Modulated Waveforms," in *Proc. IEEE 7th Int. Colloquium on Signal Proc. and Its Applications (CSPA '11)*, Penang, Malaysia, Mar. 2011, pp. 399–403.

[34] Barton, D. K., ed., *Radars, Vol. 3: Pulse Compression* (Artech Radar Library), Dedham, MA: Artech House, 1975.

[35] Schleher, D. C., *MTI and Pulsed Doppler Radar with MATLAB*, 2nd ed., Norwood, MA: Artech House, 2010.

[36] Barton, D. K., *Radar System Analysis*, Englewood Cliffs, NJ: Prentice-Hall, 1964; Dedham, MA: Artech House, 1976.

[37] Minkler, G., and J. Minkler, *CFAR*, Baltimore, MD: Magellan, 1990.

[38] Mitchell, R. L., and J. F. Walker, "Recursive Methods for Computing Detection Probabilities," *IEEE Trans. Aerosp. Electron. Syst.*, vol. 7, no. 4, Jul. 1971, pp. 671–676.

[39] Gregers-Hansen, V., "Constant False Alarm Rate Processing in Search Radars," *Radar—Present and Future*, IEE Conf. Publ. No. 105, London, Oct. 1973, pp. 325–332.

[40] Hansen, V. G., and H. R. Ward, "Detection Performance of the Cell Averaging LOG/CFAR Receiver," *IEEE Trans. Aerosp. Electron. Syst.*, vol. 8, no. 5, Sept. 1972, pp. 648–652.

[41] Nitzberg, R., "Analysis of the Arithmetic Mean CFAR Normalizer for Fluctuating Targets," *IEEE Trans. Aerosp. Electron. Syst.*, vol. 14, no. 1, Jan. 1978, pp. 44–47.

[42] Weiss, M., "Analysis of Some Modified Cell-Averaging CFAR Processors in Multiple-Target Situations," *IEEE Trans. Aerosp. Electron. Syst.*, vol. 18, no. 1, 1982, pp. 102–114.

[43] Nathanson, F. E., J. P. Reilly, and M. N. Cohen, eds, *Radar Design Principles*, 2nd ed., New York: McGraw-Hill, 1991.

[44] Barton, D. K., *Radar System Analysis and Modeling*, Norwood, MA: Artech House, 2005.

[45] Naval Air Warfare Center Weapons Division (NAWCWD), *Electronic Warfare and Radar Systems Engineering Handbook*, 4th ed., NAWCWD Technical Communications Office, Point Mugu, CA, Rep. No. NAWCWD TP 8347, Oct. 2013.

[46] Meikle, H., *Modern Radar Systems*, 2nd ed., Norwood, MA: Artech House, 2008.

[47] Skolnik, M. I., *Introduction to Radar Systems*, 3rd ed., New York: McGraw-Hill, 2001.

[48] Skolnik, M. I., ed., *Radar Handbook*, 3rd ed., New York: McGraw-Hill, 2008.

[49] Harris, F., "On the Use of Windows for Harmonic Analysis with the Discrete Fourier Transform," *Proc. IEEE*, vol. 66, no. 1, Jan. 1978, pp. 51–83.

[50] Southworth, G. C., "Hyper-Frequency Wave Guides—General Considerations and Experimental Results," *Bell Syst. Tech. J.*, vol. 15, no. 2, Apr. 1936, pp. 284–309.

[51] Barrow, W. L., "Transmission of Electromagnetic Waves in Hollow Tubes of Metal," *Proc. IRE*, vol. 24, Oct. 1936, pp. 1298–1328.

[52] Tyrrell, W. A., "Hybrid Circuits for Microwaves," *Proc. IRE*, vol. 35, no. 11, Nov. 1947, pp. 1294–1306.

[53] Montgomery, C. G., R. H. Dicke, and E. M. Purcell, eds., *Principles of Microwave Circuits*, vol. 8, New York: McGraw-Hill, 1948; Norwood, MA: Artech House (CD-ROM edition), 1999.

[54] Gardiol, F. E., *Introduction to Microwaves*, Dedham, MA: Artech House, 1984.

[55] Pozar, D. M., *Microwave Engineering*, 4th ed., New York: Wiley & Sons, 2011.

[56] Martin, W.H., "Decibel—The Name for the Transmission Unit," *Bell Syst. Tech. J.*, vol. 8, no.1, Jan. 1929, pp.1–2.

[57] Blake, L. V., "Ray Height Computation for a Continuous Nonlinear Atmospheric Refractive-Index Profile," *Radio Science,* vol. 3, no. 1, Jan. 1968, pp. 85–92.

[58] Bean, B. R., and G. D. Thayer, "Models of the Atmospheric Radio Refractive Index," *Proc. IRE,* vol. 47, no. 5, May 1959, pp. 740–755.

[59] Minzner, R.A., W.S. Ripley, and T. P. Condron, "U.S. Extension to the ICAO Standard Atmosphere," U.S. Dept. of Commerce Weather Bureau and USAF ARDC Cambridge Research Center, Geophysics Research Directorate, U.S. Government Printing Office, Washington, D.C., 1958

[60] Sissenwine, N. D., D. Grantham, and H. A. Salmela, "Humidity Up to the Mesopause," USAF Cambridge Res. Labs., Bedford, MA, Rep. No. AFCRL-68-0050, Oct. 1968.

[61] Van Vleck, J. H., "The Absoption of Microwaves by Oxygen," *Phys. Rev.,* vol. 71, no. 7, Apr. 1947, pp. 413–424.

[62] Van Vleck, J. H., "The Absorption of Microwaves by Uncondensed Water Vapor," *Phys. Rev.,* vol. 71, no. 7, Apr. 1947, pp. 425–433.

[63] Leibe, H. J., "Calculated Tropospheric Dispersion and Absorption Due to the 22-GHz Water Vapor Line," *IEEE Trans. Antennas Propag.,* vol. 17, no. 5, Sept. 1969, pp. 621–627.

[64] Meeks, M. L., and A. E. Lilley, "The Microwave Spectrum of Oxygen in the Earth's Atmosphere," *J. Geophys. Res.,* vol. 68, no. 6, Mar. 15, 1963, pp. 1683–1703.

[65] Reber, E. E., R. L. Mitchel, and C. J. Carter, "Attenuation of the 5-mm Wavelength Band in a Variable Atmosphere," *IEEE Trans. Antennas Propag.,* vol. 18, no. 4, Jul. 1970, pp. 472–479.

[66] Van Vleck, J. H., and V. F. Weisskopf, "On the Shape of Collision-Broadened Lines," *Rev. Mod. Phys.,* vol. 17, nos. 2 & 3, Apr. 1945, pp. 227–236.

[67] Ulaby, F. T., and A. W. Straiton, "Atmospheric Absorption of Radio Waves Between 150 and 350 GHz," *IEEE Trans. Antennas Propag.,* vol. 18, no. 4, Jul. 1970, pp. 479–485.

[68] Bauer, J. R., W. C. Mason, and F. A. Wilson, "Radio Refraction in a Cool Exponential Atmosphere," MIT Lincoln Lab., Cambridge, MA, Tech. Rep. No. 186, Aug. 27, 1958.

APPENDIX 5A: WAVEGUIDE ATTENUATION

Waveguide, invented by Bell Telephone Laboratories [54] and studied in parallel at MIT [55] in the 1930s, is one RF transmission line commonly used in radar because of its low loss and high power handling capability. This is especially applicable for the types of transmitters depicted in Figure 5.1, where all of the transmit power travels through a single RF path. The waveguide's dimensions (square, rectangular, circular) determine the operating frequency range and the material (gold, silver, copper, aluminum, brass) affects the loss.

Rectangular waveguide is frequently used in radar.[11] For illustration, Figure 5A.1 shows a magic T (or tee) constructed with WR-90 waveguide. The magic T is a four-port, 180°, 3-dB hybrid developed during World War II [19, 56]. A 180° hybrid is a reciprocal four-port device. When fed from the sum port (Σ), the input is split into two equal-amplitude in-phase signals exiting via ports A and B. When fed from the delta port (Δ), the input is split into two equal-amplitude 80° out-of-phase outputs exiting via ports A

---

[11] The rule of thumb for standard rectangular waveguide design is to use a 2:1 wall length ratio, which ensures that only the TE_{10} mode (dominant mode in rectangular waveguide) will propagate. In practice, wall ratios vary slightly, ~ 2:1 to 2.2:1.

and B. As such, the magic T can be used as a very low loss power combiner or a power divider [18, 24, 57], depending on the ports used. Copper and copper alloy are standard waveguide materials (solid or plating).[12]

Table 5A.1 contains the Electronic Industries Association (EIA) waveguide (WG) designations,[13] inner dimensions, frequency range, and theoretical attenuation for a number of standard rectangular copper waveguides [15]. For the EIA designation, the WR number is the internal dimension in hundredths of inches of the broad wall. Figure 5.2 contains plots of the theoretical waveguide loss versus frequency for several of the waveguides in Table 5A.1. We note that loss is inversely proportional to frequency. Also, the frequency boundaries are not coincident with radar designators. The general rule of thumb used to decide between multiple waveguide possibilities is to select the larger waveguide, which has lower loss.

Figure 5A.1 Waveguide magic T (WR-90).

---

[12] Copper waveguide provides good heat dissipation. Additional heat transfer can be achieved by brazing coolant lines directly onto the waveguide.
[13] Joint Army Navy (JAN) designators exist as well.

Table 5A.1

Standard Rectangular Waveguide Specifications (Copper)

Letter Band	EIA WG Designation	Dimensions (inches)	Frequency (GHz)	Theoretical Loss (dB/m)
L	WR-510	5.100 × 2.550	1.45–2.20	0.010–0.007
	WR-430	4.300 × 2.150	1.70–2.60	0.013–0.008
S	WR-284	2.840 × 1.340	2.60–3.95	0.024–0.016
	WR-229	2.290 × 1.145	3.30–4.90	0.031–0.022
	WR-187	1.872 × 0.872	3.95–5.85	0.045–0.031
	WR-159	1.590 × 0.795	4.90–7.05	0.050–0.038
C	WR-137	1.372 × 0.622	5.85–8.20	0.064–0.051
	WR-112	1.122 × 0.497	7.05–10.0	0.090–0.070
X	WR-90	0.900 × 0.400	8.20–12.40	0.140–0.097
	WR-75	0.750 × 0.375	10.00–15.00	0.166–0.116
Ku	WR-62	0.622 × 0.311	12.40–18.00	0.209–0.153
	WR-51	0.510 × 0.255	15.00–22.00	0.285–0.207
K	WR-42	0.420 × 0.170	18.00–26.50	0.446–0.328
	WR-34	0.340 × 0.170	22.00–33.00	0.545–0.379
Ka	WR-28	0.280 × 0.140	26.50–40.00	0.741–0.508

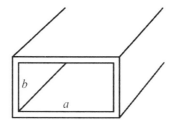

Figure 5A.2 Rectangular waveguide.

As a side note, waveguide is often pressurized, typically using dry air, nitrogen, or argon to prevent moisture buildup inside the waveguide, which can cause corrosion of the conducting surfaces, thus increasing loss.[14] In addition to using dry gas as dielectric, the slight overpressure helps keep out moisture in the event of small leaks. Microwave transparent windows are used to prevent pressure loss where the waveguide would be open (e.g., feed horn).

For example, consider a copper-plated, pressurized, rectangular WR-90 waveguide, depicted in Figure 5A.2, operating at 10 GHz (X-band) filled with nitrogen. For the

---

[14] Interestingly, extreme over-pressurization, perhaps from a pressure regulator failure, will turn rectangular waveguide into round waveguide.

dominant mode,[15] we want to determine the cutoff frequency in GHz and the attenuation due to conductor loss in dB/m.

From Table 5A.1, we see that WR-90 waveguide has interior dimensions of the broad and short walls of $a = 2.286$ cm (0.90 in) and $b = 1.016$ cm (0.4 in), respectively. Since the wall length ratio is ~ 2:1, the dominate mode of propagation is the TE_{10} mode. The cutoff frequency for the mn mode is given by [18, p. 113]

$$f_c^{mn} = \frac{k_c^{mn}}{2\pi\sqrt{\mu\varepsilon}} \quad (\text{Hz}) \tag{5A.1}$$

where

$$k_c^{mn} = \sqrt{\left(\frac{m\pi}{a}\right)^2 + \left(\frac{n\pi}{b}\right)^2} \quad \text{Hz} \tag{5A.2}$$

is the cutoff wave number. To clarify cutoff frequency as used here, for frequencies above the cutoff frequency for a given mode, the electromagnetic energy can be transmitted through the guide for that particular mode with minimal attenuation (which is backward compared to lowpass filter cutoff terminology).

For typical gaseous dielectrics used to fill waveguides (air, nitrogen, argon), the permittivity and permeability are essentially identical to those of free space (vacuum). Recall the permittivity of free space is $\mu_0 = 400\pi \approx 1256.637061$ nH/m and the permeability of free space is $\varepsilon_0 = 1/\mu_0 c^2 \approx 8.8541878176$ pF/m. For the TE_{10} mode, (5A.1) and (5A.2) simplify to

$$k_c^{10} = \frac{\pi}{a} \quad \text{Hz} \tag{5A.3}$$

and

$$f_c^{10} = \frac{1}{2a\sqrt{\mu\varepsilon}} = \frac{c}{2a}$$

$$f_c^{10} = \frac{299{,}792{,}458}{2 \cdot 0.90 \cdot 0.0254} = 6.56 \text{ GHz} \tag{5A.4}$$

The upper bound on propagation is the TE_{20} mode waveguide cutoff frequency calculated using (5A.1). Recall that above the cutoff frequency for a given mode, the electromagnetic energy will propagate through the guide for that particular mode with minimal attenuation provided that the operating frequency is appreciably above the cutoff frequency. As the operating frequency approaches the cutoff frequency from above the attenuation increases toward infinity at the cutoff frequency.

---

[15] The dominant mode is the mode with the lowest cutoff frequency, which for a rectangular waveguide ($a > b$) is the TE_{10} mode.

$$f_c^{20} = \frac{1}{a\sqrt{\mu\varepsilon}} = \frac{1}{a\sqrt{\mu_0\left(\dfrac{1}{\mu_0 c^2}\right)}} = \frac{c}{a} \tag{5A.5}$$

$$f_c^{20} = \frac{299{,}792{,}458}{0.90 \cdot 0.0254} = 13.11 \text{ GHz}$$

Therefore, TE_{10} mode will propagate at frequencies above 6.56 GHz and below 13.11 GHz.

Digressing for a moment, we note that 6.56 GHz to 13.11 GHz does not match the operating frequency range given in Table 5A.1. We illustrate the rationale for this discrepancy by comparing the waveguide loss for both the theoretical and recommended frequency ranges presented in Figure 5A.3. While the TE_{10} mode will technically propagate with up to ~3-dB loss, the amount of loss considered acceptable for waveguide is much lower.

Returning to our loss example, the attenuation due to conductor loss (loss due to the metal of the waveguide)[16] is given by [18, p. 115]

$$\alpha_c = \frac{R_s}{a^3 b \beta k \eta}\left(2b\pi^2 + a^3 k^2\right) \text{ Np/m} \tag{5A.6}$$

where we recall that nepers[17] (Np), defined in the same Bell Labs paper as dB [58], is a unit based on the natural logarithm, and is given, for voltage, by [18, p. 63]

$$\ln\left(\frac{V_1}{V_2}\right) \text{ Np} \tag{5A.7}$$

and for power by

$$\frac{1}{2}\ln\left(\frac{P_1}{P_2}\right) \text{ Np} \tag{5A.8}$$

---

[16] The attenuation due to dielectric loss (the material filling the waveguide) is negligible for typical gaseous dielectrics.
[17] Derived from the name of John Napier, who invented the natural logarithm [56].

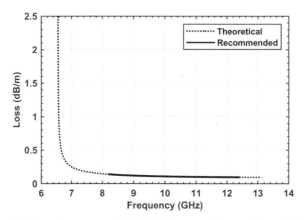

Figure 5A.3 Theoretical loss for copper-plated WR-90 waveguide over theoretical and recommended operating ranges.

The propagation constant, β, for the TE$_{10}$ mode is given by [18, p. 112]

$$\beta^{10} = \sqrt{k^2 - \left(k_c^{10}\right)^2} = \sqrt{\left(\omega\sqrt{\mu\varepsilon}\right)^2 - \left(\frac{\pi}{a}\right)^2}$$

$$\beta^{10} = \sqrt{\left(2\pi \cdot 10 \cdot 10^9 \sqrt{400\pi \cdot 10^{-9} \times 8.854187 \cdot 10^{-12}}\right)^2 - \left(\frac{\pi}{0.0229}\right)^2} \quad (5A.9)$$

$$\beta^{10} = 158.2 \quad \text{rad/m}$$

where

$$k = \omega\sqrt{\mu\varepsilon} = 209.6 \quad \text{m}^{-1} \quad (5A.10)$$

is the free space wave number. The intrinsic impedance of the dielectric is

$$\eta = \sqrt{\mu/\varepsilon} = 377 \quad \Omega \quad (5A.11)$$

The last component of (5A.6) is the surface resistivity of the metal in the waveguide, given by [55, p. 28]

$$R_s = \sqrt{\frac{\omega\mu}{2\sigma}} = \sqrt{\pi f \mu/\sigma} \quad \Omega \quad (5A.12)$$

where σ is the material conductivity. The conductivity for copper and other common waveguide materials is listed in Table 5A.2. For this example, copper is specified, which has a conductivity of 5.813×10^7 mho/m [24, p. 458]. This results in a surface resistance of $R_s = 0.0261\Omega$. Substitution into (5A.6) gives

$$\alpha_c = 0.0125 \quad \text{Np/m} \quad (5A.13)$$

Table 5A.2
Material Conductivity

Material	Conductivity (mho/m)
Aluminum	$3.816 \cdot 10^7$
Brass	$2.564 \cdot 10^7$
Copper	$5.813 \cdot 10^7$
Gold	$4.098 \cdot 10^7$
Silver	$6.173 \cdot 10^7$

Source: [54].

Converting this to dB/m gives

$$\alpha_c = 0.1083 \quad \text{dB/m} \tag{5A.14}$$

where nepers are related to dB by [18, p. 63; 19, p. 260]

$$1 \text{ Np} = 10 \log(e^2) = \frac{20}{\ln(10)} = 8.686 \text{ dB} \tag{5A.15}$$

For completeness, we convert the answer to dB/100 ft, since historically, many waveguide tables are presented in dB/100 ft.

$$\alpha_c = 3.2801 \quad \text{dB/100 ft} \tag{5A.16}$$

5A.1 EXERCISES

1. Typical waveguide plating materials are aluminum, brass, copper, gold, and silver (or alloys thereof). For these materials, calculate the theoretical loss across the recommended operating frequency for WR-90. Assume the waveguide is filled with pressurized dry nitrogen. Generate a comparison plot similar to Figure 5.2. For reference, use Table 5A.2.

2. Generate Figure 5.2 for a silver-plated waveguide. Also plot the approximation given by (5.1). How does the approximation compare?

3. Consider an air-filled rectangular waveguide operating at 2 GHz. Select an appropriate waveguide size. What are the interior dimensions? Calculate the upper and lower frequency bounds for propagation. Recall that for a wall length ratio of ~2:1, the dominate mode of propagation is the TE_{10} mode. For reference, recall the permittivity of free space is $\mu_0 = 400\pi \approx 1256.637061$ nH/m and the permeability of free space is $\varepsilon_0 = 1/\mu_0 c^2 \approx 8.8541878176$.

APPENDIX 5B: ATMOSPHERIC AND RAIN ATTENUATION

For reference, the equations and data outlined below are used to generate Figures 5.7 through 5.9 (two-way atmospheric loss) and Figure 5.10 (rain attenuation). The equations summarized in this appendix are coded in the MATLAB functions listed in Table 5B.1 and included on the CD. Note that the equations are presented in the order of execution in their associated function (e.g., terms are calculated for use in functions defined subsequently). For a complete explanation of the origins and theory for atmospheric absorption, please refer to [4, 9, 32, 59, 60].

Table 5B.1
Atmospheric and Rain Attenuation Function Summary

Function Description	Resulting Figure
[R, L, Lox, Lwv] = troploss(f, ang): This function computes the accumulated two-way tropospheric absorption loss (in dB) for an RF signal with frequency f along a refracted path that originates at the earth's surface and has an elevation angle ang. It returns the loss for oxygen, water vapor, total loss and the associated ranges.	Figure 5.7, 5.8, 5.9
[Re, h, phi] = troprefract(ang): This function computes refracted RF path through the troposphere for the elevation angle ang. It returns range, height, and the angular position PHI of the refracted path.	Called by troploss.m
[g, gox, gwv] = tropatten(f, h): Given frequency f and altitude h, this function computes tropospheric absorption coefficient versus frequency and altitude. It returns the tropospheric absorption coefficients, g, (in dB/km) as well as the component absorption coefficients for oxygen, gox, and water vapor, gwv.	Called by troploss.m
[K] = rainAttn2way(f, rr): Given frequency, f, in GHz and rain rate, rr, in mm/hr, this function uses the standard model for rain attenuation to compute two-way loss (dB/km) as a function of operating frequency and rain rate.	Figure 5.10

5B.1 FUNCTION TROPATTEN.M

5B.1.1 Compute International Civil Aviation Organization (ICAO) Standard Atmosphere 1964

In determining atmospheric attenuation, knowledge of the atmosphere's pressure, temperature, and water vapor density, all of which variy with altitude, is necessary. Given the varying nature of the atmosphere due to such factors as location, time of day, or season, a standard model is used [31, 61]. The standard atmosphere model[18] (based on experimental data) provides a defined variation of mean values of temperature, pressure, and water vapor density as a function of altitude. Specifically, temperature and pressure are modeled using an empirical equation, while water vapor density is determined via table lookup.

As the first step in computing the tropospheric absorption coefficient versus frequency and altitude, the function tropatten.m first calculates the geopotential altitude (based on the assumption of constant gravity at all altitude), h_g, which is related to the geometric altitude (referenced to mean sea level), h_a, by [32, p. 206; 61][19]

---

[18] The current standard atmosphere is the 1976 version, but below 32,000 km, the altitudes of interest for most ground-based radars, the models are equivalent [31, p. 227].
[19] The difference between geopotential and geometric altitude is very small for altitudes less than 30 km, but most standard atmospheric tables quote geopotential altitude.

$$h_g = \frac{r_0}{r_0 + h_a} h_a \quad \text{m} \tag{5B.1}$$

where $r_0 = 6{,}371$ km is the radius of the earth.

Using the results of (5B.1), the atmosphere temperature and pressure as a function of geopotential altitude are determined by [32, p. 205; 61]

$$\left. \begin{array}{l} T = 288.16 - 0.0065 h_g \\ p = 1{,}013.25 \left[\dfrac{T}{288.16} \right]^\alpha \end{array} \right\} \quad h_g \leq 11{,}000 \text{ m} \tag{5B.2}$$

$$\left. \begin{array}{l} T = 216.66 \\ p = 226.32 \exp\left[-\dfrac{\beta}{T}\left(h_g - 11{,}000\right) \right] \end{array} \right\} \quad 11{,}000 \leq h_g \leq 25{,}000 \text{ m} \tag{5B.3}$$

$$\left. \begin{array}{l} T = 216.66 - 0.003\left(h_g - 25{,}000\right) \\ p = 24.886 \left[\dfrac{216.66}{T} \right]^\gamma \end{array} \right\} \quad 25{,}000 \leq h_g \leq 47{,}000 \text{ m} \tag{5B.4}$$

which are equations describing the absolute atmospheric temperature, T (kelvin), and total atmospheric pressure, p (millibars), for the standard atmospheric model [61]. The coefficient values are: $\alpha = 5.2561222$, $\beta = 0.034164794$, and $\gamma = 11.388265$. The water vapor density for the U.S. standard atmosphere is provided in Table 5B.2 [32, p. 207].

The values in Table 5B.2 are the mid-latitude mean water vapor densities for a surface value ($h = 0$) of 5.947 g/m³ [62]. However, the current standard is to use a surface value of water vapor density of 7.5 g/m³. As such, we translate the vapor density values in Table 5B.2 such that the surface water vapor density is 7.5 g/m³ [32, p. 206] using

$$\rho(h) = \rho_{table}(h) \frac{7.5}{5.947} \tag{5B.5}$$

To determine water vapor, interpolate as necessary into Table 5B.2.

Table 5B.2
Mid-Latitude Mean Water Vapor Densities

Altitude (km)	Density (g/m^3)	Altitude (km)	Density (g/m^3)
0	5.947×10^0	18	4.449×10^{-4}
2	2.946×10^0	20	4.449×10^{-4}
4	1.074×10^0	22	5.230×10^{-4}
6	3.779×10^{-1}	24	6.138×10^{-4}
8	1.172×10^{-1}	26	7.191×10^{-4}
10	1.834×10^{-2}	28	5.230×10^{-4}
12	3.708×10^{-3}	30	3.778×10^{-4}
14	8.413×10^{-4}	32	2.710×10^{-4}
16	6.138×10^{-4}		

Source: [32].

5B.1.2 Absorption Coefficient for Oxygen

According to Blake [32], the original theory for determining the absorption coefficient for oxygen was presented by Van Vleck [63, 64, 65], with further refinements being made later on [66, 67]. To determine absorption, we take the summation of the contributions of several oxygen resonance lines. These resonant frequencies are listed in Table 5B.3 [32, p. 201; 9, p. 233; 66]. For each value of $N = 1, 3, \ldots 45$, there are two resonant frequencies, denoted f_{N+} and f_{N-}. N is comprised of odd integers because N is the quantum rotational number. For values of N greater than 45, the absorption contribution is negligible [32, p. 200].

Table 5B.3
Oxygen Resonance Frequencies

N	f_{N+} (GHz)	f_{N-} (GHz)	N	f_{N+} (GHz)	f_{N-} (GHz)
1	56.2648	118.7505	25	65.7626	53.5960
3	58.4466	62.4863	27	66.2978	53.0695
5	59.5910	60.3061	29	66.8313	52.5458
7	60.4348	59.1642	31	67.3627	52.0259
9	61.1506	58.3239	33	67.8923	51.5091
11	61.8002	57.6125	35	68.4205	50.9949
13	62.4112	56.9682	37	68.9478	50.4830
15	62.9980	56.3634	39	69.4741	49.9730
17	63.5685	55.7839	41	70.0000	49.4648
19	64.1272	55.2214	43	70.5249	48.9582
21	64.6779	54.6728	45	71.0497	48.4530
23	65.2240	54.1294			

Source: [66].

Using the parameters in Table 5B.3, we calculate the following values [32, p. 200]

$$\mu_{N+}^2 = \frac{N(2N+3)}{N+1} \tag{5B.6}$$

$$\mu_{N-}^2 = \frac{(N+1)(2N-1)}{N} \tag{5B.7}$$

$$\mu_{N0}^2 = \frac{2(N^2+N+1)(2N+1)}{N(N+1)} \tag{5B.8}$$

$$\frac{E_N}{k} = 2.06844 N(N+1) \tag{5B.9}$$

where (5B.6), (5B.7), (5B.8) and (5B.9) are terms used in the model (5B14).

$$\Delta f = g(h) \frac{p}{p_0} \frac{T_0}{T} \tag{5B.10}$$

where $p_0 = 1013.25$ mbar (760 torr) is the pressure at sea level, and $T_0 = 360$ K[20] and [32, 67]

$$g(h) = \begin{cases} 0.640 & 0 \le h \le 8 \text{ km} \\ 0.640 + 0.04218(h-8) & 8 \le h \le 25 \text{ km} \\ 1.357 & h > 25 \text{ km} \end{cases} \tag{5B.11}$$

where h is the height above mean sea level in km.

The values calculated above are components of [32, p. 200; 68]

$$F_{N\pm}(f) = \frac{\Delta f}{(f_{N\pm}-f)^2 + (\Delta f)^2} + \frac{\Delta f}{(f_{N\pm}+f)^2 + (\Delta f)^2} \tag{5B.12}$$

Equation (5B.12) is the Van Vleck-Weisskopf formula, which provides the shapes of the resonance lines [32]. The nonresonant contribution is of the form [32, p. 200]

$$F_0 = \frac{\Delta f}{f^2 + (\Delta f)^2} \tag{5B.13}$$

Next, compute the terms given by [32, p. 200]

$$A_N = \left(F_{N+} \mu_{N+}^2 + F_{N-} \mu_{N-}^2 + F_0 \mu_{N0}^2 \right) e^{-E_N/kT} \tag{5B.14}$$

Finally, the complete expression for absorption coefficient due to oxygen is given by the summation [32, p. 201]

---

[20] P_0, T_0, and ρ_0 define the standard atmosphere [9].

$$\gamma_{ox}(f,p,T) = CpT^{-3}f^2 \sum_N A_N \qquad (5B.15)$$

where f is frequency, p is atmospheric pressure, T is absolute temperature, and $C = 2.0058$ for γ in decibels per kilometer [32].

5B.1.3 Absorption Coefficient for Water Vapor

There are two primary components to water vapor absorption (below 100 GHz). First, we will handle water vapor absorption due to the resonance at 22.235 GHz. To do this, start by computing the water vapor partial pressure (in Torr), which is a function of water vapor density and temperature. The partial pressure of water vapor (in Torr) is given by [32, p. 203]

$$p_w = \frac{\rho T}{288.75} \qquad (5B.16)$$

where ρ is water vapor density and T is temperature. Recall that 1 mbar $\sim$ 0.75 Torr. Converting to Torr we then use [32, p. 203]

$$p_t = 0.75 p_{mb} \qquad (5B.17)$$

Next, we use the equation provided by Liebe for Δf [32, p. 203; 67]

$$\Delta f = (17.99 \times 10^{-3}) \left[p_w \left(\frac{300}{T} \right) + 0.20846(p_t - p_w) \left(\frac{300}{T} \right)^{0.63} \right] \qquad (5B.18)$$

Similarly to what was done for oxygen (except for the additional factor f/f_r), we now use the Van Vleck-Weisskopf formula again to determine F [32, p. 203; 68]

$$F = \frac{f}{f_r} \left[\frac{\Delta f}{(f_r - f)^2 + (\Delta f)^2} + \frac{\Delta f}{(f_r + f)^2 + (\Delta f)^2} \right] \qquad (5B.19)$$

where $f_r = 22.235$ GHz. Finally, using the terms determined above, calculate the absorption coefficient due to vapor resonance at 22.235 GHz using [32, p. 203]

$$\gamma_{wv22} = (2.535 \times 10^{-3}) \left\{ fp_w \left(\frac{300}{T} \right)^{7/2} \exp\left[2.144 \left(1 - \frac{300}{T} \right) \right] F \right\} \text{ dB/km} \qquad (5B.20)$$

Second, compute the residual effect of water vapor absorption lines above 100 GHz, using the simpler expression [32, p. 204; 69]

$$\gamma_{wv100} = (7.347 \times 10^{-3}) \rho p T^{-5/2} f^2 \text{ dB/km} \qquad (5B.21)$$

Now finish off the water vapor result, which is given by

$$\gamma_{wv} = \gamma_{wv22} + \gamma_{wv100} \text{ dB/km} \qquad (5B.22)$$

For the total absorption, we sum the oxygen and the water vapor absorption, using

$$\gamma = \gamma_{ox} + \gamma_{wv} \quad \text{dB/km} \tag{5B.23}$$

5B.2 FUNCTION TROPREFRACT.M

When RF waves travel through the atmosphere, their path is bent, or refracted. This is because the atmosphere is a stratified medium whose refractive index varies with altitude. To determine atmospheric absorption properly, which is a function of distance, we must calculate the actual path traveled versus the straight line path.

For our calculations, we use the exponential model of refractive index [32, p. 182]. More specifically, Table 5B.4 provides the parameters that define the Central Radio Propagation Laboratory (CRPL)[21] exponential reference atmosphere [32, 60]. The values associated with index $k = 5$ ($N_s = 313$, $c_e = 0.1439$) are representative of the average values over the United States [32] and will be used for our calculations.

The values in Table 5B.4 are c_e, which is a constant related to refractive index gradient (per km), h_s, which corresponds to altitude above sea level in ft and N_s, the surface refractivity in ppm [32, p. 182].

First, convert to h_s to km

$$h_{s_{km}} = h_{s_{ft}} \times 12 \times 2.54 / 100 / 1{,}000 \tag{5B.24}$$

Next, convert surface refractivity to n_0, which is the exponential refractive index at the Earth's surface ($h = 0$). To do this, use the relationship between N_s and n_0 of [32, p. 183]

$$N_s(k) = (n_0 - 1) \times 10^6 \tag{5B.25}$$

or

$$n_0 = 1 + N_s(k) \times 10^{-6} \tag{5B.26}$$

Table 5B.4
Value of Parameters of CRPL Exponential Reference Atmosphere

K	N_s	c_e (per km)	h_s (ft)
1	200	0.118400	10,000
2	250.0	0.125625	—
3	252.9294	0.126255	5,000
4	301.0	0.139632	1,000
5	313.0	0.143859	700
6	344.5	0.156805	0
7	350.0	0.159336	0
8	377.2	0.173233	0
9	400.0	0.186720	0

---

[21] CPRL is now the National Oceanic and Atmospheric Administration (NOAA).

K	N_s	c_e (per km)	h_s (ft)
10	404.9	0.189829	0
11	450.0	0.223256	0

Using (5B.26) and the parameters from Table 5B.4, the exponential model for refractive index as function of altitude is [32, p. 182; 60; 70]

$$n(h) = 1 + (n_0 - 1)e^{-c_e(k)h} \tag{5B.27}$$

Finally, to determine the total distance over the refracted path, we compute the ray tracing integral given by [32, p. 182; 9, p. 232; 59]

$$R(h_1, \theta_0) = \int_0^{h_1} \frac{n(h)dh}{\sqrt{1 - \{(n_0 \cos(\theta_0))/[n(h)(1 + h/r_0)]\}^2}} \tag{5B.28}$$

5B.3 FUNCTION TROPLOSS.M

Using the above functions, we are ready to compute atmospheric loss as follows:

- First do the ray tracing to determine the refracted RF path
  - Call $[R, h]$ = troprefract(*angle*, M)
- Now compute the absorption coefficients versus h at the desired frequencies
  - Call $[\gamma, \gamma_{ox}, \gamma_{wv}]$ = tropatten(f, h)
- The total atmospheric loss is then determined by taking the integral over the RF path [32, p. 199]

$$L = L_{ox} + L_{wv}$$

$$L = \int_{r_1}^{r_2} [\gamma_{ox}(r) + \gamma_{wv}(r)] dr \tag{5B.29}$$

$$L = \int_{r_1}^{r_2} \gamma(r) dr$$

As indicated by (5B.29) the two components of atmospheric loss are the atmospheric loss due to oxygen, L_{ox}, and the atmospheric loss due to water vapor, L_{wv}.

5B.4 FUNCTION RAINATTN2WAY.M

The standard model for rain attenuation coefficient k_{ar} in dB/km given a rainfall rate r_r in mm/h takes the form [24, p. 246; 32, p. 215]

$$k_{ar}(r_r) = ar_r^b \quad \text{dB/km} \tag{5B.30}$$

The terms a and b are a multiplicative factor and an exponent, respectively, both of which are dependent on frequency.

Barton provides the following empirical expression for a and b (an updated version of the expression presented by Blake [32, p. 217]) that applies for a temperature of ~ 291 K [9, p. 246]:

$$a(f_0) = \frac{C_0 f_0^2 \left(1 + f_0^2/f_1^2\right)^{1/2}}{\left(1 + f_0^2/f_2^2\right)^{1/2} \left(1 + f_0^2/f_3^2\right)^{1/2} \left(1 + f_0^2/f_4^2\right)^{0.65}} \quad (5B.31)$$

where f_0 is the operting frequency. The exponent b in (5B.31) is [9, p. 246]

$$b(f_0) = 1.30 + 0.0372\left[1 - \left(1 + x_f^2\right)\right]^{1/2} \quad (5B.32)$$

and $C_0 = 3.1 \times 10^{-5}$ is a frequency parameter. The following are break frequencies (determined empirically to match various theoretical computations published by a number of authors[22]) in GHz:

$$\begin{aligned} f_1 &= 3 \\ f_2 &= 35 \\ f_3 &= 50 \\ f_4 &= 110 \end{aligned} \quad (5B.33)$$

and

$$x_f = 16.67 \log(0.13 f_0) \quad (5B.34)$$

---

[22] Blake acknowledges Wayne Rivers, a senior scientist at Technology Service Corporation, for coming up with the original expressions for (5B.31) through (5B.34) [31, p. 215].

Chapter 6

Detection Theory

6.1 INTRODUCTION

In the radar range equation exercises of Chapter 2, we considered an example of computing detection range based on SNRs of 13 and 20 dB. We now want to develop some theory explaining the use of these particular SNR values. More specifically, we will examine the concept of detection probability, P_d. Our need to study detection from a probabilistic perspective stems from our dealings with signals that are noise-like. From our studies of RCS, we found that, in practice, the signal return looks random. In fact, Peter Swerling has convinced us to use statistical models to represent target signals [1]. In addition to these target signals, we found that the signals in the radar contain a noise component, which also needs to be dealt with using the concepts of random variables, random processes, and probabilities.

The early work in detection theory, as applies to radar, was published by Stephen Oswald Rice in the *Bell System Technical Journal* [2]. Rice considered the problem of detecting a constant amplitude signal in the presence of noise, based on a single sample of the signal plus noise. A SW0/SW5 target (see Chapter 3) produces such a signal. In his 1947 paper, J. I. Marcum extended Rice's work to the case of detection after the integration of a number of signal-plus-noise samples [3, 4]. In 1954, Swerling introduced his concepts of noise-like signals caused by a target with a fluctuating RCS [1]. He developed equations for determining detection probability for single and multiple sample cases. Since then, other authors have extended Swerling's work to other target fluctuation models [5–10]. However, the standards are still the Rice model and the Swerling models.

In this chapter, we will be concerned with detection based on returns from a *single pulse*. In Chapter 9, we will extend the results to the case where detection is based on returns from several pulses. Since we are considering a single pulse, the detection equations we develop are termed *single pulse*, *single sample*, or *single hit* detection probabilities. We will develop detection equations for the five target RCS types discussed in Chapter 3: SW0/SW5, SW1, SW2, SW3, and SW4. We will also derive the "detection" equation for noise, which we term false alarm probability.

Table 6.1
Single-Pulse Detection Probability Equations for SW0 through SW5 Targets

Target Type	P_d/P_{fa} Equation
SW0/SW5	$P_d = Q_1\left(\sqrt{2(SNR)}, \sqrt{-2\ln P_{fa}}\right)$
SW1/SW2	$P_d = \exp\left(\dfrac{\ln P_{fa}}{SNR+1}\right)$
SW3/SW4	$P_d = \left[1 - \dfrac{2(SNR)\ln P_{fa}}{(2+SNR)^2}\right] e^{2\ln P_{fa}/(2+SNR)}$
Noise	$P_{fa} = e^{-TNR}$

Target Type	Signal-Plus-Noise/Noise Density Function
SW0/SW5	$f_V(V) = \dfrac{V}{\sigma^2} I_0\left(\dfrac{VS}{\sigma^2}\right) e^{-(V^2+S^2)/2\sigma^2} U(V)$
SW1/SW2	$f_V(V) = \dfrac{V}{P_S+\sigma^2} e^{-V^2/2(P_S+\sigma^2)} U(V)$
SW3/SW4	$f_V(V) = \dfrac{2V}{(2\sigma^2+P_S)^2}\left[2\sigma^2 + \dfrac{P_S V^2}{(2\sigma^2+P_S)}\right] e^{-V^2/(2\sigma^2+P_S)} U(V)$
Noise	$f_N(N) = \dfrac{N}{\sigma^2} e^{-N^2/2\sigma^2} U(N)$

The various signal models and probability derivations presented in this chapter are not new. As indicated above, they have been carried out by Rice, Swerling, and many others [11]. We include them in this book because we feel it is very important to understand the origin of the detection and false alarm probability equations, along with the limitations on when and where they can be applied. For those readers who are interested only in the final results, Table 6.1 contains a summary of the detection and false alarm probability equations derived in this chapter, along with the noise and signal-plus-noise density function equations on which probability equations are based.

In the table,

- P_d is the single pulse detection probability.
- SNR is the single pulse SNR (see (2.1), Chapter 2).
- P_{fa} is the probability of false alarm.
- Q_1 is the Marcum Q-function.
- TNR is the threshold-to-noise ratio.
- I_0 is the modified Bessel function of the first kind, order zero.

- $S = \sqrt{2P_S}$ is the amplitude of the signal return for a SW0/SW5 target.
- P_S is the signal power from the radar range equation (see Chapter 2).
- σ^2 is the noise power at the output of the matched filter.
- $U(x)$ is the unit step function.
- V and N are dummy variables (V represents signal plus noise and N represents only noise).

These parameters are defined more fully in the discussions that follow.

To develop the requisite detection probability equations, we need to develop a mathematical characterization of the target signal, the noise signal, and the target-plus-noise signal at various points in the radar. We start with a characterization of noise and then progress to the target and target-plus-noise signals.

6.2 NOISE IN RECEIVERS

We characterize noise for the two most common types of receiver implementations. The first is illustrated in Figure 6.1 and is termed the *IF representation* [12][1]. In this representation, the matched filter is implemented at some intermediate frequency, or IF. The second receiver configuration is illustrated in Figure 6.2 and is termed the *baseband representation* [12]. In this configuration, the signal is converted to a baseband signal, a complex signal centered at a frequency of zero, instead of ω_{IF}. The IF configuration is common in older radars, and the baseband representation is common in modern radars, especially those using digital signal processing.

Both the IF and baseband representations contain a matched filter, which serves as the signal processor for the case where the radar bases detection decisions on a single pulse. As we will see in Chapter 8, the matched filter is a necessary component because it maximizes SNR, which is a requirement for maximizing detection probability.

6.2.1 IF Configuration

In the IF configuration, we represent the noise by

$$\mathbf{n}_{IF}(t) = \mathbf{N}(t)\cos\left[\omega_{IF}t - \varphi(t)\right] \tag{6.1}$$

where $\mathbf{n}_{IF}(t)$, $\mathbf{N}(t)$, and $\varphi(t)$ are random processes. $\mathbf{N}(t)$ is the magnitude of the noise and $\varphi(t)$ is its associated phase. Expanding (6.1) using trigonometric identities gives

$$\begin{aligned}\mathbf{n}_{IF}(t) &= \mathbf{N}(t)\cos\varphi(t)\cos\omega_{IF}t + \mathbf{N}(t)\sin\varphi(t)\sin\omega_{IF}t \\ &= \mathbf{n}_I(t)\cos\omega_{IF}t + \mathbf{n}_Q(t)\sin\omega_{IF}t\end{aligned} \tag{6.2}$$

---

[1] In Figure 6.1, ω_o is the carrier frequency and ω_{IF} is the IF, or intermediate frequency.

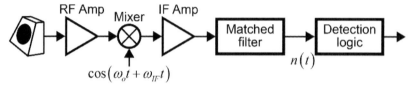

Figure 6.1 IF receiver representation.

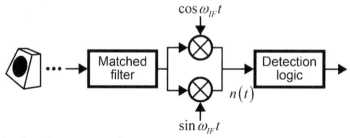

Figure 6.2 Baseband receiver representation.

In (6.2), $\mathbf{n}_I(t)$ and $\mathbf{n}_Q(t)$ are also random processes. We stipulate $\mathbf{n}_I(t)$ and $\mathbf{n}_Q(t)$ as joint, wide-sense stationary (WSS), zero-mean, equal variance, Gaussian random processes. They are also such that the *random variables* $\mathbf{n}_I = \mathbf{n}_I(t)|_{t=t1}$ and $\mathbf{n}_Q = \mathbf{n}_Q(t)|_{t=t1}$ are independent. The variance of $\mathbf{n}_I(t)$ and $\mathbf{n}_Q(t)$ is σ^2. Under these conditions, the density functions of $\mathbf{n}_I(t)$ and $\mathbf{n}_Q(t)$ are given by

$$f_{\mathbf{n}_I}(n) = f_{\mathbf{n}_Q}(n) = \frac{1}{\sigma\sqrt{2\pi}} e^{-n^2/2\sigma^2} \qquad (6.3)$$

We now show that $\mathbf{N}(t)$ is Rayleigh and $\varphi(t)$ is uniform on $(-\pi, \pi]$. We will further show that the random variables $\mathbf{N} = \mathbf{N}(t)|_{t=t1}$ and $\varphi = \varphi(t)|_{t=t1}$ are independent.

From random variable theory [13], if $\mathbf{x}$ and $\mathbf{y}$ are real random variables,

$$r = \sqrt{x^2 + y^2} \qquad (6.4)$$

and

$$\varphi = \tan^{-1}\left(\frac{y}{x}\right) \qquad (6.5)$$

where $\tan^{-1}(y/x)$ denotes the four-quadrant arctangent, then the joint density of $\mathbf{r}$ and φ can be written in terms of the joint density of $\mathbf{x}$ and $\mathbf{y}$ as

$$f_{r\varphi}(r,\phi) = r f_{xy}(r\cos\phi, r\sin\phi) U(r) \operatorname{rect}\left[\frac{\phi}{2\pi}\right] \qquad (6.6)$$

where, as a reminder, rect[x] is[2]

---

[2] We modified the definition of the rect[x] function definition of Chapter 2 slightly by making the upper bound closed.

$$\text{rect}[x] = \begin{cases} 1 & -1/2 < x \leq 1/2 \\ 0 & \text{elsewhere} \end{cases} \tag{6.7}$$

$U(x)$ is the unit step function,[3] which is defined as

$$U(x) = \begin{cases} 1 & x \geq 0 \\ 0 & 0 < x \end{cases} \tag{6.8}$$

In our case, $\mathbf{x} = \mathbf{n}_I$, $\mathbf{y} = \mathbf{n}_Q$, $\mathbf{r} = \mathbf{N}$, and $\boldsymbol{\varphi} = \boldsymbol{\varphi}$. Thus, we have

$$N = \sqrt{n_I^2 + n_Q^2} \tag{6.9}$$

$$\varphi = \tan^{-1}\left(\frac{n_Q}{n_I}\right) \tag{6.10}$$

and

$$f_{N\varphi}(N,\phi) = N f_{n_I n_Q}(N\cos\phi, N\sin\phi) U(N) \text{rect}\left(\frac{\phi}{2\pi}\right) \tag{6.11}$$

Since $\mathbf{n}_I$ and $\mathbf{n}_Q$ are independent

$$\begin{aligned} f_{n_I n_Q}(n_I, n_Q) &= f_{n_I}(n_I) f_{n_Q}(n_Q) \\ &= \left(\frac{1}{\sigma\sqrt{2\pi}} e^{-n_I^2/2\sigma^2}\right)\left(\frac{1}{\sigma\sqrt{2\pi}} e^{-n_Q^2/2\sigma^2}\right) = \frac{1}{2\pi\sigma^2} e^{-(n_I^2+n_Q^2)/2\sigma^2} \end{aligned} \tag{6.12}$$

Using this result in (6.11) with $n_I = N\cos(\phi)$ and $n_Q = N\sin(\phi)$, we get

$$\begin{aligned} f_{N\varphi}(N,\phi) &= \frac{N}{2\pi\sigma^2} e^{-(N^2\cos^2\phi + N^2\sin^2\phi)/2\sigma^2} U(N) \text{rect}\left[\frac{\phi}{2\pi}\right] \\ &= \frac{N}{2\pi\sigma^2} e^{-N^2/2\sigma^2} U(N) \text{rect}\left[\frac{\phi}{2\pi}\right] \end{aligned} \tag{6.13}$$

From random variable theory, we can find the marginal density from the joint density by integrating with respect to the variable we want to eliminate. Thus,

$$f_N(N) = \int_{-\infty}^{\infty} f_{N\varphi}(N,\phi) d\phi = \frac{N}{\sigma^2} e^{-N^2/2\sigma^2} U(N) \tag{6.14}$$

and

$$f_\varphi(\phi) = \int_{-\infty}^{\infty} f_{N\varphi}(N,\phi) dN = \frac{1}{2\pi} \text{rect}\left[\frac{\phi}{2\pi}\right] \tag{6.15}$$

---

[3] Also known as the Heaviside step function.

This proves the assertion that $\mathbf{N}(t)$ is Rayleigh and $\varphi(t)$ is uniform on $(-\pi,\pi]$. To prove the random variables $\mathbf{N} = \mathbf{N}(t)|_{t=t1}$ and $\varphi = \varphi(t)|_{t=t1}$ are independent, we note from (6.13), (6.14), and (6.15) that

$$f_{N\varphi}(N,\phi) = f_N(N) f_\varphi(\phi) \tag{6.16}$$

which means $\mathbf{N}$ and φ are independent.

Since we will need it later, we want to find an equation for the noise power out of the matched filter, and into the detection logic. Since $\mathbf{n}_{IF}(t)$ is WSS, we use (6.2) to write

$$\begin{aligned}
P_{nIF} &= E\{\mathbf{n}_{IF}^2(t)\} = E\left\{\left(\mathbf{n}_I(t)\cos\omega_{IF}t + \mathbf{n}_Q(t)\sin\omega_{IF}t\right)^2\right\} \\
&= E\{\mathbf{n}_I^2(t)\cos^2\omega_{IF}t\} + E\{\mathbf{n}_Q^2(t)\sin^2\omega_{IF}t\} \\
&\quad + 2E\{\mathbf{n}_I(t)\mathbf{n}_Q(t)\cos\omega_{IF}t\sin\omega_{IF}t\} \\
&= \sigma^2
\end{aligned} \tag{6.17}$$

In (6.17), the term on the third line is zero because $\mathbf{n}_I = \mathbf{n}_I(t)|_{t=t1}$ and $\mathbf{n}_Q = \mathbf{n}_Q(t)|_{t=t1}$ are independent and zero-mean.

6.2.2 Baseband Configuration

In the baseband configuration of Figure 6.2, we represent the noise into the detection logic as a complex random process of the form

$$\mathbf{n}_B(t) = \frac{1}{\sqrt{2}}\left[\mathbf{n}_I(t) + j\mathbf{n}_Q(t)\right] \tag{6.18}$$

where $\mathbf{n}_I(t)$ and $\mathbf{n}_Q(t)$ are joint, WSS, zero-mean, equal variance, Gaussian random processes. They are also such that the random variables $\mathbf{n}_I = \mathbf{n}_I(t)|_{t=t1}$ and $\mathbf{n}_Q = \mathbf{n}_Q(t)|_{t=t1}$ are independent. The variance of $\mathbf{n}_I(t)$ and $\mathbf{n}_Q(t)$ is σ^2. The constant of $1/\sqrt{2}$ is included to provide consistency between the noise powers in the baseband and IF receiver configurations. The power in $\mathbf{n}_B(t)$ [making use of the properties of $\mathbf{n}_I(t)$ and $\mathbf{n}_Q(t)$] is given by

$$\begin{aligned}
P_{nB} &= E\{\mathbf{n}_B(t)\mathbf{n}_B^*(t)\} \\
&= E\left\{\frac{1}{\sqrt{2}}[\mathbf{n}_I(t) + j\mathbf{n}_Q(t)]\frac{1}{\sqrt{2}}[\mathbf{n}_I(t) - j\mathbf{n}_Q(t)]\right\} \\
&= \frac{1}{2}E\{\mathbf{n}_I^2(t)\} + \frac{1}{2}E\{\mathbf{n}_Q^2(t)\} \\
&= \sigma^2 = P_{nIF}
\end{aligned} \tag{6.19}$$

We can write $\mathbf{n}_B(t)$ in polar form as

$$n_B(t) = \frac{N(t)}{\sqrt{2}} e^{j\varphi(t)} \tag{6.20}$$

where

$$N(t) = \sqrt{n_I^2(t) + n_Q^2(t)} \tag{6.21}$$

and

$$\varphi(t) = \tan^{-1}\left[\frac{n_Q(t)}{n_I(t)}\right] \tag{6.22}$$

We note that the definitions of $\mathbf{n}_I(t)$, $\mathbf{n}_Q(t)$, $\mathbf{N}(t)$, and $\varphi(t)$ are consistent between the IF and baseband representations. This means the two representations are equivalent in terms of the statistical properties of the noise. We will reach the same conclusion for the signal. As a result, the detection and false alarm performances of both types of receiver configurations are the same. Thus, the detection and false alarm probability equations we derive in the future will apply to either receiver configuration.

If the receiver being analyzed is not of one of the two forms indicated above, the detection and false alarm probability equations derived herein may not apply. A particular example is the case where the receiver uses only the I or Q channel in baseband processing. While this is not a common receiver configuration, it is sometimes used. In that case, one would need to derive a different set of detection and false alarm probability equations specifically applicable to the configuration.

6.3 SIGNAL IN RECEIVERS

6.3.1 Introduction and Background

We now turn our attention to developing a representation of the signals at the input to the detection logic. Consistent with the noise case, we consider both IF and baseband receiver configurations. Thus, we will use Figures 6.1 and 6.2, but replace $\mathbf{n}(t)$ with $\mathbf{s}(t)$, $\mathbf{N}(t)$ with $\mathbf{S}(t)$, $\varphi(t)$ with $\theta(t)$, $\mathbf{n}_I(t)$ with $\mathbf{s}_I(t)$, and $\mathbf{n}_Q(t)$ with $\mathbf{s}_Q(t)$.

We will develop three signal representations: one for SW0/SW5 targets, one for SW1/SW2 targets, and one for SW3/SW4 targets. We have already acknowledged that the SW1 through SW4 target RCS models are random process models. To maintain consistency with this idea, and consistency with what happens in an actual radar, we also use a random process model for the SW0/SW5 target.

In Chapter 3, we learned the SW1 and SW2 targets share one RCS fluctuation model and the SW3 and SW4 targets share a second RCS fluctuation model. The difference between SWodd (SW1, SW3) and SWeven (SW2, SW4) was in how their RCS varies with time. SWodd targets have an RCS that is constant from pulse to pulse, but varies from scan to scan. SWeven targets have an RCS that varies from pulse to pulse. All cases assumed the RCS did not vary during a PRI. Because of this assumption, the statistics for SW1 and SW2 targets are the same on any one pulse. Likewise, the statistics for SW3 and SW4 targets are the same on any one pulse. Consequently, in terms of single-pulse probabilities, we can combine SW1 and SW2 targets and we can combine SW3 and SW4 targets. This accounts for our use of the terminology "SW1/SW2 targets" and "SW3/SW4 targets" when discussing single-pulse detection probability. In Chapter 9, we will develop separate equations for each of the Swerling target types, since we will base detection decisions on the results from processing several pulses.

Since the target RCS models are random processes, we also represent the target *voltage* signals in the radar (henceforth termed the *target* signal) as random processes. To that end, the IF representation of the target signal is

$$s_{IF}(t) = S(t)\cos[\omega_{IF}t - \theta(t)] = s_I(t)\cos\omega_{IF}t + s_Q(t)\sin\omega_{IF}t \qquad (6.23)$$

where

$$s_I(t) = S(t)\cos\theta(t) \qquad (6.24)$$

and

$$s_Q(t) = S(t)\sin\theta(t) \qquad (6.25)$$

The baseband signal model is

$$s_B(t) = \frac{1}{\sqrt{2}}[s_I(t) + js_Q(t)] \qquad (6.26)$$

We note that both of the signal models are consistent with the noise model of the previous sections. We assume $\mathbf{S} = S(t)|_{t=t1}$ and $\mathbf{\theta} = \theta(t)|_{t=t1}$ are independent. $S(t)$ is the signal magnitude and $\theta(t)$ is its phase.

We have made many assumptions concerning the statistical properties of the signal and noise. A natural question is: Are the assumptions reasonable? The answer is that radars are usually designed so that the assumptions are satisfied. In particular, designers endeavor to make the receiver and matched filter linear. Because of this and the central limit theorem, we can reasonably assume $\mathbf{n}_I(t)$ and $\mathbf{n}_Q(t)$ are Gaussian. Further, if we enforce reasonable constraints on the bandwidth of receiver components, we can reasonably assume the validity of the independence requirements. The stationarity requirements are easily satisfied if we assume the receiver gains and noise figures do not change with time. We enforce the zero-mean assumption by using AC coupling and bandpass filters (BPFs) to eliminate DC components. For signals, we will not need the Gaussian requirement. However, we will need the stationarity, zero-mean, and other requirements. These constraints are usually satisfied for signals by using the same assumptions as for noise, by requiring a WSS random process for the target RCS, and by requiring $\theta(t)$ be wide sense stationary and uniform on $(-\pi,\pi]$. The latter two assumptions are valid for practical radars and targets.

At this point, we need to develop separate signal models for the different types of targets because each signal amplitude fluctuation, $\mathbf{S}(t)$, is governed by a different model.

6.3.2 Signal Model for SW0/SW5 Targets

For the SW0/SW5 target case, we assume a constant target RCS. This means the target power, and thus the target signal amplitude, is constant. With this assumption, we let

$$S(t) = S \tag{6.27}$$

The IF signal model becomes

$$\begin{aligned} s_{IF}(t) &= S\cos(\omega_{IF}t - \theta) = S\cos\theta\cos\omega_{IF}t + S\sin\theta\sin\omega_{IF}t \\ &= s_I \cos\omega_{IF}t + s_Q \sin\omega_{IF}t \end{aligned} \tag{6.28}$$

We introduce the random variable θ to force $s_{IF}(t)$ to be a random process. We specifically choose θ to be uniform on $(-\pi, \pi]$. This makes s_I and s_Q random variables, rather than random processes. $s_{IF}(t)$ is a random process because of the presence of the $\omega_{IF}t$ term. This model is actually consistent with what happens in an actual radar. Specifically, the phase of the signal is random for any particular target return.

The density functions of s_I and s_Q are the same and are given by [13]:

$$f_{s_I}(s) = f_{s_Q}(s) = \frac{1}{\pi\sqrt{S^2 - s^2}} \operatorname{rect}\left[\frac{s}{2S}\right] \tag{6.29}$$

We cannot assert the independence of random variables s_I and s_Q because we have no means of showing $f_{s_I s_Q}(s_I, s_Q) = f_{s_I}(s_I) f_{s_Q}(s_Q)$.

The signal power is given by

$$\begin{aligned} P_{sIF} &= E\{s_{IF}^2(t)\} = E\{(S\cos\theta\cos\omega_{IF}t + S\sin\theta\sin\omega_{IF}t)^2\} \\ &= S^2 E\{\cos^2\theta\}\cos^2\omega_{IF}t + S^2 E\{\sin^2\theta\}\sin^2\omega_{IF}t \\ &\quad + 2SE\{\cos\theta\sin\theta\}\cos\omega_{IF}t\sin\omega_{IF}t \end{aligned} \tag{6.30}$$

In the above, we can write

$$\begin{aligned} E\{\cos^2\theta\} &= \int_{-\infty}^{\infty} \cos^2\theta f_\theta(\theta) d\theta = \int_{-\infty}^{\infty} \cos^2\theta \left[\frac{1}{2\pi}\operatorname{rect}\left(\frac{\theta}{2\pi}\right)\right] d\theta \\ &= \frac{1}{2\pi} \int_{-\pi}^{\pi} \cos^2\theta d\theta = \frac{1}{2} \end{aligned} \tag{6.31}$$

Similarly, we get

$$E\{\sin^2\theta\} = \frac{1}{2} \tag{6.32}$$

and

$$E\{\cos\theta\sin\theta\} = 0 \tag{6.33}$$

Substituting (6.31), (6.32), and (6.33) into (6.30) results in

$$P_{sIF} = S^2\left(\frac{1}{2}\right)\cos^2\omega_{IF}t + S^2\left(\frac{1}{2}\right)\sin^2\omega_{IF}t + 2(0)S\cos\omega_{IF}t\sin\omega_{IF}t \quad (6.34)$$

$$= \frac{S^2}{2}$$

From (6.26), the baseband signal model is

$$s_B(t) = s_B = \frac{1}{\sqrt{2}}(s_I + js_Q) = \frac{S}{\sqrt{2}}(\cos\theta + j\sin\theta) \quad (6.35)$$

and the signal power is

$$P_{sB} = E\{s_B s_B^*\} = E\left\{\frac{S}{\sqrt{2}}(\cos\theta + j\sin\theta)\frac{S}{\sqrt{2}}(\cos\theta - j\sin\theta)\right\} \quad (6.36)$$

$$= \frac{S^2}{2} = P_{sIF}$$

6.3.3 Signal Model for SW1/SW2 Targets

For the SW1/SW2 target case, we have already stated that the target RCS is governed by the density function (see Chapter 3)

$$f_\sigma(\sigma) = \frac{1}{\sigma_{AV}} e^{-\sigma/\sigma_{AV}} U(\sigma) \quad (6.37)$$

where σ is the target RCS and σ_{AV} is the average RCS (see Chapter 3).

Since the power is a direct function of the RCS (from the radar range equation), the signal power at the detection logic input has a density function identical in form to (6.37). That is,

$$f_P(p) = \frac{1}{P_S} e^{-p/P_S} U(p) \quad (6.38)$$

where

$$P_S = \frac{P_T G_T G_R \lambda^2}{(4\pi)^3 R^4 L}\sigma_{AV} \quad [4] \quad (6.39)$$

Random variable theory shows the signal amplitude, $S(t)$, governed by the density function,

---

[4] In (6.39) we normalized the receiver gain, G, to unity. This is acceptable as long as we use the same normalization for noise, which we do (see Chapter 2).

$$f_S(S) = \frac{S}{P_S} e^{-S^2/2P_S} U(S) \qquad (6.40)$$

which is recognized as a Rayleigh density function [13]. This, combined with the fact that $\theta(t)$ in (6.21) is uniform, and the assumption of the independence of random variables $\mathbf{S} = \mathbf{S}(t)|_{t=t1}$ and $\boldsymbol{\theta} = \boldsymbol{\theta}(t)|_{t=t1}$, leads to the interesting observation that the signal model for a SW1/SW2 target takes the same form as the noise model. That is, the IF signal model for a SW1/SW2 target takes the form

$$s_{IF}(t) = S(t)\cos[\omega_{IF}t - \theta(t)] = s_I(t)\cos\omega_{IF}t + s_Q(t)\sin\omega_{IF}t \qquad (6.41)$$

where $S(t)$ is Rayleigh and $\theta(t)$ is uniform on $(-\pi, \pi]$. If we adapt the results from our noise study, we conclude that $s_I(t)$ and $s_Q(t)$ are Gaussian with the density functions

$$f_{s_I}(s) = f_{s_Q}(s) = \frac{1}{\sqrt{2\pi P_S}} e^{-s^2/2P_S} \qquad (6.42)$$

Furthermore, $\mathbf{s}_I = \mathbf{s}_I(t)|_{t=t1}$ and $\mathbf{s}_Q = \mathbf{s}_Q(t)|_{t=t1}$, are independent.

The signal power is given by

$$\begin{aligned} P_{sIF} &= E\{s_{IF}^2(t)\} = E\left\{[s_I(t)\cos\omega_{IF}t + s_Q(t)\sin\omega_{IF}t]^2\right\} \\ &= E\{s_I^2(t)\}\cos^2\omega_{IF}t + E\{s_Q^2(t)\}\sin^2\omega_{IF}t \\ &\quad + E\{s_I(t)s_Q(t)\}\cos\omega_{IF}t\sin\omega_{IF}t \end{aligned} \qquad (6.43)$$

Invoking the independence of, $\mathbf{s}_I = \mathbf{s}_I(t)|_{t=t1}$ and $\mathbf{s}_Q = \mathbf{s}_Q(t)|_{t=t1}$, and the fact that $s_I(t)$ and $s_Q(t)$ are zero mean and have equal variances of P_S, leads to the conclusion that

$$P_{sIF} = P_S \qquad (6.44)$$

The baseband representation of the signal is

$$s_B(t) = \frac{1}{\sqrt{2}}[s_I(t) + js_Q(t)] = \frac{S(t)}{\sqrt{2}} e^{j\theta(t)} \qquad (6.45)$$

where the various terms are as defined above. The power in the baseband signal representation can be written as

$$\begin{aligned} P_{sB} &= E\{s_B(t)s_B^*(t)\} = E\left\{\frac{1}{\sqrt{2}}[s_I(t) + js_Q(t)]\frac{1}{\sqrt{2}}[s_I(t) - js_Q(t)]\right\} \\ &= \frac{1}{2}(E\{s_I^2(t)\} + E\{s_Q^2(t)\}) = P_S \end{aligned} \qquad (6.46)$$

as expected.

6.3.4 Signal Model for SW3/SW4 Targets

For the SW3/SW4 target case, we have already stated that the target RCS is governed by the density function (see Chapter 3)

$$f_\sigma(\sigma) = \frac{4\sigma}{\sigma_{AV}^2} e^{-2\sigma/\sigma_{AV}} U(\sigma) \tag{6.47}$$

Since the power is a direct function of the RCS (from the radar range equation), the signal power at the signal processor output has a density function that takes the same form as (6.47). That is,

$$f_P(p) = \frac{4p}{P_S^2} e^{-2p/P_S} U(p) \tag{6.48}$$

where P_S is defined earlier in (6.39). From random variable theory it can be shown that the signal amplitude, $S(t)$, is governed by the density function

$$f_S(S) = \frac{2S^3}{P_S^2} e^{-S^2/P_S} U(S) \tag{6.49}$$

Unfortunately, this is about as far as we can carry the signal model development for the SW3/SW4 case. We can invoke the previous statements and write

$$s_{IF}(t) = S(t)\cos[\omega_{IF} t - \theta(t)] = s_I(t)\cos\omega_{IF} t + s_Q(t)\sin\omega_{IF} t \tag{6.50}$$

and

$$s_B(t) = \frac{1}{\sqrt{2}}[s_I(t) + js_Q(t)] = \frac{S(t)}{\sqrt{2}} e^{j\theta(t)} \tag{6.51}$$

However, we do not know the form of the densities of $s_I(t)$ and $s_Q(t)$. Furthermore, deriving their form has proven very laborious and elusive.

We can find the power in the signal from

$$\begin{aligned}P_{sIF} &= E\{s_{IF}^2(t)\} = E\left\{[S(t)\cos(\omega_{IF} t - \theta(t))]^2\right\} \\ &= P_{sB} = E\{s_B(t)s_B^*(t)\} = \frac{1}{2} E\{S^2(t)\} = P_S \end{aligned} \tag{6.52}$$

We will need to deal with the inability to characterize $s_I(t)$ and $s_Q(t)$ when we consider the characterization of signal-plus-noise.

6.4 SIGNAL-PLUS-NOISE IN RECEIVERS

6.4.1 General Formulation

Now that we have characterizations for the signal and noise, we want to develop characterizations for the sum of signal and noise. That is, we want to develop the appropriate density functions for

$$v(t) = s(t) + n(t) \qquad (6.53)$$

If we are using the IF representation, we write

$$\begin{aligned} v_{IF}(t) &= s_{IF}(t) + n_{IF}(t) \\ &= S(t)\cos[\omega_{IF}t - \theta(t)] + N(t)\cos[\omega_{IF}t - \varphi(t)] \\ &= V(t)\cos[\omega_{IF}t - \psi(t)] \end{aligned} \qquad (6.54)$$

and if we are using the baseband representation, we write

$$\begin{aligned} v_B(t) &= [s_I(t) + n_I(t)] + j[s_Q(t) + n_Q(t)] \\ &= v_I(t) + jv_Q(t) \\ &= \frac{V(t)}{\sqrt{2}} e^{\psi(t)} \end{aligned} \qquad (6.55)$$

In either representation, the primary variable of interest is the magnitude of the signal-plus-noise voltage, $V(t)$, since this quantity is used in computing detection probability. We will compute the other quantities as needed, and as we are able.

We begin the development with the easiest case—the SW1/SW2 case—and progress through the SW0/SW5 case to the most difficult—the SW3/SW4 case.

6.4.2 Signal-Plus-Noise Model for SW1/SW2 Targets

For the SW1/SW2 case, we found the real and imaginary parts of both signal and noise were zero-mean, Gaussian random processes. We will use the baseband representation to derive the density function of $V(t)$. Since $s_I(t)$ and $n_I(t)$ are Gaussian, $v_I(t)$ will also be Gaussian. Since $s_I(t)$ and $n_I(t)$ are zero-mean, $v_I(t)$ will also be zero-mean. Finally, since $s_I(t)$ and $n_I(t)$ are independent, the variance of $v_I(t)$ will equal to the sum of the variances of $s_I(t)$ and $n_I(t)$. That is,

$$\sigma_v^2 = P_S + \sigma^2 \qquad (6.56)$$

With this we get

$$f_{v_I}(v) = \frac{1}{\sqrt{2\pi(P_S + \sigma^2)}} e^{-v^2/2(P_S + \sigma^2)} \qquad (6.57)$$

By similar reasoning, we get

$$f_{V_Q}(v) = f_{V_I}(v) = \frac{1}{\sqrt{2\pi(P_S + \sigma^2)}} e^{-v^2/2(P_S + \sigma^2)} \qquad (6.58)$$

Since $s_I = s_I(t)|_{t=t1}$, $s_Q = s_Q(t)|_{t=t1}$, $n_I = n_I(t)|_{t=t1}$ and $n_Q = n_Q(t)|_{t=t1}$ are mutually independent, $v_I = v_I(t)|_{t=t1}$ and $v_Q = v_Q(t)|_{t=t1}$ are independent. This, coupled with our reasoning above and our previous discussions of noise and the SW1/SW2 signal model, leads to the observation that $V(t)$ is Rayleigh with density

$$f_V(V) = \frac{V}{P_S + \sigma^2} e^{-V^2/2(P_S + \sigma^2)} U(V) \qquad (6.59)$$

6.4.3 Signal-Plus-Noise Model for SW0/SW5 Targets

Since $s_I(t)$ and $s_Q(t)$ are not Gaussian for the SW0/SW5 case, when we add them to $n_I(t)$ and $n_Q(t)$, the resulting $v_I(t)$ and $v_Q(t)$ are not Gaussian. This means that directly manipulating $v_I(t)$ and $v_Q(t)$ to obtain the density function of $V(t)$ will be difficult. Therefore, we will take a different tack and invoke some properties of joint and marginal density functions [13, 14]. Specifically, we use

$$f_{V\psi\theta}(V, \psi, \theta) = f_{V\psi}(V, \psi | \theta = \theta) f_\theta(\theta) \qquad (6.60)$$

We then use

$$f_V(V) = \int_{-\infty}^{\infty} \int_{-\infty}^{\infty} f_{V\psi\theta}(V, \psi, \theta) d\psi d\theta \qquad (6.61)$$

to get the density function of $V(t)$. This procedure involves some tedious math, but it is math that can be found in many books on random variable theory [13–17].

To execute the derivation, we start with the IF representation and write

$$v_{IF}(t) = S\cos(\omega_{IF}t - \theta) + N(t)\cos[\omega_{IF}t - \varphi(t)] \qquad (6.62)$$

where we made use of (6.28). When we expand (6.62) and group terms, we get

$$v_{IF}(t) = [S\cos\theta + n_I(t)]\cos\omega_{IF}t + [S\sin\theta + n_Q(t)]\sin\omega_{IF}t \qquad (6.63)$$

According to the conditional density of (6.60), we want to consider (6.63) for the specific value of $\theta = \theta$. Doing this, we get

$$\begin{aligned} v_{IF}(t)|_{\theta=\theta} &= [S\cos\theta + n_I(t)]\cos\omega_{IF}t + [S\sin\theta + n_Q(t)]\sin\omega_{IF}t \\ &= v_I(t)\cos\omega_{IF}t + v_Q(t)\sin\omega_{IF}t \\ &= V(t)\cos[\omega_{IF}t - \psi(t)] \end{aligned} \qquad (6.64)$$

With this we note $[S\cos\theta + n_I(t)]$ and $[S\sin\theta + n_Q(t)]$ are Gaussian random variables with means of $S\cos\theta$ and $S\sin\theta$. These variables also have the same variance of σ^2.

Furthermore, since $\mathbf{n}_I = \mathbf{n}_I(t)|_{t=t1}$ and $\mathbf{n}_Q = \mathbf{n}_Q(t)|_{t=t1}$ are independent, $(S\cos\theta + \mathbf{n}_I)$ and $(S\sin\theta + \mathbf{n}_Q)$ are independent. With this we can write

$$f_{v_I v_Q}(v_I, v_Q | \theta = \theta) = \frac{1}{2\pi\sigma^2} e^{-\left[(v_I - S\cos\theta)^2 + (v_Q - S\sin\theta)^2\right]/2\sigma^2} \tag{6.65}$$

Invoking the discussions related to (6.4), (6.5), and (6.6), we get

$$f_{V\psi}(V,\psi|\theta=\theta) = V f_{v_I v_Q}(V\cos\psi, V\sin\psi | \theta = \theta) U(V) \text{rect}\left[\frac{\psi}{2\pi}\right] \tag{6.66}$$

If we substitute from (6.65), we get

$$f_{V\psi}(V,\psi|\theta=\theta) = \frac{V}{2\pi\sigma^2} e^{-\left[(V\cos\psi - S\cos\theta)^2 + (V\sin\psi - S\sin\theta)^2\right]/2\sigma^2} U(V) \text{rect}\left[\frac{\psi}{2\pi}\right] \tag{6.67}$$

and manipulate the exponent to yield

$$f_{V\psi}(V,\psi|\theta=\theta) = \frac{V}{2\pi\sigma^2} e^{-\left[V^2 + S^2 - 2VS\cos(\psi-\theta)\right]/2\sigma^2} U(V) \text{rect}\left[\frac{\psi}{2\pi}\right] \tag{6.68}$$

Finally, we use

$$f_\theta(\theta) = \frac{1}{2\pi} \text{rect}\left[\frac{\theta}{2\pi}\right] \tag{6.69}$$

along with (6.60), to write

$$f_{V\psi\theta}(V,\psi,\theta) = \frac{V}{(2\pi)^2 \sigma^2} e^{-\left[V^2 + S^2 - 2VS\cos(\psi-\theta)\right]/2\sigma^2} \cdot U(V) \text{rect}\left[\frac{\psi}{2\pi}\right] \text{rect}\left[\frac{\theta}{2\pi}\right] \tag{6.70}$$

For the next step, we integrate $f_{V\psi\theta}(V,\psi,\theta)$ with respect to ψ and θ to derive the desired marginal density, $f_V(V)$. That is (after a little manipulation),

$$f_V(V) = \left(\frac{V}{\sigma^2} e^{-(V^2+S^2)/2\sigma^2} U(V)\right) \times \frac{1}{(2\pi)^2} \int_{-\infty}^{\infty}\int_{-\infty}^{\infty} e^{2VS\cos(\psi-\theta)/2\sigma^2} \text{rect}\left[\frac{\psi}{2\pi}\right] \text{rect}\left[\frac{\theta}{2\pi}\right] d\psi d\theta \tag{6.71}$$

We first consider the integral with respect to ψ, or

$$\Upsilon_\psi(S,V) = \frac{1}{2\pi} \int_{-\infty}^{\infty} e^{2VS\cos(\psi-\theta)/2\sigma^2} \text{rect}\left[\frac{\psi}{2\pi}\right] d\psi$$
$$= \frac{1}{2\pi} \int_{-\pi}^{\pi} e^{-2VS\cos(\psi-\theta)/2\sigma^2} d\psi \tag{6.72}$$

We recognize the integrand is periodic with a period of 2π and that the integral is performed over one period. This means we can evaluate the integral over any period. Specifically, we choose the period from θ to $2\pi + \theta$ and get

$$\Upsilon_\psi(S,V) = \frac{1}{2\pi} \int_{\theta}^{2\pi+\theta} e^{2VS\cos(\psi-\theta)/2\sigma^2} d\psi \tag{6.73}$$

With the change of variables $\alpha = \psi - \theta$, the integral becomes [18]:

$$\Upsilon_\psi(S,V) = \frac{1}{2\pi} \int_{0}^{2\pi} e^{VS\cos\alpha/\sigma^2} d\alpha = I_0\left(\frac{VS}{\sigma^2}\right) \tag{6.74}$$

where $I_0(x)$ is a modified Bessel function of the first kind and order zero [19].

Substituting (6.74) into (6.71) yields

$$f_V(V) = \left[\frac{V}{\sigma^2} e^{-(V^2+S^2)/2\sigma^2} U(V)\right] I_0\left(\frac{VS}{\sigma^2}\right) \left[\frac{1}{2\pi} \int_{-\infty}^{\infty} \text{rect}\left(\frac{\theta}{2\pi}\right) d\theta\right]$$
$$= \frac{V}{\sigma^2} I_0\left(\frac{VS}{\sigma^2}\right) e^{-(V^2+S^2)/2\sigma^2} U(V) \tag{6.75}$$

where the last step derives from the fact that the integral with respect to θ is equal to 1. Equation (6.75) is the desired result, which is the density function of $V(t)$.

6.4.4 Signal-Plus-Noise Model for SW3/SW4 Targets

As with the SW0/SW5 case, $s_I(t)$ and $s_Q(t)$ are not Gaussian for the SW3/SW4 case. Thus, when we add them to $n_I(t)$ and $n_Q(t)$, the resulting $v_I(t)$ and $v_Q(t)$ are not Gaussian and directly manipulating them to obtain the density function of $V(t)$ is difficult. Based on our experience with the SW0/SW5 case, we again use the joint/conditional density approach. We note that the IF signal-plus-noise voltage is given by

$$v_{IF}(t) = S(t)\cos[\omega_{IF}t - \theta(t)] + N(t)\cos[\omega_{IF}t - \varphi(t)]$$
$$= V(t)\cos[\omega_{IF}t - \psi(t)] \tag{6.76}$$

In this case, we find the joint density of $V(t)$, $S(t)$, $\psi(t)$, and $\theta(t)$, and perform the appropriate integration to obtain the marginal density of $V(t)$. More specifically, we will find

$$f_{VS\psi\theta}(V,S,\psi,\theta) = f_{V\psi}(V,\psi|\theta=\theta,S=S) f_{S\theta}(S,\theta) \tag{6.77}$$

and

$$f_V(V) = \int_{-\infty}^{\infty}\int_{-\infty}^{\infty}\int_{-\infty}^{\infty} f_{VS\psi\theta}(V,S,\psi,\theta) d\psi d\theta dS \qquad (6.78)$$

Drawing on our work from the SW0/SW5 case, we write

$$f_{V\psi}(V,\psi|S=S,\theta=\theta) = \frac{V}{2\pi\sigma^2} e^{-\left[V^2+S^2-2VS\cos(\psi-\theta)\right]/2\sigma^2} U(V)\text{rect}\left[\frac{\psi}{2\pi}\right] \qquad (6.79)$$

Further, since $\mathbf{S}(t)$ and $\boldsymbol{\theta}(t)$ are, by definition, independent, we write

$$f_{S\theta}(S,\theta) = f_S(S)f_\theta(\theta) = \left[\frac{2S^3}{P_S^2} e^{-S^2/P_S} U(S)\right]\left[\frac{1}{2\pi}\text{rect}\left(\frac{\theta}{2\pi}\right)\right] \qquad (6.80)$$

Substituting (6.79) and (6.80) into (6.77) results in

$$\begin{aligned}f_{VS\psi\theta}(V,S,\psi,\theta) &= \frac{V}{2\pi\sigma^2} e^{-\left[V^2+S^2-2VS\cos(\psi-\theta)\right]/2\sigma^2} U(V)\text{rect}\left[\frac{\psi}{2\pi}\right] \\ &\times \left[\frac{2S^3}{P_S^2} e^{-S^2/P_S} U(S)\right]\left[\frac{1}{2\pi}\text{rect}\left(\frac{\theta}{2\pi}\right)\right]\end{aligned} \qquad (6.81)$$

From (6.78), with some manipulation, we write

$$f_V(V) = \frac{V}{\sigma^2} e^{-V^2/2\sigma^2} \left[\int_{-\infty}^{\infty} \frac{2S^3}{P_S^2} e^{-\alpha S^2} \Upsilon_{\psi\theta}(S,V) U(S) dS\right] U(V) \qquad (6.82)$$

where

$$\alpha = \frac{1}{P_S} + \frac{1}{2\sigma^2} \qquad (6.83)$$

and

$$\Upsilon_{\psi\theta}(S,V) = \frac{1}{(2\pi)^2} \int_{-\infty}^{\infty}\int_{-\infty}^{\infty} e^{2VS\cos(\psi-\theta)/2\sigma^2} \text{rect}\left[\frac{\psi}{2\pi}\right]\text{rect}\left[\frac{\theta}{2\pi}\right] d\psi d\theta \qquad (6.84)$$

We recognize (6.84) as the same double integral of (6.71). Thus, using discussions related to (6.74), we get

$$\Upsilon_{\psi\theta}(S,V) = \Upsilon_\psi(S,V) = I_0\left(\frac{VS}{\sigma^2}\right) \qquad (6.85)$$

and

$$f_V(V) = \frac{V}{\sigma^2} e^{-V^2/2\sigma^2} \left[\int_{-\infty}^{\infty} \frac{2S^3}{P_S^2} e^{-\alpha S^2} I_0\left(\frac{VS}{\sigma^2}\right) U(S) dS \right] U(V)$$
$$= \frac{V}{\sigma^2} e^{-V^2/2\sigma^2} \left[\int_{0}^{\infty} \frac{2S^3}{P_S^2} e^{-\alpha S^2} I_0\left(\frac{VS}{\sigma^2}\right) dS \right] U(V) \quad (6.86)$$

To complete the calculation of $f_V(V)$, we compute the integral

$$\Upsilon = 2\int_{0}^{\infty} S^3 e^{-\alpha S^2} I_0(\beta S) dS \quad (6.87)$$

where

$$\beta = V/\sigma^2 \quad (6.88)$$

Using a symbolic mathematics software package to compute the integral, we get

$$\Upsilon = \frac{1}{\alpha^2}\left(1 + \frac{\beta^2}{4\alpha}\right) e^{\beta^2/4\alpha} \quad (6.89)$$

With this result, $f_V(V)$ becomes

$$f_V(V) = \frac{V}{P_S^2 \sigma^2} e^{-V^2/2\sigma^2} \left[\frac{1}{\alpha^2}\left(1 + \frac{\beta^2}{4\alpha}\right) e^{\beta^2/4\alpha} \right] U(V) \quad (6.90)$$

which, after manipulation, can be written as

$$f_V(V) = \frac{2V}{(2\sigma^2 + P_S)^2} \left[2\sigma^2 + \frac{P_S V^2}{(2\sigma^2 + P_S)} \right] e^{-V^2/(2\sigma^2 + P_S)} U(V) \quad (6.91)$$

Now that we have completed the characterization of noise, signal, and signal-plus-noise, we are ready to attack the detection problem.

6.5 DETECTION PROBABILITY

6.5.1 Introduction

A functional block diagram of the detection process is illustrated in Figure 6.3. This process consists of an amplitude detector and a threshold device. The amplitude detector determines the magnitude of the signal coming from the matched filter (MF), and the threshold device—a binary decision device—outputs a detection declaration if the signal magnitude is above some threshold, or a no-detection declaration if the signal magnitude is below the threshold.

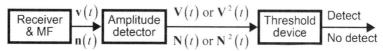

Figure 6.3 Block diagram of the detector and threshold device.

6.5.2 Amplitude Detector Types

The amplitude detector can be a *square-law* detector or a *linear* detector. Figure 6.4 provides a functional illustration of both variants for the IF implementation and the baseband implementation. In the IF implementation, the detector consists, functionally, of a diode followed by a lowpass filter (LPF). If the circuit design uses small voltage levels, the diode will be operating in its small signal region and will result in a square-law detector. If the circuit design uses large voltage levels, the diode will be operating in its large signal region and will result in a linear detector.

For the baseband case, the digital hardware (which we assume in the baseband case) actually forms the square of the magnitude of the complex signal out of the receiver/matched filter by squaring the real and imaginary components of the receiver/matched filter output and then adding them. This operation results in a square-law detector. In some instances, the detector also performs a square root to form the magnitude.

In either the IF or baseband representation, the square-law detector outputs $\mathbf{N}^2(t)$ when only noise is present at the receiver/matched filter output and $\mathbf{V}^2(t)$ when signal-plus-noise is present at the receiver/matched filter output. The linear detector outputs $\mathbf{N}(t)$ when only noise is present at the receiver/matched filter output and $\mathbf{V}(t)$ when signal-plus-noise is present at the receiver/matched filter output.

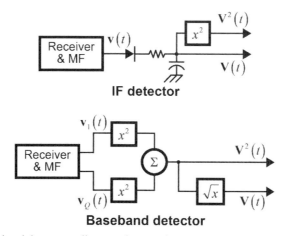

Figure 6.4 IF and baseband detectors—linear and square law.

6.5.3 Detection Logic

Since both $\mathbf{N}(t)$ and $\mathbf{V}(t)$ are random processes, we must use concepts from random processes theory to characterize the detection logic performance. In particular, we use probabilities.

Since we have two signal conditions (noise only and signal-plus-noise) and two outcomes from the threshold check, we have four possible events to consider:

1. Signal-plus-noise $\geq$ threshold $\Rightarrow$ detection
2. Signal-plus-noise $<$ threshold $\Rightarrow$ missed detection
3. Noise $\geq$ threshold $\Rightarrow$ false alarm
4. Noise $<$ threshold $\Rightarrow$ no false alarm

Of the above examples, the two desired events are 1 and 4. That is, we want to detect targets when they are present, and we do not want to detect noise when targets are not present. Since events 1 and 2 are related and events 3 and 4 are related, we need only find probabilities associated with events 1 and 3. We term the probability of the first event occurring the *detection* probability, and the probability of the third event occurring the *false alarm* probability. In equation form

$$P_d: \text{detection probability} = P(\mathbf{V} \geq T | \text{target present}) \quad (6.92)$$

and

$$P_{fa}: \text{false alarm probability} = P(\mathbf{N} \geq T | \text{target not present}) \quad (6.93)$$

where $\mathbf{V} = \mathbf{V}(t)|_{t=t1}$ indicates signal-plus-noise voltage evaluated at a specific time, and $\mathbf{N} = \mathbf{N}(t)|_{t=t1}$ indicates noise voltage evaluated at a specific time.

The definition above carries some subtle implications. First, when one finds detection probability, it is tacitly assumed that the target return is present at the time the output of the threshold device is checked. Likewise, when one finds false alarm probability, it is tacitly assumed that the target return is not present at the time the output of the threshold device is checked.

In practical applications, it is more appropriate to say: that at the time the output of the threshold device is checked, the probability of a threshold crossing equals P_d *if the signal contains a target signal* and P_{fa} *if the signal does not contain a target signal*.

Note that the above probabilities are conditional probabilities. In normal practice, we do not explicitly use the conditional notation, and write

$$P_d = P(\mathbf{V} \geq T) \quad (6.94)$$

and

$$P_{fa} = P(\mathbf{N} \geq T) \quad (6.95)$$

Further, we recognize that we should use signal-plus-noise when we assume the target is present and only noise when we assume the target is not present, and that the probabilities are conditional.

The discussion above relates to a linear detector. If the detector is square law, the appropriate equations would be

$$P_d = P(V^2 \geq T^2) \tag{6.96}$$

and

$$P_{fa} = P(N^2 \geq T^2) \tag{6.97}$$

6.5.4 Calculation of Pd and Pfa

From probability theory, we can write [13]

$$P_d = \int_T^\infty f_V(v)\,dv \text{ or } P_d = \int_{T^2}^\infty f_{V^2}(v)\,dv \tag{6.98}$$

and

$$P_{fa} = \int_T^\infty f_N(n)\,dn \text{ or } P_{fa} = \int_{T^2}^\infty f_{N^2}(n)\,dn \tag{6.99}$$

In the expression above, T denotes the threshold voltage level and T^2 denotes the threshold expressed as normalized power.

To avoid having to use two sets of P_d and P_{fa} equations, we will digress to show how we can compute P_d and P_{fa} using either of the integrals of (6.98) and (6.99).

It can be shown [13] that if $x = \sqrt{y}$ and $y \geq 0$, the densities of **x** and **y** are related by

$$f_x(x) = 2x f_y(x^2) \tag{6.100}$$

If we write

$$P_d = \int_T^\infty f_V(v)\,dv \tag{6.101}$$

we can use (6.100) to write

$$P_d = \int_T^\infty f_V(v)\,dv = \int_T^\infty 2v f_{V^2}(v^2)\,dv \tag{6.102}$$

With the change of variables $x = v^2$, we have

$$P_d = \int_T^\infty f_V(v)dv = \int_{T^2}^\infty f_{V^2}(x)dx \qquad (6.103)$$

Similar results apply to P_{fa} and indicate we can use either form to compute detection and false alarm probability.

We note that the integrals for P_d and P_{fa} are over the same limits. Figure 6.5 provides an illustration of this. Notice P_d and P_{fa} are areas under their respective density functions, to the right of the threshold value. Increasing the threshold decreases the areas, and thus the probabilities, and decreasing the threshold increases the areas, and thus probabilities. This is not exactly what we want. Ideally, we want to select the threshold so that we have $P_{fa} = 0$ and $P_d = 1$. Because this is not possible, we usually choose the threshold as some sort of trade-off between P_d and P_{fa}. In fact, we choose the threshold to achieve a certain P_{fa} and find other means of increasing P_d (see Chapter 9).

Referring to (6.12), the only parameter that affects $f_N(n)$ is the noise power, σ^2. While we have some control over this via noise figure, executing that control can be very expensive. On the other hand, $f_V(v)$ depends on both P_S and σ^2. This gives us some degree of control. In fact, we usually try to affect both $f_N(n)$ and $f_V(v)$ by increasing P_S and decreasing σ^2. The net result of this is that we try to maximize SNR.

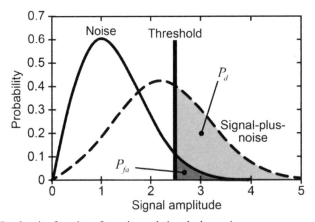

Figure 6.5 Probability density functions for noise and signal-plus-noise.

6.5.4.1 False Alarm Probability

Using (6.14) in (6.99), we can derive an equation for false alarm probability as

$$P_{fa} = \int_T^\infty f_N(n)\,dn = \int_T^\infty \frac{n}{\sigma^2} e^{-n^2/2\sigma^2}\,dn = e^{-T^2/2\sigma^2} = e^{-TNR} \qquad (6.104)$$

In (6.104), we define

$$TNR = \frac{T^2}{2\sigma^2} \qquad (6.105)$$

as the *threshold-to-noise ratio* (TNR). We usually select a desired P_{fa} and, from this, derive the required TNR as

$$TNR = -\ln P_{fa} \qquad (6.106)$$

6.5.4.2 Detection Probability

We compute the detection probability for the three target classes by substituting (6.59), (6.75), and (6.91) into (6.103).

SW0/SW5 Target

For the SW0/SW5 case, we have

$$\begin{aligned} P_d &= \int_T^\infty f_V(V)\,dV \\ &= \int_T^\infty \frac{V}{\sigma^2} I_0\!\left(\frac{VS}{\sigma^2}\right) e^{-(V^2+S^2)/2\sigma^2}\,dV \end{aligned} \qquad (6.107)$$

where we took advantage of $T > 0$ to eliminate $U(V)$ from the integrand.

Equation (6.107) is in the form of the Marcum Q function [3, 4], which has the general form

$$Q_M(a,b) = \int_b^\infty x\!\left(\frac{x}{a}\right)^{M-1} I_{M-1}(ax)\,e^{-(x^2+a^2)/2}\,dx \qquad (6.108)$$

In (6.107), we make the change of variables $x = V/\sigma$ and get

$$P_d = \int_{T/\sigma}^\infty x I_0\!\left(x \frac{S}{\sigma}\right) e^{-(x^2+S^2/\sigma^2)/2}\,dx \qquad (6.109)$$

This is of the form of (6.108), with $a = S/\sigma$, $b = T/\sigma$, and $M = 1$. Thus, we have

$$P_d = Q_1\left(\frac{S}{\sigma}, \frac{T}{\sigma}\right) \tag{6.110}$$

Since we are interested in finding P_d as a function of *SNR* and P_{fa}, we want to manipulate (6.110) so it is a function of these variables. From (6.105) and (6.106), we have

$$T/\sigma = \sqrt{-2\ln P_{fa}} \tag{6.111}$$

From (6.17) or (6.19), we have

$$P_{nIF} = P_{nB} = \sigma^2 \tag{6.112}$$

and from (6.34) or (6.36), we have

$$P_S = S^2/2 \tag{6.113}$$

We note that

$$SNR = \frac{P_S}{P_N} = \frac{S^2}{2\sigma^2} \tag{6.114}$$

which leads to

$$S/\sigma = \sqrt{2(SNR)} \tag{6.115}$$

Substituting (6.111) and (6.115) into (6.110) results in

$$P_d = Q_1\left(\sqrt{2(SNR)}, \sqrt{-2\ln P_{fa}}\right) \tag{6.116}$$

Unfortunately, $Q_1(a,b)$ has no simple form. However, Steen Parl has developed an algorithm that appears to work quite well [20]. Parl's algorithm is described in Appendix 8B. The Marcum Q function is also a standard function in MATLAB, and most likely other languages.

Skolnik presents the approximation [21, p. 27]

$$P_d = \frac{1}{2}\left[1 - \text{erf}\left(\sqrt{TNR} - \sqrt{SNR}\right)\right] \\ + \frac{e^{-(\sqrt{TNR}-\sqrt{SNR})^2}}{4\sqrt{\pi}\sqrt{SNR}}\left[1 - \frac{\sqrt{TNR}-\sqrt{SNR}}{4\sqrt{SNR}} + \frac{1+2(\sqrt{TNR}-\sqrt{SNR})^2}{16 SNR} - \cdots\right] \tag{6.117}$$

where

$$TNR = -\ln P_{fa} \tag{6.118}$$

and

$$\text{erf}(x) = \frac{2}{\sqrt{\pi}} \int_0^x e^{-u^2} du \qquad (6.119)$$

is one form of the error function. It has been the authors' experience that Skolnik's approximation degrades as *SNR* approaches and falls below *TNR*.

A recent paper by Barton [9] presents an equation attributed to David A. Shnidman [22]. In this equation,

$$P_d = \frac{1}{2}\text{erfc}\left(\sqrt{-0.8\ln\left[4P_{fa}(1-P_{fa})\right]} - \sqrt{SNR}\right) \qquad (6.120)$$

where erfc denotes the complementary error function and is defined as $\text{erfc}(x) = 1 - \text{erf}(x)$.

SW1/SW2 Target

For the SW1/SW2 case, we substitute (6.59) into (6.103) and write

$$P_d = \int_T^\infty f_V(V) dV = \int_T^\infty \frac{V}{P_S + \sigma^2} e^{-V^2/2(P_S + \sigma^2)} dV \qquad (6.121)$$

With the change of variables, $x = V^2/2(P_s + \sigma^2)$ we get

$$\begin{aligned} P_d &= \int_{T^2/2(P_S+\sigma^2)}^\infty e^{-x} dx = e^{-T^2/2(P_S+\sigma^2)} \\ &= \exp\left(\frac{-T^2/2\sigma^2}{P_S/\sigma^2 + 1}\right) \end{aligned} \qquad (6.122)$$

Using $\ln P_{fa} = -T^2/2\sigma^2$ and $SNR = P_S/\sigma^2$, we can write (6.122) as

$$P_d = \exp\left(\frac{\ln P_{fa}}{SNR + 1}\right) \qquad (6.123)$$

SW3/SW4 Target

For the SW3/SW4 case, we substitute (6.91) into (6.103), and write

$$\begin{aligned} P_d &= \int_T^\infty f_V(V) dV \\ &= \int_T^\infty \frac{2V}{(P_S + 2\sigma^2)^2}\left[2\sigma^2 + \frac{P_S V^2}{(P_S + 2\sigma^2)}\right] e^{-V^2/(P_S + 2\sigma^2)} dV \end{aligned} \qquad (6.124)$$

With the change of variables, $x = V^2/2(P_s + 2\sigma^2)$ we get

$$P_d = \int_{T^2/(P_S+2\sigma^2)}^{\infty} \frac{1}{P_S + 2\sigma^2}\left[P_s x + 2\sigma^2\right]e^{-x}dx$$

$$= \frac{2\sigma^2}{P_S + 2\sigma^2}\int_{T^2/(P_S+2\sigma^2)}^{\infty} e^{-x}dx + \frac{P_S}{P_S + 2\sigma^2}\int_{T^2/(P_S+2\sigma^2)}^{\infty} xe^{-x}dx \quad (6.125)$$

$$= \frac{2\sigma^2}{P_S + 2\sigma^2}e^{-T^2/(P_S+2\sigma^2)} + \frac{P_S}{P_S + 2\sigma^2}e^{-T^2/(P_S+2\sigma^2)}\left(\frac{T^2}{P_S + 2\sigma^2} + 1\right)$$

Substituting $TNR = T^2/2\sigma^2$ and $SNR = P_S/\sigma^2$ and manipulating yields

$$P_d = \left[1 + \frac{2(SNR)(TNR)}{(2+SNR)^2}\right]e^{-2TNR/(2+SNR)} \quad (6.126)$$

Finally, with $TNR = -\ln P_{fa}$

$$P_d = \left[1 - \frac{2(SNR)\ln P_{fa}}{(2+SNR)^2}\right]e^{2\ln P_{fa}/(2+SNR)} \quad (6.127)$$

As a reminder, for all the P_d equations, SNR denotes the signal-to-noise ratio computed from the radar range equation (see Chapter 2).

6.5.5 Behavior versus Target Type

Figure 6.6 contains plots of P_d versus SNR for the three target types and $P_{fa} = 10^{-6}$, a typical value [23, p. 45]. It is interesting to note the P_d behavior for the three target types. In general, the SW0/SW5 target provides the largest P_d for a given SNR; the SW1/SW2 target provides the lowest P_d; and the SW3/SW4 falls somewhere between the other two. With some thought, this makes sense. For the SW0/SW5 target model, only the noise affects a threshold crossing (since the target RCS is constant). For the SW1/SW2, the target RCS can fluctuate considerably; thus, both noise and RCS fluctuation affect the threshold crossing. The standard assumption for the SW3/SW4 model is that it consists of a predominant (presumably constant RCS) scatterer and several smaller scatterers. Thus, RCS fluctuation affects the threshold crossing for the SW3/SW4 target somewhat, but not to the extent of the SW1/SW2 target.

It is interesting to note that a SW1/SW2 target requires an SNR of about 13 dB for $P_d = 0.5$, with $P_{fa} = 10^{-6}$. This same SNR gives a $P_d = 0.9$ on a SW0/SW5 target. A $P_d = 0.9$ on a SW1/SW2 target requires an SNR of about 21 dB. These are the origins of the 13-dB and 20-dB SNR numbers used in the radar range equation examples of Chapter 2.

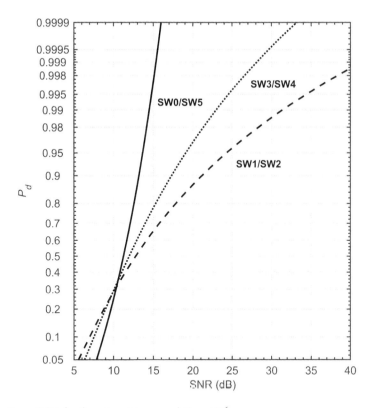

Figure 6.6 P_d vs. SNR for three target types and $P_{fa} = 10^{-6}$.

To reiterate an earlier statement, we term the P_{fa} and P_d variables given above the single pulse, single sample, or single hit P_{fa} and P_d. This term derives from the fact that the threshold check (i.e., check for a target detection) is based on target returns from a single pulse. If the signal contains both a target and noise component [i.e., $\mathbf{s}(t)$ and $\mathbf{n}(t)$], we are computing P_d. If the sample contains only noise, we are computing P_{fa}. Figure 6.7 illustrates this concept.

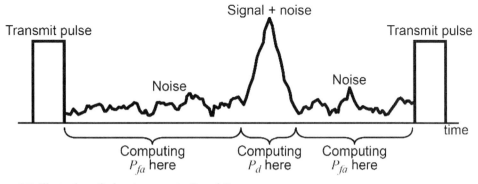

Figure 6.7 Illustration of when to compute P_d and P_{fa}.

6.6 DETERMINATION OF FALSE ALARM PROBABILITY

One parameter included in the detection probability equations is the threshold-to-noise ratio, *TNR*. As indicated in (6.105), $TNR = -\ln P_{fa}$, where P_{fa} is the false alarm probability. System requirements set false alarm probability.

In a radar, false alarms result in wasted radar resources (energy, timeline, and hardware), because every time a false alarm occurs, the radar must expend resources determining whether the alarm was the result of a random noise peak or of an actual target that can be redetected at or near that location. Said another way, every time the output of the amplitude detector exceeds the threshold, *T*, a detection is recorded. The radar data processor does not know, a priori, whether the detection is a target detection or the result of noise (i.e., a false alarm). Therefore, the radar must verify each detection, a process that usually requires transmission of another pulse and another threshold check (an expenditure of time and energy). Further, until the detection is verified, it must be carried in the computer as a valid target detection (an expenditure of hardware or software).

To minimize wasted radar resources, we want to minimize the probability of false alarm. Said another way, we want to minimize P_{fa}. However, we cannot set P_{fa} to an arbitrarily small value because this increases *TNR* and reduces detection probability, P_d. As a result, we set P_{fa} to provide an acceptable number of false alarms within a given time period. This last statement provides the criterion normally used to compute P_{fa}. Specifically, P_{fa} is chosen to provide an average of one false alarm within a time period termed the *false alarm time*, T_{fa}. T_{fa} is usually set by some criterion driven by radar resource limitations.

The classical method of determining P_{fa} is based strictly on timing [24]. Figure 6.8, which contains a plot of noise at the output of the amplitude detector, helps illustrate this concept. The horizontal line labeled "Threshold, T" represents the detection threshold voltage level. Note that the noise voltage is above the threshold during four time intervals of length t_1, through t_4. Further, the spacings between threshold crossings are T_1, T_2, and T_3. Since a threshold crossing constitutes a false alarm, one can say that over the interval T_1, false alarms occur for a period of t_1. Likewise, over the interval T_2, false alarms occur for a period of t_2, and so forth. Averaging all t_k produces an average time, $\bar{t}_k$, when the noise is above the threshold. Likewise, averaging all of T_k produces the average time between false alarms (i.e., the false alarm time, T_{fa}). To determine the false alarm probability, we find the ratio of $\bar{t}_k$ to T_{fa}; that is,

$$P_{fa} = \frac{\bar{t}_k}{T_{fa}} \quad (6.128)$$

While T_{fa} is reasonably easy to specify, the specification of $\bar{t}_k$ is not obvious. The standard assumption sets $\bar{t}_k$ to the range resolution expressed as time, $\tau_{\Delta R}$. For an unmodulated pulse, $\tau_{\Delta R}$ is the pulsewidth. For a modulated pulse, $\tau_{\Delta R}$ is the reciprocal of the modulation bandwidth.

It has been the authors' experience that the above method of determining P_{fa} is not very accurate. While it would be possible to place the requisite number of caveats on (6.128) to make it more accurate, with modern radars, this is not necessary.

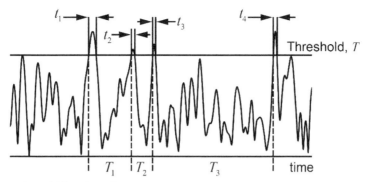

Figure 6.8 Illustration of false alarm time.

The previously described method of determining P_{fa} relies on the assumption that hardware operating on a continuous-time signal records the detections. Modern radars base detection on the examination of signals that have been converted to the discrete-time domain by sampling or by an analog-to-digital converter. This makes determination of P_{fa} easier, and more intuitively appealing, in that we can deal with discrete events. With modern radars, we compute the number of false alarm chances, N_{fa}, within the desired false alarm time, T_{fa}, and compute the probability of false alarm from

$$P_{fa} = \frac{1}{N_{fa}} \tag{6.129}$$

Computing N_{fa} requires us to know certain things about the radar's operation. We will outline some such thoughts.

A typical radar samples the return signal from each pulse with a period equal to the range resolution, $\tau_{\Delta R}$, of the pulse. As indicated above, this would equal the pulsewidth for an unmodulated pulse and the reciprocal of the modulation bandwidth for a modulated pulse. These *range samples* are usually taken over the instrumented range, ΔT. In a search radar, ΔT might be only slightly less than the PRI, T. However, for a track radar, ΔT may be significantly less than T. With the above, we compute the number of range samples per PRI as

$$N_R = \frac{\Delta T}{\tau_{\Delta R}} \tag{6.130}$$

Each range sample provides the opportunity for a false alarm.

In a time period of T_{fa}, the radar transmits

$$N_{pulse} = \frac{T_{fa}}{T} \tag{6.131}$$

pulses. Thus, the number of range samples (and, thus, chances for false alarm) over the time period of T_{fa} is

$$N_{fa} = N_R N_{pulse} \tag{6.132}$$

In some radars, the receiver contains several (N_{Dop}) parallel Doppler channels. Such radars also contain N_{Dop} amplitude detectors. Each amplitude detector generates N_R range samples per PRI (actually, coherent dwell – see Chapters 9, 19, and 20). Thus, in this case, the total number of range samples in the time period T_{fa} would be

$$N_{fa} = N_R N_{pulse} N_{Dop} \tag{6.133}$$

In either case, (6.129) gives the false alarm probability.

6.6.1 *Pfa* Computation Example

To illustrate the discussion above, we consider the simple example of a search radar with a PRI of $T = 400$ μs. This radar uses a 50-μs pulse with LFM and a bandwidth of 1 MHz. With this we get $\tau_{\Delta R} = 1$ μs. We assume the radar starts its range samples one pulsewidth after the transmit pulse and stops taking range samples one pulsewidth before the succeeding transmit pulse. From these parameters, we get $\Delta T = 300$ μs (see Chapter 1). The signal processor is not a multichannel Doppler processor. The radar has a search scan time of $T_{scan} = 1$ s, and we want no more than one false alarm every two scans. With this we get $T_{fa} = 2T_{scan} = 2$ s. If we combine this with the PRI, we get

$$N_{pulse} = \frac{T_{fa}}{T} = \frac{2}{400 \times 10^{-6}} = 5{,}000 \tag{6.134}$$

From ΔT and $\tau_{\Delta R}$, we get

$$N_R = \frac{\Delta T}{\tau_{\Delta R}} = \frac{300\ \mu s}{1\ \mu s} = 300 \tag{6.135}$$

This results in

$$N_{fa} = N_R N_{pulse} = 300 \times 5{,}000 = 1.5 \times 10^6 \tag{6.136}$$

and

$$P_{fa} = \frac{1}{N_{fa}} = \frac{1}{1.5 \times 10^6} = 6.667 \times 10^{-7} \tag{6.137}$$

6.6.2 Detection Contour Example

For this example, we combine the radar range equation discussions of Chapter 2 with the P_d and P_{fa} discussions of this chapter to plot a detection contour for a search radar. As used here, a detection contour is a boundary, in altitude versus downrange space, on which a radar achieves a given P_d. For all points outside the area bounded by the contour, P_d is less than the desired value. For all points inside the boundary, P_d is greater than the desired value.

Table 6.2 lists the radar parameters. The table also includes other parameters we will need. As implied by Table 6.2, the antenna constantly rotates in azimuth and the antenna's directivity varies with elevation angle. This directivity variation can be represented by the equation

$$G(\varepsilon) = 900 \sum_{k=1}^{4} a_k \text{sinc}^2 \left[N_k \left(\sin \varepsilon - \sin \varepsilon_k \right) \right] \qquad (6.138)$$

where Table 6.3 (next page) provides the values of a_k, N_k, and ε_k. Figure 6.9 (next page) contains a plot of $G(\varepsilon)$.

We acknowledge that the directivity also varies with azimuth and, as a result, the SNR varies as the antenna sweeps by the target. We accounted for this variation by including a scan loss in the total losses of Table 6.2.

Table 6.2
Radar and Other Parameters Used for Example 2

Parameter	Value
Peak transmit power at the power amp output	50 kW
Operating frequency	2 GHz
PRF	1,000 Hz
Pulsewidth, τ_p	100 μs
Pulse modulation bandwidth	1 MHz
Antenna directivity (transmit and receive)	See function
Total losses	13 dB
System noise figure	5 dB
Target type and RCS	SW1, 6 dBsm
Antenna rotation rate	6 rpm
Instrumented range	PRI – $2\tau_p$ – 50 μs
False alarm criterion	No more than one false alarm per 360° rotation
Detection probability	0.5

Table 6.3
Parameter Values for (6.138)

a_k	N_k	ε_k (deg)
1	45	1
1	25	3
1	13	7
0.5	7	15

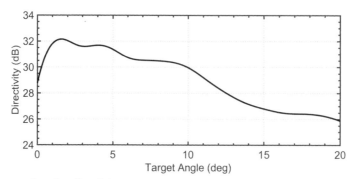

Figure 6.9 Antenna elevation directivity pattern.

The elevation angle is computed from

$$\varepsilon = \tan^{-1}(h/R_d) \qquad (6.139)$$

where h denotes the target height, or altitude, and R_d denotes the downrange (or horizontal or ground range) position of the target. We relate h and R_d to slant range, R, and ε by

$$R_d = R\cos\varepsilon \quad \text{and} \quad h = R\sin\varepsilon \qquad (6.140)$$

To generate the detection contour, we solve the radar range equation for R in terms of the other parameters. That is,

$$R(\varepsilon) = \left[\frac{P_T G_T(\varepsilon) G_R(\varepsilon) \lambda^2 \sigma \tau_p}{(4\pi)^3 kT_0 F_n L (SNR)}\right]^{1/4} \qquad (6.141)$$

We then use (6.140) to obtain $h(\varepsilon)$ and $R_d(\varepsilon)$ as we vary ε from 0° to 90°. In (6.141) we used $T_S = T_0 F_n$ for convenience (see Chapter 2). This means we are assuming $T_a = T_0$, which is okay since we are performing a preliminary analysis.

Most of the required parameters are in Table 6.2, or can be easily computed from the parameters in the table. The exception is the required SNR. We will use the specified P_d and the false alarm specification, along with the target type, to compute SNR.

Since the table specifies a SW1 target, we can solve (6.123) for SNR and get

$$SNR = \frac{\ln P_{fa}}{\ln P_d} - 1 \qquad (6.142)$$

We already have P_d, but need to compute P_{fa}. To do this, we must first compute N_{fa}.

Since the antenna rotates at a rate of f_{rot} revolutions per minute (rpm), the time to complete a rotation is

$$T_{rot} = \frac{1}{(f_{rot} \text{ rev/min}) \times (1/60 \text{ min/s})} = \frac{60}{6} = 10 \text{ s/rev} \qquad (6.143)$$

Since the false alarm criterion is one false alarm per revolution (see Table 6.2), the false alarm time is

$$T_{fa} = T_{rot} = 10 \text{ s} \tag{6.144}$$

Had we specified no more than one false alarm every three rotations, we would have had $T_{fa} = 3T_{rot} = 30$ s.

Given the PRF of 1,000 Hz, we can compute the PRI as (see Chapter 1):

$$T = 1/PRF = 10^{-3} \text{ s} \tag{6.145}$$

From (6.134), we compute the number of pulses in T_{fa} (as a reminder, we are considering single-pulse detection) as

$$N_{pulse} = T_{fa}/T = 10/10^{-3} = 10,000 \tag{6.146}$$

If we sample the return at the reciprocal of the modulation bandwidth, we get

$$\tau_{\Delta R} = \frac{1}{1 \text{ MHz}} = 1 \text{ μs} \tag{6.147}$$

From the instrumented range specified in Table 6.2, we get

$$\Delta T = T - \tau_p - 50 \text{ μs} = 750 \text{ μs} \tag{6.148}$$

and from (6.135)

$$N_R = \frac{\Delta T}{\tau_{\Delta R}} = \frac{750 \text{ μs}}{1 \text{ μs}} = 750 \tag{6.149}$$

We assume the radar does not implement any type of Doppler processing. Therefore, we can use (6.132) to compute

$$N_{fa} = N_R N_{pulse} = 750 \times 10,000 = 7.5 \times 10^6 \tag{6.150}$$

and find

$$P_{fa} = 1/N_{fa} = 1.33 \times 10^{-7} \tag{6.151}$$

Finally, we use (6.142) to compute

$$SNR = \frac{\ln P_{fa}}{\ln P_d} - 1 = 21.8 \text{ W/W or } 13.4 \text{ dB} \tag{6.152}$$

Substituting the appropriate values into (6.141) gives

$$R(\varepsilon) = \left[\frac{50 \times 10^3 G^2(\varepsilon)(0.15)^2 (10^{6/10})(100 \times 10^{-6})}{(4\pi)^3 (4 \times 10^{-21})(10^{5/10})(10^{13/10})(21.8)} \right]^{1/4} \tag{6.153}$$

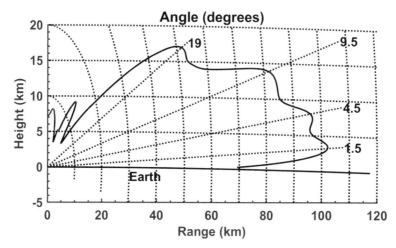

Figure 6.10 Detection contour.

Figure 6.10 contains a plot of $h(\varepsilon)$ versus $R_d(\varepsilon)$. The curved grid lines on the figure result from the round earth model used to plot the detection contour. The solid curved line represents the Earth's surface, and the slanted, numbered lines represent elevation lines.

6.7 SUMMARY

The major results of this chapter are the three detection probability equations for the three different target types and the methodology for computing P_{fa}. The three P_d equations are

- SW0/SW5 targets— $P_d = Q_1\left(\sqrt{2(SNR)}, \sqrt{-2\ln P_{fa}}\right)$

- SW1/SW2 targets— $P_d = \exp\left(\dfrac{\ln P_{fa}}{SNR+1}\right)$

- SW3/SW4 targets— $P_d = \left[1 - \dfrac{2(SNR)\ln P_{fa}}{(2+SNR)^2}\right] e^{2\ln P_{fa}/(2+SNR)}$

As yet another reminder, the P_d and P_{fa} discussed in this chapter are single-sample, or single-pulse, values. In Chapter 9, we consider the problem of computing P_d and P_{fa} based on processing several samples of signal-plus-noise and noise.

6.8 EXERCISES

1. A phased array radar searches a volume of space with a search raster containing 400 beams. The dwell time per beam is 10 ms. and the radar uses a pulsed Doppler waveform. The signal processor has 10 range gates with a 64-point fast Fourier transformer (FFT) on each range gate. This produces a 10-by-64 range-Doppler map on each dwell. The detection logic checks each range-Doppler cell once per beam dwell. We want the radar to support a 20-s time between false alarms. For purposes

of computing P_d and P_{fa}, we consider a dwell a single sample or single pulse. Thus, the P_d and P_{fa} equations of this chapter apply to this problem.

a) What false alarm probability, Pfa, is necessary to support the specified false alarm rate?
b) What SNR, in dB, is required at the signal processor output for the radar to provide a single-sample detection probability of 0.95 on a SW0/SW5 target?
c) What SNR, in dB, is required at the signal processor output for the radar to provide a single-sample detection probability of 0.95 on a SW1 target?
d) What SNR, in dB, is required at the signal processor output for the radar to provide a single-sample detection probability of 0.95 on a SW3 target?

2. A monostatic radar has the following parameters:

- Peak transmit power at power amplifier output—100 kW
- Transmit losses—2 dB
- Operating frequency—10 GHz
- PRF—2,000 Hz
- Antenna diameter—1.5 m (circular aperture)
- Antenna efficiency—60%
- Other losses—10 dB
- System noise figure—6 dB
- Antenna temperature (T_a)—100 K
- The radar transmits a 10-μs rectangular pulse.
- The radar maintains a P_{fa} of 10^{-9}

Plot P_d versus target range, in km, for a 6-dBsm, SW1 target. Let the range vary from 5 km to the unambiguous range of the radar.

3. Derive (6.14) and (6.15).

4. Derive (6.17) and (6.19).

5. Show that (6.71) can be obtained by manipulating (6.70).

6. Show that (6.91) follows from (6.90).

7. Show that (6.100) is correct.

8. Show that (6.126) follows from (6.125).

References

[1] Swerling, P., "Probability of Detection for Fluctuating Targets," RAND Corp., Santa Monica, CA, Res. Memo. RM-1217, Mar. 17, 1954. Reprinted: *IRE Trans. Inf. Theory,* vol. 6, no. 2, Apr. 1960, pp. 269–308.

[2] Rice, S. O., "Mathematical Analysis of Random Noise," *Bell Syst. Tech. J.,* vol. 23, no. 3, Jul. 1944, pp. 282–332; vol. 24, no. 1, Jan. 1945, pp. 461–556.

[3] Marcum, J. I., "A Statistical Theory of Target Detection by Pulsed Radar," RAND Corp., Santa Monica, CA, Res. Memo. RM-754-PR, Dec. 1, 1947. Reprinted: *IRE Trans. Inf. Theory,* vol. 6, no.

[4] Marcum, J. I., "A Statistical Theory of Target Detection by Pulsed Radar (Mathematical Appendix)," RAND Corp., Santa Monica, CA, Res. Memo. RM-753-PR, Jul. 1, 1948. Reprinted: *IRE Trans. Inf. Theory,* vol. 6, no. 2, Apr. 1960, pp. 145–268.

[5] Swerling, P., "Radar Probability of Detection for Some Additional Fluctuating Target Cases," *IEEE Trans. Aerosp. Electron. Syst.,* vol. 33, no. 2, Apr. 1997, pp. 698–709.

[6] Shnidman, D. A., "Expanded Swerling Target Models," *IEEE Trans. Aerosp. Electron. Syst.,* vol. 39, no. 3, Jul. 2003, pp. 1059–1069.

[7] Heidbreder, G. R., and R. L. Mitchell, "Detection Probabilities for Log-Normally Distributed Signals," *IEEE Trans. Aerosp. Electron. Syst.,* vol. 3, no. 1, Jan. 1967, pp. 5–13.

[8] Barton, D. K., "Simple Procedures for Radar Detection Calculations," *IEEE Trans. Aerosp. Electron. Syst.,* vol. 5, no. 5, Sept. 1969, pp. 837–846. Reprinted: Barton, D. K., ed., *Radars, Vol. 2: The Radar Range Equation* (Artech Radar Library), Dedham, MA: Artech House, 1974, pp. 113–122.

[9] Barton, D. K., "Universal Equations for Radar Target Detection," *IEEE Trans. Aerosp. Electron. Syst.,* vol. 41, no. 3, Jul. 2005, pp. 1049–1052.

[10] Weinstock, W. W., *Target Cross Section Models for Radar Systems Analysis,* Ph.D. Dissertation in Electrical Engineering, University of Pennsylvania, 1964.

[11] DiFranco, J. V., and W. L. Rubin, *Radar Detection,* Prentice-Hall, 1968. Reprinted: Dedham, MA: Artech House, 1980.

[12] Budge, M. C., Jr., "EE 619: Intro to Radar Systems," www.ece.uah.edu/courses/material/EE619/index.htm.

[13] Papoulis, A., *Probability, Random Variables, and Stochastic Processes*, 3rd ed., New York: McGraw-Hill, 1991.

[14] Ross, S. M., *A First Course in Probability,* 5th ed., Upper Saddle River, NJ: Prentice-Hall, 1998.

[15] Cooper, G. R., and C. D. McGillem, *Probabilistic Methods of Signal and System Analysis,* 2nd ed., Fort Worth, TX: Holt, Rinehart and Winston, 1986.

[16] Stark, H., and J. W. Woods, *Probability, Random Processes and Estimation Theory for Engineers,* 2nd ed., Upper Saddle River, NJ: Prentice-Hall, 1994.

[17] Peebles, P. Z., Jr., *Probability, Random Variables, and Random Signal Principles*, 3rd ed., New York: McGraw-Hill, 1993.

[18] Gradshteyn, I. S., and I. M. Ryzhik, *Table of Integrals, Series, and Products,* 8th ed., D. Zwillinger and V. Moll, eds., New York: Academic Press, 2015. Translated from Russian by Scripta Technica, Inc.

[19] Abramowitz, M., and I. A. Stegun, eds., *Handbook of Mathematical Functions with Formulas, Graphs, and Mathematical Tables,* National Bureau of Standards Applied Mathematics Series 55, Washington, DC: U.S. Government Printing Office, 1964; New York: Dover, 1965.

[20] Parl, S., "A New Method of Calculating the Generalized Q Function," *IEEE Trans. Inf. Theory,* vol. 26, no. 1, Jan. 1980, pp. 121–124.

[21] Skolnik, M. I., *Introduction to Radar Systems,* 2nd ed., New York: McGraw-Hill, 1980.

[22] Shnidman, D. A., "Determination of Required SNR Values," *IEEE Trans. Aerosp. Electron. Syst.,* vol. 38, no. 3, Jul. 2002, pp. 1059–1064.

[23] Barton, D. K., *Radar System Analysis and Modeling,* Norwood, MA: Artech House, 2005.

[24] Meyer, D. P., and H. A. Mayer, *Radar Target Detection,* New York: Academic Press, 1973.

Chapter 7

CFAR Processing

7.1 INTRODUCTION

Maintaining a *Constant False Alarm Rate* (CFAR) is critical for predictable target detection performance. The probability of false alarm, P_{fa}, and thus false-alarm rate (FAR), is very sensitive to changes in noise, clutter, or other interference. FAR variations present a significant problem affecting a radar's detection performance, which can increase probability of false alarm drastically, or reduce the probability of detection, P_d. In this chapter, we introduce the basics of CFAR processing, which is a standard component in most modern radars. CFAR processing is sometimes referred to synonymously as CFAR detection, automatic detection, adaptive detection, or simply CFAR.

CFAR processing refers to various signal processing techniques, algorithms, and procedures used to automatically detect targets against a background of varying noise, clutter, or other interference. The term *clutter* refers to backscatter from unwanted objects in the environment, examples of which (for ground-based radars) include echoes from the ground, water, buildings, hills, rain, snow, birds, insects, and chaff [1, p. 166; 2, p. 107]. Examples of noise and interference sources include receiver noise, atmospheric noise, signals from other radars, and jammers. In the context of CFAR processing, the term *clutter* is sometimes used to refer to all unwanted noise, clutter, and other interference signals, collectively.

In Chapter 6, we developed design equations, relating SNR, P_d, and P_{fa}, which enable us to set up a fixed threshold detector, operating on a matched filter output, to provide a specified P_{fa} and P_d. The target signal is in a background of Gaussian-distributed thermal receiver noise [3, 4]. The relationships between threshold-to-noise ratio, *TNR*; detection threshold voltage, T; average noise power, σ^2; and P_{fa} for single-pulse detection were shown to be [see (6.104), (6.105), (6.106), and (6.111)]

$$P_{fa} = e^{-T^2/2\sigma^2} = e^{-TNR} \qquad (7.1)$$

$$TNR = \frac{T^2}{2\sigma^2} = \frac{\text{rms sinewave power}}{\text{rms noise power}} = -\ln\left(P_{fa}\right) \qquad (7.2)$$

$$\frac{T}{\sigma} = \frac{\text{peak sinewave amplitude}}{\text{rms noise voltage}} = \sqrt{-2\ln\left(P_{fa}\right)} \qquad (7.3)$$

The noise power in (7.1) and (7.2), and the noise voltage in (7.3) are assumed to be constant and known a priori, allowing fixed threshold detection to provide for predictable P_{fa} and P_d (see Chapter 6). In practical radar environments, noise, clutter, and interference power are not generally known a priori, nor do they remain at a constant level.

For example, drifts in receiver temperature and component aging can cause minor deviations in receiver noise power. Clutter that includes ground returns can exhibit power levels that vary greatly with terrain type and weather conditions. Likewise, interference may include noise jamming, which can occur at high levels compared to thermal receiver noise. This noise, clutter, and interference power variability causes difficulties when detecting targets using fixed threshold detection, motivating the invention of CFAR processing.

Additionally, clutter statistics may be governed by some other probability density function (PDF) than Gaussian. In order to evaluate a CFAR processor's ability to establish a desired P_{fa}, we need knowledge of the clutter characteristics (statistics). To this end, a model that describes the clutter properties is usually assumed. Clutter, or more precisely its amplitude out of a detector, is usually modelled as a random process described by a specific PDF with particular parameters/statistics. One (or more) parameters of the chosen density function used to model interference is assumed unknown.

Examples of PDFs used to describe ground clutter, sea clutter, and weather clutter include Rayleigh [5–10], Weibull [11–16], log-normal [17, 18], log-Weibull [19], Gamma [20], and K-distribution [21, 22]. Clutter distributions vary as a function of several parameters, including clutter type, grazing angle, radar resolution, frequency, and polarization. Given the complexity of clutter returns, the PDF that best represents a particular clutter environment is generally determined experimentally.

In addition to a CFAR processor's ability to establish a desired P_{fa}, another figure of merit is the P_d achieved, usually evaluated for various signal-to-clutter ratios (SCRs). Determining the P_d for a CFAR adds another level of complexity when analyzing CFAR performance. This is because calculating the P_d requires a target RCS model, such as the Swerling models presented in Chapter 3, be selected and incorporated into the CFAR analysis. While not strictly necessary to set up a CFAR processor, P_d is one factor used to compare different CFAR processors, typically when selecting which CFAR technique to adopt.

While there are various approaches used to achieve CFAR operation, this chapter focuses on *adaptive CFAR processing*, which is the most prevalent CFAR method used in radars. Adaptive CFAR processing suppresses fluctuations in FAR (stabilizes Pfa) caused by a varying clutter level by constantly estimating the clutter level and adjusting the detection threshold accordingly. Estimates of clutter amplitude (linear detector) or power (square-law detector) are generated using an estimation method tailored to specified clutter amplitude characteristics. The adaptive threshold is formed by continually scaling the clutter estimate by a constant that sets the P_{fa}.

Our goal is a fundamental understanding of basic CFAR processor topologies and design choices necessary to design a CFAR processor that will establish a specified P_{fa}. Specifically, the CFAR multiplier, which is used to control false alarm probability, and the number of reference cells, which controls CFAR loss. Larger reference windows result in lower CFAR loss (see Section 5.6), at the expense of responding more slowly to changes in clutter power. Typical CFAR processor shortcomings, such as self-masking, target masking, and clutter edge effects, are illustrated.

There are numerous adaptive CFAR processing variations, analyzed for different combinations of clutter and target RCS models, present in literature, some of which are fairly complex and difficult to implement and analyze. Examples of some of the analysis of the various CFAR techniques can be found in [5, 10, 23–32].

In the interest of simplicity, the only clutter considered in this chapter is represented as a zero-mean Gaussian random process with a variance (power) equal to σ^2. This is the simplest clutter model, referred to as the Rayleigh-Envelope or Gaussian clutter model, and allows clutter to be analyzed the same way as thermal noise. Using the Rayleigh clutter model is generally sufficient when considering varying levels of ground clutter, weather clutter, sea clutter, and receiver noise for low to moderate resolution radars.

As a matter of specifying CFAR parameters, P_d generally plays a negligible role. As such, this chapter will not focus on P_d. For simplicity, only SW1 or SW2 target RCS models are used where necessary for illustration in the discussion. A SW1 or SW2 target is Rayleigh-distributed at the output of a linear detector and exponentially distributed at the output of a square-law detector. This allows SW1 and SW2 targets to be treated like thermal noise as well.

For more detailed CFAR analysis, we recommend Minkler & Minkler [5], Nitzberg [26, 32], Levanon [28], Barkat [24], Kang [31] and Richards [25].

At this point, we emphasize the clutter modelling and analysis methods typically used to evaluate CFAR performance differ from the clutter modelling and analysis to be presented in Chapters 15 through 21. While typical figures of merit for CFAR performance include P_{fa}, P_d, CFAR loss, and implementation efficiency, the material in Chapters 15 through 21 focus on clutter attenuation and signal-to-clutter ratio improvement, for a given signal processor. To avoid possible confusion with how clutter is defined and modeled in Chapters 15 through 21, we will discuss clutter using the term interference for the remainder of this Chapter 7. Given the stipulation of using only Rayleigh clutter in this chapter, we will at times use the terms noise and interference, synonymously.

Example – P_{fa} sensitivity

When the noise power, σ^2, is a known constant, the fixed threshold equations of Chapter 6 hold true. In this example, we want to investigate the change in P_{fa} (and thus false alarm rate) when the actual noise power differs from the design noise power used to tune a fixed threshold detector. Solving (7.2), the design threshold is set to

$$T_d^2 = 2\sigma_d^2 \ln\left(P_{fad}^{-1}\right) \quad (7.4)$$

where the subscript d denotes design value. Substituting (7.4) into (7.1), the actual P_{fa} is then [25, 28]

$$P_{fa} = e^{-T_d^2/2\sigma^2} = e^{-2\sigma_d^2 \ln\left(P_{fad}^{-1}\right)/2\sigma^2}$$
$$= e^{\left(\sigma_d^2/\sigma^2\right)\ln\left(P_{fad}\right)} = P_{fad}^{\left(\sigma_d^2/\sigma^2\right)} \quad (7.5)$$

where we made use of

$$e^{x\ln(y)} = y^x \quad (7.6)$$

The change in P_{fa} and P_{fad} is then [25]

$$\frac{P_{fa}}{P_{fad}} = P_{fad}^{-1} P_{fad}^{\left(\sigma_d^2/\sigma^2\right)} = P_{fad}^{\left(\sigma_d^2/\sigma^2 - 1\right)} \quad (7.7)$$

The resultant P_{fa} versus change in noise power, $10\log(\sigma_d^2/\sigma^2)$, for several values of P_{fa}, are presented in Figure 7.1. The plots in Figure 7.1 illustrate relatively small changes in noise power can result in P_{fa} changing by orders of magnitude. For example, as depicted in Figure 7.1, for a radar designed to have a P_{fad} of 10^{-8}, an increase in noise power of only 3 dB increases P_{fa} by 4 orders of magnitude to 10^{-4}. This increase in P_{fa} by a factor of 10,000 significantly decreases false alarm time.

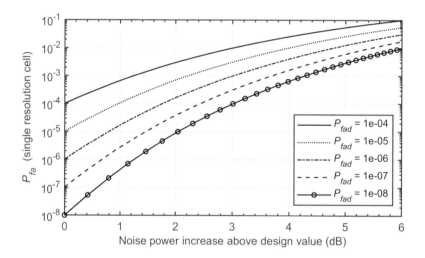

Figure 7.1 Resulting P_{fa} versus noise power increase from design level (After [32]).

This increase in noise power increases the false alarm rate, resulting in an increased demand on radar or signal processor resources to confirm target detections and reject false alarms. At best, too many false alarms can increase noise presented on operator displays, "cluttering" the display and fatiguing radar operators, or overwhelm subsequent processing, such as target parameter extraction. At worst, erroneous tracks could be initiated. Similarly, a decrease in noise power from the design level results in fewer false alarms. This is not necessarily a good thing, since now the threshold is set higher than necessary, reducing detection probability. In short, a constant false alarm rate is necessary for optimal, predictable, and consistent radar performance.

7.2 CELL-AVERAGING CFAR

In Chapter 6 we analyzed detection of a single target against a background of thermal receiver noise. In Section 6.2, we stipulated thermal noise as being a zero-mean Gaussian random process with a variance (power) equal to σ^2. We note that the Gaussian PDF describing receiver noise is governed by a single parameter, σ^2. In practice, σ^2 is not known in advance and can vary over time. To address this problem, we can instead use estimates for the unknown statistical parameter, which in this case is σ^2, to establish a detection threshold.

One such solution, depicted in Figure 7.2, is the classic cell averaging constant false alarm rate (CA-CFAR) processor, which is specifically designed for a single target signal competing with Gaussian noise of unknown power[1]. CA-CFAR was developed in 1968 by Finn and Johnson who proposed a mean-level estimator using N samples taken from cells near the CUT to estimate noise power. [32].

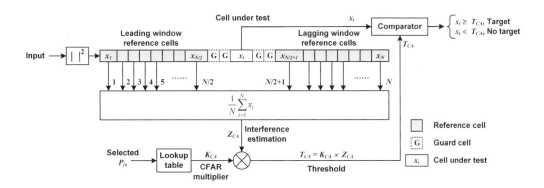

Figure 7.2 Block diagram of CA-CFAR detector.

---

[1] Even though Finn and Johnson stipulated Gaussian interference when analyzing their CA-CFAR processor design, this should not be interpreted to mean CA-CFAR cannot be analyzed and used against non-Gaussian interference.

Complex samples enter the CA-CFAR processor from the matched filter, or signal processor output, and are square–law detected. When considering CFAR processing, samples are typically referred to as *cells*. The detector output feeds into shift registers used to form a *reference window* or *CFAR window* consisting of N reference cells comprised of leading and lagging cells collected in the vicinity of the cell under test (CUT), x_t. The statistics of the CFAR window are used to estimate the statistics in the CUT (noise power).

For the configuration depicted in Figure 7.2, the number of reference cells, N, is usually an even number, allowing an equal number of leading and lagging reference cells. The location of the reference cells depicted in Figure 7.2 is typical of CFAR detection literature, implying CFAR across range samples (hence the terms early and late). While this is customary for CFAR illustration and discussion purposes, the location of the reference cells can vary according to the type of waveform, so long as they are in the vicinity of the CUT.

The noise power estimate is scaled by a *CFAR multiplier*, also referred to as threshold multiplier, or CFAR constant, to form an *adaptive detection threshold*, sometimes referred to as the *CFAR threshold*. The CFAR constant is a function of the desired P_{fa}. Selection of the CFAR constant for a given P_{fa} is often implemented as a table lookup to avoid implementing potentially complicated equations (see Section 7.7.2).

Noise power estimation is carried out continuously as a sliding window averaging process, incremented cell by cell. The CFAR threshold is recalculated after each cell increment. The resulting CFAR threshold essentially "rides" above the estimated noise power. This allows targets exceeding this adaptive CFAR threshold to be detected, while stabilizing the P_{fa}.

For CFAR analysis, noise is assumed independent and identically distributed (i.i.d.). The means that the noise power is the same for all cells, which is referred to as homogeneous noise. Homogeneity ensures that cells near the CUT have identical noise power (statistics) as the CUT [24, p. 642]. Also, Finn and Johnson stipulated only one cell contains a target signal to be detected. As such, each cell can only contain either noise or signal plus noise. Additionally, when the target is in the CUT, reference window samples can contain only noise. These constraints are sometimes referred to as the CFAR assumptions, of which violation of any causes a degradation in CFAR performance from both a P_d and P_{fa} perspective.

Stipulating CFAR processing across range, these assumptions equate to range samples spaced at least one waveform resolution apart (digital system), or adjacent, non-overlapping range gates (analog system), with the target being exactly aligned with a single range sample or range gate. In practice, range samples are sometimes separated by less than a resolution and range gates are sometimes overlapped in digital and analog signal processors, respectively. This overlap (oversampling) is intended to reduce straddle loss. Additionally, targets are not always aligned exactly with a range sample or gate. Large target returns can also bleed into adjacent cells. As a result, a single target can potentially occupy more than one cell, violating the single target cell assumption.

To compensate, the *Guard cells* shown in Figure 7.2 bracket (surround) the CUT, and are excluded from serving as reference cells. The typical number of guard cells is one or two on either side of (around) the CUT. The use of guard cells is intended to prevent target returns in the CUT from leaking into the reference window, which would cause an inaccurately high noise power estimate.

For example, a typical range gate overlap is 1/3 to 1/2 of a resolution, resulting in a single target usually occupying two or three adjacent cells, which would require one or two guard cells on either side of the CUT, respectively. This is why the number of guard cells on either side of the CUT is usually only one or two. A very large target might bleed into five or six adjacent cells (depending on level).

While Figure 7.2 shows a square-law detector, implementation and analysis of CFAR processing can use either linear or square-law detection. It turns out that the detector law used is usually not critical. In most practical CFAR processors, detection performance is identical, or very nearly so, irrespective of detector law used. Analysis is usually carried out assuming square-law detection, which generally results in more tractable math. Implementation often favors linear detection because it requires less dynamic range, is less likely to introduce distortion, and is easy to implement digitally (see Exercise 2).

7.2.1 Estimation of Interference Power

For CA-CFAR, the particular statistic of interest, power level, is estimated by the mean value of N reference cells, which can be written as[2]

$$Z_{CA} = \widehat{\sigma^2} = \frac{1}{N}\sum_{i=1}^{N} x_i \tag{7.8}$$

where Z_{CA} is the noise power estimate, N is the reference window size, and σ^2 is the actual noise power. The $\wedge$ symbol over σ^2 in (7.8) denotes approximation. The subscript CA denotes cell averaging.

While selecting the arithmetic mean to estimate the noise power might seem intuitive, this choice is not as arbitrary as it might seem. The estimate expressed by (7.8) is the maximum likelihood estimate (MLE) for, and also an unbiased estimation of, the unknown noise power (see Appendix 7A). One implication of Z_{CA} being an unbiased estimate of σ^2, is the noise power estimate is approximately equal to the actual noise power, which can be expressed as

$$Z_{CA} = \widehat{\sigma^2} \approx \sigma^2 \tag{7.9}$$

The noise power estimate is then multiplied by a CFAR multiplier, to form an adaptive CFAR threshold. The threshold for a CA-CFAR, can be expressed as

$$T_{CA} = K_{CA} Z_{CA} = K_{CA}\widehat{\sigma^2} = K_{CA}\frac{1}{N}\sum_{i=1}^{N} x_i \tag{7.10}$$

where K_{CA} is a scale factor and T_{CA} is the CFAR threshold. The factor $1/N$ is sometimes absorbed into the CFAR multiplier to avoid implementing a division in hardware.

The sample in the CUT, denoted by x_t, is compared to the CFAR threshold to provide a detection decision. A target is declared if the value of the sample in the CUT exceeds the detection threshold; otherwise, no detection is declared. This decision rule can be expressed as

$$x_t \geq T_{CA}, \quad \text{Target detected}$$
$$x_t < T_{CA}, \quad \text{No target detected} \quad (7.11)$$

The CA-CFAR process is performed in a sliding window fashion, incremented on a cell-by-cell basis. For the case of a stationary, homogeneous noise background, where reference cells contain i.i.d. observations with an exponential distribution (square-law detector), CA-CFAR is the optimum CFAR detector in the sense that is results in the maximum P_d for a given SNR [5, p. 165, 33–40].

7.2.2 CA-CFAR Analysis

Looking at Figure 7.2, we see that we need to specify the number of reference cells and the threshold multiplier. The threshold multiplier is usually determined based on the specified P_d and P_{fa} for a CA-CFAR. Selection of N is based on an acceptable CFAR loss. In this CFAR analysis, we will generally follow the approach of Finn and Johnson [32] and Kang [31, p. 281].

We can take advantage of the techniques presented in Chapter 6 for determining P_d and P_{fa} for a CA-CFAR. There is, however, one important difference. In Chapter 6, P_d and P_{fa} are related to SNR and σ^2 which were assumed constant. In this analysis, we are using an estimate of σ^2, which is a random variable characterized by a PDF. The resultant P_{fa} and P_d derived using an estimate of σ^2 are also random variables. To accommodate this, we use expected values of P_d and P_{fa}.

Design and analysis of a CFAR processor usually includes the following steps (not necessarily in this order):

- Select an interference model that best characterizes interference statistics
- Find optimal estimator(s) for unknown parameter(s) of interference statistics
- Determine the probability density function(s) of estimated parameter(s)
- Derive expression for P_{fa}
- Find threshold multiplier, which results in specified P_{fa}
- Select desired target type(s)
- Evaluate P_d for various SCRs

One standard a CFAR processor is usually compared to is the fixed threshold detection of Chapter 6. As mentioned in the introduction, we will not emphasis the last two steps listed above concerning P_d, though for illustration, a SW1 or SW2 target RCS model is mentioned.

In this chapter we are primarily interested in selecting CFAR parameters necessary to design a CFAR processor, which will establish a specified P_{fa}. Specifically, this is the CFAR multiplier, which is used to control false alarm probability, and the number of reference cells, which controls CFAR loss. In specifying CFAR parameters, detection probability generally plays a negligible role.

To begin this analysis, we need to select an interference model that best characterizes interference amplitude statistics. We will use the *Rayleigh-envelope model* (Gaussian model), which corresponds to Finn and Johnson's specifying a background of Gaussian noise where the unknown parameter is noise power [32]. After selecting the Rayleigh-envelope model, we need to find an optimal estimator for the unknown parameter σ^2, noise power. Finn and Johnson selected a mean-level estimate to determine noise power, which is an unbiased MLE of σ^2 [32].

Next, we need to choose a model to represent the amplitude statistics for a target in the CUT, called the *primary target* [24, p. 637]. We pick a SW1 target model (see Section 3.3) [4]. The Gaussian quadrature components of a SW1 target are zero-mean, Gaussian random processes. The amplitude for a SW1 target can be expressed as[2]

$$f_S(S) = \begin{cases} \dfrac{S}{\sigma^2} e^{-S^2/2\sigma^2}, & \text{linear} \\ \dfrac{1}{2\sigma^2} e^{-S/2\sigma^2}, & \text{square-law} \end{cases} \quad (7.12)$$

As indicated by (7.12), a SW1 target is Rayleigh-distributed at the output of a linear detector and exponentially distributed at the output of a square-law detector. As such, we can treat the SW1 target like the Rayleigh interference, simplifying analysis. It should be noted that while a SW1 target model is selected, any target RCS model could be chosen.

Reference cells are assumed to contain only noise. Given a complex valued Gaussian random process at the detector input, the linear or square-law detected noise voltage can be shown to be governed by the *Rayleigh* PDF or by the *exponential* (*Rayleigh power*) PDF, respectively, both with a single parameter σ^2 [41]. We can express this as [42–44]

$$f_N(N) = \begin{cases} \dfrac{N}{\sigma^2} e^{-N^2/2\sigma^2}, & \text{linear} \\ \dfrac{1}{2\sigma^2} e^{-N/2\sigma^2}, & \text{square-law} \end{cases} \quad (7.13)$$

The CUT is assumed to contain either noise or a SW1 target plus noise. Going forward, we opt to use a square-law detector as did Finn and Johnson [32]. The resulting conditional PDF of a square-law detected output for the CUT can be expressed as [4, 24, 45]

$$f_{X_t|H_i}(x_t|H_i) = \begin{cases} \dfrac{1}{2(\sigma^2 + P_S)} e^{-x_t/2(\sigma^2 + P_S)}, & \text{for } H_1 \\ \dfrac{1}{2\sigma^2} e^{-x_t/2\sigma^2}, & \text{for } H_0 \end{cases} \quad (7.14)$$

where hypothesis H_0 is the case where only noise is present, and hypothesis H_1 is the case where signal and noise are present. We can express (7.14) in terms of SNR as [24, p. 637]

---

[2] Using a substitution of $\sigma/2 = \sigma_{AV}$ into (7.12) results in the form presented in (3.9).

$$f_{\mathbf{X}_t|H_i}(x_t|H_i) = \begin{cases} \dfrac{1}{2\sigma^2(1+SNR)} e^{-x_t/2\sigma^2(1+SNR)}, & \text{for } H_1 \\ \dfrac{1}{2\sigma^2} e^{-x_t/2\sigma^2}, & \text{for } H_0 \end{cases} \quad (7.15)$$

The mean-level estimation function used for CA-CFAR is given by (7.8), where x_i are reference cells assumed to contain samples of square-law detected Gaussian noise. As such, the PDF at the output of a square-law detector is exponential, which can be expressed by [25, p. 338]

$$f_{\mathbf{X}_t|H_t}(x_t|H_0) = \dfrac{1}{2\sigma^2} e^{-x_t/2\sigma^2} \quad (7.16)$$

Using (7.16) we can determine P_{fa}, which is given by

$$\begin{aligned} P_{fa} &= \int_{T^2}^{\infty} \dfrac{1}{2\sigma^2} e^{-x_t/2\sigma^2} dx_t = \left[-e^{-x_t/2\sigma^2} \right]_{T^2}^{\infty} \\ &= -e^{-\infty/2\sigma^2} - \left(-e^{-T^2/2\sigma^2} \right) = e^{-T^2/2\sigma^2} \end{aligned} \quad (7.17)$$

Letting $Th = T^2$, after Kang's notation, the P_{fa} for the CUT, given it contains only noise, is then [5, p. 171; 31, p. 283]

$$P_{fa} = \int_{Z_{CA}=0}^{\infty} \left[\int_{Th}^{\infty} f_{\mathbf{X}_t|H_0}(x_t|H_0) dx_t \right] f_{\mathbf{Z_{CA}}}(Z_{CA}) dZ_{CA} \quad (7.18)$$

and $f_Z(z)$ is the PDF of the sum of reference cells.

For N random variables governed by the exponential PDF, the sum of these random variables has a Gamma PDF with N degrees of freedom, which can be expressed as [46, p. 210]

$$\begin{aligned} f_{\mathbf{Z_{CA}}}(Z_{CA}) &= \dfrac{1}{\Gamma(N)} \left(\dfrac{1}{2\sigma^2} \right)^N Z_{CA}^{N-1} e^{-Z_{CA}/2\sigma^2} \\ &= \dfrac{1}{(2\sigma^2)^N (N-1)!} Z_{CA}^{N-1} e^{-Z_{CA}/2\sigma^2} \end{aligned} \quad (7.19)$$

Substituting (7.16) and (7.19) into (7.18), absorbing $1/N$ into the limit of integration, we obtain [31, p. 284]

$$P_{fa} = \int_0^\infty \left[\int_{K_{CA}Z_{CA}/N}^\infty \frac{1}{2\sigma^2} e^{-x_t/2\sigma^2} dx \right] \frac{1}{(N-1)!} \left(\frac{1}{\sigma}\right)^N Z_{CA}^{N-1} e^{-Z_{CA}/\sigma} dZ_{CA}$$

$$= \frac{1}{\sigma^N (N-1)!} \int_0^\infty Z_{CA}^{N-1} e^{-(1+K_{CA}/N)Z_{CA}/\sigma} dZ_{CA} \qquad (7.20)$$

$$= \frac{1}{(1+K_{CA}/N)^N}$$

The important aspect of (7.20) is that P_{fa} is independent of interference power, indicating that the circuitry shown in Figure 7.2 results in a CFAR.

Solving (7.20), for the CFAR multiplier, we obtain the expression [31]

$$K_{CA} = N\left(e^{-\ln(P_{fa})/N} - 1\right) = N\left(P_{fa}^{-1/N} - 1\right) \qquad (7.21)$$

For illustration, plots of the CFAR multiplier K_{CA} versus the number of reference cells for several P_{fa} are plotted in Figure 7.3.

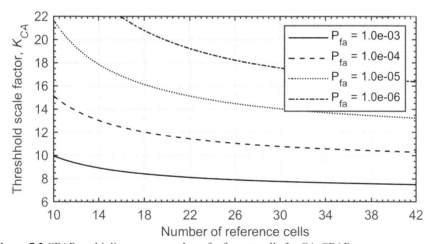

Figure 7.3 CFAR multiplier versus number of reference cells for CA-CFAR.

7.2.3 CA-CFAR Example

For this example, we want to design a CA-CFAR processor to provide a false alarm probabilty of 10^{-6} and a detection probability of 0.8, with a CFAR loss less than 1 dB. We will use two guard cells on either side of the CUT. The interference is zero-mean Gaussian noise with a nominal power level of -120 dBm. We will test this CA-CFAR using four typical test cases, the parameters of which are summarized in Table 7.2. These test cases are selected to demonstrate potential CFAR processor shortcomings.

For the single-target scenario, there is a target present in cell 80, with an SNR of 20 dB. The noise level is constant at -120 dBm, or homogeneous. This scenario represents a single target signal competing with Gaussian noise of unknown power, as specified for CA-CFAR by Finn and Johnson [32]. This is the nominal scenario, and does not violate any of the CFAR assumptions presented in Section 7.2.

For the two target scenario, there is a target present in cells 80 and 90, with SNRs of 20 dB and 15 dB, respectively. The noise level is constant at -120 dBm. This scenario represents a single target signal competing with Gaussian noise and an interfering target. This scenario violates the single target CFAR assumption presented in Section 7.2.

For the extended target scenario, there is a single target occupying cells 80 to 86, with a SNR of 20 dB. The noise level is constant at -120 dBm. This scenario represents a single extended target signal competing with Gaussian noise. The extended target leaks into the reference window. This scenario violates the assumption that only noise be present in the reference window presented in Section 7.2.

For the interference boundary edge scenario, there are no targets present. The nominal noise level is at -120 dBm, with an elevated Gaussian interference region at a level of -110 dBm extending from cell 50 to cell 110. This scenario represents non-homogeneous interference, violating the identically distributed interference CFAR assumption presented in Section 7.2.

Table 7.2
CFAR test scenarios

Scenario	Range cell(s) with target	SNR	Range cells with elevated interference	SNR
Single target	80	20	–	–
Two target	80, 90	20, 15	–	–
Extended target	80 – 86	15	–	–
Interference boundary	–	–	50 – 110	10

To start off this CFAR design, we first select the number of reference cells to use in the reference window that meets the specified CFAR loss. CFAR loss can be estimated using (5.5), repeated here [47][3]

$$L_{cfar} = \frac{P_{fa}^{-1/N} - P_d^{-1/N}}{P_{fa}^{-1/N} - 1} \cdot \frac{\ln(P_d)}{\ln(P_{fa}) - \ln(P_d)} \cong P_{fa}^{-1/2N} \qquad (7.22)$$

As we can see from the approximation in (7.22), P_d plays a trivial role in this calculation. For example, the difference between the full equation in (7.22) and the approximation, for $P_{fa} = 10\text{-}6$, $P_d = 0.7$, and $N = 32$, is 0.0319 dB.

Solving the approximation (7.22) for N, we obtain the minimum number of reference cells for a given P_{fa} and L_{cfar}, given by

$$N \cong \text{ceil}\left\{-\ln(P_{fa}) / \left[2 \cdot \ln(L_{cfar})\right]\right\} \qquad (7.23)$$

where ceil[·] represents rounding up to the nearest integer towards infinity. We also stipulate that N be an even number, so the reference window can be divided evenly into equal length leading and lagging windows.

Evaluating (7.23) for a false alarm probability of $P_{fa} = 10^{-6}$ and CFAR loss of $L_{cfar} = 1$ dB = 1.2589, results in

$$N \cong \text{ceil}\left\{-\ln(10^{-6}) / \left[2 \cdot \ln(1.2589)\right]\right\} = \text{ceil}[30.0] = 30 \qquad (7.24)$$

as the minimum number of reference cells. Ironincally, (7.24) evaluates to exactly 30. Since a CFAR loss of less than 1 dB is specified, we select $N = 32$, which corresponds to $L_{cfar} = 0.97$ dB.

Next, we need to calculate the CA-CFAR multiplier. Using (7.21) results in

$$K_{CA} = N\left(P_{fa}^{-1/N} - 1\right) = 32\left[\left(10^{-6}\right)^{-1/32} - 1\right] = 17.3 = 12.4 \text{ dB} \qquad (7.25)$$

The simulation results for this CA-CFAR processor example are presented in Figure 7.4 and Figure 7.5. There are plots in both figures corresponding to each of the four test scenarios: single target, dual target, extended target, and interference boundary edges, which are summarized in Table 7.2.

---

[3] Technically, CFAR loss varies slightly depending on CFAR type, but (7.22) is a good starting point.

176 Basic Radar Analysis

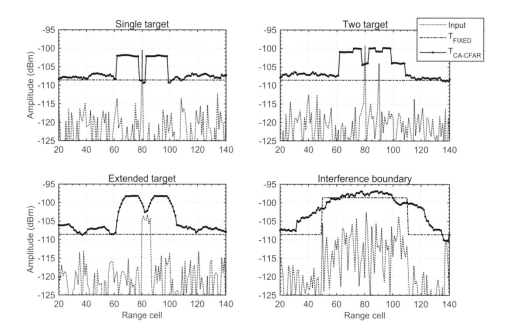

Figure 7.4 CA-CFAR example – single run.

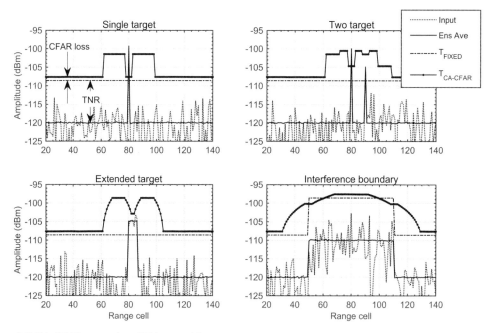

Figure 7.5 CA-CFAR example – 2000 ensemble average.

The plots in Figure 7.4 present a single simulation run to illustrate the adaptive detection threshold during radar operation. The plots in Figure 7.4 include the CFAR input, the theoretical fixed detection threshold, and the adaptive threshold generated by the CA-CFAR for a single run. The Monte Carlo method, averaging 2000 simulation runs (ensembles), was used to evaluate the median value of the adaptive detection threshold, the results of which are shown in Figure 7.5. The plots in Figure 7.5 include one input ensemble, an ensemble average of the input, the theoretical fixed threshold, and the threshold generated by the CA-CFAR.

Examining the four plots in Figure 7.5, let's first compare the adaptive CA-CFAR threshold to the fixed detection threshold in regions where the cells only contain the nominal -120 dB level of interference. For example, region from cell 20 to cell 50 in the single target plot. Using (7.1), the fixed threshold is $10 \cdot \log[-\log(10^{-6})] = 11.4$ dB above the noise level, or -108.6 dBm. The ensemble average of the CA-CFAR threshold is ~107.6 dBm. Using (7.25), this equates to $10 \cdot \log 10(K_{CA}) = 12.4$ dB above the noise. The ~ 1 dB difference is the result of CFAR loss.

CFAR loss, compared to fixed threshold detection, is the result of using an approximate interference power instead of the actual interference power. The larger the reference window, the lower the CFAR loss (see Section 5.2). The tradeoff is that the larger a reference window, the more slowly a CA-CFAR responds to changes in interference power. If the number of reference cells is allowed to approach infinity, the P_d approaches that of fixed threshold detection (assuming homogeneous interference) [24, p. 639].

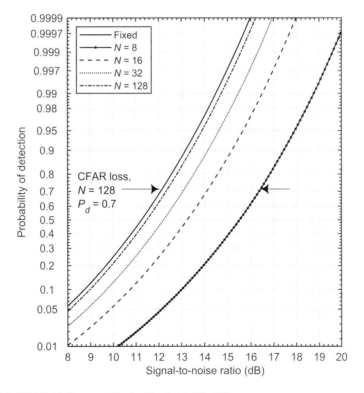

Figure 7.6 CA-CFAR, SNR versus P_d, for N = 8, 16, 32, 128.

To illustrate the relationship between the size of the reference window, N, and CFAR loss, plots of SNR versus P_d, for $N = 8$, 16, 32, and 28 are presented in Figure 7.5. As N increases, the SNR vs P_d approaches that of a fixed threshold detector, theoretically converging as N approaches infinity. Annotated in Figure 7.6 is the CFAR loss for $P_d = 0.7$, and $N = 8$, which is 4.25 dB.

Returning to Figure 7.5, now consider the CA-CFAR threshold for the single target case. On either side of the target in cell 80, there are two 16-cell windows where the threshold is elevated. The spacing between these two windows is five samples, corresponding to two guard cells, one CUT, and two more guard cells. The CA-CFAR threshold is elevated because the target in cell 80 was included in the noise estimate during the sliding window process, violating the CFAR assumption that reference window samples contain only interference. So long as there is only a single target, this is not a problem.

Now consider the plot in Figure 7.5 for the two target case. There is a target present in cell 80, and another target present in cell 90, with SNRs of 20 and 15 dB, respectively. Portions of the CA-CFAR threshold are elevated, compared to the single-target case. This is the result of one target, the other target, or both being included in the noise estimate. The CA-CFAR threshold is higher than the target signal in cell 90, resulting in a very low P_d for that target. This phenomenon of one target overshadowing a nearby target, is referred to as *target masking*.

This is an example of violating the CFAR assumption that only one cell contains a target signal to be detected. This situation is generally referred to as a *multiple target environment*, in the context of CFAR performance. As the number of interfering targets increases, CA-CFAR detection performance deteriorates rapidly. In Section 7.4, we present a basic CA-CFAR modification aimed at minimizing the effects of interfering targets. The modification is referred to as smallest of CFAR (SO-CFAR).

For the extended target case of Figure 7.5, the target return is present in 7 adjacent cells (80–86). This could be the result of a very large target bleeding into seven adjacent cells or a large interference patch occupying multiple resolution cells. Examining Figure 7.5, we see the CA-CFAR threshold is elevated such that the extended target falls below the threshold. This is because samples from the extended target spanned across the guard cells, elevating the noise estimate erroneously. This demonstrates a shortcoming of CA-CFAR refered to as *self-masking*.

The lower right plot in Figure 7.4 contains an elevated interference region in cells 50–110, and illustrates what we term interference boundaries (changes in interference power) or interference *edge effects* [5, p. 215]. For this scenario, the CA-CFAR threshold increases leading into, and decreases exiting, the elevated interference region, as the sliding window process progresses.

The increasing CA-CFAR threshold leading into the elevated interference region exhibits a 37-cell transient. The rising threshold from cells 32 to 47 is the result of 16 leading reference cells. The subsequent 5-cell flat portion of the CFAR threshold from cells 48 to 52 is due to the CUT and 4 bracketing guard cells. The rise in the CA-CFAR threshold, from cells 53 to 68, is due to 16 lagging reference cells. The flat portion of the CA-CFAR threshold, from cell 69 to cell 91, occurs when the elevated interference fills all 37 cells used by the CA-CFAR. A similar 37-cell threshold transient occurs from cell 92 to cell 128 when leaving the elevated interference region.

To help understand the structure of the CA-CFAR threshold exhibited in the lower right plot of Figure 7.4, the sliding window averaging process of a CA-CFAR can be thought of (and implemented) as an FIR filter, with changes in threshold being filter transients (see Section 7.2.4).

These CFAR threshold transients are the results of failing to meet the CFAR assumption of a constant interference level (homogeneity). Changes in interference level can potentially skew the interference power estimate near interference boundaries, resulting in false alarm rate, and thereby P_{fa}, being significantly higher than designed.

These CFAR threshold transients illustrate how long a CA-CFAR threshold can take to react to changes in interference power. Using fewer reference cells results in a more responsive CFAR. The tradeoff is between threshold responsiveness (reference window length) and CFAR loss. A CFAR using a longer reference window incurs less CFAR loss, but responds more slowly to an interference boundary. A general rule of thumb is to chose a reference window length that results in ~ 1 dB CFAR loss.

The elevated CFAR threshold results in degraded detection performance near interference boundaries. Masking of weak targets near the interference boundary is another potential problem that can occur near interference boundaries. In Section 7.3, we present a basic CA-CFAR modification aimed at addressing interference edged effects. The modification is referred to as greatest of CFAR (GO-CFAR).

7.2.4 CA-CFAR FIR Implementation

One simple method of implementing a CA-CFAR is to consider the sliding window averaging process of a CA-CFAR in terms of a finite inpulse response (FIR) digital filter[4]. The mean-level estimator used by a CA-CFAR to form interference power estimations can be represented as a causal moving (running) average FIR filter.

This CA-CFAR FIR filter can be written in the form of a difference equation as

$$y[n] = \frac{1}{N_R} \sum_{k=0}^{M=N_{Total}-1} b_k x[n-k] \qquad (7.26)$$

where $x[n]$ is the input signal, $y[n]$ is the output signal, M is the filter order, b_k is the filter impulse response, and N_R is the size of the reference window.

N_{Total} is the total number of cells involved in the CA-CFAR process, given by

$$N_{Total} = N_R + 2N_G + 1 \qquad (7.27)$$

where N_G is the number of guard cells on either side of the CUT. The FIR filter's impulse response, b_k, is formed by using 1's for reference cells and 0's for guard cells and the CUT.

---

[4] FIR filter delay needs to be accounted for to align input signal with CFAR threshold.

As an example, consider a CA-CFAR operating across range that contains $N_R = 20$ reference cells (10 early, 10 late), and $N_G = 4$ guard cells (2 on either side of the CUT). We want the CA-CFAR to estabilsh a false alarm probability of 10^{-6}. Let the interference be zero-mean Gaussian noise with a nominal power level of -120 dBm.

The total number of cells is therefore equal to $N_{Total} = 20+2\cdot2+1 = 25$. The associated impulse response, b_k, consisting of 20 ones and 5 zeros, is given by

$$b_k = [1\ 1\ 1\ 1\ 1\ 1\ 1\ 1\ 1\ 1\ 0\ 0\ 0\ 0\ 0\ 1\ 1\ 1\ 1\ 1\ 1\ 1\ 1\ 1\ 1] \quad (7.28)$$

The CFAR input, fixed detection threshold, CA-CFAR detection threshold, and average detection threshold (ADT) for this CA-CFAR FIR filter example are presented in Figure 7.7.

The total set of range cell data for which this example CA-CFAR processor operates across equates to some instrumented range (see Section 1.5). At the start of CFAR processing, the early and late reference cells that form the reference window are initially zero, and need to be populated with data before a valid estimate of interference power can be calculated. Specifically, the CA-CFAR threshold for the first N_{Total} cells shown in Figure 7.7 is the result of FIR filter transients. This transient behavior results in interference power estimates and in turn, the CA-CFAR threshold being too low in the transient region, which causes an elevated false alarm rate during CA-CFAR processing.

For this example, the 10 early and 10 late reference cells are populated with input data, forming a useable CA-CFAR threshold, when range cell index $n \geq N_{Total} = 25$. Looking at the ADT plotted in Figure 7.7, we note that cells 1 through 10 correspond to the leading refernce window; cells 11 through 15 correspond to two guard cells, the CUT, followed by 2 more guard cells; and cells 16 through 25 correspond to the lagging reference window. The CA-CFAR threshold plotted in Figure 7.7 spanning from cell 25 to cell 50 is valid.

While using a FIR filter to implement a CA-CFAR is simple, the approach lacks flexibility. We can use Toeplitz matrices to help speed CFAR computations (see Appendix 7B). Using Toeplitz matrices versus FIR filtering is more flexible, and is specifically applicable to CA-CFAR, GO-CFAR, or SO-CFAR. It is suspected that using Toeplitz matrices can help in implementing other types of CFAR.

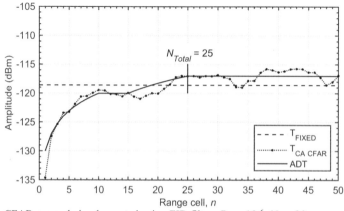

Figure 7.7 CA-CFAR example implemented using FIR filter, $P_{fa} = 10^{-6}$, $N_R = 20$.

7.2.5 CFAR Processing at the Edges of Instrumentation

Various methods can be used to address the problem of CFAR processing at the edges of instrumentation, as discussed in Section 7.2.4. One approach used in some radars is simply to accept the transient, with unpopulated cells in the reference window remaining zero. This approach results in low threshold estimates in the transient region, causing an elevated false alarm rate during these CFAR threshold transients. According to Lok, the effect of CFAR transients is generally negligible unless weather interference is present during the transients [48]. A similar tactic is to discard the transient portion of CFAR processing altogether, though this may be impractical for small instrumented regions. Both of these approaches have the advantage of simplicity.

Two other means of addressing CFAR processing at the end of instrumentation are presented by Lok. Lok's first approach is to *clamp the threshold*, which freezes the last valid threshold, extending it to the instrumented boundary. In the second approach, the size of the sliding window of reference cells is allowed to shrink, as it is limited by processing boundaries. Lok's conclusion is that the clamped threshold approach is preferable to shrinking the sliding window [48]. Clamping the threshold and shrinking the reference window size both exhibit good P_{fa}. However, the clamped threshold approach has a higher P_d and is easier to implement than shrinking the reference window [48].

7.3 CA-CFAR WITH GREATEST-OF SELECTION

One of the basic modifications to CA-CFAR, depicted in Figure 7.8, was first proposed by Hansen and Sawyers and is referred to as cell averaging CFAR with greatest of (GO) selection (GOCA-CFAR), or simply GO-CFAR [47, 49, 50]. The greatest of logic, $Z_{GO} = \max(U, V)$, presented by Vilhelm Gregers-Hansen and James Sawyers, is intended to mitigate an increase in P_{fa} that CA-CFAR processing exhibits at interference edges (boundaries) [34]. Specifically, the leading and lagging reference windows are averaged seperately, and the larger average is the parameter used to estimate interference power. Each reference window consists of $N/2$ cells that bracket the CUT (N even).

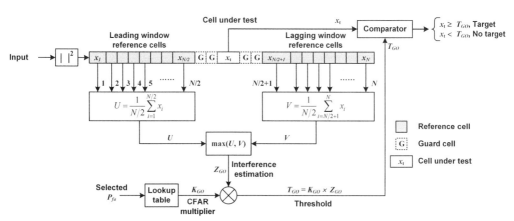

Figure 7.8 Block diagram of a GO-CFAR detector.

Like CA-CFAR processing, GO-CFAR processing was analyzed and designed for a single target competing with Gaussian noise of unknown power with reference window data being used to estimate the statistics in the CUT[5]. Noise power estimation is carried out in a sliding window fashion, incremented cell by cell. The CFAR threshold is recalculated after each cell increment. GO-CFAR analysis generally follows the same process summarized in Section 7.2.2.

For GO-CFAR, the statistic used for interference power estimation is the greater of the averages of the leading and lagging reference cells, which can be expressed as

$$Z_{GO} = \widehat{\sigma^2} = \max(U,V) = \begin{cases} U, & U > V \\ V, & U \leq V \end{cases} \quad (7.29)$$

$$U = \frac{1}{M}\sum_{i=1}^{N/2} x_i, \quad V = \frac{1}{M}\sum_{i=N/2+1}^{N} x_i \quad (7.30)$$

where Z_{GO} is the interference power estimate, N is the total number of reference cells, and $M = N/2$ is the number of reference cells in the leading and lagging windows.

The interference power estimate is then multiplied by a CFAR multiplier to form a detection threshold. The detection threshold for a GO-CFAR, can be written as

$$T_{GO} = K_{GO}Z_{GO} = K_{GO}\widehat{\sigma^2} = K_{GO}\max\left(\frac{1}{M}\sum_{i=1}^{N/2} x_i, \frac{1}{M}\sum_{i=N/2+1}^{N} x_i\right) \quad (7.31)$$

where K_{GO} is a CFAR multiplier and T_{GO} is the detection threshold.

According to Papoulis, the probability density function of the maximum of two continuous random variables, $\max(U,V)$, used to form the estimate (7.29), can be expressed as [42, p. 194]

$$f_Z(z) = F_U(z)f_V(z) + f_U(z)F_V(z) \quad (7.32)$$

where $f_U(z)$ and $f_V(z)$ are PDFs and $F_U(z)$ and $F_V(z)$ are cumulative distribution functions (CDFs) of the leading and lagging windows, respectively.

The expected value of the P_{fa} for a GO-CFAR processor can be shown to be [25, p. 350; 40]

$$\frac{\overline{P}_{fa}}{2} = \left(1 + \frac{K_{GO}}{N/2}\right)^{-N/2} - \left(2 + \frac{K_{GO}}{N/2}\right)^{-N/2} \times \left\{\sum_{k=0}^{N/2-1}\binom{\frac{N}{2}-1+k}{k}\left(2 + \frac{K_{GO}}{N/2}\right)^{-k}\right\} \quad (7.33)$$

To determine the GO-CFAR multiplier, (7.33) is usually solved iteratively for several values of P_{fa} and stored in a lookup table (see Exercise 5).

---

[5] As with CA-CFAR, even though Hansen and Sawyers stipulated Gaussian interference when analyzing their GO-CFAR processor design, this should not be interpreted to mean GO-CFAR can't be analyzed and used against non-Gaussian interference.

7.3.1 GO-CFAR Example

For this example, we design a GO-CFAR processor to provide a $P_{fa} = 10^{-6}$, a $P_d = 0.8$, and have a CFAR loss of no more than 1 dB. We will use two guard cells on either side of the CUT. The interference is zero-mean Gaussian noise with a nominal power level of -120 dBm. We will test this GO-CFAR using the four typical test cases from the CA-CFAR example, the parameters of which are summarized in Table 7.2.

First, we select the number reference cells to use that meet the specified CFAR loss. In this case, it will be the same as for the CA-CFAR example. As such, we choose to use $N = 32$ reference cells. Using (7.22), for a P_{fa} of 10^{-6} and 32 reference cells, the CFAR loss is 0.97 dB. Iterating on (7.33), the GO-CFAR multiplier is $K_{GO} = 15.7 = 12$ dB.

The simulation results for this GO-CFAR processor for a single run and the average of 2000 ensembles are plotted in Figure 7.9 and Figure 7.10, respectively. There are plots in both figures for each of the four test cases: single target, dual target, extended target, and interference edges. The plots in Figure 7.9 include the input, the theoretical fixed threshold for a matched filter output, and the threshold generated by the GO-CFAR. The plots in Figure 7.10 include one input ensemble, an ensemble average of the input, the theoretical fixed threshold, and the threshold generated by the GO-CFAR.

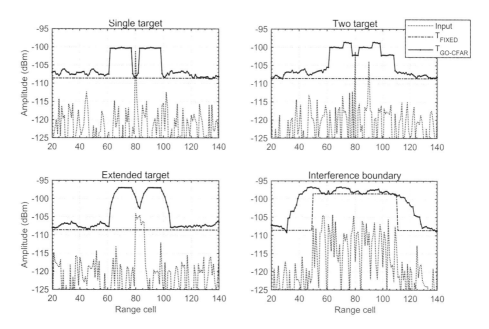

Figure 7.9 GO-CFAR example – single run.

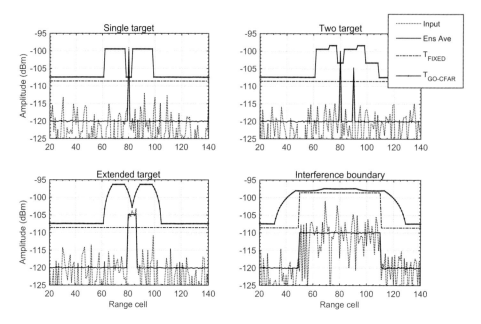

Figure 7.10 GO-CFAR example – 2000 ensemble average.

GO-CFAR processing is intended to mitigate the increase in P_{fa} that CA-CFAR processing exhibits at interference edges (boundaries). Looking at the interference boundary example in Figure 7.10, we can see that this is indeed the case. Specifically, the GO-CFAR threshold is always above the fixed detection threshold. However, examining the other three scenarios, we can see that, like CA-CFAR, GO-CFAR has no problems against single targets, but is susceptible to target masking and self-masking. For GO-CFAR, P_d decreases markedly if an interfering target of equal amplitude to the CUT is present [40].

7.4 CA-CFAR WITH SMALLEST OF SELECTION

Depicted in Figure 7.1 is the smallest of CA-CFAR (SOCA-CFAR, or simply SO-CFAR) detector, which is another basic modification of CA-CFAR. CFAR processing using smallest of logic, $Z_{so} = \min(U,V)$, was first proposed by Gerald V. Trunk while studying the target resolution of some adaptive threshold detectors [51, 52]. Using smallest of selection logic is intended to reduce the effects of interfering targets exhibited by CA-CFAR.

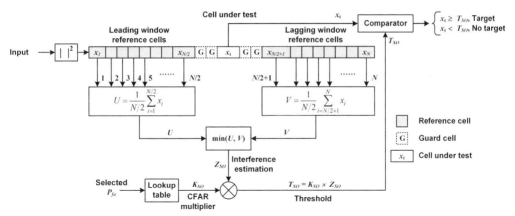

Figure 7.11 Block diagram of an SO-CFAR detector.

In Trunk's SO-CFAR, the leading and lagging reference windows are averaged seperately, with the smaller average used as the parameter to estimate interference power. Each reference window consists of $N/2$ cells that bracket the CUT.

SO-CFAR processing and analysis parallels that of CA-CFAR. Likewise, the SO-CFAR detector is designed for a single target competing with Gaussian noise, where noise power is unknown. Reference window samples are used to estimate the statistics in the CUT. For SO-CFAR, the statistic used for interference power estimation is the smaller of the averages of the leading and lagging reference cells, which can be written as

$$Z_{SO} = \widehat{\sigma^2} = \min(U, V) = \begin{cases} V, & U > V \\ U, & U \leq V \end{cases} \quad (7.34)$$

where Z_{SO} is the interference power estimate. U and V are averages of the early and late reference cells, respectively, which can be expressed as

$$U = \frac{1}{M}\sum_{i=1}^{M} x_i, \quad V = \frac{1}{M}\sum_{i=M+1}^{N} x_i \quad (7.35)$$

where $M = N/2$ is the number of reference cells in the leading and lagging windows.

The interference power estimate Z_{SO} is then multiplied by a CFAR multiplier to form an adaptive detection threshold. The adaptive detection threshold for a SO-CFAR can be expressed as

$$T_{SO} = K_{GS} Z_{SO} = K_{SO}\widehat{\sigma^2} = K_{SO} \min\left(\frac{1}{M}\sum_{i=1}^{N/2} x_i, \frac{1}{M}\sum_{i=N/2+1}^{N} x_i\right) \quad (7.36)$$

where K_{SO} is the CFAR multiplier and T_{SO} is the adaptive threshold.

According to Papoulis, the probability density function of the minimum of two continuous random variables, min(U, V), used to form the estimate (7.34), can be expressed as [42, p. 195]

$$f_Z(z) = f_U(z) + f_V(z) - f_U(z)F_V(z) - F_U(z)f_V(z) \tag{7.37}$$

where $f_U(z)$ and $f_V(z)$ are PDFs and $F_U(z)$ and $F_V(z)$ are CDFs of the leading and lagging windows, respectively.

The expected value of the P_{fa} for a SO-CFAR processor can be shown to be [25, p. 350; 40]

$$\frac{\bar{P}_{fa}}{2} = \left(2 + \frac{K_{SO}}{N/2}\right)^{-N/2} \left\{ \sum_{k=0}^{N/2-1} \binom{\frac{N}{2}-1+k}{k} \left(2 + \frac{K_{SO}}{N/2}\right)^{-k} \right\} \tag{7.38}$$

As with GO-CFAR, to determine the SO-CFAR multiplier, (7.38) is usually solved iteratively for several values of P_{fa} and stored in a lookup table (see Exercise 6).

7.4.1 SO-CFAR Example

For this example, we design a SO-CFAR processor to provide a $P_{fa} = 10^{-6}$, a $P_d = 0.8$, and have a CFAR loss of no more than 1 dB. We will use two guard cells on either side of the CUT. The interference is zero-mean Gaussian noise with a nominal power level of -120 dBm. We will test this SO-CFAR using the four typical test cases from the CA-CFAR example, the parameters of which are summarized in Table 7.2.

First, we select the number reference cells to use that meet the specified CFAR loss. As with the CA- and GO-CFAR, we choose to use $N = 32$ reference cells. Using (7.22), for a P_{fa} of 10^{-6} and 32 reference cells, the CFAR loss is 0.97 dB. Iterating on (7.38), the SO-CFAR multiplier is $K_{SO} = 20 = 13$ dB.

The simulation results for this SO-CFAR processor for a single run and the average of 2000 ensembles are plotted in Figure 7.12 and Figure 7.13, respectively. There are plots for each of the four test cases: single target, dual target, extended target, and interference edges. The plots in Figure 7.12 include the input, the theoretical fixed threshold for a matched filter output, and the threshold generated by the SO-CFAR. The plots in Figure 7.13 include one input ensemble, an ensemble average of the input, the theoretical fixed threshold for a matched filter output, and the threshold generated by the SO-CFAR.

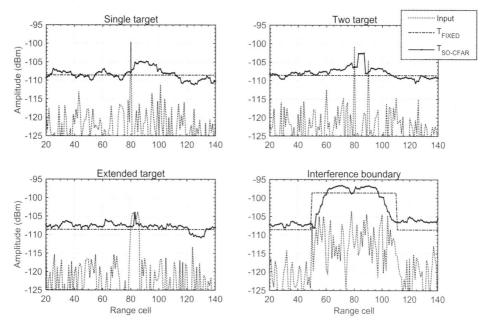

Figure 7.12 SO-CFAR example – single run.

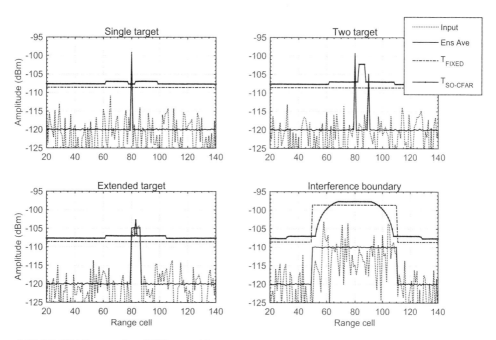

Figure 7.13 SO-CFAR example – 2000 ensemble average.

SO-CFAR is intended to reduce the effects of interfering targets exhibited by CA-CFAR. Looking at the two-target and extended examples in Figure 7.13, we can see that SO-CFAR exhibits neither self-masking nor target masking, which was demonstrated by CA-CFAR. Examining the other three scenarios plotted in Figure 7.13, we can see that SO-CFAR has no problems against single targets, but is susceptible to interference edge effects. Compared to CA-CFAR and GO-CFAR, SO-CFAR's ability to control false alarm rate (P_{fa}) is degraded significantly due to interference-edge effects [5, p. 222].

7.5 ORDERED STATISTIC CFAR

A block diagram of a CFAR processor based on the theory of ordered statistics, referred to as an ordered statistic (OS) CFAR (OS-CFAR) processor, is presented in Figure 7.14. Using ordered statistics to generate an interference power estimate was first introduced by Hermann Rohling as a means of mitigating the negative impact of both interference edges and interfering targets (target masking) [52–54]. OS-CFAR does not restrict the interference power level to be a constant (homogeneous). Additionally, the guard cells shown in Figure 7.14 are not strictly necessary, but are sometimes included [55, p. 219].

When using OS-CFAR processing, the amplitude data from N reference cells $[x_1, x_2, \ldots, x_N]$ is first rank ordered (sorted) by decreasing magnitude to form a sequence that occurrs in ascending order. The kth *ordered statistic* is the kth element or rank of the ordered list. The kth rank of the ordered statistic (magnitude of the kth element of the sorted data) is then selected as the interference power estimate rather than the mean-level used by CA-CFAR [55, p. 218]. As a general rule of thumb, $k \approx 0.75 \cdot N$. The rank ordering of the data in the reference window according to amplitude can be represented by [25]

$$x_{(1)} \leq x_{(2)} \leq \cdots \leq x_{(i)} \leq \cdots \leq x_{(k)} \leq \cdots x_{(N)} \tag{7.39}$$

where $x_{(1)}$ contains the smallest magnitude and $x_{(N)}$ is the largest magnitude. The OS-CFAR threshold is formed by scaling the kth ordered statistic from this ranking, which can be written as

$$T_{OS} = K_{OS} \cdot x_{(k)} \tag{7.40}$$

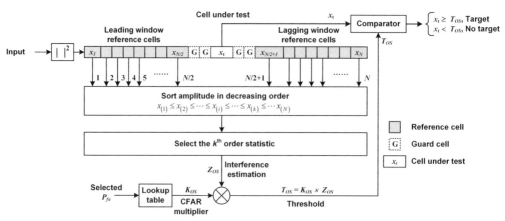

Figure 7.14 Block diagram of an OS-CFAR detector.

If the random variable x has a PDF of $f_X(x)$ with CDF $F_X(x)$, the PDF of the kth ranked sample of N i.i.d. random variables selected from (7.36) is [25, p. 354]

$$f_K(x) = k \binom{N}{k} \left[F_X(x)\right]^{k-1} \left[1 - F_X(x)\right]^{N-k} f_X(x) \tag{7.41}$$

For Rayleigh interference, out of a square-law detector

$$f_X(x) = \frac{1}{2\sigma^2} e^{-x/2\sigma^2} \tag{7.42}$$

Following Levanon [28], we normalize (7.39) to simplify notation, which results in

$$f_X(x) = e^{-x} \tag{7.43}$$

The CDF can be written as

$$F_X(x) = \int_0^x f_X(x)dx = \int_0^x e^{-x}dx = 1 - e^{-x} \tag{7.44}$$

Substitution of (7.40) and (7.41) into (7.38), we obtain the PDF of the kth ranked sample:

$$\begin{aligned} f_K(x) &= k\binom{N}{k}\left[1-e^{-x}\right]^{k-1}\left[1-\left(1-e^{-x}\right)\right]^{N-k} e^{-x} \\ &= k\binom{N}{k}\left[1-e^{-x}\right]^{k-1}\left(-e^{-x}\right)^{N-k} e^{-x} \\ &= k\binom{N}{k}\left[1-e^{-x}\right]^{k-1}\left(-e^{-x}\right)^{N-k+1} \end{aligned} \tag{7.45}$$

Using the above information, the expected value of the P_{fa} for an OS-CFAR processor can be shown to be [25, p. 355; 53]

$$\bar{P}_{fa} = k \frac{N!}{k!(N-k)!} \frac{(k-1)!(K_{OS}+N-k)!}{(K_{OS}+N)!} = \frac{N!(K_{OS}+N-k)!}{(N-k)!(K_{OS}+N)!} \tag{7.46}$$

Compared to CA-CFAR loss, OS-CFAR loss can be ~ 0.5 dB higher when using $k \approx 0.75 \cdot N$. When using median values of k, the difference in OS-CFAR loss, compared to CA-CFAR, is negligible [5, p. 223]. Because of the sorting of the reference window data, OS-CFAR processing is computationally expensive compared to CA-, GO-, and SO-CFAR. Real-time implementation of OS-CFAR can be challenging, though less so as field programmable gate array (FPGA) technology continues to improve.

7.5.1 OS-CFAR Example

For this example, we design an OS-CFAR processor to provide a $P_{fa} = 10^{-6}$, a $P_d = 0.8$, and have a CFAR loss of no more than 1 dB. We will use two guard cells on either side of the CUT. The interference is zero-mean Gaussian noise with a nominal power level of

-120 dBm. We will test this OS-CFAR using the four typical test cases from the CA-CFAR example, the parameters of which are summarized in Table 7.2.

First, we select the number reference cells to use that meet the specified CFAR loss. As with the CA- and GO-CFAR, we choose to use $N = 32$ reference cells. Technically, CFAR loss for an OS-CFAR depends on the value used for k, but this is a good starting point. Using (7.22) for a P_{fa} of 10^{-6} and 32 reference cells, the CFAR loss is 0.97 dB.

To determine the CFAR constant, we can iterate on (7.46). Alternatively, we can generate Figure 7.15, which is a graph of (7.46) for various values of N, and pick the CFAR constant off the chart. For example, using Figure 7.15, given a P_{fa} of 10^{-6}, $N = 32$, $k \approx 0.75 \times 32 = 24$, the OS-CFAR multipler is $K_{OS} = 14.4 = 11.6$ dB.

The simulation results for this OS-CFAR processor for a single run and the average of 2000 ensembles are plotted in Figure 7.16 and Figure 7.17, respectively. There are plots for each of the four test cases: single target, dual target, extended target, and interference edges. The plots in Figure 7.16 include the input, the theoretical fixed threshold for a matched filter output, and the threshold generated by the SO-CFAR. The plots in Figure 7.17 include one input ensemble, an ensemble average of the input, the theoretical fixed threshold for a matched filter output, and the threshold generated by the OS-CFAR.

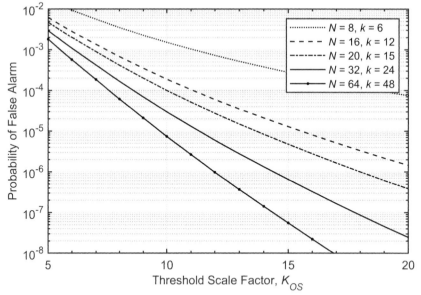

Figure 7.15 OS-CFAR multiplier versus P_{fa}.

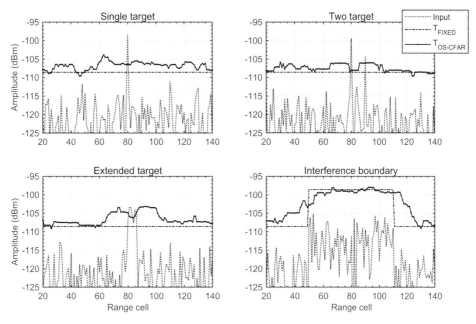

Figure 7.16 OS-CFAR example, k = 0.75·N – single run.

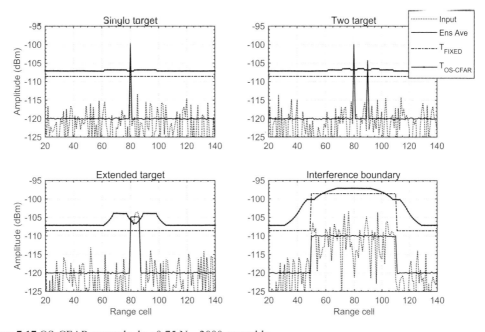

Figure 7.17 OS-CFAR example, k = 0.75·N – 2000 ensemble.

Looking at Figure 7.17, we can see that OS-CFAR did not have any problems with the single target, the double target, or the extended target. The behavior for the interference case resembles that of the CA-CFAR case. More specifically, the OS-CFAR threshold case would experience an increase in false alarms near the interference boundwary compared to a fixed threshold case. The OS-CFAR was designed to mitigate both multiple target and interference edge effects. To alter the behavior of the OS-CFAR threshold, we can tune the order k. For example, median or higher values of k result in the OS-CFAR providing false alarm rate control similar to that of SO-CFAR or GO-CFAR, respectively [5, p. 222].

7.6 MINIMUM SELECTED CA-CFAR

A simplified block diagram of a minimum selected cell averaging (MSCA) CFAR detector presented by Jin Mo Yang and Whan Woo Kim is shown in Figure 7.18 [56]. The description, analysis, and notation presented here follow that of Yang and Kim closely. This CFAR detector is based upon the minimum selected cell in a sub-reference window (SRW) [56–58].

First, the end samples in each sliding SRW are compared, keeping the sample with the lower magnitude. The SRW is long enough such that a single target should not appear in both SRW end samples. The premise is that, given homogeneous interference and a single target present, the smaller sample has to be interference only. For the case where there is only interference present, either end sample of the SRW contains only interference.

While the functionality of guard cells is present in MSCA-CFAR processing due to the operation of the SRW, guard cells are not indicated explicitly in Figure 7.18 (though they could be). For example, CA-CFARs typically use two guard cells on either side of the CUT. An SRW might be five or so cells long for the same reason, serving the same purpose as guard cells.

Selecting the minimum of the SRW end cells prior to cell averaging is used in an attempt to reject strong interference, to eliminate target masking, and to alleviate interference edge effects. The outputs of the minimum selection process are then used as reference cells feeding a subsequent classical CA-CFAR used for target detection [5, 32].

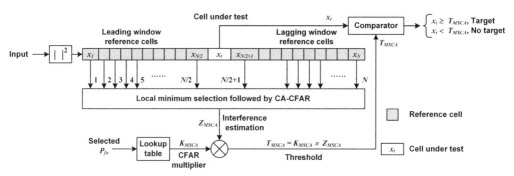

Figure 7.18 Block diagram of a MSCA-CFAR detector (After [56]).

7.6.1 MSCA-CFAR Algorithm

A block diagram of the MSCA-CFAR detector algorithm is shown in Figure 7.19. While not emphasized in this chapter, the upper and lower reference windows, which bracket the CUT, consist of a total of N cells. The sliding SRW is used for selecting the minimum cell between the leftmost and rightmost boundaries (ends) of the SRW. In theory, the output of the minimum selection process contains samples of interference only. These interference-only samples are then used by a subsequent CA-CFAR.

As with the CA-CFAR analysis, the background interference is stipulated to be zero-mean Gaussian noise (Rayleigh-envelope model). As such, the noise amplitude in each cell, after square-law detection, is modelled as an i.i.d. random variable X with an exponential PDF, which can be expressed as [56][6]

$$p_x(x) = \frac{1}{\lambda} e^{-x/\lambda} \quad (7.47)$$

where $x \geq 0$ and $\lambda > 0$. In (7.47), λ is normalized with respect to the thermal noise. Therefore, $\lambda = 1$ if the cell contains thermal noise; $\lambda = 1 + S$ if the cell contains a target return with an average signal-to-noise ratio (SNR) of S, and $\lambda = 1 + C$ if the cell contains an interference return with an average interference-to-noise ratio (INR) of C. Yang and Kim stipulate SW2 targets for their analysis [34, 56].

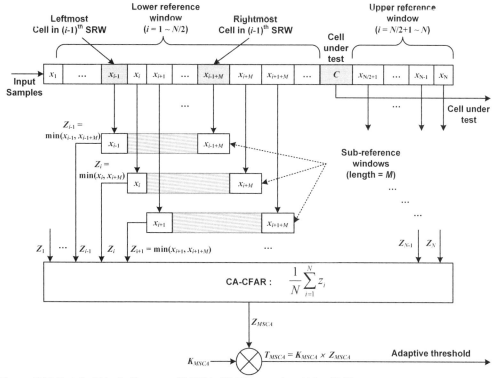

Figure 7.19 Detailed block diagram of MSCA-CFAR operation (After [56]).

---

[6] Compared to (7.13), in (7.44), $\lambda = 2\sigma^2$, which is further normalized w.r.t. thermal noise.

The MSCA-CFAR processor implements the following detection steps [56].

Step 1) Reference samples collected near the CUT form an N-sample reference window. The reference window is partitioned into a sliding SRW of length M. Like guard cells, the parameter M is used to exclude target samples from the averaging used to estimate interference power. The value for M should exceed the main lobe width of the matched-filter response. More specifically, it is convenient to specify M as the nearest even integer that exceeds main-lobe width [56].

Step 2) The cells in the leftmost and rightmost boundaries of the SRW are compared with the minimum of these two cells being selected. Let Z_i be a new random variable defined as [56]

$$Z_i \equiv \min(Z_i, Z_{i+M}) = \min(V, W) \qquad (7.48)$$

where $Z_i = V$ is the leftmost cell in the ith SRW and $Z_{i+M} = W$ is the rightmost cell. The simplified block diagram of the minimum selected cell-averaging operation in the ith SRW is shown in Figure 7.18 [56]. According to Papoulis, the PDF of the minimum of two random variables, $Z_i = \min(V, W)$, used in (7.48), can be expressed as [42, p.195; 56]

$$p_{Z_i}(Z_i) = p_V(Z_i) + p_W(Z_i) - p_V(Z_i) P_W(Z_i) - P_V(Z_i) p_W(Z_i) \qquad (7.49)$$

where $p_V(Z_i)$ and $p_W(Z_i)$ are PDFs and $P_V(Z_i)$ and $P_W(Z_i)$ are CDFs of the first and last cell in the SRW, respectively.

Step 3) After finishing the minimum selection operation for all the SRWs, the results are averaged according to a typical CA-CFAR process. A target detection is declared in the CUT if [56]

$$x_t \geq T_{MSCA} = K_{MSCA} \frac{1}{N} \sum_{i=1}^{N} Z_i \qquad (7.50)$$

where T_{MSCA} is an adaptive detection threshold, N represents the number of cells in the CA-CFAR reference window, and K_{MSCA} is a scale factor used to achieve the desired P_{fa}. The cells after minimum cell selection, Z_i, are averaged to obtain an estimate of the interference power level. In (7.50), the detection threshold, T_{MSCA}, which is itself a random variable, is determined by multiplying the interference power estimate by the scale factor K_{MSCA} [56].

7.6.2 MSCA-CFAR Analysis

To simplify this analysis, (and in accordance with the stipulated Rayleigh interference model), we let random variable V (leftmost cell in the SRW) and random variable W (rightmost cell in the SRW) have a Rayleigh PDF. After square-law detection, V and W can be expressed as an exponential PDF,

$$p_A(a) = \frac{1}{\lambda} e^{-a/\lambda} \qquad (7.51)$$

with a CDF of

$$P_A(a) = \int_0^a p_A(a)\,da = \int_0^a \frac{1}{\lambda} e^{-w/\lambda}\,dw = \left[e^{-w/\lambda}\right]_0^a = e^{-0/\lambda} - e^{-a/\lambda} \tag{7.52}$$
$$= 1 - e^{-a/\lambda}$$

Using (7.49), the PDF of random variable $Z_i = \min(V, W)$ can be expressed as

$$\begin{aligned}p_{Z_i}(Z_i) &= p_V(Z_i) + p_W(Z_i) - p_V(Z_i)P_W(Z_i) - P_V(Z_i)p_W(Z_i) \\ &= \frac{1}{\lambda}e^{-Z_i/\lambda} + \frac{1}{\lambda}e^{-Z_i/\lambda} - \frac{1}{\lambda}e^{-Z_i/\lambda}\left(1 - e^{-Z_i/\lambda}\right) - \left(1 - e^{-Z_i/\lambda}\right)\frac{1}{\lambda}e^{-Z_i/\lambda} \\ &= \frac{2}{\lambda}e^{-Z_i/\lambda} - \frac{1}{\lambda}e^{-Z_i/\lambda} + \frac{1}{\lambda}e^{-Z_i/\lambda}e^{-Z_i/\lambda} - \frac{1}{\lambda}e^{-Z_i/\lambda} + \frac{1}{\lambda}e^{-Z_i/\lambda}e^{-Z_i/\lambda} \\ &= \frac{2}{\lambda}e^{-2Z_i/\lambda}\end{aligned} \tag{7.53}$$

After the minimum cell selection process, the detection threshold, T_{MSCA}, is determined by multiplying the interference power estimate, formed by averaging N cells of Z_i, by the scale factor K_{MSCA}, used to establish the desired P_{fa}, which results in

$$T_{MSCA} = K_{MSCA} \frac{1}{N} \sum_{i=1}^{N} Z_i \tag{7.54}$$

The PDF of T_{MSCA}, which is itself a random variable, can be calculated by convolving the PDFs of Z_i. When the PDF of Z_i is (7.53), the PDF of T_{MSCA} is Erlang [56; 59, p. 79].

$$p_{T_{MSCA}}(x) = p_{Z_1}(x) * p_{Z_2}(x) * \cdots * p_{Z_N}(x) = 2^N e^{-2x} \frac{x^{N-1}}{(N-1)!} \tag{7.55}$$

The P_d for a SW2 target can be expressed as [56]

$$P_d(S) = \int_0^\infty P_d(S|T_{MSCA}) p_{T_{MSCA}}(T_{MSCA})\,dT_{MSCA} = 2^N \left(2 + \frac{K_{MSCA}}{N(1+S)}\right)^{-N} \tag{7.56}$$

where $P_d(S|T_{MSCA})$ is equal to $e^{(-T_{MSCA}/\lambda)}$ for SW2, and S is the signal-to-noise ratio (SNR) at the CFAR input. To calculate P_{fa}, set $S = 0$ in (7.56), corresponding to an input that is noise only. The P_{fa} for a MSCA-CFAR is given by [56]

$$P_{fa} = 2^N \left(2 + \frac{K_{MSCA}}{N}\right)^{-N} \tag{7.57}$$

Equation (7.57) shows that the MSCA-CFAR algorithm is a CFAR process because the P_{fa} depends only on the scalar factor, K_{MSCA}, and the reference window length, N, and not on the estimated interference power [56].

Since the MSCA-CFAR is a combination of taking the minimum of two samples, followed by CA-CFAR processing, one might think the threshold multipliers K_{MSCA} and K_{CA} would be related in some fashion. This is indeed the case. Manipulating (7.57), we can solve for K_{MSCA} as follows:

$$\frac{P_{fa}}{2^N} = \left(2 + \frac{K_{MSCA}}{N}\right)^{-N}$$

$$\left(2^{-N} P_{fa}\right)^{-1/N} = 2 + \frac{K_{MSCA}}{N} \qquad (7.58)$$

$$K_{MSCA} = 2\left[N\left(P_{fa}^{-1/N} - 1\right)\right]$$

We note that the bracketed portion of (7.58) is the threshold multiplier for a CA-CFAR, given by (7.21). The relationship between the MSCA- and CA-CFAR threshold multipliers can therefore be expressed as

$$K_{MSCA} = 2 K_{CA} \qquad (7.59)$$

With a little more manipulation, we can express the MSCA-CFAR multiplier as

$$K_{MSCA} = -2N \cdot P_{fa}^{-1/N}\left(\sqrt[N]{P_{fa}} - 1\right) = 2N\left(P_{fa}^{-1/N} - 1\right) \qquad (7.60)$$

The proof of (7.60) is left as an exercise. A plot of (7.60) for several values of N is shown in Figure 7.20.

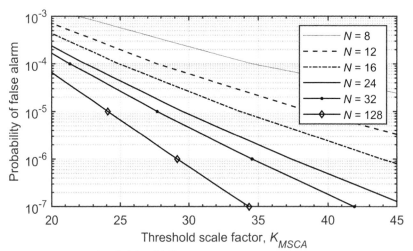

Figure 7.20 MSCA-CFAR threshold multiplier.

7.6.3 MSCA-CFAR Example

For this example, we want to design a MSCA-CFAR processor similar to the CA-CFAR example. The MSCA-CFAR processor is specified to provide a $P_{fa} = 10^{-6}$ and a $P_d = 0.8$, and have a CFAR loss of no more than 1 dB. The interference is zero-mean Gaussian noise with a nominal power level of -120 dBm. We will test this MSCA-CFAR using the four typical test cases from the CA-CFAR example, the parameters of which are summarized in Table 7.2.

First, we select the number of reference cells to use that meets the specified CFAR loss. As with the CA-CFAR, we choose to use $N = 32$ reference cells. Using (7.22) for a P_{fa} of 10^{-6} and 32 reference cells, the CFAR loss is 0.97 dB. Using (7.60), the MSCA-CFAR multiplier is $K_{MSCA} = 34.6 = 15.4$ dB.

The simulation results for this MSCA-CFAR processor for a single run and the average of 2000 ensembles are plotted in Figure 7.21 and Figure 7.22, respectively. There are plots for each of the four test cases: single target, dual target, extended target, and interference edges. The plots in Figure 7.21 include the input, the theoretical fixed threshold for a matched filter output, and the threshold generated by the MSCA-CFAR. The plots in Figure 7.22 include one input ensemble, an ensemble average of the input, the theoretical fixed threshold for a matched filter output, and the threshold generated by the MSCA-CFAR.

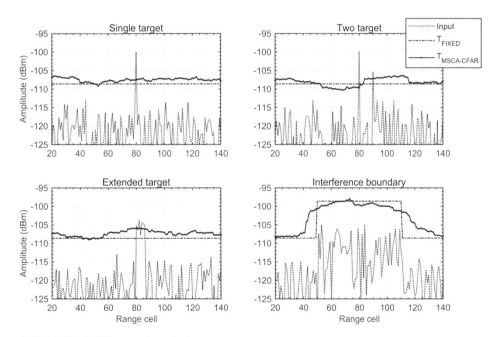

Figure 7.21 MSCA-CFAR example – single run.

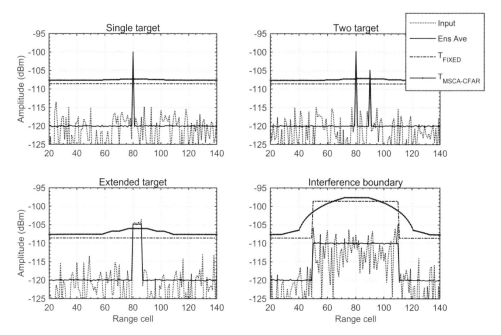

Figure 7.22 MSCA-CFAR example – 2000 ensembles.

MSCA-CFAR detection has good target detection performance in adjacent target situations compared with CA-CFAR. The detection performance of MSCA-CFAR is similar to that of OS-CFAR in a nonhomogeneous environment with interfering targets [56]. However, OS-CFAR is relatively processing intensive, since sorting the cells in the reference window is necessary. Because MSCA-CFAR takes the minimum cell within the SRW, it requires less processing time than OS-CFAR. By getting rid of the sorting process in OS-CFAR, the MSCA-CFAR scheme can be easily implemented [56].

In the never-ending CFAR development process, a further evolution of the MSCA-CFAR is the local minimum selected cell averaging (LMSCA) CFAR is presented in [58]. The authors state that LMSCA-CFAR provides more robust detection performance than CA- and MSCA-CFAR in homogeneous background with strong interfering targets [58].

7.7 SUMMARY

7.7.1 CFAR Problems and Remedies

Violating any of the CFAR assumptions in Section 7.2 causes a degradation in CFAR performance (P_d and P_{fa} behavior). Several problems resulting from violating the CFAR assumptions and possible remedies are summarized in Table 7.3.

Table 7.3
CFAR problems and solutions

CFAR type	Notes	Problem(s)
CA		Interference edges, interfering targets
GO	Designed to address interference edges effects	Interfering targets
SO	Designed to address interfering targets effects	Interference edges
OS	Designed to address interference boundaries and interfering target effects	Computationally intensive algorithm because of the ordered sorting required
MSCA	Similar performance to OS-CFAR, given interference edges and interfering targets	Easily implemented because of the minimum selection process

7.7.2 CFAR Scale Factors

The CFAR multipliers presented in this chapter are summarized in Table 7.4 [25, 31, 40, 49, 53, 54, 56].

Table 7.4
CFAR multipliers

CFAR Type	Scale Factor Equation	Notes
CA	$K_{CA} = N\left(\overline{P}_{fa}^{-1/N} - 1\right)$	Direct calculation
SO	$\dfrac{\overline{P}_{fa}}{2} = \left(2 + \dfrac{K_{SO}}{N/2}\right)^{-N/2} \left\{ \sum_{k=0}^{N/2-1} \binom{N/2 - 1 + k}{k} \left(2 + \dfrac{K_{SO}}{N/2}\right)^{-k} \right\}$	Iterative solution
GO	$\dfrac{\overline{P}_{fa}}{2} = \left(1 + \dfrac{K_{GO}}{N/2}\right)^{-N/2} - \left(2 + \dfrac{K_{GO}}{N/2}\right)^{-N/2}$ $\times \left\{ \sum_{k=0}^{N/2-1} \binom{N/2 - 1 + k}{k} \left(2 + \dfrac{K_{GO}}{N/2}\right)^{-k} \right\}$	Iterative solution
OS	$\overline{P}_{fa} = k \dfrac{N!}{k!(N-k)!} \dfrac{(k-1)!(K_{OS} + N - k)!}{(K_{OS} + N)!} = \dfrac{N!(K_{OS} + N - k)!}{(N-k)!(K_{OS} + N)!}$	Table lookup, K_{OS} integer
MSCA	$K_{MSCA} = -2N \cdot \overline{P}_{fa}^{-1/N}\left(\sqrt[N]{\overline{P}_{fa}} - 1\right)$	Direct calculation

7.8 EXERCISES

1. Design a CA-CFAR to provide a false alarm probabilty of 10^{-4} and a detection probability of 0.9, with a CFAR loss less than 1.5 dB using two guard cells on either side of the CUT. The noise floor at the matched filter output is -125 dBm. Use the Monte Carlo method [60], averaging 2000 ensembles. Exercise this CFAR using four typical test cases, the parameters of which are summarized in Table 7.2. Generate plots like those in Figure 7.4 and Figure 7.5.

2. A task performed frequently by a radar signal processor is the magnitude operation, or linear detector, which can be time intensive. Argentina Filip presented 13 digital magnitude approximations [61]. Implement the binary fraction, single region, digital linear magnitude approximation No. 6, which can be expressed as

$$\hat{R} = \frac{31}{32}\max(|I|,|Q|) + \frac{3}{8}\min(|I|,|Q|) \tag{7.61}$$

This approximation is geared toward simplicity. Determine the SNR loss resulting from using the approximation. Plot I, Q, root sum of the squares (RSS).

3. Using (7.7), and given a design P_{fa} of 10^{-4}, how many orders of magnitude will P_{fa} increase as the result of a 3-dB increase in interference power compared to the design value?

4. Prove (7.53) is equal to

$$p_{\mathbf{Z}_i}(Z_i) = 2 \cdot p_{\mathbf{W}}(Z_i) \cdot \left[1 - P_{\mathbf{W}}(Z_i)\right] \tag{7.62}$$

5. 5. Using (7.33), generate a CFAR multiplier versus number of reference cells graph for GO-CFAR similar to Figure 7.20.

6. 6. Using (7.38), generate a CFAR multiplier versus number of reference cells graph for SO-CFAR similar to Figure 7.20.

References

[1] IEEE Standard 100, IEEE Standard Dictionary of Electrical and Electronics Terms, 6th ed., 1996.

[2] Barton, D. K., Radar System Analysis and Modeling, Boston, MA: Artech House, 2005.

[3] Marcum, J., "A statistical theory of target detection by pulsed radar," *IRE Transactions on Information Theory*, vol. 6, no. 2, April 1960, p. 59–267.

[4] Swerling, P., 'Probability of Detection for Fluctuating Targets," *IRE Transactions on Information Theory*, Vols. IT-6, no. 2, April 1960, p. 269–308.

[5] Minkler, G. and J. Minkler, *CFAR: The Principles of Automatic Detection in Clutter*, Baltimore, MD: Magellan Book Company, 1990.

[6] Skolnik, M. I., *Introduction to Radar Systems*, Third ed., New York: McGraw-Hill, 2001.

[7] Ulaby, F. T. and M. C. Dobson, *Handbook of Radar Scattering Statistics for Terrain*, Norwood, MA: Artech House, 1989.

[8] Billingsley, J. B., "Ground Clutter Measurements for Surface-Sited Radar, Tech. Rep. 786, Revision 1," MIT Lincoln Laboratory, Lexington, February 1, 1993.

[9] Barton, D. K., *Radars*, Volume 5, Radar Clutter, Dedham, MA: Artech House, 1975.

[10] Белецкий, Ю. С., Методы и Алгоритмы Контрастного Обнаружения Сигналов на Фоне Помех с Априори Неизвестными Характеристиками (*Methods and Algorithms for Contrast Detection of Signals on a Background Interference with A priori Unknown Characteristics*), Москва: Радиотехника, 2011.

[11] Schleher, D. C., "Radar Detection in Weibull Clutter," *IEEE Transactions on Aerospace and Electronic Systems,* Vols. AES-12, no. 6, November 1976, p. 736–743.

[12] Booth, R. R., "The Weibull Distribution Applied to the Ground Clutter Backscatter Coefficient," U. S. Army Missile Command, Report No. RE-TR-69-15, AD 691–109, June, 1969.

[13] Farina, A., A. Russo, F. Scannapieco and S. Barbarossa, "Theory of Radar Detection in Coherent Weibull Clutter," *IEE Proc.,* no. 134, Pt. F, 1987, pp. 174–190.

[14] Sekine, M. and Y. H. Mao, *Weibull Radar Clutter*, London: Peter Peregrinus Ltd., 1990.

[15] Sekine, M., T. Musha, Y. Tomita, T. Hagisawa, T. Irabu and E. Kiuichi, "Weibull-distributed sea clutter," *IEE Proc. F.,* vol. 130, 1983, p. 476.

[16] Ogawa, H., M. Sekine, T. Musha, M. Aota, M. Ohi and H. Fukushi, "Weibull-distributed radar clutter reflected from sea ice," *Trans. IEICE Japan,* vol. E70, 1987, pp. 116–120.

[17] Schleher, D. C., "Radar Detection in Log-Normal Clutter," in *IEEE International Radar Conference*, Washington, D.C., 1975.

[18] Farina, A., A. Russo and F. A. Studer, "Coherent Radar Detection in Log-Normal Clutter," in *IEE Proc.*, 1986.

[19] Sekine, M., T. Musha, Y. Tomita, T. Hagisawa, T. Irabu and E. Kiuchi, "Log-Weibull Distributed Sea Clutter," *IEE Proceedings,* Vols. 127, Pt. F, no. 3, June 1980, p. 225–228.

[20] Oliver, C. J., "Representation of Radar Sea Clutter," in *IEE Proc.*, 1988.

[21] Pentini, F. A., A. Farina and F. Zirilli, "Radar Detection of Targets Located in a Coherent K Distributed Clutter Background," in *IEE Proc. 139, Pt. F.*, 1992.

[22] Sangston, K. J., "Coherent Detection of Radar Targets in K-Distributed, Correlated Clutter, Report 9130," Naval Research Laboratory, Washington, D.C., August, 1988.

[23] Schleher, D. C., *Automatic Detection and Radar Data Processing*, Dedham, MA: Artech House, Inc., 1980.

[24] Barkat, M., *Signal Detection and Estimation*, Second ed., Norwood, MA: Artech House, 2005.

[25] Richards, M. A., *Fundamentals of Radar Signal Processing*, Second ed., New York: McGraw-Hill Education, 2014.

[26] Nitzberg, R., *Adaptive Signal Processing for Radar*, Norwood, MA: Artech House, 1992.

[27] Weinberg, G. V., *Radar Detection Theory of Sliding Window Processes*, New York: CRC Press, 2017.

[28] Levanon, N., *Radar Principles*, New York: John Wiley & Sons, Inc., 1988.

[29] You, H., G. Jian and M. Xiangwei, *Automatic Radar Detection and CFAR Processing*, 2nd ed., Bejing: Tsinghua University Press, 2011.

[30] 关健, 何友, and 孟祥伟, 雷达目标检测与恒虚警处理 (*Radar Target Detection and Constant False Alarm Processing*), 北京: 清华大学出版社, 2011.

[31] Kang, E. W., *Radar System Analysis, Design, and Simulation*, Norwood, MA: Artech House, 2008.

[32] Finn, H. M. and R. S. Johnson, "Adaptive Detection Model with Threshold Control as a Function of Spatially Sampled Clutter-Level Estimates," *RCA Review*, vol. 29, Sept. 1968, pp. 414–464.

[33] Gandhi, P. P. and S. A. Kassam, "Optimality of the Cell Averaging CFAR Detector," *IEEE Transactions on Information Theory*, vol. 40, no. 4, July 1994, pp. 1226–1228.

[34] Gandhi, P. P. and S. A. Kassam, "Analysis of CFAR Processors in Nonhomogenious Background," *IEEE Transactions on AES*, vol. 24, no. 4, July 1988.

[35] Steenson, B. O., "Detection performance of a mean-level threshold," *IEEE Transactions on Aerospace and Electronic Systems*, Vols. AES-4, July 1968, pp. 529–534.

[36] Dillard, G. M., "Mean-level detection of nonfluctuating signals," *IEEE Transactions on Aerospace and Electronic Systems*, Vols. AES-10, November 1974, pp. 795–799.

[37] Mitchell, R. L. and J. F. Walker, "Recursive Methods for Computing Detection Probabilities," *IEEE Transactions on Aerospace and Electronic Systems*, Vols. AES-7, no. 4, July 1971, pp. 671–676.

[38] Nitzberg, R., "Analysis of the Arithmetic Mean CFAR Normalizer for Fluctuating Targets," *IEEE Transactions on Aerospace and Electronic Systems*, Vols. AES-14, no. 1, January 1978, pp. 44–47.

[39] Moore, J. D. and N. B. Lawrence, "Analysis of the Arithmetic Mean CFAR Normalizer for Fluctuating Targets," *IEEE Transactions on Aerospace and Electronic Systems*, Vols. AES-14, January 1978, pp. 44–47.

[40] Weiss, M., "Analysis of Some Modified Cell-Averaging CFAR Processors in Multiple-Target Situations," *IEEE Transactions on Aerospace and Electronic Systems*, Vols. AES-18, no. 1, 1982, pp. 102–114.

[41] Urkowitz, H., *Signal Theory and Random Processes*, Dedham, MA: Artech House, 1983.

[42] Papoulis, A. and S. U. Pillai, *Probability, Random Vairables, and Stochastic Processes*, Fourth ed., New York: McGraw-Hill, 2002.

[43] Burdic, W. S., *Radar Signal Analysis*, Englewood Cliff, NJ: Prentice-Hall, 1968.

[44] McDonough, R. N. and A. D. Whalen, *Detection of Signals in Noise*, 2nd ed., New York: Academic Press, 1995.

[45] Barkat, M. and P. K. Varshney, "On Adaptive Cell-Averaging CFAR Radar Signal Detection," Rome Air Development Center, Griffise Air Force Base, 1987.

[46] Yao, K., F. Lorenzelli and C.-E. Chen, *Detection and Estimation for Communication and Radar Systems*, Cambridge,: Cambridge University Press, 2013.

[47] Hansen, V. G. and J. H. Sawyers, "Detectability Loss Due to "Greatest Of" Selection in a Cell-Averaging CFAR," *IEEE Transactions on Aerospace and Electronic Systems*, Vols. AES-16, no. 1, January 1980, pp. 115–118.

[48] Lok, Y. F., "CFAR at The end of Instrumented Range," *IEEE*, 1999, pp. 250–255.

[49] Hansen, V. G., "Constant False Alarm Rate Processing in Search Radars," in *Proceedings of IEEE Conference on 'Radar - Present and Future', IEEE Conf. Publ. No. 105, October*, London, 1973.

[50] Moore, J. D. and N. B. Lawrence, "Comparison of Two CFAR Methods Used with Square Law Detection of Swerling 1 Targets." in *IEEE International Radar Conference*, Washington, D.C., 1980.

[51] Trunk, G. V., "Range Resolution of Targets Using Automatic Detectors," *IEEE Transactions on Aerospace and Electronic Systems,* Vols. AES-14, no. 5, September 1978, pp. 750-755.

[52] Rohling, H., "Radar CFAR Thresholding in Clutter and Multiple Target Situations," *IEEE Transactions on Aerospace and Electronic Systems,* Vols. AES-19, no. 4, July 1983, pp. 608–621.

[53] Rohling, H., "New CFAR Processor Based on an Ordered Statistic," in *Proceedins of the IEEE 1984 International Radar Conference*, Paris, 1984.

[54] Rohling, H., "Ordered statistic CFAR technique - an overview," in *12th International Radar Symposium*, Leipzig, Germany, 2011.

[55] Harrison, L. A., *Introduction to Radar Using Python and MATLAB*, Norwood, MA: Artech House, 2019.

[56] Yang, J. M. and W. W. Kim, "Performance analysis of a minimum selected cell-averaging CFAR detection," in *2008 11th IEEE International Conference on Communication Technology*, Hangzhou, 2008.

[57] Bacallao-Vidal, J. d. l. C., J. R. Machado-Fernández and N. Mojena-Hernández, "Evaluation of CFAR detectors performance," *ITECKNE,* vol. 14, no. 2, December 2017, pp. 170–178.

[58] Hezarkhani, A. and A. Kashaninia, "Performance analysis of a CA-CFAR detector in the interfering target and homogeneous background," in *2011 International Conference on Electronics, Communications and Control (ICECC)*, Ningbo, 2011.

[59] Papoulis, A., *Probability, Random Variables, and Stochastic Processes*, 3rd Edition ed., New York: McGraw-Hill, Inc., 1991.

[60] Metropolis, N., A. W. Rosenbluth, M. N. Rosenbluth and A. H. Teller, "Equation of State Calculations by Fast Computing Machines," *Journal of Chemical Physics,* vol. 21, 1953, p. 1087.

[61] Filip, A. E., "A Baker's Dozen Magnitude Approximations and Their Detection Statistics," *IEEE Transactions on Aerospace and Electronic Systems*, January 1976, pp. 86–89.

[62] Berkowitz, R. S., *Modern Radar: Analysis, Evaluation, and System Design*, New York: John Wiley & Sons, Inc., 1965.

[63] Woodward, P. M., *Probability and Information Theory with Applications to Radar*, 2nd ed., New York: Pergamon Press, 1953, Artech House, 1980.

[64] Blahut, R. E., *Algebraic Methods for Signal Processing and Communications Coding*, NY: Springer-Verlag, 1992.

[65] Boyd, S. and L. Vandenberghe, *Introduction to Applied Linear Algebra*, NY: Cambridge University Press, 2018.

APPENDIX 7A: MAXIMUM LIKELIHOOD ESTIMATION

Consider an N-dimension vector of samples or observations of a random variable. We stipulate that the samples are independent and identically distributed, as is the case for CA-CFAR processing. The problem at hand is to determine the best or most likely value for the unknown parameter.

To optimally estimate the unknown parameter, let us look at the problem in terms of MLE. We start with the likelihood function, L, in terms of a set of N observations, which is defined by [62, p. 150]

$$L(x_1, x_2, \cdots x_N; \alpha) \equiv f(x_1; \alpha) f(x_2; \alpha) \cdots f(x_N; \alpha) \tag{7A.1}$$

where $x_1, x_2, \ldots x_N$ are the observations, $f(x_i; \alpha)$ is a PDF. Parameter α is unknown to the observer. If the results can be described as a joint distribution, then the likelihood function can be expressed as

$$L(x_1, x_2, \cdots x_N; \alpha) \equiv f(x_1, x_2, \cdots x_N; \alpha) \tag{7A.2}$$

The MLE, $\hat{\alpha}$, is the value for α that maximizes the likelihood function, using N observations. To determine $\hat{\alpha}$, we solve the likelihood function

$$\frac{\partial L}{\partial \alpha} = 0 \tag{7A.3}$$

or, equivalently, the log likelihood function

$$\frac{\partial \ln(L)}{\partial \alpha} = 0 \tag{7A.4}$$

It is sometimes easier to solve the natural logarithm of the likelihood function (7A.4) rather than the likelihood function (7A.3).

Finn and Johnson stipulated Gaussian noise, where the unknown is the noise power, σ^2. This results in the Rayleigh clutter model described in Section 7.4.1. The output of a square-law detector presented with Gaussian noise is characterized by the exponential density function given by [41, pp. 207, 452]

$$f_X(x) = \frac{1}{\sigma^2} e^{-x/\sigma^2} \tag{7A.5}$$

where variance (power) σ^2 is unknown. Specifically, we have N independent observations of random variable X governed by the exponential PDF. We would like to estimate σ^2. Substituting (7A.5) into (7A.1), the likelihood function becomes

$$\begin{aligned} L(\sigma^2) &\equiv \left(\frac{1}{\sigma^2} e^{-x_1/\sigma^2}\right)\left(\frac{1}{\sigma^2} e^{-x_2/\sigma^2}\right) \cdots \left(\frac{1}{\sigma^2} e^{-x_N/\sigma^2}\right) \\ &= \frac{1}{(\sigma^2)^N} \prod_{i=1}^{N} e^{-x_i/\sigma^2} = (\sigma^2)^{-N} e^{-\frac{1}{\sigma^2} \sum_{i=1}^{N} x_i} \end{aligned} \tag{7A.6}$$

The corresponding log likelihood function is therefore

$$\ln\left[L(\sigma^2)\right] = \ln\left[(\sigma^2)^{-N} e^{-\frac{1}{\sigma^2}\sum_{i=1}^{N} x_i}\right] = \ln\left\{(\sigma^2)^{-N}\right\} + \ln\left(e^{-\frac{1}{\sigma^2}\sum_{i=1}^{N} x_i}\right) \quad (7A.7)$$

$$= -N\ln(\sigma^2) - \frac{1}{\sigma^2}\sum_{i=1}^{N} x_i$$

To find the MLE of σ^2, we need to maximize (7A.6) or (7A.7) with respect to σ^2. We do this by setting (7A.6) or (7A.7) equal to zero and taking the derivative with respect to σ^2. Comparing (7A.6) and (7A.7), it looks like the simpler approach is to solve the log likelihood function (7A.7). First, we take the partial derivative of (7A.7) with respect to σ^2, which results in

$$\frac{\partial \ln(L)}{\partial(\sigma^2)} = \frac{\partial\left\{-N\ln(\sigma^2) - \frac{1}{\sigma^2}\sum_{i=1}^{N} x_i\right\}}{\partial(\sigma^2)} = -N\frac{\partial \ln(\sigma^2)}{\partial(\sigma^2)} + \frac{\partial(\sigma^2)^{-1}}{\partial(\sigma^2)}\sum_{i=1}^{N} x_i$$

$$= \frac{1}{(\sigma^2)^2}\sum_{i=1}^{N} x_i - \frac{N}{\sigma^2} = -\frac{1}{(\sigma^2)^2}\left(\sigma^2 N - \sum_{i=1}^{N} x_i\right) \quad (7A.8)$$

Setting (7A.8) equal to zero and solving for σ^2, we can express the MLE of σ^2 used for CA-CFAR as [52, 63]

$$\widehat{\sigma^2} = \frac{1}{N}\sum_{i=1}^{N} x_i \approx \sigma^2 \quad (7A.9)$$

which is the noise power estimate proposed by Finn and Johnson [32]. As a point of interest, the arithmetic mean of (7A.9) is also an unbiased estimator. This means that the expected value of the estimate is equal to the parameter being estimated, which can be expressed as

$$E\left\{\widehat{\sigma^2}\right\} = E\left\{\frac{1}{N}\sum_{i=1}^{N} x_n\right\} = \sigma^2 \quad (7A.10)$$

APPENDIX 7B: TOEPLITZ MATRIX AND CFAR

We want to discuss the use of Toeplitz matrices to help speed CFAR computations in MATLAB. We are particularly interested in the cell averaging, greatest-of or smallest-of CFAR, although it is suspected that the Toeplitz matrix will help in implementing other types of CFAR. For reference, a Toeplitz matrix is an $n \times n$ matrix of the form [64, p. 3]

$$A = \begin{bmatrix} a_0 & a_1 & a_2 & \cdots & a_{n-1} \\ a_{-1} & a_0 & a_1 & \cdots & a_{n-2} \\ a_{-2} & a_{-1} & a_0 & \cdots & a_{n-3} \\ \vdots & & & \ddots & \vdots \\ a_{-n+1} & & & \cdots & a_0 \end{bmatrix} \qquad (7B.1)$$

where the elements running down each minor diagonal (parallel with the main diagonal) are equal. Toeplitz matrices are named after mathematician Otto Toeplitz [65].

A block diagram that we will use to discuss the implementation is contained in Figure 7B.1. For a given CUT, the CFAR averages the powers in N_E and N_L cells that constitute the early and late gates. The early and late gates are separated from the CUT by G_E and G_L guard cells, respectively. The averaging is performed for all CUT that lie within a range window of interest.

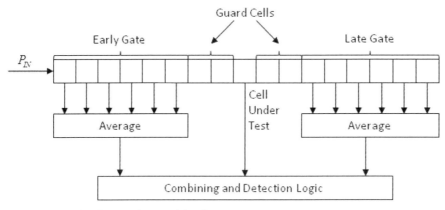

Figure 7B.1 CFAR implementation considered.

For the first CUT, the CFAR averages the powers in cells 1 through N_E and in cells $N_E + G_E + 1 + G_L + 1$ through $N_E + G_E + 1 + G_L + N_L$. For the second CUT, all of the above numbers are increased by one. Thus, the CFAR averages the powers in cells 2 through $N_E + 1$ and in cells $N_E + G_E + 1 + G_L + 2$ through $N_E + G_E + 1 + G_L + N_L + 1$. This continues until the upper limit on the late gate sum is equal to the number of cells in the range window, N_{cell}. In equation form, the process continues until

$$N_E + G_E + 1 + G_L + N_L + M - 1 = N_{cell} \qquad (7B.2)$$

where M is the Mth CUT.

The above may be clearer with the example of Figure 7B.2. In this figure, $N_E = 4$, $G_E = 2$, $G_L = 1$, $N_L = 5$, $M = 6$, and $N_{cell} = 18$[7]. For CUT – 1, the CFAR averages the powers in cells 1 to $N_E = 1$ to 4 to form the early gate average and in cells $N_E + G_E + 1 + G_L + 1 = N_{TOT} - N_L + 1$ to $N_{TOT} = 9$ to 13 to form the late gate average. In this expression we note that

---

[7] The asymmetry in this example is somewhat arbitrary. Most CFAR processors use symmetric gates.

$$N_{TOT} = N_E + G_E + 1 + G_L + N_L \qquad (7B.2)$$

is the total number of range cells involved in the CFAR process.

For CUT – 2, all of the limits are increased by one. Thus the CFAR averages the powers in cells 2 to $N_E + 1 = 2$ to 5 to form the early gate average and in cells $N_{TOT} - N_L + 2$ to $N_{TOT} + 1 = 10$ to 14 to form the late gate average. For the kth CUT, the early average is the average of power in cells k to $N_E + k - 1$, and the late average is the average of powers in cells $N_{TOT} - N_L + k$ to $N_{TOT} + k - 1$.

If we arrange the indices for the cells used in the early and late averages into matrices we get, for this example, the tables shown in Figure 7B.3. Recall that $N_E = 4$, $G_E = 2$, $G_L = 1$, $N_L = 5$, $M = 6$, and $N_{cell} = 18$, and that $N_{TOT} = N_E + G_E + 1 + G_L + N_L = 13$.

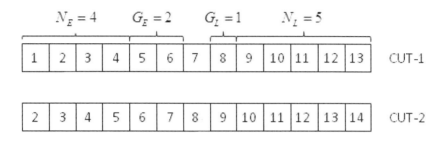

Figure 7B.2 Cells in the various gates as a function of CUT.

EARLY CELLS						LATE CELLS					
1	2	3	4		(N_E)	$(N_{TOT} - N_L + M)$ 9	10	11	12	13	(N_{TOT})
2	3	4	5				10	11	12	13	14
3	4	5	6				11	12	13	14	15
4	5	6	7				12	13	14	15	16
5	6	7	8				13	14	15	16	17
6	7	8	9	$(N_E + M - 1)$			14	15	16	17	18 (N_{cell})

Figure 7B.3 Matrices of indices for the early and late Cells involved in the early and late gate averages.

If we reverse the ordering of the columns of the two matrices, we get the matrices in Figure 7B.4.

	EARLY CELLS					LATE CELLS				
(N_E)	4	3	2	1	(N_{TOT})	13	12	11	10	9	$(N_{TOT} - N_L + M)$
	5	4	3	2		14	13	12	11	10
	6	5	4	3		15	14	13	12	11
	7	6	5	4		16	15	14	13	12
	8	7	6	5		17	16	15	14	13
$(N_E + M - 1)$	9	8	7	6	(N_{cell})	18	17	16	15	14

Figure 7B.4 Matrices of Figure 7B.2 with the columns reversed.

In this form, we recognize that the two matrices are Toeplitz matrices.

If we define a column vector as $C = [c_1 \ c_2 \ \ldots \ c_M]^T$ and a row vector as $R = [r_1 \ r_2 \ \ldots \ r_N]$ with $r_1 = c_1$, their associated Toeplitz matrix is of the form

$$T = toeplitz(C, R) = \begin{bmatrix} r_1(c_1) & r_2 & \cdots & r_N \\ c_2 & r_1 & \cdots & r_{N-1} \\ c_3 & c_2 & \cdots & r_{N-2} \\ c_4 & c_3 & \cdots & r_{N-3} \\ \vdots & \vdots & \ddots & \vdots \end{bmatrix} \quad (7B.3)$$

For our case, we have for the early gate, $C_E = [N_E : N_E + M + 1]^T$, $R_E = [N_E : -1 : 1]$, and $EARLY = toeplitz(C_E, R_E)$, where we have used MATLAB notation. For the late gate, we have $C_L = [N_{TOT} : N_{cell}]^T$, $R_L = [N_{TOT} : -1 : N_{TOT} - N_L + 1]$, and $LATE = toeplitz(C_L, R_L)$.

The *EARLY* and *LATE* Toeplitz matrices are used as the indices into the array of powers into the CFAR (P_{IN} in Figure 7B.1) needed to form the average powers in the early and late gates. The MATLAB commands used to compute the vector of average powers in the early and late gates are

$$P_E = sum(P_{IN}(EARLY), 2)/N_E \quad (7B.4)$$

and

$$P_L = sum(P_{IN}(LATE), 2)/N_L \quad (7B.5)$$

where P_E and P_L are M element vectors of early and late gate averages associated with the M CUT of the CFAR. In the above statements, $P_{IN}(EARLY)$ and $P_{IN}(LATE)$ causes MATLAB to, effectively, reorganize P_{IN} into two-dimensional arrays. The sum with a second parameter of 2 causes the sum to be executed across the rows of $P_{IN}(EARLY)$ and $P_{IN}(LATE)$.

Chapter 8

Matched Filter

8.1 INTRODUCTION

In the detection probability equations [(6.116), (6.123), and (6.127)], we noted that P_d depends directly on SNR. That is, P_d increases as SNR increases. Because of this, we want to try to ensure the receiver is designed to maximize SNR by including a *matched filter* in the receiver. In most radars, the matched filter is included immediately before the signal processor, and in some, the matched filter is the signal processor.

J. H. Van Vleck and David Middleton coined the term "matched filter" in a 1946 *Journal of Applied Physics* article [1]. They credited D. O. North with arriving at the same formulation for the matched filter, but by a different approach based on calculus of variations, instead of the Cauchy-Schwarz inequality they used. North's development first appeared in a classified report, which was later published in a 1963 journal article [2]. Van Vleck and Middleton indicated the matched filter equations were also developed by Henry Wallman as a specific case of a more general theory developed by Norbert Wiener.

8.2 PROBLEM DEFINITION

The statements in the first paragraph of this chapter provide the design requirement for the matched filter. Specifically, given some signal, $s(t)$, and noise, $\mathbf{n}(t)$, we find a filter impulse response, $h(t)$, that maximizes SNR at the filter output. For purposes of this discussion, we assume the signal is not a random process. Actually, we assume the *form* (e.g., a linear frequency modulated (LFM) pulse) of the signal is deterministic; its amplitude and phase can be random variables. As a note, we are using complex signal notation (see Chapter 1) in this chapter. This is consistent with the notation used in Chapter 6 and applies to both the IF and baseband representation. Thus, for example, $\mathbf{n}(t)$ could be a representation of $\mathbf{n}_{IF}(t)$ or $\mathbf{n}_B(t)$ as appropriate.

As indicated in Figure 8.1, if the input to the matched filter is $s(t)$, the output will be $s_o(t)$, and if the input is $\mathbf{n}(t)$, the output will be $\mathbf{n}_o(t)$. The *output, instantaneous, normalized, signal power* is

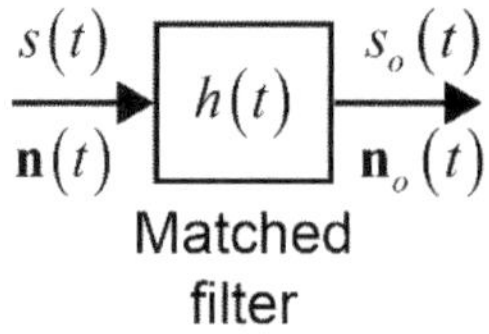

Basic Radar Analysis

```
    s(t)  ┌──────┐  s_o(t)
    ────▶ │ h(t) │ ────▶
    n(t)  └──────┘  n_o(t)
          Matched
           filter
```

Figure 8.1 Matched filter block diagram.

$$P_{so}(t) = |s_o(t)|^2 = s_o(t)s_o^*(t) \tag{8.1}$$

For purposes of the matched filter design, we define the *normalized, peak signal power* at the matched filter output as[1]

$$P_S = \max_t P_{so}(t) = P_{so}(t_o) = |s_o(t_o)|^2 \tag{8.2}$$

where t_o is the time of the peak.

Since $\mathbf{n}_o(t)$ is a random process that we assume is WSS, we must work with its average power. Thus, the *normalized average noise power* at the output of the matched filter is

$$P_N = E\{|\mathbf{n}_o(t)|^2\} \tag{8.3}$$

where we use expected values ($E\{\mathbf{x}\}$) because we are dealing with random processes [3].

With the above, we can define the design criterion for the matched filter. Specifically, we choose the matched filter to maximize the ratio of peak signal power to average noise power at the output of the matched filter. In equation form,

$$h(t): \max_{h(t)} \frac{P_S}{P_N} \tag{8.4}$$

8.3 PROBLEM SOLUTION

Equation (8.4) states we must first write the ratio of P_S and P_N in terms of $h(t)$ and then maximize it with respect to $h(t)$.

We assume $h(t)$ is linear and write

$$s_o(t) = s(t) * h(t) \tag{8.5}$$

and

$$\mathbf{n}_o(t) = \mathbf{n}(t) * h(t) \tag{8.6}$$

---

[1] The P_S and P_N in this chapter are the same as in Chapter 6.

where ∗ denotes convolution. We choose to solve the optimization problem in the frequency domain through the use of Fourier transforms. To this end, we write

$$H(f) = \int_{-\infty}^{\infty} h(t) e^{-j2\pi f t} dt = \Im[h(t)] \qquad (8.7)$$

$$S(f) = \Im[s(t)] \qquad (8.8)$$

and

$$S_o(f) = \Im[s_o(t)] \qquad (8.9)$$

and recall that

$$S_o(f) = H(f) S(f) \qquad (8.10)$$

As a note, $S(f)$ is the signal spectrum (the signal voltage spectral density) at the input to the matched filter, as such it will experience the receiver gain of $\sqrt{G}$ discussed in Chapter 2. (We will again encounter the factor of G when we consider noise.)

Since $\mathbf{n}(t)$ and $\mathbf{n}_o(t)$ are random processes, we must deal with them as such, which means we write [3]

$$N(f) = \Im[R_n(\tau)] = \Im\left[E\{\mathbf{n}(t+\tau)\mathbf{n}^*(t)\}\right] \qquad (8.11)$$

$$N_o(f) = \Im[R_{no}(\tau)] = \Im\left[E\{\mathbf{n}_o(t+\tau)\mathbf{n}_o^*(t)\}\right] \qquad (8.12)$$

and

$$N_o(f) = |H(f)|^2 N(f) \qquad (8.13)$$

In the above, $R_n(\tau)$ and $R_{no}(\tau)$ are the autocorrelation functions of $\mathbf{n}(t)$ and $\mathbf{n}_o(t)$, respectively. As a reminder, we note that $\mathbf{n}(t)$ and $\mathbf{n}_o(t)$ are WSS, which implies the autocorrelation is a function of time difference, τ, only, not absolute time, t.

We recognize $N_o(f)$ is a power spectral density (noise energy). Thus, the noise power at the output of the matched filter is

$$P_N = \int_{-\infty}^{\infty} N_o(f) df = \int_{-\infty}^{\infty} |H(f)|^2 N(f) df \qquad (8.14)$$

From (8.2), the normalized peak signal power at the matched filter output is given by

$$P_S = |s_o(t_o)|^2 \qquad (8.15)$$

However, we can write

$$s_o(t_o) = \mathfrak{F}^{-1}\left[S_o(f)\right]\Big|_{t=t_o} = \int_{-\infty}^{\infty} S(f)H(f)e^{j2\pi f t_o} df \qquad (8.16)$$

If we combine (8.16), (8.15), (8.14), and (8.4), we get [4]

$$h(t): \max_{h(t)} \frac{P_S}{P_N} = \max_{h(t)} \frac{\left|\int_{-\infty}^{\infty} S(f)H(f)e^{j2\pi f t_o} df\right|^2}{\int_{-\infty}^{\infty} |H(f)|^2 N(f) df} \qquad (8.17)$$

At this point, we make the assumption that the noise power spectral density at the input to the matched filter is

$$N(f) = kT_s G \qquad (8.18)$$

where G is the receiver gain. Equation (8.18) implies the noise into the matched filter is white. In Chapter 4, we justified that the noise into the radar is white. Radar designers design radar receivers so as to assure the spectrum of noise into the matched filter is flat over an extent that is larger than the bandwidth of the pulse. This is sufficient to satisfy the white noise assumption. As a note, the assumption of white noise is critical to the derivation of the matched filter. If the noise cannot be assumed white, the equations to follow will not apply.

Substituting (8.18) into (8.17) leads to

$$h(t): \max_{h(t)} \frac{\left|\int_{-\infty}^{\infty} S(f)H(f)e^{j2\pi f t_o} df\right|^2}{kT_s G \int_{-\infty}^{\infty} |H(f)|^2 df} \qquad (8.19)$$

We perform the maximization process by applying one of the Cauchy-Schwarz inequalities to the numerator of (8.19) [5], specifically

$$\left|\int_a^b A(f)B(f)df\right|^2 \leq \left(\int_a^b |A(f)|^2 df\right)\left(\int_a^b |B(f)|^2 df\right) \qquad (8.20)$$

with the equality valid only when $A(f)$ is proportional to the complex conjugate of $B(f)$ [5]. That is, when

$$A(f) = KB^*(f) \qquad (8.21)$$

where K is an arbitrary (complex) constant. If we apply (8.20) to the ratio of (8.19) with the associations

$$A(f) = H(f) \qquad (8.22)$$

and

$$B(f) = S(f)e^{j2\pi f t_o} \tag{8.23}$$

we get

$$\frac{\left|\int_{-\infty}^{\infty} S(f)H(f)e^{j2\pi f t_o} df\right|^2}{kT_s G \int_{-\infty}^{\infty} |H(f)|^2 df} \leq \frac{\left(\int_{-\infty}^{\infty} |H(f)|^2 df\right)\left(\int_{-\infty}^{\infty} |S(f)|^2 df\right)}{kT_s G \int_{-\infty}^{\infty} |H(f)|^2 df} \tag{8.24}$$

where we made use of

$$\left|S(f)e^{j2\pi f t_o}\right| = |S(f)| \tag{8.25}$$

We note that (8.24) reduces to

$$\frac{\left|\int_{-\infty}^{\infty} S(f)H(f)e^{j2\pi f t_o} df\right|^2}{kT_s G \int_{-\infty}^{\infty} |H(f)|^2 df} \leq \frac{\int_{-\infty}^{\infty} |S(f)|^2 df}{kT_s G} \tag{8.26}$$

Equation (8.26) tells us that *for all H(f)*, the upper bound on the left side is equal to that on the right side. That is, we have found the maximum value of P_S/P_N (the ratio of peak signal power to average noise power at the matched filter output) over all $h(t)$ and have solved part of the maximization problem. To find the $h(t)$ that yields the maximum P_S/P_N, we invoke the second part of the Cauchy-Schwarz inequality given in (8.21). Specifically, we say

$$\max_{h(t)} \frac{P_S}{P_N} = \frac{\int_{-\infty}^{\infty} |S(f)|^2 df}{kT_s G} \tag{8.27}$$

when we choose $H(f)$ as [see (8.21)]

$$H(f) = KS^*(f)e^{-j2\pi f t_o} \tag{8.28}$$

Thus, we have found the Fourier transform of the filter impulse response that maximizes peak signal power to average noise power at the filter output. Furthermore, we have an equation for the maximum in the form of (8.27) and have determined that the maximum occurs at $t = t_o$.

We note from the form of (8.28) that

$$|H(f)| = |KS(f)| \tag{8.29}$$

In other words, the matched filter frequency response has the same shape as the frequency spectrum of the signal. They simply differ by a scaling factor $|K|$. This is the reason Van Vleck and Middleton termed $H(f)$ a *matched filter*.

We now want to look at the specific form of $h(t)$ relative to $s(t)$. We can write

$$h(t) = \mathfrak{I}^{-1}\left[KS^*(f)e^{-j2\pi f t_o}\right] = \int_{-\infty}^{\infty} KS^*(f)e^{-j2\pi f t_o}e^{j2\pi f t}df$$

$$= K\int_{-\infty}^{\infty} S^*(f)e^{-j2\pi f(t_o-t)}df = K\left[\int_{-\infty}^{\infty} S(f)e^{j2\pi f(t_o-t)}df\right]^* = Ks^*(t_o - t) \quad (8.30)$$

Thus, $h(t)$ is the conjugate of a scaled (by K), time reversed (because of the $-t$), and shifted (by t_o) version of the signal, $s(t)$ at the input to the matched filter. This operation is illustrated in Figure 8.2. The left sketch of this figure is $s(t)$, while the center figure is a sketch of $s^*(-t)$. Finally, the right figure is $Ks^*(t_o - t)$, or $h(t)$. We normally assume that $s(t)$ has the same shape as the signal generated in the transmitter (i.e., a pulse with a rect[x] envelope). That is, we ignore any distortion that occurs in the transmit, propagation, and receive paths. We account for the distortion by incorporating a mismatch loss in the radar range equation (see Chapter 5).

Now that we have established the equation for the maximum value of the SNR at the output of the matched filter and have a filter that can provide the maximum SNR, we want to determine the value of the SNR. Specifically, we want to relate the maximum SNR to the value of SNR we compute from the radar range equation.

From (8.27), we have

$$SNR_{MAX} = \frac{\int_{-\infty}^{\infty} |S(f)|^2 df}{kT_s G} \quad (8.31)$$

Recalling Parseval's theorem (also known as Rayleigh's energy theorem) [6], which can be expressed as

$$\int_{-\infty}^{\infty} |x(t)|^2 dt = \int_{-\infty}^{\infty} |X(f)|^2 df \quad (8.32)$$

and noting $s(t)$ has finite energy and power, we write

$$SNR_{MAX} = \frac{\int_{-\infty}^{\infty} |s(t)|^2 dt}{kT_s G} \quad (8.33)$$

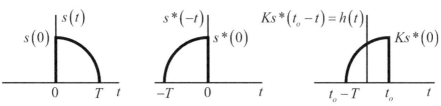

Figure 8.2 Evolution of $h(t)$.

We recognize the numerator of (8.33) as the *energy* in the signal at the input to the matched filter. From Chapter 2, we found this to be

$$E_S = \frac{P_T G_T G_R \lambda^2 \sigma \tau_p}{(4\pi)^3 R^4 L} G \quad (8.34)$$

With this we get

$$SNR_{MAX} = \frac{E_S}{kT_s G} = \frac{P_T G_T G_R \lambda^2 \sigma \tau_p}{(4\pi)^3 R^4 kT_s L} \quad (8.35)$$

We recognize (8.35) as the SNR given by the radar range equation (see Chapter 2). This tells us the peak value of SNR (the peak *power* ratio) at the output of the matched filter is the SNR (the *energy* ratio) we obtain from the radar range equation. In essence, the matched filter ekes out the maximum possible SNR from the signal and noise the radar must deal with. For the case where the interference is due to white noise at the input to the matched filter, there is no other linear filter that will give a larger value of SNR for the transmitted signal. If the interference is other than white noise (e.g., clutter), there are other filters that will provide larger values of signal-to-*interference* power ratio (SIR) than the filter defined by (8.30). This is discussed further in Chapters 15 through 20.

In Chapter 6, we found that P_d depended on the SNR power ratio. The results above say that the maximum SNR *power* ratio is equal to the SNR *energy* ratio derived in Chapter 2. Thus, the SNR provided by the radar range equation will provide the maximum P_d for a given P_{fa}. Further, this maximum P_d can be achieved if the radar includes a matched filter. As a re-reminder: here we are dealing with single-pulse, or single-sample, P_d and P_{fa}, and with the SNR for a single transmitted (and received) pulse. We will consider how to handle multiple pulses in Chapter 9.

8.4 MATCHED FILTER EXAMPLES

We now want to consider two matched filter examples and derive equations for the output of the matched filter. The specific examples we consider are matched filters for an unmodulated pulse and a pulse with LFM. We consider these two cases because the equations for the matched filter output are reasonably simple. We consider more complicated waveforms and associated matched filters in Chapter 11.

8.4.1 General Formulation

From (8.30), we have

$$h(t) = Ks^*(t_o - t) \tag{8.36}$$

where K is an arbitrary (complex) constant and t_o is the value of t at which the matched filter response to $s(t)$ will reach its peak.

Since K and t_o can be anything we want, without loss of generality we let $K = 1$ and $t_o = 0$. The latter statement says that the output of the matched filter will reach its peak at a relative time of zero. With this we get

$$h(t) = s^*(-t) \tag{8.37}$$

The response of $h(t)$ to $s(t)$ is given by

$$s_o(t) = h(t) * s(t) = \int_{-\infty}^{\infty} s(\gamma) h(t-\gamma) d\gamma \tag{8.38}$$

But $h(t) = s^*(-t)$ so $h(t - \gamma) = s^*[-(t - \gamma)] = s^*(\gamma - t)$ and

$$s_o(t) = \int_{-\infty}^{\infty} s(\gamma) s^*(\gamma - t) d\gamma \tag{8.39}$$

We note that this integral is the complex, time autocorrelation of $s(t)$ [5].

8.4.2 Response for an Unmodulated Pulse

For an unmodulated (rectangular) baseband pulse,

$$s(t) = Ae^{j\theta} \text{rect}\left[\frac{t - \tau_p/2}{\tau_p}\right] \tag{8.40}$$

where A is the amplitude of the pulse and θ is the phase. With this we have

$$s^*(t) = Ae^{-j\theta} \text{rect}\left[\frac{t - \tau_p/2}{\tau_p}\right] \tag{8.41}$$

A plot of $s(t)$ is shown notionally in Figure 8.3. The plot of $s^*(t)$ would look the same except the "height" would be $Ae^{-j\theta}$ rather than $Ae^{j\theta}$.

In the $s_o(t)$ integral of (8.39), we note that t is the separation between $s(\gamma)$ and $s^*(\gamma - t)$, as shown in Figure 8.4. Figure 8.4 corresponds to the case where $t \geq 0$.

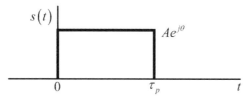

Figure 8.3 Unmodulated pulse.

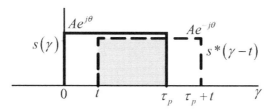

Figure 8.4 Plot of $s(\gamma)$ and $s^*(\gamma - t)$ for $t \geq 0$.

When $t \geq \tau_p$, $s(\gamma)$ and $s^*(\gamma - t)$ do not overlap and we have $s(\gamma) s^*(\gamma - t) = 0$. Thus,

$$s_o(t) = \int_{-\infty}^{\infty} s(\gamma) s^*(\gamma - t) d\gamma = 0 \quad t \geq \tau_p \tag{8.42}$$

For $0 \leq t < \tau_p$, the overlap region of $s(\gamma)$ and $s^*(\gamma - t)$ is $t \leq \gamma < \tau_p$. In the overlap region, $s(\gamma) s^*(\gamma - t) = A e^{j\theta} A e^{-j\theta} = A^2$ and thus,

$$s_o(t) = \int_t^{\tau_p} A^2 d\gamma = A^2 (\tau_p - t) \tag{8.43}$$

Since $t \geq 0$, we can use the substitution $|t| = t$ (we are doing this because we will need it to compare the form of (8.43) to the case where $t < 0$) and write (8.42) as

$$s_o(t) = A^2 (\tau_p - |t|) \quad 0 \leq t < \tau_p \text{ or } |t| < \tau_p \tag{8.44}$$

With (8.42), we get

$$s_o(t) = \begin{cases} A^2 (\tau_p - |t|) & 0 \leq t < \tau_p \text{ or } |t| < \tau_p \\ 0 & t \geq \tau_p \text{ or } |t| \geq \tau_p \end{cases} \tag{8.45}$$

The arrangement of $s(\gamma)$ and $s^*(\gamma - t)$ for $t < 0$ is shown in Figure 8.5. It should be clear that if $t + \tau_p \leq 0$ or $t \leq -\tau_p$, $s_o(t) = 0$. If we multiply both sides of the inequality by -1, we get $-t \geq \tau_p$ and, since t is negative (i.e., $t < 0$), we can write $-t = |t|$ and

$$s_o(t) = 0 \quad |t| \geq \tau_p \tag{8.46}$$

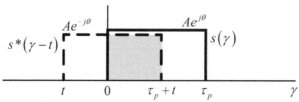

Figure 8.5 Plot of $s(\gamma)$ and $s^*(\gamma - t)$ for $t < 0$.

The overlap region is $0 \leq \gamma < t + \tau_p$, which yields

$$s_o(t) = \int_0^{t+\tau_p} A^2 d\gamma = A^2(\tau_p + t) \tag{8.47}$$

Since $t < 0$, $t = -|t|$ and we replace t with $-|t|$ to get

$$s_o(t) = A^2(\tau_p - |t|) \quad -\tau_p < t < 0 \tag{8.48}$$

We can multiply the terms of the inequality by -1 to get $\tau_p > -t > 0$ or $\tau_p > |t| > 0$ since $-t = |t|$. This leads to

$$s_o(t) = A^2(\tau_p - |t|) \quad |t| < \tau_p \tag{8.49}$$

If we combine this with (8.46) we have

$$s_o(t) = \begin{cases} 0 & |t| \geq \tau_p \\ A^2(\tau_p - |t|) & |t| < \tau_p \end{cases} \tag{8.50}$$

We note that this is the same form as (8.45). Thus, (8.50) and (8.45) apply for all t.

We can combine the two parts of (8.50) and use the rect[x] function to write $s_o(t)$ in a more compact form as

$$s_o(t) = A^2(\tau_p - |t|)\text{rect}\left[\frac{t}{2\tau_p}\right] \tag{8.51}$$

A plot of $s_o(t)$ is shown in Figure 8.6. Note that $s_o(t)$ is a triangle with a height of $A^2 \tau_p$ and a base width of $2\tau_p$. This height and base width property is common to the matched filter response of all pulses that have a constant amplitude of A and width of τ_p. That is, the peak value of the matched filter output is $A^2 \tau_p$ and a base width is $2\tau_p$.

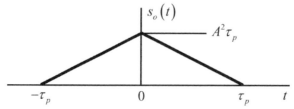

Figure 8.6 Plot of matched filter output for an unmodulated pulse.

8.4.3 Response for an LFM Pulse

There are very few practical pulses that lead to simple expressions for $s_o(t)$. One is the unmodulated pulse of the previous example, and another is a pulse with LFM across the pulse. For an LFM pulse, the form of $s(t)$ is

$$s(t) = Ae^{j\pi\alpha t^2} \text{rect}\left[\frac{t - \tau_p/2}{\tau_p}\right] \quad (8.52)$$

We note the difference between (8.52) and (8.40) is that we replaced the constant phase, θ, with a time-varying phase,

$$\theta(t) = \pi\alpha t^2 \quad (8.53)$$

If we take the derivative of $\theta(t)$, we get the frequency modulation

$$f(t) = \frac{1}{2\pi}\frac{d\theta(t)}{dt} = \alpha t \quad (8.54)$$

and note the frequency changes linearly across the pulse. This is the origin of the term *linear frequency modulation*.

The parameter α is termed the *LFM slope*. If $\alpha > 0$, we say we have increasing LFM because the frequency increases across the pulse. If $\alpha < 0$, we have decreasing LFM. An LFM waveform is also termed a *chirp* waveform because of the sound it makes at audio frequencies. Increasing LFM is termed *up chirp,* and decreasing LFM is termed *down chirp*.

The frequency, $f(t)$, starts at zero at the beginning of the pulse and increases (decreases) to $\alpha\tau_p$ ($-\alpha\tau_p$) at the end of the pulse. Thus, the total frequency extent is $|\alpha\tau_p|$. This is termed the *LFM bandwidth*. As we will see, the width of the central lobe of $s_o(t)$ is approximately $1/|\alpha\tau_p|$, or the reciprocal of the LFM bandwidth.

To compute the matched filter output for an LFM pulse, we start with (8.39) and consider the $t \geq 0$ and $t < 0$ intervals as before. Like the unmodulated pulse, we note $s_o(t) = 0$ for $|t| \geq \tau_p$.

Similar to (8.43) for $0 \leq t < \tau_p$, we have

$$\begin{aligned}
s_o(t) &= \int_t^{\tau_p} s(\gamma)s^*(\gamma - t)d\gamma = \int_t^{\tau_p}\left(Ae^{j\pi\alpha\gamma^2}\right)\left(Ae^{-j\pi\alpha(\gamma-t)^2}\right)d\gamma \\
&= A^2\int_t^{\tau_p}\exp\left\{j\pi\alpha\left[\gamma^2 - \left(\gamma^2 - 2\gamma t + t^2\right)\right]\right\}d\gamma \\
&= A^2 e^{-j\pi\alpha t^2}\int_t^{\tau_p} e^{j2\pi\alpha t\gamma}d\gamma \\
&= A^2 e^{-j\pi\alpha t^2}\frac{1}{j2\pi\alpha t}\left[e^{j2\pi\alpha t\tau_p} - e^{j2\pi\alpha t^2}\right]
\end{aligned} \quad (8.55)$$

If we factor $\exp[j\pi\alpha t(\tau_p + t)]$ from the bracketed term, we get

$$s_0(t) = A^2 e^{j\pi\alpha t \tau_p} \frac{1}{\pi\alpha t} \left[\frac{e^{j\pi\alpha t(\tau_p - t)} - e^{-j\pi\alpha t(\tau_p - t)}}{2j} \right]$$

$$= A^2 e^{j\pi\alpha t \tau_p} \frac{\sin\left[\pi\alpha t(\tau_p - t)\right]}{\pi\alpha t} \tag{8.56}$$

We note that $s_o(t)$ is complex. Since we are concerned only with the shape of $s_o(t)$, we can use $|s_o(t)|$ and write

$$|s_0(t)| = A^2 \left| \frac{\sin\left[\pi\alpha t(\tau_p - t)\right]}{\pi\alpha t} \right| \tag{8.57}$$

Multiplying by $|\tau_p - t|/|\tau_p - t|$ gives

$$|s_0(t)| = A^2 \left| (\tau_p - t) \frac{\sin\left[\pi\alpha t(\tau_p - t)\right]}{\pi\alpha t(\tau_p - t)} \right| = A^2 (\tau_p - t) \left| \mathrm{sinc}\left(\alpha t[\tau_p + t]\right) \right| \tag{8.58}$$

where we were able to remove the absolute value from $\tau_p - t$ because we are only considering $0 \le t \le t_p$.

If we perform similar math for $-\tau_p < t < 0$, we get (see Exercise 1)

$$|s_0(t)| = A^2 (\tau_p + t) \left| \mathrm{sinc}\left[\alpha t(\tau_p + t)\right] \right| \tag{8.59}$$

As with the unmodulated pulse, we can use $t = |t|$ in (8.58) and $t = -|t|$ in (8.59) to get

$$|s_0(t)| = A^2 (\tau_p - |t|) \left| \mathrm{sinc}\left(\alpha t[\tau_p - |t|]\right) \right| \tag{8.60}$$

for both (8.58) and (8.59). The even property of the sinc function is the reason we use t instead of $|t|$ in the first appearance of t in the argument of the sinc function (see Exercise 1).

If we combine (8.60) with $s_o(t) = 0$ for $|t| \ge \tau_p$ and make use of the $\mathrm{rect}[x]$ function, we get

$$|s_0(t)| = A^2 (\tau_p - |t|) \left| \mathrm{sinc}\left(\alpha t[\tau_p - |t|]\right) \right| \mathrm{rect}\left[\frac{t}{2\tau_p}\right] \tag{8.61}$$

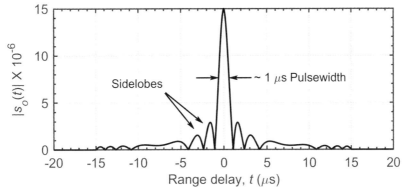

Figure 8.7 Plot of matched filter output for an LFM pulse with $\tau_p = 15$ μs, $B = |\alpha\tau_p| = 1$ MHz and $A = 1$.

Figure 8.7 contains a plot of $|s_o(t)|$ for an example case of a $\tau_p = 15$ μs pulse width, an LFM bandwidth of $B = |\alpha\tau_p| = 1$ MHz, $\alpha > 0$, and $A = 1$. The spacing between the points where the response is $A^2\tau_p/2$ is approximately 1 μs or $1/B$. Also, the height is 15×10^{-6}, or τ_p, since $A = 1$. The total extent of the response is 30 μs or twice the pulse width of 15 μs. As a note, for modulated waveforms, τ_p is termed the *uncompressed pulse width,* and the aforementioned spacing between $A^2/2$ points is termed the *compressed pulse width*. The details of the sidelobe structure of $|s_o(t)|$ depends on $B\tau_p$, which is termed the *time-bandwidth product* or *BT product* of the waveform. Plots for other BT products are considered in the exercises.

8.5 SUMMARY

We summarize this chapter by repeating that the impulse response of a matched filter for some signal, $s(t)$, is given by

$$h(t) = Ks^*(t_o - t) \tag{8.62}$$

where K and t_o are arbitrary. The sole function of a matched filter is to maximize SNR. The matched filter is under no constraint to preserve the shape of the signal. We also note that:

- The SNR at the output to the matched filter is peak signal to average noise power and is equal to ratio of the signal energy to the noise energy at the input to the matched filter, as given by the radar range equation.
- The matched filter impulse response is the time-reversed impulse response of the waveform to which it is matched.
- The frequency response of the matched filter is the spectrum (Fourier transform) of the complex conjugate of the waveform to which it is matched.
- The matched filter output is the autocorrelation of the waveform to which it is matched.

In this chapter, we developed matched filter responses for an unmodulated rectangular pulse and a rectangular pulse with LFM. The equations for these responses are

Unmodulated:
$$s_o(t) = A^2(\tau_p - |t|)\text{rect}\left[\frac{t}{2\tau_p}\right] \quad (8.63)$$

and

LFM:
$$|s_o(t)| = A^2(\tau_p - |t|)\left|\text{sinc}(\alpha t[\tau_p - |t|])\right|\text{rect}\left[\frac{t}{2\tau_p}\right] \quad (8.64)$$

In Chapter 11, we will consider other types of waveform modulation and how to compute their matched filter responses.

8.6 CLOSING COMMENTS

In deriving the equation for the matched filter, we rationalized the suitability of the white noise assumption. However, we did not account for the effects the transmitter, environment, antenna, target, and receiver would have on $s(t)$, the signal input to the matched filter. In fact, in the two examples, we assumed $s(t)$ was a rectangular pulse. This is an idealization that is not completely satisfied in practice. If we tried to account for distortion effects caused by the radar, environment, and target, the math associated with the matched filter derivation would very quickly become untenable. Our justification for considering only the ideal case is that, in practice, if the radar waveform generator creates an ideal rectangular pulse, and the target is not too large relative to the reciprocal of the pulse modulation bandwidth, the signal at the input to the matched filter will be close to an ideal rectangular pulse. Deviations in matched filter output shape due to a nonideal input are generally small, and losses due to the fact that the matched filter is not perfectly matched to the signal at its input are accounted for via the matched filter loss discussed in Chapter 5.

As an example, Figure 8.8 contains the envelope of a nonideal, 1.5-V, 1-µs, unmodulated pulse at the input to a matched filter that is matched to an ideal, $A = 1.5$-V, $\tau_p = 1$-µs, unmodulated pulse. The nonideal pulse was generated by passing an ideal pulse through a receiver we represented by a filter with a bandwidth of 4 MHz. We chose the bandwidth to be representative of the relation between receiver bandwidth and the reciprocal of the duration of a 1-µs, unmodulated pulse, which is taken to be the modulation bandwidth of the pulse. As can be seen, the envelope of the nonideal pulse is not perfectly rectangular, but is reasonably close.

Figure 8.9 contains plots of the output of the matched filter for the ideal and nonideal pulses. For the ideal pulse, the matched filter has the expected triangular shape, with a peak amplitude of $A^2\tau_p = (1.5)^2 \times 10^{-6}$. For the non-ideal pulse, the matched filter does not have a perfect triangular shape. Also, the peak amplitude is not the ideal value of $A^2\tau_p$. There is also a delay between the peak of the ideal output and the nonideal output. The delay is caused by the delay through the 4-MHz filter and is easily accounted for in the range measurement logic. The difference in peak amplitude, relative to the ideal case, is accounted for by the matched filter mismatch loss discussed in Chapter 5.

For LFM pulses, the response of Figure 8.7 is representative of an actual response, except that it is likely it will not be as perfectly symmetric as shown in that figure, and the peak will be smaller than the ideal value. Again, these are deviations that are easily accounted for.

For cases where the target is large relative to the range resolution, the peak of the matched filter output may broaden or could exhibit several peaks. This will also translate to an SNR loss relative to the case of a point target with the same RCS, since the target RCS will be essentially distributed in range. It is something that must be taken into account when analyzing the output of the detection logic. This problem arises in radars that have very narrow compressed pulse widths (less than about 0.05 to 0.1 μs depending on target size) or very large targets such as ships or very large aircraft (e.g., blimps).

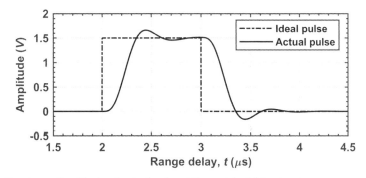

Figure 8.8 Envelope of an ideal and actual pulse at the matched filter input.

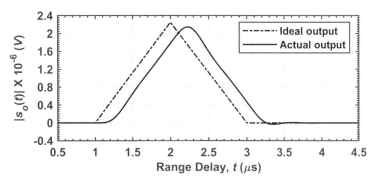

Figure 8.9 Plot of matched filter output for an unmodulated pulse.

In some radars, the designers intentionally use filters that are not matched to the transmit pulse. The most common case is an LFM pulse where the filter is intentionally mismatched to reduce the range (time) sidelobes (the lobes around the main lobe—see Figure 8.7) [7]. In such cases, the designer is concerned with interference (from other targets or clutter) and is willing to accept the loss in SNR caused by using a mismatched filter.

Radars that use unmodulated pulses and analog processing may not include a matched filter per se. Instead, they may use a narrowband filter that will pass most of the pulse power and minimize the noise power to some extent. This approach is usually taken as a cost savings where the designer is willing to accept the potential 1 or 2 dB loss in SNR associated with such an implementation.

In analog radars that use LFM pulses, the matched filter can be implemented with surface or bulk wave acoustic devices with piezoelectric transducers at the input and output [8, 9] and, in some instances, lumped parameter filters. In modern radars that use digital signal processing, the matched filter could be implemented using a fast convolver based on FFTs, FIR filters, or some other digital processing methodology.

8.7 EXERCISES

1. Derive (8.59) and show that it can be combined with (8.58) and $s_o(t) = 0$ for $|t| > \tau_p$ to arrive at (8.60).

    2. Derive $s_o(t)$ for the unmodulated pulse of Section 8.4.2 when the phase is $\theta(t) = 2\pi f_{IF} t + \phi$ instead of a constant. You will note that $s_o(t)$ is not real as was the case of the example. Because of this, one would plot $|s_o(t)|$ versus t instead of $s_o(t)$ versus t.

    3. Plot Re[$s(t)$] versus t for -5 μs $\leq t \leq 20$ μs, a pulse width $\tau_p = 15$ μs, a chirp bandwidth $B = \alpha \tau_p = 1$ MHz, and an amplitude $A = 1$. Use $\theta(t) = 2\pi f_{IF} t + \pi \alpha t^2$ instead of $\theta(t) = \pi \alpha t^2$. Let $f_{IF} = 0.5$ MHz. This plot illustrates the increasing frequency behavior of an LFM pulse.

    4. Repeat Exercise 3 for the case where α is negative. This plot illustrates the decreasing frequency behavior of an LFM pulse for a negative chirp slope.

    5. Plot $|s_o(t)|$ for an LFM pulse with amplitude $A = 1$, pulse width $\tau_p = 15$ μs, and LFM bandwidths of 0.2, 0.5, 2.0, and 5.0 MHz. Note the difference in the sidelobe structure.

    6. Find an equation for the impulse response, $h(t)$, of a matched filter for a pulse defined by

    $$s(t) = e^{\beta t} \text{rect}\left[\frac{t - \tau_p/2}{\tau_p}\right] \quad (8.65)$$

    where $\beta < 0$. Sketch $s(t)$ and $h(t)$. Find and sketch the matched filter output, $s_o(t)$. This is an example of a pulse that does not have a rectangular shape. As a note, there is nothing in matched filter theory that says a transmit pulse must have a rectangular envelope, or shape. However, most radar transmitters are not capable of transmitting pulses that do not have a rectangular envelope. This is due to the fact that the transmitter must operate at full power to achieve maximum efficiency. Modern radars that use active electronically scanned antennas (AESAs) with gallium-nitride power amplifiers will most likely be able to transmit variable amplitude pulses at some time in the near future.

References

[1] Van Vleck, J. H., and D. Middleton, "A Theoretical Comparison of the Visual, Aural, and Meter Reception of Pulsed Signals in the Presence of Noise," *J. Appl. Phys.*, vol. 17, no. 11, Nov. 1946, pp. 940–971.

[2] North, D. O., "An Analysis of the Factors which Determine Signal/Noise Discrimination in Pulsed Carrier Systems," RCA Labs, Princeton, NJ, Tech. Rep. PTR-6C, Jun. 25, 1943. Reprinted: *Proc. IEEE*, vol. 51, no. 7, Jul. 1963, pp. 1016–1027. Reprinted: *Detection and Estimation* (S. S. Haykin, ed.), Stroudsburg, PA: Halstad Press, 1976, pp. 10–21.

[3] Papoulis, A., *Probability, Random Variables, and Stochastic Processes*, 3rd ed., New York: McGraw-Hill, 1991.

[4] Budge, M. C., Jr., "EE 619: Intro to Radar Systems," www.ece.uah.edu/courses/material/EE619/index.htm.

[5] Urkowitz, H., *Signal Theory and Random Processes*, Dedham, MA: Artech House, 1983.

[6] Barkat, M., *Signal Detection and Estimation*, 2nd ed., Norwood, MA: Artech House, 2005.

[7] Klauder, J. R. et al., "The Theory and Design of Chirp Radars," *Bell Syst. Tech. J.*, vol. 39, no. 4, Jul. 1960, pp. 745–808.

[8] Brookner, E., *Radar Technology*, Dedham, MA: Artech House, 1977.

[9] Skolnik, M. I., ed., *Radar Handbook*, New York: McGraw-Hill, 1970.

Chapter 9

Detection Probability Improvement Techniques

9.1 INTRODUCTION

In Chapters 6 and 8, we derived equations for single-pulse detection probability and showed that the use of a matched filter provides the maximum SNR and P_d that can be obtained for a given set of radar parameters and a given, single transmitted pulse. We termed the resultant SNR and P_d *single-pulse* SNR and P_d. We now want to address the improvement in P_d that we can obtain by using multiple transmit pulses. We will examine four techniques:

- Coherent integration;
- Noncoherent integration;
- m-of-n detection;
- Cumulative probability.

According to a correspondence from David K. Barton,[1] the earliest published mention of coherent integration was in a paper by D. O. North [1] where he discussed coherent integration and some of the problems (that he saw at the time) of implementation. In a 1950 book by Lawson and Uhlenbeck [2], the authors referenced a 1944 MIT Radiation Laboratories report by Emslie titled "Coherent Integration" [3]. In the early 1950s, Lincoln Laboratories developed a pulsed-Doppler radar called Porcupine [4], and in 1956, Westinghouse built an airborne intercept radar using coherent integration in the form of a Doppler processor. By 1957, coherent integration was being widely discussed in the literature [5–8].

Noncoherent integration has apparently been used since the early days of radar since this is the type of integration performed by radar displays such as A-scopes, plan position indicators (PPIs), and the like [2, 9]. However, it appears that the first rigorous treatment of noncoherent integration was presented by Marcum in his seminal paper [10]. Shortly after

---

[1] David K. Barton, private communication to author containing historical notes on coherent integration, cumulative integration, and binary integration, September 15, 2014. Portions are paraphrased in this introduction.

Marcum published his paper, Swerling expanded on Marcum's analyses and considered noncoherent integration of signal returns for his four target fluctuation models [11].

The first publication of a paper on m-of-n detection, which is also known as binary integration, coincidence detection, and dual threshold detection, appears to have been the 1955 paper by Harrington [12]. Papers by Dinneen and Reed in 1956 [13] and Schwartz [14] followed.

Marcum discusses cumulative detection in his 1947 paper [10]. Hall also discusses cumulative detection probability in a 1956 paper [15].

9.2 COHERENT INTEGRATION

With coherent integration, we insert a coherent integrator (a type of signal processor) between the matched filter and amplitude detector, as shown in Figure 9.1. This coherent integrator adds returns (thus the word integrator) from n pulses. After accumulating the n-pulse sum, amplitude detection and the threshold check are performed.

In practice, the process of forming the n-pulse sum is somewhat complicated. In one implementation, the coherent integrator samples the return from each transmit pulse at a spacing equal to the range resolution of the radar. Thus, for example, if we are interested in a range window from 5 to 80 km and have a range resolution of 150 m, the signal processor forms 75,000/150 or 500 complex samples for each pulse return. The coherent integrator stores the 500 samples for each pulse. After it has stored n sets of 500 samples, it sums across n to form 500 sums. In modern, phased array radars with digital signal processors, the summation is accomplished by summers or FFTs. In older, analog radars, the summation (integration) is performed by filters [16] or integrate-and-quench circuits similar to those used in communications receivers.

We will first consider the effects of coherent integration on SNR and then discuss its effect on P_d. As we did in previous chapters, we will separately consider the signal and noise for the SNR analysis and noise and signal-plus-noise for the detection analyses.

9.2.1 SNR Analysis

For the signal, we assume the complex amplitude of the signal on pulse k at the matched filter output is given by

$$s(k) = Se^{j\theta} \tag{9.1}$$

where $S > 0$ is the signal amplitude and θ is the phase. We assume we are looking at the specific range cell—out of the 500 discussed in the above example—that contains the target return. Further, the sample timing corresponds to matched range so that we are sampling the output of the matched filter at its peak (see Chapter 8).

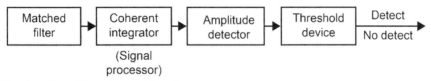

Figure 9.1 Location of the coherent integrator.

The formulation of $s(k)$ in (9.1) carries several assumptions about the target. It implies that the amplitude and phase of the signal returned from the target is constant, at least over the n pulses that are to be integrated. This means we are assuming the target is SW0/SW5, SW1, or SW3. It does not admit SW2 or SW4 targets. As we will show later, coherent integration offers no SNR benefit for SW2 and SW4 targets.

The formulation also implies there is nothing in the radar or environment that would cause the signal amplitude or phase to vary across the n pulses. In particular, the radar and environment must be such that all of the parameters of the radar range equation remain constant across the n pulses. Thus, for example, the antenna beam must be stationary, the transmit power must be constant, the target must be stationary, the radar frequency must be constant, the parameters of the radar receiver must not change, and the environment between the radar and the target must not change.

Another implication of (9.1) is that there is no Doppler on the target return. If the target is moving, its return will have a Doppler frequency and thus a changing phase. This Doppler frequency must be removed by the coherent integrator before the summation takes place. In digital signal processors that use FFTs, Doppler removal is effectively accomplished by the FFT. In analog processors, Doppler is removed through the use of bandpass filters tuned to various Doppler frequencies that cover the range of expected Doppler frequencies, or by mixers before the integrate-and-quench circuits.

It should be noted that not all of the aforementioned constraints can be perfectly satisfied. We account for the fact that some will be violated by including loss terms in the radar range equation. These were discussed in Chapter 5 and will be reviewed later in this chapter.

If we sum over n pulses, the output of the summer will be (for the range cell or sample being investigated)

$$s_{out} = \sum_{k=1}^{n} s(k) = nSe^{j\theta} \tag{9.2}$$

If the signal power at the input to the summer is

$$P_{sin} = S^2 = P_S \tag{9.3}$$

the signal power at the output of the summer will be

$$P_{sout} = n^2 S^2 = n^2 P_S \tag{9.4}$$

In these equations, P_S is the single-pulse signal power from the radar range equation.

We can write the noise at the input to the coherent integrator on the kth pulse as[2]

$$n(k) = \frac{1}{\sqrt{2}}\left[n_I(k) + jn_Q(k)\right] \tag{9.5}$$

Consistent with our previous noise discussions (see Chapters 6 and 8), we assume $\mathbf{n}_I(k)$ and $\mathbf{n}_Q(k)$ are wide sense stationary (WSS), zero mean, and independent. They each have a variance of σ^2. We also assume they are Gaussian.

If we sum the n pulses, the noise at the output of the summer will be

$$n_{out} = \sum_{k=1}^{n} n(k) = \frac{1}{\sqrt{2}}\left[\sum_{k=1}^{n} n_I(k) + j\sum_{k=1}^{n} n_Q(k)\right] = n_{outI} + jn_{outQ} \tag{9.6}$$

The noise power at the output of the summer will be

$$P_{nout} = E\{n_{out}n_{out}^*\} = E\{n_{outI}^2\} + E\{n_{outQ}^2\} \tag{9.7}$$

In (9.7), we made use of the fact that $\mathbf{n}_I(k)$ and $\mathbf{n}_Q(k)$ being independent and zero mean implies that $\mathbf{n}_{outI}$ and $\mathbf{n}_{outQ}$ are independent and zero mean.

We can write

$$E\{n_{outI}^2\} = E\left\{\left[\frac{1}{\sqrt{2}}\sum_{k=1}^{n} n_I(k)\right]\left[\frac{1}{\sqrt{2}}\sum_{l=1}^{n} n_I(l)\right]\right\}$$
$$= \frac{1}{2}\sum_{k=1}^{n} E\{n_I^2(k)\} + \frac{1}{2}\sum_{\substack{l,k\in[1,n] \\ k\neq l}} E\{n_I(k)n_I(l)\} \tag{9.8}$$

Since $\mathbf{n}_I(k)$ is WSS and zero mean,

$$E\{n_I^2(k)\} = \sigma^2 \;\forall k \tag{9.9}$$

We also assume the noise samples are uncorrelated from pulse to pulse.[3] This means $\mathbf{n}_I(k)$ and $\mathbf{n}_I(l)$ are uncorrelated $\forall k \neq l$. Since $\mathbf{n}_I(k)$ and $\mathbf{n}_I(l)$ are also zero mean, we get

$$E\{n_I(k)n_I(l)\} = 0 \;\forall\; k \neq l \tag{9.10}$$

If we use (9.9) and (9.10) in (9.8), we get

---

[2] Recall that the factor of square root of two is used to provide consistent notation between the noise power in IF and baseband signals.
[3] This assumption carries implications about the spacing between pulses relative to the impulse response of the matched filter. Specifically, the spacing between noise samples must be greater than the length of the impulse response of the matched filter. If the matched filter is matched to a rectangular pulse with a width of τ_p, the spacing between noise samples must be greater than τ_p. Since the noise (in a particular range cell—of the 500 of the previous example) is sampled once per pulse, the pulses must be spaced more than one pulsewidth apart. This is easily satisfied in pulsed radars since the pulses can never be spaced by less than one pulsewidth. This is discussed in more detail in Appendix 9A.

$$E\{n_{outI}^2\} = \frac{n\sigma^2}{2} = \frac{nP_{nin}}{2} \tag{9.11}$$

where P_{nin} is the noise power at the output of the matched filter (the "single-pulse" noise term from the radar range equation with $B=1/\tau_p$; see Chapters 2 and 4).

By similar reasoning, we have

$$E\{n_{outQ}^2\} = \frac{n\sigma^2}{2} = \frac{nP_{nin}}{2} \tag{9.12}$$

and, from (9.7),

$$P_{nout} = E\{n_{outI}^2\} + E\{n_{outQ}^2\} = nP_{nin} \tag{9.13}$$

If we combine (9.4) and (9.13), we find that the SNR at the output of the coherent integrator is

$$SNR_{out} = \frac{P_{sout}}{P_{nout}} = \frac{n^2 P_S}{nP_{nin}} = n(SNR) \tag{9.14}$$

or n times the SNR at the output of the matched filter (the SNR given by the radar range equation). With this we conclude the coherent integrator provides an SNR *gain*, or SNR *improvement*, of n.

If the target is SW2 or SW4, coherent integration does not increase SNR. This stems from the fact that, for SW2 and SW4 targets, the signal is not constant from pulse to pulse but, instead, behaves like noise. This means we must treat the target signal the same as we do noise. Thus, in place of (9.2), we would write

$$s_{out} = \sum_{k=1}^{n} s(k) = \frac{1}{\sqrt{2}}\left[\sum_{k=1}^{n} s_I(k) + j\sum_{k=1}^{n} s_Q(k)\right] = s_{outI} + js_{outQ} \tag{9.15}$$

Following the procedure we used for the noise case, we have

$$E\{s_{outI}^2\} = E\{s_{outQ}^2\} = \frac{nP_S}{2} \tag{9.16}$$

and

$$P_{sout} = E\{s_{outI}^2\} + E\{s_{outQ}^2\} = nP_S \tag{9.17}$$

This leads to the result

$$SNR_{out} = \frac{P_{sout}}{P_{nout}} = \frac{nP_S}{nP_{nin}} = SNR \tag{9.18}$$

In other words, the SNR at the coherent integrator output would be the same as the SNR at the matched filter output.

9.2.2 Detection Analysis

We have addressed the signal power, the noise power, and the SNR at the output of the coherent integrator. In order to compute P_d, we need to consider the forms of the density functions of the noise and signal plus noise at the output of the signal processor. We address the noise first.

From (9.6), we have

$$\frac{1}{\sqrt{2}}\left[\sum_{k=1}^{n}\mathbf{n}_I(k) + j\sum_{k=1}^{n}\mathbf{n}_Q(k)\right] = \mathbf{n}_{outI} + j\mathbf{n}_{outQ} \qquad (9.19)$$

We already made the assumption that the $\mathbf{n}_I(k)$ and $\mathbf{n}_Q(k)$ are independent, zero-mean, Gaussian random variables with equal variances of σ^2. This means $\mathbf{n}_{outI}$ and $\mathbf{n}_{outQ}$ are zero-mean, Gaussian random variables and have variances of $n\sigma^2/2$. They are also independent. This is exactly the same as the conditions we had on the I and Q components of noise in the single-pulse case. This means the density of the noise magnitude, $\mathbf{N}_{out}$, at the detector output will be of the form of (6.14) (Chapter 6), and the P_{fa} equation is given by (6.104). They will differ in that the σ^2 in these two equations will be replaced by $n\sigma^2$. The specific equations are

$$f_N(N) = \frac{N}{n\sigma^2}e^{-N^2/2n\sigma^2}U(N) \qquad (9.20)$$

and

$$P_{fa} = e^{-T^2/2n\sigma^2} = e^{-TNR} \qquad (9.21)$$

where TNR is the threshold to noise ratio used in the detection logic (see Chapter 6).

We now turn our attention to signal plus noise. For the SW0/SW5 target, we can write the signal-plus-noise voltage at the coherent integrator output as

$$\mathbf{v}_{out} = \frac{1}{\sqrt{2}}\left[\sum_{k=1}^{n}\mathbf{v}_I(k) + j\sum_{k=1}^{n}\mathbf{v}_Q(k)\right] = \mathbf{v}_{outI} + j\mathbf{v}_{outQ} \qquad (9.22)$$

where each of the $\mathbf{v}_I(k)$ and $\mathbf{v}_Q(k)$ are independent, Gaussian random variables with equal variances of σ^2. The mean of $\mathbf{v}_I(k)$ is $S\cos\theta$, and the mean of $\mathbf{v}_Q(k)$ is $S\sin\theta$ (see Section 6.4 of Chapter 6). With this $\mathbf{v}_{outI}$ and $\mathbf{v}_{outQ}$ are also Gaussian. Their variances are equal to $n\sigma^2$ and their means are $nS\cos\theta$ and $nS\sin\theta$. They are also independent. In this case, the density of the signal-plus-noise magnitude, $\mathbf{v}_{out}$, at the detector output is of the form given in (6.75) with S replaced by nS and σ^2 replaced by $n\sigma^2$. With this we conclude P_d is given by (6.116) with SNR replaced by

$$SNR_{out} = \frac{(nS)^2}{2n\sigma^2} = n(SNR) \qquad (9.23)$$

where SNR is the single-pulse SNR given by the radar range equation. Specifically, we have

$$P_d = Q_1\left(\sqrt{2n(SNR)}, \sqrt{-2\ln P_{fa}}\right) \tag{9.24}$$

where, from Chapter 6, $Q_1(a,b)$ is the Marcum Q function.

For the SW1 and SW3 target, we need to take an approach similar to that used in Chapter 6 for SW3 targets. For SW1 and SW3 targets, the signal amplitude, S, and phase, θ, are constant across the n pulses that are coherently integrated. However, the amplitude of the group of pulses, termed the coherent dwell, is governed by the SW1 or SW3 amplitude fluctuation density [see (6.40) and (6.49)]. The phase of the group of pulses is governed by the uniform probability density function as discussed in Chapter 6. This means that, during the n pulses, the signal plus noise for SW1 and SW3 targets is the same form as for the SW0/SW5 target. That is, $v_I(k)$ and $v_Q(k)$ are independent, Gaussian random variables with variances of σ^2 and means of $S\cos\theta$ and $S\sin\theta$. This implies that the densities of v_{outI} and v_{outQ}, given that S and θ are fixed, are also Gaussian, but with variances of $n\sigma^2/2$ and means of $nS\cos\theta$ and $nS\sin\theta$. This was the same form of the conditional density presented in Chapter 6. If we follow this argument through and follow the procedure of Chapter 6, we can derive the density function of the magnitude of v_{out} as

$$f_V(V) = \frac{V}{n^2 P_S + n\sigma^2} e^{-V^2/2(n^2 P_S + n\sigma^2)} U(V) \tag{9.25}$$

for the SW1 target and

$$f_V(V) = \frac{2V}{\left(2n\sigma^2 + n^2 P_S\right)^2}\left[2n\sigma^2 + \frac{n^2 P_S V^2}{\left(2n\sigma^2 + n^2 P_S\right)}\right]e^{-V^2/\left(2n\sigma^2 + n^2 P_S\right)} U(V) \tag{9.26}$$

for the SW3 target.

By performing the appropriate integrations, we can show the equations for P_d are of the same form as (6.123) and (6.127) with SNR replaced by $nSNR$. In particular,

$$P_d = \exp\left(\frac{\ln P_{fa}}{n(SNR)+1}\right) \tag{9.27}$$

for SW1 targets and

$$P_d = \left(1 - \frac{2n(SNR)\ln P_{fa}}{\left[2+n(SNR)\right]^2}\right) e^{2\ln P_{fa}/\left[2+n(SNR)\right]} \tag{9.28}$$

for SW3 targets.

For a SW2 target, the signal-plus-noise, $[v_I(k)+jv_Q(k)]\sqrt{2}$, is independent from pulse to pulse (across the n pulses). Further, $v_I(k)$ and $v_Q(k)$ are zero mean and Gaussian with variances of $P_s + \sigma^2$ [see (6.57)]. Their sums are also zero mean and Gaussian, but have variances of $n(P_s + \sigma^2)/2$. This means the magnitude of v_{out} has the density

$$f_V(V) = \frac{V}{n(P_S + \sigma^2)} e^{-V^2/2n(P_S+\sigma^2)} U(V) \qquad (9.29)$$

By performing the appropriate integration, we find P_d is as given by (6.123), with SNR equal to the single-pulse SNR. In other words, the coherent integrator does not improve detection probability.

Derivation of a similar result for SW4 targets is not as easy as for SW2 targets because we cannot claim that $\mathbf{v}_I(k)$ and $\mathbf{v}_Q(k)$ are Gaussian for SW4 targets. This means we cannot easily find the density functions of the coherent integrator output, $\mathbf{v}_{outI}$ and $\mathbf{v}_{outQ}$, for the SW4 target. Without these density functions, we cannot compute P_d. As a consequence, we have no rigorous mathematical basis for claiming that coherent integration will or will not improve P_d for a SW4 target. The standard assumption appears to be that, like SW2 targets, coherent integration offers no P_d improvement for SW4 targets.

In the above development, we made some ideal assumptions concerning the target, radar, and environment based on the fact that we were collecting and summing returns from a sequence of n pulses. In particular, we assumed the target amplitude was constant from pulse to pulse. Further, we assumed that we sampled the output of the matched filter at its peak. In practice, neither of these is strictly true. First, we really cannot expect to sample the matched filter output at its peak. Because of this, the SNR in the P_d equations will not be the peak SNR at the matched filter output (the SNR given by the radar range equation). It will be some smaller value. We usually account for this by degrading SNR by a factor we call *range straddling loss* [17, p. 236] (see Chapter 5). If the samples (the 500 samples of the aforementioned example) are spaced one range resolution cell apart, the range straddling loss is usually taken to be 3 dB.

There are other reasons that the signal into the coherent integrator will vary. One is target motion. This will create a Doppler frequency, which will cause phase variations from pulse to pulse (which translate to amplitude variations in the I and Q components). If the Doppler frequency is large enough to cause large phase variations, the gain of the coherent integrator will be nullified. In general, if the Doppler frequency is greater than about PRF/n, the coherent integration gain will be nullified. In fact, the coherent integration could result in an SNR *reduction*. Doppler frequency offsets can be circumvented by using banks of coherent integrators that are tuned to different Doppler frequencies. This is usually accomplished by FFTs in digital signal processors and bandpass filters in analog processors. However, even in this case, the SNR at the output of the coherent integrator will not be the peak SNR because the coherent integrator cannot usually be perfectly matched to the target Doppler frequency. We account for this reduction in SNR by including a *Doppler straddle loss* (see Chapter 5). Doppler straddle loss usually ranges from 0.3 to 1 dB.

Another degradation related to Doppler is termed *range gate walk*. Because of the nonzero range rate, the target signal will move relative to the time location of the various samples fed to the coherent integrator. This means that, over the n pulses, the signal amplitude will change. As indicated above, this could result in a degradation of SNR at the output of the coherent integrator. In practical radars, designers take steps to avoid range walk by not integrating too many pulses. Unavoidable range walk is usually accounted for by including a small (0.1 to 0.2 dB) SNR degradation (SNR loss). Also, if the radar computer has some knowledge of target range rate, it can adjust the range samples to account for range walk. This is reasonably easy to accomplish when the radar is tracking. It may be more difficult during search.

Still another factor that causes the signal amplitude to vary is the fact that the coherent integration may take place while the radar scans its beam across the target. The scanning beam will cause the G_T and G_R terms in the radar range equation to vary across the n pulses that are coherently integrated. As before, this will degrade the SNR, and its effects are included in what is termed a *beamshape loss* [17, p. 493] (Chapter 5). This loss, or degradation, is usually 1 to 3 dB in a well-designed radar.

Phased array radars have a similar problem. For phased array radars, the beam does not move continuously (in most cases), but in discrete steps. This means the phased array radar may not point the beam directly at the target. In turn, the G_T and G_R of the radar range equation will not be their maximum values. As with the other cases, this degradation is accommodated through the inclusion of a beamshape loss term (see Chapter 5).

9.3 NONCOHERENT INTEGRATION

We now want to discuss noncoherent, video, or post-detection integration. The term *post-detection integration* derives from the fact that the integrator, or summer, is placed after the amplitude or square-law detector, as shown in Figure 9.2. The term *noncoherent integration* derives from the fact that since the signal has undergone amplitude or square law detection, the phase information is lost. The synonym *video* appears to be a carryover from older radars and refers to the video displayed on PPIs, A-scopes, and the like. The noncoherent integrator operates in the same fashion as the coherent integrator in that it sums the returns from n pulses before performing the threshold check. However, where the coherent integrator operates on the output of the matched filter, the noncoherent integrator operates on the output of the amplitude detector.

A noncoherent integrator can be implemented in several ways. In older radars, it was implemented via the persistence on displays plus the integrating capability of a human operator. These types of noncoherent integrators are very difficult to analyze and will not be considered here. The reader is referred to [2, 9, 16].

A second implementation is termed an *m-of-n* detector and uses more of a logic circuit rather than a device that integrates. Simply stated, the radar examines the output of the threshold device for n pulses. If a DETECT is declared on any m or more of those n pulses, the radar declares a target detection. This type of implementation is also termed a *dual threshold detector* or a *binary integrator* [18–20]. We will consider this type of noncoherent integrator later in this chapter.

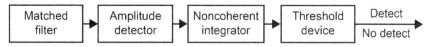

Figure 9.2 Location of the noncoherent integrator.

The third type of noncoherent integrator is implemented as a summer or integrator. In older radars, lowpass filters were used to implement them. In newer radars, they are implemented in special-purpose hardware or the radar computer as digital summers.

In a fashion similar to coherent integration, the noncoherent integrator samples the (amplitude detected) return from each transmit pulse at a spacing equal to the range resolution of the radar. Repeating the previous example, if we are interested in a range window from 5 to 80 km and have a range resolution of 150 m, the noncoherent integrator forms 75,000/150 or 500 samples for each pulse return. The noncoherent integrator stores the 500 samples for each pulse. After it has stored n sets of 500 samples, it sums across n to form 500 sums.

For SW0/SW5, SW1, and SW3 targets, the main advantage of a noncoherent integrator over a coherent integrator is hardware simplicity. As indicated in earlier discussions, coherent integrators must contend with the effects of target Doppler. In terms of hardware implementation, this usually translates to increased complexity of the coherent integrator. Specifically, it is usually necessary to implement a bank of coherent integrators that are tuned to various Doppler frequencies. Because of this, the radar will need a number of integrators equal to the number of range cells in the search window multiplied by the number of Doppler bands needed to cover the Doppler frequency range of interest. Although not directly stated earlier, this will also require a larger number of amplitude (or square-law) detectors and threshold devices, since one will be required for each Doppler sample of each range cell.

Since the noncoherent integrator is placed after the amplitude detector, it does not need to accommodate multiple Doppler frequencies. This lies in the fact that the amplitude detection process recovers the signal (plus noise) amplitude without regard to phase (i.e., Doppler). Because of this, the number of integrators is reduced; it is equal to the number of range cells in the search window.

Recall that coherent integration offers no improvement in detection probability for SW2 or SW4 targets. In fact, it can degrade detection probability relative to that which can be obtained from a single pulse. In contrast, noncoherent integration can offer significant improvement in detection probability relative to a single pulse. It is interesting to note that some radar designers use various schemes, such as frequency hopping, to force targets to exhibit SW2 or SW4 characteristics and exploit the significant detection probability improvement offered by noncoherent integration [21, 22].

Analysis of noncoherent integrators is much more complicated than analysis of coherent integrators because the integration takes place after the nonlinear process of amplitude or square-law detection. From our previous work in Chapter 6, we note that the density functions of the magnitude of noise and signal-plus-noise are somewhat complicated. More importantly, they are not Gaussian. Therefore, when we sum the outputs from successive pulses, we cannot conclude that the density function of the sum of signals will be Gaussian (as we can if the density function of each term in the sum was Gaussian). In fact, the density functions become very complicated. This has the further ramification that the computation of P_{fa} and P_d becomes very complicated. Analysts such as DiFranco and Rubin, Marcum, Swerling, and Meyer and Mayer have devoted considerable energy to analyzing noncoherent integrators and documenting the results of these analyses [10, 11, 23, 24]. We will not attempt to duplicate the analyses here; instead, we present the results of their labor.

An equation for P_{fa} at the output of an *n*-pulse, noncoherent integrator is

$$P_{fa} = 1 - \Gamma(n, TNR) \tag{9.30}$$

where $\Gamma(n, TNR)$ is the incomplete gamma function [25, p. 112] defined by[4]

$$\Gamma(a, x) = \frac{1}{\Gamma(a)} \int_0^x e^{-t} t^{a-1} dt \tag{9.31}$$

For $a = n$, where *n* is a positive integer, $\Gamma(n)$ becomes the factorial operation [26, p. 98]. That is

$$\Gamma(n) = (n-1)! = (n-1)(n-2)\cdots \times 2 \times 1 \tag{9.32}$$

Many modern software packages, such as MATLAB and Mathcad®, include the incomplete gamma function in their standard library. These software packages also have the inverse incomplete gamma function, which is necessary for determining *TNR* for a given P_{fa}. Specifically,

$$TNR = \Gamma^{-1}(n, 1 - P_{fa}) \tag{9.33}$$

where $\Gamma^{-1}(n, 1 - P_{fa})$ is the inverse of the incomplete gamma function.

The P_d equations for the five target types we have studied are

SW0/SW5:

$$P_d = Q_1\left(\sqrt{2n(SNR)}, \sqrt{2(TNR)}\right) \\ + e^{-TNR - n(SNR)} \sum_{r=2}^{n} \left(\frac{TNR}{n(SNR)}\right)^{(r-1)/2} I_{r-1}\left(2\sqrt{(TNR)n(SNR)}\right) \tag{9.34}$$

SW1:

$$P_d = 1 - \Gamma(n-1, TNR) \\ + \left(1 + \frac{1}{n(SNR)}\right)^{n-1} \Gamma\left(n-1, \frac{TNR}{1 + 1/[n(SNR)]}\right) e^{-TNR/[1+n(SNR)]} \tag{9.35}$$

SW2:

$$P_d = 1 - \Gamma\left(n, \frac{TNR}{1+SNR}\right) \tag{9.36}$$

SW3:

$$P_d \approx \left(1 + \frac{2}{n(SNR)}\right)^{n-2} \left[1 + \frac{TNR}{1 + n(SNR)/2} - \frac{2(n-2)}{n(SNR)}\right] e^{-TNR/[1+n(SNR)/2]} \tag{9.37}$$

---

[4] Some forms/implementations of the incomplete gamma function omit the $1/\Gamma(a)$ term.

SW4:

$$P_d = 1 - \left(\frac{SNR}{SNR+2}\right)^n \sum_{k=0}^{n} \frac{n!}{k!(n-k)!} \left(\frac{SNR}{2}\right)^{-k} \Gamma\left(2n-k, \frac{2TNR}{SNR+2}\right) \quad (9.38)$$

In the above, $Q_1(a,b)$ is the Marcum Q function, $I_r(x)$ is the modified Bessel function of the first kind and order r [26, p. 104], and $\Gamma(a,x)$ is the aforementioned incomplete gamma function. TNR is the threshold-to-noise ratio and is computed from (9.33).

SNR in the above equations is the *single-pulse* SNR defined by the radar range equation (see Chapter 2).

With the exception of (9.37), (9.34) through (9.38) are exact equations. Equation (9.37) is usually taken to be an exact equation, but is actually an approximation, as indicated by the use of ≈ instead of =. An exact equation for the SW3 case can be found in the appendix of [24][5].

In a paper [27] and his recent books [17, 25], Barton provides a set of "universal" equations for the SW1 through SW4 cases. He attributes the original formulation of these equations to the Russian author, P. A. Bakut [28]. In an internal memo, the universal equation for P_d, as modified by Hardaker[4] and the authors, is

$$P_d \approx K_m\left(2\left[\frac{TNR - n + n_e}{(n/n_e)(SNR) + 1}\right], 2n_e\right) \quad (9.39)$$

where

$$K_m(2x, 2k) = 1 - \Gamma(k, x) \quad (9.40)$$

and TNR is computed from (9.33). SNR is the single-pulse SNR.

The integer, n_e is the number of degrees of freedom associated with the different Swerling target types (see Chapter 3). This stems from Swerling's definition of his four target types or, more accurately, signal fluctuation models [11, 27]. Specifically, he defined four signal fluctuation models whose amplitude statistics are governed by a chi-square density function having $2n_e$ degrees of freedom (DOF) (see Chapter 3). The four values of n_e associated with the four Swerling target types are:

- SW1, $n_e = 1$
- SW2, $n_e = n$
- SW3, $n_e = 2$
- SW4, $n_e = 2n$

where n is the number of pulses noncoherently integrated.

Barton also gives a universal equation for determining the single-pulse SNR required to provide a desired P_d. This equation is quite useful. Before its introduction, the single-pulse SNR was found by using a root solver in conjunction with the exact equations of

---

[5] We have not tested the conditions under which (9.37) is a valid approximation. That would be an interesting exercise.

(9.35) to (9.38). The "inverse" universal equation, again as modified by Hardaker and the authors, is

$$SNR \approx \left[\frac{2(TNR) - 2(n - n_e)}{K_m^{-1}(P_d, 2n_e)} - 1\right]\left(\frac{n_e}{n}\right) \quad (9.41)$$

where

$$K_m^{-1}(x, 2k) = 2\Gamma^{-1}(k, (1-x)) \quad (9.42)$$

and $\Gamma^{-1}(k,z)$ is the inverse of the incomplete gamma function.

Barton compared the universal equations to the exact equations for several values of P_{fa} and a range of SNRs and n [27]. His results indicate that the universal equations are quite accurate for P_d greater than about 0.2 and P_{fa} less than about 10^{-4}. As an interesting note, the universal equation is exact for SW2 targets.

The universal equations are not recommended for SW0/SW5 targets. However, in his 2005 book [17, pp. 42–53], Barton provides an approximation to the exact equation of (9.34), along with its inverse. Those equations are, using Barton's notation

$$P_d(S, P_{fa}, n) = Q\left[Q^{-1}(P_{fa}) - \sqrt{\frac{2nS^2}{S + 2.3}}\right] \quad (9.43)$$

and

$$S(P_d, P_{fa}, n) = \frac{S_1(P_d, P_{fa})}{2n}\left[1 + \sqrt{1 + \frac{9.2n}{S_1(P_d, P_{fa})}}\right] \quad (9.44)$$

where

$$S_1(P_d, P_{fa}) = \frac{1}{2}\left[Q^{-1}(P_{fa}) - Q^{-1}(P_d)\right]^2 \quad (9.45)$$

and

$$Q(x) = \frac{1}{\sqrt{2\pi}}\int_x^\infty e^{-t^2/2}dt = \frac{1}{2}\text{erfc}\left(\frac{x}{\sqrt{2}}\right) \quad (9.46)$$

erfc(x) is the complementary error function [29, p. 214] and $Q^{-1}(x)$ is its inverse, erfc$^{-1}(x)$. Both of these functions are included as standard functions in software packages such as MATLAB and Mathcad.

In an internal memo,[6] Hardaker recasts (9.43) through (9.45) in a form directly in terms of erfc(x) and erfc$^{-1}(x)$. These are

---

[6] David A. Hardaker, Dynetics Inc. internal memo, *Application of Barton's Universal Equations for Radar Target Detection*, September 15, 2014.

$$P_d = \frac{1}{2}\text{erfc}\left[\text{erfc}^{-1}(2P_{fa}) - \sqrt{\frac{n(SNR)^2}{SNR + 2.3}}\right] \quad (9.47)$$

for finding P_d in terms of *SNR*, and

$$SNR = \frac{S_1(P_d, P_{fa})}{2n}\left[1 + \sqrt{1 + \frac{9.2n}{S_1(P_d, P_{fa})}}\right] \quad (9.48)$$

for determining the required single-pulse *SNR* for a given P_d. In (9.48),

$$S_1(P_d, P_{fa}) = \left[\text{erfc}^{-1}(2P_{fa}) - \text{erfc}^{-1}(2P_d)\right]^2 \quad (9.49)$$

The noncoherent P_d equations discussed herein are based on the assumption that the amplitude detector of Figure 9.2 is a square-law detector.[7] According to Meyer and Mayer [24], Marcum [30] considered the effect on P_d of using an amplitude (linear) detector instead of a square-law detector. Marcum showed that the P_d performance using either detector was very similar (~0.2-dB difference) for a constant RCS target (SW0/SW5 target). It is not clear whether Swerling or other analysts have performed such a comparison for other Swerling target types. However, it is commonly accepted that the P_d equations developed for the square law detector also apply to the case where the radar use a linear detector.

9.3.1 Coherent and Noncoherent Integration Comparison

Figures 9.3 through 9.7 contain plots that provide a comparison of coherent integration, noncoherent integration and single-pulse operation for the five Swerling target models. The figures contain plots of P_d versus required single-pulse SNR. Figures 9.3, 9.4, and 9.6 contain plots for the single-pulse case and two sets of two plots for the cases of coherent and noncoherent integration of 10 and 100 pulses, for the SW0/SW5, SW1 and SW3 cases. Figures 9.5 and 9.7 contain plots for the single-pulse case and noncoherent integration of 10 and 100 pulses for the SW2 and SW4 cases. Coherent integration was not considered for the SW2 and SW4 targets since we already concluded that coherent integration offers no improvement in P_d for these two target types.

All of the plots were generated for a P_{fa} of 10^{-6}. As a reminder, for the coherent and noncoherent integrators, this is the P_{fa} at the output of the detector that follows the integrators. For the single-pulse case, it is the P_{fa} for a single detection attempt (i.e., no integration). The "required single-pulse SNR" label on the horizontal axis means that this is the SNR required at the matched filter output to achieve the indicated P_d at the output of the threshold device that follows the coherent or noncoherent integrator.

---

[7] For analysis, a square-law detector is typically assumed because the resulting mathematical analysis tends to be more tractable.

As expected, Figures 9.3, 9.4, and 9.6 show that, with coherent integration of 10 pulses, the required single-pulse SNR is 10 dB lower than when only a single pulse is used. For coherent integration of 100 pulses, the required single-pulse SNR is 20 dB lower.

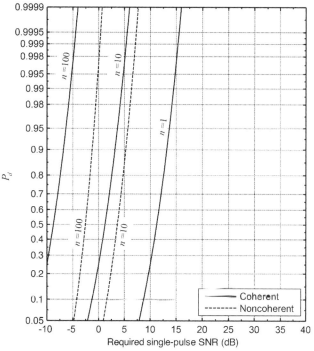

Figure 9.3 Plots of desired P_d vs. required single-pulse SNR for a SW0/SW5 target and coherent and noncoherent integration of 10 and 100 pulses—$P_{fa} = 10^{-6}$.

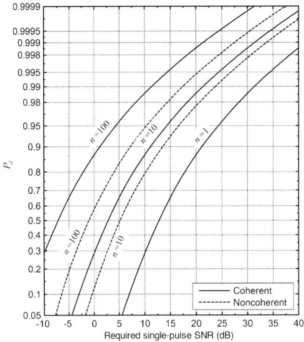

Figure 9.4 Plots of desired P_d vs. required single-pulse SNR for a SW1 target and coherent and noncoherent integration of 10 and 100 pulses—$P_{fa} = 10^{-6}$.

Table 9.1
Reduction in Required Single-Pulse SNR with Noncoherent Integration

	P_d	Reduction in Required Single-Pulse SNR	
		10 Pulses Integrated	100 Pulses Integrated
SW0/SW5	0.5	7.6	13.9
	0.9	7.9	14.4
	0.99	8.1	14.8
SW1	0.5	7.6	13.8
	0.9	7.7	13.9
	0.99	7.2	14.8
SW3	0.5	7.6	13.8
	0.9	7.7	13.9
	0.99	7.6	13.9

For noncoherent integration, the reduction in required single-pulse SNR depends on the number of pulses noncoherently integrated and the desired P_d after integration. Examples of the reduction for the three target types (SW0/SW5, SW1, and SW3) are contained in Table 9.1. As indicated, the values range from 7 to 8 dB for noncoherent integration of 10 pulses and 14 to 15 dB for noncoherent integration of 100 pulses. This relation leads to a useful rule of thumb for the reduction in required single-pulse SNR for noncoherent integration. Specifically, the reduction is

$$I(n) \approx 7.5\log(n) \quad \text{dB} \tag{9.50}$$

Some authors term $I(n)$ noncoherent integration gain [31, 32]. For preliminary calculation of P_d, they suggest adding $I(n)$ to the single-pulse SNR (from the radar range equation) and using it in the single-pulse P_d equation to compute P_d at the output of the noncoherent integrator (for SW0/SW5, SW1, and SW3 targets).

The curves for the SW2 (Figure 9.5) and SW4 (Figure 9.7) indicate that noncoherent integration can offer significant reductions in single-pulse SNR requirements when compared to basing detection on only a single pulse. For example, for a SW2 target and 10 pulses integrated, the reduction is about 15 dB for a desired P_d of 0.9. This increases to 23 dB for a desired P_d of 0.99. For 100 pulses integrated, the reductions are 22 and 31 dB for the two P_d cases. This is a significant reduction in single-pulse SNR requirements and is a reason for radar designers to try to arrange for aircraft targets to appear as SW2 targets to the radar, which they can do by changing RF (see Chapter 3).

The reduction in single-pulse SNR requirements is not as dramatic for the SW4 case, but they are still significant, as indicated by Figure 9.7.

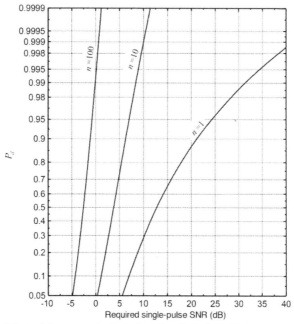

Figure 9.5 Plots of desired P_d vs. required single-pulse SNR for a SW2 target and noncoherent integration of 10 and 100 pulses—$P_{fa} = 10^{-6}$.

244 Basic Radar Analysis

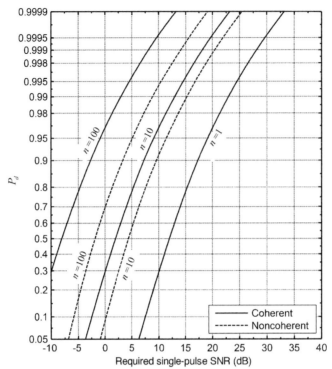

Figure 9.6 Plots of desired P_d vs. required single-pulse SNR for a SW3 target and coherent and noncoherent integration of 10 and 100 pulses—$P_{fa} = 10^{-6}$.

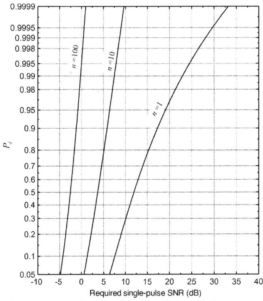

Figure 9.7 Plots of desired P_d vs. required single-pulse SNR for a SW4 target and noncoherent integration of 10 and 100 pulses—$P_{fa} = 10^{-6}$.

9.3.2 Detection Example with Coherent and Noncoherent Integration

For this example, we consider the radar of the example in Section 6.6.2. The radar parameters are listed in Table 2.2 of Chapter 2 and are repeated in Table 9.2. We have added some specific antenna parameters since we will need to use them in this example. We assume the radar has a fan beam (see Figure 6.9) with a peak directivity of 32 dB. The azimuth beamwidth is 1.3°.

Table 9.2
Radar Parameters for Coherent and Noncoherent Integration Example

Parameter	Value
Peak transmit power at the power amp output	50 kW
Operating frequency	2 GHz
PRF	1,000 Hz
Pulsewidth—τ_p	100 µs
Pulse modulation bandwidth	1 MHz
Antenna directivity	32 dB
Elevation beamwidth	Fan beam
Azimuth beamwidth	1.3°
Total losses—excluding beamshape loss	13 dB
Noise figure—referenced to the antenna feed	5 dB
Antenna rotation rate	6 rpm
Instrumented range	PRI $- 2\tau_p - 50$ µs
False alarm criterion	No more than one false alarm per 360° rotation

We want to generate a plot of detection probability versus target range for a 0.1 m², SW1 target. We will assume cases where the radar coherently and noncoherently integrates the number of pulses received as the beam scans by the target. We will assume the elevation of the target is such that it is at the peak of the antenna beam in elevation.

To start, we need to find the single-pulse SNR for the case where the radar beam is pointed directly at the target. We use the radar range equation of Chapter 2 to obtain this. That is,

$$SNR = \frac{P_t G_T G_R \lambda^2 \sigma \tau_p}{(4\pi)^3 R^4 k T_0 F L} = \frac{(50 \times 10^3)(10^{3.2})(10^{3.2})(0.15)^2 (0.1)(100 \times 10^{-6})}{(4\pi)^3 R^4 (4 \times 10^{-21})(10^{0.5})(10^{1.3})}$$

$$= \frac{5.64 \times 10^{19}}{R^4} \qquad (9.51)$$

We next need to compute the number of pulses that can be coherently or noncoherently integrated. We said this would be the number of pulses received as the beam scans (in azimuth) across the target. The standard way to compute this is to see how many pulses are in the 3-dB azimuth beamwidth, which is 1.3° in this example.

The antenna rotation rate is 6 rpm or 6 × 360° per minute. This gives a scan rate of

$$F_{scan} = (6 \text{ rev/min} \times 360°/\text{rev})/60 \text{ sec/min} = 36°/\text{sec} \qquad (9.52)$$

The time for the antenna to travel one beamwidth is

$$\begin{aligned} T_{beam} &= (\theta_{BW}°/\text{beam})/F_{scan} = (1.3°/\text{beam})/(36°/\text{sec}) \\ &= 0.036 \text{ sec/beam} \end{aligned} \qquad (9.53)$$

The waveform PRF is 1,000 Hz, which means the radar transmits (and receives) 1,000 pulses per second. We can use this to compute the number of pulses per beam as

$$n = T_{beam} \times PRF = (0.036 \text{ sec/beam})(1,000 \text{ pulses/sec}) = 36 \text{ pulses/beam} \qquad (9.54)$$

This tells us that we can coherently or noncoherently integrate about 36 pulses as the beam scans by the target. Thus, this is the n we need to include in the appropriate P_d equation.

We note that, as the beam scans by the target, the SNR associated with the 36 pulses will not be constant. As we discussed earlier, we will account for this by incorporating a beamshape loss in the computation of single-pulse SNR. Since we assumed the target was on the peak of the antenna pattern in elevation, we need only account for the variation of SNR due to azimuth scanning. This means that we need to include an additional 1.24 or 1.6 dB loss in the single-pulse SNR calculation [33].[8] We will use 1.6 dB. This reduces the single-pulse SNR we use in the detection calculations to

$$SNR = \frac{5.64 \times 10^{19}}{R^4} \bigg/ 10^{0.16} = \frac{3.9 \times 10^{19}}{R^4} \qquad (9.55)$$

As a note, we will assume the coherent integrator has been tuned to the Doppler frequency of the target. This is an idealization, since we do not know the target Doppler frequency in a search radar. In practice, the coherent integrator would actually consist of many coherent integrators tuned to different Doppler frequencies. As indicated earlier, this complicates the design of the coherent integrator. Also, because of the multiple Doppler channels, with their associated detection circuits, we should adjust the P_{fa} to account for multiple Doppler channels (see Chapter 6). We assume the range straddle loss is already included in the total losses.

The other term we need is the P_{fa} at the output of the detector that follows the integrator and associated amplitude detector. We will assume that the integrator performs a running sum, or integration, and makes a detection decision on every pulse. Thus, we can use the P_{fa} we computed in the detection contour example of Chapter 6. That P_{fa} was 1.33×10^{-7}.

To create plots of P_d versus range, we use the SNR from (9.55) along with $n = 36$ and $P_{fa} = 1.33 \times 10^{-7}$ in (9.27) and (9.35). The specific equations are

$$P_d = \exp\left(\frac{\ln P_{fa}}{n(SNR)+1}\right) \qquad (9.56)$$

for coherent integration and

---

[8] Blake suggests 1.6 dB for 1-D scanning and 3.2 dB for 2-D scanning. Barton suggests 1.24 dB and 2.48 dB, respectively, for a typical radar beam (see Chapter 5 for a discussion of this).

$$P_d = 1 - \Gamma(n-1, TNR)$$
$$+ \left(1 + \frac{1}{n(SNR)}\right)^{n-1} \Gamma\left(n-1, \frac{TNR}{1 + 1/[n(SNR)]}\right) e^{-TNR/[1+n(SNR)]} \quad (9.57)$$

for noncoherent integration.

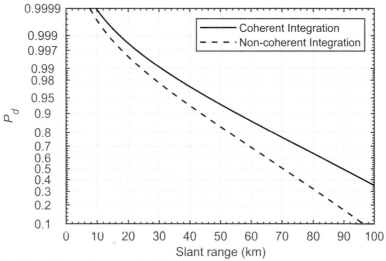

Figure 9.8 Plots of P_d vs. slant range for Example 2.

Figure 9.8 contains plots of P_d versus R for the two integration cases. As expected, the coherent integrator achieves a given P_d at longer ranges. As a specific example, coherent integration gives $P_d = 50\%$ at 90 km, which is not achievable with noncoherent integration until 70 km.

9.4 CUMULATIVE DETECTION PROBABILITY

The third technique we examine for increasing detection probability is the use of multiple detection attempts. The premise behind using multiple detection attempts is that if we attempt to detect the target several times, we will increase the overall detection probability. The question we pose in the cumulative detection problem is

> If we check for a threshold crossing on several occasions, what is the probability that the signal-plus-noise voltage will cross the threshold *at least once*?

Thus, suppose for example we check for a threshold crossing on three occasions. We want to determine the probability of a threshold crossing on any one, two, or three of the occasions. To compute the appropriate probabilities, we must use probability theory. The details are somewhat involved and are presented in Appendix 9C. The main results are as follows.

We assume we have N detection opportunities (i.e., the chance to detect the target on N tries) and that they are independent. This limits when we can use cumulative detection concepts. Specifically, we should use cumulative detection concepts only on a scan-to-scan basis. If we do, we will satisfy the constraints on all of the Swerling target types. Specifically, for SW1 and SW3 targets, the signal-plus-noise samples are, by definition, independent from scan to scan. For SW0/SW5, SW2, and SW4 targets, the signal-plus-noise samples are independent from pulse to pulse and will thus also be independent from scan to scan. Having said this, we must also assure that the coherent or noncoherent integrator does not cause the independence restriction to be violated. The restriction will not be violated if the time between target illuminations is significantly larger than the coherent or noncoherent processing time.

If the detection probability on the kth detection try is P_{dk}, the probability of detecting the target on at least one of the tries is

$$P_{dcum} = 1 - \prod_{k=1}^{N}(1 - P_{dk}) \qquad (9.58)$$

where P_{dcum} is the *cumulative detection probability* over N tries.

In addition to increasing detection probability, the use of cumulative detection techniques also increases false alarm probability. In fact, if we consider the false alarm case, we can express (9.58) as

$$P_{facum} = 1 - \prod_{k=1}^{N}(1 - P_{fak}) \qquad (9.59)$$

In the case of false alarm probability, we usually have that $P_{fak} = P_{fa} \ \forall k \in [1,N]$ and write

$$P_{facum} = 1 - \left(1 - P_{fa}\right)^N \tag{9.60}$$

If we further recall that $P_{fa} \ll 1$, and N is not too large, we can write

$$P_{facum} \approx NP_{fa} \tag{9.61}$$

Equation (9.61) tells us that when we use cumulative detection concepts, we should compute the individual P_{dk} detection probabilities using $P_{fak} = P_{facum}/N$ where P_{facum} is the desired false alarm probability.

As a rule of thumb, one should be careful about invoking cumulative detection concepts in a fashion that allows any P_{dk} to be such that the SNR per scan falls below 10 to 13 dB. If the SNR is below 10 to 13 dB, the radar may not be able to establish track on a target it has detected. If this is stated in terms of P_{dk}, P_{dk} should not be allowed to fall below about 0.5. A reviewer of this book pointed out that this rule of thumb may not be "applicable to a multifunction radar or a system in which a multiple-target phased array tracker is assigned to validate a single detection from an associated search radar." It is assumed that the rationale behind this statement is that even if the detection probability is low (because of losses associated with computing SNR during detection), the SNR on verify may be sufficient to establish track since it may be possible to devote more radar resources to verify, and thus increase SNR.

9.4.1 Cumulative Detection Probability Example

Suppose we have a phased array radar that is performing search. It illuminates the target with a search beam every 20 seconds and transmits a single pulse. We will assume an aircraft-type target, which means we can assume it is a SW1 target. Because of this and the 20 seconds between search illuminations, we can safely assume the signal-plus-noise samples will be independent from look to look.

We will assume the radar achieves detection probabilities of 0.5, 0.51, 0.52, 0.53, and 0.54 on five consecutive search beams (looks). With this we get a cumulative detection probability of

$$\begin{aligned} P_{dcum} &= 1 - \prod_{k=1}^{5}\left(1 - P_{dk}\right) = 1 - (1-0.5)(1-0.51)(1-0.52)(1-0.53)(1-0.54) \\ &= 1 - 0.025 = 0.975 \end{aligned} \tag{9.62}$$

Suppose we want to compute the required SNRs on each look to achieve the various P_{dk} and obtain a P_{facum} of 10^{-6} over the five looks. From (9.61), we get

$$P_{facum} = 10^{-6} \approx NP_{fa} = 5P_{fa} \tag{9.63}$$

This says we must set the detection threshold for each threshold check such that we obtain

$$P_{fa} = \frac{10^{-6}}{5} = 0.2 \times 10^{-6} \tag{9.64}$$

We can then use (6.123) to determine the required SNR values. Specifically, we would have

$$SNR_k = \frac{\ln P_{fa}}{\ln P_{dk}} - 1 \tag{9.65}$$

This would give required SNR values of 13.3, 13.4, 13.5, 13.7, and 13.8 dB.

9.5 M-OF-N DETECTION

We can think of m-of-n detection as an extension of cumulative detection where, instead of requiring one or more detections on n tries, we require m or more detections on the n tries. This is the origin of the term m-of-n detection [34–36].

Since m-of-n detection usually operates on target returns that are closely spaced in time, we cannot necessarily assume independent detection events. However, for the same reason, we can reasonably assume the detection probability will be the same on each detection attempt. That is, $P_{dk} = P_d$.

For the case of SW0/SW5, SW2, and SW4 targets, we can assume the detection events are independent. For SW0/SW5 targets, the randomness in the signal-plus-noise is due only to noise since the signal amplitude is a known constant across the n detection tries. Since we assume the noise is independent from pulse to pulse, the detection events will be independent from try to try.

For SW2 and SW4 targets, we assume both the signal amplitude and noise are random, and independent, from pulse to pulse. Thus, again, the detection events will be independent from try to try.

Based on the discussions of the previous two paragraphs, we can directly extend the cumulative detection discussions to m-of-n detection for SW0/SW5, SW2, and SW4 targets.

For SW1 and SW3 targets, the signal amplitude is constant across the n detection tries. However, it is not a known constant, as was the case for SW0/SW5 targets. Instead, it is a random variable that is governed by (6.40) and (6.49).

To accommodate the fact that the signal amplitude is random, we will approach the SW1 and SW3 m-of-n detection problem by using the approach we used in Chapter 6 to find P_d for SW3 targets. Specifically, we will determine the m-of-n detection probability by assuming a constant, known, signal amplitude across the n detection attempts. We will then form a weighted average across all possible signal amplitudes using the appropriate density function for the Swerling target type being considered. In equation form, we have

$$P_{mofn} = \int_{-\infty}^{\infty} P_{mofn0}(S) f_S(S) dS \tag{9.66}$$

where $P_{mofn0}(S)$ is the m-of-n detection probability for a given signal amplitude and $f_S(S)$ is the amplitude density function associated with the particular Swerling target type of interest. The analyses leading to (9.66) is included in Appendix 9B.

Once we fix the signal amplitude, the density function of signal-plus-noise is the same as the density function of signal-plus-noise for a SW0/SW5 target.

We will first develop the m-of-n detection probability equations applicable to SW0/SW5, SW2, and SW4 targets and then extend the results to SW1 and SW3 targets.

If D_k is the detection event on any one try, the detection event on exactly m of n tries will be (in the following, $\cap$ and $\cup$ are set and probability theory notations for intersection and union – $\cap$ is interpreted as "and" and $\cup$ is interpreted as "or"—see Appendix 9C)

$$\mathcal{D}_{mn} = \left(\bigcap_{m} \mathcal{D}_k\right) \cap \left(\bigcap_{n-m} \bar{\mathcal{D}}_k\right) \tag{9.67}$$

where the first term is the intersection across the m detection events, and the second term is the intersection over the $n-m$ events of a missed detection. As an example, we consider the case of exactly two of three detections. Let the three detection events be D_1, D_2, and D_3 and their corresponding missed detection events be $\bar{D}_1$, $\bar{D}_2$, and $\bar{D}_3$. The event of exactly two of three detections can be either

$$\begin{aligned}\mathcal{D}_a &= (\mathcal{D}_1 \cap \mathcal{D}_2) \cap \bar{\mathcal{D}}_3, \\ \mathcal{D}_b &= (\mathcal{D}_1 \cap \mathcal{D}_3) \cap \bar{\mathcal{D}}_2, \\ \text{or } \mathcal{D}_c &= (\mathcal{D}_2 \cap \mathcal{D}_3) \cap \bar{\mathcal{D}}_1\end{aligned} \tag{9.68}$$

That is, we can have detections on tries 1 and 2, but not 3, and so forth. In this case, we see that there are three ways we can have exactly two of three detection events.

The event consisting of any two of three detection events is the union of D_a, D_b and D_c, or

$$\mathcal{D}_{23} = \mathcal{D}_a \cup \mathcal{D}_b \cup \mathcal{D}_c \tag{9.69}$$

We want to compute

$$P(\mathcal{D}_{23}) = P(\mathcal{D}_a \cup \mathcal{D}_b \cup \mathcal{D}_c) \tag{9.70}$$

We note that the events D_a, D_b, and D_c are mutually exclusive since, for example, if there are detections on tries 1 and 2, but not 3, we cannot have the possibility of detections on tries 1 and 3, but not 2. In other words, the occurrence of any one of D_a, D_b, or D_c precludes the occurrence of any of the others. Since D_a, D_b, and D_c are mutually exclusive, the probability of their union is equal to the sum of their individual probabilities. Thus,

$$P(\mathcal{D}_{23}) = P(\mathcal{D}_a) + P(\mathcal{D}_b) + P(\mathcal{D}_c) \tag{9.71}$$

We now want to examine the individual probabilities on the right side of (9.71). Recall that we assumed the probability of each of the n detection events was the same. For our 2 of 3 example, this means

$$P(\mathcal{D}_1) = P(\mathcal{D}_2) = P(\mathcal{D}_3) = P_d \tag{9.72}$$

We also note that

$$P(\bar{\mathcal{D}}_k) = 1 - P(\mathcal{D}_k) = 1 - P_d \tag{9.73}$$

Given the assumption that D_1, D_2, and D_3 are independent, we have, as an example,

$$P(\mathcal{D}_a) = P(\mathcal{D}_1 \cap \mathcal{D}_2 \cap \bar{\mathcal{D}}_3) = P(\mathcal{D}_1)P(\mathcal{D}_2)P(\bar{\mathcal{D}}_3)$$
$$= P_d P_d (1 - P_d) = P_d^2 (1 - P_d) \tag{9.74}$$

Extending this further, we have that

$$P(\mathcal{D}_a) = P(\mathcal{D}_b) = P(\mathcal{D}_c) = P_d^2 (1 - P_d) \tag{9.75}$$

If we use this in (9.71), we have

$$P_{23} = P(\mathcal{D}_{23}) = 3P_d^2 (1 - P_d) \tag{9.76}$$

If we extend our 2 of 3 example to the general case, we can write the probability of a particular combination of m of n detections occurring as

$$P_a = P(\mathcal{D}_a) = P\left[\left(\bigcap_m \mathcal{D}_k\right) \cap \left(\bigcap_{n-m} \bar{\mathcal{D}}_k\right)\right] = P_d^m (1 - P_d)^{n-m} \tag{9.77}$$

To get all possible combinations of m detections and $n - m$ missed detections we need to ask how many ways we can combine the m D_k detection events and the $(n - m)$ $\bar{D}_k$ missed detection events. For the 2 of 3 case, this was three. For the general case, we turn to combinatorial theory [29] and ask how many ways can m objects be arranged in a string of n objects. The answer is

$$C_m^n = \frac{n!}{m!(n-m)!} \tag{9.78}$$

For our 2 of 3 example, we have

$$C_2^3 = \frac{3!}{2!(3-2)!} = 3 \tag{9.79}$$

Given (9.78), we find that the probability of having exactly m detections and $n - m$ missed detections in n tries is

$$P_{mn} = C_m^n P_d^m (1 - P_d)^{n-m} = \frac{n!}{m!(n-m)!} P_d^m (1 - P_d)^{n-m} \tag{9.80}$$

In our original problem statement, we said we wanted the probability of obtaining detections on at least m of n tries. Said another way, we want to find the probability of obtaining detections on m or $m+1$ or $m+2$, ... of n tries. Thus, we want to find the probability of

$$\mathcal{D}_{mofn} = \mathcal{D}_{mn} \cup \mathcal{D}_{(m+1)n} \cup \mathcal{D}_{(m+2)n} \cdots \cup \mathcal{D}_{nn} \tag{9.81}$$

Since the event of having exactly m detections and $n-m$ missed detections precludes the possibility of having, say, exactly r detections and $n-r$ missed detections, all of the events of (9.81) are mutually exclusive. Thus,

$$P(\mathcal{D}_{mofn}) = \sum_{k=m}^{n} P(\mathcal{D}_{kn}) \tag{9.82}$$

or

$$P_{mofn} = \sum_{k=m}^{n} P_{kn} \tag{9.83}$$

Substituting (9.80) into (9.83) gives our final answer of

$$P_{mofn} = \sum_{k=m}^{n} \frac{n!}{k!(n-k)!} P_d^k (1-P_d)^{n-k} \tag{9.84}$$

As indicated earlier, (9.84) does not directly apply to SW1 and SW3 targets. For these targets, the appropriate equation is

$$P_{mofn} = \int_{-\infty}^{\infty} P_{mofn\_SW0}(S) f_{\mathbf{S}}(S) dS \tag{9.85}$$

In this equation, $P_{mofn\_SW0}(S)$ is the m-of-n detection probability for a SW0/SW5 target as a function of signal amplitude, S, and $f_S(S)$ is the density function of the signal amplitude for a SW1 or SW3 target. Specifically, for a SW1 target

$$f_{\mathbf{S}}(S) = \frac{S}{P_s} e^{-S^2/2P_s} U(S) \tag{9.86}$$

and for the SW3 target

$$f_{\mathbf{S}}(S) = \frac{2S^3}{P_s^2} e^{-S^2/P_s} U(S) \tag{9.87}$$

Substituting (9.84) into (9.85) gives

$$P_{mofn} = \sum_{k=m}^{n} \frac{n!}{k!(n-k)!} \int_{-\infty}^{\infty} P_{d0}^k(S) \left[1 - P_{d0}(S)\right]^{n-k} f_{\mathbf{S}}(S) dS \tag{9.88}$$

or with

$$P_{d0}(S) = Q_1\left(\frac{S}{\sigma}, \sqrt{2(TNR)}\right) \tag{9.89}$$

$$P_{mofn} = \sum_{k=m}^{n} \frac{n!}{k!(n-k)!} \int_{-\infty}^{\infty} Q_1^k\left(\frac{S}{\sigma}, \sqrt{2(TNR)}\right)$$
$$\times \left[1 - Q_1\left(\frac{S}{\sigma}, \sqrt{2(TNR)}\right)\right]^{n-k} f_S(S) dS \tag{9.90}$$

For the specific cases of SW1 and SW3 targets we get, with some manipulation (see Appendix 9B),

$$P_{mofn} = \sum_{k=m}^{n} \frac{n!}{k!(n-k)!} \int_0^{\infty} Q_1^k\left(\sqrt{2x}, \sqrt{2(TNR)}\right)$$
$$\times \left[1 - Q_1\left(\sqrt{2x}, \sqrt{2(TNR)}\right)\right]^{n-k} \frac{e^{-x/SNR}}{SNR} dx \tag{9.91}$$

for SW1 targets and

$$P_{mofn} = \sum_{k=m}^{n} \frac{n!}{k!(n-k)!} \int_0^{\infty} Q_1^k\left(\sqrt{2x}, \sqrt{2(TNR)}\right)$$
$$\times \left[1 - Q_1\left(\sqrt{2x}, \sqrt{2(TNR)}\right)\right]^{n-k} \frac{4x}{(SNR)^2} e^{-2x/SNR} dx \tag{9.92}$$

for SW3 targets. The integrals of (9.91) and (9.92) must be computed numerically.

The next subject we want to address is how to handle false alarm probability. Specifically, to compute the P_{mofn} of (9.84), (9.91), and (9.92) we need $TNR = -\ln P_{fa}$ where P_{fa} is the probability of false alarm on any one of the n detection tries. However, the false alarm probability of an m-of-n detector, P_{fa_mofn}, is usually specified as a requirement. Thus, we need a way to compute P_{fa}, given P_{fa_mofn}.

Since we assume noise samples are independent, we can directly use (9.84) to write

$$P_{fa\_mofn} = \sum_{k=m}^{n} \frac{n!}{k!(n-k)!} P_{fa}^k (1 - P_{fa})^{n-k} \tag{9.93}$$

To find P_{fa} we need to solve (9.93) for P_{fa}. This can be done using an iterative approach. To obtain the initial guess of P_{fa} we take advantage of the fact that P_{fa} is small to simplify (9.93). Specifically, we note that $(P_{fa})^m$ will be larger than $(P_{fa})^{m+1}$, usually by several orders of magnitude. Also, $(1 - P_{fa})^{n-m}$ will be approximately one since $P_{fa} \ll 1$. With this we claim that the first term of the sum will be much larger than the subsequent terms. We can thus drop all but the first term of the sum. Further, we can replace the $(1 - P_{fa})^{n-m}$ term by one. With this we get a first-order approximation of (9.93) as

$$P_{fa\_mofn} \approx \frac{n!}{m!(n-m)!} P_{fa}^m \tag{9.94}$$

which we can use to initialize an iterative algorithm to find a more accurate value of P_{fa}.

To illustrate this, we consider the case where $m = 3$ and $n = 5$ (3-of-5 detection) and $P_{fa\_mofn} = 10^{-8}$. From (9.94) we have

$$P_{fa\_mofn} = 10^{-8} \approx \frac{5!}{3!(5-3)!} P_{fa}^3 = 10 P_{fa}^3 \qquad (9.95)$$

which yields $P_{fa} = \sqrt[3]{10^{-9}} = 10^{-3}$. If we use this to seed the iterative solution of (9.93), we get a final value of $P_{fa} = 1.0005 \times 10^{-3}$, which is very close to the initial guess. It must be noted that this is not always the case. As P_{fa_mofn} becomes larger, the initial guess of P_{fa} will become poorer. However, it still provides a good starting point for the iterative routine.

9.5.1 m-of-n Detection Example for SW0/SW5, SW2 and SW4 Targets

Figure 9.9 contains a plot of P_{mofn} versus P_d for $m = 3$ and $n = 5$, using (9.84). This curve demonstrates an interesting feature of the m-of-n detector. Specifically, for P_d above a certain value, 0.5 for the 3-of-5 detector, P_{mofn} will be larger than P_d. However, for P_d below this value, P_{mofn} will be less than P_d. This feature tells us m-of-n detectors tend to increase detection probability while simultaneously decreasing false alarm probability relative to their single sample values.

Equation (9.84) and the plot of Figure 9.9 applies to the case where the detection events were independent. This will be the case for noise, SW0/SW5, SW2 and SW4 targets. For those cases, the signal-plus-noise, $\mathbf{v}_{out}$ at the output of the matched filter will be independent from pulse to pulse. For SW1 and SW3 targets, (9.91) or (9.92) must be used to generate a plot like Figure 9.9.

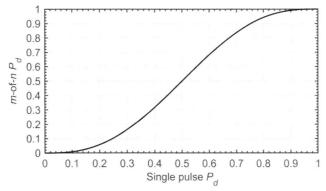

Figure 9.9 P_{mofn} vs. P_d for a 3-of-5 detector, SW0/SW5, SW2, SW4 targets.

9.5.2 m-of-n and Noncoherent Comparison for SW1 and SW2 Targets

Figures 9.10 and 9.11 contain plots of detection probability for SW1 (Figure 9.10) and SW2 (Figure 9.11) targets for the case of single-pulse detection, 5-pulse noncoherent integration and a 3-of-5 detector. The noncoherent integration curves were generated using (9.35) and (9.36) and the m-of-n curves were generated using (9.84) for the SW2 target and (9.91) for the SW1 target. The *TNR* used in (9.35) and (9.36) was computed from (9.33), and the *TNR* used in (9.84) and (9.91) was computed using the aforementioned iterative approach. The detection thresholds (i.e., the *TNR*) was adjusted to provide a P_{fa} of 10^{-6} at the output of the detection process.

For the SW1 target, P_{fa} and the *m* and *n* considered in this example, the performances of the noncoherent integrator and m-of-n detector are very similar, with the plots of P_d versus required single-pulse SNR in Figure 9.10 having the same shape and differing by only about 1 dB. For the SW2 target, the difference in required SNR slightly larger, at 3 to 5 dB.

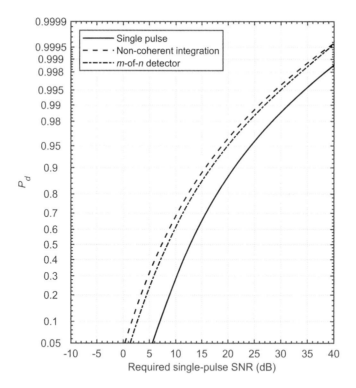

Figure 9.10 P_d vs. required single-pulse SNR for 5-pulse noncoherent integration and 3-of-5 detection—SW1 target.

The slight loss in detection performance (1 dB for SW1 and 3 to 5 dB for SW2) may be an acceptable exchange for simplicity of implementation of an m-of-n detector versus a noncoherent integrator. Also, the m-of-n detector may be less susceptible to false alarms due to random pulse interference.

In these discussions, we assumed the m-of-n detector operated on single-pulse detections. An m-of-n detector can follow a coherent or noncoherent integrator if proper restrictions are placed on its use. Specifically, the signal-plus-noise samples into the m-of-n detector must be independent. This must also be true of the noise samples. These conditions can be satisfied during search if the time between target illuminations is long relative to the coherent or noncoherent processing time. This would be an example of m-of-n detection used on a scan-to-scan basis. The problem with the use of an m-of-n detector in this fashion is that the P_{dk} on each scan will most likely not be the same. Because of this, one of the basic assumptions of the m-of-n development of this section, constant P_{dk}, could be violated. Extension to the case of unequal P_{dk} is very tedious. Several authors have attempted to approach the problem through the use of Markov chains [35, 36].[9] However, they have been able to do so for only a limited number of small values of m and n.

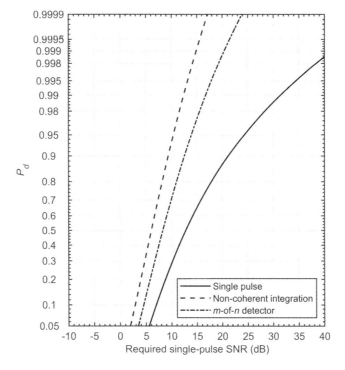

Figure 9.11 P_d vs. required single-pulse SNR for 5-pulse noncoherent integration and 3-of-5 detection—SW2 target.

---

[9] B. K. Bhagavan, Dynetics Inc. internal memo, *Markov Chain for M out of N Detection Schemes*, 2014.

9.6 EXERCISES

1. A certain radar achieves an SNR of 13 dB, with a P_{fa} of 10^{-6} on a SW1 target. What is the SNR after the coherent integration of 10 pulses? What is the P_d?

   2. Repeat Exercise 1 for a SW2 target.

   3. Repeat Exercise 1 for a SW3 target.

   4. Repeat Exercise 1 for a SW4 target.

   5. Repeat Exercise 1 for a SW0/SW5 target.

   6. What is the P_d for the conditions of Exercise 1 for the case where the 10 pulses are noncoherently integrated?

   7. Repeat Exercise 6 for a SW2 target.

   8. Repeat Exercise 6 for a SW3 target.

   9. Repeat Exercise 6 for a SW4 target.

   10. Repeat Exercise 6 for a SW0/SW5 target.

   11. A certain radar noncoherently integrates 10 pulses from a SW1 target. What single-pulse SNR is required to achieve a P_d of 0.99 and P_{fa} of 10^{-6} at the output of the noncoherent integrator?

   12. Repeat Exercise 11 for a SW2 target.

   13. Repeat Exercise 11 for a SW3 target.

   14. Repeat Exercise 11 for a SW4 target.

   15. Repeat Exercise 11 for a SW0/SW5 target.

   16. A search radar achieves an SNR of 13 dB on a SW1 target at a range of 100 km. The radar scan period is 10 s. That is, it illuminates the target every 10 seconds. The target is approaching the radar with a range rate of 500 m/s. What is the cumulative detection probability after three scans? How many scans are required to achieve a cumulative detection probability of 0.999? In both cases, the radar must maintain a cumulative false alarm probability of 10^{-6}.

   17. Repeat Exercise 16 for a SW2 target.

   18. Repeat Exercise 16 for a SW3 target.

   19. Repeat Exercise 16 for a SW4 target.

   20. Repeat Exercise 16 for a SW0/SW5 target.

21. For this exercise, we want to compare the exact and approximate equations for noncoherent integration of 10 pulses on a SW2 target. To do so, generate a plot like Figure 9.5 where the three curves correspond to: 1) single-pulse P_d, 2) P_d using the exact equation, and 3) P_d using the approximate equation.

22. Repeat Exercise 21 for a SW1 target.

23. Repeat Exercise 21 for a SW3 target.

24. Repeat Exercise 21 for a SW4 target.

25. Repeat Exercise 21 for a SW0/SW5 target.

26. Create a figure like Figure 9.9 for 5 of 10, 6 of 10, and 7 of 10 detection.

27. Create a figure like Figure 9.10 for a SW3 target.

28. Create a figure like Figure 9.10 for a SW4 target.

References

[1] North, D. O., "An Analysis of the Factors Which Determine Signal/Noise Discrimination in Pulsed Carrier Systems," RCA Labs., Princeton, NJ, Tech. Rep. No. PTR-6C, Jun. 25, 1943. Reprinted: Proc. IEEE, vol. 51, no. 7, Jul. 1963, pp. 1016–1027. Reprinted: *Detection and Estimation* (S. S. Haykin, ed.), Stroudsburg, PA: Halstad Press, 1976, pp. 10–21.

[2] Uhlenbeck, G. E., and J. L. Lawson, *Threshold Signals*, vol. 24 of MIT Radiation Lab. Series, New York: McGraw-Hill, 1950; Norwood, MA: Artech House (CD-ROM edition), 1999.

[3] Emslie, A. G., "Coherent Integration," MIT Radiation Lab. Series, Rep. 103–5, May 16, 1944. Listed as a reference in Lawson and Uhlenbeck, *Threshold Signals*, MIT Rad. Lab. Series, vol. 24, McGraw-Hill, 1950, p. 166, 331; Norwood, MA: Artech House (CD-ROM edition), 1999.

[4] Delaney, W. P., and W. W. Ward, "Radar Development at Lincoln Laboratory: An Overview of the First Fifty Years," *Lincoln Lab. J.*, vol. 12, no. 2, 2000, pp. 147–166.

[5] Miller, K. S., and Bernstein, R. I., "An Analysis of Coherent Integration and Its Application to Signal Detection," *IRE Trans. Inf. Theory*, vol. 3, no. 4, Dec. 1957, pp. 237–248.

[6] Mooney, R. K., and G. Ralston, "Performance in Clutter of Airborne Pulse, MTI, CW Doppler and Pulse Doppler Radar," *IRE Conv. Rec.*, 1961, pt. 5, pp. 55–62.

[7] Meltzer, S. A., and S. Thaler, "Detection Range Predictions for Pulse Doppler Radar," *Proc. IRE*, vol. 49, no. 8, Aug. 1961, pp. 1299–1307. Reprinted: Barton, D. K., Radars, Vol. 2: *The Radar Range Equation* (Artech Radar Library) Dedham, MA: Artech House, 1974, pp. 61–70.

[8] Steinberg, B. D., "Predetection Integration," in *Modern Radar* (R. S. Berkowitz, ed.), New York: Wiley & Sons, 1965.

[9] Soller, T., M. A. Starr, and G. E. Valley, Jr., eds., *Cathode Ray Tube Displays*, vol. 22 of MIT Radiation Lab. Series, New York: McGraw-Hill, 1948; Norwood, MA: Artech House (CD-ROM edition), 1999.

[10] Marcum, J. I., "A Statistical Theory of Target Detection by Pulsed Radar," RAND Corp., Santa Monica, CA, Res. Memo. RM-754-PR, Dec. 1, 1947. Reprinted: *IRE Trans. Inf. Theory*, vol. 6, no. 2, Apr. 1960, pp. 59–144. Reprinted: *Detection and Estimation* (S. S. Haykin, ed.), Halstad Press, 1976, pp. 57–121.

[11] Swerling, P., "Probability of Detection for Fluctuating Targets," RAND Corp., Santa Monica, CA, Res. Memo. RM-1217, Mar. 17, 1954. Reprinted: *IRE Trans. Inf. Theory*, vol. 6, no. 2, Apr. 1960, pp. 269–308.

[12] Harrington, J. V., "An Analysis of the Detection of Repeated Signals in Noise by Binary Integration," *IRE Trans.*, vol. 1, no. 1, Mar. 1955, pp. 1–9.

[13] Dinneen, G. P., and I. S. Reed, "An Analysis of Signal Detection and Location by Digital Methods," *IRE Trans. Inf. Theory*, vol. 2, no. 1, Mar. 1956, pp. 29–38.

[14] Schwartz, M., "A Coincidence Procedure for Signal Detection," *IRE Trans.*, vol. 2, no. 6, Dec. 1956, pp. 135–139.

[15] Hall, W. M., "Prediction of Pulse Radar Performance," *Proc. IRE*, vol. 44, no. 2, Feb. 1956, pp. 224–231. Reprinted: Barton, D. K., Radars, Vol. 2: The Radar Range Equation (Artech Radar Library), Dedham, MA: Artech House, 1974, pp. 31–38.

[16] Chance, B., et al., *Waveforms*, vol. 19 of MIT Radiation Lab. Series, New York: McGraw-Hill, 1949; Norwood, MA: Artech House (CD-ROM edition), 1999.

[17] Barton, D. K., *Radar System Analysis and Modeling*, Norwood, MA: Artech House, 2005.

[18] Swerling, P, "The 'Double Threshold' Method of Detection," RAND Corp., Santa Monica, CA, Res. Memo. RM-1008, Dec. 1952.

[19] Shnidman, D. A., "Binary Integration for Swerling Target Fluctuations," *IEEE Trans. Aerosp. Electron. Syst.*, vol. 34, no. 3, Jul. 1998, pp. 1043–1053.

[20] Worley, R., "Optimum Thresholds for Binary Integration (Corresp.)," *IEEE Trans. Inf. Theory*, vol. 14, no. 2, Mar. 1968, pp. 349–353.

[21] Ray, H., "Improving Radar Range and Angle Detection with Frequency Agility," Microwave J., vol. 9, no. 5, p. 63–68, May 1966. Reprinted: Barton, D. K., Radars, Vol. 6: *Frequency Agility and Diversity* (Artech Radar Library), Dedham, MA: Artech House, 1977, pp. 13–18.

[22] Fuller, J. B., "Implementation of Radar Performance Improvements by Frequency Agility," IEEE Conf. Publ. No. 105, London, Oct. 1973, pp. 56–61.

[23] DiFranco, J. V., and W. L. Rubin, *Radar Detection*, Englewood Cliffs, NJ: Prentice-Hall, 1968. Reprinted: Dedham, MA: Artech House, 1980.

[24] Meyer, D. P., and H. A. Mayer, *Radar Target Detection*, New York: Academic Press, 1973.

[25] Barton, D. K., *Radar Equations for Modern Radar*, Norwood, MA: Artech House, 2013.

[26] Barkat, M., *Signal Detection and Estimation*, 2nd ed., Norwood, MA: Artech House, 2005.

[27] Barton, D. K., "Universal Equations for Radar Target Detection," *IEEE Trans. Aerosp. Electron. Syst.*, vol. 41, no. 3, Jul. 2005, pp. 1049–1052.

[28] Bakut, P. A., et. al., *Problems in the Statistical Theory of Radar*, vol. 1, Moscow: Soviet Radio Publishing House, 1963 (in Russian; translation available from DTIC as AD608462).

[29] Urkowitz, H., *Signal Theory and Random Processes*, Dedham, MA: Artech House, 1983.

[30] Marcum, J. I., "A Statistical Theory of Target Detection by Pulsed Radar (Mathematical Appendix)," RAND Corp., Santa Monica, CA, Res. Memo. RM-753, Jul. 1, 1948. Reprinted: *IRE Trans. Inf. Theory*, vol. 6, no. 2, Apr. 1960, pp. 145–268.

[31] Skolnik, M. I., *Introduction to Radar Systems*, 3rd ed., New York: McGraw-Hill, 2001.

[32] Skolnik, M. I., ed., *Radar Handbook*, 3rd ed., New York: McGraw-Hill, 2008.

[33] Blake, L. V., "Recent Advancements in Basic Radar Range Calculation Technique," *IRE Trans. Mil. Electron.*, vol. 5, no. 2, Apr. 1961, pp. 154–164.

[34] Papoulis, A., *Probability, Random Variables, and Stochastic Processes*, 3rd ed., New York: McGraw-Hill, 1991.

[35] Castella, F. R., "Sliding Window Detection Probabilities," *IEEE Trans. Aerosp. Electron. Syst.*, vol. 12, no. 6, Nov. 1976, pp. 815–819.

[36] Williams, P., "Evaluating the State Probabilities of M Out of N Sliding Window Detectors," Aeronautical and Maritime Research Laboratory, Rep. No. AR-010-448, Feb. 1998. Approved for public release by DSTO Aeronautical and Maritime Research Laboratory, Rep. No. DSTO-TN-0132, Pyrmont, New South Wales, Australia, 2009.

[37] Parl, S., "A New Method of Calculating the Generalized Q Function," *IEEE Trans. Inf. Theory*, vol. 26, no. 1, Jan. 1980, pp. 121–124.

APPENDIX 9A: NOISE AUTOCORRELATION AT THE OUTPUT OF A MATCHED FILTER

In this appendix, we consider the correlation of the noise at the output of a matched filter. In particular, we show that noise samples are uncorrelated if they are separated by more than the duration of the matched filter impulse response.

Let the normalized impulse response of the matched filter be

$$h(t) = e^{j\phi(t)} \text{rect}\left[\frac{t - \tau_p/2}{\tau_p}\right] \tag{9A.1}$$

where $\phi(t)$ is an arbitrary phase modulation.

Let the noise into the matched filter be

$$\mathbf{n}(t) = \mathbf{n}_I(t) + j\mathbf{n}_Q(t) \tag{9A.2}$$

where $\mathbf{n}_I(t)$ and $\mathbf{n}_Q(t)$ are WSS, zero-mean, white, Gaussian random processes with equal power spectral densities of $N/2$. Further, assume that $\mathbf{n}_I(t)$ and $\mathbf{n}_Q(t)$ are uncorrelated.

The noise out of the matched filter is

$$\mathbf{n}_o(t) = \int_{-\infty}^{\infty} h(\alpha)\mathbf{n}(t-\alpha)d\alpha \tag{9A.3}$$

The autocorrelation of the noise at the output of the matched filter is

$$\begin{aligned} R_o(\tau) &= E\left\{\mathbf{n}_o(t+\tau)\mathbf{n}_o^*(t)\right\} \\ &= E\left\{\int_{-\infty}^{\infty}\int_{-\infty}^{\infty} h(\alpha)\mathbf{n}(t+\tau-\alpha)h^*(\beta)\mathbf{n}^*(t-\beta)d\beta d\alpha\right\} \\ &= \int_{-\infty}^{\infty}\int_{-\infty}^{\infty} h(\alpha)h^*(\beta)E\left\{\mathbf{n}(t+\tau-\alpha)\mathbf{n}^*(t-\beta)\right\}d\beta d\alpha \\ &= \int_{-\infty}^{\infty}\int_{-\infty}^{\infty} h(\alpha)h^*(\beta)N\delta(t+\tau-\alpha-(t-\beta))d\beta d\alpha \end{aligned} \tag{9A.4}$$

where $\delta(x)$ is the Dirac delta function.

Evaluating the β integral yields

$$R_o(\tau) = N \int_{-\infty}^{\infty} h(\alpha) h^*(\alpha - \tau) d\alpha \quad (9A.5)$$

From matched filter theory (see Chapter 8), we can write this as

$$R_o(\tau) = Nm(\tau) \text{rect}\left[\frac{\tau}{2\tau_p}\right] \quad (9A.6)$$

where $m(\tau)$ captures the fine detail of the autocorrelation. As an example, for $\phi(t) = 0$; $h(t)$ would be the impulse response of a filter matched to an unmodulated pulse and we would have

$$m(\tau) = \tau_p - |\tau| \quad (9A.7)$$

The key thing to note about (9A.6) is that

$$R_o(\tau) = 0 \text{ for } |\tau| > \tau_p \quad (9A.8)$$

Since $\mathbf{n}_o(t)$ is zero-mean (because $\mathbf{n}(t)$ is zero-mean), $R_o(\tau) = C_o(\tau)$ where $C_o(\tau)$ is the autocovariance of $\mathbf{n}_o(t)$. Since the autocovariance is zero for $|\tau| > \tau_p$, output noise samples separated, in time, by more than τ_p will be uncorrelated. Since the noise samples are Gaussian, they will also be independent. For a phase modulated pulse, $R_o(\tau)$ will be small for $|\tau| > \tau_c$ where τ_c is the chip width, or the reciprocal of the LFM bandwidth for LFM pulses. Because of this, we often assume the noise samples will be uncorrelated (and independent) if the samples are spaced by more than one compressed pulse width of the phase-coded waveform (see Chapters 10 and 11).

APPENDIX 9B: PROBABILITY OF DETECTING SW1 AND SW3 TARGETS ON *m* CLOSELY SPACED PULSES

In this appendix, we address the problem of computing the probability of detecting a SW1 or SW3 target on m closely spaced pulses. The instinctive method of computing this probability is to say that if P_{di} is the probability of detecting the target on any one pulse, the probability of detecting the target on m pulses is $P_{d1} \times P_{d2} \times P_{d3} \times \ldots \times P_{dm}$. This method makes the assumption the detection events on each pulse are independent. This is true for SW0/SW5, SW2, and SW4 targets. It is not true for SW1 and SW3 targets.

Let D_i be the event of detecting the target on the ith pulse. We can write

$$P(\mathcal{D}_i) = P(\{\mathbf{V}_i \geq T_i\}) \quad (9B.1)$$

where $\mathbf{V}_i$ is the magnitude of the signal plus noise on the ith received pulse (in a particular range cell) and T_i is the detection threshold used when the return from the ith pulse is present. We are interested in determining

$$P(\mathcal{D}) = P\left(\bigcap_{i=1}^{m}\mathcal{D}_i\right) = P\left(\bigcap_{i=1}^{m}\{\mathbf{V}_i \geq T_i\}\right)$$
$$= \int_{T_1}^{\infty}\int_{T_2}^{\infty}\cdots\int_{T_m}^{\infty} f_{\mathbf{V}_1\mathbf{V}_2\cdots\mathbf{V}_m}(V_1,V_2,\cdots,V_m)dV_1 dV_2 \cdots dV_m \quad (9B.2)$$

where $f_{\mathbf{V}_1\mathbf{V}_2\cdots\mathbf{V}_m}(V_1,V_2,\cdots,V_m)$ is the joint density of $\mathbf{V}_1, \mathbf{V}_2, \ldots, \mathbf{V}_m$. The issue becomes one of finding this joint density, and then performing the integrations.

For SW0/SW5, SW2, and SW4 targets, the random variables $\mathbf{V}_1, \mathbf{V}_2, \ldots, \mathbf{V}_m$ are independent and

$$f_{\mathbf{V}_1\mathbf{V}_2\cdots\mathbf{V}_m}(V_1,V_2,\cdots,V_m) = \prod_{i=1}^{m} f_{\mathbf{V}_i}(V_i) \quad (9B.3)$$

This leads to the aforementioned statement that

$$P(\mathcal{D}) = \prod_{i=1}^{m} P(\mathcal{D}_i) \quad (9B.4)$$

For SW1 and SW3 targets, we cannot assume that $\mathbf{V}_1, \mathbf{V}_2, \ldots, \mathbf{V}_m$ are independent. This stems from the fact that they all depend on the target RCS, which is a random variable governed by the SW1 and SW3 target models. To compute the joint density, we resort to conditional density functions and write

$$f_{\mathbf{V}_1\mathbf{V}_2\cdots\mathbf{V}_m}(V_1,V_2,\cdots,V_m) = \int_{-\infty}^{\infty}\int_{-\infty}^{\infty} f_{\mathbf{V}_1\mathbf{V}_2\cdots\mathbf{V}_m\mathbf{S}\Theta}(V_1,V_2,\cdots,V_m,S,\theta)dSd\theta$$
$$= \int_{-\infty}^{\infty}\int_{-\infty}^{\infty} f_{\mathbf{V}_1\mathbf{V}_2\cdots\mathbf{V}_m}(V_1,V_2,\cdots,V_m|\mathbf{S}=S,\Theta=\theta)f_{\mathbf{S}}(S)f_{\Theta}(\theta)dSd\theta \quad (9B.5)$$

We note that once we fix $\mathbf{S}$ to S and Θ to θ, the random variables $\mathbf{V}_i|\mathbf{S}=S, \Theta=\theta$ are no longer dependent on the random variables $\mathbf{S}$ and Θ. They are dependent on the noise component of the signal-plus-noise. However, the noise is independent from pulse to pulse. Therefore, the random variables $\mathbf{V}_i|\mathbf{S}=S, \Theta=\theta$ are independent from pulse to pulse. With this we conclude that

$$f_{\mathbf{V}_1\mathbf{V}_2\cdots\mathbf{V}_m}(V_1,V_2,\cdots,V_m|\mathbf{S}=S,\Theta=\theta) = \prod_{i=1}^{m} f_{\mathbf{V}_i}(V_i|\mathbf{S}=S,\Theta=\theta) \quad (9B.6)$$

To allow for the case where the SNR can be different on each of the pulses, we represent $\mathbf{V}_i|\mathbf{S}=S, \Theta=\theta$ as $\mathbf{V}_i|\mathbf{S}=S, \Theta=\theta = \rho_i S + \mathbf{n}_i$ where ρ_i is a voltage scaling factor.

From our experience with determining the density functions of the magnitude of signal-plus-noise, the density functions of the in-phase and quadrature components of the conditioned signal-plus-noises are Gaussian with variances of σ^2 and means of $\rho_i S\cos\theta$ and $\rho_i S\sin\theta$ (see Chapter 6). With this (see Section 6.4) we have

$$f_{\mathbf{V}_i}(V_i|\mathbf{S}=S,\Theta=\theta) = \frac{V_i}{\sigma^2} I_0\left(\frac{V_i \rho_i S}{\sigma^2}\right) \exp\left[\frac{-\left(V_i^2 + (\rho_i S)^2\right)}{2\sigma^2}\right] U(V_i) \qquad (9B.7)$$

For a SW1 target

$$f_\mathbf{S}(S) = \frac{S}{P_S} e^{-S^2/2P_S} U(S) \qquad (9B.8)$$

and

$$f_\Theta(\theta) = \frac{1}{2\pi} \operatorname{rect}\left[\frac{\theta}{2\pi}\right] \qquad (9B.9)$$

For a SW3 target

$$f_\mathbf{S}(S) = \frac{2S^3}{P_S^2} e^{-S^2/P_S} U(S) \qquad (9B.10)$$

and $f_\Theta(\theta)$ is as in (9B.9). In the above, σ^2 is the noise power and P_S is a signal power.

We can use (9B.9) to eliminate the random variable Θ through the appropriate integration to yield

$$f_{\mathbf{V}_1 \mathbf{V}_2 \cdots \mathbf{V}_m}(V_1, V_2, \cdots, V_m) = \int_{-\infty}^{\infty} f_{\mathbf{V}_1 \mathbf{V}_2 \cdots \mathbf{V}_m}(V_1, V_2, \cdots, V_m | \mathbf{S}=S) f_\mathbf{S}(S) dS \qquad (9B.11)$$

If we use (9B.11) in (9B.2), we get

$$P(\mathcal{D}) = \int_{T_1}^{\infty}\int_{T_2}^{\infty} \cdots \int_{T_m}^{\infty} \int_{-\infty}^{\infty} f_{\mathbf{V}_1 \mathbf{V}_2 \cdots \mathbf{V}_m}(V_1, V_2, \cdots, V_m | \mathbf{S}=S) f_\mathbf{S}(S) dS dV_1 dV_2 \cdots dV_m \qquad (9B.12)$$

We can now use (9B.6) and write

$$P(\mathrm{D}) = \int_{-\infty}^{\infty} \left(\prod_{i=1}^{m} \int_{T_i}^{\infty} f_{\mathbf{V}_i}(V_i | \mathbf{S}=S) dV_i \right) f_\mathbf{S}(S) dS \qquad (9B.13)$$

If we use (9B.7), we recognize the inner integrals of (9B.13) as the detection probabilities for a SW0/SW5 target. Thus, we can rewrite (9B.13) as

$$P(\mathcal{D}) = \int_{-\infty}^{\infty} \left(\prod_{i=1}^{m} P_{d0}(\rho_i S, \sigma^2, T_i) \right) f_\mathbf{S}(S) dS$$

$$= \int_{-\infty}^{\infty} \left(\prod_{i=1}^{m} \int_{T_i}^{\infty} \frac{V}{\sigma^2} I_0\left(\frac{V \rho_i S}{\sigma^2}\right) \exp\left[\frac{-\left(V^2 + (\rho_i S)^2\right)}{2\sigma^2}\right] dV \right) f_\mathbf{S}(S) dS \qquad (9B.14)$$

In the inner integral, we let $u = V/\sigma$ and write

$$P(\mathcal{D}) = \int_{-\infty}^{\infty} \left(\prod_{i=1}^{m} \int_{T_i/\sigma}^{\infty} u I_0 \left(\frac{u \rho_i S}{\sigma} \right) \exp\left[\frac{-(u^2 + \rho_i^2 S^2/\sigma^2)}{2} \right] du \right) f_\mathbf{S}(S) dS \quad (9\text{B}.15)$$

We can use the Marcum Q function [10] to evaluate the inner integral and write

$$P(\mathcal{D}) = \int_{-\infty}^{\infty} \left[\prod_{i=1}^{m} Q_1 \left(\frac{\rho_i S}{\sigma}, \frac{T_i}{\sigma} \right) \right] f_\mathbf{S}(S) dS \quad (9\text{B}.16)$$

Substituting for $f_\mathbf{S}(S)$ from (9B.9) results in

$$P(\mathcal{D}) = \int_{0}^{\infty} \left[\prod_{i=1}^{m} Q_1 \left(\frac{\rho_i S}{\sigma}, \frac{T_i}{\sigma} \right) \right] \frac{S}{P_S} e^{-S^2/2P_S} dS \quad (9\text{B}.17)$$

for the SW1 case. Substitution for $f_\mathbf{S}(S)$ from (9B.10) results in

$$P(\mathcal{D}) = \int_{0}^{\infty} \left[\prod_{i=1}^{m} Q_1 \left(\frac{\rho_i S}{\sigma}, \frac{T_i}{\sigma} \right) \right] \frac{2S^3}{P_S^2} e^{-S^2/P_S} dS \quad (9\text{B}.18)$$

for the SW3 case. In (9B.17) and (9B.18), we make the change of variables $x = S^2/2\sigma^2$. Manipulation of some of the arguments, with the change of variables, yields

$$\frac{S^2}{2\rho_i^2 P_S} = \frac{x}{\rho_i^2 (SNR)} \quad (9\text{B}.19)$$

$$\frac{S dS}{\rho_i^2 P_S} = \frac{dx}{\rho_i^2 (SNR)} \quad (9\text{B}.20)$$

$$\frac{T_i}{\sigma} = \sqrt{2(TNR)_i} \quad (9\text{B}.21)$$

$$\frac{S}{\sigma} = \sqrt{2x} \quad (9\text{B}.22)$$

$$SNR = \frac{P_S}{\sigma^2} \quad (9\text{B}.23)$$

$$(TNR)_i = \frac{T_i^2}{2\sigma^2} \quad (9\text{B}.24)$$

In the above, *SNR* is a *reference* SNR. We would compute it using the radar range equation using a set of reference parameters. The term $\rho_i^2(SNR)$ is the SNR with a set of parameters specific to the ith pulse. We would compute ρ_i^2 from $[\rho_i^2(SNR)]/SNR$. For example, we might compute *SNR* from

$$SNR = \frac{P_T G_T G_R \lambda^2 \sigma \tau_{p0}}{(4\pi)^3 R^4 k T_S L} \tag{9B.25}$$

where τ_{p0} is the pulse width of one of the pulses. For another pulse, with a different width, we would compute

$$\rho_i^2 (SNR) = \frac{P_T G_T G_R \lambda^2 \sigma \tau_{pi}}{(4\pi)^3 R^4 k T_S L} \tag{9B.26}$$

If we combine (9B.25) and (9B.26), we have

$$\rho_i^2 = \frac{\tau_{pi}}{\tau_{p0}} \tag{9B.27}$$

It would be reasonable to expect that the pulse width, peak power, antenna directivities, and noise temperature and losses could change from pulse to pulse. However, it would not be reasonable to expect that wavelength or range could change because this would cause decorrelation of the target RCS, which would convert the RCS from Swerling odd to Swerling even, and violate the assumptions of this appendix.

$(TNR)_i = -\ln P_{fai}$, is the threshold-to-noise ratio used on the ith pulse, and P_{fai} is the desired false alarm probability for that pulse.

With the above substitutions, the equation that must be implemented for SW1 targets is

$$P(\mathcal{D}) = \int_0^\infty \left[\prod_{i=1}^m Q_1\left(\rho_i \sqrt{2x}, \sqrt{2(TNR)_i}\right) \right] \frac{1}{SNR} e^{-x/SNR} dx \tag{9B.28}$$

The equation that must be implemented for SW3 targets is

$$P(\mathcal{D}) = \int_0^\infty \left[\prod_{i=1}^m Q_1\left(\rho_i \sqrt{2x}, \sqrt{2(TNR)_i}\right) \right] \frac{4x}{(SNR)^2} e^{-2x/SNR} dx \tag{9B.29}$$

These are the integrals that we need to evaluate. We will need to do so via numerical integration.

For the case where the SNR and detection threshold is the same on every detection attempt, (9B.28) and (9B.29) reduce to

$$P(\mathcal{D}) = \int_0^\infty Q_1^m\left(\sqrt{2x}, \sqrt{2TNR}\right) \frac{1}{SNR} e^{-x/SNR} dx \tag{9B.30}$$

and

$$P(\mathcal{D}) = \int_0^\infty Q_1^m\left(\sqrt{2x}, \sqrt{2TNR}\right) \frac{4x}{(SNR)^2} e^{-2x/SNR} dx \tag{9B.31}$$

9B.1 Marcum Q Function

The Marcum Q-function is defined as

$$Q_M(a,b) = \frac{1}{a^{M-1}} \int_b^\infty x^M I_{M-1}(ax) e^{-\frac{x^2+a^2}{2}} dx \qquad (9B.30)$$

An efficient and accurate algorithm for computing $Q_1(a,b)$ was developed by Steen Parl [37].

The Parl algorithm for computing $Q_1(a,b)$ is:

- Initialization

$$a_{-1} = 0, \; \beta_0 = 0.5, \; \beta_{-1} = 0$$

$$\text{if } a < b, \; a_0 = 1, \; d_1 = a/b$$

$$\text{if } a \geq b, \; a_0 = 0, \; d_1 = b/a$$

- Iteration, $n = 1, 2, \cdots$

$$\alpha_n = d_n + \frac{2n}{ab}\alpha_{n-1} + \alpha_{n-2}$$

$$\beta_n = 1 + \frac{2n}{ab}\beta_{n-1} + \beta_{n-2}$$

$$d_{n+1} = d_n d_1$$

terminate when $\beta_n > 10^p$

- Final Step

$$\text{if } a < b, \; Q_1(a,b) = \frac{\alpha_n}{2\beta_n} e^{-\frac{(a-b)^2}{2}}$$

$$\text{if } a \geq b, \; Q_1(a,b) = 1 - \frac{\alpha_n}{2\beta_n} e^{-\frac{(a-b)^2}{2}}$$

Typical values of p in the termination criterion are $p = 3$ to 9. A reasonable value seems to be $p = 6$.

APPENDIX 9C: CUMULATIVE DETECTION PROBABILITY

This appendix contains a derivation of the cumulative detection probability equations enumerated in Section 9.4.

To develop the technique, we start by considering the *events* [26, 29, 34] of the occurrence of a threshold crossing on two occasions. We denote these two events as

- D_1: Threshold crossing on occasion 1;
- D_2: Threshold crossing on occasion 2.

If we form the event

$$D = D_1 \cup D_2 \tag{9C.1}$$

where $\cup$ denotes the *union* operation [26], then D is the event consisting of a threshold crossing on occasion 1, or occasion 2, or occasions 1 and 2. Since D is the event of interest to us, we want to find the probability that it will occur. That is, we want

$$P(D) = P(D_1 \cup D_2) \tag{9C.2}$$

From probability theory, we can write

$$P(D) = P(D_1 \cup D_2) = P(D_1) + P(D_2) - P(D_1 \cap D_2) \tag{9C.3}$$

where $D_1 \cap D_2$ represents the *intersection* of D_1 and D_2 and is the event consisting of a threshold crossing on occasion 1 *and* occasion 2. The first two probability terms on the right side, $P(D_1)$ and $P(D_2)$, are computed using the appropriate single or n pulse probability equation discussed in Chapter 6 and Sections 9.2 and 9.3, depending on the target type and whether or not coherent or noncoherent integration is used.

To compute the third term, $P(D_1 \cap D_2)$, we need to make an assumption about the events D_1 and D_2. Specifically, we assume they are independent. This, in turn, limits when we can use cumulative detection concepts. Specifically, we should use cumulative detection concepts only on a scan-to-scan basis. If we do, we will satisfy the constraints on all of the Swerling target types. Specifically, for SW1 and SW3 targets, the signal-plus-noise samples are, by definition, independent from scan to scan. For SW0/SW5, SW2, and SW4 targets, the signal-plus-noise samples from pulse to pulse and will thus also be independent from scan to scan. Having said this, we must also assure that the coherent or noncoherent integrator does not cause the independence restriction to be violated. The restriction will not be violated if the time between target illuminations is significantly larger than the coherent or noncoherent processing time (e.g., scan to scan).

If D_1 and D_2 are independent, we can write

$$P(D_1 \cap D_2) = P(D_1)P(D_2) \tag{9C.4}$$

and

$$P(D) = P(D_1) + P(D_2) - P(D_1)P(D_2) \tag{9C.5}$$

As an example, suppose $P(D_1) = P(D_2) = 0.9$. Using (9C.5), we would obtain

$$\begin{aligned} P(D) &= P(D_1) + P(D_2) - P(D_1)P(D_2) \\ &= 0.9 + 0.9 - 0.9 \times 0.9 = 1.8 - 0.81 = 0.99 \end{aligned} \tag{9C.6}$$

While (9C.5) is reasonably easy to implement for two events, its direct extension to many events is tedious. In order to set the stage for a simpler extension, we consider a different means of determining $P(D)$. We begin by observing that

$$S = D_i \cup \bar{D}_i \tag{9C.7}$$

where S is the *universe* and $\bar{D}_i$ is the *complement* of D_i. $\bar{D}_i$ contains all elements that are in S but not in D_i. By the definition of $\bar{D}_i$ we note that D_i and $\bar{D}_i$ are *mutually exclusive*. We also note that $P(S) = 1$. With this we get

$$P(S) = 1 = P(D_i \cup \bar{D}_i) = P(D_i) + P(\bar{D}_i) \tag{9C.8}$$

and

$$P(D_i) = 1 - P(\bar{D}_i) \tag{9C.9}$$

To proceed with the derivation, we let

$$D = D_1 \cup D_2 \tag{9C.10}$$

and

$$S = D \cup \bar{D} = (D_1 \cup D_2) \cap \overline{(D_1 \cup D_2)} \tag{9C.11}$$

From (9C.8), we get

$$1 = P(D_1 \cup D_2) + P(\overline{D_1 \cup D_2}) \tag{9C.12}$$

By DeMorgan's Law [29], we can write

$$\overline{(D_1 \cup D_2)} = \bar{D}_1 \cap \bar{D}_2 \tag{9C.13}$$

and

$$1 = P(D_1 \cup D_2) + P(\bar{D}_1 \cap \bar{D}_2) \tag{9C.14}$$

Now, since D_1 and D_2 are independent, so are $\bar{D}_1$ and $\bar{D}_2$. If we use this along with (9C.10), we can write

$$P(D) = 1 - P(\bar{D}_1) P(\bar{D}_2) \tag{9C.15}$$

Finally, making use of (9C.9), we obtain

$$P(D) = 1 - [1 - P(D_1)][1 - P(D_2)] = 1 - \prod_{k=1}^{2} [1 - P(D_k)] \tag{9C.16}$$

We can now generalize (9C.16) to any number of events. Specifically, if

$$D = D_1 \cup D_2 \cup D_3 \cdots \cup D_N \tag{9C.17}$$

where $D_1, D_2, D_3 \ldots D_N$ are independent, then

$$P(D) = 1 - \prod_{k=1}^{N} [1 - P(D_k)] \tag{9C.18}$$

As an example of the use of (9C.16) or (9C.18), we consider the previous example wherein $P(D_1) = P(D_2) = 0.9$. With this we get

$$P(D) = 1 - \prod_{k=1}^{2}\left[1 - P(D_k)\right] = 1 - \left[1 - P(D_1)\right]\left[1 - P(D_2)\right]$$
$$= 1 - (1 - 0.9)(1 - 0.9) = 1 - 0.1 \times 0.1 = 0.99$$
(9C.19)

We now want to restate (9C.19) in terminology more directly related to detection probability. To that end, we write

$$P_{dcum} = 1 - \prod_{k=1}^{N}(1 - P_{dk})$$
(9C.20)

where P_{dcum} is the *cumulative detection probability* over N scans, and P_{dk} is the detection probability on the kth scan.

Chapter 10

Ambiguity Function

10.1 INTRODUCTION

The ambiguity function, which is denoted as $|\chi(\tau, f)|^2$, is primarily used to gain an understanding of how a signal processor responds, or reacts, to a given returned signal. As indicated in the notation, the independent variables of the ambiguity function are time (τ) and frequency (f). The time variable is normally associated with target range, and the frequency variable is normally associated with target Doppler frequency. The magnitude square (i.e., $|\ |^2$) is used to indicate we are characterizing the amplitude squared of the signal processor output.

In a strict sense, when one uses the phrase "ambiguity function," there is an underlying assumption that the signal processor is matched to the transmitted waveform. If the signal processor is not matched to the transmitted waveform, the proper terminology, in the ambiguity function context, would be to refer to the "cross ambiguity function." In practice, however, we do not always make the distinction and simply use the phrase "ambiguity function."

We will derive a general equation for the (cross) ambiguity function and then derive specific ambiguity function expressions for two simple waveforms and signal processors. We will then discuss a representation that is suitable for numerically computing the ambiguity function using the FFT. This will allow generation of ambiguity functions of advanced waveforms and signal processors.

The ambiguity function was first developed by a British mathematician named Philip M. Woodward [1]. As such, the ambiguity function is sometimes referred to as the Woodward ambiguity function. In 2009, Woodward received the IEEE Dennis J. Picard Medal for radar technologies and applications "[f]or pioneering work of fundamental importance in radar waveform design, including the Woodward Ambiguity Function, the standard tool for waveform and matched filter analysis" [2].

10.2 AMBIGUITY FUNCTION DEVELOPMENT

Since the ambiguity function can be thought of as the response of a signal processor to a received radar waveform, this is the approach we take in deriving the ambiguity function.

Let the normalized, transmitted waveform be represented by the baseband signal, $u(t)$.[1] The normalized (baseband) signal received from a (constant range rate) target at a range R and range rate $\dot{R}$ is given by

$$v_R(t) = u(t - \tau_R) e^{j2\pi f_d t} \tag{10.1}$$

where $\tau_R = 2R/c$ is the range delay and $f_d = -2\dot{R}/\lambda$ is the target Doppler frequency. λ is the wavelength of the transmitted signal, and c is the speed of light.

The signal processor configuration we use in deriving the ambiguity function is shown in Figure 10.1. In this figure, $h(t)$ is a lowpass function and f_s is thought of as the frequency to which the signal processor is tuned. Thus, the overall signal processor is a bandpass filter centered at f_s. This is indicated in Figure 10.2, where the left plot is the frequency response of $h(t)$ and the right plot is the frequency response of the signal processor.

In keeping with the concept of matched filters, we normally define $h(t)$ in terms of the waveform to which it is matched. Specifically, we use

$$h(t) = K v^*(t_0 - t) \tag{10.2}$$

and set $K = 1$ and $t_0 = 0$ to get

$$h(t) = v^*(-t) \tag{10.3}$$

In (10.3), $v(t)$ is the waveform to which the signal processor is matched and the star (*) superscript denotes the complex conjugate.

With the above, the impulse response of the signal processor is

$$h_{SP}(t) = e^{j2\pi f_s t} h(t) = e^{j2\pi f_s t} v^*(-t) \tag{10.4}$$

The output of the signal processor is the convolution of $v_R(t)$ and $h_{SP}(t)$ or

$$v_o(\gamma) = \int_{-\infty}^{\infty} v_R(\alpha) h_{SP}(\gamma - \alpha) d\alpha \tag{10.5}$$

where we temporarily replaced the time variable, t, by γ.

---

[1] Recall that if an actual signal is $v_{RF}(t) = A(t)\cos[\omega_{RF}t - \phi(t)]$, the baseband representation of that signal is $v_B(t) = A(t)e^{j\phi(t)}$. That is, the baseband signal captures the amplitude and phase modulation of the actual signal in the form of a complex variable. Baseband notation is a special case of *complex signal notation* (see Chapter 1).

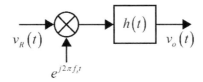

Figure 10.1 Signal processor.

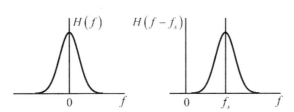

Figure 10.2 Signal processor frequency response.

Using (10.1) and (10.4) in (10.5) yields

$$\begin{aligned}v_o(\gamma) &= \int_{-\infty}^{\infty} u(\alpha - \tau_R) e^{j2\pi f_d \alpha} e^{j2\pi f_s(\gamma - \alpha)} v^*(\alpha - \gamma) d\alpha \\ &= e^{j2\pi f_s \gamma} \int_{-\infty}^{\infty} u(\alpha - \tau_R) v^*(\alpha - \gamma) e^{j2\pi (f_d - f_s)\alpha} d\alpha\end{aligned} \quad (10.6)$$

If we make the change of variables, $t = \alpha - \tau_R$, (10.6) can be rewritten as

$$\begin{aligned}v_o(\gamma) &= e^{j2\pi f_s \gamma} \int_{-\infty}^{\infty} u(t) v^*(t + \tau_R - \gamma) e^{j2\pi (f_d - f_s)(t + \tau_R)} dt \\ &= e^{-j2\pi f_s \tau} e^{j2\pi f_d \tau_R} \int_{-\infty}^{\infty} u(t) v^*(t + \tau) e^{j2\pi f t} dt\end{aligned} \quad (10.7)$$

where we made the substitutions $\tau = \tau_R - \gamma$ and $f = f_d - f_s$.

The variables τ and f are often termed the mismatched range and Doppler of the ambiguity function. More specifically, τ is the difference between the target range delay and the time we look at the signal processor output. If $\tau = 0$, we say that we are at matched range. That is, we are looking at the signal processor output at a time equal to the time delay of the target. f is the difference between the target Doppler frequency and the frequency to which the signal processor is matched. If $f = 0$, we say the signal processor is matched to the target Doppler frequency, or vice versa. In this case, we say we are at matched Doppler.

Since τ and f are the variables of interest, we rewrite (10.7) in terms of them and change the dependent variable from $v_o(\gamma)$ to $\chi(\tau, f)$. Thus, we get

$$\chi(\tau,f) = e^{-j2\pi f_s \tau} e^{j2\pi f_d \tau_R} \int_{-\infty}^{\infty} u(t) v^*(t+\tau) e^{j2\pi f t} dt \qquad (10.8)$$

Finally, if we take the magnitude squared of (10.8), we get the ambiguity function or

$$|\chi(\tau,f)|^2 = \left| \int_{-\infty}^{\infty} u(t) v^*(t+\tau) e^{j2\pi f t} dt \right|^2 \qquad (10.9)$$

We often attribute special names to plots of $|\chi(\tau, f)|^2$ for specific values of τ and f. In particular:

- If we let $f = 0$ to yield $|\chi(\tau, 0)|^2$, we have the *matched-Doppler, range cut* of the ambiguity function. This is what we normally think of as the output of the classical matched filter.
- If we let $\tau = 0$ to yield $|\chi(0, f)|^2$, we have the *matched-range, Doppler cut* of the ambiguity function.
- If we let $f = f_k$ to yield $|\chi(\tau, f_k)|^2$, we have a range cut at some mismatched Doppler of f_k.
- If we let $\tau = \tau_k$ to yield $|\chi(\tau_k, f)|^2$, we have a Doppler cut at some mismatched range of τ_k.

As we will discuss in Chapter 11, the ambiguity function provides a wealth of information about radar waveforms and how they interact with the environment and the radar signal processor. For now, as examples, we want to derive equations for the ambiguity function of an unmodulated pulse and a pulse with LFM.

10.3 EXAMPLE 1: UNMODULATED PULSE

We want to derive the equation for the ambiguity function of an unmodulated pulse of width τ_p. We will assume the signal processor is matched to the transmitted pulse. With this we can write[2]

$$v(t) = u(t) = \text{rect}\left[\frac{t - \tau_p/2}{\tau_p}\right] \qquad (10.10)$$

If we substitute this into (10.9), we get

$$|\chi(\tau,f)|^2 = \left| \int_{-\infty}^{\infty} \text{rect}\left[\frac{t-\tau_p/2}{\tau_p}\right] \text{rect}\left[\frac{t-\tau_p/2+\tau}{\tau_p}\right] e^{j2\pi f t} dt \right|^2 \qquad (10.11)$$

As we did for the matched filter derivation (see Chapter 8), we need to consider several regions of τ. To see this, refer to Figures 10.3 and 10.4.

---

[2] We are using a normalized pulse amplitude of unity. For the more general case, we would need to multiply $u(t)$ and $v(t)$ by appropriate amplitudes.

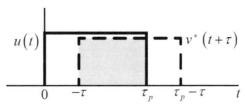

Figure 10.3 Plot of $u(t)$ and $v^*(t + \tau)$ for $\tau < 0$.

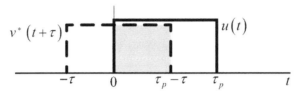

Figure 10.4 Plot of $u(t)$ and $v^*(t + \tau)$ for $\tau \geq 0$.

In these figures, $u(t) = \text{rect}[(t - \tau_p/2)/\tau_p]$ and $v^*(t + \tau) = \text{rect}[(t - \tau_p/2 + \tau)/\tau_p]$. Note that for $|\tau| \geq \tau_p$, the rect functions will not overlap and the integral of (10.11) will be zero. This leads to the observation that $|\chi(\tau, f)|^2 = 0$ for $|\tau| \geq \tau_p$. We will account for this by multiplying $|\chi(\tau, f)|^2$ by $\text{rect}[\tau/(2\tau_p)]$.

For $-\tau_p < \tau < 0$, the rect functions overlap from $-\tau$ to τ_p (see Figure 10.3). Thus, (10.11) becomes

$$|\chi(\tau, f)|^2 = \left| \int_{-\tau}^{\tau_p} e^{j2\pi ft} dt \right|^2 = \left| \frac{e^{j2\pi f\tau_p} - e^{-j2\pi f\tau}}{j2\pi f} \right|^2 \tag{10.12}$$

If we factor $e^{j\pi f(\tau_p - \tau)}$ from both terms on the far right side of (10.12) and then multiply by $|\tau_p + \tau|/|\tau_p + \tau|$, we get

$$|\chi(\tau, f)|^2 = \left| e^{j\pi f(\tau_p - \tau)} \left(\frac{e^{j\pi f(\tau_p + \tau)} - e^{-j\pi f(\tau_p + \tau)}}{j2\pi f} \right) \right|^2 \tag{10.13}$$

$$= |\tau_p + \tau|^2 \left| \text{sinc}\left(f[\tau_p + \tau] \right) \right|^2$$

Finally, if we recognize $\tau < 0$, we can use $|\tau| = -\tau$, and rewrite (10.13) as

$$|\chi(\tau, f)|^2 = |\tau_p - |\tau||^2 \left| \text{sinc}\left(f[\tau_p - |\tau|] \right) \right|^2 \tag{10.14}$$

For $\tau_p > \tau \geq 0$, the rect functions overlap from 0 to $\tau_p - \tau$ (see Figure 10.4). Thus, (10.11) becomes

$$|\chi(\tau, f)|^2 = \left| \int_0^{\tau_p - \tau} e^{j2\pi ft} dt \right|^2 = \left| \frac{e^{j2\pi f(\tau_p - \tau)} - 1}{j2\pi f} \right|^2 \tag{10.15}$$

If we factor $e^{j\pi f(\tau_p - \tau)}$ from both terms on the far right side (10.15) and then multiply by $|\tau_p - \tau|/|\tau_p - \tau|$, we get

$$|\chi(\tau,f)|^2 = \left| e^{j\pi f(\tau_p - \tau)} \left(\frac{e^{j\pi f(\tau_p - \tau)} - e^{-j\pi f(\tau_p - \tau)}}{j2\pi f} \right) \right|^2 \quad (10.16)$$

$$= |\tau_p - \tau|^2 \left| \text{sinc}\left(f[\tau_p - \tau] \right) \right|^2$$

If we observe that $\tau \geq 0$, we can use $|\tau| = \tau$, and rewrite (10.16) as

$$|\chi(\tau,f)|^2 = |\tau_p - |\tau||^2 \left| \text{sinc}\left(f[\tau_p - |\tau|] \right) \right|^2 \quad (10.17)$$

which is the same result we obtained for $-\tau_p < \tau < 0$.

If we combine all of the above, we arrive at our final answer of

$$|\chi(\tau,f)|^2 = (\tau_p - |\tau|)^2 \left| \text{sinc}\left(f[\tau_p - |\tau|] \right) \right|^2 \text{rect}\left[\frac{\tau}{2\tau_p} \right] \quad (10.18)$$

We note that the square root of the matched-Doppler, range cut, which we obtain by setting $f = 0$, is the same form as the matched filter output we found in Chapter 8. Specifically,

$$|\chi(\tau,0)| = (\tau_p - |\tau|)\text{rect}\left[\frac{\tau}{2\tau_p} \right] \quad (10.19)$$

A sketch of $|\chi(\tau, 0)|$ is shown in Figure 10.5.[3]

The square root of the matched-range, Doppler cut, which we obtain by setting $\tau = 0$, is

$$|\chi(0,f)| = \tau_p \left| \text{sinc}(f\tau_p) \right| \quad (10.20)$$

A plot of this function is shown in Figure 10.6.

Finally, a plot of the center portion of $|\chi(\tau,f)|$ is shown in Figure 10.7 for the specific case where $\tau_p = 1$ μs. The plot has been normalized to a height of unity. Its actual height is τ_p, or 1 μs.

---

[3] The standard convention is to plot $|\chi(\tau,f)|$ since it usually provides more detail about the structure of the matched filter response, especially in the sidelobes.

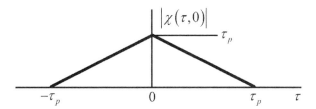

Figure 10.5 Matched-Doppler, range cut for an unmodulated pulse.

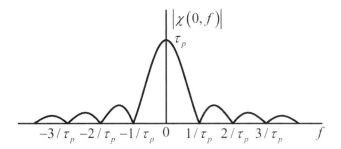

Figure 10.6 Matched-range, Doppler cut of an unmodulated pulse.

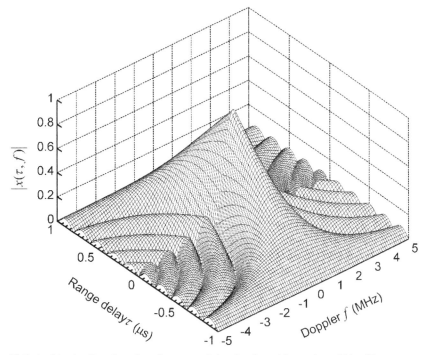

Figure 10.7 Ambiguity function plot of an unmodulated pulse with a pulse width of 1 μs.

10.4 EXAMPLE 2: LFM PULSE

For the LFM pulse, we let

$$v(t) = u(t) = e^{j\pi\alpha t^2} \operatorname{rect}\left[\frac{t - \tau_p/2}{\tau_p}\right] \quad (10.21)$$

where α is the LFM slope. Recall that the LFM slope is related to the LFM bandwidth by (see Chapter 8)

$$B = |\alpha \tau_p| \quad (10.22)$$

$\alpha > 0$ means the modulation frequency increases across the pulse, and $\alpha < 0$ means the modulation frequency decreases across the pulse.

If we substitute (10.21) into (10.9), we get

$$|\chi(\tau,f)|^2 = \left|\int_{-\infty}^{\infty} e^{j\pi\alpha t^2} \operatorname{rect}\left[\frac{t - \tau_p/2}{\tau_p}\right] e^{-j\pi\alpha(t+\tau_p)^2} \operatorname{rect}\left[\frac{t - \tau_p/2 + \tau}{\tau_p}\right] \cdot e^{j2\pi f t} dt\right|^2 \quad (10.23)$$

which, after manipulation, becomes

$$|\chi(\tau,f)|^2 = \left|e^{-j\pi\alpha\tau^2}\int_{-\infty}^{\infty} \operatorname{rect}\left[\frac{t - \tau_p/2}{\tau_p}\right] \operatorname{rect}\left[\frac{t - \tau_p/2 + \tau}{\tau_p}\right] e^{j2\pi(f - \alpha\tau)t} dt\right|^2$$

$$= \left|\int_{-\infty}^{\infty} \operatorname{rect}\left[\frac{t - \tau_p/2}{\tau_p}\right] \operatorname{rect}\left[\frac{t - \tau_p/2 + \tau}{\tau_p}\right] e^{j2\pi(f - \alpha\tau)t} dt\right|^2 \quad (10.24)$$

where we made use of $\left|e^{-j\pi\alpha\tau^2}\right| = 1$ to eliminate it from the absolute value.

If we follow the integration steps of Example 1, we get $|\chi(\tau,f)|^2 = 0$ for $|\tau| \geq \tau_p$. For $-\tau_p < \tau < 0$, we get [see (10.12)]

$$|\chi(\tau,f)|^2 = \left|\int_{-\tau}^{\tau_p} e^{j2\pi(f - \alpha\tau)t} dt\right|^2 = \left|\frac{e^{j2\pi(f - \alpha\tau)\tau_p} - e^{-j2\pi(f - \alpha\tau)\tau}}{j2\pi(f - \alpha\tau)}\right|^2 \quad (10.25)$$

If we factor $e^{j\pi(f - \alpha\tau)(\tau_p - \tau)}$ from the third term of (10.25), we get

$$|\chi(\tau,f)|^2 = \left|e^{j\pi(f - \alpha\tau)(\tau_p - \tau)}\left(\frac{e^{j\pi(f - \alpha\tau)(\tau_p + \tau)} - e^{-j\pi(f - \alpha\tau)(\tau_p + \tau)}}{j2\pi(f - \alpha\tau)}\right)\right|^2 \quad (10.26)$$

As with Example 1, we multiply (10.26) by $|\tau_p + \tau|/|\tau_p + \tau|$ and manipulate it to get

$$|\chi(\tau,f)|^2 = |\tau_p + \tau|^2 \left|\operatorname{sinc}\left([f-\alpha\tau][\tau_p+\tau]\right)\right|^2 \tag{10.27}$$

For $\tau_p > \tau \geq 0$ we get [see (10.15)]

$$|\chi(\tau,f)|^2 = \left|\int_0^{\tau_p-\tau} e^{j2\pi(f-\alpha\tau)t} dt\right|^2 = \left|\frac{e^{j2\pi(f-\alpha\tau)(\tau_p-\tau)} - 1}{j2\pi(f-\alpha\tau)}\right|^2 \tag{10.28}$$

If we factor $e^{j\pi(f-\alpha\tau)(\tau_p-\tau)}$ from both terms on the far right side (10.28) and then multiply by $|\tau_p - \tau|/|\tau_p - \tau|$, we get

$$|\chi(\tau,f)|^2 = \left|e^{j\pi(f-\alpha\tau)(\tau_p-\tau)}\left(\frac{e^{j\pi(f-\alpha\tau)(\tau_p-\tau)} - e^{-j\pi(f-\alpha\tau)(\tau_p-\tau)}}{j2\pi(f-\alpha\tau)}\right)\right|^2 \tag{10.29}$$

$$= |\tau_p - \tau|^2 \left|\operatorname{sinc}\left([f-\alpha\tau][\tau_p-\tau]\right)\right|^2$$

Since $\tau < 0$ in (10.27), we can replace $\tau_p + \tau$ by $\tau_p - |\tau|$ in that equation. In (10.29) $\tau \geq 0$ and we can replace $\tau_p - \tau$ with $\tau_p - |\tau|$. When we do this, (10.27) and (10.29) have the same form. Finally, if we combine this with the $|\tau| \geq \tau_p$ condition, we have

$$|\chi(\tau,f)|^2 = (\tau_p - |\tau|)^2 \left|\operatorname{sinc}\left([f-\alpha\tau][\tau_p-|\tau|]\right)\right|^2 \operatorname{rect}\left[\frac{\tau}{2\tau_p}\right] \tag{10.30}$$

We note that (10.30) is the same form as (10.18) except f is replaced by $f - \alpha\tau$. In fact, if we compare (10.24) to (10.11), the only difference is, the f in (10.11) is replaced by $f - \alpha\tau$ in (10.24). Thus, we could have replaced the f in (10.18) to get (10.30) and avoided the various integration steps of this example. However, we did not know this in advance.

As a specific example, we consider the LFM pulse of Chapter 8. For that case we had a pulsewidth of $\tau_p = 15$ μs and an LFM bandwidth of $B = 1$ MHz. Further, we stipulated $\alpha > 0$.

The square root of the matched-Doppler range cut of the ambiguity function is obtained by setting $f = 0$. The result is

$$|\chi(\tau,0)| = (\tau_p - |\tau|) \left|\operatorname{sinc}\left([-\alpha\tau][\tau_p-|\tau|]\right)\right| \operatorname{rect}\left[\frac{\tau}{2\tau_p}\right] \tag{10.31}$$

This is plotted in Figure 10.8 and is the same as the matched filter output from Chapter 8.

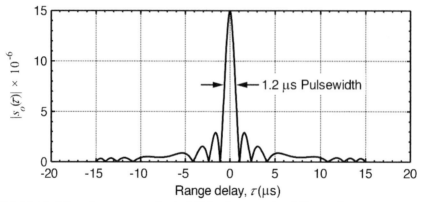

Figure 10.8 Matched-Doppler, range cut for the LFM pulse example of Chapter 8.

The square root of the matched-range, Doppler cut, which we obtain by setting $\tau = 0$, is

$$|\chi(0,f)| = \tau_p \left| \text{sinc}(f\tau_p) \right| \qquad (10.32)$$

This has exactly the same form as for the unmodulated pulse. This leads to an interesting fact of ambiguity functions: the matched-range, Doppler cut is given by the same equation [i.e., (10.20) and (10.32)] for any $u(t)$ and $v(t)$ that satisfies

$$u(t) = v(t) = e^{j\phi(t)} \text{rect}\left[\frac{t - \tau_p/2}{\tau_p}\right] \qquad (10.33)$$

In this equation, $\phi(t)$ is an arbitrary phase function. The proof of this is left as an exercise.

Figure 10.9 contains a plot of $|\chi(\tau, f)|$ for the case where $-\tau_p \leq \tau \leq \tau_p$ and $-B \leq f \leq B$. As with Figure 10.7, the height has been normalized to unity. The actual height is τ_p, or 15 μs. This exhibits the same triangle ridge as does the plot of $|\chi(\tau, f)|$ for an unmodulated pulse, except that the ridge slants across range-Doppler space for the LFM pulse while it is concentrated along $f = 0$ for the unmodulated pulse. This slanting of the LFM ridge makes the LFM waveform useful in search radars, when compared to an unmodulated pulse of the same duration.

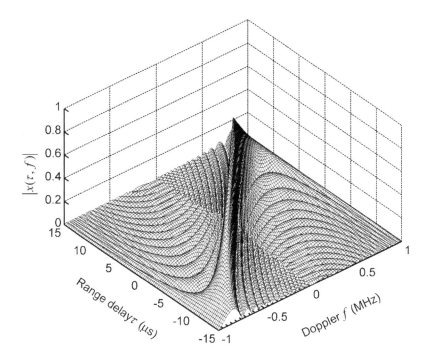

Figure 10.9 Ambiguity function plot of an LFM pulse for a pulse width of 15 μs and a bandwidth of 1 MHz.

10.5 NUMERICAL TECHNIQUES

While the analytical approach discussed above is suitable for deriving the ambiguity function of simple waveforms and signal processors, it becomes very tedious for complex waveforms and signal processors. Because of this, we present a numerical technique for generating the ambiguity function. This technique relies on the fact that modern computers are very fast and that efficient software algorithms are available—specifically, FFT algorithms.

With this technique, we develop the ambiguity function as a sequence of range cuts at different Doppler frequencies. If we recall the general representation of the ambiguity function, we note that $\chi(\tau,f)$ can be interpreted as the correlation of $u(\tau)e^{j2\pi ft}$ with $v(\tau)$. That is,

$$\chi(\tau,f) = \int_{-\infty}^{\infty} u(t)v^*(t+\tau)e^{j2\pi ft}dt = \left[u(\tau)e^{j2\pi f\tau}\right] \otimes v(\tau) \qquad (10.34)$$

where $\otimes$ denotes correlation.

From Fourier transform theory, we recognize that

$$\Im\{\chi(\tau,f)\} = \Im^*\{u(\tau)e^{j2\pi f\tau}\}\Im\{v(\tau)\} \qquad (10.35)$$

or

$$F_\chi(\theta) = F_u^*(\theta) F_v(\theta) \tag{10.36}$$

where the star (*) superscript denotes complex conjugation. Also, the symbology $\Im\{x\}$ denotes the Fourier transform operator and

$$F(\theta) = \Im\{x(t)\} = \int_{-\infty}^{\infty} x(t) e^{-j2\pi\theta t} dt \tag{10.37}$$

In the algorithm discussed in the next section, we use the FFT to approximate the two Fourier transforms on the right of (10.36). We then perform the indicated multiplication in the θ domain and use the inverse FFT (IFFT) to determine $\chi(\tau, f)$. We do this for the values $f = f_k$ of interest. Thus, every FFT-multiply-IFFT will result in a range cut of the ambiguity function at $f = f_k$. This is why we say that we develop the ambiguity function as a sequence of range cuts.

10.6 AMBIGUITY FUNCTION GENERATION USING THE FFT

We have already discussed how we can use Fourier transforms to compute the ambiguity function by computing individual range cuts. Specifically, let $u(t)$ be the baseband transmitted signal and $v(t)$ be the signal to which the matched filter (or signal processor) is matched. Let f be the Doppler mismatch at which we want to generate a range cut. The algorithm used to generate the range cut, $\chi(\tau, f)$, is

- Find $\Im[u(t)e^{j2\pi ft}] = F_u(\theta)$.
- Find $\Im[v(t)] = F_v(\theta)$.
- Find $F_\chi(\theta) = F_u^*(\theta)F_v(\theta) = \Im[\chi(\tau, f)]$.
- Find $\chi(\tau, f) = \Im^{-1}[F_\chi(\theta)]$.

In the above, we use the FFT and inverse FFT to approximate $\Im[\bullet]$ and $\Im^{-1}[\bullet]$. When using the FFT, we need to be sure we satisfy the Shannon sampling theorem[4] and account for some time properties of the FFT and ambiguity function [3, 4]. The algorithm is as follows:

1. Create a time array, t, whose length, N, is an integer power of 2 (i.e., $N = 2^M$ where M is a positive integer)[5] and extends from 0 to T, where T is equal to, or greater than, the sum of the durations of $u(t)$ and $v(t)$. This assures the resulting range cut will cover all time values over which the range cut is not zero. Some restrictions on N are:
    - Choose N so that N/T satisfies Shannon's sampling theorem. Specifically, choose N such that $N/T > 2F$ where F is the highest frequency of $u(t)$ and/or $v(t)$ that is of interest. F should be chosen so that it is more than twice the waveform bandwidth. Five times the waveform bandwidth works fairly well.

---

[4] Also known as the Nyquist sampling theorem (perhaps erroneously) after Harry Nyquist [5]. In Russia, the equivalent theorem is known by the name Kotel'nikov after Vladimir Aleksandrovich Kotel'nikov (Владимир Александрович Котельников) [6].

[5] This restriction on N is not a necessity with modern computers and FFT algorithms. It turns out that if N is a product of relatively small prime numbers, modern FFT routines are quite fast. Even if this is not the case, they are reasonably fast. The restriction stated here is convenient, but could result in the computation of range-cut values at unnecessary points.

- The larger N is, the better each range cut will look.
2. Compute v = $v(t)$. Note: $v(t)$ (and v) will contain several zero values at the end.
3. Compute FFT(v) = F_v.
4. Select a Doppler mismatch frequency, f.
5. Compute u = $u(t)e^{j2\pi ft}$. $u(t)e^{j2\pi ft}$ (and u) will also contain several zero values at the end.
6. Compute FFT(u) = F_u.
7. Compute $|\chi(\tau,f)|$ = X = $|\text{IFFT}(F_u^* F_v)|$.
   - Because of the way the FFT and IFFT is implemented, $\tau = 0$ corresponds to the first tap, $\tau = T/N$ corresponds to the second tap, $\tau = T - T/N$ corresponds to the Nth FFT tap, and so forth.
   - To get the $\tau = 0$ point in the center of a plot of $|\chi(\tau,f)|$ versus τ, which is desired, we need to rearrange the IFFT outputs and redefine a time array using the next two steps.
8. Swap the upper and lower $N/2$ samples of X. In MATLAB, this is accomplished by using the fftshift function.
9. Create a new τ array that contains N samples and extends from $-T/2$ to $T/2 - T/N$ in steps of T/N.
   - When we plot X versus τ, we have a range cut of the square root of the ambiguity function at a mismatch Doppler of f.
10. To create another range cut at a different Doppler mismatch, repeat steps 4 through 9 with a different f.
11. To create a three-dimensional-looking depiction of $|\chi(\tau,f)|$ (as in Figures 10.7 and 10.9), assemble the series of range cuts and plot them using a 3-D plotting routine such as the mesh or surf plotting function in MATLAB. Figures 10.7 and 10.9 were plotted using the mesh plotting function.

10.7 EXERCISES

1. Show that the matched-range, Doppler cut of the ambiguity function of a rectangular pulse with a width of τ_p and an arbitrary phase modulation [see (10.33)] is given by (10.32). That is, that the matched-range, Doppler cut is a $(\text{sinc}[x])^2$ function.

   2. Implement the algorithm of Section 10.6 and use it to recreate a plot like Figure 10.7. Figure 10.7 was created for a 1-μs unmodulated pulse. The range delay axis extends from −1 μs to 1 μs and the Doppler axis extends from −5 MHz to 5 MHz.

   3. Use the algorithm from Exercise 2 to recreate a plot like Figure 10.9. The waveform corresponding to the plot of Figure 10.9 is a 15-μs LFM pulse with a bandwidth of 1 MHz and an increasing frequency (positive α). The range delay axis of the plot extends from −15 μs to 15 μs and the Doppler axis extends from −1 MHz to 1 MHz. Also reproduce the range cut of Figure 10.8.

   4. Show that if $u(t) = v(t)$, $|\chi(\tau,f)|^2 = |\chi(-\tau,-f)^2|$. This will prove useful when we consider the ambiguity function of more complicated waveforms in Chapter 11.

References

[1] Woodward, P. M., *Probability and Information Theory with Applications to Radar,* 2nd ed., New York: Pergamon Press, 1953; reprinted: Dedham, MA: Artech House, 1980.

[2] Pace, T., "From the Editor-in-Chief [Dr. Philip Woodward, Recipient of the IEEE Picard Model]," *IEEE Aerosp. Electron. Syst. Mag.,* vol. 25, no. 2, 2010, p. 3.

[3] Shannon, C. E., "A Mathematical Theory of Communication," *Bell Syst. Tech. J.,* vol. 27, no. 3, Jul. 1948, pp. 379–423.

[4] Shannon, C. E., "Communication in the Presence of Noise," *Proc. IRE,* vol. 37, no. 1, Jan. 1949, pp. 10–21. Reprinted as a classic paper in *Proc. IEEE,* vol. 86, no. 2, Feb. 1998.

[5] Nyquist, H., "Certain Topics in Telegraph Transmission Theory," *Trans. AIEE,* vol. 47, no. 2, Apr. 1928, pp. 617–644. Reprinted as a classic paper in *Proc. IEEE,* vol. 90, no. 2, Feb. 2002.

[6] Kotel'nikov, V. A., "On the Capacity of the 'Ether' and Cables in Electrical Communications," in *Proc. First All-Union Conf. Technolog. Reconstruction of the Commun. Sector and Develop. of Low-Current Eng.,* Moscow: 1933. Translated by C. C. Bissell and V. E. Katsnelson. http://ict.open.ac.uk/classics/1.pd.

Chapter 11

Waveform Coding

11.1 INTRODUCTION

Waveform coding means a phase modulation is applied to the transmit pulse. Specifically, we assume the transmit pulse is of the form

$$v_T(t) = e^{j\phi(t)} \text{rect}\left[\frac{t}{\tau_p}\right] \quad (11.1)$$

where $\phi(t)$ is the phase modulation function and τ_p is the pulsewidth. The inclusion of the rect[x] function means we assume the transmit pulse has a rectangular envelope or, more specifically, a constant amplitude.[1] The assumption of a constant amplitude is consistent with current transmitter technology in that the final amplifier of most transmitters operate in saturation and thus cannot support pulses with amplitude modulation [1, 2].

Our first encounter with a phase-coded pulse was the LFM pulse (see Chapters 8 and 10), which had a $\phi(t)$ of the form $\phi(t) = \pi\alpha t^2$, a quadratic function of time. Because of this, we say the pulse has quadratic phase coding. As may be recalled, the term *linear* FM derives from the fact that the frequency variation across the pulse is a linear function of time. That is, $f(t) = \alpha t$. $f(t) = (1/2\pi)(d\phi(t)/dt) = \alpha t$

A variant of LFM that we will examine in this chapter is nonlinear FM, or NLFM. With NLFM, the frequency variation across the pulse is a nonlinear function of time. The attraction of NLFM is that the matched-Doppler, range cut of the ambiguity function of an NLFM waveform can have lower sidelobes than an equivalent bandwidth LFM waveform, without the normal degradation of SNR associated with sidelobe reduction.

---

[1] Strictly speaking, $v_T(t)$ is an idealized form of the transmit pulse. The actual pulse cannot have a true rectangular envelope because such an envelope implies the transmitter has infinite bandwidth. Practically, the envelope of the transmit pulse is close to rectangular.

With LFM and NLFM waveforms, $\phi(t)$ is a continuous function of time. Another type of phase-coded waveform is one where $\phi(t)$ is a discrete function of time. That is, $\phi(t)$ is of the form

$$\phi(t) = \sum_{k=0}^{K-1} \phi_k \operatorname{rect}\left[\frac{t - k\tau_c}{\tau_c}\right] \quad (11.2)$$

In other words, the phase is constant over some time period, τ_c, but can change from time period to time period. Examples of this type of phase-coded pulse include Frank polyphase pulses, Barker coded pulses, and pseudorandom noise (PRN) coded pulses, all of which we will consider in this chapter.

FM and the discrete phase coding just mentioned are applied to a single pulse. Another type of waveform coding we will discuss is frequency coding, or frequency hop waveforms. The frequency coded waveforms we will consider consist of a group, or burst, of pulses, where each pulse has a different carrier frequency and the pulses are spaced such that the return from pulse k is received before pulse $k+1$ is transmitted (unambiguous range operation—see Chapter 1).

A variation of step frequency waveforms are those based on Costas arrays, or Costas sequences. These waveforms, which we term Costas waveforms, are similar to the aforementioned phase-coded waveforms. However, rather than use a different phase over the interval τ_c, they use a different frequency.

The main tool we will use to analyze phase-coded waveforms is the ambiguity function or, more accurately, the square root of the ambiguity function, $|\chi(\tau, f)|$. This implies the coded pulses are processed by a matched filter, which we will assume. The exception to this will be the LFM pulse. In that case, we will consider a mismatched filter designed to reduce range sidelobes (the sidelobes of the matched-Doppler, range cut of $|\chi(\tau, f)|$).

The ambiguity function is the analysis tool of choice because we can use it to examine range resolution and range sidelobes, as well as the sidelobes in the regions off of the range cut (matched-Doppler, range cut) and Doppler cut (matched-range, Doppler cut).

It appears Robert H. Dicke developed the concept of waveform coding in the early 1940s [3]. In 1945, he applied for a patent for a system that used an LFM waveform [4]. Sidney Darlington also worked on coded waveforms during that time, but Dicke beat Darlington to print [5–7]. According to Skolnik, the first use of a coded waveform in a fielded radar occurred in the mid-1950s. That radar used a pulse with 200 discrete phase changes [$K = 200$ in (11.2)] [8]. The phases changed randomly between 0 and π (binary phase coding). Skolnik indicated the first use of LFM in a radar occurred sometime after that.

In his patent description, Dicke termed his matched filter a *compression filter*. That was most likely the origin of the term *pulse compression* that is commonly used in connection with phase-coded waveforms and their processing.

Since their introduction in the 1940s and 1950s, many different types of waveform coding have been developed or adapted from other disciplines, such as cryptography, cell phones, spread spectrum, GPS, communications, and information theory [9–16].

We will begin our discussions by revisiting LFM pulses. We will specifically investigate the use of a mismatched filter that incorporates amplitude weighting for the purpose of reducing range sidelobes. After that, we will discuss pulses with NLFM. We will present a method for synthesizing $\phi(t)$ for NLFM pulses.

We next consider discrete phased-coded pulses. We start by discussing two classic codings: Frank polyphase and Barker. With the latter, we also briefly discuss polyphase Barker codes and minimum peak sidelobe codes. We next discuss coding based on PRN sequences. PRN sequences are widely used in communications and have interesting properties that make them attractive as radar waveforms.

We next discuss step frequency waveforms. Step frequency waveforms provide a means of achieving fine range resolution without requiring the radar to have a large instantaneous bandwidth. We close the chapter with a discussion of Costas waveforms.

11.2 FM WAVEFORMS

11.2.1 LFM with Amplitude Weighting

One of the characteristics of LFM waveforms is that the first few sidelobes of the matched filter output are somewhat large. This is illustrated in the left half of Figure 11.1, which is a plot of the matched filter output for a 15-μs pulse with an LFM bandwidth of 2 MHz. As the figure shows, the first and second sidelobes are about 14 and 19 dB below the peak.[2] This ratio is fairly consistent for different BT products, where we recall that the BT product is the product of the pulsewidth, τ_p, and the LFM bandwidth, B. The waveform associated with Figure 11.1 has a BT product of 2 MHz × 15 μs = 30.

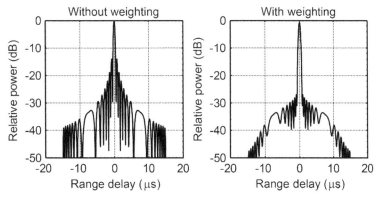

Figure 11.1 Matched filter response for an unweighted and weighted LFM pulse.

---

[2] As a note, as the BT product becomes larger, the first and second sidelobes will approach those of the first and second sidelobes of a sinc(x) function, 13.3 and 17.8 dB, respectively. See Appendix 10A for a more detailed explanation.

With LFM waveforms, it is possible to apply an amplitude weighting in the matched filter to reduce range sidelobes. The result of doing this is illustrated in the right half of Figure 11.1. In this case, the weighting function was a $\bar{n} = 6$, 30-dB Taylor window. As can be seen, the range sidelobes have been significantly reduced.

The amplitude weighting has had two other effects: the peak response is about 0.6 dB below the peak of the unweighted response and the main lobe is wider. The reduction in peak value translates to a loss in SNR, and the width increase translates to a degradation in range resolution.

An example of a weighted, mismatched filter implementation is illustrated in Figure 11.2, which contains a functional block diagram of an FFT-based (mis)matched filter processor. The processor implements (actually, approximates) the equation

$$m(\tau) = \mathcal{F}^{-1}\left[U(f)V^*(f)\right] = \int_{-\infty}^{\infty} u(t)\left[w(t+\tau)v^*(t+\tau)\right]dt \quad (11.3)$$

That is, it correlates the received signal, $u(t)$, with a weighted version of the conjugate of the transmit signal, $w(t)v^*(t)$ (see Chapter 10). In (11.3)

$$U(f) = \mathcal{F}\left[u(t)\right] = \int_{-\infty}^{\infty} u(t)e^{-j2\pi ft}dt \quad (11.4)$$

is the Fourier transform of $u(t)$ and

$$V(f) = \mathcal{F}\left[w(t)v(t)\right] = \int_{-\infty}^{\infty} w(t)v(t)e^{-j2\pi ft}dt \quad (11.5)$$

is the Fourier transform of $w(t)v(t)$. As a note, the weight (Taylor in the above example) is real, so $w^*(t) = w(t)$. $V(f)$ is precomputed and stored.

As an implementation note, in search, the received pulse could be anywhere in the instrumented range interval (see Chapter 1). Thus, the FFT, IFFT (inverse FFT), and the stored matched filter frequency response must be long enough to accommodate the number of samples of $u(t)$ that are in the instrumented range interval.

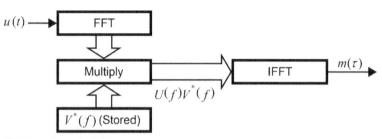

Figure 11.2 FFT-based matched filter.

The minimum sample rate necessary to satisfy the Nyquist criterion [17–19] is the waveform bandwidth, assuming complex samples. However, we have found that we should sample $u(t)$ at about twice the waveform bandwidth to avoid distortion of the range sidelobes of the processor output. This means the FFT, memory, multiplier, and IFFT lengths would need to be $2B\tau_I$, where τ_I is the instrumented range interval. Suppose the PRI associated with the 15-µs, 2-MHz pulse of the previous example was 500 µs and the instrumented range window was 400 µs. This would give $2B\tau_I = 2 \times 2 \times 400 = 1,600$ samples and indicate that the FFT, memory, multiplier, and IFFT sizes should be 2,048 or 2^{11}.

Since the target range is known reasonably well in track, a smaller FFT, memory, multiplier, and IFFT can be used during track. However, the sizes of the devices must be as long as the number of samples in a time interval of twice the pulsewidth. If this is not satisfied, there will be aliasing of $U(f)$ and $V(f)$, which will cause $m(\tau)$ to be incorrect. Thus, the minimum sizes of the components of the matched filter must be greater than $2B\tau_p$. For our example, $2B\tau_p = 2 \times 2 \times 15 = 60$ samples, so the minimum size of each component of the matched filter should be 64, or 2^6.

The implementation of Figure 11.2 was used to generate the matched filter responses and the ambiguity function plots in this book. For these cases, the signals were sampled at 5 to 10 times the pulse bandwidth to produce smooth matched filter plots and plots of $|\chi(\tau,f)|$. To compute the various range cuts of $|\chi(\tau,f)|$, $u(t)$ was offset in frequency by an amount equal to the Doppler mismatch, f, of the range cut.

11.2.2 Nonlinear FM

An alternate method of reducing range sidelobes of FM waveforms is through the use of a nonlinear frequency variation across the pulse. This idea was originally conceived by Key et al. [20, 21] and Watters [22], according to statements by Fowle [23]. The technique has also been discussed by other authors [24–28]. In this chapter, we outline the technique presented by Fowle in his 1964 paper.

Fowle developed his technique through the use of stationary phase integration [29, 30]. We will not attempt to repeat Fowle's development here. Instead, we present the results needed to use the technique to design NLFM waveforms.

In his paper, Fowle addressed the following problem:

Given a function of the form

$$u(t) = u_m(t)e^{j\phi(t)} \tag{11.6}$$

and its Fourier transform

$$U(f) = U_m(f)e^{j\theta(f)} \tag{11.7}$$

how does one determine the phase function $\phi(t)$ so that $u_m^2(t)$ and $U_m^2(f)$ closely approximate desired functions? As a note, $u_m(t)$ and $U_m(f)$ are real and usually positive.

The idea that this approach will provide low range sidelobes stems from the results of the previous section where we found that applying an amplitude weighting in the LFM-matched filter can reduce range sidelobes. Amplitude weighting changes the magnitude of the matched filter frequency response. The question addressed in Fowle's paper is whether changing $\phi(t)$ from a quadratic function of time (which it is for LFM) to some other function of time will similarly change the waveform and matched filter spectrums to produce low range sidelobes. Stated another way, will changing the frequency deviation from a linear function of time to a nonlinear function of time result in lower range sidelobes?

Fowle's technique is as follows: given a desired time function, $u_m(t)$, and a desired frequency function, $U_m(f)$, evaluate the integrals

$$\int_{-\infty}^{t} u_m^2(\gamma)d\gamma = \int_{-\infty}^{f(t)} U_m^2(\zeta)d\zeta \qquad (11.8)$$

to obtain

$$P(t) = Q[f(t)] \qquad (11.9)$$

Solve (11.9) to obtain

$$f(t) = Q^{-1}[P(t)] \qquad (11.10)$$

Next, use

$$\phi(t) = 2\pi \int f(t)dt \qquad (11.11)$$

to determine the desired phase function for $u(t)$. $U_m(\zeta)$ and $u_m(\gamma)$ must be chosen to satisfy Parseval's theorem [13]. That is, they must be chosen so that

$$\int_{-\infty}^{\infty} u_m^2(\gamma)d\gamma = \int_{-\infty}^{\infty} U_m^2(\zeta)d\zeta \qquad (11.12)$$

This can be accomplished through the use of scaling factors, assuming the integrals of (11.12) exist.

The algorithm is applicable to any $U_m(\zeta)$ and $u_m(\gamma)$. However, the case of interest to us is where $u_m(\gamma)$ is a rectangular pulse. That is, where

$$u_m(\gamma) = \frac{1}{\sqrt{\tau_p}} \text{rect}\left[\frac{\gamma}{\tau_p}\right] \qquad (11.13)$$

where, for convenience, we scaled $u_m(\gamma)$ so that the left integral of (11.12) will be unity. With this we have

$$u(\gamma) = \frac{e^{j\phi(\gamma)}}{\sqrt{\tau_p}} \text{rect}\left[\frac{\gamma}{\tau_p}\right] \qquad (11.14)$$

which is a phase modulated, rectangular pulse. With the assumption of (11.13), (11.12) becomes

$$\int_{-\infty}^{f(t)} U_m^2(\zeta)d\zeta = \int_{-\infty}^{t} u_m^2(\gamma)d\gamma = \frac{1}{\tau_p}\int_{-\infty}^{t}\text{rect}\left[\frac{\gamma}{\tau_p}\right]d\gamma$$

$$= \begin{cases} 0 & t < -\tau_p/2 \\ \dfrac{t+\tau_p/2}{\tau_p} & |t| \leq \tau_p/2 \\ 1 & t > \tau_p/2 \end{cases} \qquad (11.15)$$

11.2.2.1 Fowle Example with Uniform $U_m(f)$

As a first example of Fowle's method, we derive a pulse with LFM. It can be shown that the magnitude of the spectrum of an LFM pulse with a bandwidth of B is close to a rectangle function (see Figure 12.1, Chapter 12). Thus, we choose $U_m^2(\zeta)$ as

$$U_m^2(\zeta) = \frac{1}{B}\text{rect}\left[\frac{\zeta}{B}\right] \qquad (11.16)$$

where we chose the scale factor, $1/B$, so that the right side of (11.12) was unity [since the left side is unity for the $u_m(\gamma)$ of (11.13)].

Using this we get, over the interval $|t| \leq \tau_p/2$,

$$\frac{t+\tau_p/2}{\tau_p} = \int_{-\infty}^{f(t)} U_m^2(\zeta)d\zeta = \int_{-\infty}^{f(t)} \frac{1}{B}\text{rect}\left[\frac{\zeta}{B}\right]d\zeta$$

$$= \begin{cases} 0 & f(t) < -B/2 \\ \dfrac{f(t)+B/2}{B} & |f(t)| \leq B/2 \\ 1 & f(t) > B/2 \end{cases} \qquad (11.17)$$

Using the region $|f(t)| \leq B/2$, we get

$$\frac{t+\tau_p/2}{\tau_p} = \frac{f(t)+B/2}{B} \qquad (11.18)$$

or

$$f(t) = (B/\tau_p)t = \alpha t \qquad (11.19)$$

and

$$\phi(t) = 2\pi\int f(t)dt = \pi\alpha t^2 \qquad (11.20)$$

In other words, the method results in an LFM pulse

$$u(t) = \frac{e^{j\pi\alpha t^2}}{\sqrt{\tau_p}} \text{rect}\left[\frac{t}{\tau_p}\right] \tag{11.21}$$

11.2.2.2 Fowle Example with Cosine on a Pedestal $U_m(f)$

As another example, we consider the case where $u_m(\gamma)$ is as in (11.13) and $U_m^2(\zeta)$ is a cosine on a pedestal function (e.g., Hamming, Hanning, and so forth.). With this we get

$$U_m^2(\zeta) = K\left[a + b\cos\left(\frac{\pi\zeta}{B}\right)\right]\text{rect}\left[\frac{\zeta}{B}\right] \tag{11.22}$$

where $b \leq 1$, $a = 1 - b$, and $K = \pi/[B(\pi a + 2b)]$ is chosen so that the right side of (11.12) is equal to unity. The derivation of K is left as an exercise. With this we get

$$\frac{t + \tau_p/2}{\tau_p} = \int_{-\infty}^{f(t)} U_m^2(\zeta)d\zeta = \frac{K}{B}\int_{-\infty}^{f(t)}\left[a + b\cos\left(\frac{\pi\zeta}{B}\right)\right]\text{rect}\left[\frac{\zeta}{B}\right]d\zeta$$

$$= \begin{cases} 0 & f(t) < -B/2 \\ Ka(f(t) + B/2) + Kb\dfrac{B}{\pi}\left[\sin\left(\dfrac{\pi f(t)}{B}\right) + 1\right] & |f(t)| \leq B/2 \\ 1 & f(t) > B/2 \end{cases} \tag{11.23}$$

The above leads to

$$\frac{t + \tau_p/2}{\tau_p} = Ka(f(t) + B/2) + Kb\frac{B}{\pi}\left[\sin\left(\frac{\pi f(t)}{B}\right) + 1\right] \tag{11.24}$$

which we must solve for $f(t)$. Herein lies one of the difficulties with Fowle's method: numerical techniques are often needed to find $f(t)$ for $U_m^2(\zeta)$ functions of interest.

For this particular example, assuming $B = 2$, we can get a closed-form solution of

$$f(t) = \frac{2}{\pi}\sin^{-1}\left(\frac{2t}{\tau_p}\right) \tag{11.25}$$

and

$$\phi(t) = 4\left[t\sin^{-1}\left(\frac{2t}{\tau_p}\right) + \sqrt{(\tau_p/2)^2 - t^2}\right] \tag{11.26}$$

for the case where $b = 1$ (and, thus, $a = 0$). The details are left as an exercise.

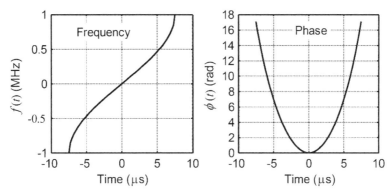

Figure 11.3 NLFM frequency and phase plots.

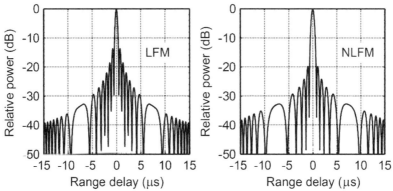

Figure 11.4 Matched filter response for an NLFM pulse and LFM pulse.

Figure 11.3 contains plots of $f(t)$ and $\phi(t)$, and Figure 11.4 contains plots of the matched-Doppler, range cut of the ambiguity function for an LFM waveform and the nonlinear FM waveform we obtained in this example. As the figures show, the frequency modulation is quite nonlinear. Also, the first range sidelobe has been reduced from −14 dB to −20 dB.

11.2.2.3 NLFM Design Procedures

In the examples above, we made some assumptions that allowed us to develop closed-form expressions for $f(t)$ and $\phi(t)$. In general, this is not possible and we must resort to numerical techniques. To that end, we outline a procedure for deriving $f(t)$ and $\phi(t)$. The technique assumes the steps are performed using numerical methods. However, some of them could be completed using analytical techniques, if the various functions are conducive to analytical methods.

1. Select a desired $U_m(\zeta)$ or $U_m^2(\zeta)$ and compute it for several values of ζ over the interval of $-B/2$ to $B/2$, where B is the desired NLFM bandwidth. The values of ζ should be chosen close enough to capture the shape $U_m(\zeta)$. A rule of thumb is to space them less than $1/\tau_p$ apart.

2. Square $U_m(\zeta)$ to get $U_m^2(\zeta)$. This step can be omitted if one starts with $U_m^2(\zeta)$. We did this in the two examples.
3. Numerically compute the integral

$$P(f) = \int_{-B/2}^{f} U_m^2(\zeta) d\zeta \quad (11.27)$$

for f between $-B/2$ and $B/2$. Scale $P(f)$ so that $P(B/2) = 1$. This is needed to satisfy (11.12). These three steps result in

$$\frac{t + \tau_p/2}{\tau_p} = P(f) \quad |f| \leq B/2 \quad (11.28)$$

or

$$t = \tau_p [P(f) - 1/2] \quad (11.29)$$

4. Use the results of Step 3 to generate a tabulation of t versus f and use interpolation to find f as a function of t for values of t between $-\tau_p/2$ and $\tau_p/2$ with a spacing of $\Delta t < 1/B$. A rule of thumb is to start with $\Delta t = 1/(10B)$. The result of this will be $f(t)$.
5. To find $\phi(t)$, numerically compute the integral

$$\phi(t) = \int_{-\tau_p/2}^{t} f(\gamma) d\gamma \quad (11.30)$$

11.3 PHASE-CODED PULSES

In this section, we consider phase coding where the phase changes in discrete steps, rather than continuously. We consider a single pulse that is subdivided into a series of subpulses, or chips, where the durations of the chips are equal. This is not a requirement, but a convenience for our purposes. We then assign a different phase to each chip according to some rule defined by a phase-coding algorithm. We assume the amplitudes of all chips are equal. This is a "semi-requirement" of most phase coding schemes in that they were developed under the assumption that the amplitudes of the chips are the same. In most applications, the chips are adjacent. That is, the waveform has a 100% duty cycle. Again, this is not a hard requirement but is a standard to which waveform designers generally adhere.

We can use this definition to write the normalized, baseband equation of a phase-coded pulse as

$$v(t) = \sum_{k=0}^{K-1} e^{j\phi_k} \text{rect}\left[\frac{t - k\tau_c}{\tau_c}\right] \quad (11.31)$$

where τ_c is the chip width and the pulse consists of K chips. The phases, ϕ_k are assigned according to some phase-coding algorithm.

As indicated earlier, the first phase-coding algorithm used in a radar was based on a random selection of 0 or π phase shifts across 200 chips. Since that time, analysts have developed a wide variety, and a large number, of phase-coding algorithms [9, 31–41].

In this book, we consider only a few phase-coding algorithms. Two of these are Frank polyphase and Barker codes, which are classical phase codes discussed in many radar books [9, 42–48]. As an extension of Barker codes, we briefly discuss minimum peak sidelobe codes [31, 32, 34, 49] and polyphase Barker codes [50–53].

The other phase-coding algorithm we consider is derived from PRN codes. These are also called maximal length codes, shifter register codes, shift register sequences, linear shift register (LSR) codes, and a host of other names [44, 54]. PRN codes are used in many applications including digital television, Global Positioning System (GPS) cell phones, spread spectrum communications, and deep-space communications. They are attractive for radars because they exhibit "good" range sidelobes and "good" off-axis sidelobes. They are also useful in multiple radar applications, such as multiple input, multiple output (MIMO) radars, [55, 56] because there are PRN codes of the same length that are almost orthogonal.

11.3.1 Frank Polyphase Coding

Frank polyphase coding is a digital representation of a quadratic phase shift, the phase shift exhibited by LFM. Frank polyphase codes have lengths that are perfect squares, that is, $K = L^2$ where L is an integer. The code can be formed by first creating an $L \times L$ matrix of the form

$$F_L = \begin{bmatrix} 0 & 0 & 0 & \cdots & 0 \\ 0 & 1 & 2 & \cdots & L-1 \\ 0 & 2 & 4 & \cdots & 2(L-1) \\ \vdots & \vdots & \vdots & \ddots & \vdots \\ 0 & (L-1) & 2(L-1) & \cdots & (L-1)^2 \end{bmatrix} \qquad (11.32)$$

Next, the rows or columns are concatenated to form a vector of length $K = L^2$. Finally, the phase is determined by multiplying each element of the vector by

$$\Delta \phi = 2\pi / L \qquad (11.33)$$

We illustrate this by an example. We consider $L = 4$, which produces a $K = L^2 = 16$ element Frank polyphase code. The Frank polyphase matrix is

$$F_4 = \begin{bmatrix} 0 & 0 & 0 & 0 \\ 0 & 1 & 2 & 3 \\ 0 & 2 & 4 & 6 \\ 0 & 3 & 6 & 9 \end{bmatrix} \qquad (11.34)$$

and

$$\Delta \phi = 2\pi / 4 = \pi / 2 \qquad (11.35)$$

The vector of phase shifts is

$$\phi_k = [0\ 0\ 0\ 0\ 0\ 1\ 2\ 3\ 0\ 2\ 4\ 6\ 0\ 3\ 6\ 9] \times \pi/2 \qquad (11.36)$$

The resulting Frank polyphase-coded pulse is

$$v_{FP}(t) = \sum_{k=0}^{K-1} e^{j\phi_k} \operatorname{rect}\left[\frac{t - k\tau_c}{\tau_c}\right] \qquad (11.37)$$

A plot of $|\chi(\tau, f)|$ for this example is shown in Figure 11.5. In the plot, Doppler ranges from 0 to $1/\tau_c$ and range delay goes from $-16\tau_c$ to $16\tau_c$, where $16\tau_c$ is the total duration of the pulse.

The plot of Figure 11.5 allows us to visualize the structure of the overall $|\chi(\tau, f)|$ function while still being able to visualize the matched-Doppler range cut.[3]

Note that the plot of Figure 11.5 exhibits some semblance of the ridge that is characteristic of LFM waveforms. We might have expected this since the Frank polyphase waveform is a discrete version of an LFM waveform. For reference, a plot of $|\chi(\tau, f)|$ for an LFM waveform with a BT product of 16 is contained in Figure 11.6.[4]

Figure 11.7 contains a plot of ϕ_k (with appropriate phase unwrapping) for the 16-chip example above. It also contains a plot of the phase shift of an LFM pulse that has a BT product of 16, the same as the BT product of the 16-chip Frank polyphase pulse. As the figure illustrates, the Frank polyphase pulse has approximately the same quadratic phase characteristic as an equivalent LFM pulse.

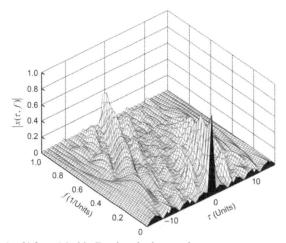

Figure 11.5 Plot of $|\chi(\tau, f)|$ for a 16-chip Frank polyphase pulse.

---

[3] This plotting methodology was adapted from that used in Levanon and Mozeson [9].
[4] The BT product of a K-chip phase coded pulse is normally equal to K. This derives from the observation that the pulse bandwidth is $B = 1/\tau_c$ and the duration of the pulse is $\tau_p = K\tau_c$. Thus BT $= B\tau_p = (1/\tau_c)(K\tau_c) = K$.

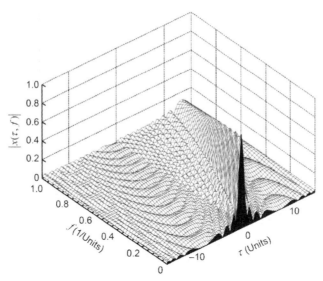

Figure 11.6 Plot of $|\chi(\tau,f)|$ for an LFM pulse with a BT product of 16.

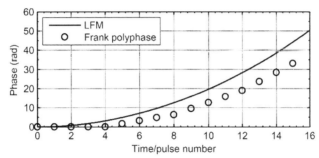

Figure 11.7 Phases for 16-chip Frank polyphase and equivalent LFM pulse.

Several other digital approximations of LFM waveforms have been developed over the years. Two of these are the Zadoff and Chu codes discussed in [9, 57–59]. Others are the Lewis-Kretschmer P1 through P4 codes [43].

11.3.2 Barker-coded Waveforms

A simplification of polyphase-coded pulses are those that use only two phase shifts, usually separated by π (e.g., 0 and π, or $-\pi/2$ and $\pi/2$). These are termed binary phase, or biphase, codes. A common set of binary phase codes found in radar texts are the Barker codes [60, 61]. Barker codes have the interesting property that the peak level of the range sidelobes is $1/K$, assuming the peak of $|\chi(\tau,f)|$ is normalized to unity. Although Barker codes have low range sidelobes, the sidelobe levels of $|\chi(\tau,f)|$ off of matched Doppler can be high, as shown in Figures 11.8 and 11.9.

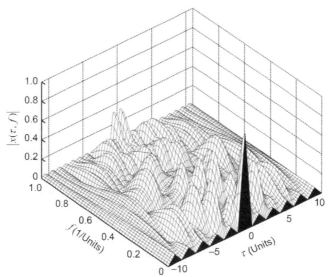

Figure 11.8 Plot of $|\chi(\tau,f)|$ for an 11-chip Barker-coded pulse.

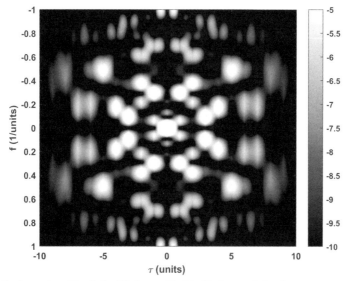

Figure 11.9 Contour plot of $\log(|\chi(\tau,f)|)$ for an 11-chip Barker-coded pulse.

Figure 11.8 is a plot of $|\chi(\tau,f)|$, similar to Figures 11.5 and 11.6. Figure 11.9 is a contour plot of $|\chi(\tau,f)|$, showing f versus τ with $10\log(|\chi(\tau,f)|)$ shown in grayscale. The bar to the right of the plot provides the relation between the values of $10\log(|\chi(\tau,f)|)$ and gray level.

As the plots of Figures 11.8 and 11.9 illustrate, the region near $f=0$ has low-amplitude sidelobes. However, the off-axis regions exhibit several ancillary lobes. This is a property of all of the Barker-coded pulses.

The off-axis behavior of Barker-coded pulses is an illustration of a property of the ambiguity function proved by Woodward, the inventor of the ambiguity function [62]. Specifically, the volume under the ambiguity function is constant and equal to its peak value. That is,

$$|\chi(0,0)|^2 = \int_{-\infty}^{\infty} \int_{-\infty}^{\infty} |\chi(\tau,f)|^2 \, d\tau df \qquad (11.38)$$

This says that if a coding reduces the ambiguity function in one region, the volume in that region is distributed, in some fashion, to other regions. Sometimes it results in ancillary lobes, as in Figures 11.8 and 11.9, and in other cases, it spreads out somewhat uniformly over the τ-f region, as is the case with PRN-coded pulses. As a note, this property applies to the matched filter case, not to the mismatched filter.

There are only seven known Barker codes. They have lengths of 2, 3, 4, 5, 7, 11, and 13. The phase shifts for the seven codes are shown in Table 11.1. Although not "officially" Barker codes, time-reversed versions of Barker codes are Barker codes, in the sense they exhibit the same sidelobe properties.

The low-range sidelobe characteristics of Barker-coded pulses has motivated researchers to find other, longer binary phase-coded pulses that exhibit low range sidelobes. One of these is a class of pulses termed *minimum peak sidelobe* coded pulses. According to Levanon and Mozeson [9], sets of these have been developed by Linder [32], Cohen et al. [31, 39], and Coxson et al. [33, 49]. The peak range sidelobes are not $1/K$ as with Barker-coded pulses; however, they are quite small. Table 11.2 contains a list of minimum peak sidelobe codes of lengths 15 to 25. Other lists can be found in [31, 33, 34, 36]. In Table 11.2, 0 equates to a phase shift of 0 and 1 equates to a phase shift if π.

Figure 11.10 contains range cuts of a 25-chip, minimum peak sidelobe pulse and a 25-chip, Frank polyphase pulse. Note the peak sidelobes of the minimum peak sidelobe pulse are considerably lower than those of the Frank polyphase pulse.

Another extension of Barker codes are *generalized Barker codes* or *polyphase Barker codes* [50]. As the second name implies, these are not binary phase codes but polyphase codes. The range sidelobes of pulses with these codes are below $1/K$. Listings of polyphase Barker codes can be found in [9, 35, 37, 38, 50–53], covering lengths of 4 to 45.

Table 11.1
Phase Shifts for Barker Codes

Code Length	Phase Shifts
2	0 0 or 0 π
3	0 0 π
4	0 0 0 π or 0 0 π 0
5	0 0 0 π 0
7	0 0 0 π π 0 π
11	0 0 0 π π π 0 π π 0 π
13	0 0 0 0 0 π π 0 0 π 0 π 0

Table 11.2
Partial List of Minimum Peak Sidelobe Codes

Length	Code
15	001100000101011
16	0110100001110111
17	00111011101001011
18	011001000011110101
19	1011011101110001111
20	01010001100000011011
21	101101011101110000011
22	0011100110110101011111
23	01110001111110101001001
24	011001001010111111100011
25	1001001010100000011100111

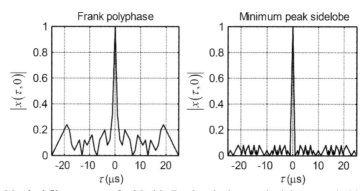

Figure 11.10 Matched filter response for 25-chip Frank polyphase and minimum peak sidelobe pulses.

11.3.3 PRN-coded Pulses

PRN-coded pulses are another class of pulses that use binary phase coding. In this case, the coding is based on PRN codes, which consist of sequences of 0s and 1s and most often have lengths of $K = 2^M - 1$, where M is an integer. The sequences of 0s and 1s are generated by *feedback shift register* devices [54] a functional block diagram of a feedback shift register is contained in Figure 11.11. The boxes with Stage 1, Stage 2, and so forth, represent shift register elements (flip-flops) and the adder is a modulo-2 adder. The block with z^{-1} is a delay, or buffer, that holds the result of the modulo-2 addition before it is loaded in the first shift register. The feedback configuration is usually chosen such that the sequence of 0s and 1s at the output repeats only after $K = 2^M - 1$ samples. Such a sequence of 0s and 1s is termed a *maximal length sequence* or *m-sequence* [16, 42, 63–66]. The phase codes used on the chips of the PRN-coded pulse is the PRN sequence multiplied by π. Solomon Wolf Golomb is generally credited with developing and characterizing maximal length sequences [15, 54]. However, in his book [16], Golomb gives credit to James Singer as the actual inventor of maximal length sequences.[5]

Table 11.3 contains a partial list of feedback configurations that can be used to generate maximal length sequences for M between 3 and 10. The numbers in the table denote the shift register outputs that are added and fed back to the input. The tap numbering corresponds to the stage number in Figure 11.11. For example, the $M = 4$ case shown in the table is (4, 3) and indicates that the output of shift registers 3 and 4 would be added and fed back to the first shift register input. This specific example is illustrated in Figure 11.12. The entries in Table 11.3 were obtained from a website hosted by New Wave Instruments [70], which has a much more complete list. Other sources include [63, 64, 71].

As pointed out in [70], the entries in Table 11.3 represent only half of the possible feedback configurations. If one of the entries in the table is (M, a, b, c), the companion to that entry would be (M, M − a, M − b, M − c). For example, one of the entries for $M = 6$ is (6, 5, 4, 1), so its companion would be (6, 6 − 5, 6 − 4, 6 − 1) = (6, 1, 2, 5).

Table 11.3

Feedback Tap Configurations for M = 3 to 10

M	Feedback Taps
3	3, 2
4	4, 3
5	(5, 3), (5, 4, 3, 2), (5, 4, 3, 1)
6	(6, 5), (6, 5, 4, 1), (6, 5, 3, 2)
7	(7, 6), (7, 4), (7, 6, 5, 4), (7, 6, 5, 2), (7, 6, 4, 2), (7, 6, 4, 1), (7, 5, 4, 3), (7, 6, 5, 4, 3, 2), (7, 6, 5, 4, 2, 1)
8	(8, 7, 6, 1), (8, 7, 5, 3), (8, 7, 3, 2), (8, 6, 5, 4), (8, 6, 5, 3), (8, 6, 5, 2), (8, 7, 6, 5, 4, 2), (8, 7, 6, 5, 2, 1)
9	(9, 5), (9, 8, 7, 2), (9, 8, 6, 5), (9, 8, 5, 4), (9, 8, 5, 1), (9, 8, 4, 2), (9, 7, 6, 4), (9, 7, 5, 2), (9, 6, 5, 3), (9, 8, 7, 6, 5, 3), (9, 8, 7, 6, 5, 1), (9, 8, 7, 6, 4, 3), (9, 8, 7, 6, 4, 2), (9, 8, 7, 6, 3, 2), (9, 8, 7, 6, 3, 1), (9, 8, 7, 6, 2, 1), (9, 8, 7, 5, 4, 3), (9, 8, 7, 5, 4, 2), (9, 8, 6, 5, 4, 1), (9, 8, 6, 5, 3, 2), (9, 8, 6, 5, 3, 1), (9, 7, 6, 5, 4, 3), (9, 7, 6, 5, 4, 2), (9, 8, 7, 6, 5, 4, 3, 1)
10	(10, 7), (10, 9, 8, 5), (10, 9, 7, 6), (10, 9, 7, 3), (10, 9, 6, 1), (10, 9, 5, 2), (10, 9, 4, 2), (10, 8, 7, 5), (10, 8, 7, 2), (10, 8, 5, 4), (10, 8, 4, 3), (10, 9, 8, 7, 5, 4), (10, 9, 8, 7, 4, 1), (10, 9, 8, 7, 3, 2), (10, 9, 8, 6, 5, 1), (10, 9, 8, 6, 4, 3), (10, 9, 8, 6, 4, 2), (10, 9, 8, 6, 3, 2), (10, 9, 8, 6, 2, 1), (10, 9, 8, 5, 4, 3), (10, 9, 8, 4, 3, 2), (10, 9, 7, 6, 4, 1), (10, 9, 7, 5, 4, 2), (10, 9, 6, 5, 4, 3), (10, 8, 7, 6, 5, 2), (10, 9, 8, 7, 6, 5, 4, 3), (10, 9, 8, 7, 6, 5, 4, 1), (10, 9, 8, 7, 6, 4, 3, 1), (10, 9, 8, 6, 5, 4, 3, 2), (10, 9, 7, 6, 5, 4, 3, 2)

---

[5] Among other awards for his contributions to information theory and shift register sequence theory and their application in digital communications, Golomb was awarded the IEEE Shannon Award in 1985 [67], IEEE Richard W. Hamming Medal in 2000 [68], and National Medal of Science in 2011 [69].

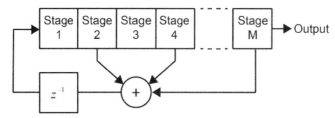

Figure 11.11 *M*-stage feedback shift register.

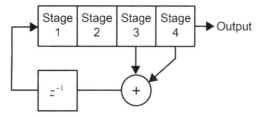

Figure 11.12 Shift register configuration of the (4, 3) entry of Table 11.3.

A particular maximal length sequence is generated by initializing the shift register with any M digit binary number except zero. As an interesting note, the sequence generated with any one shift register initialization (initial load) is not a unique sequence, but a circular shift of a sequence that results from some other initial load (see Exercise 12). Unique sequences are generated by choosing different feedback configurations. Although changing the initial load does not produce a unique sequence, it can have a significant effect of the range sidelobes of the PRN-coded pulse.

The operation of the shift register generator is as follows:

1. The modulo 2 addition is performed and the result is loaded into the z^{-1} buffer.
2. The shift register contents are shifted one bit to the right, and the output of shift register element M is shifted into an output buffer.
3. The result stored in the z^{-1} buffer is loaded into shift register stage 1.
4. Steps 1 through 3 are repeated until the output buffer contains $2^M - 1$ elements.

An interesting feature of PRN-coded pulses is that codes based on different feedback configurations will be almost orthogonal. By "almost orthogonal" we mean that if the PRN pulse based on one feedback configuration is processed through a matched filter matched to a PRN pulse based on a different feedback configuration, the output will not have a peak, but will look like noise (see Exercise 13).

Figure 11.13 contains a plot of $|\chi(\tau, f)|$ for a 15-chip PRN-coded pulse where the PRN code was generated with the feedback configuration of Figure 11.12 and an initial load of 1111. The range cut does not have sidelobes that are as low as comparable length Frank polyphase or Barker pulses. However, $|\chi(\tau, f)|$ does not have the ridge or peaks that the other two codings exhibit. This is a characteristic of PRN-coded waveforms: their sidelobe levels are generally "okay" but not extremely low or high. Long PRN-coded waveforms have $|\chi(\tau, f)|$ that approach the ideal "thumbtack" function [17].

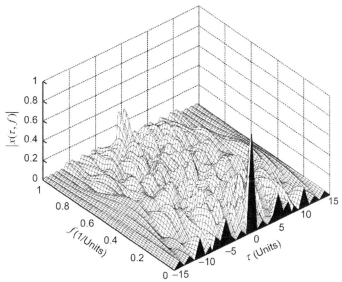

Figure 11.13 Plot of $|\chi(\tau,f)|$ for a 15-chip PRN-coded pulse.

The particular initial load used to generate Figure 11.13 resulted in low sidelobe levels near the central peak. It turns out that the initial load can have a fairly significant impact on the matched-Doppler, range sidelobes of the ambiguity function (see Exercise 15). It also has a lesser impact on the other range-Doppler sidelobes. The only known way to choose an initial load that provides the desired sidelobe characteristics is to experiment.

11.3.3.1 Mismatched PRN Processing

We now want to investigate a special type of processing of PRN-coded pulses that takes advantage of an interesting property of PRN codes. The property we refer to is that the *circular* (or periodic) autocorrelation of a PRN sequence has a value of either K or -1. With a circular correlation, when we shift the sequence to the right by k chips, we take the k chips that "fall off" the end of the shifted sequence and place them at the beginning of the shifted sequence. This is illustrated in Figure 11.14. In this figure, we used the 7-bit PRN code of 1001110 to generate the PRN-coded sequence of $-111-1-1-11$ (in Figure 11.14, and 11.15, the 1 is due to a phase shift of 0 and the -1 is due to a phase shift of π).

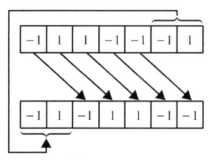

Figure 11.14 Illustration of a 2-bit circular shift.

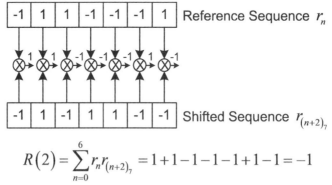

$$R(2) = \sum_{n=0}^{6} r_n r_{(n+2)_7} = 1+1-1-1-1+1-1 = -1$$

Figure 11.15 Illustration of circular correlation for $k = 2$.

To perform the circular autocorrelation, we make a copy of the sequence to produce two sequences. We then circularly shift one sequence by k chips, multiply the result in the K-chip pairs, and form a sum across the K-chip result. This is illustrated in Figure 11.15. Mathematically, we can write the circular correlation as

$$R(k) = \sum_{n=0}^{K-1} r_n r_{(n+k)_K} \qquad (11.39)$$

where $(m)_K$ denotes evaluation of m modulo K. The interesting property of PRN sequences is that

$$R(k) = \begin{cases} K & k = 0, K, 2K, \cdots \\ -1 & \text{otherwise} \end{cases} \qquad (11.40)$$

We now apply this property to examine a special type of PRN-coded waveforms. We will assume that we encode a 0 of the PRN code to a phase of 0 and a 1 to a phase shift of β instead of π. Thus, a PRN-coded pulse corresponding to the 7-bit PRN code of Figures 11.14 and 11.15 (i.e., 1001110) would be as shown in Figure 11.16.

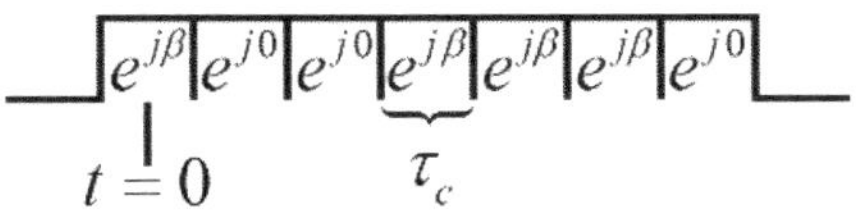

Figure 11.16 Seven-chip, PRN-coded waveform.

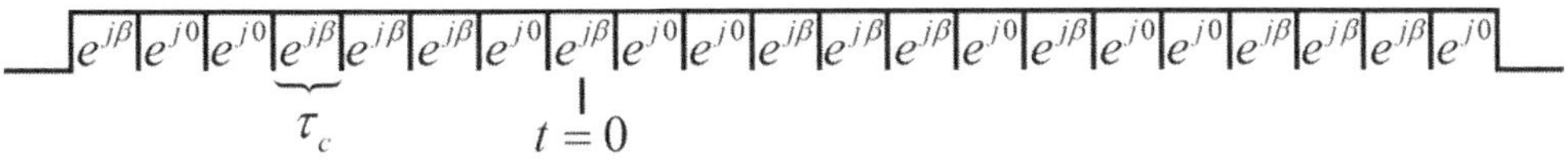

Figure 11.17 Waveform to which the matched filter is matched.

We assume the transmit waveform, $u(t)$, is as shown in Figure 11.16. We define a matched filter that is matched to a signal $v(t)$, where $v(t)$ is a concatenation of three $u(t)$'s. Thus, $v(t)$ would be as shown in Figure 11.17. The $t = 0$ reference points in Figures 11.16 and 11.17 are used to denote the time alignment for matched range. Thus, when the received signal (a scaled version of Figure 11.16) is aligned with the center of the three PRN-coded pulses of Figure 11.17, the matched filter is matched in range to the received pulse.

The matched filter output is

$$v_{MF}(\tau) = \int_{-\infty}^{\infty} u(t) h(\tau - t) dt = \chi(\tau, f)\big|_{f=0} = \int_{-\infty}^{\infty} u(t) v^*(t + \tau) e^{j2\pi ft} dt \bigg|_{f=0} \quad (11.41)$$

We want to examine $v_{MF}(t)$ for $\tau = n\tau_c$ where n is an integer between $-(N-1)$ and $N-1$. We particularly want to examine the form of $u(t)v^*(t + n\tau_c)$. This is illustrated in Figure 11.18 for the 7-chip PRN-coded waveform and $n = 3$.

When we form $u(t)v^*(t + n\tau_c)$, we get

$$u(t)v^*(t + 3\tau_c) = \sum_{k=0}^{K-1} e^{j\phi_k} e^{j\phi_{(k+3)_K}} \operatorname{rect}\left[\frac{t - k\tau_c}{\tau_c}\right] \quad (11.42)$$

and

$$v_{MF}(3\tau_c) = \int_{-\infty}^{\infty} \sum_{k=0}^{K-1} e^{j\phi_k} e^{-j\phi_{(k+3)_K}} \operatorname{rect}\left[\frac{t - k\tau_c}{\tau_c}\right] dt = \tau_c \sum_{k=0}^{K-1} e^{j\left(\phi_k - \phi_{(k+3)_K}\right)} \quad (11.43)$$

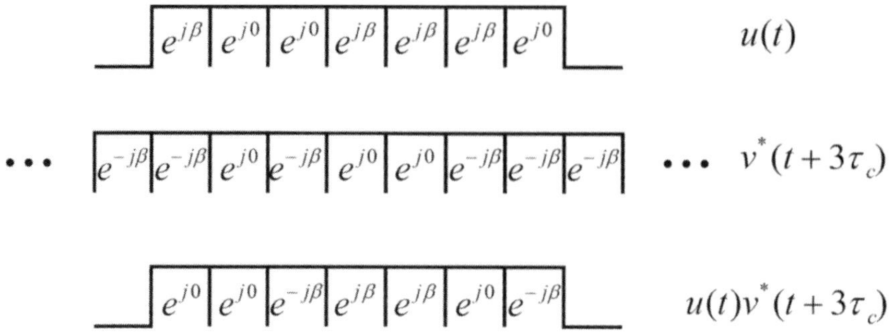

Figure 11.18 Formation of $u(t)v^*(t + n\tau_c)$.

We note from Figure 11.18 that $\phi_k - \phi_{(k+3)K}$ is equal to either 0, β, or $-\beta$. We note further that there are three cases where $\phi_k - \phi_{(k+3)K} = 0$, two cases where $\phi_k - \phi_{(k+3)K} = \beta$, and two cases where $\phi_k - \phi_{(k+3)K} = -\beta$. With this we get

$$v_{MF}(3\tau_c) = 3e^{j0} + 2e^{j\beta} + 2e^{-j\beta} = 3 + 4\cos\beta \qquad (11.44)$$

It turns out that for all $\tau_c < |\tau| < (N-1)\tau_c$

$$v_{MF}(\tau) = 3 + 4\cos\beta \qquad (11.45)$$

In fact, for any K-chip ($K = 2^M - 1$) PRN-coded waveform with $u(t)$ and $v(t)$ chosen by the above rule,

$$v_{MF}(\tau) = \frac{K-1}{2} + \frac{K+1}{2}\cos\beta \qquad \tau_c < |\tau| < (K-1)\tau_c \qquad (11.46)$$

Stated in words, the range sidelobes within $K - 1$ chips of the mainlobe have a constant value as given by (11.46).

As an interesting extension of the above, if we choose β such that

$$\frac{K-1}{2} + \frac{K+1}{2}\cos\beta = 0 \qquad (11.47)$$

or

$$\beta = \cos^{-1}\left(\frac{1-K}{1+K}\right) \qquad (11.48)$$

we get

$$v_{MF}(\tau) = 0 \qquad \tau_c < |\tau| < (K-1)\tau_c \qquad (11.49)$$

That is, range sidelobes within $K-1$ chips of the mainlobe are *zero*. This has the potential of being useful when a radar must be able to detect or track a very small target in the presence of a very large target, provided both targets are at the same Doppler frequency.

Figure 11.19 contains a plot of $|\chi(\tau,f)|$ for the 7-chip PRN example above for the case where β was chosen to be

$$\beta = \cos^{-1}\left(\frac{1-7}{1+7}\right) = \cos^{-1}(-0.75) = 138.59° \qquad (11.50)$$

As predicted, the range sidelobes around the central peak are zero. However, the sidelobes off of matched Doppler rise significantly. Also, the range cut contains two extra peaks. These peaks are range ambiguities and are due to the fact that $u(t)$ correlates with each of the other two end PRN-coded segments of $v(t)$. The range sidelobes adjacent to these range ambiguities are the normal range sidelobes associated with PRN-coded waveforms.

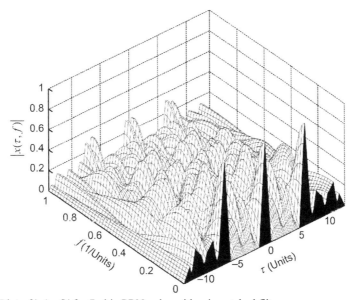

Figure 11.19 Plot of $|\chi(\tau,f)|$ for 7-chip PRN pulse with mismatched filter.

11.4 STEP FREQUENCY WAVEFORMS

With step frequency waveforms the carrier frequency is changed from pulse to pulse. The specific case we consider is shown in Figure 11.20. For this analysis, and in most practical applications, we assume the radar operates unambiguously in range. That is, the signal from pulse k is received before pulse $k + 1$ is transmitted. Thus, we can think of processing one pulse at a time and saving the results for later, further, processing.

We assume the frequency, f_k, of the kth pulse is given by

$$f_k = f_0 + k\Delta f \qquad (11.51)$$

where f_0 is the carrier frequency and Δf is the frequency step. To simplify the development, we assume the individual pulses are unmodulated, though this is not necessary, or desired, in practical applications. We can write the normalized transmit signal for the kth pulse as

$$v_{Tk}(t) = e^{j2\pi(f_0 + k\Delta f)t} \text{rect}\left[\frac{t}{\tau_p}\right] \qquad (11.52)$$

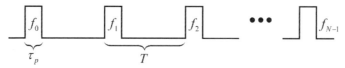

Figure 11.20 Step frequency waveform.

The normalized signal returned from a target at a range delay of τ_R is

$$v_{Rk}(t) = e^{j2\pi(f_0 + k\Delta f)(t - \tau_R)} \text{rect}\left[\frac{t - \tau_R}{\tau_p}\right] \quad (11.53)$$

We assume we know τ_R well enough to be able to sample the matched filter output near its peak. A more accurate measurement of τ_R will be obtained from the output of the step waveform signal processor. For now, we assume the radar and target are fixed so that τ_R is constant.

The heterodyne signal is given by

$$h_k(t) = e^{j2\pi(f_0 + k\Delta f)t} \text{rect}\left[\frac{t - \tau'_R}{\tau_h}\right] \quad (11.54)$$

where τ'_R is close to τ_R and τ_h is large enough for the rect[x] of (11.54) to overlap the rect[x] of (11.53). The frequency of the heterodyne signal is different for every pulse and $h_k(t)$ is perfectly coherent with $v_{Tk}(t)$. The output of the heterodyne operation is

$$v_{Hk}(t) = v_{Rk}(t) h_k^*(t) = e^{-j2\pi f_0 \tau_R} e^{-j2\pi k\Delta f \tau_R} \text{rect}\left[\frac{t - \tau_R}{\tau_p}\right] \quad (11.55)$$

The first term is a constant phase shift that is common to all pulses. We will lump it into some constant that we normalize to unity.

For the next step, we process $v_{Hk}(t)$ by a matched filter matched to rect[t/τ_p] to produce a normalized output of

$$v_{Mk}(t) = e^{-j2\pi k\Delta f \tau_R} \text{tri}\left[\frac{t - \tau_R}{\tau_p}\right] \quad (11.56)$$

where tri[x] is a triangle centered at $x = 0$ with a base width of 2 and a height of unity.

Finally, we sample $v_{Mk}(t)$ at some τ, close to τ_R, to obtain

$$v_{Mk}(\tau) = e^{-j2\pi k\Delta f \tau_R} \text{tri}\left[\frac{\tau - \tau_R}{\tau_p}\right] \quad (11.57)$$

After we obtain $v_{Mk}(\tau)$ from N pulses, we form the sum

$$V(\tau - \tau_R) = \sum_{k=0}^{N-1} a_k v_{Mk}(\tau) = \text{tri}\left[\frac{\tau - \tau_R}{\tau_p}\right] \sum_{k=0}^{N-1} a_k e^{-j2\pi k \Delta f \tau_R} \quad (11.58)$$

where the a_k are complex weight coefficients that we want to choose to maximize $|V(\tau - \tau_R)|$. We recognize (11.58) as the form of the sum we encounter in antenna and stretch processing analyses. We can use this knowledge to postulate that the forms of the a_k that will maximize $|V(\tau - \tau_R)|$ are

$$a_k = e^{j2\pi k \Delta f \tau} \quad (11.59)$$

With this we write $V(\tau - \tau_R)$ as

$$V(\tau - \tau_R) = \text{tri}\left[\frac{\tau - \tau_R}{\tau_p}\right] \sum_{k=0}^{N-1} e^{j2\pi k \Delta f (\tau - \tau_R)} \quad (11.60)$$

which we evaluate and normalize to yield

$$|V(\tau - \tau_R)| = \text{tri}\left[\frac{\tau - \tau_R}{\tau_p}\right] \left|\frac{\sin\left[N\pi\Delta f (\tau - \tau_R)\right]}{\sin\left[\pi\Delta f (\tau - \tau_R)\right]}\right| \quad (11.61)$$

A plot of $|V(\tau - \tau_R)|$ versus $(\tau - \tau_R)\Delta f$ is shown in Figure 11.21 for $N = 10$ and without the tri[x] function. The central peak occurs at $(\tau - \tau_R)\Delta f = 0$ and the first null occurs at $|(\tau - \tau_R)\Delta f| = 1/N = 0.1$. The other peaks, which are range ambiguities, are located at integer values of $(\tau - \tau_R)\Delta f$. This tells us the range resolution of the waveform is

$$\Delta\tau = \frac{1}{N\Delta f} \quad (11.62)$$

and the range ambiguities are located at

$$\tau_{amb} = \frac{k}{\Delta f} \quad (11.63)$$

In the above development, we ignored the tri[x] function to emphasize the location of range ambiguities. If we now include it, we can quantify the effect of the single-pulse matched filter on $|V(\tau - \tau_R)|$. We will add the extra step of recognizing that $V(\tau - \tau_R)$ is the matched-Doppler, range cut $|\chi(\tau,f)|$ for $v_{Tk}(t)$ and, in future references, use $V(\tau - \tau_R) = \chi(\tau - \tau_R, 0)$. With this Figure 11.22 contains plots of $|\chi(\tau - \tau_R, 0)|$ for $\Delta f \tau_p = 0.5, 1$ and 2. The top plot corresponds to the case of $\Delta f \tau_p = 0.5$ and the bottom plot corresponds to the case of $\Delta f \tau_p = 2$. The dashed triangles are the single-pulse matched filter responses.

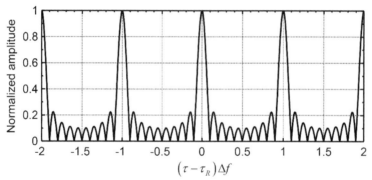

Figure 11.21 Plot of $|V(\tau - \tau_R)|$ without tri$[x]$.

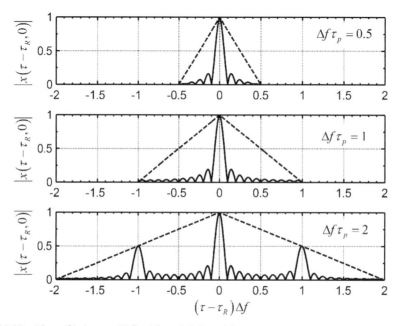

Figure 11.22 Plots of $|\chi(\tau - \tau_R, 0)|$ for $\Delta f \tau_p = 0.5$, 1, and 2.

For $\Delta f \tau_p = 0.5$ and 1, the single-pulse matched filter response nullifies the range ambiguities. However, when $\Delta f \tau_p = 2$, the range ambiguities are present. This interaction between the single-pulse matched filter response and the presence of range ambiguities is a limitation that must be considered when designing step frequency waveforms.

Equation (11.62) tells us we can improve resolution by increasing either Δf or N. If we increase Δf we must consider the properties indicated in Figure 11.22 and ensure that

$$\Delta f \tau_p \leq 1 \tag{11.64}$$

We note that (11.64) can also be satisfied by decreasing τ_p. Thus, we could increase Δf to improve resolution if we can effectively reduce τ_p to satisfy (11.64).

A means of effectively reducing τ_p is to phase code the individual pulses. In that case, τ_p would be the compressed pulsewidth, τ_c.

Increasing the number of pulses to improve resolution must be done with care because it can have negative consequences in terms of timelines and the potential impact of target motion.

11.4.1 Doppler Effects

We now consider the effects of target motion. For now, we will be concerned only with target Doppler frequency. To include target Doppler frequency, we write the target range as

$$R(t) = R_0 + \dot{R}t \tag{11.65}$$

and the target range delay as

$$\tau_R(t) = \tau_{R0} + (2\dot{R}/c)t = \tau_{R0} - (f_d/f_0)t \tag{11.66}$$

We can write the target range delay at the time of the kth transmit pulse as

$$\tau_R(kT) = \tau_{Rk} = \tau_{R0} - (f_d/f_0)kT \tag{11.67}$$

where T is the PRI. In these equations, $f_d = -2\dot{R}/\lambda$ is the Doppler frequency.

If the transmit signal as defined in (11.52), the received signal is

$$\begin{aligned} v_{Rk}(t) &= e^{j2\pi(f_0+k\Delta f)(t-\tau_{Rk})} \text{rect}\left[\frac{t-\tau_{Rk}}{\tau_p}\right] \\ &= e^{j2\pi(f_0+k\Delta f)(t-\tau_{R0}+(f_d/f_0)kT)} \text{rect}\left[\frac{t-\tau_{Rk}}{\tau_p}\right] \end{aligned} \tag{11.68}$$

Manipulating the above using $(f_0 + k\Delta f)(f_d/f_0) \approx f_d$ and $\tau_{Rk} \approx \tau_{R0}$ (in the rect[x] function), we get

$$v_{Rk}(t) = e^{j2\pi(f_0+k\Delta f)t} e^{-j2\pi k\Delta f \tau_{R0}} e^{j2\pi f_d kT} \text{rect}\left[\frac{t-\tau_{R0}}{\tau_p}\right] \tag{11.69}$$

We note that the approximation of $(f_0 + k\Delta f)(f_d/f_0) \approx f_d$ may not be very good because the $k\Delta f$ term could cause a degradation in range resolution. However, the approximation allows us to focus on the effects of target Doppler, and not the potential resolution degradation. This would need to be considered in a more complete analysis.

If we compare (11.69) to (11.53), we note the only difference is the appearance of the term related to Doppler. Thus, if we repeat the heterodyning and matched filtering math from above, we get

$$v_{Mk}(\tau - \tau_{R0}) = e^{-j2\pi k \Delta f \tau_{R0}} e^{j2\pi f_d kT} \text{tri}\left[\frac{\tau - \tau_{R0}}{\tau_p}\right] \quad (11.70)$$

Forming the weighted sum of the $v_{Mk}(\tau - \tau_{R0})$ yields

$$V(\tau - \tau_{R0}) = \sum_{k=0}^{N-1} v_{Mk}(\tau) = \sum_{k=0}^{N-1} b_k e^{-j2\pi(\Delta f \tau_{R0} - f_d T)k} \text{tri}\left[\frac{\tau - \tau_{R0}}{\tau_p}\right] \quad (11.71)$$

We choose b_k as

$$b_k = e^{j2\pi(\Delta f \tau - fT)k} \quad (11.72)$$

which yields

$$\chi(\tau - \tau_{R0}, f - f_d) = \text{tri}\left[\frac{\tau - \tau_{R0}}{\tau_p}\right] \sum_{k=0}^{N-1} e^{j2\pi[(\tau - \tau_{R0})\Delta f - (f - f_d)T]k} \quad (11.73)$$

or, evaluating and normalizing the sum,

$$|\chi(\tau - \tau_{R0}, f - f_d)| = \text{tri}\left[\frac{\tau - \tau_{R0}}{\tau_p}\right] \left|\frac{\sin\{N\pi[(\tau - \tau_{R0})\Delta f - (f - f_d)T]\}}{\sin\{\pi[(\tau - \tau_{R0})\Delta f - (f - f_d)T]\}}\right| \quad (11.74)$$

Figure 11.23 contains a matched-range, Doppler cut (a plot of $|\chi(0, f - f_d)|$ versus $f - f_d$,) of the 10-pulse waveform discussed earlier. In this case, we needed specific values for the parameters and thus chose a PRI of 500 μs, $\tau_p = 1$ μs, and $\Delta f = 1$ MHz. Note that the Doppler resolution of this waveform is 200 Hz or $1/NT = 1/(10 \times 500$ μs), as expected.

Figure 11.24 contains range cuts at matched Doppler and at a Doppler offset of one Doppler resolution cell (200 Hz). Note that a Doppler offset of one Doppler resolution cell causes a range error of one range resolution cell. This indicates the step frequency waveform is very sensitive to Doppler and that, if we want accurate absolute range *measurement,* the range shift due to target Doppler must be removed.

This can be done if the target is in track and the relative velocity between the radar and target is known with reasonable accuracy.

If the step frequency waveform is used in its more common role of target imaging, the various scatterers of the target should be moving at about the same range rate so that range errors due to Doppler differences of the scatterers should be small. One would still want to remove the gross Doppler to minimize losses due to Doppler mismatch. Note that the range cut at $f = 1/NT$ shown in Figure 11.24 is down about $-20\log(0.9) \approx 1$ dB.

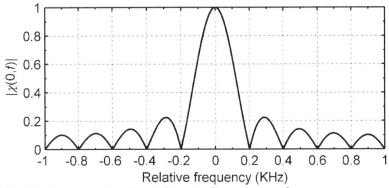

Figure 11.23 Matched-range, Doppler cut for a step frequency waveform.

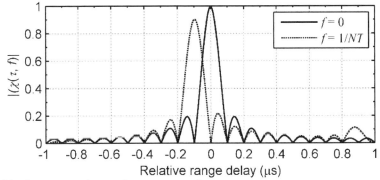

Figure 11.24 Range cuts of a step frequency waveform.

11.5 COSTAS WAVEFORMS

A variation of step frequency waveforms is a waveform based on Costas sequences. These waveforms, which we term Costas waveforms, are similar to the phase-coded pulses discussed in Section 11.3 in that the pulse consists of K adjacent chips or subpulses. However, like the step frequency waveforms of Section 11.4, instead of phase, the frequency is changed from chip to chip. The specific frequency assignments on the chips are based on a Costas sequence.

According to [78], it appears Dr. Costas conceived of his waveforms as a means of mitigating multipath in sonar. He appears to have been searching for a frequency-coded waveform whose ambiguity function approached the ideal, thumbtack ambiguity function. With such a waveform, multipath returns could be separated from direct path returns. The antithesis of such a waveform is a discrete version of an LFM waveform—one where the frequency changes in a linear, monotonic fashion from chip to chip. Costas' idea was that if the frequency changed nonmonotonically, the ambiguity function of the waveform would not exhibit the ridge associated with LFM pulses. He went further to speculate that specific, nonmonotonic frequency changes might be better than others in terms of producing something close to the ideal, thumbtack ambiguity function. The result of his conjectures and studies led to what is now termed Costas sequences.

In his original attempt at deriving Costas sequences [78], Costas used what he termed a permutation array, which is now termed a Costas array. Such an array is a K-by-K matrix whose rows correspond to frequency index and columns correspond to chip number. He placed K 1's in the array such that each row and column contained only one 1. He then shifted a duplicate of the array in all possible row and column directions and compared it to the original array. If the arrangement of 1's in the shifted and original array were always such that they had only one, or no, 1's in common, the frequency index-to-chip number assignment used to create the array was a Costas sequence. According to [78], while this approach was easy to conceptualize, it was difficult to implement. Because of this, Costas devised another technique based on what he termed a difference triangle [9, 79]. He indicates he could use the difference triangle approach to manually derive Costas sequences for orders up to 12. Beyond 12, finding Costas sequences became too tedious by manual means.

In a quest to find longer Costas sequences, Costas contacted Solomon W. Golomb, who apparently provided valuable insight on construction of Costas sequences. According to [78], after contacting Golomb "Notable progress has since been made by Golomb, Taylor, Welch, and Lempel" in developing Costas sequences. He also cites several other developers for contributing to the field [80–84].

Interestingly, according to Aaron Sterling [85], the same year Costas' paper was published, the mathematician Edgar Nelson Gilbert published a paper entitled "Latin Squares which Contain No Repeated Diagrams" [86]. In it he described a technique that is the same as one known as the "logarithmic Welch" technique.

More modern research in the areas of Costas arrays is being conducted by James K. Beard, Scott Rickard, K. Drakakis, F. Iorio, R. Caballero, G. O'brien, and John Walsh [87–91]. It appears that two of the leaders in the field are currently Beard and Rickard. According to a website hosted by Dr. Beard [92], Costas sequences up to order 1030 have been developed to date.

An example of a Costas array for an 8-element Costas sequence is illustrated in Figure 11.25. The numbers along the left side are frequency indices, and the numbers along the bottom denote chip number. The eight boxes that contain 1's denote frequency indices associated with the various chips. For example, frequency 2 is assigned to chip 1, frequency 6 is assigned to chip 2, and so forth. The resulting Costas sequence defined by the Costas array of Figure 11.25 is

$$a = \begin{bmatrix} 2 & 6 & 3 & 8 & 7 & 5 & 1 & 4 \end{bmatrix} \quad (11.75)$$

Table 11.4 contains example Costas sequences of length 3 through 20 that were obtained from Dr. Beard's website.

Waveform Coding 315

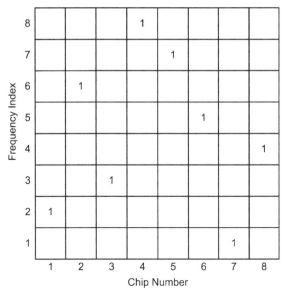

Figure 11.25 Costas array for an 8-element Costas sequence.

Table 11.4
Costas Sequences for K = 3 to 20

K	Costas Sequence
3	1 3 2
4	1 2 4 3
5	1 3 4 2 5
6	1 2 5 4 6 3
7	1 2 6 4 7 3 5
8	1 2 5 7 6 4 8 3
9	1 2 6 4 9 8 5 7 3
10	1 2 4 8 5 10 9 7 3 6
11	1 2 5 9 4 11 10 7 3 8 6
12	1 2 4 8 3 6 12 11 9 5 10 7
13	1 2 4 9 13 6 12 11 7 5 8 3 10
14	1 2 5 7 14 8 12 11 6 4 13 10 3 9
15	1 2 6 14 9 3 15 13 5 10 12 11 8 4 7
16	1 2 6 11 5 13 8 4 15 14 16 9 12 3 10 7
17	1 2 9 13 11 16 8 17 6 12 5 4 14 10 7 15 3
18	1 2 4 8 16 13 7 14 8 18 17 15 11 3 6 12 5 10
19	1 2 13 7 11 4 18 10 17 19 6 5 14 9 15 12 3 8 16
20	1 2 6 11 20 18 15 7 13 9 12 19 5 4 17 8 3 14 16 10

Given the Costas sequence

$$a = \begin{bmatrix} a_1 & a_2 & \cdots & a_K \end{bmatrix} \qquad (11.76)$$

an equation for a Costas waveform can be written as

$$u(t) = \sum_{k=1}^{K} e^{j2\pi f_k(t-(k-1)\tau_c)} \mathrm{rect}\left[\frac{t-(k-1)\tau_c}{\tau_c}\right] \qquad (11.77)$$

where

$$f_k = \frac{a_k}{\tau_c} \qquad (11.78)$$

and τ_c is the chip width.

Equation (11.78) indicates the frequencies of the chips of a Costas waveform must be integer multiples of the reciprocal of the chip width. If this condition is satisfied, and $|\chi(\tau,f)|$ is normalized so that $|\chi(0,0)|=1$, the maximum value of $|\chi(\tau,f)|$ for $|\tau| > \tau_c$ will be mostly less than or equal to $1/K$. That is, the peak sidelobe level more than one chip removed from matched range will be mostly less than or equal to $1/K$. If the frequency steps are not chosen in accordance with (11.78), the peak sidelobe levels will rise above $1/K$. As f_k deviates further away from the values defined by (11.78), the sidelobe structure can deteriorate significantly.

If (11.78) is satisfied, there will be k cycles of a sinusoid on chip k. This means there will be no phase change between chips. It also means the signals on each chip will be orthogonal. The impact of this is that the correlation of chip l with chip m will be zero. This means the matched-Doppler, range cut of the ambiguity function will be zero at integer multiples of τ_c. If (11.78) is not satisfied, there will not be an integer number of cycles of a sinusoid on each chip. Because of this, the signals on the chips will not be orthogonal, and the matched-Doppler, range cut of the ambiguity will not be zero at integer multiples of the chip width. Also, the peak sidelobe levels of the ambiguity function, both at matched Doppler and elsewhere, will increase [see Exercise 18].

Because of (11.78) the frequency extent, and thus bandwidth, of the waveform will be $(K-1)/\tau_c$. However, the range resolution of the waveform will be approximately τ_c/K. This apparent disagreement between waveform bandwidth and range resolution is due to the discrete frequency change from chip to chip.

The overall pulse width will be $\tau_p = K\tau_c$. These relations between frequency deviation, pulse width, and the number of chips in the pulse is different than for the LFM or phase-coded pulses discussed earlier. With LFM, the range resolution was determined by the frequency extent, independent of pulse width. For phase-coded waveforms, the range resolution was determined by the chip width, independent of the pulse width or number of chips on the pulse. The interrelation between frequency deviation, pulse width, and number of chips in the pulse is something that must be considered when designing Costas waveforms.

11.5.1 Costas Waveform Example

Figure 11.26 contains a plot of $|\chi(\tau,f)|$ for a $K = 10$ chip, $\tau_p = 100$-μs Costas pulse, based on the sequence

$$a = \begin{bmatrix} 1 & 2 & 4 & 8 & 5 & 10 & 9 & 7 & 3 & 6 \end{bmatrix} \qquad (11.79)$$

Since $K = 10$ and $\tau_p = 100$ μs, $\tau_c = \tau_p/K = 100/10 = 10$ μs. The associated frequencies on the chips are

$$f_k = a_k/\tau_c = 0.1 a_k \text{ MHz}. \tag{11.80}$$

The waveform bandwidth is BW = $(K-1)/\tau_c$ = 0.9 MHz, and the expected range resolution is τ_c/K = 10/10 = 1 μs.

Although not obvious from Figure 11.26, the sidelobe levels in the regions defined by $|\tau| > \tau_c$ = 10 μs have peak values that are mostly less than $1/K$ = 1/10. This is more evident in Figure 11.27, which is a contour plot where the lines define values of τ and f (range and Doppler mismatch) where $|\chi(\tau,f)|$ = 0.1. Regions inside of the lines represent values of τ and f where $|\chi(\tau,f)|$ exceeds $1/K$ = 0.1. In the rest of τ-f space, $|\chi(\tau,f)|$ is less than 0.1. In regions where $|\tau| > 10$ μs and $|f| < 0.1$ MHz, the Doppler resolution of the waveform, we note that $|\chi(\tau,f)|$ is, indeed, mostly less than $1/K$ = 0.1.

Figure 11.28 contains a plot like Figure 11.27, but with contour lines at $|\chi(\tau,f)|$ = 0.2. In this case, there are no regions outside of $|\tau| = \tau_c$ = 10 μs where $|\chi(\tau,f)|$ is above 0.2.

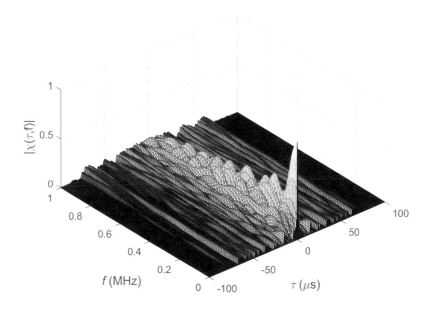

Figure 11.26 Plot of $|\chi(\tau,f)|$ for a 10-chip Costas Waveform.

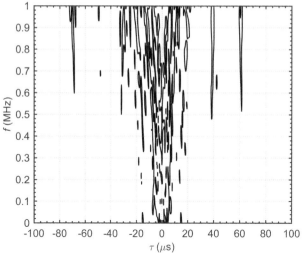

Figure 11.27 Contour plot of $|\chi(\tau,f)|$ for a 10-chip Costas Waveform—regions with $|\chi(\tau,f)|>0.1$.

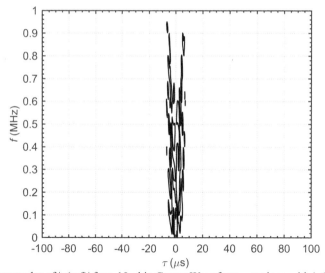

Figure 11.28 Contour plot of $|\chi(\tau,f)|$ for a 10-chip Costas Waveform—regions with $|\chi(\tau,f)|>0.2$.

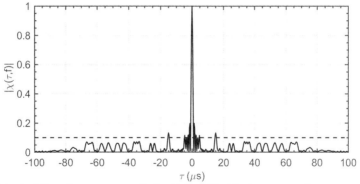

Figure 11.29 Matched-Doppler, range cut of $|\chi(\tau,f)|$ for a 10-chip Costas waveform.

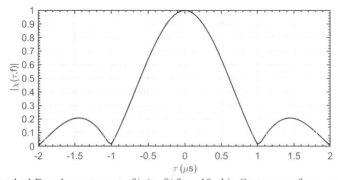

Figure 11.30 Matched-Doppler, range cut of $|\chi(\tau,f)|$ for a 10-chip Costas waveform—zoomed.

Figure 11.29 contains a matched-Doppler, range cut of $|\chi(\tau,f)|$. The horizontal, dashed line represents the level of $1/K = 0.1$. Except for the region near matched range and the peaks at about $\tau = \pm 15$ µs, the sidelobe levels are below 0.1. Also, as expected $|\chi(\tau,f)| = 0$ at integer multiples of $\tau_c = 10$ µs.

Figure 11.30 contains a zoomed-in version of Figure 11.29. The first nulls in the range cut occur at ± 1 µs. This implies a range resolution of 1 µs, which is consistent with the earlier statement that the range resolution of Costas waveforms is τ_c/K.

11.6 CLOSING COMMENTS

With the exception of LFM pulses, Barker-coded pulses, and some short PRN-coded pulses, waveforms of the type mentioned in this chapter were very difficult to implement in older radars. This is because the more complicated waveforms discussed in this chapter were difficult to generate and process with older, analog hardware. However, the advent of direct digital synthesizers [72–74], fast FIR [93–96] filter hardware, and FFT-based signal processors [75–77] has essentially removed the hardware constraint. This means we can expect to see waveforms such as those discussed in this chapter, and even more complicated waveforms, come into fairly common use.

11.7 EXERCISES

1. We did not discuss matched-range, Doppler cuts of $|\chi(\tau,f)|$. This is because the matched-range, Doppler cut does not depend on the phase modulation on the pulse, $\phi(t)$. Prove that this is a correct statement by showing that the matched-range, Doppler cut of the waveform of (11.1) is given by $|\chi(0,f)| = K|\text{sinc}(f\tau_p)|$.

2. Reproduce the plots of Figure 11.1.

3. Show that the scaling constant, K, of (11.22) is $K = \pi/[B(\pi a + 2b)]$.

4. Derive (11.25) and (11.26). Assume $B = 2$.

5. Reproduce the plots of Figure 11.4 ($\tau_p = 15$-µs, LFM bandwidth $B = 2$ MHz).

6. Use the numerical technique of Section 11.2.2.3 to design an NLFM waveform for the case where $U_m^2(\zeta)$ is a Hamming weighting function (see Example 2 of Section 11.2.2.2).

7. Reproduce the plot of Figure 11.3. Produce a similar plot for a 25-chip pulse with Frank polyphase coding.

8. Generate a plot like Figure 11.8 for a 25-chip pulse with minimum peak sidelobe coding.

9. Generate a plot like Figure 11.8 for a 13-chip Barker-coded pulse.

10. Generate a plot like Figure 11.13 for a 63-chip PRN-coded pulse.

11. Generate a plot like Figure 11.19 where the base pulse is a 15-chip PRN-coded pulse.

12. Show, by example, that two PRN codes generated by two different loads of a shift register generator are circular shifts of each other. Use a 7- or 15-element code to simplify the problem.

13. It was stated that PRN-coded pulses based on different shift register feedback configurations were "almost orthogonal." That is, if a PRN-coded pulse is based on one feedback configuration and the matched filter is based on a pulse derived with a different feedback configuration, the output of the matched filter would not exhibit a predominant peak as it would if the matched filter was matched to the input pulse. To verify this assertion, generate a PRN-coded pulse, $u(t)$, using one of the feedback configurations for $M = 6$ in Table 11.3. Match the matched filter to another pulse, $v(t)$, based on a different feedback tap configuration for $M = 6$. Process $u(t)$ through the matched filter matched to $v(t)$ and plot the output. Does it behave as claimed?

14. Reproduce the plots of Figures 11.21 and 11.22.

15. Generate plots like Figure 11.13 for different initial loads of the feedback shift register. Make a note of the range sidelobe levels close to the main peak as you change initial loads.

16. Derive a Costas array for the 4- and 6-element Costas sequence of Table 11.4.

17. Reproduce Figures 11.26 through 11.30.

18. Generate a figure like Figure 11.29 for the case where $f_k = \alpha a_k/\tau_c$, instead of $f_k = a_k/\tau_c$. Use $\alpha = 0.8, 0.9, 1.1,$ and 1.2. Note the effect on when/if the range cut goes to zero, and on the peak side levels. Also generate figures like Figures 11.27 and 11.28 to note the effect on overall sidelobe levels.

19. Generate figures like Figures 11.26 through 11.30 for a 6-chip Costas waveform based on the 6-element Costas sequence of Table 11.4.

20. Generate figures like Figures 11.26 through 11.30 for a 20-chip Costas waveform based on the 20-element Costas sequence of Table 11.4.

References

[1] Ewell, G. W., *Radar Transmitters,* New York: McGraw-Hill, 1981.

[2] Ostroff, E. D., et al., *Solid-State Radar Transmitters,* Dedham, MA: Artech House, 1985.

[3] Cooke, C.E., "The Early History of Pulse Compression Radar: The History of Pulse Compression at Sperry Gyroscope Company," *IEEE Trans. Aerosp. Electron. Syst.,* vol. 24, no. 6, Nov. 1988, pp. 825–833.

[4] Dicke, R. H., "Object Detection System," U.S. Patent 2,624,876, Jan. 6, 1953.

[5] Darlington, S., "Pulse Transmission," U.S. Patent 2,678,997. May 18, 1954.

[6] Cook, C. E., "Pulse Compression-Key to More Efficient Radar Transmission," *Proc. IRE,* vol. 48, no. 3 Mar. 1960, pp. 310–316. Reprinted in: Barton, D. K., *Radars, Vol. 3: Pulse Compression* (Artech Radar Library), Dedham, MA: Artech House, 1975.

[7] Klauder, J. R. et al., "The Theory and Design of Chirp Radars," *Bell Syst. Tech. J.,* vol. 39, no. 4, Jul. 1960, pp. 745–808.

[8] Skolnik, M. I., "Fifty Years of Radar," *Proc. IEEE,* vol. 73, no. 2, Feb. 1985, pp. 182–197.

[9] Levanon, N., and E. Mozeson, *Radar Signals,* New York: Wiley-Interscience, 2004.

[10] Rihaczek, A. W., *Principles of High-Resolution Radar,* New York: McGraw-Hill, 1969; Norwood, MA: Artech House, 1995.

[11] Ipatov, V. P., *Spread Spectrum and CDMA: Principles and Applications*, Chichester, UK: Wiley & Sons, 2005.

[12] Holmes, J. K., *Spread Spectrum Systems for GNSS and Wireless Communications*, Norwood, MA: Artech House, 2007.

[13] Ziemer, R. E., and W. H. Tranter, *Principles of Communications,* 3rd ed., Hoboken, NJ: Houghton Mifflin, 1990.

[14] Glisic, S., and B. Vucetic, *Spread Spectrum CDMA Systems for Wireless Communications,* Norwood, MA: Artech House, 1997.

[15] Golomb, S. W., ed., *Digital Communications with Space Applications,* Englewood Cliffs, NJ: Prentice-Hall, 1964.

[16] Golomb, S. W., and G. Gong, *Signal Design for Good Correlation: For Wireless Communication, Cryptography, and Radar*, New York: Cambridge University Press, 2005.

[17] Nyquist, H., "Certain Topics in Telegraph Transmission Theory," *Trans. AIEE,* vol. 47, no. 2, Apr. 1928, pp. 617–644. Reprinted as a classic paper in Proc. IEEE, vol. 90, no. 2, Feb. 2002.

[18] Shannon, C. E., "Communication in the Presence of Noise," *Proc. IRE,* vol. 37, no. 1, Jan. 1949, pp. 10–21. Reprinted as a classic paper in *Proc. IEEE,* vol. 86, no. 2, Feb. 1998.

[19] Kotel'nikov, V. A., "On the Capacity of the 'Ether' and Cables in Electrical Communications," in *Proc. First All-Union Conf. Technolog. Reconstruction of the Commun. Sector and Develop. of Low-Current Eng.,* Moscow: 1933. Translated by C. C. Bissell and V. E. Katsnelson. http://ict.open.ac.uk/classics/1.pdf.

[20] Key, E. L., E. N. Fowle, and R. D. Haggarty, "A Method of Pulse Compression Employing Non-Linear Frequency Modulation," MIT Lincoln Lab., Lexington, MA., Tech. Rep. No. 207, Aug. 1959.

[21] Key, E. L., E. N. Fowle, and R. D. Haggarty, "A Method of Designing Signals of Large Time-Bandwidth Product," in *IRE Int. Conv. Rec.,* pt. 4, 1961, pp. 146–154.

[22] Watters, E. C. "A Note on the Design of Coded Pulses," in *Proc. Pulse Compression Symp.,* Rome Air Develop. Ctr., Rep. No. TR–59–161, Griffis AFB, NY, Defense Tech. Inf. Ctr., Sept. 1959.

[23] Fowle, E., "The Design of FM Pulse Compression Signals," *IEEE Trans. Inf. Theory,* vol. 10, no. 1, Jan. 1964, pp. 61–67.

[24] Labitt, M., "Obtaining Low Sidelobes Using Non-Linear FM Pulse Compression," MIT Lincoln Lab., Lexington, MA, Rep. No. ATC-223, Nov. 4, 1994. Made available by Nat. Tech. Inf. Service, Springfield, VA.

[25] Doerry, A. W., "Generating Nonlinear FM Chirp Waveforms for Radar," Sandia Nat. Labs., Albuquerque, NM, Rep. No. SAND2006–5856, Sept. 2006.

[26] Collins, T., and P. Atkins, "Nonlinear Frequency Modulation Chirps for Active Sonar," *IEEE Proc. Radar, Sonar and Navigation,* vol. 146, no. 6, Dec. 1999, pp. 312–316.

[27] Varshney, L. R., and D. Thomas, "Sidelobe Reduction for Matched Filter Range Processing," in *Proc. 2003 IEEE Radar Conf.,* Huntsville, AL, May 5–8, 2003, pp. 446–451.

[28] Yichun, P., et al., "Optimization Design of NLFM Signal and Its Pulse Compression Simulation," *2005 IEEE Int. Radar Conf.,* Arlington, VA, May 9–12, 2005, pp. 383–386.

[29] Erdelyi, A., *Asymptotic Expansions,* New York: Dover, 1956.

[30] Bleistein, N., and R. A. Handelsman, *Asymptotic Expansions of Integrals,* New York: Holt, Rinehart and Winston, 1975.

[31] Cohen, M. N., M. R. Fox, and J. M. Baden, "Minimum Peak Sidelobe Pulse Compression Codes," in *Proc. 1990 IEEE Int. Radar Conf.,* Arlington, VA, May 7–10, 1990, pp. 633–638.

[32] Lindner, J. "Binary Sequences Up to Length 40 with Best Possible Autocorrelation Function," *Proc. IEEE,* vol. 11, no. 21 (Oct. 1975); *Electron. Lett.,* vol. 11, no. 21, Oct. 16, 1975, p. 507.

[33] Coxson, G., and J. Russo, "Efficient Exhaustive Search for Optimal-Peak-Sidelobe Binary Codes," *IEEE Trans. Aerosp. Electron. Syst.,* vol. 41, no. 1, Jan. 2005, pp. 302–308.

[34] Leukhin, A. N., and E. N. Potekhin, "Optimal Peak Sidelobe Level Sequences Up to Length 74," *2013 European Radar Conf. (EuRAD),* Nuremberg, Germany, Oct. 9–11, 2013, pp. 495–498.

[35] Gabbay, S. "Properties of Even-Length Barker Codes and Specific Polyphase Codes with Barker Type Autocorrelation Functions," Naval Research Laboratory, Washington, DC, Rep. 8586, Jul. 12, 1982. Available from DTIC as AD A117415.

[36] Nunn, C. J., and G. E. Coxson, "Best-Known Autocorrelation Peak Sidelobe Levels for Binary Codes of Length 71 to 105," *IEEE Trans. Aerosp. Electron. Syst.,* vol. 44, no. 1, Jan. 2008, pp. 392–395.

[37] Nunn, C. J., and G. E. Coxson, "Polyphase Pulse Compression Codes with Optimal Peak and Integrated Sidelobes," *IEEE Trans. Aerosp. Electron. Syst.,* vol. 45, no. 2, Apr. 2009, pp. 775–781.

[38] Borwein, P., and R. Ferguson, "Polyphase Sequences with Low Autocorrelation," *IEEE Trans. Inf. Theory,* vol. 51, no. 4, Apr. 2005, pp. 1564–1567.

[39] Ipatov, V. P., "Ternary Sequences with Ideal Autocorrelation Properties," *Radio Eng. Electron. Phys.,* vol. 24, no. 10, Oct. 1979, pp. 75–79.

[40] Ipatov, V. P., "Contribution to the Theory of Sequences with Periodic Autocorrelation Properties," *Radio Eng. Electron. Phys.,* vol. 25, no. 4, Apr. 1980, pp. 31–34.

[41] Hoholdt, T., and Justesen, J., "Ternary Sequences with Perfect Periodic Autocorrelation," *IEEE Trans. Info. Theory,* vol. 29, no. 4, Jul. 1983, pp. 597–600.

[42] Cook, C. E., and M. Bernfeld, *Radar Signals: An Introduction to Theory and Application,* New York: Academic Press, 1967; Norwood, MA: Artech House, 1993.

[43] Lewis, B. L., et al., *Aspects of Radar Signal Processing,* Norwood, MA: Artech House, 1986.

[44] Skolnik, M. I., *Introduction to Radar Systems,* 3rd ed., New York: McGraw-Hill, 2001.

[45] Meikle, H., *Modern Radar Systems,* 2nd ed., Norwood, MA: Artech House, 2008.

[46] Hovanessian, S. A., *Radar System Design and Analysis,* Norwood, MA: Artech House, 1984.

[47] Eaves, J. L., and E. K. Reedy, *Principles of Modern Radar,* New York: Van Nostrand Reinhold, 1987.

[48] Nathanson, F. E., J. P. Reilly, and M. N. Cohen, eds., *Radar Design Principles,* 2nd ed., New York: McGraw-Hill, 1991.

[49] Coxson, G. E., A. Hirschel, and M. N. Cohen, "New Results on Minimum-PSL Binary Codes," *Proc. 2001 IEEE Nat. Radar Conf.,* Atlanta, GA, May 2001, pp. 153–156.

[50] Bömer, L., and M. Antweiler, "Polyphase Barker Sequences," *Electron. Lett.,* vol. 25, no. 23, Nov. 1989, pp. 1577–1579.

[51] Friese, M., and H. Zottmann, "Polyphase Barker Sequences Up to Length 31," *Electron. Lett.,* vol. 30, no. 23, Nov. 1994, pp. 1930–1931.

[52] Friese, M., "Polyphase Barker Sequences Up to Length 36," *IEEE Trans. Inf. Theory,* vol. 42, no. 4, Jul. 1996, pp. 1248–1250.

[53] Brenner, A. R., "Polyphase Barker Sequences Up to Length 45 with Small Alphabets," *Electron. Lett.,* vol. 34, no. 16, Aug. 1998, pp. 1576–1577.

[54] Golomb, S. W., *Shift Register Sequences,* San Francisco, CA: Holden-Day, 1967.

[55] Guerci, J. R., *Cognitive Radar: The Knowledge-Aided Fully Adaptive Approach*, Norwood, MA: Artech House, 2010.

[56] Li, J., and P. Stoica, eds., *MIMO Radar Signal Processing,* New York: Wiley & Sons, 2009.

[57] Zadoff, S. A., "Phase-coded Communication System," U. S. Patent 3,099,796, Jul. 30, 1963.

[58] Chu, D. C., "Polyphase Codes with Good Periodic Correlation Properties," *IEEE Trans. Inf. Theory,* vol. 18, no. 4, Jul. 1972, pp. 531–532.

[59] Antweiler, M., and L. Bömer, "Merit Factor of Chu and Frank Sequences," *Electron. Lett.,* vol. 26, no. 25, Dec. 6, 1990, pp. 2068–2070.

[60] Barker, R. H., "Group Synchronization of Binary Digital Systems," in *Communication Theory* (W. Jackson, ed.), London: Academic Press, 1953, pp. 273–287.

[61] Turyn, R., "On Barker Codes of Even Length," *Proc. IEEE,* vol. 51, no. 9, Sept. 1963, p. 1256.

[62] Woodward, P. M., *Probability and Information Theory with Applications to Radar,* 2nd ed., New York: Pergamon Press, 1953; Dedham, MA: Artech House, 1980.

[63] Dixon, R. C., *Spread Spectrum Systems with Commercial Applications,* 3rd ed., New York: Wiley & Sons, 1994.

[64] Michelson, A. M., and A. H. Levesque, *Error-Control Techniques for Digital Communication,* New York: Wiley & Sons, 1985.

[65] Berkowitz, R. S., *Modern Radar: Analysis, Evaluation, and System Design*, New York: Wiley & Sons, 1965.

[66] Peterson, W. W., and E. J. Weldon, *Error-Correcting Codes,* Cambridge, MA: MIT Press, 1972.

[67] IEEE Information Theory Society (ITSOC), "Claude E. Shannon Award." http://www.itsoc.org/honors/ claude-e-shannon-award.

[68] IEEE, "IEEE Richard W. Hamming Medal Recipients." http://www.ieee.org/documents/hamming_rl.pdf.

[69] National Science Foundation (NSF), "US NSF–About Awards." http://www.nsf.gov.

[70] New Wave Instruments, "Linear Feedback Shift Registers—Implementation, M-Sequence Properties, Feedback Tables," Apr. 5, 2005. www.newwaveinstruments.com.

[71] Skolnik, M. I., ed., *Radar Handbook,* 2nd ed., New York: McGraw-Hill, 1990.

[72] Cushing, R., "A Technical Tutorial on Digital Signal Synthesis," 1999. www.analog.com.

[73] Analog Devices, "AD9858 1 GSPS Direct Digital Synthesizer," 2003. www.analog.com.

[74] Kester, W., ed., *The Data Conversion Handbook,* New York: Newnes, 2005.

[75] Rabiner, L. R., and B. Gold, *Theory and Application of Digital Signal Processing,* Englewood Cliffs, NJ: Prentice-Hall, 1975.

[76] Martinson, L., and R. Smith, "Digital Matched Filtering with Pipelined Floating Point Fast Fourier Transforms (FFT's)," *IEEE Trans. Acoust., Speech, Signal Process.,* vol. 23, no. 2, Apr. 1975, pp. 222234.

[77] Blankenship, P., and E. M. Hofstetter, "Digital Pulse Compression Via Fast Convolution," *IEEE Trans. Acoust., Speech, Signal Process.,* vol. 23, no. 2, Apr. 1975, p. 189–201.

[78] Costas, J. P., "A Study of a Class of Detection Waveforms Having Nearly Ideal Range—Doppler Ambiguity Properties," in *Proceedings of the IEEE*, 1984.

[79] Levanon, N., *Radar Principles*, New York: John Wiley & Sons, Inc., 1988.

[80] Golomb, S. W. and H. Taylor, "Two-dimensional Synchronization Patterns for Minimum Ambiguity," *IEEE Trans. Inform. Theory,* Vols. IT-28, no. 4, July 1982, pp. 600–604.

[81] Colomb, S. W., "Algebraic Constructions for Costas Arrays," *Journal of Combinatorial Theory,* vol. 37, no. 1, 1984, pp. 13–21.

[82] Cooper, G. R. and R. D. Yate, "Design of Large Signal Sets with Good Aperiodic Correlation Properties," Purdue University, Tech. Rep. TR-EE66-13, 1966.

[83] Merserau, R. M. and T. S. Kay, "Multiple Access Frequency Hopping Patterns with Low Ambiguity," *IEEE Transactions on Aerospace and Electronic Systems,* Vols. AES-17, no. 4, July 1981, pp. 571–578.

[84] Sites, M. J., "Coded Frequency Shift Keyed Sequences with Applications to Low Data Rate Communication and Radar," Stanford Electronics Lab. Rep. 3606-5 (AD 702063), Sept. 1969.

[85] Sterling, A., "An Independent Discovery of Costas Arrays," [online]. Available: https://nanoexplanations.wordpress.com/2011/10/09/an-independent-discovery-of-costas-arrays/. [accessed 3 December 2018].

[86]   Gilbert, E. N., "Latin Squares which Contain No Repeated Digrams," *Society for Industrial and Applied Mathematics Review,* vol. 7, no. 2, April 1965, pp. 189–198.

[87]   Beard, J. K., "Generating Costas Arrays to Order 200," in 2006 40th Annual Conference on Information Sciences and Systems, (CISS) 2006, 2006.

[88]   Drakakis, K., S. Rickard, J. K. Beard, R. Caballero, F. Iorio, G. O'Brien and J. Walsh, "Results of the Enumeration of Costas Arrays of Order 27," *IEEE Transactions on Information Theory,* vol. 54, no. 10, October 2008, pp. 4684–4687.

[89]   Drakakis, K., F. Iorio and S. Rickard, "The Enumeration of Costas Arrays of Order 28 and its Consequences," *Advances in Mathematics of Communications,* vol. 5, no. 2, February 2011, pp. 69–86.

[90]   Drakakis, K., F. Iorio, S. Rickard and J. Walsh, "Results of the Enumeration of Costas Arrays of Order 29," *Advances in Mathematics of Communications,* vol. 5, no. 3, 2011, pp. 547–553.

[91]   Beard, J. K., "Costas Array Generator Polynomials in Finite Fields," in *42nd Annual Conference on Information Sciences and Systems (CISS 2008),* April, 2008.

[92]   Beard, J. K., "Costas Arrays," [Online]. Available: http://jameskbeard.com/jameskbeard/Costas_Arrays.html. [Accessed 3 December 2018].

[93]   Hamming, R. W., *Digital Filters,* 2nd ed., Englewood Cliffs, NJ: Prentice-Hall, Inc., 1983.

[94]   Lyons, R. G., *Understanding Digital Signal Processing,* 3rd ed., New York: Prentice Hall, 2011.

[95]   Mitra, S. K., *Digital Signal Processing: A Computer-Based Approach,* 2nd ed., NY: McGraw-Hill, 2001.

[96]   Proakis, J. G. and D. G. Manolakis, *Digital Signal Processing,* 4th ed., Upper Saddle River, New Jersey: Pearson Prentice Hall, 2007.

APPENDIX 11A: LFM AND THE sinc²(x) FUNCTION

In footnote 2 of Section 11.2.1, we noted that the matched filter output for an unweighted LFM pulse had a shape similar to a sinc²(x) function, but the match was not exact. In particular, we noted the first two sidelobe levels were about −14 and −19 dB instead of −13.2 and −17.8 dB. This can be explained by examining the equation for the matched-Doppler range cut of the ambiguity function we derived in Chapter 10 [see (10.30)]. That equation is

$$|\chi(\tau,0)|^2 = (\tau_p - |\tau|)^2 \left|\text{sinc}\left[(\alpha\tau)(\tau_p - |\tau|)\right]\right|^2 \text{rect}\left[\frac{\tau}{2\tau_p}\right] \quad (11A.1)$$

We note the matched-Doppler range cut does contain a sinc²(x) function, but with the added term $(\tau_p - |\tau|)$ in the argument. It is the presence of this term that causes the sidelobes to be lower than those of the sinc²(x) function. As the BT product of the waveform is increased, the $(\tau_p - |\tau|)$ term has less of an effect on the first few sidelobes, which means they would approach those of a sinc²(x) function.

We can also explain this from a frequency domain perspective. To that end, Figure 11A.1 contains a plot of the frequency spectrum of the 15-μs, 2-MHz LFM pulse considered in Section 11.2. The figure also contains an ideal spectrum with the same bandwidth. If the ideal spectrum was that of some hypothetical pulse, the matched-Doppler range cut of the pulse would be a sinc²(x) function. The nature of the matched-Doppler range cut of the LFM ambiguity function is due to the ripples (which are termed Fresnel ripples) and skirts of the LFM pulse spectrum. The ripples and skirts of the LFM spectrum are also what caused the sidelobes of the matched filter output of the weighted LFM pulse (Figure 10.1) to be other than the expected −30 dB normally associated with 30-dB Taylor weighting.

As a comparison, Figure 11A.2 contains the matched-Doppler range cut and spectrum for a LFM pulse with a duration of 150 μs and a bandwidth of 2 MHz (a BT product of 300 instead of 30). The full extent of the matched-Doppler range cut is not shown so we could more easily see the first few sidelobes. Note that the spectrum more closely approximates the ideal spectrum and the first two sidelobe of the matched-Doppler range cut are closer to −13.2 and −17.8 dB.

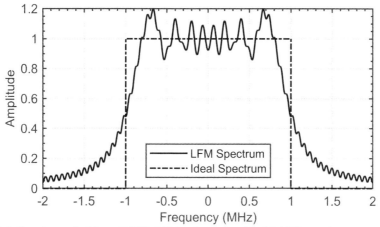

Figure 11A.1 Spectrum of a 15-μs, 2-MHz LFM pulse and an ideal 2-MHz spectrum.

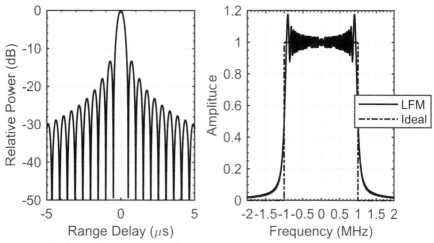

Figure 11A.2 Matched-Doppler range cut (left) and spectrum (right) of a 150-μs, 2-MHz LFM pulse.

Chapter 12

Stretch Processing

12.1 INTRODUCTION

Stretch processing is an alternative to matched filtering that is used on large bandwidth waveforms. It is most often applied to waveforms with linear frequency modulation (LFM). The concepts of stretch processing also appear in other applications such as frequency modulated continuous wave (FMCW) radar [1] and, as we will see in Chapter 23, synthetic aperture radar (SAR). With stretch processing the frequency modulation (LFM) of the return signal is removed by the first mixer of the radar. This process of removing the frequency modulation results in a constant frequency signal whose frequency is proportional to target range. The origin of the term stretch processing is not clear. It is possible it derives from the fact that the waveform processing is moved from a single device, the matched filter, at the end of the receiver chain and distributed between the first mixer and a fast Fourier transformer (FFT) at the end of the receiver chain. Thus, one could think of the waveform processor as being "stretched" over the entire receiver chain.

Stretch processing was developed by Dr. William J. Caputi, Jr. [2]. In recognition of this and other efforts in SAR, Dr. Caputi was awarded the IEEE Dennis Picard Medal "for conception and development of innovative range and Doppler bandwidth reduction techniques used in wideband radars and high resolution synthetic aperture radars" [3].

We consider a normalized, LFM, transmit waveform of the form

$$v(t) = e^{j\pi\alpha t^2} \operatorname{rect}\left[\frac{t}{\tau_p}\right] \tag{12.1}$$

where

$$\operatorname{rect}[x] = \begin{cases} 1 & |x| \leq 1/2 \\ 0 & |x| > 1/2 \end{cases} \tag{12.2}$$

α is the LFM slope, and τ_p is the uncompressed pulse width. The instantaneous phase of $v(t)$ is

$$\phi(t) = \pi \alpha t^2 \tag{12.3}$$

and the instantaneous frequency is

$$f(t) = \frac{1}{2\pi}\frac{d\phi(t)}{dt} = \alpha t \tag{12.4}$$

Over the duration of the pulse, $f(t)$ varies from $-\alpha\tau_p/2$ to $\alpha\tau_p/2$. Thus, the bandwidth of the LFM signal, $v(t)$, is

$$B = |\alpha\tau_p| \tag{12.5}$$

We can also determine the bandwidth of $v(t)$ by finding and plotting its Fourier transform. Specifically [see Exercise 1],

$$\begin{aligned}V(f) &= \int_{-\infty}^{\infty} v(t) e^{-j2\pi ft} dt \\ &= \sqrt{\tfrac{1}{2\alpha}} e^{-j\pi f^2/\alpha} \left\{ F\left[\sqrt{\tfrac{2}{\alpha}}(f+\alpha\tau_p/2)\right] - F\left[\sqrt{\tfrac{2}{\alpha}}(f-\alpha\tau_p/2)\right] \right\}\end{aligned} \tag{12.6}$$

where

$$F(x) = C(x) + jS(x) \tag{12.7}$$

is the Fresnel integral, and $C(x)$ and $S(x)$ are the cosine and sine Fresnel integrals, respectively, defined by [4, p. 296]

$$C(x) = \int_0^x \cos\left(\frac{\pi t^2}{2}\right) dt \tag{12.8}$$

and

$$S(x) = \int_0^x \sin\left(\frac{\pi t^2}{2}\right) dt \tag{12.9}$$

A normalized plot of $|V(f)|$ for an LFM bandwidth of $B = 500$ MHz and a pulsewidth of $\tau_p = 100$ µs is shown in Figure 12.1. Note that the spectrum width is equal to the LFM bandwidth of 500 MHz.

If we were to process the LFM pulse using a matched filter, the normalized impulse response of the matched filter would be

$$h(t) = v^*(-t) = e^{-j\pi\alpha t^2} \text{rect}\left[\frac{t}{\tau_p}\right] \tag{12.10}$$

where we have made use of the fact that $\text{rect}[x]$ is an even function.

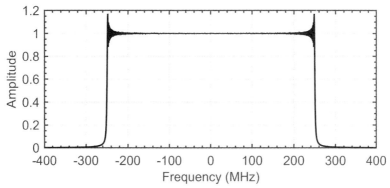

Figure 12.1 Baseband spectrum of an LFM pulse with B = 500 MHz and τ_p = 100 μs.

The form of $h(t)$ means the matched filter would need to have a bandwidth of $B = |\alpha \tau_p|$. Herein lies a problem: large bandwidth matched filters are still difficult and costly to build. Two methods of building LFM matched filters (LFM pulse compressors, LFM signal processors) are surface acoustic wave (SAW) devices and digital signal processors [5–7]. A cursory survey of manufacturer literature and other sources indicates that the current state of SAW technology limits these types of processors to 1,000 MHz bandwidth and BT products on the order of 10,000.

The bandwidth of digital signal processors is usually limited by the sample rate of the analog-to-digital converters (ADCs) needed to convert the analog signal to a digital signal. Although the technology is progressing rapidly, the current limit on ADC rates is 1,000 MHz or so [5]. If an upper limit on ADC sample rate is 1,000 MHz, then the maximum bandwidth of an LFM signal processor would also be 1,000 MHz (assuming complex signals and processors).

Still another impact of wide bandwidth signals is that the RF and IF amplifiers of the receiver must have a bandwidth wider than the waveform bandwidth. Given that the center frequency of bandpass amplifiers should be four to five times the bandwidth (see Chapter 22), the various amplifiers must be centered in the GHz region, rather than the MHz region. This can impose stringent requirements of the design of the receiver.

Stretch processing relieves the receiver and signal processor bandwidth requirement by sacrificing all-range processing to obtain a narrowband receiver and signal processor. If we were to use a classic matched filter approach, we could look for targets over the entire waveform pulse repetition interval (PRI). With stretch processing, we are limited to a range extent that is usually smaller than an uncompressed pulse width. Thus, we could not use stretch processing for search because search requires looking for targets over a large range extent, usually many pulse widths long. We could use stretch processing for track because we already know range fairly well but want a more accurate measurement of it. However, we point out that, in general, wide bandwidth waveforms, and thus the need for stretch processing, is "overkill" for tracking. Generally speaking, bandwidths of 1s to 10s of MHz are sufficient for tracking.

One of the most common uses of wide bandwidth waveforms and stretch processing is in discrimination, where we need to distinguish individual scatterers along a target.

Another use is in SAR. In that application, we only try to map a small range extent of the ground but want very good range resolution to distinguish the individual scatterers that constitute the scene.

In the above discussion, we focused on the receiver and signal processor and have argued, without proof at this point, that we can use stretch processing to ease the bandwidth requirements on a signal processor used to compress wide bandwidth waveforms. Stretch processing does not relieve the bandwidth requirements on other components of the radar. Specifically, the transmitter must be capable of generating and amplifying the wide bandwidth signal, the antenna must be capable of radiating the transmit signal and capturing the return signal, and the RF portion of the receiver must be capable of amplifying and heterodyning the wide bandwidth signal. This places stringent requirements on the transmitter, antenna, and receiver front end, but current technology has advanced to cope with the requirements [8–11].

12.2 STRETCH PROCESSOR CONFIGURATION

Figure 12.2 contains a functional block diagram of a stretch processor. Stretch processing is technically distributed across multiple radar subsystems; with processing occurring in what is commonly defined as the receiver (i.e., mixing), exciter, track filters, and the digital signal processor. For our purposes, the functional block diagram is sufficient to guide us through the math of stretch processing. Processing consists of a mixer, an LFM generator, timing circuitry, and a spectrum analyzer.

We can write the normalized (idealized) signal returned from a stationary point scatterer, at a range delay of τ_R, as

$$r(t) = \sqrt{P_S}\, v(t-\tau_R) = \sqrt{P_S}\, e^{j\pi\alpha(t-\tau_R)^2} \operatorname{rect}\left[\frac{t-\tau_R}{\tau_p}\right] \qquad (12.11)$$

where $(P_S)^{1/2}$ is a scaling factor that we will use when we address signal-to-noise ratio (SNR). P_S can be interpreted as the peak signal power at the matched filter output and comes from the radar range equation (see Chapters 2 and 8).

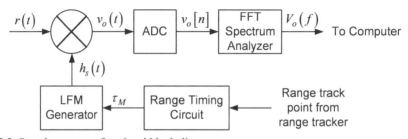

Figure 12.2 Stretch processor functional block diagram.

The normalized heterodyne signal generated by the LFM generator is[1]

$$h_s(t) = e^{j\pi\alpha(t-\tau_M)^2} \text{rect}\left[\frac{t-\tau_M}{\tau_h}\right] \qquad (12.12)$$

In the above, τ_M is the range delay to which the stretch processor is "matched" and is usually close to τ_R. Actually, *usually* is not the correct word. A more precise statement is that τ_M *must be close to* the τ_R of the scatterers we wish to resolve. τ_h is the duration of the heterodyne signal and, as we will show, must satisfy $\tau_h > \tau_p$.

Notional sketches of the frequency behavior of $r(t)$ and $h_s(t)$ are shown in Figure 12.3. The horizontal axis is time and the vertical axis is frequency. The frequency of each signal is shown only over the time the signal itself is not zero. Since $r(t)$ and $h_s(t)$ are LFM signals, we note that their frequencies increase linearly over their respective durations. Furthermore, by design, both frequency versus time plots have the same slope of α. The top plot corresponds to the case where the target range delay, τ_R, is greater than τ_M and the lower plot corresponds to the case where the range delay is less than τ_M. Note that when $\tau_R > \tau_M$, the frequency of $h_s(t)$ is greater than the frequency of $r(t)$. When $\tau_R < \tau_M$, the frequency of $h_s(t)$ is less than the frequency of $r(t)$. Further, the size of the frequency difference between $r(t)$ and $h_s(t)$ depends on the difference between τ_R and τ_M.

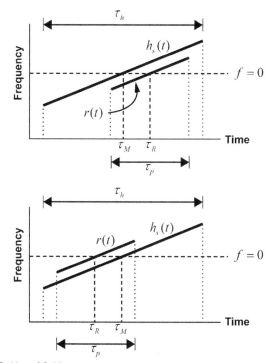

Figure 12.3 Sketches of $r(t)$ and $h_s(t)$.

---

[1] Consistent with previous analyses, we are using complex, baseband signal notation. As such, carrier and intermediate frequencies are zero.

Figure 12.3 also tells us how to select the value of τ_h, the duration of the heterodyne signal. Specifically, we want to choose τ_h so that $r(t)$ is completely contained within $h_s(t)$ for all expected values of τ_R relative to τ_M. From the bottom plot of Figure 12.3, we conclude we want to choose τ_h such that

$$\tau_{RMIN} - \tau_p/2 \geq \tau_M - \tau_h/2 \qquad (12.13)$$

From the top plot we want to choose it such that

$$\tau_{RMAX} + \tau_p/2 \leq \tau_M + \tau_h/2 \qquad (12.14)$$

In (12.13) and (12.14), τ_{RMIN} and τ_{RMAX} are the minimum and maximum expected values of τ_R. Equations (12.13) and (12.14) lead to the requirement on τ_h that it satisfy

$$\tau_h \geq \Delta\tau_R + \tau_p \qquad (12.15)$$

where

$$\Delta\tau_R = \tau_{RMAX} - \tau_{RMIN} \qquad (12.16)$$

is the range delay extent over which we want to use stretch processing. If τ_h satisfies the above constraint and

$$\tau_{RMIN} \leq \tau_R \leq \tau_{RMAX} \qquad (12.17)$$

then $h_s(t)$ will completely overlap $r(t)$, and the stretch processor will offer almost the same SNR performance as a matched filter. If the various timing parameters are such that $h_s(t)$ does not completely overlap $r(t)$, the stretch processor will experience an SNR loss proportional to the extent of $r(t)$ that does not lie within the extent of $h_s(t)$. It could also suffer a loss in resolution ability.

12.3 STRETCH PROCESSOR OPERATION

Given that $r(t)$ and $h_s(t)$ satisfy the above requirements, we can write the output of the mixer (of Figure 12.2) as

$$\begin{aligned} v_o(t) &= h_s(t) r^*(t) \\ &= \sqrt{P_S}\, e^{j\pi\alpha(t-\tau_M)^2} \mathrm{rect}\left[\frac{t-\tau_M}{\tau_h}\right] e^{-j\pi\alpha(t-\tau_R)^2} \mathrm{rect}\left[\frac{t-\tau_R}{\tau_p}\right] \end{aligned} \qquad (12.18)$$

or

$$v_o(t) = \sqrt{P_S}\, e^{j\pi\alpha\left(\tau_M^2 - \tau_R^2\right)} e^{j2\pi\alpha(\tau_R - \tau_M)t} \mathrm{rect}\left[\frac{t-\tau_R}{\tau_p}\right] \qquad (12.19)$$

The first exponential term of (12.19) is simply a phase term. However, the second exponential term tells us the output of the mixer is a constant frequency signal with a frequency that depends on the difference between the target range delay, τ_R, and the range delay to which the stretch processor is tuned, τ_M. Thus, if we can determine the frequency of the signal out of the mixer, we can determine the target range. Specifically, if we define the frequency out of the mixer as

$$f_m = \alpha(\tau_R - \tau_M) \qquad (12.20)$$

we get

$$\tau_R = f_m/\alpha + \tau_M \qquad (12.21)$$

The spectrum analyzer of Figure 12.2 is used to measure f_m. Ideally, the spectrum analyzer computes the Fourier transform of $v_o(t)$. With this in mind, we can write

$$V_o(f) = \int_{-\infty}^{\infty} v_o(t) e^{-j2\pi ft} dt = \sqrt{P_S} e^{j\phi} \int_{-\infty}^{\infty} e^{j2\pi f_m t} \text{rect}\left[\frac{t - \tau_R}{\tau_p}\right] e^{-j2\pi ft} dt \qquad (12.22)$$

or

$$V_o(f) = \tau_p \sqrt{P_S} e^{j\phi} e^{j2\pi(f_m - f)\tau_R} \text{sinc}\left[(f - f_m)\tau_p\right] \qquad (12.23)$$

where

$$\phi = \pi\alpha\left(\tau_M^2 - \tau_R^2\right) \qquad (12.24)$$

The information of interest is contained in $|V_o(f)|$, a normalized plot of which is contained in Figure 12.4. As we would expect, the sinc[x] function is centered at f_m and has a nominal width of $1/\tau_p$. Thus, we can measure f_m, but not with perfect accuracy. This is consistent with the result we would get with a classical matched filter. That is, the range measurement accuracy is related to the width of the main lobe of the output of a matched filter. For an LFM signal with a bandwidth of B, the nominal width of the main lobe is $1/B$ (see Chapter 8).

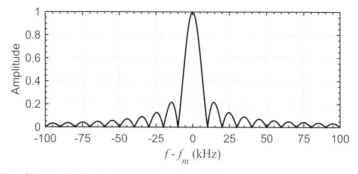

Figure 12.4 Plot of $|V_o(f - f_m)|$ for $B = 500$ MHz and $\tau_p = 100$ μs.

We now want to examine the range resolution of the stretch processor. Since the nominal width of the central lobe of the sinc[x] function is $1/\tau_p$, we normally say that the frequency resolution at the output of the spectrum analyzer is also $1/\tau_p$. Suppose we have a target at a range of τ_{R1} and a second target at a range of $\tau_{R2} > \tau_{R1}$. The mixer output frequencies associated with the two targets will be

$$f_{m1} = \alpha(\tau_{R1} - \tau_M) \quad (12.25)$$

and

$$f_{m2} = \alpha(\tau_{R2} - \tau_M) \quad (12.26)$$

Suppose further that τ_{R1} and τ_{R2} are such that

$$\delta f_m = f_{m2} - f_{m1} = 1/\tau_p \quad (12.27)$$

That is, the frequencies are separated by a (frequency) resolution cell of the stretch processor. With this we can write

$$\delta f_m = f_{m2} - f_{m1} = 1/\tau_p = \alpha(\tau_{R2} - \tau_M) - \alpha(\tau_{R1} - \tau_M) \quad (12.28)$$

or

$$\tau_{Rres} = \tau_{R2} - \tau_{R1} = \delta f_m / \alpha = 1/\alpha\tau_p = 1/B \quad (12.29)$$

or that the stretch processor has the same range resolution as a matched filter.

With LFM we can use an amplitude taper, implemented by a filter at the input or output of the matched filter, to reduce the range sidelobes at the matched filter output [12–15, Chapter 11]. We can apply a similar taper to a stretch processor by applying an amplitude taper to $v_o(t)$ before sending it to the spectrum analyzer.

12.4 STRETCH PROCESSOR SNR

At this point, we want to compare the SNR at the output of a matched filter to the SNR at the output of a stretch processor. Since neither processor includes nonlinearities, we can invoke superposition and treat the signal and noise separately.

12.4.1 Matched Filter

For the matched filter case, we can write the (normalized) signal voltage at the output of the matched filter as (see Chapter 10)

$$v_{sm}(\tau_{mm}, f_{mm}) = \chi(\tau_{mm}, f_{mm}) = \sqrt{P_S} e^{j\phi_{MF}} \int_{-\infty}^{\infty} v(t) v^*(t + \tau_{mm}) e^{j2\pi f_{mm} t} dt \quad (12.30)$$

where τ_{mm} and f_{mm} are the range delay and Doppler frequency mismatch, respectively, between the target return and matched filter. $v(t)$ is given by (12.1). We are interested in the power[2] out of the matched filter at matched range and Doppler. That is, we want

$$P_{sm} = |v_{sm}(0,0)|^2 \tag{12.31}$$

Substituting (12.30) into (12.31) yields

$$P_{sm} = \left| \sqrt{P_S} e^{j\phi_{MF}} \int_{-\infty}^{\infty} v(t) v^*(t) dt \right|^2 = P_S \tau_p^2 \tag{12.32}$$

The noise voltage at the output of the matched filter is given by

$$v_{nm}(t) = \int_{-\infty}^{\infty} h(x) n(t-x) dx \tag{12.33}$$

where $n(t)$ is zero-mean, wide-sense stationary, white noise with

$$E\{n(t) n^*(\tau)\} = N_o \delta(t-\tau) \tag{12.34}$$

$N_o = kT_s$ is the noise power spectral density at the antenna face (see Chapter 2, and $\delta(x)$ is the Dirac delta function[3] [16, p. 41]. Since $n(t)$ is a random process, so is $v_{nm}(t)$. Thus, the average noise power out of the matched filter is given by

$$P_{nm} = E\{|v_{nm}(t)|^2\} = E\left\{ \left| \int_{-\infty}^{\infty} h(x) n(t-x) dx \right|^2 \right\} = N_o \tau_p \tag{12.35}$$

where we have made use of (12.33). [The relation of (12.35) was derived in Appendix 9A.] With this we get the SNR at the matched filter output as

$$SNR_m = \frac{P_{sm}}{P_{nm}} = \frac{P_S \tau_p}{N_o} \tag{12.36}$$

which we recognize from radar range equation theory (see Chapter 2).

12.4.2 Stretch Processor

For the stretch processor, we are interested in the signal power at the target range delay, τ_R. Thus, we are interested in the output of the spectrum analyzer at $f = f_m$. With this we get, using (12.23),

$$P_{ss} = |V_o(f_m)|^2 = P_S \tau_p^2 \tag{12.37}$$

---

[2] P_{sm} is the power in a 1-ohm resistor. The right side of (12.32) is the *value* of P_{sm} in terms of P_S, the received signal power. The τ_p^2 term is a scaling factor that is caused by the math.
[3] Introduced in quantum mechanics by Paul Dirac.

If the noise into the mixer part of the stretch processor is $n(t)$, the noise out of the mixer is

$$v_{nM}(t) = n^*(t) h_s(t) \tag{12.38}$$

Recall that the signal power was computed at the spectrum analyzer output where $f = f_m$. The noise signal at the same frequency tap of the spectrum analyzer output is

$$V_{ns}(f_m) = \mathcal{F}[v_{mN}(t)]\big|_{f=f_m} = \int_{-\infty}^{\infty} n^*(t) h_s(t) e^{-j2\pi f_m t} dt \tag{12.39}$$

where $\mathcal{F}[\bullet]$ denotes the Fourier transform. The average power at the spectrum analyzer output is

$$P_{ns} = E\left\{|V_{ns}(f_m)|^2\right\} = N_0 \tau_h \tag{12.40}$$

where we have made use of (12.34) and (12.12).

From (12.37) and (12.40), the SNR at the output of the stretch processor is

$$SNR_s = \frac{P_{ss}}{P_{ns}} = \frac{P_s \tau_p^2}{N_0 \tau_h} \tag{12.41}$$

Combining (12.36) and (12.41), we get

$$\frac{SNR_s}{SNR_m} = \frac{\tau_p}{\tau_h} \tag{12.42}$$

Thus, the stretch processor encounters an SNR loss of τ_h/τ_p relative to the matched filter. This means we should be careful about using stretch processing for range extents that are significantly longer than the transmit pulse width.

At first inspection, it appears as if stretch processing could offer *better* SNR than a matched filter, which would contradict the fact that the matched filter maximizes SNR (see Chapter 8). This apparent contradiction is resolved by the stretch processor constraint imposed by (12.15). Specifically, $\tau_h > \tau_p$. Equation (12.42) also demonstrates another reason why stretch processing should not be used in a search function: it would be too lossy, as τ_h would need to be significantly larger than τ_p.

12.5 PRACTICAL IMPLEMENTATION ISSUES

Next, we turn our attention to practical implementation issues. The mixer, timing, and heterodyne generation are reasonably straightforward. However, we want to address how to implement the spectrum analyzer. The most obvious method of implementing the spectrum analyzer is to use an FFT. To do so, we need to determine the required ADC sample rate and the number of points to use in the FFT. To determine the ADC rate, we need to know the expected frequency limits of the signal out of the mixer.[4]

If τ_{RMIN} and τ_{RMAX} are the minimum and maximum range delays over which stretch processing is performed, the corresponding minimum and maximum frequencies out of the mixer are [see (12.25) and (12.26)]

$$f_{mMIN} = \alpha(\tau_{RMIN} - \tau_M) \tag{12.43}$$

and

$$f_{mMAX} = \alpha(\tau_{RMAX} - \tau_M) \tag{12.44}$$

Thus, the expected range of frequencies out of the mixer is

$$\Delta f_m = f_{mMAX} - f_{mMIN} = \alpha(\tau_{RMAX} - \tau_{RMIN}) = \alpha \Delta \tau_R \tag{12.45}$$

Thus, the ADC sample rate should be at least Δf_m.

The FFT will need to operate on data samples taken between $\tau_{RMIN} - \tau_p/2$ and $\tau_{RMAX} + \tau_p/2$ or over a time window of at least [see (12.15)]

$$\tau_{RMAX} - \tau_{RMIN} + \tau_p = \Delta \tau_R + \tau_p = \tau_h \tag{12.46}$$

The total number of data samples processed by the FFT will be

$$N_{samp} = \Delta f_m \tau_h \tag{12.47}$$

This means the FFT length will need to be some power of 2 that is greater than N_{samp}.

12.5.1 Stretch Processor Example

As an example of the above calculations, we consider the following parameters:

- $\tau_p = 100$ μs
- $B = 500$ MHz
- Stretch processing performed over 1,500 m

With this we get

$$\tau_{RMAX} - \tau_{RMIN} = 10 \text{ μs} \tag{12.48}$$

and

$$\tau_h = \tau_{RMAX} - \tau_{RMIN} + \tau_p = 110 \text{ μs} \tag{12.49}$$

To compute Δf_m, we first need to compute α as

$$\alpha = B/\tau_p = 5 \text{ MHz/μs} \tag{12.50}$$

---

[4] Again, we will assume baseband processing in these discussions. In practice, the mixer output will be at some intermediate frequency (IF). The signal could be brought to baseband using a synchronous detector or, as in some modern radars, by using IF sampling (i.e., a digital receiver). In either case, the effective ADC rate (the sample rate of the complex, digital, baseband signal) will be as derived here.

With this we get

$$\Delta f_m = \alpha(\tau_{RMAX} - \tau_{RMIN}) = 50 \text{ MHz} \quad (12.51)$$

Thus, the minimum required ADC sample rate is 50 MHz (since we are using complex samples). The number of samples to be processed by the FFT is

$$N_{samp} = \Delta f_m \tau_h = 5,500 \quad (12.52)$$

This means that we would want to use an 8,192-point FFT. One method of getting to 8,192 samples would be to increase the size of the range window. This would cause both Δf_m and τ_h to increase. An alternative would be to zero-pad the FFT by filling the last 8,192–5,500 taps with zero.

If we continue the calculations, we find that the time extent of the heterodyne window is

$$\tau_h \geq \Delta \tau_R + \tau_p = 110 \text{ μs} \quad (12.53)$$

The SNR loss associated with the use of stretch processing, relative to a matched filter, is $\tau_h/\tau_p = 110/100$ or about 0.4 dB.

12.6 RANGE-RATE EFFECTS

In the previous discussions we considered a stationary target. We now want to examine the effects of nonzero range-rate on the output of the stretch processor. We start by extending the definition of $r(t)$ in (12.11) to include a carrier term. We also let the range delay, τ_R, be a function of time, t. We then specifically examine how range-rate affects the returned signal, $r(t)$. After this we examine the stretch processor response to $r(t)$ from the specific perspectives of range resolution degradation and range error due to nonzero range-rate.

12.6.1 Expanded Transmit and Receive Signal Models

We need to extend the previous definition of the transmitted LFM pulse to include the carrier term. Specifically, we write

$$v(t) = e^{j2\pi f_c t} e^{j\pi\alpha t^2} \text{rect}\left[\frac{t}{\tau_p}\right] \quad (12.54)$$

where the first exponential is the carrier term, and f_c is the carrier frequency.

The signal returned from the target is

$$r(t) = \sqrt{P_S}\, v[t - \tau_R(t)] = \sqrt{P_S}\, e^{j2\pi f_c[t-\tau_R(t)]} e^{j\pi\alpha[t-\tau_R(t)]^2} \text{rect}\left[\frac{t-\tau_R(t)}{\tau_p}\right] \quad (12.55)$$

where we now show range delay, $\tau_R(t)$, as a function of time to account for the fact that range changes with time because the range-rate is not zero. We assume the target range-rate is constant. With this we can write

$$\tau_R(t) = \frac{2R(t)}{c} = \frac{2R_0}{c} + \frac{2\dot{R}}{c}t \qquad (12.56)$$

where R_0 is the range at $t = 0$ (the center of the transmit pulse in this case), and $\dot{R}$ is the (constant) range-rate.

Substituting (12.56) into (12.55) gives

$$r(t) = \sqrt{P_S}\, e^{j2\pi f_c\left[t - 2R_0/c - (2\dot{R}/c)t\right]} e^{j\pi\alpha\left[t - 2R_0/c - (2\dot{R}/c)t\right]^2} \mathrm{rect}\left[\frac{t - 2R_0/c - (2\dot{R}/c)t}{\tau_p}\right]$$

$$= \sqrt{P_S}\, e^{j\phi_1(t)} e^{j\phi_2(t)} \mathrm{rect}\left[\frac{t - \tau_{R\dot{R}}}{\tau_{p\dot{R}}}\right] \qquad (12.57)$$

where $\phi_1(t)$ is a phase term we associate with the interaction of the target range-rate with the carrier and $\phi_2(t)$ is a phase term we associate with the interaction of range-rate with the LFM. We also changed the rect[x] function to denote the effect of the nonzero range-rate on pulsewidth and perceived target range.

Expanding $\phi_1(t)$ gives

$$\phi_1(t) = 2\pi f_c t - 4\pi R_0/\lambda - 2\pi\left(2\dot{R}/\lambda\right)t = 2\pi f_c t - 4\pi R_0/\lambda + 2\pi f_d t \qquad (12.58)$$

where the first term on the right is the carrier component, the second term is a phase shift associated with the initial target position, and the third term is the Doppler frequency term. This term [$\phi_1(t)$] is the same as we developed in Chapter 1 when we discussed Doppler frequency.

If we expand the second phase term, we get

$$\phi_2(t) = \pi\alpha\left[t - 2R_0/c - (2\dot{R}/c)t\right]^2 = \pi\alpha\left[(1 - 2\dot{R}/c)t - 2R_0/c\right]^2$$

$$= \pi\alpha(1 - 2\dot{R}/c)^2\left[t - \frac{2R_0/c}{1 - 2\dot{R}/c}\right]^2 = \pi\alpha_r\left(t - \tau_{R\dot{R}}\right)^2 \qquad (12.59)$$

From (12.59) we see that range-rate has two effects on the signal returned from the target. The first is that the LFM slope changes from α to

$$\alpha_r = \alpha\left(1 - 2\dot{R}/c\right)^2 \qquad (12.60)$$

The change in LFM slope is very small—fractions of a percent. However, as we will see, for high speed targets (e.g., ballistic targets going 1,000s m/s) and large BT product waveforms (BT $= B\tau_p = \alpha\tau_p^2$), if not accounted for, the slope change due to nonzero range-rate can have a significant impact on range resolution and SNR at the stretch processor output.

The second impact of nonzero range-rate is that the range delay of the target moves from $\tau_R = 2R_0/c$ to

$$\tau_{R\dot{R}} = \frac{\tau_R}{1-2\dot{R}/c} = \frac{2R_0/c}{1-2\dot{R}/c} \quad (12.61)$$

In other words, range-rate causes a bias in the range measurement, relative to R_0. For high-speed, long-range targets, this bias can be a significant multiple of the range resolution of the waveform. However, the only time the bias will have an impact on range measurement, from a system perspective, if not accounted for, is when trying to reconcile a range measurement from a wideband waveform with a range measurement from a low bandwidth waveform. If the wideband waveform is used to resolve scatterers on a target (i.e., in discrimination) the bias will have virtually no effect since all the scatterers have the same range-rate, and would thus experience the same bias.

The last term in (12.57), the rect[x] function, captures another effect caused by nonzero range-rate targets: It changes the width of the return pulse from τ_p to

$$\tau_{p\dot{R}} = \frac{\tau_p}{1-2\dot{R}/c} \quad (12.62)$$

It is this change in pulse width that causes the change in LFM slope and the apparent change in target range. This can be heuristically explained as follows: For a stationary point target, the trailing edge of the transmitted pulse will reach the target τ_p seconds after the leading edge. However, if the target is moving, it will change position as the pulse is interacting with the target. As a result, the trailing edge of the pulse will not reach the target in τ_p seconds after the leading edge. For an approaching target, the trailing edge will reach the target in less than τ_p seconds after the leading edge reaches the target. As a result, the return pulse will be shorter than the transmit pulse. The reverse is true for a receding target: the return pulse will be longer than the transmit pulse. Even though the pulse width changes, the frequency still varies by B across the return pulse. Since the pulse width changes, but B remains the same, the LFM slope must change.

A similar argument explains the change in apparent target range, given that range is measured at the center of the return pulse. Because of target motion, the center of the return pulse will not be at $\tau_p/2$ relative to the leading edge. Thus, the apparent range will be something other than R_0.

As a summary of the above discussions, we can now write the return signal (12.11) as

$$r(t) = \sqrt{P_S}\, e^{j2\pi f_c t} e^{j2\pi f_d t} e^{-j4\pi R_0/\lambda} e^{j\pi\alpha_r[t-\tau_{R\dot{R}}]^2} \mathrm{rect}\left[\frac{t-\tau_{R\dot{R}}}{\tau_{p\dot{R}}}\right]. \quad (12.63)$$

12.6.2 Stretch Processor Modification

The form of (12.63) tells us how we must modify the stretch processor of Figure 12.2 to accommodate the modified received signal and the nonzero range-rate. First, the processor must contain mixers to remove the carrier frequency term and the estimated Doppler frequency term. We don't need to worry about the third exponential because it is a phase term we temporarily ignore.

To remove the carrier and Doppler frequency terms, we include a pair of mixers whose inputs are $\exp[-j2\pi f_c t]$ and $\exp[-j2\pi f_{Md} t]$. We set the frequency of the first mixer to f_c because the carrier frequency is accurately known (the same oscillator generates the transmit carrier and the carrier term used on receive). We used a frequency of f_{Md} for the second mixer to account for the possibility the radar will have only an estimate of f_d rather than the actual target Doppler frequency. This will allow us to examine the impact of having an approximate measurement of range-rate. As a note, the measurement of range-rate, and R_0, would most often be provided by the radar tracker.

We must also modify the stretch processor heterodyne signal, $h_s(t)$, to be of the form

$$h_{s\dot{R}}(t) = e^{j\pi\alpha_{h\dot{R}}\left(t-\tau_{M\dot{R}}\right)^2} \text{rect}\left[\frac{t-\tau_{M\dot{R}}}{\tau_h}\right] \quad (12.64)$$

Specifically, the LFM slope of the heterodyne signal must be changed so that it is close to α_r. That is

$$\alpha_{h\dot{R}} \approx \alpha_r = \alpha\left(1-2\dot{R}/c\right)^2 \quad (12.65)$$

The reason for stating $\alpha_{h\dot{R}}$ can be only approximately set to α_r is that it is unlikely range-rate will be known well enough for $\alpha_{h\dot{R}}$ to exactly equal α_r. This does not present serious problems since resolution and SNR degradation due to range-rate error will be small unless the range-rate error is very large and/or the waveform BT product is large. We will examine this further via examples.

The matched range setting, $\tau_{M\dot{R}}$, of the stretch processor heterodyne signal should be close to $\tau_{R\dot{R}}$. That is

$$\tau_{M\dot{R}} \approx \tau_{R\dot{R}} = \frac{2R_0/c}{1-2\dot{R}/c} \quad (12.66)$$

The reason $\tau_{M\dot{R}}$ can be only approximately set to $\tau_{R\dot{R}}$ is that it is unlikely exact values of $\dot{R}$ and R_0 can be measured.

In Section 12.2, we found we could accommodate inaccuracies in estimates of R_0 by proper selection of τ_h, the width of the heterodyne signal. Lack of accurate $\dot{R}$ and R_0 values could require τ_h be increased somewhat. However, this will only be the case for large uncertainties in $\dot{R}$ and long-range targets (large R_0). We will quantify some of these impacts in upcoming discussions.

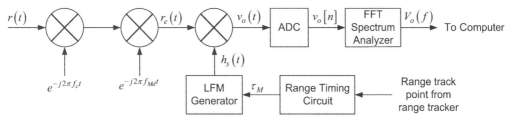

Figure 12.5 Modified stretch processor.

Figure 12.5 contains a block diagram of the stretch processor with the modifications just discussed. With the modifications, the output of the third mixer will be

$$v_o(t) = \left(r(t)e^{-j2\pi f_c t}e^{-j2\pi f_{Md} t}\right)^* e^{j\pi\alpha_{h\dot{R}}(t-\tau_{M\dot{R}})^2} \text{rect}\left[\frac{t-\tau_{M\dot{R}}}{\tau_h}\right]$$
$$= \sqrt{P_S}\, e^{j\theta} e^{j2\pi\Delta f_d t} e^{-j\pi\alpha_r(t-\tau_{R\dot{R}})^2} e^{j\pi\alpha_{h\dot{R}}(t-\tau_{M\dot{R}})^2} \text{rect}\left[\frac{t-\tau_{R\dot{R}}}{\tau_{p\dot{R}}}\right] \quad (12.67)$$

where $\Delta f_d = f_d - f_{Md}$ is the difference between the actual and estimated Doppler frequency. θ is a phase that accounts for the constant phase term of (12.63) and any other phase shifts encountered up to the output of the second mixer of Figure 12.5. This phase shift can be ignored, which we will do by setting it to zero.

The fact that the last term is rect$[(t - \tau_{R\dot{R}})/\tau_{R\dot{R}}]$ instead of rect$[(t - \tau_{M\dot{R}})/\tau_h]$ means we assume τ_h is chosen such that the heterodyne pulse (i.e., $h_{s\dot{R}}(t)$) completely overlaps the received pulse (i.e., $r(t)$). This is consistent with our earlier discussions. It also carries the implicit assumption that we know $\dot{R}$ and R_0 well enough for $\tau_{M\dot{R}}$ to be close to $\tau_{R\dot{R}}$.

Now that we have an equation for the output of the third mixer, we want to examine the effect of range-rate errors. We first examine the effect of slope mismatch caused by inaccurate knowledge of range-rate. After that we consider the effect of inaccurate range-rate knowledge on range measurement bias.

12.6.3 Slope Mismatch Effects

To examine the effect of LFM slope mismatch (the difference between α_r and $\alpha_{h\dot{R}}$), we consider only the last three terms of (12.67). That is,

$$v_{o\alpha}(t) = e^{-j\pi\alpha_r(t-\tau_{R\dot{R}})^2} e^{j\pi\alpha_{h\dot{R}}(t-\tau_{M\dot{R}})^2} \text{rect}\left[\frac{t-\tau_{R\dot{R}}}{\tau_{p\dot{R}}}\right] = e^{j\phi_3(t)} \text{rect}\left[\frac{t-\tau_{R\dot{R}}}{\tau_{p\dot{R}}}\right] \quad (12.68)$$

Specifically, we assume $\Delta f_d = 0$ and $P_S = 1$ (we already assumed $\theta = 0$). If we expand the exponent, we get

$$\phi_3(t) = \pi\left(\alpha_{h\dot{R}} - \alpha_r\right)t^2 + 2\pi\left(\alpha_r\tau_{R\dot{R}} - \alpha_{h\dot{R}}\tau_{M\dot{R}}\right)t + \pi\left(\alpha_{h\dot{R}}\tau_{M\dot{R}}^2 - \alpha_r\tau_{R\dot{R}}^2\right) \quad (12.69)$$

The first term of (12.69) is a quadratic phase, or LFM, term; the second term is a constant frequency term; and the third term is a phase term we ignore.

For now, we assume we are able to choose $\tau_{M\dot{R}}$ such that

$$f_m = \alpha_r\tau_{R\dot{R}} - \alpha_{h\dot{R}}\tau_{M\dot{R}} = 0 \quad (12.70)$$

With this and the fact we ignored the constant phase term of (12.69), we have

$$v_{o\alpha}(t)\Big|_{f_m=0} = e^{j\pi(\alpha_{h\dot{R}} - \alpha_r)t^2} \text{rect}\left[\frac{t-\tau_{R\dot{R}}}{\tau_{p\dot{R}}}\right] \quad (12.71)$$

We now want to consider a few specific slope mismatch cases.

12.6.3.1 Slope Mismatch Case 1: $\alpha_{h\dot{R}} = \alpha$ – no compensation

When we considered a fixed target (zero range-rate), we had $\alpha_r = \alpha$, and we chose $\alpha_{h\dot{R}} = \alpha$. With that (12.71) reduces to

$$v_{o\alpha}(t)\big|_{f_m=0,\text{matched}} = \text{rect}\left[\frac{t-\tau_{R\dot{R}}}{\tau_{p\dot{R}}}\right] \tag{12.72}$$

When we process this through the spectrum analyzer, we get a sinc[x] function, as was illustrated in Figure 12.4.

For the non-zero range-rate case we have

$$\alpha_r = \alpha\left(1 - 2\dot{R}/c\right)^2 \tag{12.73}$$

If we choose $\alpha_{h\dot{R}} = \alpha$ (i.e., ignore the effect of range-rate) we get

$$\Delta\alpha = \alpha_{h\dot{R}} - \alpha_r = \alpha - \alpha\left(1 - 2\dot{R}/c\right)^2 = \alpha\left(4\dot{R}/c\right)\left(1 - \dot{R}/c\right) \tag{12.74}$$

and

$$v_{o\alpha}(t)\big|_{f_m=0} = e^{j\pi\Delta\alpha t^2}\text{rect}\left[\frac{t-\tau_{R\dot{R}}}{\tau_{p\dot{R}}}\right] = e^{j\pi\alpha(4\dot{R}/c)(1-\dot{R}/c)t^2}\text{rect}\left[\frac{t-\tau_{R\dot{R}}}{\tau_{p\dot{R}}}\right] \tag{12.75}$$

Because of the slope mismatch, we no longer have a simple rectangular pulse at the output of the third mixer of Figure 12.5. Instead, we have an LFM pulse with an LFM slope of $\alpha(4\dot{R}/c)(1-\dot{R}/c)$. As a result, the output of the spectrum analyzer will not be a sinc[x] function. Instead, it will look like an LFM frequency response (see Section 12.1). More importantly, the width of the main lobe of the spectrum analyzer output will be broader than shown in Figure 12.4. This increase in width translates to a degradation in range resolution. It also leads to a reduction in SNR.

Slope Mismatch Example 1

To quantify the effect of slope mismatch, we consider a specific example. Since $v_{o\alpha}(t)|_{f_m=0}$ is an LFM pulse with a width of $\tau_{p\dot{R}}$ and an LFM slope of $\alpha(4\dot{R}/c)(1-\dot{R}/c)$, we can use (12.6) to write the magnitude of the response of the spectrum analyzer as

$$\begin{aligned}|V_o(f)| &= \left|\int_{-\infty}^{\infty} v_{o\alpha}(t)\big|_{f_m=0} e^{-j2\pi ft}dt\right| \\ &= \left|F\left[\sqrt{\tfrac{2}{\Delta\alpha}}\left(f + \Delta\alpha\tau_{p\dot{R}}/2\right)\right] - F\left[\sqrt{\tfrac{2}{\Delta\alpha}}\left(f - \Delta\alpha\tau_{p\dot{R}}/2\right)\right]\right|\end{aligned} \tag{12.76}$$

As a specific example, we consider the waveform in Section 12.3: a 100-µs pulse with an LFM bandwidth of 500 MHz. We also consider range-rates of −300 m/s, −3,000 m/s, and –6,000 m/s. The range-rate of –300 m/s would be representative of a target flying at about Mach 1, and –6000 m/s would be representative of the range-rate of an intercontinental ballistic missile (ICBM). The range-rate of –3,000 m/s is an intermediate value between –300 m/s and –6000 m/s.

Figure 12.6 contains plots of $|V_o(f)|$ for these three range-rates. The heights of the curves are scaled relative to a plot of $|V_o(f)|$ for a zero range-rate target. As such, the heights represent the square root of the relative SNRs for the three range-rates. If we compare the plots of Figure 12.6 to those of Figure 12.4, we note that the Figure 12.6 curve for the –300 m/s target is virtually the same as the plot for a zero range-rate target.

For the –3,000 m/s target, the width of the main lobe is broadened slightly (from 8.8 kHz to 9.4 kHz, at the –3-dB points) and the peak amplitude is reduced from one to about 0.9. The broadening represents about a 6% reduction in range resolution (recall that range resolution is related to frequency by $\delta\tau = \delta f/\alpha$—see Section 12.3) and the reduced amplitude equates to about a 1 dB loss in SNR.

The plot of $|V_o(f)|$ for the –6,000-m/s range-rate is quite distorted. In this case, the –3 dB width is about 16.6 kHz, which translates to a degradation in range resolution by a factor of about two. The peak amplitude is 0.63, which translates to about a 4 dB SNR loss.

Slope Mismatch Example 2

In the above discussion, we focused on the effect of range-rate for a given transmit waveform (100 µs, 500-MHz LFM pulse). We now consider the effect of waveform properties, namely BT product, on resolution and SNR for a given range-rate. Figure 12.7 contains a plot of $|V_o(f)|$ for the case where the waveform is a 100-µs pulse with a 250-MHz LFM bandwidth, instead of 500 MHz. This means the BT product of the waveform is 25,000 rather than the 50,000 of the 100-µs, 500-MHz LFM pulse. The responses are about the same for the –300-m/s and –3,000-m/s targets, and for the zero range-rate target.

Further, the response for the –6,000-m/s target shows much less degradation than in Figure 12.6.

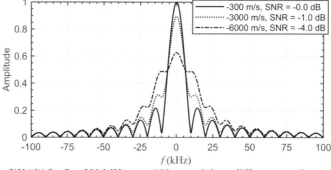

Figure 12.6 Plot of $|V_o(f)|$ for $B = 500$ MHz, $\tau_p = 100$ µs and three different speed targets.

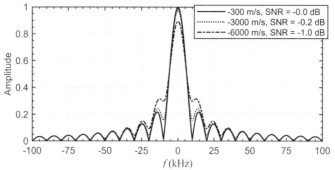

Figure 12.7 Plot of $|V_o(f)|$ for $B = 250$ MHz, $\tau_p = 100$ μs, and three different target speeds.

12.6.3.2 Slope Mismatch Case 2: $\alpha_{h\dot{R}} = \alpha_r$ – Perfect Compensation

In the above discussions, we assumed the LFM slope of the heterodyne signal was α, the slope of the transmitted LFM pulse. Had we used a slope of

$$\alpha_{h\dot{R}} = \alpha_r = \alpha\left(1 - 2\dot{R}/c\right)^2 \tag{12.77}$$

instead of α, (perfect compensation), we would have

$$\Delta\alpha = \alpha_{h\dot{R}} - \alpha_r = 0 \tag{12.78}$$

and

$$v_{o\alpha}(t)\big|_{f_m=0} = e^{j\pi\Delta\alpha t^2}\mathrm{rect}\left[\frac{t - \tau_{R\dot{R}}}{\tau_{p\dot{R}}}\right] = \mathrm{rect}\left[\frac{t - \tau_{R\dot{R}}}{\tau_{p\dot{R}}}\right] \tag{12.79}$$

Because of this, $|V_o(f)|$ would be the ideal sinc[x] function we obtained for the zero range-rate target.

12.6.3.3 Slope Mismatch Case 3: $\alpha_{h\dot{R}} = \alpha(1 - 2\dot{R}_h/c)^2$ – Partial Compensation

The ability to achieve $\alpha_{h\dot{R}} = \alpha_r$ requires perfect knowledge of $\dot{R}$. However, perfect knowledge is not necessary. Suppose we have an estimate of $\dot{R}$ as $\dot{R}_h$. With this we can write (partial compensation)

$$\alpha_{h\dot{R}} = \alpha\left(1 - 2\dot{R}_h/c\right)^2 \tag{12.80}$$

which would lead to

$$\begin{aligned}\Delta\alpha = \alpha_{h\dot{R}} - \alpha_r &= \alpha\left(1 - 2\dot{R}_h/c\right)^2 - \alpha\left(1 - 2\dot{R}/c\right)^2 \\ &= 4\alpha\left(\frac{\dot{R} - \dot{R}_h}{c}\right)\left(1 - \left(\frac{\dot{R} + \dot{R}_h}{c}\right)\right)\end{aligned} \tag{12.81}$$

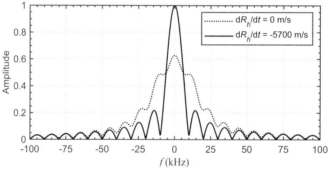

Figure 12.8 Plot of $|V_o(f)|$ for $B = 500$ MHz, $\tau_p = 100$-μs, $\dot{R} = -6{,}000$-m/s and $\dot{R}_h = -5{,}700$-m/s.

Although not obvious, the $\Delta\alpha$ of (12.81) can be made considerably smaller than the $\Delta\alpha$ of (12.74). This, in turn, means the $v_{o\alpha}(t)|_{fm=0}$ of (12.75) would approach the $v_{o\alpha}(t)|_{fm=0}$ of (12.79), and $|V_o(f)|$ would approach a sinc[x] function. To illustrate this, we consider the case where we have a 100-μs, 500-MHz LFM pulse and a −6,000-m/s target. From Figure 12.6 we found this led to a significant degradation in resolution and SNR if we used $\alpha_{h\dot{R}} = \alpha$. We now assume we are able to measure $\dot{R}$ to within 300 m/s. Specifically, we assume we have $\dot{R}_h = -5{,}700$ m/s. If we use this to compute $\Delta\alpha$ from (12.81) and use this in (12.76), we get the solid curve of Figure 12.8. The dotted curve is the dash-dot curve of Figure 12.6. As expected, the estimate of range-rate used to compute the LFM slope of the heterodyne signal is close enough to provide a sinc-like response at the spectrum analyzer output.

In the above discussions, we assumed the LFM slope of the stretch processor, $\alpha_{h\dot{R}}$, was changed to match, or partially match, the slope of the received signal, α_r. An alternate approach would be to alter the transmit slope so that the LFM slope of the received waveform was close to α, the desired LFM slope. In that case, the LFM slope of the stretch processor would be set to α.

12.6.4 Range-rate Effects on Range Bias

The next subject we want to consider is the effect of range-rate on measurement bias. To do so, we examine the frequency term of $v_{o\alpha}(t)$. Specifically [see (12.70)],

$$f_m = \alpha_r \tau_{R\dot{R}} - \alpha_{h\dot{R}} \tau_{M\dot{R}} = 0 \tag{12.82}$$

For the zero range-rate target, we had $\alpha_r = \alpha_{h\dot{R}} = \alpha$ and $\tau_{R\dot{R}} = \tau_R = 2R_0/c$ and $\tau_{M\dot{R}} = \tau_M$. With this (12.82) reduces to

$$f_m = \alpha(\tau_R - \tau_M) \tag{12.83}$$

from which we arrive at [see (12.21)]

$$\tau_R = f_m/\alpha + \tau_M \tag{12.84}$$

In other words, given a τ_M of the stretch processor we could measure f_m and use it to determine τ_R. If we coincidently choose τ_M such that $f_m = 0$, we would have $\tau_R = \tau_M$, or $R_0 = R_M = 2\tau_M/c$.

The question we now ask is: How does a nonzero range-rate affect the equality between $\tau_{R\dot{R}}$ and $\tau_{M\dot{R}}$ when $f_m = 0$, and $\alpha_{h\dot{R}} \neq \alpha_r$? We consider two cases:

1. Where we ignore the nonzero range-rate and set $\alpha_{h\dot{R}} = \alpha$
2. Where we estimate $\dot{R}$ and use (12.80) to compute $\alpha_{h\dot{R}}$

12.6.4.1 Case 1 – $\alpha_{h\dot{R}} = \alpha$

For the first case we have

$$f_m|_{\alpha_{h\dot{R}}=\alpha} = \alpha_r \tau_{R\dot{R}} - \alpha \tau_{M\dot{R}} = \alpha\left(1 - 2\dot{R}/c\right)^2 \tau_{R\dot{R}} - \alpha \tau_{M\dot{R}} \quad (12.85)$$

If we set $f_m|_{\alpha_{h\dot{R}}=\alpha} = 0$ we get

$$\tau_{M\dot{R}} = \left(1 - 2\dot{R}/c\right)^2 \tau_{R\dot{R}} \quad (12.86)$$

or, in terms of R_M and R_0,

$$\frac{2R_M}{c} = \left(1 - 2\dot{R}/c\right)^2 \frac{2R_0/c}{\left(1 - 2\dot{R}/c\right)} \quad (12.87)$$

or

$$R_M = \left(1 - 2\dot{R}/c\right) R_0 \quad (12.88)$$

We see that because of the nonzero range-rate, R_M differs from R_0 by the bias

$$\Delta R = |R_M - R_0| = \frac{2|\dot{R}|}{c} R_0 \quad (12.89)$$

This is an interesting equation in that it says the range bias depends on the target range, R_0.

As a specific example, we consider the –6,000-m/s target at a range of 300 km. In that case, we would have

$$\Delta R = \frac{2(6000)}{3 \times 10^8}\left(3 \times 10^5\right) = 12 \text{ m} \quad (12.90)$$

While 12 m is quite small relative to $R_0 = 300$ km, it could be quite large relative to the waveform range resolution. For example, it would be 40 times the 0.3 m resolution of the 500-MHz bandwidth LFM pulse.

Before we consider the second case, we consider the other extreme of Case 1 and assume we have perfect knowledge of $\dot{R}$ so that we can use

$$\alpha_{h\dot{R}} = \alpha_r = \alpha\left(1 - 2\dot{R}/c\right)^2 \qquad (12.91)$$

We suppose further that we choose $\tau_{M\dot{R}}$ so that $f_m = 0$. With this we would deduce that

$$\tau_{M\dot{R}} = \tau_{R\dot{R}} = \frac{2R_0/c}{1 - 2\dot{R}/c} \qquad (12.92)$$

We carry the concept of knowing $\dot{R}$ a step further and assume $\tau_{M\dot{R}}$ is the same form as $\tau_{R\dot{R}}$. Specifically, we write

$$\tau_{M\dot{R}} = \frac{2R_M/c}{1 - 2\dot{R}/c} \qquad (12.93)$$

Considering (12.93) and (12.92) leads to the conclusion that

$$R_M = R_0 \qquad (12.94)$$

In other words, the stretch processor provides a measurement of R_0, the same as for the zero range-rate case. Keep in mind that we assumed $f_m = 0$ in these examples. This implies we (actually the stretch processor) are concluding $\tau_{M\dot{R}}$ is equal to the actual range delay, $\tau_{R\dot{R}}$.

12.6.4.2 Case 2: Imperfect Estimate of $\dot{R}$

We next consider Case 2, where we have an estimate of $\dot{R}$, which we previously denoted as $\dot{R}_h$. From (12.80) we use the estimate of $\dot{R}_h$ to compute $\alpha_{h\dot{R}}$ as

$$\alpha_{h\dot{R}} = \alpha\left(1 - 2\dot{R}_h/c\right)^2 \qquad (12.95)$$

We take a hint from our earlier work and write $\tau_{M\dot{R}}$ as

$$\tau_{M\dot{R}} = \frac{2R_M/c}{1 - 2\dot{R}_h/c} \qquad (12.96)$$

We then use (12.96) with (12.82) to write

$$\begin{aligned}
f_m &= \alpha_r \tau_{R\dot{R}} - \alpha_{h\dot{R}} \tau_{M\dot{R}} \\
&= \alpha\left(1 - 2\dot{R}/c\right)^2 \left(\frac{2R_0/c}{(1 - 2\dot{R}/c)}\right) - \alpha\left(1 - 2\dot{R}_h/c\right)^2 \left(\frac{2R_M/c}{(1 - 2\dot{R}_h/c)}\right) \\
&= \alpha\left[R_0\left(1 - 2\dot{R}/c\right) - R_M\left(1 - 2\dot{R}_h/c\right)\right]
\end{aligned} \qquad (12.97)$$

As before, we assume we (actually, the stretch processor logic) choose $\tau_{M\dot{R}}$ so that $f_m = 0$. With this we get

$$R_0\left(1 - 2\dot{R}/c\right) - R_M\left(1 - 2\dot{R}_h/c\right) = 0 \qquad (12.98)$$

which we can solve for R_M as

$$R_M = R_0 \frac{\left(1 - 2\dot{R}/c\right)}{\left(1 - 2\dot{R}_h/c\right)} \tag{12.99}$$

We note that if $\dot{R}_h = \dot{R}$, then $R_M = R_0$, which is what we found earlier. If $\dot{R}_h = \dot{R} + \Delta\dot{R}$, and $\Delta\dot{R} \ll \dot{R}$, then we have ($\Delta\dot{R}$ is the range-rate estimation error)

$$R_M = R_0 + R_0 \frac{2\Delta\dot{R}}{c} \tag{12.100}$$

or

$$\Delta R = R_0 \frac{2\Delta\dot{R}}{c} \tag{12.101}$$

In other words, the use of R_M as the target range, via the selection of $\tau_{M\dot{R}}$, produceD a range estimate that was off by ΔR.

As a specific example, we revisit the previous example of a $-6{,}000$-m/s target at a range of $R_0 = 300$ km. Suppose we can measure $\dot{R}$ to within $\Delta\dot{R} = 150$ m/s. In that case, we would have

$$\Delta R = \left(3 \times 10^5\right) \frac{2 \times 150}{3 \times 10^8} = 0.3 \text{ m} \tag{12.102}$$

which is quite small. For a 500-MHz LFM pulse, this value of ΔR is equal to the range resolution of 0.3 m. As a note, the range bias due to range-rate error does not depend on the actual range-rate, but on the range-rate error. Thus, (12.101) applies to both slow and fast targets.

12.6.5 Doppler Frequency Measurement Effects

Thus far, we have assumed the Doppler frequency estimate used in the second mixer of Figure 12.5 is the true Doppler frequency, and that the frequency of the signal out of the second mixer was $\Delta f_d = f_d - f_{Md}$. In practice, f_{Md} will be related to the estimated range-rate, $\dot{R}_h$, so that

$$\Delta f_d = \frac{-2\Delta\dot{R}}{\lambda} \tag{12.103}$$

If we carry the Δf_d term of (12.67) through to (12.75) we would have

$$v_{o\alpha}(t)\big|_{f_m = 0, \Delta f_d \neq 0} = e^{j2\pi\Delta f_d t} e^{j\pi\Delta\alpha t^2} \text{rect}\left[\frac{t - \tau_{R\dot{R}}}{\tau_{p\dot{R}}}\right] \tag{12.104}$$

and (12.76) would become

$$\left|V_o(f)\right|_{\Delta f_d = 0} = \left|F\left[\sqrt{\tfrac{2}{\Delta\alpha}}\left(f - \Delta f_d + \Delta\alpha\tau_{p\dot{R}}/2\right)\right]\right. \quad (12.105)$$
$$\left. - F\left[\sqrt{\tfrac{2}{\Delta\alpha}}\left(f - \Delta f_d - \Delta\alpha\tau_{p\dot{R}}/2\right)\right]\right|$$

In words, the peak of the spectrum analyzer output would be shifted from zero to Δf_d. For the case where $f_m \neq 0$, the peak would be located at $f_m + \Delta f_d$. This shift in the peak of the spectrum analyzer output will translate to a range measurement error of $\Delta\tau_R = \Delta f_d / \alpha$, or $\Delta R = (\Delta f_d / \alpha)(c/2)$. As a note, this relation between Doppler measurement error and range measurement error is the normal range-Doppler coupling relation associated with the ambiguity function of LFM waveforms.

To get an idea of the magnitude of the range measurement error caused by a Doppler frequency error, we revisit a combination of our previous examples. Specifically, we consider a 100-μs, 500-MHz LFM pulse. For this case we have

$$\alpha = B/\tau_p = 500 \times 10^6 / 100 \times 10^{-6} = 5 \times 10^{12} \text{ Hz/s} \quad (12.106)$$

We also consider the most recent example where we assumed the range-rate was estimated to within 150 m/s of the true range-rate. Finally, we assume the radar is operating at X-band with $f_c = 10$ GHz. This results in a wavelength of $\lambda = c/f_c = 0.03$ m. With these parameters we have

$$\Delta f_d = \frac{2\Delta\dot{R}}{\lambda} = \frac{2 \times 150}{0.03} = 10 \text{ kHz} \quad (12.107)$$

which is the resolution of the spectrum analyzer. This Doppler error translates to a range measurement error of

$$\Delta R = \frac{\Delta f_d}{\alpha}\frac{c}{2} = \frac{10 \times 10^3}{5 \times 10^{12}}\frac{3 \times 10^8}{2} \quad (12.108)$$
$$= (2 \times 10^{-9}) \times (1.5 \times 10^8) = (2 \text{ ns}) \times (1.5 \times 10^8) = 0.3 \text{ m}$$

which is the range resolution of the LFM pulse. This is also expected since the Doppler resolution of the spectrum analyzer, and the LFM pulse, is 10 kHz.

12.6.6 A Matched Filter Perspective

Earlier in this chapter we established a relation between stretch processing and matched filter processing in terms of resolution and SNR. We now want to see how this relation carries over to range-rate effects. We assume the input to the matched filter is the output of the second mixer of Figure 12.5. That is, the matched filter replaces the stretch processor portion of Figure 12.5 (see Figure 12.1) The input to the matched filter is [see (12.63)]

$$r_e(t) = r(t)e^{-j2\pi f_c t}e^{-j2\pi f_{Md}t}$$
$$= \sqrt{P_S}\, e^{j2\pi \Delta f_d t} e^{-j4\pi R_0/c} e^{j\pi \alpha_r [t-\tau_{R\dot{R}}]^2} \mathrm{rect}\left[\frac{t-\tau_{R\dot{R}}}{\tau_{p\dot{R}}}\right] \quad (12.109)$$
$$= K_1 e^{j2\pi \Delta f_d t} e^{j\pi \alpha_r [t-\tau_{R\dot{R}}]^2} \mathrm{rect}\left[\frac{t-\tau_{R\dot{R}}}{\tau_{p\dot{R}}}\right]$$

where we combined the terms that are not a function of t into the constant K_1.

We can write the matched filter impulse response as

$$h(t) = e^{-j\pi \alpha_{h\dot{R}} t^2} \mathrm{rect}\left[\frac{t}{\tau_h}\right] \quad (12.110)$$

The response of the matched filter to the received signal is the convolution of $r_e(t)$ and $h(t)$. That is,

$$v_o(t) = r_e(t) * h(t) = \int_{-\infty}^{\infty} r_e(\gamma) h(t-\gamma) d\gamma \quad (12.111)$$

Substituting for $r_e(t)$ and $h(t)$ results in

$$v_o(t) = \int_{-\infty}^{\infty} K_1 e^{j2\pi \Delta f_d t} e^{j\pi \alpha_r [\gamma-\tau_{R\dot{R}}]^2} \mathrm{rect}\left[\frac{\gamma-\tau_{R\dot{R}}}{\tau_{p\dot{R}}}\right] e^{-j\pi \alpha_{h\dot{R}}(t-\gamma)^2} \mathrm{rect}\left[\frac{t-\gamma}{\tau_p}\right] d\gamma \quad (12.112)$$

If we expand the two quadratic exponentials and group terms, (12.112) becomes

$$v_o(t) = K_2 \int_{-\infty}^{\infty} e^{j2\pi f \gamma} e^{j\pi \Delta \alpha \gamma^2} \mathrm{rect}\left[\frac{\gamma-\tau_{R\dot{R}}}{\tau_{p\dot{R}}}\right] \mathrm{rect}\left[\frac{t-\gamma}{\tau_p}\right] d\gamma \quad (12.113)$$

where

$$K_2 = K_1 e^{j\pi \alpha_r \tau_{R\dot{R}}^2} e^{-j\pi \alpha_{h\dot{R}} t^2} \quad (12.114)$$

$$\Delta \alpha = \alpha_r - \alpha_{h\dot{R}} \quad (12.115)$$

and

$$f = \Delta f_d - \alpha_r \tau_{R\dot{R}} + \alpha_{h\dot{R}} t \quad (12.116)$$

K_2 is a complex constant we will absorb into another complex constant that we normalize away (K_3 below). $\Delta \alpha$ is the difference between the LFM slope of the received signal and the LFM slope of the waveform to which the matched filter is matched. This is the same $\Delta \alpha$ as in (12.105) and the equations that preceded it. f is a frequency that depends on the Doppler frequency mismatch, Δf_d, between the target Doppler frequency and the frequency of the LO signal into the second mixer of Figure 12.5. f also depends on the LFM slopes and the target range. When we considered the stretch processor, these frequency terms were used to locate the target via the spectrum analyzer response. In the case of the matched filter, it is a range-related, Doppler frequency mismatch that will cause a range bias because of the range-Doppler coupling of LFM waveforms.

After considerable manipulation (see Exercise 6), it can be shown that

$$|v_o(t)| = \begin{cases} \left\| K_3 \begin{Bmatrix} F\left[\sqrt{2}\left(\sqrt{\Delta\alpha}U + f/\sqrt{\Delta\alpha}\right)\right] - \\ F\left[\sqrt{2}\left(\sqrt{\Delta\alpha}L + f/\sqrt{\Delta\alpha}\right)\right] \end{Bmatrix} \right\| & U \geq L \\ 0 & U < L \end{cases} \quad (12.117)$$

In (12.117):

- F is the Fresnel integral.
- $U = \min(t + \tau_h/2,\ \tau_{R\dot{R}} + \tau_{p\dot{R}}/2)$.
- $L = \max(t - \tau_h/2,\ \tau_{R\dot{R}} - \tau_{p\dot{R}}/2)$.
- K_3 is a complex constant we normalize away

Equation (12.117) applies only to the case where the $\Delta\alpha \neq 0$ and the rect[x] function widths are different (the mismatched case). If $\Delta\alpha = 0$ and the rect[x] function widths are the same (the matched case), $|v_o(t)|$ can be derived from the ambiguity function of $v(t)$ and is

$$|v_o(t)| = |\chi(t,\Delta f)| = K_1 |\tau_p - |t|| \operatorname{sinc}\left[(\Delta f - \alpha t)(\tau_p - |t|)\right] \operatorname{rect}\left[\frac{t}{2\tau_p}\right] \quad (12.118)$$

In (12.66), K_1 is a complex constant that we normalize away.

The form of (12.117) is basically the same as (12.105), except that (12.117) is a function of range delay, t; and (12.105) is a function of spectrum analyzer frequency, f, which can be converted to range delay by the relation $t = f/\alpha$. Thus, we expect the matched filter response to exhibit shapes similar to those of Figures 12.6, 12.7, and 12.8.

As an example of this, and to illustrate the bias error discussed earlier, we consider the example where we had a target at 300 km with a range rate of −6,000 m/s. We use a pulse width of 100 μs and an LFM bandwidth of 500 MHz. We set the LFM slope of the matched filter to the transmit slope of $\alpha = B/\tau_p = 5\times10^{12}$ Hz/s. A plot of the matched filter output, relative to the target range of 300 km, is contained in Figure 12.9. As expected, it exhibits the same distortion as the −6,000-m/s case of Figure 12.6. Also, the peak exhibits the 12-m bias discussed in Section 12.6.3.2. Thus, based on this one example, we surmise that the distortions and biases associated with stretch processing are also present in matched filter processing. Other examples that demonstrate this are left as an exercise.

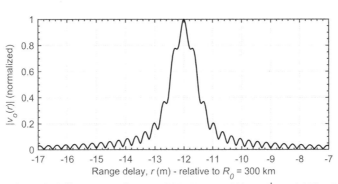

Figure 12.9 Plot of matched filter output for $B = 500$ MHz, $\tau_p = 100$ μs, $\dot{R} = -6{,}000$ m/s, and $R_0 = 300$ km, matched filter matched to transmit waveform.

12.7 EXERCISES

1. Derive (12.6) and generate a plot like Figure 12.1. Hint: think "complete the square" when performing the calculations.

2. Implement a stretch processor as discussed in Section 12.5. In your implementation, use an (unrealistic) $2^{16} = 65{,}536$-point FFT to provide a smooth output plot for visualization purposes. Zero pad the input to the FFT by loading the last 65,536–5,500 input taps with zero. Generate a plot like Figure 12.4 by plotting the magnitude of the FFT output. You will need to appropriately assign ranges to the FFT output taps. In this exercise, you will need to actually generate the received LFM pulse using (12.1).

3. Apply a window to your stretch processor from Exercise 2 to reduce range sidelobes. Use a Hamming window function. Apply the Hamming window across the 5,500-sample output of your simulated ADC, not across the 65,536-FFT input taps.

4. Use your stretch processor from Exercise 2 to produce plots like Figure 12.6. For this exercise, you will need to re-create the input LFM pulse with $\alpha = \alpha_r$ [see (12.60)] and a pulse width of $\tau_{p\dot{R}}$ [see (12.61)]. You will also need to artificially provide the scale factors needed to give the peak values of Figure 12.6.

5. Use your stretch processor from Exercise 2 to produce plots like Figure 12.6. For this exercise, you will need to re-create the input LFM pulse with $\alpha = \alpha_r$ [see (12.60)] and a pulse width of $\tau_{p\dot{R}}$ [see (12.61)]. You will also need to artificially provide the scale factors needed to give the peak values of Figure 12.7.

6. Use your stretch processor from Exercise 2 to produce plots like Figure 12.6. For this exercise, you will need to re-create the input LFM pulse with $\alpha = \alpha_r$ [see (12.60)] and a pulse width of $\tau_{p\dot{R}}$ [see (12.61)]. You will also need to artificially provide the scale factors needed to give the peak values of Figure 12.8. This will require you modify your stretch processor so that $h_s(t)$ has an LFM slope of $\alpha_{h\dot{R}}$. You will need to compute the value of $\tau_{M\dot{R}}$ that make $f_m = 0$ [see (12.82)].

7. Derive (12.76).

8. Derive (12.32), (12.35), (12.36), (12.37), (12.40), (12.41), and (12.42).
9. Derive (12.117).
10. Generate the plot of Figure 12.9.

References

[1] Cook, C. E., and M. Bernfeld, *Radar Signals: An Introduction to Theory and Application*, New York: Academic Press, 1967. Reprinted: Norwood, MA: Artech House, 1993.

[2] Caputi, W. J., "Stretch: A Time-Transformation Technique," *IEEE Trans. Aerosp. Electron. Syst.*, vol. 7, no. 2, Mar. 1971, pp. 269–278. Reprinted: Barton, D. K., ed., *Radars*, Vol. 3: Pulse Compression (Artech Radar Library), Dedham, MA: Artech House, 1975.

[3] IEEE document, "IEEE Dennis J. Picard Medal for Radar Technologies and Applications," www.ieee.org/documents/picard_rl.pdf.

[4] Abramowitz, M., and I. A. Stegun, eds., *Handbook of Mathematical Functions with Formulas, Graphs, and Mathematical Tables, National Bureau of Standards Applied Mathematics Series 55*, Washington, DC: U.S. Government Printing Office, 1964; New York: Dover, 1965.

[5] Analog Devices, "AD9680 Data Sheet: 14-Bit, 1 GSPS JESD204B, Dual Analog-to-Digital Converter," 2014. www.analog.com.

[6] Dufilie, P., C. Valerio, and T. Martin, "Improved SAW Slanted Array Compressor Structure for Achieving >20,000 Time-Bandwidth Product," 2014 IEEE Int. Ultrasonics Symp. (IUS), Chicago, IL, Sept. 3–6, 2014, pp. 2019–2022.

[7] Skolnik, M. I., ed., *Radar Handbook*, 3rd ed., New York: McGraw-Hill, 2008.

[8] Yu, J., et al., "An X-band Radar Transceiver MMIC with Bandwidth Reduction in 0.13 μm SiGe Technology," IEEE J. Solid-State Circuits, vol. 49, no. 9, Sept. 2014, pp. 1905–1915.

[9] Baturov, B. B., et al., "An S-band High-Power Broadband Transmitter," *2000 IEEE MTT-S Int. Microwave Symp. Dig.*, vol. 1, Boston, MA, Jun. 11–16, 2000, pp. 557–559.

[10] Abe, D. K., et al., "Multiple-Beam Klystron Development at the Naval Research Laboratory," 2009 IEEE Radar Conf., Pasadena, CA, May 4–8, 2009, pp. 1–5.

[11] Ender, J. H. G., and A. R. Brenner, "PAMIR—A Wideband-Phased Array SAR/MTI System," *IEEE Proc. Radar, Sonar and Navigation*, vol. 150, no. 3, Jun. 2003, pp. 165–172.

[12] Klauder, J. R. et al., "The Theory and Design of Chirp Radars," *Bell Syst. Tech. J.*, vol. 39, no. 4, Jul. 1960, pp. 745–808.

[13] Powell, T. H. J., and A. Sinsky, "A Time Sidelobe Reduction Technique for Small Time-Bandwidth Chirp," *IEEE Trans. Aerosp. Electron. Syst.*, vol. 10, no. 3, May 1974, pp. 390–392.

[14] Barton, D. K., ed., *Radars, Vol. 3: Pulse Compression* (Artech Radar Library), Dedham, MA: Artech House, 1975.

[15] Wehner, D. R., *High Resolution Radar*, Norwood, MA: Artech House, 1987.

[16] Picinbono, B., *Principles of Signals and Systems: Deterministic Signals*, Norwood, MA: Artech House, 1988.

Chapter 13

Phased Array Antenna Basics

13.1 INTRODUCTION

In this chapter, we discuss the basics of phased array antennas. We specifically develop equations and techniques to find antenna radiation and directive gain patterns. That is, we develop equations for $G(\alpha, \varepsilon)$, where α and ε are orthogonal angles such as azimuth and elevation or angles relative to a normal to the antenna face. We develop equations and algorithms to produce plots similar to the plot shown in Figure 13.1. We also discuss beamwidth, directive gain, sidelobes, and grating lobes and how these relate to antenna dimensions and other factors.

We begin with a simple two-element array antenna to illustrate some of the basic aspects of computing antenna radiation patterns and some of the properties of antennas. We then progress to linear arrays and planar phased arrays. After that, we discuss polarization and how phased array analysis methods can be used to generate antenna patterns for simple reflector antennas. We close with derivations of antenna parameter approximations that are useful in preliminary radar analyses, such as those in Chapter 2.

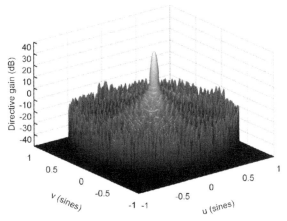

Figure 13.1 Sample antenna pattern.

It appears that the first use of array antennas was by Guglielmo Marconi in a communication experiment in 1901 [1, 2]. According to Mailloux [3], Friis and Feldman reported the use of a "fully electromechanically scanned array" in a 1937 paper [4]. In a 1947 paper [5], Friis and Lewis described a phased array antenna for the Navy Mark 8 shipboard radar. Since the 1950s, the development of phased array antennas has progressed steadily, both from a theory and hardware perspective to the point where such antennas are becoming the norm rather than the exception [6].

13.2 TWO-ELEMENT ARRAY ANTENNA

13.2.1 Transmit Perspective

Assume we have two isotropic radiators, or isotropes [7], separated by a distance, d, as shown in Figure 13.2. In Figure 13.2, the arc represents part of a sphere located at a distance of r relative to the center of the radiators. For these studies, we assume that $r \gg d$, which is termed the *far-field condition*. The sinusoids represent the electric fields (E-fields) generated by each radiator.

Since the radiators are isotropic, the power each radiates is uniformly distributed over a sphere with some radius r. The power density at any point of the sphere (at the far-field point) is

$$S(r) = \frac{P_{rad}}{4\pi r^2} = \frac{1}{2} \frac{|E(r)|^2}{Z_0} \tag{13.1}$$

where P_{rad} is the power delivered to the radiator, $|E(r)|$ is the electric field intensity at all points on the sphere and $Z_0 = 377\ \Omega$ is the characteristic impedance of free space [7, p. 12]. We can solve (13.1) for $|E(r)|$ as

$$|E(r)| = \frac{1}{r}\sqrt{\frac{P_{rad} Z_0}{2\pi}} \tag{13.2}$$

Since the signal is a sinusoid at a carrier frequency of ω_o, we can write the E-field at the far-field point as

$$E = |E(r)| e^{j\omega_o \tau_r} \tag{13.3}$$

where τ_r is the time required for the E-field to propagate from the source to the far-field point.

For the next step, we invoke the relations $\tau_r = r/c$, $\omega_o = 2\pi f_o$, and $f_o = c/\lambda$, where c is the speed of light and λ denotes wavelength. With this we can write the E-field at the far field point as

$$E = |E(r)| e^{j2\pi r/\lambda} \tag{13.4}$$

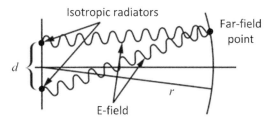

Figure 13.2 Two-element array antenna.

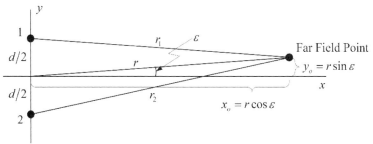

Figure 13.3 Geometry for two-element radiator problem.

We now derive an equation for the E-field at the far-field point when we have the two radiators of Figure 13.2. We use the geometry of Figure 13.3 to aid the derivation. We denote the upper radiator (point source) of Figure 13.3 as radiator 1 and the lower radiator as radiator 2. The distances from the individual radiators to the far-field point are r_1 and r_2, and each radiator radiates a power of $P_{rad}/2$. The factor of 2 is included to denote the fact that the power is split evenly between the radiators (uniform weighting).

The E-fields of the two radiators at the far-field point are

$$E_1 = |E(r_1)| = \frac{1}{r_1}\sqrt{\frac{P_{rad}Z_0}{4\pi}}e^{j2\pi r_1/\lambda} \tag{13.5}$$

and

$$E_2 = |E(r_2)| = \frac{1}{r_2}\sqrt{\frac{P_{rad}Z_0}{4\pi}}e^{j2\pi r_2/\lambda} \tag{13.6}$$

From Figure 13.3,

$$r_1 = \sqrt{x_o^2 + (y_o - d/2)^2} = \sqrt{r^2 + d^2/4 - rd\sin\varepsilon} \tag{13.7}$$

and

$$r_2 = \sqrt{x_o^2 + (y_o + d/2)^2} = \sqrt{r^2 + d^2/4 + rd\sin\varepsilon} \tag{13.8}$$

As indicated earlier, we assume the far-field condition and use $r \gg d$. With this we get

$$r_1 \approx \sqrt{r^2 - rd\sin\varepsilon} \approx r\left(1 - \frac{d}{2r}\sin\varepsilon\right) \tag{13.9}$$

and

$$r_2 \approx \sqrt{r^2 + rd\sin\varepsilon} \approx r\left(1 + \frac{d}{2r}\sin\varepsilon\right) \tag{13.10}$$

where we have used the relation

$$(1 \pm x)^N \approx 1 \pm Nx \quad \text{for} \quad x \ll 1 \tag{13.11}$$

Since r_1 and r_2 are functions of ε, the E-fields are also functions of ε. With this we get

$$E_1(\varepsilon) = \sqrt{\frac{P_{rad}Z_0}{4\pi}} \frac{1}{r\left(1 - \frac{d}{2r}\sin\varepsilon\right)} \exp\left[j2\pi\left(r - \tfrac{d}{2}\sin\varepsilon\right)/\lambda\right] \tag{13.12}$$

and

$$E_2(\varepsilon) = \sqrt{\frac{P_{rad}Z_0}{4\pi}} \frac{1}{r\left(1 + \frac{d}{2r}\sin\varepsilon\right)} \exp\left[j2\pi\left(r + \tfrac{d}{2}\sin\varepsilon\right)/\lambda\right] \tag{13.13}$$

In (13.12) and (13.13), we can set the denominator terms to r since $d/2r \ll 1$. We cannot do this in the exponential terms because phase is measured modulo 2π.

The total E-field at the far-field point is

$$E(\varepsilon) = E_1(\varepsilon) + E_2(\varepsilon) \tag{13.14}$$

or

$$\begin{aligned}
E(\varepsilon) &= \frac{1}{r}\sqrt{\frac{P_{rad}Z_0}{4\pi}} \exp\left[j2\pi\left(r - \frac{d}{2}\sin\varepsilon\right)/\lambda\right] \\
&\quad + \frac{1}{r}\sqrt{\frac{P_{rad}Z_0}{4\pi}} \exp\left[j2\pi\left(r + \frac{d}{2}\sin\varepsilon\right)/\lambda\right] \\
&= \frac{1}{r}\sqrt{\frac{P_{rad}Z_0}{4\pi}} e^{j2\pi r/\lambda} \left(e^{-j\pi d\sin\varepsilon/\lambda} + e^{j\pi d\sin\varepsilon/\lambda}\right) \\
&= \frac{1}{r}\sqrt{\frac{P_{rad}Z_0}{4\pi}} e^{j2\pi r/\lambda} \left[2\cos\left(\frac{\pi d}{\lambda}\sin\varepsilon\right)\right] \\
&= \frac{|E(r)|}{\sqrt{2}} e^{j2\pi r/\lambda} \left[2\cos\left(\frac{\pi d}{\lambda}\sin\varepsilon\right)\right]
\end{aligned} \tag{13.15}$$

We define an antenna radiation pattern as

$$R(\varepsilon) = \frac{|E(\varepsilon)|^2}{|E(r)|^2} \tag{13.16}$$

The radiation pattern for the two-element, isotropic radiator antenna is thus

$$R(\varepsilon) = 2\cos^2\left(\frac{\pi d}{\lambda}\sin\varepsilon\right) \tag{13.17}$$

We are interested in $R(\varepsilon)$ for $|\varepsilon| < 90°$. We call the region $|\varepsilon| < 90°$ *visible space*.

Figure 13.4 contains plots of $R(\varepsilon)$ for $d = \lambda$, $\lambda/2$, and $\lambda/4$. For $d = \lambda$, the radiation pattern has peaks at 0, 90°, and –90°. The peaks at ±90° are termed *grating* lobes and are undesirable in for radar applications. For $d = \lambda/4$, the radiation pattern does not return to zero, and the width of the central region is broad. This is also a generally undesirable characteristic because it decreases directivity (see Chapter 2 and later discussions). The case of $d = \lambda/2$ is a good compromise that leads to a reasonably narrow center peak and levels that go to zero at ±90°. In the design of phased array antennas, we find that $d \approx \lambda/2$ is usually a desirable design criterion.

The central region of the plots in Figure 13.4 is termed the *main beam,* and the angle spacing between the 3-dB points (the points where the radiation pattern is down 3 dB from its peak value) is termed the *beamwidth*. From Figure 13.4, we conclude that, for our two radiator example, the beamwidth is inversely proportional to the spacing between the radiators. A more accurate statement is that the beamwidth is inversely proportional to the length of the array, or the dimensions of the array for a planar array. We will investigate this relation in Section 13.13.

13.2.2 Receive Perspective

We just solved the transmit problem. That is, we supplied power to the radiators and determined how it was distributed on a sphere. We now want to consider the reverse problem and examine the receive antenna. The results of that analysis will illustrate an important property known as *reciprocity*. Reciprocity says we can analyze an antenna from a transmit or receive perspective and obtain the same radiation pattern.

For this case, we consider the two "radiators" of Figure 13.2 as receive antennas that are isotropic. Here, we call them *receive elements*. We assume an E-field radiates from an isotropic radiator located at a range r from the center of the two receive elements. The receive elements are separated by a distance of d. Figure 13.5 shows the required geometry.

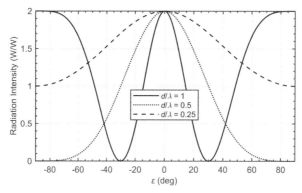

Figure 13.4 Radiation pattern for a two-element array with various element spacings.

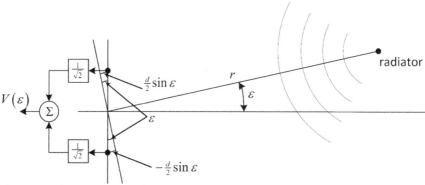

Figure 13.5 Two-element array, receive geometry.

Outputs of the receive elements are multiplied by $1/\sqrt{2}$ and summed. The voltage out of each element is proportional to the E-field at each element and is represented as a complex number to account for the fact that the actual signal, which is a sinusoid, is characterized by amplitude and phase.

The E-field of an isotropic radiator at all points on a circle (or a sphere in three dimensions) has the same amplitude and phase. Also, since $d \ll r$, the circle becomes a line at the location of the receive elements. The line is oriented at an angle of ε relative to the vertical plane of elements, and is termed the *constant E-field line*. ε is also the angle between the horizontal line and the point from which the E-field radiates. We term the horizontal line the *array normal* or *antenna broadside*. In more general terms, the antenna broadside is normal to the plane containing the elements.

The distance from the constant E-field line to the elements is $(d/2)\sin \varepsilon$. If we define the E-field at the center point between the elements as

$$E = E_r e^{j2\pi r/\lambda} \qquad (13.18)$$

then the E-field at the elements is

$$E_1(\varepsilon) = E_r e^{j2\pi(r+(d/2)\sin \varepsilon)/\lambda} \qquad (13.19)$$

and

$$E_2(\varepsilon) = E_r e^{j2\pi(r-(d/2)\sin\varepsilon)/\lambda} \quad (13.20)$$

where we made use of the approximation in (13.11).

Since the voltage out of each element is proportional to the E-field at each element, the voltages out of the elements are

$$V_1(\varepsilon) = V_r e^{j2\pi[r+(d/2)\sin\varepsilon]/\lambda} \quad (13.21)$$

and

$$V_2(\varepsilon) = V_r e^{j2\pi[r-(d/2)\sin\varepsilon]/\lambda} \quad (13.22)$$

With this the voltage at the summer output is

$$\begin{aligned} V(\varepsilon) &= \frac{1}{\sqrt{2}}\left[V_1(\varepsilon) + V_2(\varepsilon)\right] \\ &= \frac{V_r}{\sqrt{2}} e^{j2\pi r/\lambda} 2\cos\left(\frac{\pi d}{\lambda}\sin\varepsilon\right) \end{aligned} \quad (13.23)$$

We define the radiation pattern as

$$R(\varepsilon) = \frac{|V(\varepsilon)|^2}{V_r^2} \quad (13.24)$$

which yields

$$R(\varepsilon) = 2\cos^2\left(\frac{\pi d}{\lambda}\sin\varepsilon\right) \quad (13.25)$$

This is the same result we obtained for the transmit case described by (13.17) and demonstrates that reciprocity applies to this antenna. This allows us to use either the receive or transmit approach when analyzing more complex antennas.

13.3 N-ELEMENT LINEAR ARRAY

We now extend the results of the previous section to a linear array of elements shown in Figure 13.6. As Figure 13.6 implies, we use the receive approach to derive the radiation pattern. The array consists of N elements (the sideways "v" symbol on the right of each block) with a spacing of d between the elements. The output of each element is multiplied, or weighted, by a factor of a_n, and the results summed to form the signal out of the antenna. In general, the weights, a_n, are complex. In fact, we find that we move, or *steer*, the antenna beam by assigning appropriate phases to a_n. We assume each element is an isotropic radiator.

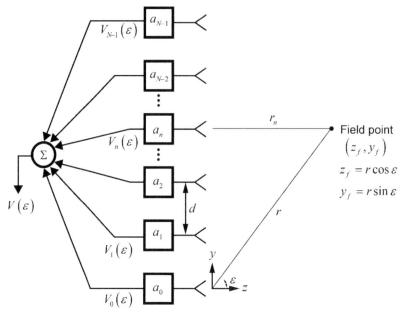

Figure 13.6 Geometry for *N*-element linear array.

We placed the origin of a coordinate system at the lower element. The axes labels, z and y, were chosen to be consistent with the planar array geometry discussed in Section 13.10.

The distance from the nth element to the (far-)field point is

$$r_n = \sqrt{z_f^2 + (y_f - nd)^2} = \sqrt{r^2 \cos^2 \varepsilon + (r \sin \varepsilon - nd)^2}$$
$$= \sqrt{r^2 - 2rnd \sin \varepsilon + n^2 d^2} \qquad (13.26)$$

Since r is much greater than the array length, due to our far-field condition, we can drop the last term of the radical and factor r^2 from the square root to get

$$r_n \approx r\sqrt{1 - \frac{2}{r} nd \sin \varepsilon} \qquad (13.27)$$

Since the magnitude of the second term of the radical is much less than 1, we can invoke (13.11) and write

$$r_n \approx r - nd \sin \varepsilon \qquad (13.28)$$

This means the E-field at the nth element is

$$E_n(\varepsilon) = E_r e^{j2\pi r_n/\lambda}$$
$$= E_r e^{j2\pi r/\lambda} e^{-j(2\pi nd \sin \varepsilon)/\lambda} \qquad (13.29)$$

and the voltage out of the nth element is

$$V_n(\varepsilon) = a_n V_r e^{j2\pi r/\lambda} e^{-j(2\pi nd \sin\varepsilon)/\lambda} \qquad (13.30)$$

where V_r is the magnitude of the voltage out of each element.

The voltage out of the summer is[1]

$$\begin{aligned} V(\varepsilon) &= \sum_{n=0}^{N-1} V_n(\varepsilon) = \sum_{n=0}^{N-1} a_n V_r e^{j2\pi r/\lambda} e^{-j(2\pi nd \sin\varepsilon)/\lambda} \\ &= V_r e^{j2\pi r/\lambda} \sum_{n=0}^{N-1} a_n e^{-j(2\pi nd \sin\varepsilon)/\lambda} \\ &= V_r e^{j2\pi r/\lambda} A(\varepsilon) \end{aligned} \qquad (13.31)$$

where $A(\varepsilon)$ is termed the *array factor* of the array.

We let $x = 2\pi d \sin(\varepsilon)/\lambda$, and write

$$V(\varepsilon) = V_r e^{j2\pi r/\lambda} \sum_{n=0}^{N-1} a_n e^{-jnx} \qquad (13.32)$$

As before, we define the radiation pattern as

$$R(\varepsilon) = \frac{|V(\varepsilon)|^2}{V_r^2} \qquad (13.33)$$

which yields

$$R(\varepsilon) = \left| \sum_{n=0}^{N-1} a_n e^{-jnx} \right|^2 = |A(\varepsilon)|^2 \qquad (13.34)$$

We now consider the special case of a linear array with constant, or *uniform*, weighting of $a_n = (N)^{-1/2}$. For the sum term, we write

$$\begin{aligned} A(\varepsilon) &= \sum_{n=0}^{N-1} a_n e^{-jnx} = \sum_{n=0}^{N-1} \frac{1}{\sqrt{N}} e^{-jnx} \\ &= \frac{1}{\sqrt{N}} \sum_{n=0}^{N-1} e^{-jnx} \end{aligned} \qquad (13.35)$$

We invoke the relation [8]

$$\sum_{n=0}^{N-1} z^n = \frac{1-z^N}{1-z} \qquad (13.36)$$

to write

---

[1] Sometimes the wave number, $k = 2\pi/\lambda$, is used to simplify notation.

$$A(\varepsilon) = \frac{1}{\sqrt{N}} \sum_{n=0}^{N-1} e^{-jnx} = \frac{1}{\sqrt{N}} \frac{1-e^{-jNx}}{1-e^{-jx}} = \frac{1}{\sqrt{N}} \frac{e^{-jNx/2}}{e^{-jx/2}} \left(\frac{e^{jNx/2} - e^{-jNx/2}}{e^{jx/2} - e^{-jx/2}} \right)$$
$$= \frac{1}{\sqrt{N}} e^{-j(N-1)x/2} \frac{\sin(Nx/2)}{\sin(x/2)}$$
(13.37)

Finally, we get

$$R(\varepsilon) = |A(\varepsilon)|^2 = \frac{1}{N} \left[\frac{\sin(Nx/2)}{\sin(x/2)} \right]^2 = \frac{1}{N} \left[\frac{\sin\left(\frac{N\pi d}{\lambda} \sin \varepsilon\right)}{\sin\left(\frac{\pi d}{\lambda} \sin \varepsilon\right)} \right]^2$$
(13.38)

Figure 13.7 contains plots of $R(\varepsilon)$ versus ε for $N = 20$ and $d = \lambda$, $\lambda/2$, and $\lambda/4$. As with the two-element example, grating lobes appear for the case of $d = \lambda$. Also, the width of the main lobe varies inversely with element spacing. Since N is fixed, the larger element spacing implies a larger antenna, which leads to the observation that, as with the two-element array, the beamwidth varies inversely with array length. The peak value of $R(\varepsilon)$ is 20, or N, and occurs at $\varepsilon = 0$. This value can also be derived by taking the limit of $R(\varepsilon)$ as $\varepsilon \to 0$, or by evaluating $A(\varepsilon)$ at $\varepsilon = 0$ and squaring it.

For the general case where a_n is not constant, we directly compute $R(\varepsilon)$ using

$$R(\varepsilon) = |A(\varepsilon)|^2 = \left| \sum_{n=0}^{N-1} a_n e^{-jnx} \right|^2, \quad x = 2\pi d \sin \varepsilon / \lambda$$
(13.39)

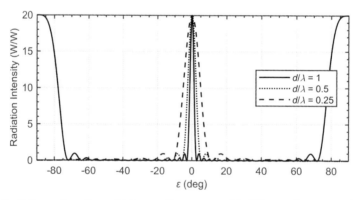

Figure 13.7 Radiation pattern for an N-element linear array with different element spacings and uniform weighting.

13.4 DIRECTIVE GAIN PATTERN (ANTENNA PATTERN)

The radiation pattern is useful when determining antenna properties such as beamwidth, grating lobes, and sidelobe levels. However, it does not provide an indication of antenna directivity. To obtain this, we define a *directive gain pattern*. The directive gain pattern indicates antenna directivity, or directive gain, as a function of angle.[2] This is the antenna gain term we use in the radar range equation.

The directive gain pattern is defined as [7, p. 125; 9]

$$G(\alpha,\varepsilon) = \frac{\text{Radiation intensity on a sphere of radius } r \text{ at an angle } (\alpha,\varepsilon)}{\text{Average radiation intensity over a sphere of radius } r} \quad (13.40)$$

or

$$G(\alpha,\varepsilon) = \frac{R(\alpha,\varepsilon)}{\dfrac{1}{4\pi r^2} \displaystyle\int_{sphere} R(\alpha,\varepsilon)\, dA} = \frac{R(\alpha,\varepsilon)}{\bar{R}} \quad (13.41)$$

where $d\Omega$ is a differential area on the sphere.

To compute the denominator integral, we consider the geometry of Figure 13.8, where the vertical row of dots represents the linear array. The differential area can be written as

$$\begin{aligned} dA &= (du)\,ds \\ &= r^2 \cos\varepsilon\, d\varepsilon\, d\alpha \end{aligned} \quad (13.42)$$

and the integral becomes

$$\bar{R} = \frac{1}{4\pi r^2} \int_{\alpha=-\pi}^{\pi} \int_{\varepsilon=-\pi/2}^{\pi/2} R(\alpha,\varepsilon) r^2 \cos\varepsilon\, d\varepsilon\, d\alpha \quad (13.43)$$

---

[2] In this book, we use the terms *directive gain pattern* and *antenna pattern* synonymously. We also use *directivity* and *directive gain* synonymously.

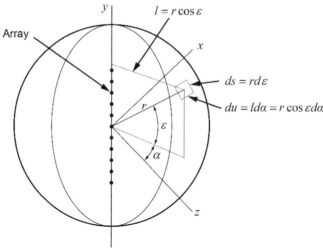

Figure 13.8 Geometry used to compute $\bar{R}$.

For the linear array, we have $R(\alpha,\varepsilon) = R(\varepsilon)$ and

$$\bar{R} = \frac{1}{2} \int_{-\pi/2}^{\pi/2} R(\varepsilon)\cos\varepsilon \, d\varepsilon \tag{13.44}$$

For the special case of a linear array with uniform weighting, we get

$$\bar{R} = \frac{1}{2} \int_{-\pi/2}^{\pi/2} \frac{1}{N}\left[\frac{\sin\left(\frac{N\pi}{\lambda}\sin\varepsilon\right)}{\sin\left(\frac{\pi}{\lambda}\sin\varepsilon\right)}\right]^2 \cos\varepsilon \, d\varepsilon \tag{13.45}$$

After considerable manipulation, it can be shown that (see Exercise 4)

$$\bar{R} = 1 + \frac{2}{N}\sum_{n=1}^{N-1}\sum_{k=1}^{n}\operatorname{sinc}(2kd/\lambda) \tag{13.46}$$

As a "sanity check," we consider a point-source (isotropic) radiator. This can be considered a special case of an N-element linear array with uniform illumination, and an element spacing of $d = 0$. For this case, we get $\operatorname{sinc}(2kd/\lambda) = 1$,

$$\bar{R} = 1 + \frac{2}{N}\sum_{n=1}^{N-1}\sum_{k=1}^{n} 1 = 1 + \frac{2}{N}\sum_{n=1}^{N-1} n = N \tag{13.47}$$

and

$$G(\varepsilon) = \frac{R(\varepsilon)}{\bar{R}} = \frac{N}{N} = 1 \tag{13.48}$$

It can also be shown that (see Exercise 6), for a general N-element, uniformly illuminated linear array with an element spacing of $d = \lambda/2$, and weights of $a_n = (N)^{-1/2}$, that $\bar{R} = 1$ and

$$G(\varepsilon) = R(\varepsilon) = \frac{1}{N}\left[\frac{\sin\left(\frac{N\pi}{2}\sin\varepsilon\right)}{\sin\left(\frac{\pi}{2}\sin\varepsilon\right)}\right]^2 \qquad (13.49)$$

For the case of a general, nonuniformly illuminated linear array, $\bar{R}$ must be computed numerically from (13.44).

Directive gain, G used in Chapter 2, is defined as the maximum value of $G(\varepsilon)$ [9, 10]. For the example of (13.49), $G = G(0)$. Figure 13.9 contains a plot of G, normalized by N, (i.e., G/N) versus d/λ for several values of N.

The shapes of the curves in Figure 13.9 are interesting, especially around integer values of d/λ. For example, for d/λ slightly less than 1, G/N is between about 1.7 and 1.9, whereas when d/λ is slightly greater than 1, G/N is about 0.7. In other words, a small change in element spacing relative to wavelength causes the directive gain to vary by a factor of about 1.8/0.7 or 4 dB. The reason for this is illustrated in Figure 13.10, which contains a plot of $R(\varepsilon)$ for d/λ values of 0.9, 1.0, and 1.1. In this case, $R(\varepsilon)$ is plotted versus $\sin\varepsilon$ to better illustrate the widths of the grating lobes (the lobes not at $\varepsilon = 0$).

For the case where d/λ is 0.9 (top plot of Figure 13.10), the radiation pattern does not contain grating lobes. This means most of the transmitted power is focused in the main beam. For the cases where d/λ is either 1.0 or 1.1, the radiation pattern contains grating lobes, and some of the transmitted power is transferred from the main lobe to the grating lobes. This reduces the directive gain of the antenna relative to the case where d/λ is 0.9. Furthermore, since there are two grating lobes for $d/\lambda = 1.1$ and only one grating lobe for $d/\lambda = 1.0$ (½ lobe at $\sin\varepsilon = 1$ and ½ lobe at $\sin\varepsilon = -1$), the directive gain is less when d/λ is 1.1 than when it is 1.0.

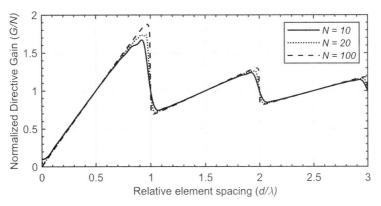

Figure 13.9 Normalized directive gain vs. element spacing.

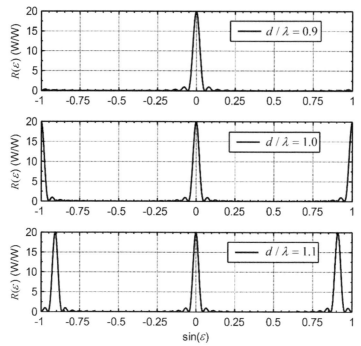

Figure 13.10 Radiation patterns for d/λ close to 1.0.

Figure 13.9 illustrates the same behavior of directive gain values of d/λ near other integer values. Directive gain decreases as the integer value of d/λ increases from below to above integer values. This is due to the number of grating lobes that appear in visible space. As d/λ becomes larger, the number of grating lobes increases. Therefore, the addition of one grating lobe for integer values of d/λ, and the two full grating lobes for d/λ slightly larger than an integer, has an increasingly smaller impact on the overall directive gain variation. Since the number of elements is fixed at 20, the length of the array increases as d/λ increases, thus causing the beamwidth to decrease and the directive gain to increase. This increase is offset by the increase in the number of grating lobes. This is what causes the curve of Figure 13.9 to vary about the nominal value of 1.

13.5 BEAMWIDTH, SIDELOBES, AND AMPLITUDE WEIGHTING

Figure 13.11 contains a plot of $G(\varepsilon)$ for a 20-element array with an element spacing of $d/\lambda = 1/2$ and uniform weighting. In this case, the units on the vertical scale are in dBi. The unit notation, dBi, stands for *decibel relative to an isotropic radiator* and indicates the directive gain is referenced to the directive gain of an isotropic radiator, which is unity.

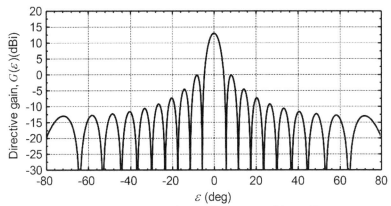

Figure 13.11 Directive gain pattern for a 20-element linear array with a uniform taper.

As discussed earlier, the lobe near $\varepsilon = 0$ is termed the *main beam*. The lobes surrounding the main beam are the *sidelobes*. The first couple of sidelobes on either side of the main beam are termed the *near-in* sidelobes, and the remaining sidelobes are termed the *far-out* sidelobes. For this antenna, the directive gain is $10\log(20) = 13$ dB, and the near-in sidelobes are about 13 dB below the peak of the main beam (13 dB below the main beam). The far-out sidelobes are greater than 20 dB below the main beam.

The *beamwidth* is defined as the width of the main beam between the 3-dB points of the main beam. For the pattern of Figure 13.11, the beamwidth is 5°.

The near-in sidelobe level of 13 dB is often considered undesirably high. To reduce this level, antenna designers usually apply an *amplitude taper* to the array by setting magnitudes of the a_n to different values. Generally, the values of a_n are varied symmetrically across the elements so that the elements on opposite sides of the center of the array have the same value of a_n. Designers usually try to choose the a_n so that they achieve a desired sidelobe level while minimizing the beamwidth increase and directive gain decrease usually engendered by weighting.

The optimum amplitude weighting in this regard is Chebyshev weighting [11, 12]. Up until recently, Chebyshev weights were difficult to generate. However, during the past 10 or so years, standard algorithms have become available. Chebyshev weights can be chosen to provide a specified sidelobe level.

While Chebyshev weighting is an optimum weighting, it is often difficult to implement in antennas. This is because Chebyshev weighting can cause the amplitude taper near the edges of the array to increase. This is difficult to achieve in practice. An alternative to Chebyshev weighting is Taylor weighting [13–16]. Like Chebyshev, it allows specification of sidelobe levels, but does not have the increase in taper amplitude near the edges of the array. An algorithm for computing Taylor weights is given in Appendix 13A. Another popular weighting is the $\cos^n$ weighting discussed in Chapter 5 [14, 15].

In *space-fed* phased arrays and reflector antennas, the amplitude taper is created by the feed. As a result, the type of taper is limited by the design of the feed. Unless the feed is complicated, it tends to provide a $\cos^n$ taper.

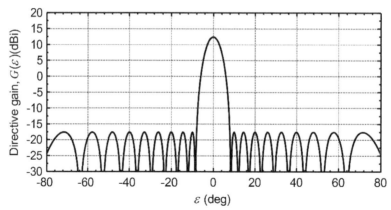

Figure 13.12 Directive gain for a 20-element linear array with Chebyshev weighting.

In *constrained-feed* phased arrays, the taper is controlled by the way power is delivered to, or combined from, the various elements. Again, this limits the type of amplitude taper that can be obtained. In solid-state phased arrays, considerable flexibility exists in controlling the amplitude taper on receive. However, it is currently difficult to obtain an amplitude taper on transmit because all of the transmit/receive (T/R) modules are typically operated at full power to maximize efficiency [17].

Figure 13.12 contains a plot of $G(\varepsilon)$ for a 20-element linear array with $d/\lambda = 1/2$ and Chebyshev weighting. The Chebyshev weighting was chosen to provide a sidelobe level of -30 dB relative to the main beam. The directive gain is about 12.4 dB rather than the 13-dB gain associated with a 20-element linear array with uniform weighting. Thus, the amplitude taper has reduced the directive gain by about 0.6 dB. Also, the beamwidth of the antenna has increased to 6.32° (a broadening factor of 1.26).

13.6 STEERING

Thus far, the antenna patterns we have generated have their main beams located at 0°. We now want to address the problem of placing the main beam at some desired angle. This is termed *beam steering*. First, we address the general problem of *time-delay steering*, and then we develop the degenerate case of *phase steering*.

13.6.1 Time-delay Steering

To address the problem of time-delay steering, we refer to the N-element linear receive array geometry of Figure 13.6. Let the idealized, normalized E-field from the far-field point source be

$$E_{pt}(t) = \text{rect}\left[\frac{t}{\tau_p}\right] e^{j2\pi f_o t} \qquad (13.50)$$

where τ_p is the pulse width, f_o is the carrier frequency, and rect[x] is the rectangle function. We assume the point-source radiator is stationary and located at some range, r.

The idealized, normalized voltage out of the nth antenna element (before the weighing, a_n) is

$$v_n(t) = \text{rect}\left[\frac{t-\tau_n}{\tau_p}\right] e^{j2\pi f_o(t-\tau_n)} \quad (13.51)$$

where τ_n is the time delay from the point-source radiator to the nth element and is given by

$$\tau_n = \frac{r_n}{c} = \frac{r + nd\sin\varepsilon}{c} = \tau_r + n\tau_{d\varepsilon} \quad (13.52)$$

Instead of treating the weights, a_n, as multiplication factors, we treat them as operators on the voltages at the output of the antenna elements. With this we write the voltage out of the summer as

$$V(\varepsilon) = \sum_{n=0}^{N-1} V_n(t) = \sum_{n=0}^{N-1} a(v_n(t), n) \quad (13.53)$$

We want to determine how the *weighting functions*, $a(v_n(t),n)$, must be chosen to focus, or collimate, the beam at some angle ε_o.

Figure 13.13 contains a notional sketch of the envelopes of the various $v_n(t)$. The main point illustrated by Figure 13.13 is that the pulses out of the various antenna elements are not aligned in time. This means the weighting functions, $a(v_n(t),n)$, must effect some desired alignment of the signals. More specifically, the $a(v_n(t),n)$ must be chosen so that the signals out of the weighting functions are aligned (and in-phase) at some desired ε_o. To accomplish this, the $a(v_n(t),n)$ must introduce appropriate time delays (and possibly phase shifts) to the various $v_n(t)$. The $a(v_n(t),n)$ must also appropriately scale the amplitudes of the various $v_n(t)$. This use of time delays to focus the beam at some angle ε_o is termed *time-delay steering*.

Substituting for τ_n into the general $v_n(t)$ expression, we get

$$v_n(t) = \text{rect}\left[\frac{t-\tau_n}{\tau_p}\right] e^{j2\pi f_o(t-\tau_n)}$$

$$= \text{rect}\left[\frac{t-\tau_r - n\tau_{d\varepsilon}}{\tau_p}\right] e^{j2\pi f_o(t-\tau_r - n\tau_{d\varepsilon})} \quad (13.54)$$

where

$$\tau_{d\varepsilon} = d(\sin\varepsilon)/c \quad (13.55)$$

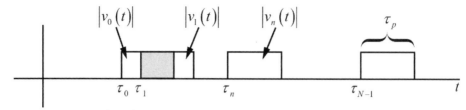

Figure 13.13 Sketch of $|v_n(t)|$.

To time align all of the pulses out of the weighting functions, the weighting function must introduce a time delay that cancels the $n\tau_{d\varepsilon}$ term in $v_n(t)$. Specifically, $a(v_n(t),n)$ must be chosen such that the voltage out of the nth element is

$$V_n(t) = a(v_n(t),n) = |a_n| v_n(t + n\tau_{do}) \tag{13.56}$$

where $\tau_{do} = d(\sin \varepsilon_o)/c$. Using this, the $V_n(t)$ are

$$V_n(t) = |a_n| \operatorname{rect}\left[\frac{t - \tau_r - n(\tau_{d\varepsilon} - \tau_{do})}{\tau_p}\right] e^{j2\pi f_o [t - \tau_r - n(\tau_{d\varepsilon} - \tau_{do})]} \tag{13.57}$$

We note that, at $\varepsilon = \varepsilon_o$, $\tau_{d\varepsilon} = \tau_{do}$, and

$$V_n(t) = |a_n| \operatorname{rect}\left[\frac{t - \tau_r}{\tau_p}\right] e^{j2\pi f_o (t - \tau_r)} \tag{13.58}$$

In other words, the pulses out of the weighting functions are time aligned, in phase, and properly amplitude-weighted.

Time-delay steering is expensive and not easy to implement. It is needed in radars that use compressed pulse widths that are small relative to antenna dimensions. This can be seen from examining Figure 13.13. If τ_p is small relative to $(N-1)\tau_{d\varepsilon}$, then, for some ε, not all of the pulses align. Stated another way, the pulse out of the first element is not aligned with the pulse out of the nth element. However, this implies either a very small τ_p or a very large antenna (large $(N-1)d_\varepsilon$). For example, if τ_p was 1 ns and the antenna was 2-m wide, we would have $(N-1)\tau d_\varepsilon = 6.7$ ns $> \tau_p$, and time-delay steering would be needed. If τ_p was 1 µs, the pulses would not be aligned, but the misalignment would be much less than the pulse width. This means that the pulses can be summed, and time-delay steering is not necessary.

The relation between array diameter and minimum compressed pulse width depends on the angle to which the beam is pointed, the scan angle (assuming the beam is pointed toward the target). If the scan angle is zero, all of the pulses will reach the array at the same time. This means time-delay steering would not be needed for any (reasonable) pulse width.

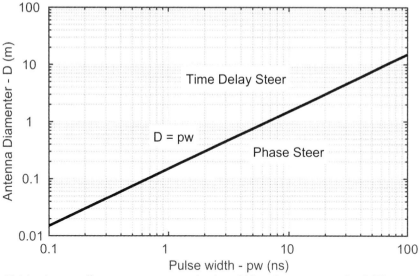

Figure 13.14 Antenna diameter vs. compressed pulse width trade at a scan angle of 60°.

A plot of the boundary where time-delay steering would and would not be needed (on both receive and transmit) is shown in Figure 13.14 for the case where the beam is scanned to 60 degrees off of array normal (off of broadside). For combinations of compressed pulse width and antenna diameter to the left of the line, time-delay steering would be needed. For combinations to the right, phase steering would be adequate. The boundary line of Figure 13.14 was based on the requirement of minimizing distortion of the compressed pulse at the output of the matched filter. This is discussed in detail in Chapter 14.

13.6.2 Phase Steering

The two regions of Figure 13.14 indicate that the alternative to time-delay steering is phase steering. Indeed, if we assume the pulses are aligned, we can write

$$\begin{aligned}V_n(t) &= a(v_n(t), n) \\ &= |a_n| \text{rect}\left[\frac{t - \tau_r}{\tau_p}\right] e^{j2\pi f_o (t - \tau_r - n\tau_{d\varepsilon} + n\tau_{do})} \\ &= |a_n| e^{j2\pi f_o n \tau_{do}} v_n(t) \\ &= |a_n| e^{j2\pi n d (\sin \varepsilon_o)/\lambda} v_n(t) \end{aligned} \quad (13.59)$$

or $a_n = |a_n|\exp(j2\pi nd(\sin\varepsilon_o)/\lambda)$. That is, the weights, a_n, modify the amplitudes and *phases* of the various $v_n(t)$. Therefore, this technique is called *phase steering*. In (13.59) we made use of (13.55), $\varepsilon = \varepsilon_o$ and $\lambda = c/f_o$.

13.6.3 Phase Shifters

In the above discussions, a tacit assumption is that the phase of each weight, a_n, can take on a continuum of values. In practice, the phase can only be adjusted in discrete steps because the devices that implement the phase shift, the *phase shifters*, are digital. Typical phase shifters use 3 to 6 bits to set the phase shift. If B_ϕ is the number of bits used in the phase shifter, then the number of phases is $N_\phi = 2^{B_\phi}$. As an example, a 3-bit phase shifter has 8 phases that range from 0 to $2\pi - 2\pi/8$ in steps of $2\pi/8$. As shown in Exercise 11, the *phase quantization* caused by the phase shifters can have a deleterious effect on the sidelobes when the beam is steered to other than broadside. Discussions of various types of phase shifters can be found in several texts [14, 15, 18–20].

13.7 ELEMENT PATTERN

In the equations above, it was assumed that all of the elements of the antenna were isotropic radiators. In practice, antenna elements are not isotropic but have their own radiation pattern. This means the voltage (amplitude and phase) out of each element depends on ε, independent of the phase shift caused by the element spacing. If all of the elements are the same, and oriented the same relative to array normal, the dependence voltage on ε is the same for each element (again, ignoring the phase shift caused by the element spacing). In equation form, the voltage out of each element is

$$v_n(t) = A_{elt}(\varepsilon) \text{rect}\left[\frac{t-\tau_n}{\tau_p}\right] e^{j2\pi f_o(t-\tau_n)} \tag{13.60}$$

where $A_{elt}(\varepsilon)$ is the array factor of the elements. The voltage out of the summer (assuming phase steering) is

$$V(\varepsilon) = A_{elt}(\varepsilon) e^{j2\pi r/\lambda} \sum_{n=0}^{N-1} a_n e^{jnd\sin\varepsilon/\lambda} = A_{elt}(\varepsilon)\left[e^{j2\pi r/\lambda} A_{array}(\varepsilon)\right] \tag{13.61}$$

The resulting radiation pattern is

$$R(\varepsilon) = |V(\varepsilon)|^2 = R_{elt}(\varepsilon) R_{array}(\varepsilon) \tag{13.62}$$

In other words, to get the radiation pattern of an antenna with nonisotropic elements, we multiply the array radiation pattern (found by the aforementioned techniques) by the radiation pattern of the element.

Although $R(\varepsilon)$ is the product of $R_{elt}(\varepsilon)$ and $R_{array}(\varepsilon)$, we can't say that $G(\varepsilon)$ is the product of $G_{elt}(\varepsilon)$ and $G_{array}(\varepsilon)$. We can see this from the following. From (13.41) we have

$$G_{elt}(\varepsilon) = \frac{R_{elt}(\varepsilon)}{\bar{R}_{elt}} \tag{13.63}$$

$$G_{array}(\varepsilon) = \frac{R_{array}(\varepsilon)}{\bar{R}_{array}} \qquad (13.64)$$

and

$$G_{array}(\varepsilon) = \frac{R(\varepsilon)}{\bar{R}} = \frac{R_{elt}(\varepsilon) R_{array}(\varepsilon)}{\bar{R}} . \qquad (13.65)$$

The only way we can say $G(\varepsilon) = G_{elt}(\varepsilon) G_{array}(\varepsilon)$ is if $\bar{R} = \bar{R}_{elt} \bar{R}_{array}$, which is not generally the case. Indeed, we have

$$\bar{R} = \frac{1}{2}\int_{-\pi/2}^{\pi/2} R(\varepsilon)\cos\varepsilon \, d\varepsilon = \frac{1}{2}\int_{-\pi/2}^{\pi/2} R_{elt}(\varepsilon) R_{array}(\varepsilon) \cos\varepsilon \, d\varepsilon \qquad (13.66)$$

which is clearly not equal to

$$\bar{R}_{elt}\bar{R}_{array} = \left(\frac{1}{2}\int_{-\pi/2}^{\pi/2} R_{elt}(\varepsilon)\cos\varepsilon \, d\varepsilon\right)\left(\frac{1}{2}\int_{-\pi/2}^{\pi/2} R_{array}(\varepsilon)\cos\varepsilon \, d\varepsilon\right) \qquad (13.67)$$

The exception to the above is the case where the elements are isotropic radiators. In that case $R_{elt}(\varepsilon) = 1$ and $\bar{R}_{elt} = 1$, thus $G(\varepsilon) = G_{elt}(\varepsilon) G_{array}(\varepsilon)$. The proof of this is left as an exercise.

As a closing note, in general, the element pattern is not steered. Also, as a note, element patterns are typically designed to have $\cos^n$ radiation patterns to allow the 3-dB beamwidth of the element to fall near the design scan limits of the overall array. This tends to minimize the deleterious effect of mutual coupling at large scan angles. Usual values of n are 1.5 to 2 for antennas that scan to ±60° relative to array normal.

13.8 ARRAY FACTOR RELATION TO THE DISCRETE-TIME FOURIER TRANSFORM

Suppose we have a discrete-time function, $x(nT)$, where T is the spacing between samples. We can write its Fourier transform (discrete time Fourier transform – DTFT) [79] as

$$X(f) = \sum_{n=-\infty}^{\infty} x(nT) e^{-j2\pi f(nT)} \qquad (13.68)$$

If we assume $x(nT)$ is finite duration and such that

$$x(nT) = \begin{bmatrix} a_n & 0 \leq n \leq N-1 \\ 0 & \text{elsewhere} \end{bmatrix} \qquad (13.69)$$

we have

$$A(f) = \sum_{n=0}^{N-1} a_n e^{-j2\pi nTf} \qquad (13.70)$$

From (13.31), we can write the array factor of a linear array as

$$A(\varepsilon) = \sum_{n=0}^{N-1} a_n e^{-j(2\pi nd \sin\varepsilon)/\lambda} = \sum_{n=0}^{N-1} a_n e^{-j2\pi n \frac{d}{\lambda}\sin\varepsilon} \tag{13.71}$$

If we let $v = \sin\varepsilon$, we get[3]

$$A(v) = \sum_{n=0}^{N-1} a_n e^{-j2\pi n \frac{d}{\lambda} v} \tag{13.72}$$

which is the same form as (13.70). Thus, we can interpret the array factor, $A(v)$, as the DTFT of the element weights where d/λ is analogous to the sample spacing in time and v is analogous to frequency.

We know we can use the FFT to compute an approximation of $A(f)$. By analogy, we can also use the FFT to compute an approximation to $A(v)$. For a time-frequency FFT, the frequency extent of the FFT output is

$$\Delta f = 1/T \tag{13.73}$$

and the frequency spacing between FFT taps is

$$\delta f = 1/(MT) \tag{13.74}$$

where M is the number of FFT taps, or the length of the FFT.[4] For a response centered at 0 Hz, the frequencies associated with the FFT taps are, assuming M is even,

$$f = \left[-\frac{M}{2}\left(\frac{1}{MT}\right), \left(\frac{M}{2}-1\right)\left(\frac{1}{MT}\right) \right] \tag{13.75}$$

By analogy, the total extent of the FFT output for the antenna case is

$$\Delta v = \Delta \sin\varepsilon = \lambda/d \tag{13.76}$$

and the spacing between FFT taps is

$$\delta v = \frac{1}{M(\lambda/d)} \tag{13.77}$$

The sine space angle associated with the FFT taps is

$$v = \left[-\frac{M}{2}\left(\frac{d}{M\lambda}\right), \left(\frac{M}{2}-1\right)\left(\frac{d}{M\lambda}\right) \right] \tag{13.78}$$

---

[3] v is termed a *sine space* angle equivalent of the *elevation angle*, ε. We will discuss this further when we discuss planar arrays in the next section. The main reason for using sine space angles is that the array factor, $A(v)$, is linear in v, but not in ε. That feature facilitates the use of the Fourier transform in antenna theory.

[4] In the past, M had to be a power of two to maximize FFT speed. However, with today's faster computers and more efficient FFT methodologies, this is no longer a strict requirement. However, power-of-two FFTs are still the fastest. In this book, we assume M is an even number.

Equation (13.77) tells us the angle spacing, δv (and thus angle spacing between samples of the resulting radiation pattern), depends on M, the length of the FFT. It does not depend on the length of the array, N. If we want to obtain a smooth radiation pattern plot we would choose $M > N$. To make use of the longer FFT, and thus finer angle spacing, we would load the element weights, a_n, in the first N input taps of the FFT and set the remaining $M-N$ taps to zero. That is, we would zero-pad the FFT.

Equation (13.78) tells us there could be some restrictions on interpreting the FFT output. If the spacing between elements is $d = \lambda/2$ (half-wavelength spacing), the FFT output taps will span the range of v of -1 to $+1$. Given the relation $v = \sin\varepsilon$, this says the FFT output will span the angles of $-\pi/2$ to $\pi/2$, or all of visible space.

If $d < \lambda/2$, the FFT output taps would span a range of v from less than -1 to greater than $+1$. Since $|v| > 1$ does not correspond to visible space (because visible space does not include $|v| = |\sin\varepsilon| > 1$), we would ignore FFT taps where $|v| > 1$.

A more difficult problem arises when $d > \lambda/2$. In this case the FFT output taps span a v region from greater than -1 to less than $+1$. That is, the FFT output taps do not span all of visible space. There are two ways to correct this limitation:

1. Add fake array elements.
2. Take advantage of the periodic property of the DTFT.

The first method derives from the fact that the v extent of the FFT output depends on the spacing, λ/d_{FFT}, between the FFT input samples, not on the element spacing. By inserting fake elements with a weight of $a_n = 0$, we can reduce d_{FFT} while maintaining the element spacing of d. The number of fake elements needed between each pair of real elements depends on the actual element spacing. For example, if $d = \lambda$ only one fake element is needed between each pair of actual elements. If $d = 1.5\lambda$, two fake elements would be needed between each pair of actual elements. The objective is to add enough fake elements to achieve $d_{FFT} \leq \lambda/2$.

The second approach makes use of the fact that the DTFT is periodic, and the FFT provides an approximation to one of the cycles of the DTFT. We can demonstrate the periodic nature of the DTFT by some straightforward math. Indeed, we have

$$A(v) = \sum_{n=0}^{N-1} a_n e^{-j2\pi n \left(\frac{d}{\lambda}\right) v} \qquad (13.79)$$

and

$$A\left(v+K\frac{\lambda}{d}\right) = \sum_{n=0}^{N-1} a_n e^{-j2\pi n\left(\frac{d}{\lambda}\right)\left(v+K\frac{\lambda}{d}\right)}$$

$$= \sum_{n=0}^{N-1} a_n e^{-j2\pi n\left(\frac{d}{\lambda}\right)v} e^{-j2\pi nK} \quad (13.80)$$

$$= \sum_{n=0}^{N-1} a_n e^{-j2\pi n\left(\frac{d}{\lambda}\right)v} = A(v)$$

In (13.80) we made use of the fact that $e^{-j2\pi nK} = 1$, since n and K are integers. We also made use of the fact that the period of the DTFT is $\Delta v = \lambda/d$ [see (13.76)]. From (13.78), we have that the v range of the FFT output taps is from about $(-\lambda/d)/2$ to $(\lambda/d)/2$. Thus, the FFT output taps span one period of the DTFT.

To implement this method, we would compute the M-point FFT of the weights, a_n. We assume the span is given by (13.78). Call this $A_1(v)$. We would append the upper $M/2$ values of $A_1(v)$ to the front of $A_1(v)$ and the lower $M/2$ values of $A_1(v)$ to the rear of this result. Call the result of these two operations $A_2(v)$. $A_2(v)$ will have a v span of $2\lambda/d$, or twice that of $A_1(v)$. If $2\lambda/d \geq 2$, the span of v would be from at least -1 to $+1$, or over all of visible space. If $2\lambda/d < 2$, the process would be repeated by appending the upper half of $A_1(v)$ to the front of $A_2(v)$ and the lower half of $A_1(v)$ to the back of $A_2(v)$. We would repeat this process until v spans at least -1 to $+1$.

A premise of the above discussion is that the first $M/2$ FFT taps correspond to negative v and the last $M/2$ taps correspond to positive v. This situation needs to be forced in most FFT algorithms because these algorithms place the first tap at $v = 0$. To get the negative-positive v configuration, the upper and lower $M/2$ FFT outputs must be swapped. This is accomplished in MATLAB by the fftshift function.

The first edition of this book presented another method of computing $A(v)$ that took advantage of the sophisticated, vectorized algorithms contained in several software languages. That algorithm is discussed in Appendix 13B of this book.

13.9 PLANAR ARRAYS

We now want to extend the linear array development to planar arrays. In a planar array, the antenna elements are located on some type of grid in a plane. Generally, the grid pattern, or lattice, is rectangular or triangular (this is discussed further in Section 13.9.2). Figure 13.15 shows an example that would apply to a rectangular grid.

The array lies in the x-y plane, and the array broadside is the z-axis. The dots with the numbers by them are the elements. The line located at the angles α and ε point to the (far-)field point (the target on transmit or the source, which could also be the target, on receive). The field point is located at a range of r that is large relative to the dimensions of the array (far-field assumption).

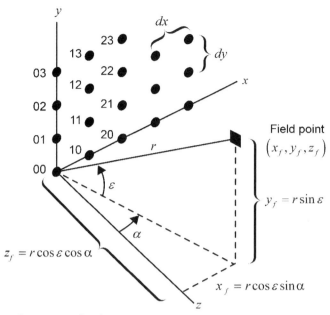

Figure 13.15 Example geometry for planar arrays.

The array shown in Figure 13.15 is oriented vertically. With this orientation, ε is elevation angle to the field point and is measured from the x-z plane, which would be the local ground plane for a ground-based radar. The angle, α, is azimuth and is measured in the x-z plane, relative to the z axis. If the array is tilted back from vertical, as is typical, ε and α are still often thought of as elevation and azimuth angles; although strictly speaking, they are not.

The angles we are using are not the angles traditionally used to develop the radiation pattern for planar arrays [21–24]. The traditional angles are those associated with a standard spherical coordinate system [25]. These angles are θ and ϕ, where θ is measured from the z axis, and ϕ is measured from the x axis in the x-y plane. We are using α and ε because they generally correspond to azimuth and elevation, and they simplify the derivation of $\bar{R}$ for planar arrays (see Section 13.9.7).

In the coordinate system of Figure 13.15, the field point is located at

$$(x_f, y_f, z_f) = (r\cos\varepsilon\sin\alpha,\ r\sin\varepsilon,\ r\cos\varepsilon\cos\alpha) \qquad (13.81)$$

The 00 element is located at the origin, and the mn^{th} element is located at (md_x, nd_y), where d_x is the spacing between elements in the x direction, and d_y is the spacing between elements in the y direction. With this and (13.81), the range from the mn^{th} element to the field point is

$$r_{mn} = \sqrt{(x_f - md_x)^2 + (y_f - nd_y)^2 + z_f^2}$$
$$= \sqrt{(r\cos\varepsilon\sin\alpha - md_x)^2 + (r\sin\varepsilon - nd_y)^2 + (r\cos\varepsilon\cos\alpha)^2} \quad (13.82)$$
$$= \sqrt{r^2 - 2rmd_x\cos\varepsilon\sin\alpha - 2rnd_y\sin\varepsilon + m^2d_x^2 + n^2d_y^2}$$

Since r is much larger than the array width and height, we can drop the last two terms of the radical, and factor r^2 from the square root to give

$$r_{mn} \approx r\sqrt{1 - \frac{2}{r}(md_x\cos\varepsilon\sin\alpha + nd_y\sin\varepsilon)} \quad (13.83)$$

Since the second term of the radical of (13.83) is small relative to 1, we can use (13.11) to write

$$r_{mn} \approx r - md_x\cos\varepsilon\sin\alpha - nd_y\sin\varepsilon \quad (13.84)$$

We invoke reciprocity and consider the receive case to write the voltage out of the mn^{th} element as

$$V_{mn}(\alpha,\varepsilon) = V_r a_{mn} e^{j2\pi r_{mn}/\lambda}$$
$$= V_r a_{mn} \exp\left(\frac{j2\pi r}{\lambda}\right)\exp\left[-\frac{j2\pi}{\lambda}(md_x\sin\alpha\cos\varepsilon + nd_y\sin\varepsilon)\right] \quad (13.85)$$

where a_{mn} is the weight applied to the mn^{th} element. Summing the outputs of all elements gives

$$V(\alpha,\varepsilon) = \sum_{n=0}^{N-1}\sum_{m=0}^{M-1} V_{mn}(\alpha,\varepsilon)$$
$$= V_r \exp\left(\frac{j2\pi r}{\lambda}\right)\sum_{n=0}^{N-1}\sum_{m=0}^{M-1} a_{mn}\exp\left[-\frac{j2\pi}{\lambda}(md_x\sin\alpha\cos\varepsilon + nd_y\sin\varepsilon)\right] \quad (13.86)$$

where M is the number of elements in the x direction, and N is the number of elements in the y direction. Dividing by V_r and ignoring the first exponential term (which disappears when we form the radiation pattern), we get the array factor

$$A(\alpha,\varepsilon) = \sum_{n=0}^{N-1}\sum_{m=0}^{M-1} a_{mn}\exp\left[-\frac{j2\pi}{\lambda}(md_x\sin\alpha\cos\varepsilon + nd_y\sin\varepsilon)\right] \quad (13.87)$$

At this point, we adopt a notation that is common in phased array antennas: *sine space*. We define

$$u = \sin\alpha\cos\varepsilon \quad (13.88)$$

and

$$v = \sin\varepsilon \quad (13.89)$$

and write

$$A(u,v) = \sum_{n=0}^{N-1} \sum_{m=0}^{M-1} a_{mn} \exp\left[-\frac{j2\pi}{\lambda}(md_x u + nd_y v)\right] \quad (13.90)$$

Consistent with our work on linear arrays, we write the radiation pattern (assuming isotropic radiators) as

$$R(u,v) = |A(u,v)|^2 \quad (13.91)$$

and the directive gain pattern as

$$G(u,v) = R(u,v)/\bar{R} \quad (13.92)$$

We will consider $\bar{R}$ shortly.

When we plot $R(u,v)$ or $G(u,v)$, we are plotting the radiation or directive gain pattern in sine space. When we plot $R(\alpha,\varepsilon)$ or $G(\alpha,\varepsilon)$, we are plotting the radiation or directive gain pattern in angle space.

From (13.88) and (13.89), it can be shown (see Exercise 18) that u and v satisfy the constraint, $u^2 + v^2 \leq 1$. These u and v constitute *visible* space. We will use this when we discuss how to generate and plot radiation and directive gain patterns.

13.9.1 Weights for Beam Steering

In (13.90) [the equation for $A(u,v)$], a_{mn} are the weights used to provide a proper taper and steer the beam; these are of the general form

$$a_{mn} = |a_{mn}| \exp\left[\frac{j2\pi}{\lambda}(md_x u_0 + nd_y v_0)\right] \quad (13.93)$$

where u_0, v_0 are the desired steering angles in sine space. Equation (13.93) assumes phase steering.

13.9.2 Array Shapes and Element Locations (Element Packing)

The development of Section 13.10 is applicable to rectangular arrays with the elements placed on a rectangular lattice. Many antennas are nonrectangular (e.g., circular or elliptical), and their elements are not placed on a rectangular lattice (i.e., rectangular packing). In both cases, the deviations from rectangular shape and/or rectangular packing are usually made to conserve array elements and increase the efficiency of the antenna. The elements at the corners of rectangular arrays do not contribute much to the directive gain and can cause the ridges in the radiation pattern. Also, the peak sidelobes of a rectangular array, with uniform weighting, are 13 dB below the main beam peak. For a circular array, they are about 17 dB below the main beam peak. If the corner elements of a rectangular array are removed, the peak sidelobe will be somewhere between 13 and 17 dB below the main beam peak.

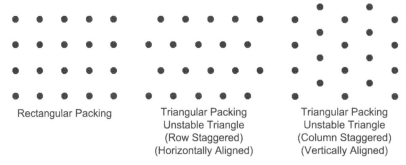

Figure 13.16 Illustration of rectangular and triangular element packing.

The most common element packing scheme, besides rectangular packing, is *triangular* packing. Figure 13.16 shows sections of a planar array with rectangular and triangular packing. With triangular packing, the elements are arranged in a triangular pattern, and there are two variants of triangular packing. We, the authors, do not know if there are "formal" names for the two variants. Some of our colleagues term them vertically or horizontally aligned triangular lattices while others term them unstable or stable triangular lattices. Our reviewer suggested the terms row or column staggered triangular lattices. These names are noted below the two triangular lattices. The main impact of using one or the other two types of triangular packing (triangular lattices) is the location of grating lobes. This is discussed further in Section 13.9.7.

13.9.3 Feeds

An antenna feed is the mechanism by which the energy from the transmitter is conveyed to the array so that it can be radiated into space. On receive, it is used to collect the energy from the array elements. Two broad classes of feed types are used in phased arrays: space feed and constrained feed. These two types of feed mechanisms are illustrated notionally in Figures 13.17 and 13.18.

In a space-fed array, the feed is some type of small antenna that radiates the energy to the array, through space. The feed could be a horn antenna or even another, smaller, phased array. On transmit the feed radiates energy that is captured by small antennas on the feed side of the array. These are represented by the sideways v-shaped symbols on the left side of the array of Figure 13.17. The voltages out of the small antennas undergo a phase shift (represented by the circles with ϕ in them). The phase-shifted voltages are converted to E-fields by the elements on the right side of the array and are radiated into space. On receive, the reverse of the above occurs:

- Antennas on the right of the array capture energy from the source.
- Phase shifters apply appropriate phase shifts.
- Antennas on the left of the array radiate the energy to the feed.
- Feed sends the energy to the receiver.

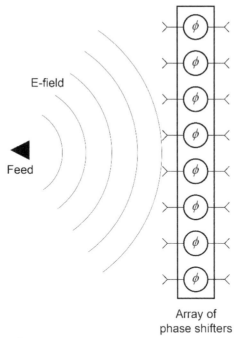

Figure 13.17 Space-fed phased array.

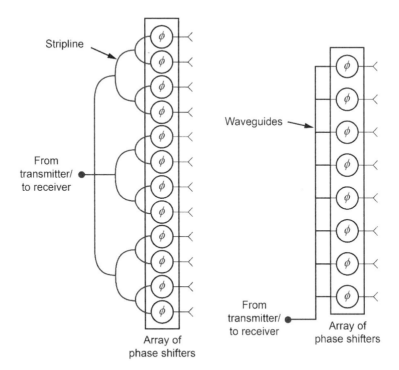

Figure 13.18 Constrained-feed phased array.

The phase shifters provide the beam steering and perform what is called a *spherical correction* (or *collimation* on transmit and *focusing* on receive). The E-field radiated from the feed nominally has constant phase on a sphere, which is represented by the arcs in Figure 13.17. This means the phase will not vary linearly across the array, which is necessary to form a beam at a desired angle. This must be accounted for in the setting of the phase shifters. The process of adjusting the phase to account for the spherical wave front is spherical correction.

The feed produces its own directive gain pattern. This means the signals entering each of the phase shifters are at different amplitudes. Thus, the feed is applying the amplitude weighting, $|a_{mn}|$, to the array. The feed is usually designed so that its directive gain pattern provides a desired sidelobe level for the overall antenna. Feed patterns typically approximate a $\cos^n$ function. To obtain a good trade-off between directive gain, sidelobe levels and efficiency for the overall antenna, the feed pattern is such that the level at the edge of the array is between 10 and 20 dB below the peak value, termed an *edge taper*. A feed that provides a 20-dB edge taper results in lower array sidelobes than a feed that provides a 10-dB edge taper (see Exercise 17). However, a space-fed phased array with a 10-dB edge taper feed has higher directive gain than a space-fed phased array with a 20-dB edge taper feed. An 8- to 12-dB edge taper provides optimized aperture efficiency [15, p. 12.9].

In a constrained-feed phased array, the energy is routed from the transmitter, and to the receiver, by a waveguide network. This is represented by the network of connections to the left of the arrays in Figure 13.18. The drawing on the left depicts a parallel or corporate feed network, and the drawing on the right depicts a series or traveling-wave feed network. Some antennas use both feed types. For example, the rows of an antenna might be fed by a series network, and the individual elements in each row might be fed by a parallel network. In some applications, the waveguide network can be structured to provide an amplitude taper.

The phase shifters in a constrained-feed array must include phase shifts to account for the different path lengths of the various legs of the waveguide network.

Space-fed phased arrays are less expensive to build than constrained-feed arrays because they do not require the waveguide network of the constrained-feed phased array. However, the constrained-feed phased array is smaller than the space-fed phased array. The space-fed phased array is generally as deep as it is tall or wide to allow proper positioning of the feed. The depth of a constrained-feed phased array is only about twice the depth of the array portion of a space-fed phased array. The extra depth is needed to accommodate the waveguide network. Finally, the constrained-feed phased array is more rugged than the space-fed array since almost all hardware is on the array structure.

A variant of the constrained-feed phased array is the solid-state phased array. For this array, the phase shifters of Figure 13.18 are replaced by solid-state T/R modules. The waveguide network can be replaced by cables or stripline since they carry only low-power signals. The transmitters in each of the T/R modules are low power (typically 10 to 1,000 W). However, a solid-state phased array can contain thousands of T/R modules so that the total transmit power is comparable to that of a space-fed or constrained-feed phased array.

13.9.4 Amplitude Weighting

As with linear arrays, planar phased arrays use amplitude weighting to reduce sidelobes. The type of weighting (Taylor, Chebyshev, or $\cos^n$, for example) is the same as in linear arrays. The difference is that the weights are applied in two dimensions. Weights can be applied in two basic ways:

1. Multiplicative weighting[5]
2. Circularly (or elliptically) symmetric weighting

For multiplicative weighting, we would write the magnitudes of the weights as

$$|a_{mn}| = |a_m||a_n| \tag{13.94}$$

This type of weighting is sometimes used in constrained-feed phased arrays because of the way the feed structures are designed.

Approximations to circularly and elliptically symmetric weighting occur in space-fed phased arrays because the weights are created by the feed pattern. This type of weighting usually provides sidelobe levels that are circularly or elliptically symmetric over the u-v plane.

The following procedure can be used to generate elliptically symmetric weights for antenna modeling purposes:

1. Generate a set of appropriate weights that has a number of terms equal to $N_{wt} \geq 2L_{max}/d_{min}$, where $2L_{max}$ is the maximum antenna dimension, and d_{min} is the minimum element spacing. Create an N_{wt} element array of numbers, x_w, evenly spaced between –1 and 1. Compute an N_{wt} array of weights, W_d, based on the desired weighting function (e.g., Chebyshev, Taylor, $\cos^n$).

2. Find the location of all of the antenna elements relative to the center of the array. Let d_{xmn} and d_{ymn} be the x and y locations of the mn^{th} element relative to the center of the array. Let $2L_x$ and $2L_y$ be the antenna widths in the x and y directions. Find the normalized distance from the center of the array to the mn^{th} element using

$$D_{mn} = \sqrt{\left(\frac{d_{xmn}}{L_x}\right)^2 + \left(\frac{d_{ymn}}{L_y}\right)^2} \tag{13.95}$$

3. Use D_{mn} to interpolate into the array W_d to get the $|a_{mn}|$.

4. If $D_{mn} > 1$, set $a_{mn} = 0$. This causes the array to have the shape of an ellipse (or a circle if $L_x = L_y$).

13.9.5 Computing Antenna Patterns for Planar Arrays

As with linear arrays, the FFT can be used to generate the antenna patterns of planar arrays. The main difference is that a two-dimensional FFT is needed for planar arrays.

---

[5] Also referred to as separable array weighting or as separable distributions.

The specific technique used depends on the type of element packing. As with linear arrays, an alternate method of generating antenna patterns is discussed in Appendix 13B.

13.9.5.1 Planar Arrays with Rectangular Packing

For an array with rectangular packing, the input array of the two-dimensional FFT would be loaded with the element weights, a_{mn}. The resulting output taps would correspond to the two sine-space angles, u and v. As with the linear array, if the spacing between elements is equal to $\lambda/2$, the output taps of the FFT will span, $-1 \leq u \leq 1$ and $-1 \leq v \leq 1$. This means the FFT output will span a rectangular grid of sine space. Not all points in this grid will correspond to visible space. For a point to lie in visible space its (u, v) location must satisfy $u^2 + v^2 \leq 1$. To capture this effect, we would set $A(u, v) = 0$ for all (u, v) where $u^2 + v^2 > 1$.

If the spacing between elements in the x direction is less than $\lambda/2$, the span of u will be greater than ± 1. Likewise, if the spacing between elements in the y direction is less than $\lambda/2$, the span of v will be greater than ± 1. Either case can be accommodated as discussed in the previous paragraph. That is, set $A(u, v) = 0$ for all (u, v) where $u^2 + v^2 > 1$.

If the spacing between elements in the x and/or y direction is greater than $\lambda/2$, one of the techniques discussed in Section 13.9 could be used to extend the appropriate FFT taps to include plus and minus one. As an example, if the spacing between taps in the x direction is greater than $\lambda/2$, either fake elements could be added to each row of the input to the FFT, or the output could be extended as discussed in Section 13.8.

13.9.5.2 Planar Arrays with Triangular Packing

The easiest way to deal with arrays that use triangular packing is to add fake elements to each row or column of the FFT input array to make the input to the FFT look like an array with rectangular packing. This is illustrated in Figure 13.19. Since the array now has rectangular packing, the methodology discussed in Section 13.9.5.1 can be used to generate the antenna pattern.

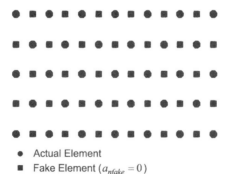

• Actual Element
■ Fake Element ($a_{nfake} = 0$)

Figure 13.19 Triangularly packed array with fake elements inserted.

13.9.6 Directive Gain Pattern

In Section 13.4, we found that the directive gain pattern was given by

$$G(\alpha,\varepsilon) = R(\alpha,\varepsilon)/\bar{R} \qquad (13.96)$$

where

$$\bar{R} = \frac{1}{4\pi} \int_{\alpha=-\pi}^{\pi} \int_{\varepsilon=-\pi/2}^{\pi/2} R(\alpha,\varepsilon) \cos\varepsilon \, d\varepsilon \, d\alpha \qquad (13.97)$$

For planar arrays, we usually compute the radiation pattern as a function of u and v instead of α and ε. Because of this, we write the directive gain pattern as

$$G(u,v) = R(u,v)/\bar{R} \qquad (13.98)$$

Since we have $R(u,v)$ and not $R(\alpha,\varepsilon)$, we want an equation for $\bar{R}$ in terms of $R(u,v)$. From (13.87), (13.88), (13.89), and (13.90), we note that $R(\alpha,\varepsilon)$ is a function of $u = \sin\alpha \cos\varepsilon$ and $v = \sin\varepsilon$. Also, we normally assume $R(\alpha,\varepsilon)$ is zero on the back of the array. Thus, we assume $R(\alpha,\varepsilon)$ is zero for α outside of the range $[-\pi/2, \pi/2]$. With this we can write (13.97) as

$$\bar{R} = \frac{1}{4\pi} \int_{\alpha=-\pi/2}^{\pi/2} \int_{\varepsilon=-\pi/2}^{\pi/2} R(\sin\alpha\cos\varepsilon, \sin\varepsilon) \cos\varepsilon \, d\varepsilon \, d\alpha \qquad (13.99)$$

We begin the derivation by making the change of variables $v = \sin\varepsilon$, and write

$$\bar{R} = \frac{1}{4\pi} \int_{\alpha=-\pi/2}^{\pi/2} \int_{v=-1}^{1} R\left(\sin\alpha\sqrt{1-v^2}, v\right) dv \, d\alpha \qquad (13.100)$$

where we made use of $\cos\varepsilon \geq 0$ for $\varepsilon \in [-\pi/2, \pi/2]$.

Next, we manipulate the α integral by making the change of variables

$$u = \left(\sqrt{1-v^2}\right)\sin\alpha$$
$$= \cos\varepsilon \sin\alpha \qquad (13.101)$$

which gives

$$du = (\cos\alpha \cos\varepsilon) d\alpha \qquad (13.102)$$

From (13.81), we have

$$\sin^2\varepsilon + \sin^2\alpha\cos^2\varepsilon + \cos^2\alpha\cos^2\varepsilon = 1 \qquad (13.103)$$

and thus,

$$\cos\alpha\cos\varepsilon = \sqrt{1 - \sin^2\varepsilon - \sin^2\alpha\cos^2\varepsilon} \qquad (13.104)$$
$$= \sqrt{1 - v^2 - u^2}$$

where we made use of $\cos\alpha \geq 0$ over the integration limits. Using this and some manipulation (see Exercise 20), we get

$$\bar{R} = \frac{1}{4\pi} \int_{v=-1}^{1} \int_{u=-\sqrt{1-v^2}}^{\sqrt{1-v^2}} \frac{R(u,v)}{\sqrt{1-v^2-u^2}}\, du\, dv \qquad (13.105)$$

When computing $\bar{R}$ by numerical integration, be careful to avoid samples on the unit circle of the u-v plane. One way to do this is to set the integrand to zero for all u,v such that $u^2 + v^2 \geq 1$. Also, it has been the authors' experience that computing $\bar{R}$ is sensitive to the integration step size. Therefore, it is recommended that care be exercised in its use.

As an interesting example, we consider the case where

$$R(u,v) = R_o \mathrm{rect}\left[\frac{u}{\delta u}\right]\mathrm{rect}\left[\frac{v}{\delta v}\right] \qquad (13.106)$$

and δu and δv are the beamwidths in the u and v directions. Equation (13.106) tells us that all of the transmit energy is concentrated in a small rectangular area centered on $u = v = 0$. For this case, we have

$$\begin{aligned}
\bar{R} &= \frac{1}{4\pi} \int_{v=-\delta v/2}^{\delta v/2} \int_{u=-\delta u/2}^{\delta u/2} \frac{R_o}{\sqrt{1-v^2-u^2}}\, du\, dv \\
&= \frac{R_o}{4\pi} \int_{-\delta v/2}^{\delta v/2} \sin^{-1}\left(\frac{u}{\sqrt{1-v^2}}\right)\Bigg|_{-\delta u/2}^{\delta u/2} dv \\
&\approx \frac{R_o \delta u}{4\pi} \int_{-\delta v/2}^{\delta/2} \frac{1}{\sqrt{1-v^2}}\, dv
\end{aligned} \qquad (13.107)$$

where we made use of the facts that δu is small and v is near zero. Performing the integration of (13.107) gives

$$\begin{aligned}
\bar{R} &= \frac{R_o \delta u}{4\pi} \sin^{-1} v\Big|_{-\delta v/2}^{\delta v/2} \\
&= \frac{R_o \delta u (\delta v)}{4\pi}
\end{aligned} \qquad (13.108)$$

where we made use of the fact that δv is small. With this we get a directive gain pattern

$$G(u,v) = \frac{R(u,v)}{\bar{R}}$$
$$= R_o \text{rect}\left[\frac{u}{\delta u}\right]\text{rect}\left[\frac{v}{\delta v}\right] \Big/ \left(R_o \delta u(\delta v)/4\pi\right) \qquad (13.109)$$
$$= \frac{4\pi}{\delta u(\delta v)} \text{rect}\left[\frac{u}{\delta u}\right]\text{rect}\left[\frac{v}{\delta v}\right]$$

The directive gain is the maximum of $G(u,v)$ or

$$G = \frac{4\pi}{\delta u(\delta v)} \qquad (13.110)$$

Saying the beam is centered on $u = v = 0$ is the same as saying it is centered on $\alpha = \varepsilon = 0$. This, with the assumption that δu and δv are small, gives $\delta u = \delta \alpha$ and $\delta v = \delta \varepsilon$ [see (13.88) and (13.89)]. This leads to

$$G = \frac{4\pi}{\delta \alpha(\delta \varepsilon)} \qquad (13.111)$$

which agrees with the form of directive gain discussed in Chapter 2.

13.9.7 Grating Lobes

We introduced the topic of grating lobes in Sections 13.2, 13.4, and 13.8, and noted that they are radiation pattern peaks (or aliases) at angles other than the location of the main beam. Grating lobes are undesirable because they take energy away from the main beam or can point toward interfering objects, such as the ground. In this section, we extend the discussion of grating lobes to planar arrays. We discuss grating lobes for arrays that use rectangular packing and arrays that use triangular packing.

13.9.7.1 Grating Lobes in Arrays with Rectangular Packing

We start by examining $A(u,v)$ from (13.90) with the a_{mn} as given by (13.93). This leads to

$$A(u,v) = \sum_{n=0}^{N-1}\sum_{m=0}^{M-1} |a_{mn}| e^{-j2\pi m d_x(u-u_0)/\lambda} e^{-j2\pi n d_y(v-v_0)/\lambda} \qquad (13.112)$$

We use $A(u,v)$ because it is easier to work with than $R(u,v)$ or $G(u,v)$. Since $R(u,v) = |A(u,v)|^2$ and $G(u,v) = R(u,v)/\bar{R}$, grating lobe observations we derive from $A(u,v)$ also apply to $R(u,v)$ and $G(u,v)$.

At the main beam location, (u_0, v_0), we have

$$A(u_0, v_0) = \sum_{n=0}^{N-1}\sum_{m=0}^{M-1} |a_{mn}| \qquad (13.113)$$

At any other $(u,v) = (u_g, v_g)$, where $2\pi d_x(u_g-u_0)/\lambda$ and $2\pi d_y(v_g-v_0)/\lambda$ are both integer multiples of 2π, the exponentials are unity for all m and n, and we have

$$A(u_g, v_g) = A(u_0, v_0) \qquad (13.114)$$

The peaks at u_g, v_g are grating lobes, and the values of u_g, v_g are the locations of the grating lobes. Thus, we say that grating lobes are located at the u_g, v_g that simultaneously satisfy

$$2\pi d_x (u_g - u_0)/\lambda = 2\pi p \qquad (13.115)$$

and

$$2\pi d_y (v_g - v_0)/\lambda = 2\pi q \qquad (13.116)$$

where p and q are integers. Solving (13.115) and (13.116) for the pair u_g, v_g, we get

$$(u_g, v_g) = \left(\frac{p\lambda}{d_x}, \frac{q\lambda}{d_y} \right) + (u_0, v_0) \qquad (13.117)$$

where p and q are integers that are not both zero (this would denote the main beam). Equation (13.117) tells us that the grating lobes are located at integer multiples of λ/d_x and λ/d_y relative to the main beam, and move with the main beam.

Figure 13.20 contains a sketch showing the locations of the main beam (the square) and the grating lobes. In this case, the main beam is steered to the left a considerable amount (the negative u direction). The unit circle of Figure 13.20 denotes the boundary of visible space. All lobes (grating or main beam) that are within the circle translate to lobes in visible space (real α, ε space), and lobes outside of the unit circle do not (they translate to lobes in imaginary α, ε space). In the example of Figure 13.20, the main beam and one grating lobe are in the unit circle. Thus, the main beam and one grating lobe are in visible space[6] since the spacing between grating lobes is λ/d_x and λ/d_y. Whether or not grating lobes enter visible space depends on the element spacing (d_x and d_y) and where the main beam is steered (u_0, v_0). This represents a trade-off that array antenna designers must face. On one hand, there is a desire to make d_x and d_y large to minimize the number of elements, and thus array cost.[7] On the other hand, d_x and d_y must be small enough to avoid grating lobes when the main beam is steered to some desired maximum angle.[8]

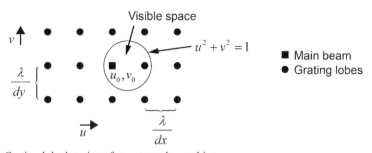

Figure 13.20 Grating lobe locations for rectangular packing.

---

[6] The main beam will always be in visible space since $(u_0, v_0) = (\sin\alpha_0 \cos\varepsilon_0, \sin\varepsilon_0)$ satisfy $u_0^2 + v_0^2 \leq 1$.

[7] Some arrays, termed limited-scan arrays, are specifically designed to use small scan angles and thus permit larger element spacings.[19]

[8] In some instances, there are other requirements on element spacing such as mutual coupling [26] and packaging.

The first grating lobes to enter visible space are the eight that surround the main beam. Of these, the four immediately adjacent to the main beam are most likely to enter visible space before the four located on the diagonals. When a grating lobe enters visible space (as the main beam is steered), it does so on the unit circle boundary of the u, v plane. We can use this to specify the maximum d_x and d_y that avoid grating lobes in visible space at some maximum steering angles of $|u_0|_{max}$ and $|v_0|_{max}$. Specifically, we get

$$\frac{\lambda}{d_{x\max}} = 1 + |u_0|_{\max} = 1 + |\sin \alpha_{0\max}| \quad (\text{when } v_0 = 0) \tag{13.118}$$

or

$$\frac{\lambda}{d_{y\max}} = 1 + |v_0|_{\max} = 1 + |\sin \varepsilon_{0\max}| \quad (\text{when } u_0 = 0) \tag{13.119}$$

The parenthetical statements mean the equations are applied when the other sine-space coordinate is zero; that is, $d_{x\max}$ is computed for $v_0 = 0$ and $d_{y\max}$ is computed for $u_0 = 0$.

Table 13.1 contains a list of $d_{x\max}/\lambda$ and $d_{y\max}/\lambda$ values for a few maximum steering angles. Note that an element spacing of $\lambda/2$ prevents grating lobes from entering visible space at all steering angles except the limiting case of $\pm 90°$.

Table 13.1

Maximum d/λ Spacing to Avoid Grating Lobes at Various Maximum Steering Angles—Rectangular Packing

Maximum Steering Angle	d/λ
30	0.67
45	0.59
60	0.54
90	0.50

13.9.7.2 Grating Lobes in Arrays with Triangular Packing

For triangular packing, we start with (13B.16) of Appendix 13B, but we write it in a slightly different form as

$$A(u,v) = A_1(u,v) + A_2(u,v) e^{-j2\pi d_x (u-u_0)/\lambda} e^{-j2\pi d_y (v-v_0)/\lambda} \tag{13.120}$$

where

$$A_i(u,v) = \sum_{n=0}^{N_i-1} \sum_{m=0}^{M_i-1} |c_{mn}| e^{-j2\pi 2md_x (u-u_0)/\lambda} e^{-j2\pi 2nd_y (v-v_0)/\lambda} \tag{13.121}$$

with $(N_1, M_1) = (M_a, N_a)$, $(N_2, M_2) = (M_b, N_b)$, and c_{mn} replaced by a_{mn} or b_{mn} as appropriate (see Appendix 13B).

At $(u, v) = (u_0, v_0)$, we have

$$A(u_0, v_0) = A_1(u_0, v_0) + A_2(u_0, v_0)$$
$$= \sum_{n=0}^{N_a-1} \sum_{m=0}^{M_a-1} |a_{mn}| + \sum_{n=0}^{N_b-1} \sum_{m=0}^{M_b-1} |b_{mn}| \qquad (13.122)$$

If $4\pi d_x(u_g - u_0)/\lambda$ and $4\pi d_y(v_g - v_0)/\lambda$ are both integer multiples of 2π, we have $A_1(u_g, v_g) = A_1(u_0, v_0)$ and $A_2(u_g, v_g) = A_2(u_0, v_0)$. This results in

$$A(u_g, v_g) = A_1(u_0, v_0) \\ + e^{-j2\pi d_x(u_g - u_0)/\lambda} e^{-j2\pi d_y(v_g - v_0)/\lambda} A_2(u_0, v_0) \qquad (13.123)$$

If $d_x(u_g - u_0)/\lambda + d_y(v_g - v_0)/\lambda$ is an integer, the product of exponentials is unity, and we have

$$A(u_g, v_g) = A_1(u_0, v_0) + A_2(u_0, v_0) \qquad (13.124)$$

If we combine the conditions that $4\pi d_x(u_g - u_0)/\lambda$ and $4\pi d_y(v_g - v_0)/\lambda$ are integer multiples of 2π and follow the logic that led to (13.117), with the added constraint $d_x(u_g - u_0)/\lambda + d_y(v_g - v_0)/\lambda$, we get that the grating lobes are located at

$$(u_g, v_g) = \left(\frac{p\lambda}{2d_x}, \frac{q\lambda}{2d_y}\right) + (u_0, v_0) \qquad (13.125)$$

with $p + q$ even.

Figures 13.21 and 13.22 contains sketches of the grating lobes for the two variants of triangular packing. As before, the square denotes the main beam, and the unit circle denotes the boundary of visible space.

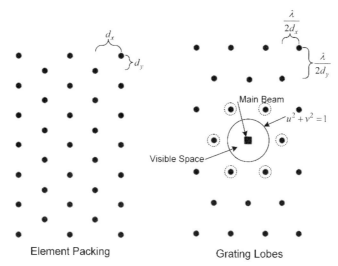

Figure 13.21 Grating lobe locations for triangular packing—stable (or vertically aligned or column staggered) triangular packing.

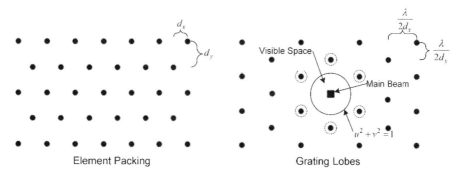

Figure 13.22 Grating lobe locations for triangular packing—unstable (or horizontally aligned or column staggered) triangular packing.

For stable (or vertically aligned or staggered column) triangular packing, the grating lobes to the left or right will enter visible space when

$$\frac{\lambda}{2d_{x\max}} = 1 + |u_0|_{\max} = 1 + |\sin \alpha_{0\max}| \quad (\text{given } v_0 = 0) \quad (13.126)$$

For unstable (or horizontally aligned or staggered row) triangular packing, the grating lobes above or below the main beam enter visible space when

$$\frac{\lambda}{2d_{y\max}} = 1 + |v_0|_{\max} = 1 + |\sin \varepsilon_{0\max}| \quad (\text{given } u_0 = 0) \quad (13.127)$$

For both types of triangle lattices, one of the grating lobes on the diagonals enters visible space when

$$\left(\frac{\lambda}{2d_x} - D_{\max}\cos\theta\right)^2 + \left(\frac{\lambda}{2d_y} - D_{\max}\sin\theta\right)^2 = 1 \qquad (13.128)$$

where

$$\theta = \tan^{-1}\left[(\lambda/2d_y)/(\lambda/2d_x)\right] = \tan^{-1}\left[d_x/d_y\right] \qquad (13.129)$$

and $D_{\max}$ is the maximum beam steering angle in the θ direction.

As a specific example, we consider the classical textbook condition [14, 18, 21, 27] where the array elements are arranged on an equilateral triangle with $2d_x$ as the width of the base and other two legs. This yields $d_y = d_x\sqrt{3}$ (see Figure 13.23). With this $\theta = 30°$ and (13.128) becomes

$$\left(\frac{\lambda}{2d_x} - D_{\max}\cos 30°\right)^2 + \left(\frac{\lambda}{2\sqrt{3}d_x} - D_{\max}\sin 30°\right)^2 = 1 \qquad (13.130)$$

With some manipulation (see Exercise 19), (13.130) can be solved to determine

$$(d_x, d_y) = \left(\frac{\lambda}{\sqrt{3}(1+D_{\max})}, \frac{\lambda}{(1+D_{\max})}\right) \qquad (13.131)$$

Table 13.2 contains a list of values for d_x and d_y for example maximum scan angles for stable and unstable triangular lattice configurations.

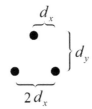

Figure 13.23 Triangular element packing parameters.

Table 13.2

Maximum d/λ Spacings to Avoid Grating Lobes as Various Maximum Steering Angles—Triangular Packing

Maximum Steering Angle	Stable (d_x/λ, d_y/λ)	Unstable (d_x/λ, d_y/λ)
30	(0.67, 0.38)	(0.38, 0.67)
45	(0.59, 0.34)	(0.34, 0.59)
60	(0.54, 0.31)	(0.31, 0.54)
90	(0.50, 0.29)	(0.29, 0.50)

13.10 POLARIZATION

Thus far in our discussions, we have played down the role of E-field orientation in antennas. We now discuss E-field orientation for the specific purpose of discussing polarization. E-fields have both direction and magnitude (and frequency). In fact, an E-field is a vector that is a function of both spatial position and time. If we consider a vector E-field that is traveling in the z direction of a rectangular coordinate system, we can express it as [28]

$$\bar{E}(t,z) = E_x(t,z)\bar{a}_x + E_y(t,z)\bar{a}_y \tag{13.132}$$

where $\bar{a}_x$ and $\bar{a}_y$ are unit vectors. Figure 13.24 contains a graphic showing the above E-field. In this drawing, the z axis is the line-of-sight (LOS) vector from the radar to the target. The x-y plane is in the neighborhood of the face of the antenna. The y axis is generally up, and the x axis is oriented to form a right-handed coordinate system. This is the configuration for propagation from the antenna to the target. When considering propagation from the target, the z axis points along the LOS from the target to the antenna, the y axis is generally up, and the x is again oriented to form a right-handed coordinate system.

When we speak of polarization, we are interested in how the E-field vector, $\bar{E}(t,z)$, behaves as a function of time (t) for a fixed z, or as a function of z for a fixed t. To proceed further, we need to write the forms of $E_x(t,z)$ and $E_y(t,z)$. We use the simplified form of sinusoidal signal. With this we get

$$\bar{E}(t,z) = E_{xo} \sin\left[2\pi(f_o t + z/\lambda)\right]\bar{a}_x \\ + E_{yo} \sin\left[2\pi(f_o t + z/\lambda) + \phi\right]\bar{a}_y \tag{13.133}$$

where, E_{xo} and E_{yo} are positive numbers and represent the E-field strength. f_o is the carrier frequency, and λ is the wavelength, which is related to f_o by $\lambda = c/f_o$. ϕ is a phase shift used to control polarization orientation.

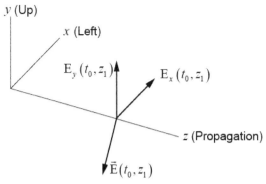

Figure 13.24 Axes convention for determining polarization.

If $\bar{E}(t,z)$ remains fixed in orientation as a function of t and z, the E-field is said to be *linearly polarized*. In particular:

- If $\phi = 0$, $E_{xo} \neq 0$, and $E_{yo} = 0$, we say the E-field is *horizontally polarized*.
- If $\phi = 0$, $E_{yo} \neq 0$, and $E_{xo} = 0$, we say the E-field is *vertically polarized*.
- If $\phi = 0$ or π, and $E_{xo} = E_{yo} \neq 0$, we say the E-field has a slant 45° polarization.
- If $\phi = 0$ or π, and $E_{xo} \neq E_{yo} \neq 0$, we say the E-field has a slant polarization at some angle other than 45°. The polarization angle is given by $\tan^{-1}(E_{yo}/E_{xo})$.

If $\phi = \pm\pi/2$ and $E_{xo} = E_{yo} \neq 0$, we say we have *circular polarization*. If $\phi = +\pi/2$, the polarization is *left-circular* because $\bar{E}(t,z)$ rotates counterclockwise, or to the left, as t or z increases. If $\phi = -\pi/2$, the polarization is *right-circular* because $\bar{E}(t,z)$ rotates clockwise, or to the right, as t or z increases.

If ϕ is any other angle besides $\pm\pi/2$, 0, or π, and/or $E_{xo} \neq E_{yo} \neq 0$, we say the polarization is *elliptical*. It can be left elliptical ($\phi > 0$) or right ($\phi < 0$) elliptical.

As a note, polarization is always measured in the direction of propagation of the E-field to/from the antenna from/to the target. When polarization of an antenna is specified, it is the polarization in the main beam. The polarization in the sidelobes can be dramatically different than the polarization in the main beam. Also, for phased array antennas, the polarization can be distorted when the beam is scanned off of array normal (off of boresight).

13.11 REFLECTOR ANTENNAS

Older radars, and some modern radars where cost is an issue, use reflector antennas rather than phased arrays. Reflector antennas are much less expensive than phased arrays (thousands to hundreds of thousands of dollars as opposed to millions or tens of millions of dollars). They are also more rugged than phased arrays and are generally easier to maintain. They can be designed to achieve good directivity and low sidelobes. The main disadvantage of reflector antennas, compared to phased array antennas, is that they must be mechanically scanned. This means radars that employ reflector antennas have limited multiple target capability. In fact, most target-tracking radars that employ reflector antennas can track only one target at a time. Search radars that employ reflector antennas can detect and track multiple targets, but the track update rate is limited by the scan time of the radar, which is usually on the order of ones to tens of seconds. This, in turn, limits the track accuracy of these radars.

Another limitation of radars that employ reflector antennas is that separate radars are needed for each function. Thus, separate radars would be needed for search, track, and missile guidance. This requirement for multiple radars leads to trade-offs in radar system design. With a phased array, it may be possible to use a single radar to perform the three aforementioned functions (referred to as a multifunction radar). Thus, while the cost of a phased array is high relative to a reflector antenna, the cost of three radars with reflector antennas may be even more expensive than a single phased array radar.

Almost all reflector antennas use some variation of a paraboloid (parabola of revolution) [14]. An example of such an antenna is shown in Figure 13.25. The feed shown in Figure 13.25 is located at the focus of the parabolic reflector (focal point). Since it is in the front, this antenna would be termed a front-fed antenna. The lines from the reflector to the feed are struts used to keep the feed in place.

A parabola is used as a reflector because of its focusing, or collimating, properties. This is illustrated in Figure 13.26. In Figure 13.26, the feed is at the focus of the parabola. From analytic geometry [29], if rays emanate from the focus and are reflected from the parabola, the reflected rays will be parallel [14; 30, p. 147]. In this way, the parabolic antenna focuses the divergent E-field from the feed into a concentrated E-field [31–33]. Stated another way, the parabolic reflector collimates the E-field of the feed.

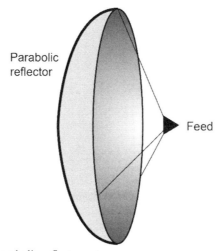

Figure 13.25 Example of a parabolic reflector antenna.

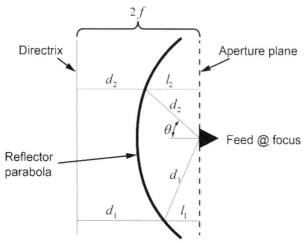

Figure 13.26 Geometry used to find reflector radiation pattern.

As with space-fed phased arrays, the feed pattern is used to control the sidelobe levels of a reflector antenna by concentrating the energy at the center of the reflector and causing it to taper toward the edge of the reflector.

The process of computing the radiation pattern for a parabolic reflector antenna, where the feed is at the focus, is reasonably straightforward. Referring to Figure 13.26, we place a hypothetical plane parallel to the face of the reflector, usually at the location of the feed. This plane is termed the *aperture plane*. We then put a grid of points in this plane. The points are typically arranged on a rectangular grid and are spaced $\lambda/2$ apart. The boundary of the points is a circle that follows the edge of the reflector. The points are used as elements in a hypothetical phased array.

We think of the points, pseudoarray elements, as being in the x-y plane whose origin is at the feed. The z axis of this coordinate system is normal to the aperture plane.

If we draw a line from the point (x, y) to the origin of the x-y plane, the angle it makes with the x axis is

$$\phi = \tan^{-1}\left(\frac{y}{x}\right) \tag{13.134}$$

where the arctangent is the four-quadrant arctangent. The distance from the origin to the point is

$$r = \sqrt{x^2 + y^2} \tag{13.135}$$

We can draw a line from the point (x, y), perpendicular to the aperture plane, to the reflector. Examples of this are the lines l_1 and l_2 in Figure 13.26.

The next step is to find the angle, θ, between the z axis and the point on the reflector. From Figure 13.26

$$d + l = 2f \tag{13.136}$$

where f is the focal length of the parabola. Also

$$r^2 + l^2 = d^2 \tag{13.137}$$

With this we can solve for d and θ to yield

$$d = \frac{r^2 + 4f^4}{4f} \tag{13.138}$$

and

$$\theta = \sin^{-1}\left(\frac{r}{d}\right) \tag{13.139}$$

Next, the angles ϕ and θ are used to find the directive gain of the feed at the point where the ray intersects the parabolic reflector. This directive gain gives the amplitude of the pseudoelement at (x, y).

The above process is repeated for all of the pseudoelements in the aperture plane. Finally, the reflector antenna radiation pattern is found by treating the pseudoelements as a planar phased array.[9]

There is no need to be concerned about the phase of each pseudoelement since the distance from the feed to the reflector to all points in the aperture plane is the same. This means the various rays from the feed take the same time to get to the aperture plane. This further implies that the E-fields along each array have the same phase in the aperture plane, and that the beam is steered to broadside.

If the feed is not located at the focus of the paraboloid, the calculations needed to find the amplitudes and phases of the E-field at the pseudoelements become considerably more complicated. It is well beyond the scope of this book.

13.12 OTHER ANTENNA PARAMETERS

In Sections 13.2 and 13.3, we showed that the beamwidth of a linear array was a function of the length of the array. We want to revisit that problem and develop a specific relation between the array length and the beam width. For an N-element, uniformly illuminated linear array steered to broadside, we have [see (13.38)]

$$R(\varepsilon) = \frac{1}{N}\left[\frac{\sin\left(\frac{N\pi d}{\lambda}\sin\varepsilon\right)}{\sin\left(\frac{\pi d}{\lambda}\sin\varepsilon\right)}\right]^2 \quad (13.140)$$

To determine the beamwidth, we need to find the value of ε_{3dB} for which $R(\varepsilon_{3dB}) = \frac{1}{2}$. With this we have that the beamwidth is $\varepsilon_B = 2\varepsilon_{3dB}$. Through a numerical search, we find that $R(\varepsilon_{3dB}) = \frac{1}{2}$ when

$$\frac{Nd}{\lambda}\sin\varepsilon_{3dB} = 0.443 \quad (13.141)$$

which gives

$$\sin\varepsilon_{3dB} = \frac{0.443\lambda}{Nd} \quad (13.142)$$

We recognize that the antenna length is $D = Nd$ and make use of the fact that D is large relative λ (which means $\sin(\varepsilon_{3dB})$ is small) to arrive at

$$\varepsilon_B = 2\varepsilon_{3dB} = \frac{0.886\lambda}{D} \text{ rad or } \frac{50.8\lambda}{D} \text{ deg} \quad (13.143)$$

If the array uses some type of amplitude weighting for sidelobe reduction, the factor of 0.886 can increase. Experimentation with a few weights indicates that the factor can be

---

[9] Technically, the feed and supporting struts result in aperture blockage, which causes perturbations in the radiation pattern (e.g., reduced gain, elevated sidelobes). This blockage can be accounted for by subtracting the antenna pattern of the blockage from the antenna pattern without blockage.

as high as about 1.4 for heavy Chebyshev weighting. The 30-dB Chebyshev weighting used to generate Figure 13.12 resulted in a factor of 1.1, which would yield a beamwidth of

$$\varepsilon_B = \frac{63\lambda}{D} \text{ deg} \qquad (13.144)$$

This is a rule of thumb. The authors use (13.144) for linear and planar arrays that employ an amplitude taper to reduce sidelobes and for reflector antennas. Our rule of thumb for arrays that use uniform illumination is (13.143).

The beamwidth equations are based on the assumption that the beam is steered to broadside. As the beam is steered off of array normal (broadside), the beamwidth increases by a factor of $1/\cos(\theta_{steer})$ where θ_{steer} is the angle to which the beam is steered, relative to array broadside.

In Chapter 2, we defined directivity in terms of effective aperture as

$$G = \frac{4\pi A_e}{\lambda^2} \qquad (13.145)$$

where A_e is the effective aperture and is related to the physical area of the antenna by $A_e = \rho_{ant} A$ where A is the physical area of the antenna and ρ_{ant} is the aperture efficiency. ρ_{ant} accounts for amplitude tapers and the fact that the beam may not be steered to broadside [the aforementioned $1/\cos(\theta_{steer})$].

The derivation of (13.145) for the general antenna is very difficult because of the difficulty in expressing $\bar{R}$ in terms of the antenna area. However, Balanis [22, Section 11.5.1] has a derivation for the case of a rectangular aperture in an infinitely conducting ground plane. The aperture has dimensions of a and b, and the electric and magnetic fields across the aperture are uniform with magnitudes of E_0 and E_0/η, respectively. η is the characteristic impedance of free space. Under these conditions, Balanis shows that the maximum radiation intensity is

$$R_{\max} = \left(\frac{ab}{\lambda}\right)^2 \frac{|E_0|^2}{2\eta} \qquad (13.146)$$

and

$$\bar{R} = \frac{1}{4\pi} P_{rad} = \frac{1}{4\pi} ab \frac{|E_0|^2}{2\eta} \qquad (13.147)$$

where P_{rad} is the total radiated power.

From (13.146) and (13.147), Balanis derives the maximum directive gain, or directivity, as

$$G_{\max} = \frac{R_{\max}}{\bar{R}} = \frac{4\pi ab}{\lambda^2} = \frac{4\pi A}{\lambda^2} \qquad (13.148)$$

where $A = ab$ is the area of the aperture. In Balanis' example, the aperture is uniformly illuminated and the main beam is pointing along the normal to the aperture (broadside).

Because of this, the aperature efficiency is $\rho_{ant} = 1$ and $A_e = A$. While this development does not prove that (13.148) holds for all antennas, experience indicates that it does. That is, the maximum directivity is directly proportional to effective aperture and inversely proportional to the square of wavelength.

13.13 EXERCISES

1. Generate the plot of Figure 13.4. What are the beamwidths for the three cases?
2. Derive (13.37).
3. Generate the plot of Figure 13.7. What are the beamwidths for the three cases?
4. Derive (13.46).
5. Derive (13.47).
6. Show that $\bar{R} = 1$ for an N-element array with $d = \lambda/2$ and $a_n = 1/\sqrt{N}$.
7. Reproduce Figure 13.9.
8. Implement the computation algorithm described in Section 13.9, and use it to generate and plot the radiation pattern for a 20-element linear array with 30-dB Chebyshev weighting. Your plot should look similar to Figure 13.12. In particular, the peak sidelobe level should be 30 dB below the main beam.
9. Compute $\bar{R}$ for the radiation pattern of Exercise 8, and use it to reproduce the directive gain pattern of Figure 13.12. Steer to 30°. What happens to the directive gain [peak of $G(\varepsilon)$] relative to the value at $\varepsilon_0 = 0°$? What happens to the beamwidth? Repeat this for $\varepsilon_0 = 60°$.
10. Repeat Exercises 8 and 9 but with $N = 50$ (50 elements) and Taylor weighting with $\bar{n} = 6$ and $SL = -30$ dB (see Appendix 13A).
11. Recompute the weights you used to steer the beam of Exercise 10 with the assumption that the phases are set by a 6-bit phase shifter. Repeat for the case where the phases are set by a 3-bit phase shifter. Repeat this for the case where the beam is steered to $\varepsilon_0 = 0°$. Can you explain the difference in the effect of quantization for this case when compared to the cases where the beam was steered to 30° and 60°?
12. Show that $G(\varepsilon) = G_{ell}(\varepsilon)G_{array}(\varepsilon)$ for the case where the array elements are isotropic radiators.
13. Implement the computation algorithm of Section 13.9.5.1 or the algorithm in Appendix 13B. Use it to generate a radiation pattern for a square array that has 51 rows and 51 columns of elements (M and $N = 51$). Assume an element spacing of $d_x/\lambda = d_y/\lambda = \frac{1}{2}$ and uniform weighting. Steer the beam to $(u_0, v_0) = (0,0)$. Generate a three-dimensional (3-D) plot of the form shown in Figure 13.1. Generate azimuth and elevation principal plane cuts for azimuth and elevation angles that range from $-5°$ to $5°$. The 3-D plot for this exercise should have the principal plane ridges indicated in Section 13.9.5.1. Change the square array of Exercise 11 to a circular array by using the method indicated in step 4 of the method for applying elliptically symmetric

weighting (Section 13.9.5). Generate the plots indicated in Exercise 11. In this case, you will note that the principal plane ridges are no longer present.

14. Repeat Exercise 13 with a circularly symmetric, Taylor, $\bar{n} = 6$, SL = −30 dB weighting applied to the array. In this case, the 3-D plot should look similar to Figure 13.1.

15. Compute $\bar{R}$ and the directive gain pattern $G(u,v)$ for the array of Exercise 11. Generate the plots indicated in Exercise 11.

16. Apply multiplicative, Taylor, $\bar{n} = 6$, SL = −30-dB weighting to the array of Exercise 11, and recompute $\bar{R}$ and $G(u,v)$. Generate the plots indicated in Exercise 11. Note how the directive gains (directivities) and beamwidths of the patterns compare to those of Exercise 14.

17. Apply an elliptically symmetric taper cos taper ($\cos^n$ taper with n = 1) to the array of Exercise 13. For the first case use a 10-dB edge taper. This means that you want the amplitude at the edge of the array to be $10^{-(10/20)}$ relative to a peak value of 1. Repeat this for a 20-dB edge taper. Discuss the difference in directive gain, beamwidth, and sidelobe levels for the two tapers.

18. In Table 13.1, we indicate that if $d_x/\lambda = d_y/\lambda$, grating lobes enter visible space when the beam is scanned more than 45° from broadside. To verify this, change d_x/λ and d_y/λ of the array of Exercise 11 to 0.59. Steer the beam to $(u_0, v_0 = 10, \sin 50°)$. Generate an elevation principal plane cut. This plot will show that the main beam is steered to 50°, but there is another lobe, the grating lobe, at a negative angle. You should also note that the width of the grating lobe is larger than the main lobe.

19. Derive (13.131).

20. Using (13.88) and (13.89), show that u and v satisfy the constraint, $u^2 + v^2 \leq 1$.

21. Derive (13.104).

References

[1] Mitra, S. K., *Digital Signal Processing: A Computer-Based Approach*, 2nd ed., New York: McGraw-Hill, 2001, p. 301.

[2] Bondyopadhyay, P. K., "The First Application of Array Antenna," *Proc. 2000 IEEE Int. Conf. on Phased Array Syst. Technol.*, Point Dana, CA, May 21–25, 2000, pp. 29–32.

[3] Brittain, J. E., "Electrical Engineering Hall of Fame: Guglielmo Marconi," *Proc. IEEE*, vol. 92, no. 9, Aug. 2004, pp. 1501–1504.

[4] Mailloux, R. J., "A Century of Scanning Array Technology," *2014 IEEE Antennas Propag. Soc. Int. Symp. (APSURSI)*, Memphis, TN, Jul. 6–11, 2014, pp. 524–525.

[5] Friis, H. T., and C. B. Feldman, "A Multiple Unit Steerable Antenna for Short-Wave Reception," IRE Proc., vol. 25, no. 7, Jul. 1937, pp. 841–917; *Bell System Tech. J.*, vol. 16, no. 3, Jul. 1937, pp. 337–419.

[6] Friis, H. T., and W. D. Lewis, "Radar Antennas," *Bell System Tech. J.*, vol. 26, no. 2, Apr. 1947, pp. 219–317.

[7] Schell, A. C., "Antenna Developments of the 1950s to the 1980s," *2001 IEEE Antennas Propag. Soc. Int. Symp.*, vol. 1, Boston, MA, Jul. 8–13, 2001, pp. 30–33.

[8] Blake, L. V., *Antennas*, Dedham, MA: Artech House, 1984.

[9] Gradshteyn, I. S., and I. M. Ryzhik, *Table of Integrals, Series, and Products*, 8th ed., (D. Zwillinger and V. Moll, eds.) New York: Academic Press, 2015. Translated from Russian by Scripta Technica, Inc.

[10] Silver, S., *Microwave Antenna Theory and Design*, vol. 12 of MIT Radiation Lab. Series, New York: McGraw-Hill, 1949; Norwood, MA: Artech House (CD-ROM edition), 1999.

[11] *IEEE Standard Dictionary of Electrical and Electronic Terms*, 6th ed., New York: IEEE, 1996.

[12] Dolph, C. L., "A Current Distribution for Broadside Arrays Which Optimizes the Relationship Between Beam Width and Side-Lobe Level," *Proc. IRE*, vol. 34, no. 6, Jun. 1946, pp. 335–348.

[13] Riblet, H. J., "Discussion of Dolph's Paper," *Proc. IRE*, vol. 35, no. 5, May 1947, pp. 489–492.

[14] Taylor, T. T., "Design of Line-Source Antennas for Narrow Beamwidth and Low Side Lobes," *Trans. IRE Prof. Group Antennas Propag.*, vol. 3, no. 1, Jan. 1955, pp. 16–28.

[15] Johnson, R. C., ed., *Antenna Engineering Handbook*, 3rd ed., New York: McGraw-Hill, 1992.

[16] Skolnik, M. I., ed., *Radar Handbook*, 3rd ed., New York: McGraw-Hill, 2008.

[17] Rhodes, D. R., "On the Taylor Distribution," *IEEE Trans. Antennas Propag.*, vol. 20, no. 2, Mar. 1972, pp. 143–145.

[18] Ostroff, E. D., et al., *Solid-State Radar Transmitters*, Dedham, MA: Artech House, 1985.

[19] Kahrilas, P. J., *Electronic Scanning Radar Systems (ESRS) Design Handbook*, Norwood, MA: Artech House, 1976.

[20] Hansen, R. C., ed., *Microwave Scanning Antennas*, Vol. III: Array Systems, New York: Academic Press, 1966.

[21] Koul, S. K., and B. Bhat, *Microwave and Millimeter Wave Phase Shifters*, Vol. I and II, Norwood, MA: Artech House, 1991.

[22] Brookner, E., ed., *Practical Phased Array Antenna Systems*, Norwood, MA: Artech House, 1991.

[23] Balanis, C. A., *Antenna Theory: Analysis and Design*, 3rd ed., New York: Wiley & Sons, 2005.

[24] Ma, M. T., *Theory and Application of Antenna Arrays*, New York: Wiley & Sons, 1974.

[25] Von Aulock, W. H., "Properties of Phased Arrays," *Proc. IRE*, vol. 48, no. 10, Oct. 1960, pp. 1715–1727.

[26] Schelkunoff, S. A., *Advanced Antenna Theory*, New York: Wiley & Sons, 1952.

[27] Mofrad, R. F., R. A. Sadeghzadeh, and S. Alidoost, "Design of a Benchmark Antenna for the Multi Function Phased Array Radar Simulation Test Bed," 2011 Loughborough Antennas and Propag. Conf. (LAPC), Nov. 14–15, 2011, pp. 1–4.

[28] Mailloux, R. J., *Phased Array Antenna Handbook*, 2nd ed., Norwood, MA: Artech House, 2005.

[29] Stutzman, W. L., *Polarization in Electromagnetic Systems*, Norwood, MA: Artech House, 1993.

[30] Leithold, L., *The Calculus with Analytic Geometry*, 5th ed., New York: Harper & Row, 1986.

[31] Barton, D. K., Radar System Analysis and Modeling, Norwood, MA: Artech House, 2005.

[32] Cutler, C. C., "Parabolic-Antenna Design for Microwave," *Proc. IRE*, vol. 35, no. 11, Nov. 1947, pp. 1284–1294.

[33] Berkowitz, B. "Antennas Fed by Horns," *Proc. IRE*, vol. 41, no. 12, Dec. 1953, pp. 1761–1765.

[34] Jones, E. M. T., "Paraboloid Reflector and Hyperboloid Lens Antennas," *Trans. IRE Prof. Group on Antennas Propag.*, vol. 2, no. 3, Jul. 1954, pp. 119–127.

APPENDIX 13A: AN EQUATION FOR TAYLOR WEIGHTS

The following are equations for calculating Taylor weights for an array antenna. It is similar to the equation on page 20-8 of the *Antenna Engineering Handbook* by Richard C. Johnson [14], with some clarifications and corrections.

The un-normalized weight for the nth element of the N-element linear array is

$$a_n = 1 + 2\sum_{n=1}^{\bar{n}-1} F(n, A, \bar{n}) \cos\left(\frac{2n\pi x_n}{N}\right) \qquad (13A.1)$$

where

$$F(n, A, \bar{n}) = \frac{\left[(\bar{n}-1)!\right]^2 \prod_{m=1}^{\bar{n}-1}\left(1 - \frac{n^2}{\sigma^2\left[A^2 + \left(m - \frac{1}{2}\right)^2\right]}\right)}{(\bar{n}-1+n)!(\bar{n}-1-n)!} \qquad (13A.2)$$

$$A = \frac{\cosh^{-1}(R)}{\pi} \qquad (13A.3)$$

$$\sigma^2 = \frac{\bar{n}^2}{A^2 + \left(\bar{n} - \frac{1}{2}\right)^2} \qquad (13A.4)$$

and

$$R = 10^{SL/20} \qquad (13A.5)$$

SL is the desired sidelobe level, in dB, relative to the peak of the main beam and is a positive number. For example, for a sidelobe level of –30 dB, $SL = 30$. This indicates that the sidelobe is 30 dB below the peak of the main beam. $\bar{n}$ is the number of sidelobes on each side of the main beam that we want to have a level of approximately SL below the main beam peak amplitude.

The x_n can be computed using the following MATLAB notation:

$$\begin{aligned} z &= [-N:N]/2 \\ x &= z(2:2:\text{end}) \end{aligned} \qquad (13A.6)$$

Finally, normalize the weights by dividing all of the a_n by $\max_n(a_n)$.

APPENDIX 13B: COMPUTATION OF ANTENNA PATTERNS

13B.1 LINEAR ARRAYS

In Section 13.3, we determined that we could compute the radiation pattern, $R(\varepsilon)$, from

$$R(\varepsilon) = |A(\varepsilon)|^2 \quad (13\text{B}.1)$$

where

$$A(\varepsilon) = \sum_{n=0}^{N-1} a_n e^{-jnx} \quad (13\text{B}.2)$$

and

$$x = 2\pi d \sin \varepsilon / \lambda \quad (13\text{B}.3)$$

The "brute force" way to compute $A(\varepsilon)$ would be to implement (13B.2) in a loop (e.g., FOR loop, DO loop) and repeat this for the ε values of interest. While this is sufficient for small values of N and few values of ε, it can be time-consuming when either or both of these are large. By recasting (13B.2) in a vector form, the computation of $A(\varepsilon)$ can be sped up when using software with efficient matrix and vector routines.

Let W_a be a row vector (a *weight* vector) of the a_n, and K_N be a column vector of integers that range from 0 to $N-1$. That is,

$$W_a = \begin{bmatrix} a_0 & a_1 & \cdots & a_{N-1} \end{bmatrix} \quad (13\text{B}.4)$$

and

$$K_N = \begin{bmatrix} 0 & 1 & \cdots & N-1 \end{bmatrix}^T \quad (13\text{B}.5)$$

where the superscript T denotes the transpose operation. Define X as

$$\begin{aligned} X &= \begin{bmatrix} x_1 & x_2 & \cdots & x_{N\varepsilon} \end{bmatrix} \\ &= \begin{bmatrix} (2\pi d \sin \varepsilon_1)/\lambda & (2\pi d \sin \varepsilon_2)/\lambda & \cdots & (2\pi d \sin \varepsilon_{N\varepsilon})/\lambda \end{bmatrix} \end{aligned} \quad (13\text{B}.6)$$

where $\varepsilon_1, \varepsilon_2, \ldots, \varepsilon_{N\varepsilon}$ are the angles at which we want to compute $A(\varepsilon)$

With the above definitions, $A(\varepsilon)$ can be written as

$$A(\varepsilon) = W_a \exp[-j K_N X] \quad (13\text{B}.7)$$

Equation (13B.7) circumvents the need for loops in higher-level languages (such as MATLAB, Mathcad, and Python®) and executes very quickly. It also results in computer code that is very concise.

13B.2 PLANAR ARRAYS

We now want to extend the method of the previous section to planar arrays. We will consider variations for rectangular and triangular packing.[10]

13B.2.1 Rectangular Packing

We start by combining (13.90) with (13.93) and write

$$A(u,v) = \sum_{n=0}^{N-1}\sum_{m=0}^{M-1} |a_{mn}| e^{-j2\pi m d_x (u-u_0)/\lambda} e^{-j2\pi n d_y (v-v_0)/\lambda} \qquad (13\text{B}.8)$$

Similar to (13B.4), we collect the $|a_{mn}|$ into a matrix

$$W_{mn} = \begin{bmatrix} |a_{00}| & |a_{01}| & \cdots & |a_{0,N-1}| \\ |a_{10}| & |a_{11}| & \cdots & |a_{1,N-1}| \\ \vdots & \vdots & \ddots & \vdots \\ |a_{M-1,0}| & |a_{M-1,1}| & \cdots & |a_{M-1,N-1}| \end{bmatrix} \qquad (13\text{B}.9)$$

and define

$$U = \begin{bmatrix} u_1 - u_0 & u_2 - u_0 & \cdots & u_{Nu} - u_0 \end{bmatrix} \qquad (13\text{B}.10)$$

$$V = \begin{bmatrix} v_1 - v_0 & v_2 - v_0 & \cdots & v_{Nv} - v_0 \end{bmatrix} \qquad (13\text{B}.11)$$

$$K_M = \begin{bmatrix} 0 & 1 & \cdots & M-1 \end{bmatrix} \qquad (13\text{B}.12)$$

and

$$K_N = \begin{bmatrix} 0 & 1 & \cdots & N-1 \end{bmatrix}^T \qquad (13\text{B}.13)$$

We combine (13B.9) through (13B.13) to write

$$A(U,V) = \exp\left(-j\frac{2\pi d_x}{\lambda} U^T K_M\right) W_{mn} \exp\left(-j\frac{2\pi d_y}{\lambda} K_N V\right) \qquad (13\text{B}.14)$$

This will produce a matrix of $A(u,v)$'s at all combinations of the u's and v's specified in the U and V vectors.

To generate an elevation principal plane pattern (elevation cut), we would replace the first exponential with a row vector containing M 1's. For an azimuth principal plane pattern (azimuth cut), we would replace the second exponential with a column vector containing N 1's.

To generate a radiation pattern over some region of the u,v plane, we would use (13B.10) and (13B.11) with the desired values of u and v and some desired steering angles u_0 and v_0. As an example, Figure 13.1 was generated using the following:

---

[10] The algorithm discussed in this section was provided by Joshua Robbins of Dynetics, Inc.

$d_x/\lambda = d_y/\lambda = 1/2$
$u_0 = v_0 = 0$
$U = [-1 \ -1+\Delta u \ \ldots \ 0 \ \ldots \ 1-\Delta u \ 1]$
$V = [-1 \ -1+\Delta v \ \ldots \ 0 \ \ldots \ 1-\Delta v \ 1]$
Δu and Δv were set to small values.

A_{mn} was chosen to provide an elliptically symmetric, Taylor weighting with $\bar{n} = 6$ and SL = 30

N and M were set to 51.

Because of the way U and V are defined, (13B.14) can have nonzero values for $u^2 + v^2 > 1$, which is not in visible space. This is taken into account by forcing the plotting routine to ignore $A(u,v)$ values for u,v pairs where $u^2 + v^2 > 1$.

13B.2.2 Triangular Packing

Calculation of $A(U,V)$ for triangular packing is more complicated in that it must be computed in two parts. Figure 13B.1 contains an illustration of triangular packing that we will use to describe the method. In Figure 13B.1, the circles and squares denote elements and form two rectangular lattices that are offset in x and y by d_x and d_y. Let the weights associated with the circles be a_{mn} and the weights associated with the squares be b_{mn}. With this we write

$$A(u,v) = A_a(u,v) + A_b(u,v)$$
$$= \sum_{n=0}^{N_a-1} \sum_{m=0}^{M_a-1} |a_{mn}| e^{-j2\pi 2md_x(u-u_0)/\lambda} e^{-j2\pi 2nd_y(v-v_0)/\lambda} \quad (13B.15)$$
$$+ \sum_{n=0}^{N_b-1} \sum_{m=0}^{M_b-1} |b_{mn}| e^{-j2\pi(2m+1)d_x(u-u_0)/\lambda} e^{-j2\pi(2n+1)d_y(v-v_0)/\lambda}$$

Equation (13B.15) tells us the overall $A(u,v)$ is the sum of $A(u,v)$'s for two offset arrays with rectangular packing. Since the equations for $A_a(u,v)$ and $A_b(u,v)$ are the same form as (13B.8), we can use the methodology of Section 13B.2.1 and write

$$A(U,V) = A_a(U,V) + A_b(U,V) \quad (13B.16)$$

To compute $A_a(U,V)$, we would use

$$W_{amn} = \begin{bmatrix} |a_{00}| & |a_{01}| & \cdots & |a_{0,Na-1}| \\ |a_{10}| & |a_{11}| & \cdots & |a_{1,Na-1}| \\ \vdots & \vdots & \ddots & \vdots \\ |a_{Ma-1,0}| & |a_{Ma-1,1}| & \cdots & |a_{Ma-1,Na-1}| \end{bmatrix} \quad (13B.17)$$

$$K_{Ma} = \begin{bmatrix} 0 & 2 & \cdots & 2(M_a-1) \end{bmatrix} \quad (13B.18)$$

and

$$K_{Na} = \begin{bmatrix} 0 & 2 & \cdots & 2(N_a - 1) \end{bmatrix}^T \tag{13B.19}$$

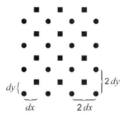

Figure 13B.1 Illustration of triangular packing used to explain calculation method.

To compute $A_b(U,V)$, we would use

$$W_{bmn} = \begin{bmatrix} |b_{00}| & |b_{01}| & \cdots & |b_{0,Nb-1}| \\ |b_{10}| & |b_{11}| & \cdots & |b_{1,Nb-1}| \\ \vdots & \vdots & \ddots & \vdots \\ |b_{Mb-1,0}| & |b_{Mb-1,1}| & \cdots & |b_{Mb-1,Nb-1}| \end{bmatrix} \tag{13B.20}$$

$$K_{Mb} = \begin{bmatrix} 1 & 3 & \cdots & 2M_b - 1 \end{bmatrix} \tag{13B.21}$$

and

$$K_{Nb} = \begin{bmatrix} 1 & 3 & \cdots & 2N_b - 1 \end{bmatrix}^T \tag{13B.22}$$

The amplitude weights can be computed using the techniques of Section 13.9.4. When using the product method, we would start with $M_a + M_b$ weights in the x direction and alternately allocate them to a_m and b_m. Likewise, for the y direction, we would start with $N_a + N_b$ weights and alternately allocate them to a_n and b_n. For the elliptically symmetric case, we recognize that the location of the m_a, n_a^{th} element is $(x_a, y_a) = (m_a d_x, n_a d_y)$ for the W_{amn} array. For the W_{bmn} array, the location of the m_b, n_b^{th} element is $(x_b, y_b) = (m_b d_x, n_b d_y)$. In these equations, m_a varies from 0 to $2M_a$ in steps of 2, n_a varies from 0 to $2N_a$ in steps of 2, m_b varies from 1 to $2M_b - 1$ in steps of 2, and n_b varies from 1 to $2N_b - 1$ in steps of 2.

Chapter 14

AESA Basics and Related Topics

14.1 INTRODUCTION

The phased array antennas we discussed in Chapter 13 are also termed *passive electronically steered arrays* (PESAs). The name derives from the fact that they are phased *arrays* where the beam is *steered electronically* and the devices used to steer the beam, the phase shifters, are *passive* devices that are part of the array structure. In this chapter, we consider the basics of *active electronically steered arrays* (*AESAs*). In an AESA, the devices used to steer the beam can still be phase shifters; however, they are part of an active device that is located in the array structure. In radars that employ a PESA, the active components of the radar, the transmitter and receiver, are not part of the array. In radars that employ an AESA, some of the active components are part of the array.

While AESAs have been used in military radars for many years, the time of appearance of the first AESA is unclear. Development of airborne AESA radars in the United States began in 1964 [1, 2]. Since then, X-band AESA radars are now the baseline in state-of-the-art combat aircraft [2]. The first volume-produced AESA for fighter and bomber aircraft is the Northrop-Grumman (formerly Westinghouse) 1,500 element, X-band, APG-77 for the F-22 Raptor [2]. The first Russian manufacturer to offer an AESA is NIIR Phazotron (Фазотрон-НИИР), which produced the Zhuk-AE (жук[1]) X-band AESA used by the MiG-35 (Микоян Миг-35) FULCRUM fighter [3].

Figures 14.1 and 14.2 contain sketches that illustrate the basic difference between a radar with a PESA and a radar with an AESA. The Figure 14.1 drawing is for a radar that contains a space-fed PESA and the Figure 14.2 drawing is for a radar that contains an AESA. In the radar that uses a PESA, the actual array (labeled Array) contains only phase shifters and radiating elements (labeled Elements). The radar body contains the active components [power amplifier, low noise amplifier (LNA), receiver, exciter, etc.]. In the radar that uses an AESA, the power amplifiers and LNAs, along with the phase shifters and radiating elements, are part of the actual array. The remaining components (receiver, exciter, etc.) are located in the radar body.

---

[1] Russian for beetle.

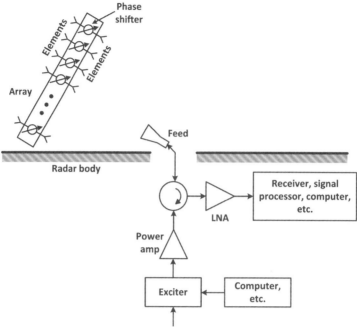

Figure 14.1 Space-fed PESA block diagram.

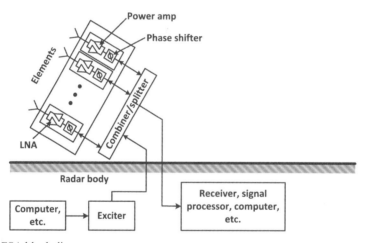

Figure 14.2 AESA block diagram.

In the space-fed PESA radar, the feed distributes transmit energy to the elements on the back (feed side) of the array and collects receive energy from those elements. In the AESA, the process of distributing energy to, and collecting energy from, the array is accomplished by the combiner/splitter. If the PESA radar were to use a corporate feed configuration, it would also have a combiner/splitter instead of a feed (see Figure 13.18).

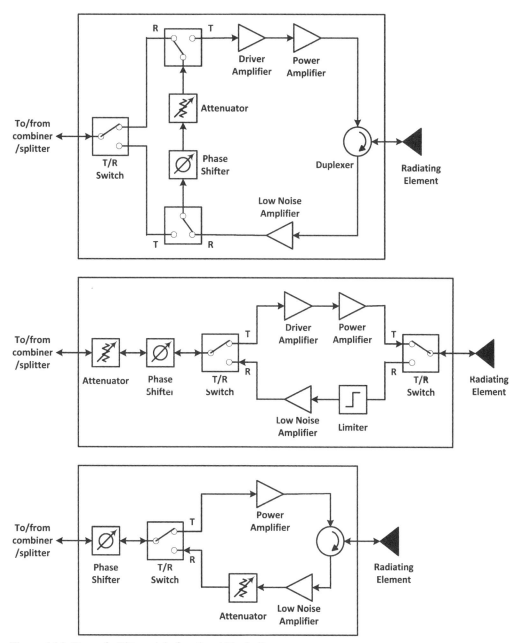

Figure 14.3 Example T/R module functional block diagrams.

14.2 T/R MODULE

In the AESA diagram, the devices that contain the power amplifier, LNA, and phase shifter are called transmit/receive modules, or T/R modules. Practical T/R modules contain more than the power amplifier, LNA, and phase shifter. Examples of these other components are indicated in the T/R module functional block diagrams of Figure 14.3. The various block diagrams represent just three of the many that appear in current literature [4–13]. Although the details of the various T/R module variants differ, all of them share common elements:

1. Devices (switches and/or circulators) to route energy to and from the radiating elements and the rest of the radar, via the combiner splitter
2. A power amplifier in the transmit path (the "T" side in Figure 14.3)
3. A LNA in the receive path (the "R" side in Figure 14.3)
4. A phase shifter that is used to steer the transmit and receive beams of the array
5. An attenuator that is used to apply amplitude weighting to shape the antenna pattern

In addition to the basic components indicated above, some T/R modules contain additional components, such as the driver amplifier and limiter indicated in Figure 14.3.

In all three of the Figure 14.3 diagrams, the phase shifter is common to the transmit and receive paths of the T/R module. This is necessary because the radar must be able to form and steer a beam on both transmit and receive.

Figure 14.4 contains a photo of a T/R module. The various components indicated in Figure 14.3 are annotated in the photo.

In all T/R module configurations of which the authors are aware, the phase shifter is on the low-power side of the T/R module on transmit, and after the LNA on receive. Thus, the phase shifter does not need to handle extremely high-power signals on transmit or extremely low-power signals on receive. This is in contrast to PESAs where the phase shifters must be able to handle high-power signals on transmit and very low-power signals on receive. As such, T/R modules in AESAs usually use diode phase shifters [14–18].

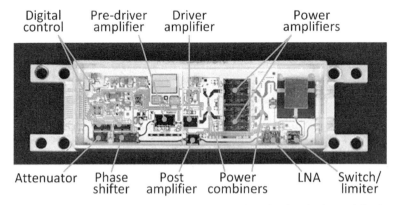

Figure 14.4 Photo of a typical T/R module. Mid-1990s MMIC-based X-band T/R Module. (Image: courtesy of Raytheon Company, Waltham, Massachusetts).

The attenuator in the top two diagrams of Figure 14.3 is common to the transmit and receive paths, while the attenuator in the bottom diagram is only in the receive path. Whether an attenuator can be used on transmit is questionable. The logic behind this statement is that the power amplifier of a T/R module is normally operated at full power output to improve efficiency [5, p. 38], and thus minimize unnecessary heating.

Many currently available T/R modules use gallium arsenide (GaAs) field-effect transistors (FETs) in their power amplifiers [4; 5, p. 62]. However, gallium nitride (GaN) technology is advancing rapidly, and T/R module manufacturers are increasingly using GaN FETs in their T/R module designs [4; 5, p. 72] . GaAs-based T/R modules currently provide power levels in the range of 1 to 10 W, depending on the design and operating frequency [6, p. 97]. GaN-based T/R modules currently provide power levels in the range of 10 to 100 W, again depending on design and operating frequency [19, p. 177]. This increase in power is a result of GaN power amplifiers being able to operate at higher power densities than GaAs power amplifiers.

Compared to GaAs power amplifiers, GaN power amplifiers have increased *power-added efficiency* (PAE), which is a figure of merit rating how well an amplifier converts DC power to RF power, expressed as [84, p. 35]

$$\eta_a = PAE = \frac{P_{out} - P_{in}}{P_{DC}} \times 100\% \qquad (14.1)$$

where P_{in} and P_{out} are RF input and output power, respectively, and P_{DC} is the DC power into the device. The more efficient an amplifier, the less DC power is required for a given output power. High PAE is critical because DC power not converted to RF must be dissipated as heat.

Even though the power generated by each T/R module is only in the range of 1 to 100 W, depending on the power amplifier, the total power radiated by an AESA can be quite high. A typical AESA can contain anywhere from 1,000 to 10,000 or more elements. This means the total radiated power of the radar can be anywhere from 1 kW to 1 MW. This is compatible with, and possibly beyond, the power levels of radars that use klystron or traveling-wave tube (TWT) power amplifiers.

The fact that an AESA uses thousands of T/R modules is the basis for one of its touted advantages: graceful degradation of performance [20–24]. In a radar that uses a single-power amplifier and RF receive amplifier (LNA), the failure of either results in failure of the radar. With AESA-based radars, a significant number of T/R modules can fail and the radar will still operate, albeit in a degraded fashion.

The LNAs of current T/R modules advertise noise figures in the range of 1 to 3 dB and gains in the range of 15 to 25 dB [5, 7, 25]. Also, since the path length between the radiating element and the LNA is short, there will be a very small increase in noise figure due to losses before the LNA. Further, since the phase shifter is after the LNA, it will not significantly add to the noise figure. Because of these T/R module characteristics, an AESA-based radar can have a fairly low system noise figure. As a note, the attenuator after the LNA, which is used to apply amplitude weighting on receive, will cause an increase in the system noise figure. However, as we will show later, it does not significantly increase the system noise figure.

In a similar fashion, because of the short path length between the power amplifier and the radiating element, and the fact that the phase shifter is before the power amplifier, the transmit losses of an AESA-based radar will be lower than for a PESA-based radar.

In the AESA block diagram of Figure 14.2, we show all of the T/R modules connected to a combiner/splitter that was subsequently connected to the exciter and receiver. This is similar to the configuration used in corporate-fed PESAs, and is probably the way early AESAs were configured. Modern AESA configurations take advantage of the fact that the signal input to the transmit side of the T/R module is low power, and that the outputs from the receive side is after the LNA. An advantage of feeding the transmit side with low-power signals is that it becomes possible to use time-delay steering to facilitate the use of wide bandwidth waveforms. Since the time-delay units (TDUs), which can be quite lossy (e.g., 15 to 30 dB) [22, 26–30], are before the power amplifier, they do not affect transmit losses from the radar range equation perspective. Similarly, TDUs can be used on receive with only a small impact on noise figure since they would be located after the LNA.

One might argue that TDUs could be used with corporate-fed PESAs. However, as implied in the previous paragraph, they would be located after the power amplifier on transmit and before the LNA on receive. In the transmit case, they would need to handle the high-power signal and would significantly increase transmit losses since they are lossy devices. On receive, they would cause a significant increase in system noise figure since they would be located before the LNA.

Another capability enabled by the use of AESAs is the ability to simultaneously form multiple receive beams and to implement space-time adaptive processing (STAP, see Chapter 24). These capabilities are supported by the availability of small, inexpensive receivers and high-speed signal and data processors. In the limit, a separate receiver would be connected to each T/R module and the digitized receiver outputs would be sent to a massively parallel signal/data processor. To the authors' knowledge, current technology is not quite able to achieve this goal in practical radars, but it is likely such a goal will be achievable in the near future.

14.3 TIME-DELAY STEERING AND WIDEBAND WAVEFORMS

The ideal implementation of time-delay steering would be to include a TDU on each T/R module. However, because of cost and packaging constraints, such an approach is not possible, except for very small AESAs. An alternative that is currently used is to divide the AESA into subarrays and use TDUs at the subarray level. An illustration of such an approach for a linear array is shown in Figure 14.5 [8, 14, 31, 32].

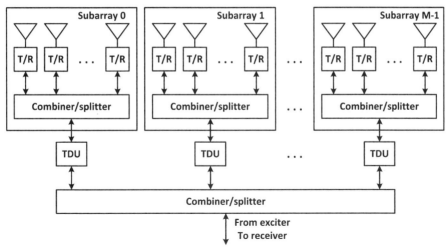

Figure 14.5 A linear array with subarrays.

Each subarray consists of N T/R modules and radiating elements whose inputs and outputs go to a combiner/splitter. The combiner/splitter sums the T/R module outputs on receive. On transmit, the splitter divides the signal from the TDU and sends the components to each of the T/R modules. Equal weighting is normally used on the summation and in the splitting. Each of the M subarrays is connected to a TDU, which is connected to another combiner/splitter. The combiner/splitters of the subarrays are termed subarray combiners (beamformers) and the combiner/splitter connected to the TDUs is termed an array combiner (beamformer) [14].

The configuration of Figure 14.5 has one level of subarraying. In large antennas, the diagram of Figure 14.5 could represent one subarray of a larger array. These subarrays would be connected to another combiner/splitter to form the overall array. In that case, the AESA would be using two levels of subarraying.

An example of two levels of subarraying is illustrated in the simple monopulse example of Figure 14.6. In the diagram, the AESA is divided into quadrants. Each quadrant contains M, one-level, subarrays like Figure 14.5. That is, each quadrant contains M subarrays. The outputs of the second-level subarrays that form the quadrants are sent to three combiner/splitters to form the sum and two difference channels of the radar. Only the combiner portion of the combiner/splitter is used on the difference channels while both the combiner and splitter portions are used on the sum channel. The exciter signal is sent to the sum combiner/splitter, which would split it four ways and send the four signals to the four quadrant subarrays.

In the subarray configuration of Figure 14.5, the phase shifters of the T/R modules are used to form and steer the antenna radiation patterns formed by the subarrays. The TDUs are then used to form and steer the radiation pattern of the overall array. The attenuators of the T/R modules would implement the amplitude taper used to control the sidelobes of the receive antenna radiation pattern. It is possible that the amplitude taper could be partitioned between the T/R modules and the amplifiers incorporated in the TDU modules.

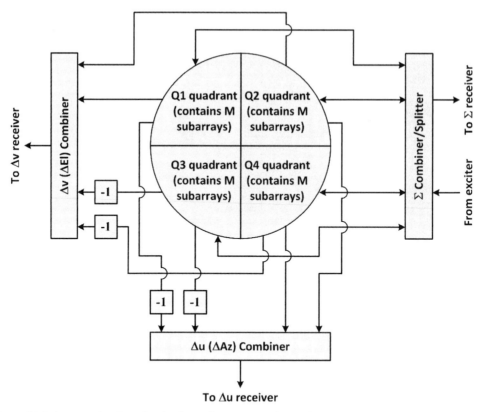

Figure 14.6 An example of two-level subarraying—monopulse AESA.

14.3.1 Subarray Size, Scan Angle, and Waveform Bandwidth

An issue with the use of subarrays and wideband waveforms is the selection of subarray size; that is, the number of elements and T/R modules in each subarray. In Chapter 13, we noted that the size of an array was a function of waveform bandwidth and maximum scan angle. We stated, without proof, that, for a maximum scan angle of 60°, the width of the array should be less than the compressed pulse width of the transmit waveform. Larger arrays could be used with smaller scan angles, and smaller arrays are needed for larger scan angles. This constraint also applies to subarrays.

The criterion used to determine the relation between array size, scan angle, and waveform bandwidth (compressed pulse width) is the degradation in resolution and signal-to-noise ratio (SNR) at the output of the matched filter (or stretch processor, since that would be the appropriate processor for wideband waveforms that use linear frequency modulation (LFM) or step frequency). To illustrate such an analysis, we consider the linear subarray geometry of Figure 14.7.

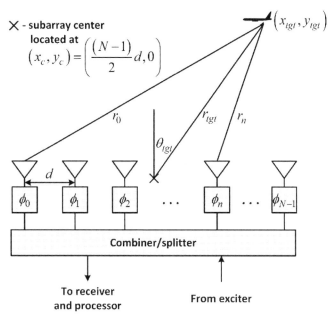

Figure 14.7 Subarray geometry used to characterize the relation between waveform bandwidth, scan angle, and subarray size.

The subarray consists of N radiating (antenna) elements and associated T/R modules that are spaced d meters apart. The radiating elements are indicated by the triangles and the T/R modules are represented by the boxes with ϕ_n in them. We assume uniform amplitude weighting and thus, do not show the attenuators of the T/R modules. We also do not show the other components of the T/R modules since they are not germane to this analysis.[2]

The target is located at (x_{tgt}, y_{tgt}) and is a stationary point target. x_{tgt} and y_{tgt} are referenced to the center of the subarray, and are related to r_{tgt} and θ_{tgt} by

$$x_{tgt} = r_{tgt} \sin \theta_{tgt} \text{ and } y_{tgt} = r_{tgt} \cos \theta_{tgt} \quad (14.2)$$

θ_{tgt} is what we normally think of as the angle to the target. It is the angle to the target relative to the normal to the array.[3]

The nth element of the subarray is located at

$$(x_n, y_n) = \left(\left(n - \frac{N-1}{2} \right) d, 0 \right) \quad n \in [0, N-1] \quad (14.3)$$

relative to array center.

The range from the nth element to the target is

---

[2] These analyses are also applicable to PESAs.
[3] Since the target is located in the far-field region, the rays from each array element to the target are essentially parallel although, for obvious reasons, they are shown differently in Figure 14.7. Thus, the angle θ_{tgt} is used for each element.

$$r_n = \sqrt{(x_n - x_{tgt})^2 + (y_n - y_{tgt})^2}$$

$$= \sqrt{\left(\left[n - \frac{N-1}{2}\right]d - r_{tgt}\sin\theta_{tgt}\right)^2 + (r_{tgt}\cos\theta_{tgt})^2} \qquad (14.4)$$

$$\approx r_{tgt}\sqrt{1 - \frac{(2n+1-N)d\sin\theta_{tgt}}{r_{tgt}}}$$

where we made use of the fact that, in the middle term, $[nd-(N-1)d/2]^2$ is much smaller than the other terms in the square root. Using the far-field assumption, where r_{tgt} becomes very large, (14.4) reduces to

$$r_n = r_{tgt} - \tfrac{1}{2}(2n+1-N)d\sin\theta_{tgt} \qquad (14.5)$$

The one-way time delay from the nth element to the target is

$$\tau_{rn} = \frac{r_n}{c} = \frac{r_{tgt}}{c} - \frac{1}{2c}(2n+1-N)d\sin\theta_{tgt} = \frac{\tau_{rtgt}}{2} - \tau_n \qquad (14.6)$$

In (14.6), $\tau_{rtgt} = 2r_{tgt}/c$ is the standard, two-way range delay to the target, and τ_n is the time-delay difference between the path length from the target to the nth element and the path length from target to array center.

For the next step in the analysis, we need to consider how the transmit signal, $v_T(t)$, travels from the exciter, through the array elements, to the target, and back through the array elements to the matched filter of the receiver. If the array contains N elements, there will be N^2 such paths. On transmit, $v_T(t)$ will be equally split into N parts and sent to each of the N phase shifters and radiating elements. The signal from the n_T^{th} radiating element will propagate a distance of $r_{nT} = r_{tgt} - (2n_T + 1 - N)(d/2)\sin\theta$ to the target. The target will then sum the N signals[4] and reradiate the summed signal back to all N elements of the array. Thus, on receive, each of the radiating elements will contain the sum of N copies of $v_T(t)$ that have been modified by the phase shifters (and amplifiers) of the T/R modules and subjected to delays due to propagation to and from the target.

The outputs of the N radiating elements will then be modified by the phase shifters (and amplifiers and attenuators, etc.) of the T/R modules, summed by the combiner splitter and sent to the matched filter via the receiver. The math that describes the above process is quite involved. As a result, it has been relegated to Appendix A. The end result is that the magnitude of the voltage out of the matched filter is

$$|V_o(\tau)| = \left|\sum_{n_R=0}^{N-1}\sum_{n_T=0}^{N-1} e^{j\psi} MF\left(\tau - \tau_{rtgt} - (n_R + n_T + 1 - N)\tfrac{d}{c}\sin\theta_{tgt}\right)\right| \qquad (14.7)$$

where[5]

---

[4] Actually, strictly speaking, the signals will be summed in space and the summed signal will be absorbed and reradiated by the target (see Chapter 3).

[5] As a note, the $MF(\tau)$ use here is different than the $MF(\tau)$ that we will use in Chapters 18 through 20.

$$\psi = 2\pi\left(n_T + n_R + 1 - N\right)\tfrac{d}{\lambda}\left(\sin\theta_{tgt} - \sin\theta_S\right) \tag{14.8}$$

d is the spacing between elements, $\lambda = c/f_c$ is the wavelength associated with the frequency f_c, θ_{tgt} is the angle to the target relative to array normal and θ_S is the angle to which the beam is steered, relative to array normal. τ is the range delay variable, and $MF(\tau)$ is the matched filter response to $v_T(t)$. If $v_T(t)$ is an LFM pulse, a normalized form of $MF(\tau)$ is, assuming no sidelobe reduction weighting,

$$MF(\tau) = e^{j\pi\tau\tau_p}\frac{\tau_p - |\tau|}{\tau_p}\operatorname{sinc}\left(\alpha\tau\left(\tau_p - \tau\right)\right)\operatorname{rect}\left[\frac{\tau}{2\tau_p}\right] \tag{14.9}$$

$\alpha = B/\tau_p$ is the LFM slope, B is the LFM bandwidth, and τ_p is the uncompressed pulse width. $\operatorname{sinc}(\tau) = \sin(\pi\tau)/(\pi\tau)$ and

$$\operatorname{rect}[x] = \begin{cases} 1 & |x| \leq 1/2 \\ 0 & |x| > 1/2 \end{cases} \tag{14.10}$$

If we assume the transmit and receive beams are steered to the target, i.e., $\theta_S = \theta_{tgt}$, we have $\psi = 0$ and (14.7) reduces to

$$|V_o(\tau)| = \left|\sum_{n_R=0}^{N-1}\sum_{n_T=0}^{N-1} MF\left(\tau - \tau_{rtgt} - \left(n_R + n_T + 1 - N\right)\tfrac{d}{c}\sin\theta_{tgt}\right)\right| \tag{14.11}$$

Each term of the double sum is the matched filter output, appropriately delayed, for a signal radiated from the $n_T{}^{th}$ element and received by the $n_R{}^{th}$ element. The sum accounts the N^2 combinations of transmit and receive elements. The individual terms of the sum are functions of path lengths from the $n_T{}^{th}$ element to the target and back to the $n_R{}^{th}$ element. The presence of N and d in (14.11) is what makes $|V_o(\tau)|$, and thus the shape of the matched filter output, a function of array size. The presence of θ_{tgt} and the fact that the beam is steered to the target are what make the shape of the matched filter output a function of scan (steering) angle. The shape of $MF(\tau)$ is what makes the shape of the matched filter output a function of waveform bandwidth, i.e., waveform range resolution.

We note that if $\theta_{tgt} = \theta_S = 0$, we would have

$$|V_o(\tau)| = \left|\sum_{n_R=0}^{N-1}\sum_{n_T=0}^{N-1} MF\left(\tau - \tau_{rtgt}\right)\right| \tag{14.12}$$

and there would be no distortion of the matched filter output. Also, if $MF(\tau)$ is a broad function of τ (low bandwidth or low resolution), the shifts due to $(n_R + n_T + 1 - N)(d/\lambda)\sin\theta_{tgt}$ would be small relative to the width of $MF(\tau)$. Thus, the distortion of the matched filter output would be almost unnoticeable. Similarly, if d and/or N are small, the shifts due to $(n_R + n_T + 1 - N)(d/\lambda)\sin\theta_{tgt}$ would be small, and thus cause little distortion of the matched filter output.

14.3.2 Subarray Pattern Distortion Examples

To illustrate the potential effect of array size, scan angle, and bandwidth on matched filter distortion, we consider some specific examples.

Suppose we have a C-band radar ($\lambda = 0.06$ m) that uses a $\tau_p = 100$-μs LFM waveform with a bandwidth of 500 MHz and a linear array (subarray) that contains 10 elements with an element spacing of $d = \lambda/2 = 0.03$ m (half wavelength element spacing). This radar has a nominal range resolution of 2 ns [1/(500 MHz)] or 0.3 m. The array length is also 0.3 m ($D = Nd = 10 \times 0.03$ m). Figure 14.8 contains a plot of the center portion of $|V_o(\tau)|/N^2$ for the case where the beam is steered to 60° relative to array normal. As a reference, it also contains a plot of $|MF(\tau)|$ from (14.9). As can be seen, the matched filter output ($|V_o(\tau)|/N^2$) is close to the reference matched filter output ($|MF(\tau)|$). The 4-dB resolution degraded by about 4% from 0.3 m to about 0.314 m, and the matched filter output suffers a SNR loss of about 1.7 dB.

If the array length is doubled to 0.6 m, or twice the waveform range resolution, the degradation is significantly worse, as evidenced by Figure 14.9. In this case, the resolution degrades by 36% to 0.41 m and the SNR loss is about 5.7 dB.

If we use the 0.6-m array length and a scan angle of 30°, the results would look similar to Figure 14.8 (this plot is left as an exercise). However, the resolution and SNR degradation would be a little worse at 5.8% and 2.2 dB.

If we use the 0.6-m array and 60° scan angle, but reduce the waveform bandwidth to 250 MHz, we would again get a plot that looks like Figure 14.8. In this case, the degradation in resolution is 3.8% and the SNR loss is again 1.7 dB.

If we were to change the pulse width, τ_p, we would find it has almost no effect on performance. In general, the performance is independent of pulse width unless very small pulse widths are used (on the order of ns).

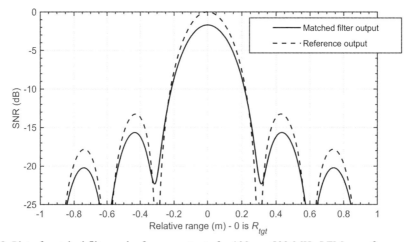

Figure 14.8 Plot of matched filter and reference outputs for 100-μs, 500-MHz LFM waveform, 10-element, C-band, array with $d = \lambda/2$ element spacing ($D = 0.3$ m), beam steered to 60°.

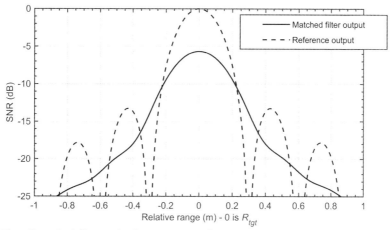

Figure 14.9 Plot of matched filter and reference outputs for 100-μs, 500-MHz LFM waveform, 20-element, C-band, array with $d = \lambda/2$ element spacing ($D = 0.6$ m), beam steered to 60°.

14.3.3 Array Beam Forming with TDUs

The final beam formation and steering of the array (see Figure 14.5) are accomplished by the TDUs connected to each subarray. The delays of the TDUs are selected to cancel the time delays to and from the target. This is illustrated in Figure 14.10. In that figure, we replaced each of the M subarrays of Figure 14.5 by a "super" element. All of the super elements have the radiation patterns of the subarray they replace and can be steered since the subarray pattern can be steered. Although not necessary for this analysis or in practice, we assume all of the subarray (and super element) radiation patterns are the same.

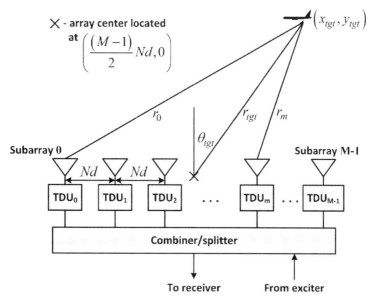

Figure 14.10 Array configuration of Figure 14.5 used to analyze how TDUs are used to steer the array radiation pattern.

The super elements are spaced Nd m apart, where N is the number of elements in the subarrays and d is the spacing between elements of the subarrays. Although not necessary for this analysis, we assume that all the subarrays are the same size.

The mth super element is located at

$$(x_m, y_m) = \left(m(Nd) - \frac{M-1}{2}(Nd), 0 \right) \quad m \in [0, M-1] \tag{14.13}$$

relative to array center. The distance from the mth super element to the target is

$$r_m = \sqrt{(x_m - x_{tgt})^2 + (y_m - y_{tgt})^2} \tag{14.14}$$

We note that (14.13) and (14.14) are the same form as (14.3) and (14.4)

If we repeat the math that led to (14.5), we get

$$r_m = r_{tgt} - \tfrac{1}{2}(2m+1-M)(Nd)\sin\theta_{tgt} \tag{14.15}$$

Thus, the one-way range delay from the mth super element (center of the mth subarray) to the target is

$$\tau_{rm} = \frac{r_m}{c} = \frac{r_{tgt}}{c} - \frac{1}{2c}(2m+1-M)(Nd)\sin\theta_{tgt} = \frac{\tau_{rtgt}}{2} - \tau_m \tag{14.16}$$

The time delay of the mth TDU is selected to remove this time delay. That is

$$\tau_{TDUm} = -\tau_{rm} = -\frac{\tau_{rtgt}}{2} + \tau_m \tag{14.17}$$

The analysis of the Figure 14.10 array is basically the same as the subarray analysis presented in Appendix A, with the phase shifters replaced by TDUs. In this case, the mth transmit signal out of the combiner/splitter and into the TDU would be [see (14A.1)]

$$v_{Tm}(t) = e^{j2\pi f_c t} p(t) \tag{14.18}$$

where $p(t)$ is defined in Appendix A.

The voltage out of the TDU, and into the super element would be

$$v_{radmT}(t) = v_T(t + \tau_{TDUm}) = e^{j2\pi f_c(t+\tau_{TDUm})} p(t + \tau_{TDUm}) \tag{14.19}$$

In other words, rather than acquire only a phase shift of $\phi_m (= 2\pi f_c \tau_{TDUm})$ from a phase shifter, the signal out of the TDU acquires a time advance (a negative time delay).[6] The time advance is, ideally, used to cancel the τ_m component of (14.16); that is, the component of range delay caused by the offset between array center and the mth super element (center of the mth subarray). Once all of the τ_m have been offset, the effective

---

[6] We note that in (14.19) and other equations, we are allowing negative time delays, or time advances. In terms of math, this is acceptable because "time is relative." In practice, the time advances would be offset by a time delay bias that is applied to all TDUs. The time delay bias would be removed in the computer when the output of the matched filter is processed to determine range.

range delay from all super elements to the target will be the same, and equal to $\tau_{rtgt}/2 = r_{tgt}/c$.

Similar to the way phase shifters are set in PESAs, the time advances of the TDUs are based on a desired scan angle, θ_S. The equation for the TDU time advance has the same form as the equation for τ_m [see (14.16)], since it ostensibly cancels τ_m. Thus,

$$\tau_{TDUm} = -\frac{1}{2c}(2m+1-M)(Nd)\sin\theta_S \quad (14.20)$$

If we repeat the analysis of Appendix A, we find that (14A.10) would become

$$V_{MF}(t) = \sum_{m_R=0}^{M-1}\sum_{m_T=0}^{M-1} e^{-j2\pi f_c(\tau_{rmT}+\tau_{rmR})} e^{j(\phi_{mT}+\phi_{mR})} \\ \times p(t + \tau_{TDUmT} - \tau_{rmT} + \tau_{TDUmR} - \tau_{rmR}) \quad (14.21)$$

where

$$\phi_{mT} = \frac{\pi}{\lambda}(2m_T + 1 - M)(Nd)\sin\theta_S \quad (14.22)$$

and

$$\phi_{mR} = \frac{\pi}{\lambda}(2m_R + 1 - M)(Nd)\sin\theta_S \quad (14.23)$$

We note that the main difference between (14A.10) and (14.21) is that $p(t)$ is advanced by $\tau_{TDUmT} + \tau_{TDUmR}$.

If we continue to follow the development steps of Appendix A, we arrive at the equation [see (14.7)]

$$|V_o(\tau)| = \left|\sum_{m_R=0}^{M-1}\sum_{m_T=0}^{M-1} e^{j\psi} MF\left(\tau - \tau_{rtgt} - (m_R + m_T + 1 - M)\frac{Nd}{c}(\sin\theta_{tgt} - \sin\theta_S)\right)\right| \quad (14.24)$$

where

$$\psi = -2\pi(m_T + m_R + 1 - M)\frac{Nd}{\lambda}(\sin\theta_{tgt} - \sin\theta_S) \quad (14.25)$$

If we select $\theta_S = \theta_{tgt}$ (14.24) reduces to

$$|V_o(\tau)| = \left|\sum_{m_R=0}^{M-1}\sum_{m_T=0}^{M-1} MF(\tau - \tau_{rtgt})\right| \quad (14.26)$$

which tells us there will be no distortion of the matched filter output.

A question related to the discussion above is: What happens if the subarray beam cannot be steered directly to the target? That is, what happens if $\theta_S \neq \theta_{tgt}$? The answer is that there can be considerable error in selecting θ_S. An example of this is illustrated in Figure 14.11. The plot was generated for a C-band radar ($\lambda = 0.06$ m) that uses a $\tau_p = 100$-

μs LFM waveform with a bandwidth of 500 MHz and a linear array that contains 10 super elements (10 subarrays) (the same parameters as used to generate Figure 14.8). In this case, the super element spacing is set to $10 \times d = 10 \times \lambda/2 = 0.3$ m (the width of the subarray used to generate Figure 14.8). The target angle was set to 60° and the subarray (super element) beam steering angle was set to 55°. The solid curve is a plot of the center portion of $|V_o(\tau)|/N^2$ and the dashed curve is a plot of the $|MF(\tau)|$ of (14.9). The subarray beam steering error caused only a small degradation in resolution (about 0.8%) and a SNR loss of only 0.5 dB. As a note, the beam width of this array is about 1°, which means the subarray beamsteering error is five beam widths, which is quite large.

The overall conclusion of this section is that, for scan and target angles of 60°, if the subarray size is approximately equal to the range resolution of the wideband waveform, and time delay steering is used at the array level (as shown in Figure 14.5), the array will cause minimal distortion of the matched filter output. If the scan and target angle is reduced, the subarrays can be larger and still cause minimal distortion in the matched filter output. If the scan, and target, angle is reduced to zero, there will be no distortion of the matched filter output for all (reasonable) array sizes.

As an interesting side note, parabolic reflector antennas will cause very little distortion in matched filter output, even if the antenna is much larger than the waveform range resolution. This is because the path lengths from the feed to the target, via the reflector, are all the same (see Section 13.25). This assumes the target is in the main beam of the antenna. Space-fed phased arrays will always experience distortion of the matched filter output, unless the array is small relative to the waveform resolution. This is because of the difference in path lengths from the feed to the back of the array and from the array elements to the target.

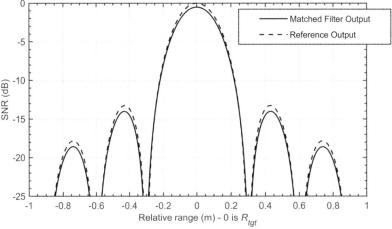

Figure 14.11 Plot of matched filter and reference outputs for 100-μs, 500-MHz LFM waveform, 10-element, C-band, array with $d = 10 \times \lambda/2$ element spacing ($D = 3$ m), beam steered to 60°.

14.4 SIMULTANEOUS MULTIPLE BEAMS

Another capability facilitated by the use of an AESA is the use of simultaneous multiple receive beams [4, 5]. To implement this capability, a separate receiver and signal processor would be connected to each subarray. The outputs of the signal processors would then be sent to a beamforming data processor, which would do the final beam formation [22, 34–36]. This is illustrated in Figure 14.12.

For now, we assume all the subarrays are the same size. We will consider different size subarrays later. We also assume the subarray size is such that we do not need to be concerned with matched filter output distortion at the subarray level when we use wide bandwidth waveforms. We already learned that if the TDUs are set to steer the overall antenna radiation pattern close to the target, the TDUs will prevent matched filter output distortion for the overall array. Thus, we do not need to consider the time delays of the TDUs, or the time delays to and from the target, in the following analyses.

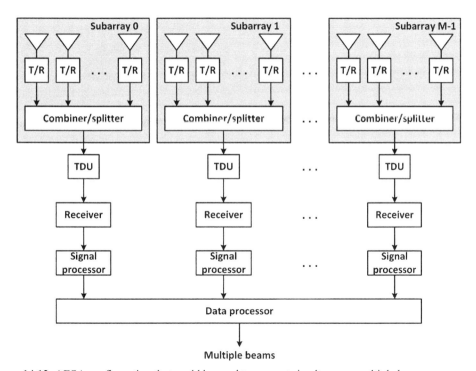

Figure 14.12 AESA configuration that could be used to support simultaneous multiple beams.

We assume the subarray and overall antenna radiation patterns are steered to the same, nominal angle, θ_S, which is close to the target angle, θ_{tgt}. The phase shifters of the subarrays are used to steer the subarray radiation patterns, and the TDUs are used to steer the overall antenna pattern. This means we set the T/R module phase shifters of all subarrays to

$$\phi_n = 2\pi n \frac{d}{\lambda} \sin\theta_S = 2\pi n \frac{d}{\lambda} u_S \quad n \in [0, N-1]^{\,7} \tag{14.27}$$

and the time delays of the TDUs to

$$\tau_{TDUm} = \frac{1}{2c}(2m+1-M)(Nd)u_S \tag{14.28}$$

Recall that the TDUs not only provide a time delay (advance) of τ_{TDUm} but also a phase shift of

$$\phi_{TDUm} = \pi(2m+1-M)\frac{Nd}{\lambda}u_S \quad m \in [0, M-1] \tag{14.29}$$

If we assume a point source located in the antenna far field at an sine-space angle of u (real angle of θ), we can write the output of the mth signal processor as

$$\begin{aligned}V_m(u) &= V_{sub}(u) e^{-j2\pi m \frac{Nd}{\lambda} u} e^{j\phi_{TDUm}} \\ &= V_{sub}(u) e^{-j2\pi m \frac{Nd}{\lambda} u} e^{-j2\pi m \frac{Nd}{\lambda} u_S} e^{-j\pi(1-M)\frac{Nd}{\lambda} u_S}\end{aligned} \tag{14.30}$$

where

$$V_{sub}(u) = \sum_{n=0}^{N-1} a_n e^{-j2\pi n \frac{d}{\lambda}(u-u_S)} \tag{14.31}$$

is the array factor (see Chapter 13) of the subarray. For now, we assume uniform weighting and let $a_n = 1/N$. The ϕ_{TDUm} term of (14.30) is the phase shift introduced by the mth TDU.

If we assume the 10-element subarray of the previous section, we get, with $a_n = 1/N$, (see Chapter 13)

$$V_{sub}(u) = \frac{1}{\sqrt{N}} e^{-j2\pi(N-1)\frac{d}{\lambda}(u-u_S)} \frac{\sin\left(2\pi N \frac{d}{\lambda}(u-u_S)\right)}{\sin\left(2\pi \frac{d}{\lambda}(u-u_S)\right)} \tag{14.32}$$

The normalized radiation intensity pattern (see Chapter 13) of the subarray is

$$R_{sub}(u) = |V_{sub}(u)|^2 = \frac{1}{N}\left(\frac{\sin\left(2\pi N \frac{d}{\lambda}(u-u_S)\right)}{\sin\left(2\pi \frac{d}{\lambda}(u-u_S)\right)}\right)^2 \tag{14.33}$$

---

[7] We have changed notation from angle space (i.e., θ) to sine-space, $u = \sin\theta$. We did this for notational convenience. See Chapter 13 for a definition of sine-space.

The beam former, which is in the beamformer data processor, forms the overall radiation pattern by forming a weighted sum of the $V_m(u)$. Specifically,

$$V_{array}(u) = \sum_{m=0}^{M-1} b_m V_m(u)$$
$$= e^{-j\pi(1-M)\frac{Nd}{\lambda}u_S} V_{sub}(u) \sum_{m=0}^{M-1} b_m e^{-j2\pi m \frac{Nd}{\lambda}(u-u_S)} \quad (14.34)$$
$$= e^{-j\pi(1-M)\frac{Nd}{\lambda}u_S} V_{sub}(u) A(u)$$

$A(u)$ is the array factor of the array discussed in Chapter 13.

The beamformer weights are of the form

$$b_m = |b_m| e^{j2\pi m \frac{Nd}{\lambda}\Delta u} \quad (14.35)$$

where Δu is the desired beam offset angle, relative to u_S. To form multiple beams, we would use multiple values of Δu. If we use $\Delta u = 0$, the beam will be steered to the nominal angle of u_S. With (14.35), we can write $A(u)$ as

$$A(u) = \sum_{m=0}^{M-1} b_m e^{-j2\pi m \frac{Nd}{\lambda}(u-u_S-\Delta u)} \quad (14.36)$$

If we use $|b_m| = (1/M)^{1/2}$, $A(u)$ has the same form as (14.31), which means we can write it as

$$A(u) = \frac{1}{\sqrt{M}} e^{-j2\pi(M-1)\frac{d}{\lambda}(u-u_S-\Delta u)} \frac{\sin\left(2\pi M \frac{Nd}{\lambda}(u-u_S-\Delta u)\right)}{\sin\left(2\pi \frac{Nd}{\lambda}(u-u_S-\Delta u)\right)} \quad (14.37)$$

The associated normalized radiation intensity pattern will be

$$R(u) = |A(u)|^2 = \frac{1}{M}\left(\frac{\sin\left(2\pi M \frac{Nd}{\lambda}(u-u_S-\Delta u)\right)}{\sin\left(2\pi \frac{Nd}{\lambda}(u-u_S-\Delta u)\right)}\right)^2 \quad (14.38)$$

When we combine (14.38) with (14.34), we get a normalized radiation intensity pattern for the entire array, with the beam steered to $u_S - \Delta u$, of

$$R_{array}(u) = |V_{array}(u)|^2 = R_{sub}(u) R(u)$$
$$= \frac{1}{MN}\left(\frac{\sin\left(2\pi N \frac{d}{\lambda}(u-u_S)\right)}{\sin\left(2\pi \frac{d}{\lambda}(u-u_S)\right)}\right)^2 \left(\frac{\sin\left(2\pi M \frac{Nd}{\lambda}(u-u_S-\Delta u)\right)}{\sin\left(2\pi \frac{Nd}{\lambda}(u-u_S-\Delta u)\right)}\right)^2 \quad (14.39)$$

We note that, if $\Delta u = 0$, $R_{array}(u)$ reduces to

$$R_{array}(u) = \frac{1}{MN}\left(\frac{\sin\left(2\pi \frac{MNd}{\lambda}(u-u_S)\right)}{\sin\left(2\pi \frac{d}{\lambda}(u-u_S)\right)}\right)^2 \quad (14.40)$$

which is the normalized radiation intensity pattern of an *MN*-element, uniformly illuminated linear array (see Chapter 13).

Figure 14.13 contains plots of the subarray pattern, $R_{sub}(u)$, (actually, the plots are $R_{sub}(\theta)$) the array factor pattern, $R(u)$, and the overall array pattern, $R_{array}(u)$, for the case where $\Delta u = 0$. That is, there is no beam offset. Because of the spacing between subarray centers (spacing between super radiators), the array factor pattern (dotted curve) has grating lobes. The grating lobes are spaced $\lambda/D = 0.2$ sines, or 11.54° for the first grating lobe (see Section 13.10.7). The other grating lobes are spaced farther apart in angle because of the nonlinear relation between sine space and angle space. The first null of the subarray pattern (the narrow-line curve) is also at 0.2 sines, and the other nulls are spaced by 0.2 sines. Thus, the grating lobes of the array factor fall in the nulls of the subarray pattern, and are canceled. Because of this, the overall array pattern does not have grating lobes (when the beam is steered to $u = \theta = 0$).

Figure 14.14 contains the same plots as Figure 14.13, but with the beam steered to $\Delta\theta = 2.3°$, or about twice the one-way beamwidth of $R_{array}(u)$. In this case, the grating lobes of $R(u)$ are not canceled by the nulls of $R_{sub}(u)$, and show up as grating lobes in the plot of $R_{array}(u)$.

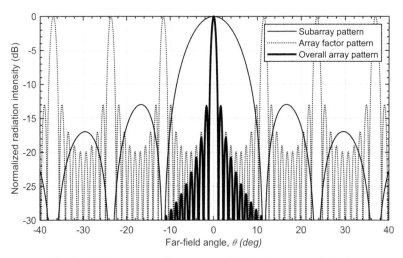

Figure 14.13 Normalized radiation patterns for an array with 10 subarrays and 10 elements per subarray—½ wavelength spacing, $\Delta\theta = 0$.

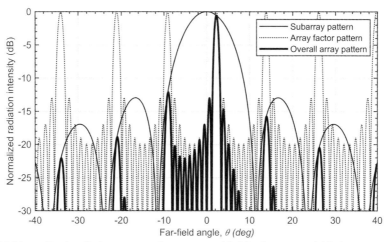

Figure 14.14 Normalized radiation patterns for an array with 10 subarrays and 10 elements per subarray—½ wavelength spacing, $\Delta\theta = 2.3°$ (two beam widths).

As indicated earlier, one way of mitigating grating lobes closest to the main beam is to use subarrays that are spaced closer together. However, to maintain the same overall array size, the size of the subarrays must be decreased and the number of subarrays must be increased. Decreasing the size of the subarrays might be desirable because it would allow the use of wider bandwidth waveforms. However, increasing the number of subarrays would mean the number of TDUs would need to be increased. This could be problematic from a packaging and cost perspective.

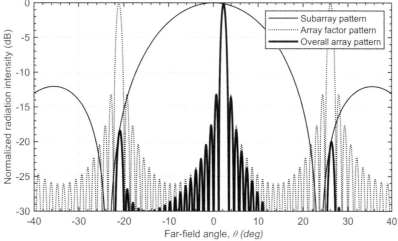

Figure 14.15 Normalized radiation patterns for an array with 20 subarrays and five elements per subarray—½ wavelength spacing, $\Delta\theta = 2.3°$ (two beamwidths).

Figure 14.15 contains plots like those in Figure 14.14, but with 20 subarrays of five elements each. While the wider subarray pattern spacing did not eliminate the grating lobes, it did move them farther apart. Also, the $R_{sub}(u)$ is smaller at the locations of the grating lobes, so the amplitudes of the grating lobes in Figure 14.15 are smaller than those in Figure 14.14.

Figures 14.14 and 14.15 contain radiation patterns for a single beam. Figures 14.16 and 14.17 contain plots like these two figures for seven simultaneous beams. In both plots, the beams were placed at 0°, and offsets of plus and minus one, two- and three beamwidths. The thin, solid curve is the subarray pattern in both cases. Also, we expanded the horizontal scale to show the multiple beams more clearly.

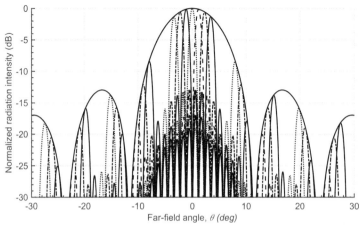

Figure 14.16 Normalized radiation patterns for an array with 10 subarrays and 10 elements per subarray—½ wavelength spacing, $\Delta\theta = 0°, \pm 1.15°, \pm 2.3°, \pm 3.45°$.

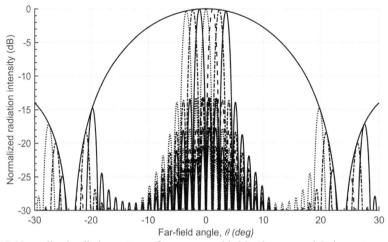

Figure 14.17 Normalized radiation patterns for an array with 20 subarrays and 5 elements per subarray—½ wavelength spacing, $\Delta\theta = 0°, \pm 1.15°, \pm 2.3°, \pm 3.45°$.

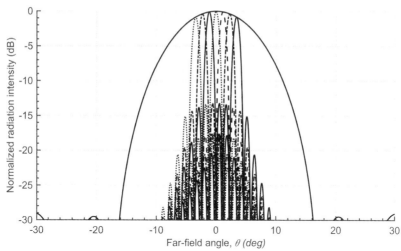

Figure 14.18 Normalized radiation patterns for an array with 10 subarrays and 10 elements per subarray – ½ wavelength spacing, $\Delta\theta = 0°, \pm 1.15°, \pm 2.3°, \pm 3.45°$, 30 dB, $\bar{n} = 6$ Taylor weighting on the subarrays.

The effect of grating lobes on the overall radiation pattern is much worse for the antenna with 10 subarrays of 10 elements each. In that case, the grating lobes for the beams steered to ±3 beamwidths are only about 8 dB below the seven main beams. The situation for the other antenna (20 subarrays of five elements each) is better. However, the grating lobes for the beams steered to ±3 beamwidths are still somewhat large.

It would be possible to suppress the grating lobes beyond about ±10° for the 10-element/10-subarray antenna by using amplitude weighting on the elements of the subarrays. This option is illustrated in Figure 14.18. In this case, we used a 30 dB, $\bar{n} = 6$, Taylor weight to the subarrays. While this suppressed the grating lobes past about 15°, it aggravated the situation for the closer-in grating lobes.

14.4.1 Overlapped Subarrays

If the subarray radiation pattern of Figure 14.18 could be used in place of the subarray pattern of Figure 14.17, the grating lobes of Figure 14.17 would be suppressed by the subarray pattern of Figure 14.18. Unfortunately, the smaller spacing between subarrays for the 20-subarray five-element case does not allow enough room to use a 10-element subarray, unless the spacing between elements is significantly reduced. Also, since the width of the main lobe of the subarray pattern is determined by the physical size of the subarray, the subarray main lobe width of Figures 14.14 and 14.16 will be the same.

To achieve the desired effect of increasing the spacing between grating lobes, and being able to suppress them with the radiation pattern of the subarrays, the distance between subarrays needs to be decreased, but the size of the subarrays needs to stay the same. This is where the concept of *overlapped subarrays* comes in [37–43]. With overlapped subarrays, the subarray combiners combine the T/R module outputs from more than one subarray. Stated another way, the output of each T/R module is split and sent to more than one subarray combiner. This concept is illustrated in Figure 14.19, which is an 18-element array with three subarrays that have six elements each. The outputs of the three subarrays go to subarray combiners 1, 3, and 5, which is the standard arrangement. However, half the outputs of each of the three subarrays also go to subarray combiners. Thus, each of the five subarray combiners is connected to six T/R modules. Because of this, the array effectively has five subarrays with six elements per subarray. However, because of the overlap, the phase centers of the five effective subarrays are separated by half of the subarray length. This closer-phase center spacing is what moves the grating lobes farther away from the main beam.

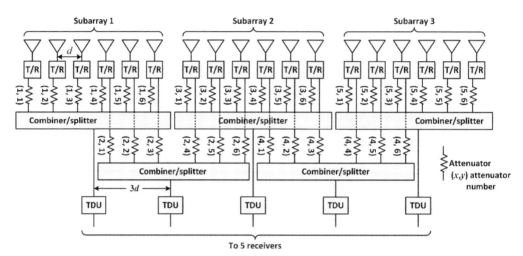

Figure 14.19 Example of an overlapped subarray configuration—2-to-1 overlap.

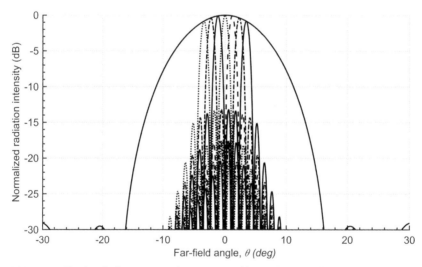

Figure 14.20 Normalized radiation patterns for an array with overlapped subarrays. Nineteen subarrays and 10 effective elements per subarray—½ wavelength spacing, $\Delta\theta = 0°, \pm1.15°, \pm2.3°, \pm3.45°$, 30 dB, $\bar{n} = 6$ Taylor weighting on the subarrays.

To enable implementation of appropriate weighting on each of the subarrays, the two outputs of each T/R module must have different attenuators to control the amplitude taper of the five effective subarrays. Referring to Figure 14.19, the amplitude tapers of the three physical subarrays are controlled by attenuators (1,1) through (1,6), (3,1) through (3,6), and (5,1) through (5,6). Attenuators (2,1) through (2,6) and (4,1) through (4,6) apply the amplitude tapers to the effective subarrays made up of the halves of physical subarrays 1 and 2 and subarrays 2 and 3, respectively. The addition of extra attenuators in the T/R modules and the use of five instead of three subarray combiners is a complication associated with overlapped subarrays.

Figure 14.20 contains a plot like Figure 14.18 for the 100-element array of the examples of this chapter. In this case, we used 19 overlapped subarrays with 10 effective elements per subarray. We also used the 30 dB, $\bar{n} = 6$ Taylor weighting to generate Figure 14.20. In this case, the grating lobes have been suppressed below the 30 dB lower limit of the plot.

In the combined radiation pattern of Figure 14.20, each of the steered beams suffers a small decrease in amplitude due to the subarray pattern. Had we included more beams farther away from θ=0, the decrease in amplitude for those beams would have been larger.

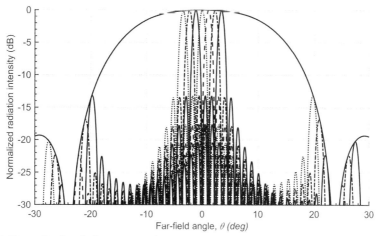

Figure 14.21 Normalized radiation patterns for an array with overlapped subarrays. 19 subarrays and 10 effective elements per subarray – ½ wavelength spacing, $\Delta\theta = 0°, \pm1.15°, \pm2.3°, \pm3.45°$, sinc weighting on the subarrays.

In an attempt to avoid the decrease in beam amplitude, several authors propose using a sinc-type of weighting on the subarrays [39, 41, 44]. Given that the array factor of an array is the discrete Fourier transform of the amplitude weighting function (see Chapter 13), a perfect sinc weighting will, ideally, produce a rectangular beam. However, a perfect sinc weighting would require an infinite number of elements in the subarray. Thus, we must use an approximate sinc weighting, which will produce an approximate rectangular beam. Figure 14.21 contains a plot like Figure 14.20 with the weighting

$$a_n = \text{sinc}\left(\frac{n-(N+1)/2}{N/2.75}\right) \quad n \in [1, N] \quad (14.41)$$

instead of the 30 dB, $\bar{n}=6$ Taylor weighting. In this case, the amplitudes of the seven beams are very close to the same value. However, the grating lobes have reappeared.

The overlapped subarray technique illustrated in Figure 14.19 is said to use 2-to-1 overlap because each physical subarray is connected to two subarray combiners. Had we changed the configuration so that each physical subarray was connected to three subarray combiners, the overlapped subarray configuration would use 3-to-1 overlap. An example of this is illustrated in Figure 14.22. The effect of using 3-to-1 overlap is to move the phase centers of the effective subarrays even closer than in the 2-to-1 overlap case. This would move the grating lobes farther apart. While this would not help much in the case where Taylor weighting was used, it may remove the close-in grating lobes of Figure 14.21, where sinc weighting was used. Examination of this is left as an exercise.

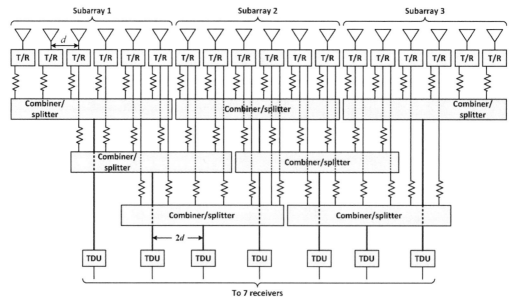

Figure 14.22 Example of an overlapped subarray configuration with 3-to-1 overlap.

14.4.2 Nonuniform Subarray Sizes

Another method of mitigating grating lobes that is suggested by some authors is the use of non-uniform subarray sizes [45, 46]. If the subarrays of the antenna are different sizes, the phase centers of the subarrays will have different spacings instead of the regular, periodic spacing that occurs when the subarrays are the same size. The result of this is that the regular, periodic nature of the radiation pattern of the array will be replaced by a radiation pattern that is somewhat erratic-looking. More importantly, if the subarray sizes are selected properly, there will be no predominant sidelobes; that is, no grating lobes. Instead, the normal sidelobes and grating lobe structure will be replaced by an erratic looking sidelobe structure.

An example of a linear array with nonuniform subarray sizes is illustrated in Figure 14.23. The figure is the normalized radiation intensity pattern of a 100-element linear array (with uniform weighting) where we replaced the ten 10-element subarrays with subarrays that had 16, 12, 10, 7, 5, 5, 7, 10, 12, and 16 elements. The beam is steered to $\Delta\theta = 2.3°$, or about twice the beam width of the array. Note that the grating lobes present in Figure 14.14 do not appear in Figure 14.23. However, the sidelobes of the nonuniform array are somewhat high. As a reminder, Figure 14.14 is an equivalent plot for a 10 subarray with 10 elements per subarray.

Figure 14.24 is a plot like Figures 14.16, 14.17, and 14.20 for the array with nonuniform subarray sizes. Again, the grating lobes are not present, but the sidelobe levels are somewhat high.

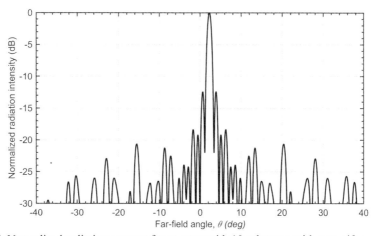

Figure 14.23 Normalized radiation patterns for an array with 10 subarrays with nonuniform subarray sizes—½ wavelength spacing, beam steered to 2.3° (two beam widths off of array normal)—subarray sizes of 16, 12, 10, 7, 5, 5, 7, 10, 12, and 16 elements, uniform weighting.

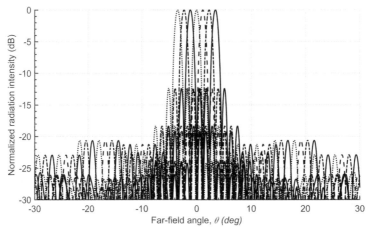

Figure 14.24 Normalized radiation patterns for an array with 10 subarrays with non-uniform subarray sizes – ½ wavelength spacing, $\Delta\theta = 0°, \pm 1.15°, \pm 2.3°, \pm 3.45°$ – subarray sizes of 16, 12, 10, 7, 5, 5, 7, 10, 12, and 16 elements, uniform weighting.

14.4.3 Transmit Array Considerations

If the radar uses multiple beams on receive, the transmit array must generate a beam whose width is the combined width of the multiple receive beams. In the examples of the previous sections, the transmit beam would need to be seven beam widths, or 8.05° wide.

Forming a wide transmit beam cannot, with today's technology, be accomplished with amplitude weighting because the T/R modules of the AESA usually operate at full power to maximize efficiency. This leaves only phase weighting. If the array uses only narrow bandwidth waveforms, it does not need to use time-delay steering. In this case, the phase weighting can be applied at the element level. For the case where the array must use time-delay steering, the phase weighting must be applied at the subarray level.

One method that can be used to form a wide transmit is to apply a quadratic phase taper across the array. The idea that such a phase taper will widen the beam is based on the fact that the Fourier transform of a signal with a quadratic phase taper, or LFM, has very close to a rectangular shape (see, for example Section 12.1). Figure 14.25 contains the plot of Figure 14.20 along with a plot of a normalized transmit radiation intensity pattern with a quadratic phase taper applied to the elements (T/R module phase shifters) of the array. While the shape of the transmit pattern is not as good as desired, and contains considerable amplitude variation, it is wide enough to encompass the seven receive beams. It could be that if the phase taper was changed somewhat, the shape of the pattern could be improved. The specific phase taper used to generate transmit pattern of Figure 14.25 is

$$\phi_{quad} = e^{j0.0035x^2} \quad x = n - (100+1)/2 \quad n \in [1,100] \qquad (14.42)$$

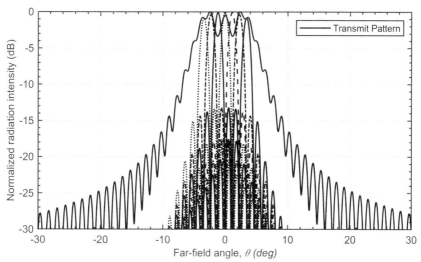

Figure 14.25 Plot of Figure 14.20 with a transmit pattern that uses the quadratic phase taper of (14.42) across the 100 T/R modules of the array.

Figure 14.26 contains the plot of Figure 14.20 along with a plot of the transmit radiation intensity pattern with a quadratic phase taper applied to the 19 subarrays, instead of the overall array. In this case, the transmit pattern is considerably wider than desired. It could be that using something other than a quadratic phase taper would work better. However, that was not pursued. The specific phase taper used to generate the transmit pattern of Figure 14.26 is

$$\phi_{quad} = e^{j0.0035x^2} \quad x = m - (19+1)/2 \quad n \in [1,19] \tag{14.43}$$

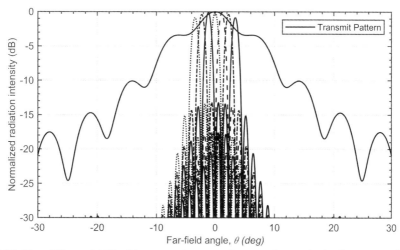

Figure 14.26 Plot of Figure 14.20 with a transmit pattern that uses the quadratic phase taper of (14.43) (14.42) across the 19 subarrays of the array.

14.5 AESA NOISE FIGURE

In radars that use reflector antennas or PESAs, we represent the antenna as a loss, L_{ant} (see Chapter 2 of [1]). As such, its noise figure is $F_{ant} = L_{ant}$ and its signal gain (not directive gain) is $G_{ant} = 1/L_{ant}$. Because the T/R modules are part of the AESA and the LNA is part of a T/R module, we cannot use this simple relation for AESAs. Instead, we need to represent the AESA as an active device that has a gain (signal gain) and noise figure. That means we need a means of computing the noise figure and gain of an AESA [5, 47–53].

Because the outputs of the T/R modules are combined in a parallel fashion—instead of a series, or cascade fashion—we cannot use the single-path cascade noise figure analysis approach discussed in Chapter 4. Instead, we need to develop a parallel cascade algorithm that is applicable to AESAs. Further, the algorithm must account for the fact that AESAs can use subarrays. To that end, we develop an equation for the noise figure and gain of a general subarray, and then use that result to develop an equation for the noise figure and gain of the overall AESA.

14.5.1 T/R Module Noise Figure

We assume the LNA of the nth T/R module of the subarray has a gain of G_{LNAn} and a noise figure of F_{LNAn}. If we include the (ohmic) loss in the radiating element, the circulator, and any other devices before the LNA (see Figure 14.2), we get a gain and noise figure through the LNA, referenced to the input to the radiating element, as

$$G_{LNA\&Front\_n} = \frac{G_{LNAn}}{L_{before\_n}} \qquad (14.44)$$

and

$$F_{LNA\&Front\_n} = F_{LNAn} L_{before\_n} \qquad (14.45)$$

where L_{before_n} is the product of the losses of all components from the radiating element up to the input to the LNA.

To include the components after the LNA and up to the T/R module output, we need to perform the cascaded device noise figure calculation discussed in Section 4.5. If we do that, we get an overall T/R module noise figure of

$$F_{TRn} = F_{LNAn} L_{before\_n} + \frac{L_{after\_n} - 1}{G_{LNAn} / L_{before\_n}} \qquad (14.46)$$

and an overall T/R module gain of

$$G_{TRn} = \frac{G_{LNAn}}{L_{before\_n} L_{after\_n}} \qquad (14.47)$$

In (14.46) and (14.47),

$$L_{after\_n} = L_{attn\_n} L_{other\_n} \qquad (14.48)$$

where L_{attn_n} is the attenuator setting and L_{other_n} is the losses in the other devices between the LNA output and the T/R module output (e.g., phase shifter, circulator, transmission lines, etc.).[8]

14.5.2 Subarray Gain and Noise Figure

As indicated in the second paragraph of this subsection, we cannot use the cascaded devices approach to finding the gain and noise figure of the overall subarray. Instead, we must return to the definition of noise figure (see Section 4.4.1) to find the subarray noise figure. From (4.18), we have that

$$F_{subarray} = \frac{E_{n\_out\_actual\,subarray}}{E_{n\_out\_ideal\,subarray}} \tag{14.49}$$

where $E_{n\_out\_actual\,subarray}$ is the noise energy (noise power spectral density) out of the actual subarray and $E_{n\_out\_ideal\,subarray}$ is the noise energy out of an ideal, or "noiseless," subarray (see Section 4.4.1).

From Section 4.4.1, the noise energies out of the actual and ideal T/R module are

$$\begin{aligned}E_{n\_out\_actual\,n} &= G_{TRn}E_{TRn\_in} + E_{TRn\_int} = G_{TRn}kT_0 + G_{TRn}kT_0(F_{TRn}-1) \\ &= G_{TRn}F_{TRn}kT_0\end{aligned} \tag{14.50}$$

and

$$E_{n\_out\_ideal\,n} = G_{TRn}E_{TRn\_in} = G_{TRn}kT_0 \tag{14.51}$$

In (14.50) and (14.51), we used the fact that the noise energy into the nth radiating element, and thus the T/R module, is $E_{TRn\_in} = kT_0$. We use T_0 here since we are computing noise figure (see Chapter 4).

We assume the noise signals into the T/R modules are independent, as are the internally generated noise signals. Because of this, the ideal and actual noise signals out of the various T/R modules are independent.

We can represent the combiner/splitter as consisting of an ideal, lossless summer followed by a lossy device that captures the combiner/splitter losses. Since the summer is treated as lossless, it does not generate internal noise. It only sums the input noises. Thus, the noise energy out of the summer is the sum of the noise energies out of the T/R modules. For the case of an ideal (i.e., lossless) T/R module, we have that the noise energy out of the summer is

$$E_{n\_out\_ideal\,\Sigma} = \sum_{n=1}^{N} E_{n\_out\_ideal\,n} = kT_0 \sum_{n=1}^{N} G_{TRn} \tag{14.52}$$

The actual noise energy out of the summer is

---

[8] If the T/R module contains an amplifier in addition to the attenuator, phase shifter, and circulator, it would need to be included in the cascaded analysis.

$$E_{n\_out\_actual\,\Sigma} = \sum_{n=1}^{N} E_{n\_out\_actual\,n} = kT_0 \sum_{n=1}^{N} G_{TRn} F_{TRn} \qquad (14.53)$$

Thus, the noise figure of the subarray from the inputs to all T/R modules up to the output of the ideal summer is

$$F_\Sigma = \frac{E_{n\_out\_actual\,\Sigma}}{E_{n\_out\_ideal\,\Sigma}} = \frac{\sum_{n=1}^{N} G_{TRn} F_{TRn}}{\sum_{n=1}^{N} G_{TRn}} \qquad (14.54)$$

That is, the noise figure of the AESA, up to the output of the ideal summer, is a weighted average of the noise figures of the T/R modules, where the weighting is the net gains of the T/R modules, up to the input to the combiner/splitter.

Since the noise energy into each of the T/R modules (actually, into each of the radiators of the T/R modules) is kT_0,[9] the total noise energy into the subarray is

$$E_{n\_in\,\Sigma} = NkT_0 \qquad (14.55)$$

The noise out of the ideal T/R modules and summer is $E_{n\_out\_ideal\,\Sigma}$. Thus, the net gain of the subarray up to the output of the summer is

$$G_\Sigma = \frac{E_{n\_out\_ideal\,\Sigma}}{E_{n\_in\,\Sigma}} = \frac{\sum_{n=1}^{N} G_{TRn}}{N} \qquad (14.56)$$

That is, the net gain of the subarray, up to the output of the ideal summer, is the average of net gains of all of the T/R modules.

To arrive at the overall noise figure and gain of the subarray, we need to account for the combiner/splitter loss, a buffer amplifier before the TDU, the actual TDU, a buffer amplifier after the TDU, and any other losses before what is designated at the subarray output. We can use the cascade noise figure technique of Chapter 4 to write the combined gain and noise figure of these devices as [see equation (4.41)]

$$F_{after} = L_{comb} + \frac{F_{B1}-1}{1/L_{comb}} + \frac{L_{TDU}-1}{G_{B1}/L_{comb}} + \frac{F_{B2}-1}{G_{B1}/(L_{comb}L_{TDU})} + \frac{L_{other}-1}{G_{B2}G_{B1}/(L_{comb}L_{TDU})} \qquad (14.57)$$

and

$$G_{after} = \frac{G_{B1}G_{B2}}{L_{other}L_{comb}L_{TDU}} \qquad (14.58)$$

In (14.57) and (14.58), L_{comb} is the combiner loss, F_{B1} is the noise figure of the buffer amplifier before the TDU, G_{B1} is the gain of the buffer amplifier before the TDU, L_{TDU} is the TDU loss (the TDU is assumed to be strictly a lossy device), F_{B2} is the noise figure of

---

[9] Recall that when computing noise figure we assume the noise energy into the device is kT_0 (see Chapter 4).

the buffer amplifier after the TDU, G_{B2} is the gain of the buffer amplifier after the TDU, and L_{other} is the loss of the devices between the second buffer amplifier and what is designated as the subarray output. If either of the buffer amplifiers is not present, we would set their gain and noise figure to one.

To get the final gain and noise figure of the subarray, from the inputs to the T/R modules to the designated subarray output, we would perform a cascade noise figure analysis of the ideal summer and the total of the components after the ideal summer. This would lead to a subarray noise figure and gain of

$$F_{subarray} = F_\Sigma + \frac{F_{after} - 1}{G_\Sigma} \tag{14.59}$$

and

$$G_{subarray} = G_\Sigma G_{after} \tag{14.60}$$

14.5.3 Array Gain and Noise Figure

If the radar is processing each of the M subarrays through the receiver and signal processor, we would consider the subarrays as separate "LNAs" and use their gains and noise figures to find the overall noise figures of the M receivers. However, if the subarray outputs are combined at RF to form the multiple receive beams, we would need to carry the noise figure to the output of the array combiner/splitter (see Figure 14.10). We would treat the array combiner/splitter the same as we treated the subarray combiner. That is, we model it as an ideal, lossless summer followed by a lossy device that represents the losses in the array combiner/splitter. Specifically, the gain of the AESA through the lossless summer would be

$$G_{\Sigma\_array} = \frac{\sum_{m=1}^{M} G_{subarray\,m}}{M} \tag{14.61}$$

and

$$F_{\Sigma\_array} = \frac{\sum_{m=1}^{M} G_{subarray\,m} F_{subarray\,m}}{\sum_{m=1}^{M} G_{subarray\,m}} \tag{14.62}$$

where $G_{subarray\,m}$ and $F_{subarray\,m}$ are the gain and noise figure of the mth subarray. If there are buffer amplifiers and/or lossy devices after the array combiner/splitter, we would perform a noise figure cascade analysis to find combined the gain and noise figure of those devices. With that, we would find the overall gain and noise figure of the AESA using

$$G_{array} = G_{\Sigma\_array} G_{after\_array} \tag{14.63}$$

and

$$F_{array} = F_{\Sigma\_array} + \frac{F_{after\_array} - 1}{G_{\Sigma\_array}} \qquad (14.64)$$

where G_{after_array} and F_{after_array} are the combined gain and noise figure of the combiner/splitter and the devices after the combiner/splitter.

14.5.4 AESA Noise Figure Example

As an example of how to compute the noise figure of a subarray and an overall array, we consider the C-band example of Figure 14.27. The array is a quantized version of a circular array that consists of four quadrants (four quadrant arrays) with 19 subarrays each. The subarrays are square, with 10 rows and 10 columns of elements, and T/R modules, that are spaced 0.03 m or ½ wavelength apart. We chose the 10×10 subarray configuration because we earlier found that a 10-element, linear subarray that was 10×0.03 = 0.3 m wide would cause minimal distortion of a 500-MHz LFM pulse when the antenna beam was scanned to a maximum angle of 60°. We chose rectangular array packing (see Chapter 13) because the math was easier than we would have had if we considered triangular packing. We chose T/R module attenuator values to provide a radial, $\bar{n} = 6$, 30 dB, Taylor weighting across the array (see Chapter 13). We assumed the outputs of the four quadrants were sent to separate receivers. For this example, we are interested in finding the noise figure and RF gain of each quadrant array. These four gains and noise figures will be the same because the four quadrant arrays are identical. We assumed the data processor used the outputs of the four receivers to form the overall sum beam, and to extract u and v angle information.

Figure 14.28 contains a schematic of the components of one leg of the array, from the radiating element to the receiver. We only show the receive side of the T/R module since the transmit side is not relevant to computing noise figure.

We assumed all of the radiating elements and the T/R module components up to the attenuator are the same for each element of the array. However, the attenuator settings of the T/R modules are different because they realize the Taylor weighting.

The outputs of each of the 100 subarray elements are sent to a subarray combiner, which we assume has a loss of 0.5 dB.

The outputs of the 19 subarrays go to the cascade of a buffer amplifier, a TDU and a second buffer amplifier. We assume both buffer amplifiers have the same gain and noise figure of 10 dB and 6 dB, respectively. We assume the TDU has a loss of 15 dB [26]. The output of each of the second buffer amplifiers goes to an array combiner, which we assume has a loss of 0.5 dB. Finally, the output of the array combiner is sent to the quadrant receiver.

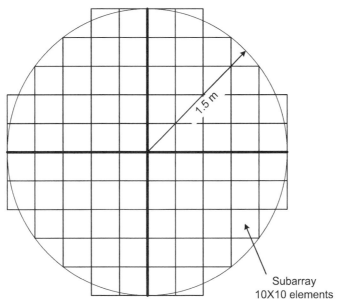

Figure 14.27 Array configuration used in the AESA noise figure example.

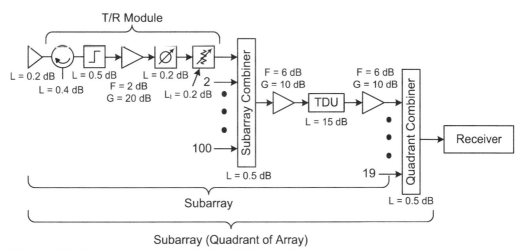

Figure 14.28 Array configuration used in the AESA noise figure example.

We computed the noise figures of the T/R modules using (14.46), (14.47), and (14.48) where L_{before_n} was the combined losses of the radiator (0.2 dB), the circulator (0.4 dB) and the limiter (0.5 dB). L_{after_n} included a phase shifter insertion loss of 0.2 dB and the loss due to the attenuator setting. It also included a 0.2-dB insertion loss for the attenuator. This would be the loss when the attenuator setting is 0 dB. The attenuator setting was determined by the Taylor weighting.

We used (14.53) and (14.55) to compute the noise figure and gain of the summing component of the subarray combiner. We then used (14.57), (14.58), (14.59), and (14.60) to find the gain and noise figure of each subarray.

Finally, we used (14.60), (14.61), (14.63), and (14.64) to find the signal gain and noise figure of the quadrant array. The gain and noise figure is from the input to the radiating elements of the array to the input to the receiver

Table 14.1 contains a list of the noise figures and gains for the 19 subarrays in each of the quadrants. The format of the table corresponds to the layout of subarrays on the upper right quadrant of Figure 14.27. That is, the lower left pair of numbers corresponds to the subarray closest to the center of the array of Figure 14.27. The top number (*slanted* text) of each entry is the noise figure ($F_{subarray}$) and the bottom entry (normal text) is the gain ($G_{subarray}$), both in dB. As expected, the noise figures of the subarrays increase and the gains decrease as the subarrays move farther from the center of the overall array. Because of the Taylor weighting, the attenuators of the T/R modules of the subarrays near the edge of the antenna are set to higher values than those of the subarrays near the center of the array. This causes a net decrease in the gains of these subarrays which, in turn, causes an increase in the noise figures of the subarrays. As indicated in the table title, the overall noise figure (F_{array}) and gain (G_{array}) of each quadrant of the array, through the quadrant combiner of Figure 14.28, is 4.62 dB and 17.31, respectively.

Had we used uniform weighting across the array, all of the attenuator settings would have been set to their minimum 0 dB, which means all of the subarrays would have noise figures and gains of 3.60 and 23 dB, respectively. The noise figures and gains of the quadrant arrays would be 3.60 dB and 22.5 dB.

Table 14.1

Noise figure and gain of the 19 subarrays of each quadrant of Figure 14.27 – $\bar{n}$=6, 30 dB Taylor weighting—Overall noise figure and gain of each quadrant is 4.62 dB and 17.31 dB, respectively

					Legend:
7.21	*7.43*				*Noise Figure*
12.08	11.73				Gain
5.38	*5.90*	*6.86*			
15.64	14.48	12.66			
4.22	*4.52*	*5.38*	*6.86*		
19.31	18.12	15.64	12.66		
3.79	*3.98*	*4.52*	*5.90*	*7.43*	
21.55	20.45	18.12	14.48	11.73	
3.64	*3.79*	*4.22*	*5.38*	*7.21*	
22.65	21.55	19.31	15.64	12.08	

14.6 EXERCISES

1. Derive (14.5).

    2. Derive (14.6). See appendix for hints.

    3. Derive (14.10).

    4. Reproduce Figures 14.8 and 14.9.

5. Derive (14.20) and (14.23).
6. Reproduce Figure 14.11.
7. Derive (14.38).
8. Reproduce Figures 14.13, 14.14, and 14.15.
9. Reproduce Figures 14.16, 14.17, and 14.18.
10. Reproduce Figures 14.20 and 14.21.
11. Reproduce Figures 14.23 and 14.24.
12. Reproduce Figures 14.25 and 14.26.
13. Derive (14.43) through (14.47).
14. Derive (14.53) and (14.55).
15. Derive (14.56) through (14.59).
16. Derive (14.62) and (14.63).
17. Repeat the example of Section 14.5.4 for the case of a $\bar{n} = 6$, 40-dB Taylor weight.
18. Repeat Exercise 17 for the case where the array uses Hamming weighting.
19. Repeat Exercise 17 for the case where the array uses cosine weighting with a −20-dB edge illumination. That is, the weight is determined by a cosine function where the amplitude is 1.0 at array center and $10^{-20/20}$ at the edges of the circle of Figure 14.27.

References

[1] W. Yue and C. Lanying, "Status in U.S. AESA Fighter Radar and Development Trends," in *IET International Radar Conference 2013*, Xi'an, 2013.

[2] C. Kopp, "Evolution of AESA Radar Technology," *Aerospace and Defense Channel, Microwave Journal,* vol. 55, no. 8, August 2012.

[3] C. Kopp, "Advances in Russian and Chinese Active Electronically steered arrays (AESAs)," in *2013 IEEE International Symposium on Phased Array Systems and Technology*, Waltham, MA, 2013.

[4] R. Sturdivant, C. Quan and E. Chang, Systems Engineering of Phased Arrays, Norwood, MA: Artech House, 2018.

[5] R. Sturdivant and M. Harris, Transmit Receive Modules for Radar and Communication Systems, Norwood, MA: Artech House, 2016.

[6] R. L. Haupt, Timed Arrays: Wideband and Time Varying Antenna Arrays, Hoboken, New Jersey: Wiley-IEEE Press, 2015.

[7] D. J. Carlson, C. Weigand and T. Boles, "MMIC Based Phased Array Radar T/R Modules," in *Proceedings of the 7th European Radar Conference*, Paris, France, 2010.

[8] 张光义 and 赵玉洁, 相控阵雷达技术 (Phased Array Radar Technology), 北京 (Beijing), Haidian District: 电子工业出版社 (Publishing House of Electronics Industry), 2006.

[9] W. Wojtasiak, D. Gryglewski, T. Morawski and E. Sedek, "Designing T/R Module for Active Phased Array Radar," in *14th International Conference on Microwaves, Radar and Wireless Communications. MIKON - 2002. Conference Proceedings*, Gdansk, Poland, 2002.

[10] A. Agrawal, R. Clark and J. Komiak, "T/R Module Architecture Tradeoffs for Phased Array Antennas," in *1996 IEEE MTT-S International Microwave Symposium Digest*, San Francisco, CA, 1996.

[11] R. Tange and R. W. Burns, "Array Technology," *Proceedings of the IEEE*, vol. 80, no. 1, 1992, pp. 173–182.

[12] M. D. R. Tench, P. A. Claridge, W. G. Gates, W. J. M. Graham, R. Young and P. Hazeldine, "Design of a Phased Array Antenna Using Solid State Transmit/Receive Modules," in *IEE Colloquium on Mechanical Aspects of Antenna Design*, London, UK, 1989.

[13] D. I. Voskresensky and A. I. Kanashchenkov, eds., *Active Phased Array Antenna Arrays*, Moscow: Radiotekhnika Publishing House, 2004.

[14] R. J. Mailloux, *Phased Array Handbook*, 3rd ed., Norwood, MA: Artech House, 2018.

[15] R. J. Mailloux, "Technology for Array Control," in *IEEE International Symposium on Phased Array Systems and Technology*, Boston, MA, 2013.

[16] D. Gryglewski, T. Morawski, E. Sedek and J. Zborowska, "Microwave Phase Shifters for Radar Applications," in *2006 International Conference on Microwaves, Radar & Wireless Communications*, Krakow, 2006.

[17] 殷连生, 相控阵雷达馈线技术 (Phased Array Radar Feedline Technology), 北京 (Beijing): 国防工业出版社 (National Defense Industry Press), 2007.

[18] J. F. White, "Diode Phase Shifters for Array Antennas," *IEEE Transactions on Microwave Theory and Techniques,* vol. 22, no. 6, 1974, pp. 658–674.

[19] C. Bowick, J. Blyler and C. Ajluni, RF Circuit Design, second ed., Burlington, MA: Newness, 2008.

[20] W.D. Wirth, Radar Techniques Using Array Antennas, London, UK: The Institution of Electrical Engineers, 2001.

[21] A. J. Fenn, Adaptive Antennas and Phased Arrays for Radar and Communications, Norwood, MA: Artech House, 2008.

[22] N. Fourikis, Advanced Array Systems, Applications and RF Technologies, New York: Academic Press, 2000.

[23] A. K. Agrawal and E. L. Holzman, "Active Phased Array Architectures for High Reliability," in *Proceedings of International Symposium on Phased Array Systems and Technology*, Boston, MA, 1996.

[24] E. D. Cohen, "Active Electronically Scanned Arrays," in *Proceedings of IEEE National Telesystems Conference - NTC '94*, San Diego, CA, 1994.

[25] L. A. Belov, S. M. Smolskiy and V. N. Kochemasov, *Handbook of RF, Microwave, and Millimeter-Wave Components*, Norwood, MA: Artech house, 2012.

[26] E. Ackerman, S. Wanuga, D. Kasemset, W. Minford, N. Thorsten and J. Watson, "Integrated 6-bit Photonic True-Time-Delay Unit for Lightweight 3-6 GHz Radar Beamformer," in *1992 IEEE MTT-S Microwave Symposium Digest*, Albuquerque, NM, 1992.

[27] M. Longbrake, "True Time-Delay Beamsteering for Radar," in *2012 IEEE National Aerospace and Electronics Conference (NAECON)*, Dayton, OH, 2012.

[28] A.V. Pham, "Development of True Time Delay Circuits," 2014. [Online]. Available: https://apps.dtic.mil/dtic/tr/fulltext/u2/a608902.pdf.

[29] X. Lan, M. Kintis, C. Hansen, W. Chan, G. Tseng and M. Tan, "A Simple DC to 110 GHz MMIC True Time Delay Line," *IEEE Microwave and Wireless Components Letters*, vol. 22, no. 7, 2012, pp. 369–371.

[30] R. Rotman, M. Tur and L. Yaron, "True Time Delay in Phased Arrays," *Proceedings of the IEEE*, vol. 104, no. 3, March 2016, pp. 504–518.

[31] 王德纯, 宽带相控阵雷达 (Wideband Phased Array Radar), 北京 (Beijing): 国防工业出版社 (National Defense Industry Press), 2010.

[32] J. J. Lee, R. Y. Loo, S. Livingston, V. I. Jones, J. B. Lewis, H.-W. Yen, G. L. Tangonan and M. Wechsberg, "Photonic Wideband Array Antennas," *IEEE Transactions on Antennas and Propagation*, vol. 42, no. 9, 1995, pp. 966–982.

[33] W. Manqing, "Digital Array Radar: Technology and Trends," *2011 IEEE CIE International Conference on Radar*, vol. 1, October 2011, pp. 1–4, 24–27.

[34] J. S. Herd and M. D. Conway, "The Evolution to Modern Phased Array Architectures," *Proceedings of the IEEE*, vol. 104, no. 3, 2016, pp. 519–529.

[35] C. Fulton, M. Yeary, D. Thompson, J. Lake and A. Mitchell, "Digital Phased Arrays: Challenges and Opportunities," *Proceedings of the IEEE*, vol. 104, no. 3, March 2016, pp. 487–503.

[36] S. H. Talisa, K. W. O'Haver, T. M. Comberiate, M. D. Sharp and O. F. Somerlock, "Benefits of Digital Phased Array Radars," *Proceedings of the IEEE*, vol. 104, no. 3, 2016, pp. 530–543.

[37] R. C. Hansen, Phased Array Antennas, New York: John Wiley & Sons, Inc., 1998.

[38] Q. M. Alfred, T. Chakravarty and S. K. Sanyal, "Overlapped Subarray Architecture of an Wideband Phased Array Antenna with Interference Suppression Capability," in *Journal of Electromagnetic Analysis and Applications*, 2013.

[39] T. Azar, "Overlapped Subarrays: Review and Update [Education Column]," *IEEE Antennas and Propagation Magazine*, vol. 55, no. 2, April 2013, pp. 228–234.

[40] R. Fante, "Sidelobe and nulling properties of overlapped, subarrayed scanning antennas," in *1980 Antennas and Propagation Society International Symposium*, Quebec, Canada, 1980.

[41] R. Fante, "Systems study of overlapped subarrayed scanning antennas," *IEEE Transactions on Antennas and Propagation*, vol. 28, no. 5, September 1980, pp. 668–679.

[42] J. S. Herd, S. M. Duffy and H. Steyskal, "Design considerations and results for an overlapped subarray radar antenna," in *2005 IEEE Aerospace Conference*, Big Sky, MT, 2005.

[43] C. T. Lin and H. Ly, "Sidelobe reduction through subarray overlapping for wideband arrays," in *Proceedings of the 2001 IEEE Radar Conference (Cat. No.01CH37200)*, Atlanta, GA, 2001.

[44] G. Borgiotti, "An antenna for limited scan in one plane: Design criteria and numerical simulation," *IEEE Transactions on Antennas and Propagation,* vol. 25, no. 2, March 1977, pp. 232–243.

[45] J. P. Basart, "Synthesis of nonuniformly spaced antenna arrays, Dissertation, 3992," 1967. [Online]. Available: https://lib.dr.iastate.edu/rtd/3992.

[46] J. F. Morris, "Design of Non-Uniform Linear Antenna Arrays with Digitized Spacing and Amplitude levels, Masters Theses. 5236.," 1965. [Online]. Available: https://scholarsmine.mst.edu/masters_theses/5236.

[47] R. V. Gatti, M. Dionigi and R. Sorrentino, "Computation of Gain, Noise Figure, and Third-Order Intercept of Active Array Antennas," *IEEE Transactions on Antennas and Propagation,* vol. 52, no. 11, Nov. 2004, pp. 3139–3143.

[48] E. L. Holzman and A. K. Agrawal, "A Comparison of Active Phased Array, Corporate Beamforming Architectures," in *Proceedings of International Symposium on Phased Array Systems and Technology*, Boston, 1996.

[49] K. F. Warnick, M. V. Ivashina, R. Maaskant and B. Woestenburg, "Unified Definitions of Efficiencies and System Noise Temperature for Receiving Antenna Arrays," *IEEE Transactions on Antennas and Propagation,* vol. 58, no. 6, June 2010, pp. 2121–2125.

[50] E. Brookner, "Active Electronically Scanned Array (AESA) System Noise Temperature," in *IEEE International Symposium on Phased Array Systems and Technology*, Waltham, MA, Oct. 15–28, 2013, pp. 760–767.

[51] J. J. Lee, "G/T and Noise Figure of Active Array Antennas," *IEEE Transactions on Antennas and Propagation,* vol. 41, no. 2, Feb. 1993, pp. 241–244.

[52] 张光义, 相控阵雷达原理 (Principle of Phased Array Radar), 北京 (Beijing): 国防工业出版社 (National Defense Industry Press), 2009.

[53] 郭崇贤, 相控 (Receive Techniques for Phased Array Radar), 北京 (Beijing): 国防工业出版社 (National Defense Industry Press), 2009.

APPENDIX 14A: DERIVATION OF THE MATCHED FILTER OUTPUT FOR AN AESA (EQUATION 14.10)

In this appendix, we present the derivation that led to (14.10). We start with the voltage pulse, $v_T(t)$, from the exciter, which will be split into N equal parts by the combiner/splitter of Figure 14.6. We represent these as

$$v_{Tn}(t) = e^{j2\pi f_c t} p(t) \qquad (14A.1)$$

where f_c is the RF, or carrier frequency, and $p(t)$ represents the transmit pulse. For an unmodulated pulse, $p(t)$ has the form

$$p_u(t) = \text{rect}\left[\frac{t}{\tau_p}\right] \qquad (14A.2)$$

and τ_p is the pulse width. For an LFM pulse, $p(t)$ has the form

$$p_{LFM}(t) = e^{j\pi\alpha t^2} \text{rect}\left[\frac{t}{\tau_p}\right] \tag{14A.3}$$

where α is the LFM slope. The associated LFM bandwidth is $B = |\alpha|\tau_p$ (see Chapter 11). Because of the form of $v_{Tn}(t)$, we are assuming the combiner/splitter imparts no delay to $v(t)$. We have also normalized $v_{Tn}(t)$ to an amplitude of one.

Each of the $v_{Tn}(t)$ signals goes from the combiner splitter, through the nth phase shifter, to the nth radiating element. The voltage at the nth radiating element is, again assuming zero delay (and no change in amplitude),

$$v_{radnT}(t) = e^{j2\pi f_c t} e^{j\phi_{nT}} p(t) \tag{14A.4}$$

where ϕ_{nT} is the phase shift imparted on $v_{Tn}(t)$ by the phase shifter.

The radiating element converts the voltage to an E field, $E_{nT}(t)$, which subsequently travels to the target. On its way to the target, the E field experiences a delay of τ_{rnT}, where we added the subscript T to the range delay, τ_{rn}, defined by (14.5). If we ignore (actually normalize away) range attenuation, the E field at that target, due to the signal radiated from the nth element, is

$$E_{nT}(t) = e^{j2\pi f_c(t-\tau_{rnT})} e^{j\phi_{nT}} p(t-\tau_{rnT}) \tag{14A.5}$$

The target sums the n E fields from the n elements to form a normalized, composite E field we can write as

$$E_{tgt}(t) = \sum_{n_T=0}^{N-1} E_{nT}(t) = \sum_{n_T=0}^{N-1} e^{j2\pi f_c(t-\tau_{rnT})} e^{j\phi_{nT}} p(t-\tau_{rnT}) \tag{14A.6}$$

As a note, we are assuming a stationary, point target.

The composite E field is radiated back to the AESA. On the way back to the nth element of the AESA, $E_{tgt}(t)$ undergoes a range delay of τ_{rnR}, where we added the subscript R to the range delay, τ_{rn}, defined by (14.5). Thus, the E field at the nth array element is

$$\begin{aligned} E_{nR}(t) &= E_{tgt}(t-\tau_{rnR}) = \sum_{n_T=0}^{N-1} E_{nT}(t) \\ &= \sum_{n_T=0}^{N-1} e^{j2\pi f_c(t-\tau_{rnT}-\tau_{rnR})} e^{j\phi_{nT}} p(t-\tau_{rnT}-\tau_{rnR}) \end{aligned} \tag{14A.7}$$

We again ignore range attenuation.

The nth array element converts the $E_{nR}(t)$ E field to a voltage, $v_{nR}(t)$, which is sent through the T/R module to the combiner/splitter to be summed with all of the other $v_{nR}(t)$ and sent to the matched filter via the receiver. The LNAs, other amplifiers, and attenuators of the T/R module will amplify $v_{nR}(t)$, and the phase shifter will apply a phase shift of

ϕ_{nR}. For purposes of this analysis, we assume the T/R modules apply the same amplification to all $v_{nR}(t)$ and normalize the amplification to one.

With the above, the signal that leaves the nth T/R module, and is sent to the combiner/splitter, is

$$V_{nR}(t) = e^{j\phi_{nR}} v_{nR}(t) = e^{j\phi_{nR}} \sum_{n_T=0}^{N-1} e^{j2\pi f_c(t-\tau_{rnT}-\tau_{rnR})} e^{j\phi_{nT}} p(t-\tau_{rnT}-\tau_{rnR}) \quad (14A.8)$$

Finally, the normalized voltage at the output of the combiner/splitter is

$$\begin{aligned} V_R(t) &= \sum_{n_R=0}^{N-1} e^{j\phi_{nR}} v_{nR}(t) \\ &= \sum_{n_R=0}^{N-1} e^{j\phi_{nR}} \sum_{n_T=0}^{N-1} e^{j2\pi f_c(t-\tau_{rnT}-\tau_{rnR})} e^{j\phi_{nT}} p(t-\tau_{rnT}-\tau_{rnR}) \\ &= \sum_{n_R=0}^{N-1} \sum_{n_T=0}^{N-1} e^{j2\pi f_c(t-\tau_{rnT}-\tau_{rnR})} e^{j(\phi_{nT}+\phi_{nR})} p(t-\tau_{rnT}-\tau_{rnR}) \end{aligned} \quad (14A.9)$$

We assume the receiver removes the carrier term and multiplies $V_R(t)$ by a normalized gain of one, so that the normalized signal into the matched filter is

$$V_{MF}(t) = \sum_{n_R=0}^{N-1} \sum_{n_T=0}^{N-1} e^{-j2\pi f_c(\tau_{rnT}+\tau_{rnR})} e^{j(\phi_{nT}+\phi_{nR})} p(t-\tau_{rnT}-\tau_{rnR}) \quad (14A.10)$$

From (14.5), we can write

$$\tau_{rnT} = \frac{\tau_{rtgt}}{2} - \frac{1}{2c}(2n_T+1-N)d\sin\theta_{tgt} \quad (14A.11)$$

and

$$\tau_{rnR} = \frac{\tau_{rtgt}}{2} - \frac{1}{2c}(2n_R+1-N)d\sin\theta_{tgt} \quad (14A.12)$$

where θ_{tgt} is the angle of the target relative to array normal (see Figure 14.6). With this we can write the argument of the first exponential as

$$\begin{aligned} -j2\pi f_c(\tau_{rnT}+\tau_{rnR}) &= -j2\pi f_c\left(\frac{\tau_{rtgt}}{2} - \frac{1}{2c}(2n_T+1-N)d\sin\theta_{tgt}\right. \\ &\quad \left. + \frac{\tau_{rtgt}}{2} - \frac{1}{2c}(2n_R+1-N)d\sin\theta_{tgt}\right) \\ &= -j2\pi f_c\left(\tau_{rtgt} - \frac{1}{c}(n_T+n_R+1-N)d\sin\theta_{tgt}\right) \\ &= j2\pi r_{tgt}/\lambda + j2\pi(n_T+n_R+1-N)\tfrac{d}{\lambda}\sin\theta_{tgt} \end{aligned} \quad (14A.13)$$

We note that the range delay of the various components of the signal to and from the target appears not only as delays in the return signal, but also as phase shifts to the various components.

From Chapter 13, we recall that if we want to steer the antenna beam to some θ_S relative to array normal, we need to select

$$\phi_{nT} = 2\pi n_T \tfrac{d}{\lambda} \sin\theta_S \text{ and } \phi_{nR} = 2\pi n_R \tfrac{d}{\lambda} \sin\theta_S \qquad (14\text{A}.14)$$

Substituting (14A.11), (14A.13), and (14A.14) into (14A.10) and manipulating, we get

$$V_{MF}(t) = e^{-j2\pi r_{tgt}/\lambda} \sum_{n_R=0}^{N-1} \sum_{n_T=0}^{N-1} e^{-j2\pi(n_T+n_R+1-N)(d/\lambda)(\sin\theta_{tgt}-\sin\theta_S)} \\ p\left(t - \tau_{rtgt} - (n_R + n_T + 1 - N)\tfrac{d}{c}\sin\theta_{tgt}\right) \qquad (14\text{A}.15)$$

for the input to the matched filter.

Finally, we assume the matched filter is matched to $p(t)$ and that its output is $MF(\tau)$, where τ is range delay. With this the output of the matched filter, when the input is $V_{MF}(t)$, is

$$V_o(\tau) = e^{-j2\pi r_{tgt}/\lambda} \sum_{n_R=0}^{N-1} \sum_{n_T=0}^{N-1} e^{-j2\pi(n_T+n_R+1-N)(d/\lambda)(\sin\theta_{tgt}-\sin\theta_S)} \\ MF\left(\tau - \tau_{rtgt} - (n_R + n_T + 1 - N)\tfrac{d}{c}\sin\theta_{tgt}\right) \qquad (14\text{A}.16)$$

If we assume the beam is steered to the target, we have $\theta_S = \theta_{tgt}$. With this the exponential inside the sum becomes 1.0, and (14A.16) reduces to

$$V_o(\tau)\big|_{\theta_S=\theta_{tgt}} = e^{-j2\pi r_{tgt}/\lambda} \sum_{n_R=0}^{N-1} \sum_{n_T=0}^{N-1} MF\left(\tau - \tau_{rtgt} - (n_R + n_T + 1 - N)\tfrac{d}{c}\sin\theta_{tgt}\right) \qquad (14\text{A}.17)$$

If we take the magnitude of $V_o(\tau)$ and drop the conditioning of $V_o(\tau)$, we get (14.10).

Chapter 15

Signal Processors

15.1 INTRODUCTION

In this and the next few chapters, we turn our attention to signal processors and the analysis of their performance. Our first encounters with signal processors were in Chapters 8 and 9, where we studied matched filters and coherent integrators. In those studies, the purpose of the signal processor was to improve SNR. In the discussions of the next few chapters, we will be concerned with signal processors whose primary function is clutter rejection, with SNR improvement as a secondary objective. By clutter, we mean returns from unwanted sources such as the ground or rain.

We will consider two types of signal processors: MTI processors and pulsed Doppler processors. MTI processors subtract returns from two or more successive pulses to provide mainly clutter rejection. They can be thought of as nonrecursive, high-pass filters. Pulsed Doppler processors combine a high-pass filter (usually recursive) and a coherent-integrator, or bank of coherent integrators tuned to different Doppler frequencies. In some applications, a pulsed Doppler processor will contain only the bank of coherent integrators. MTI processors are used strictly for clutter rejection (usually ground clutter) and pulsed Doppler processors are used to provide both clutter rejection and SNR improvement. Pulsed Doppler processors are often subdivided into high PRF (HPRF) pulsed Doppler processors, medium PRF (MPRF) pulsed Doppler processors, and low PRF (LPRF) pulsed Doppler processors. A variant of the latter is also sometimes termed moving target detector (MTD) processors [1–7]. MTI processors are discussed in Chapter 18 and pulsed Doppler and MTD processors are discussed in Chapters 19 and 20.

When we considered matched filters (see Chapter 8) we were concerned with how they processed a single pulse. The processors we consider in the next few chapters process anywhere from two pulses for MTI, to many hundreds of pulses for HPRF pulsed Doppler processors. MTI and pulsed Doppler processors also include a matched filter to maximize SNR at the input to the processor, and to reduce the SNR improvement burden on the coherent integrator.

It appears that the first use of signal processing to mitigate clutter occurred during World War II when radar operators noted they could distinguish targets from clutter by the fluttering of the clutter return on an A-scope (an oscilloscope-type display), which came to be known as the butterfly effect. This effect was incorporated into a target-in-clutter detection system that came to be known as a noncoherent MTI [8]. In the mid-1940s, Alfred G. Emslie invented coherent MTI, for which he was granted several patents [9–18].

Harry B. Smith, who was the president of the Westinghouse Defense and Electronics Systems Center for 10 years, was inducted into the Innovation Hall of Fame in 1987 "for invention of pulsed Doppler radar and other innovations in airborne electronics" [19]. Likewise, Smith, Leroy C. Perkins, and David H. Mooney were awarded the IEEE Pioneer Award in 1984 "for Contributions to the development of the high-repetition-rate Airborne Pulse Doppler Radar" [20, 21]. Smith, Perkins, and Mooney were awarded patents for a "Pulse Doppler Radar System" in 1961 and 1962 [22, 23].

It is not clear who invented the MTD processor. In 2005, Charles E. Muehe was awarded the IEEE Aerospace and Electronic Systems Society (AESS) Pioneer Award "for the invention of the Moving Target Detector (MTD) digital signal processor for aircraft surveillance radar" [24–26]. However, a 1977 MIT Lincoln Laboratories report seems to indicate that Ronald S. Bassford, William Goodchild, and Alfred de la Marche developed it to solve problems associated with air route surveillance radars (ARSRs) [27].

We begin our discussions with a description of the general type of processor we will be considering. Following that, in Chapter 16, we derive equations of the signals that are processed by the signal processor. We will be specifically interested in the spectra of these signals because we perform most signal processor analyses in the frequency domain

After we derive the signal model, we use that to derive general equations that we will use to analyze signal processors.

In preparation for the specific example processors of Chapters 18 through 20, we derive models for ground and rain clutter in Chapter 17. The clutter models are specific to ground-based radars, although they can also be used for sea-based radars and could be extended to airborne radars.

In Chapter 19 we consider pulsed Doppler signal processors. We first derive equations for clutter rejection and SNR improvement of pulsed Doppler processors. We also analyze the impact of phase noise, ADC quantization and range correlation on clutter rejection and SNR improvement. After we derive the general performance equations, we consider specific examples of HPRF, MPRF and LPRF pulsed Doppler processors. The HPRF processor uses a bandpass filter (BPF) as the coherent integrator while the MPRF processor uses a fast Fourier transformer (FFT), and the LPRF processor uses as bank of what we will term *bandpass integrator*s. This approach affords the opportunity to learn how to analyze the clutter rejection and SNR improvement of the three types of coherent integrators.

All of the processors of Chapter 19 are assumed to be implemented with digital hardware. In Chapter 20 we consider an example of a HPRF pulsed Doppler processor that is implemented with analog hardware. The analog processor is considered separately because of some subtle differences in analysis methodology for digital and analog signal processors.

To close this series of chapters on signal processors we discuss, in Chapter 21, examples of the performance of MTI and Doppler processor in a clutter environment that consists of chaff.

15.2 SIGNAL PROCESSOR STRUCTURE

Figure 15.1 contains a general block diagram of the type of processor we consider in Chapters 16 through 21. The processor consists of a matched filter, a clutter rejection filter and a coherent integrator. The main job of the clutter rejection filter is to reject clutter. It generally provides no SNR improvement. The coherent integrator provides SNR improvement, and can also provide clutter rejection.

In ground-based radars, ground clutter is centered at zero-Doppler frequency because the scattering surfaces that constitute ground clutter are, on average, not moving. Weather clutter can be centered at other than zero Doppler frequency, but at a low frequency relative to the nominal target Doppler frequency. Because of these conditions, the clutter rejection filter is normally a high-pass filter (HPF).

When we studied coherent integrators in Chapter 9, we considered them as just that – integrators or summers. We showed that they can be thought of as increasing signal power by N^2 and noise power by N, where N is the number of pulses integrated, or summed. Because of these properties, they increased SNR by $N^2/N = N$.

Another way of thinking about a coherent integrator is that it is a BPF with a center band gain of G. As such, the signal power out of the filter will equal G times the signal power into the bandpass filter. That is $P_{Sout} = GP_{Sin}$. An assumption of this statement is that the signal is a single tone, or at least is narrowband relative to the bandwidth of the bandpass filter.

While we normally think of signals as narrowband, we think of noise as a wideband signal. That is, its power is spread uniformly over a bandwidth of B_N. If the bandwidth of the bandpass filter is B_{BPF}, and less than B_N, the bandpass filter will "carve out" B_{BPF}/B_N of the input noise power and amplify it by G. In equation form we have $P_{Nout} = G(B_{BPF}/B_N)P_{Nin}$, where G is the gain of the bandpass filter. This is illustrated in Figure 15.2.

Given $P_{Sout} = GP_{Sin}$ and $P_{Nout} = G(B_{BPF}/B_N)P_{Nin}$, we have

$$SNR_{out} = \frac{P_{Sout}}{P_{Nout}} = \frac{GP_{Sin}}{GP_{Nin}(B_{BPF}/B_N)} = \frac{B_N}{B_{BPF}}\frac{P_{Sin}}{P_{Nin}} = \frac{B_N}{B_{BPF}}SNR_{in} \qquad (15.1)$$

which says that the BPF (coherent integrator) increases the SNR by B_N/B_{BPF}.

As we will see, a classical MTI processor contains only the clutter rejection filter, and not the coherent integrator.

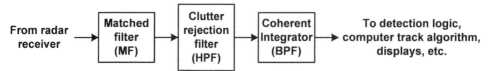

Figure 15.1 Signal processor structure.

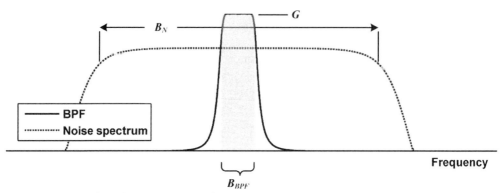

Figure 15.2 Illustration of noise power out of a BPF.

Since the clutter rejection requirements for LPRF waveforms are not as stringent as for MPRF and HPRF) waveforms, LPRF signal processors can often dispense with the clutter rejection filter and rely on the coherent integrator to provide clutter rejection, in addition to SNR improvement.

In contrast to LPRF waveforms, the clutter rejection requirements of HPRF waveforms are very stringent. As we will see, this is because targets at long ranges must compete with clutter that is at a very short range. As a result, HPRF signal processors almost always contain both clutter rejection filters and coherent integrators. Part of the clutter rejection is provided by the clutter rejection filter and part is provided by the coherent integrator.

The clutter rejection requirements of MPRF waveforms are between those of LPRF and HPRF waveforms. As a result, MPRF processors may or may not need to contain a clutter rejection filter. However, they almost always contain a coherent integrator.

Having gone through the above lengthy discussion, we need to caveat it by saying they are generalizations. The signal processor configuration for a specific application will depend on the applicable clutter and target characteristics, the radar parameters, hardware constraints, and the preferences of the radar and signal processor designers, among other factors.

In some situations, such as search, the target Doppler frequency is not known. Instead, only a range of possible Doppler frequencies is known. Furthermore, the range of Doppler frequencies is much larger than the bandwidth of the coherent integrator that is needed to provide a desired SNR improvement. As a result, a bank of coherent integrators, tuned to different frequencies, is needed. This is illustrated in Figure 15.3, where $f_{d1}, f_{d2} \ldots f_{dN}$ are the Doppler frequencies to which the coherent integrators are tuned.

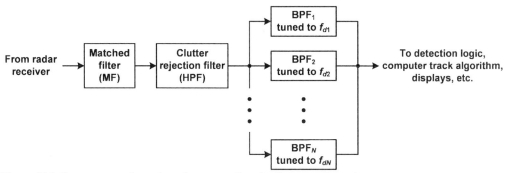

Figure 15.3 Processor configuration when target Doppler frequency is not known.

The bank of coherent integrators can be implemented as a bank of bandpass filters, a bank of tuned integrators (bandpass integrators), or with a FFT. In the case of the FFT, each output tap of the FFT is associated with a different frequency. This means the FFT acts as a bank of bandpass filters.

The processor doesn't need a bank of clutter rejection filters because the HPF frequency characteristics are set by the clutter, which (usually) has a single center frequency–zero for ground clutter and the wind velocity for rain (keep in mind we are focusing on ground-based radars).

During track, it is likely the target Doppler frequency can be estimated with a reasonable amount of accuracy. In that case, only one, or a few, coherent integrators (BPFs) would be needed.

During search, the range to the target is not known. As a result, separate signal processors (or signal processing functions) are needed for each possible target range. How this statement translates to the configuration of the overall signal processor is illustrated in Figure 15.4. The operation of this overall signal processor can be explained as follows.

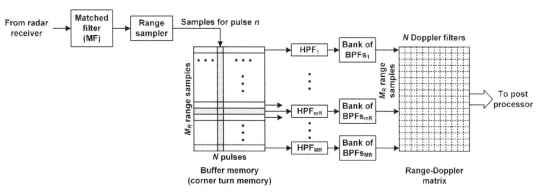

Figure 15.4 Overall signal processor configuration.

- The output of the matched filter is sampled at intervals nominally equal to the range resolution of the pulse. Sampling starts at some minimum range delay (see Chapter 2) relative to the transmit pulse and continues to some maximum range delay. This produces M_R samples. In the block diagram of Figure 15.4 the sampler is shown as an ADC. In practice it could be an ADC or some type of sample-and-hold (S&H) device in analog processors [28, Sect. 7–4].
- The M_R samples are loaded into a buffer memory for later processing. In the block diagram of Figure 15.4, the M_R samples from the nth pulse are stored in the nth column of the buffer memory. In analog processors the buffer memory is some type of analog delay line. In digital processors, the buffer memory is a digital memory device.
- This process of sampling the matched filter output and storing the samples in the columns of the buffer memory continues for N pulses.
- At the end of the N pulse interval, the contents of the buffer memory are read out across rows and sent to M_R signal processors. Analog processors could literally use M_R processors like those of Figure 15.3 (the HPFs followed by BPFs). In a digital processor, there could be M_R digital processors like Figure 15.3, or a single software processor that sequentially processes the outputs of the buffer memory. Because of the fact that the buffer memory is loaded column-wise and unloaded row-wise, it is often termed a *corner turn memory* [29, p. 267; 30, p. 149].
- The duration of the N pulses is NT, where T is the waveform PRI. NT is termed a *coherent processing interval* (CPI) [31, p. 113]. The adverb "coherent" derives from the fact that the signal processor coherently processes the N pulses.
- The signal processor generates N outputs for each of the M_R range cells. These $M_R \times N$ outputs are subsequently sent to some type of post processor. In search, this is usually some type of CFAR and detection processor (see Chapter 7). The outputs could also be sent to a display. The array of M_R range by N Doppler filter outputs is termed a *range-Doppler matrix* [32].[1]

The above discussion applies to search. During track, range will be known reasonably well. Thus, M_R can be small, as small as two in the range track channel and one in the angle track channels [33]. Also, as indicated earlier, Doppler frequency is known reasonably well so only one Doppler filter may be sufficient in the signal processor, or two for Doppler estimation and/or Doppler tracking.

Now that we have a basic understanding of signal processor structure, we return to the problem of how to analyze signal processors.

References

[1]   Barton, D. K., *Radar System Analysis and Modeling,* Norwood, MA: Artech House, 2005.

[2]   Skolnik, M. I., *Introduction to Radar Systems,* 3rd ed., New York: McGraw-Hill, 2001.

---

[1] In some instances, the clutter rejection filter (HPF) will process all N pulses but the coherent integrator (BPF) will process N_{Dop} pulses, where $N_{Dop} < N$. The result is a $M_R \times N_{Dop}$ range-Doppler matrix. The fact that the BPF does not process all N pulses has to do with filter transients. In particular, the output of the HPF cannot be sent to the BPF until the HPF transients settle, and the HPF subsequently achieves its design clutter rejection.

[3] Lijun, W., L. Feng, and W. Shunjun, "An Improved Design and Practical Application of MTD," *Fire Control Radar Technology,* vol. 34, no. 1, Mar. 2005, pp. 9–12, 25.

[4] Kun, H., et al., "Design and Implementation of Doppler Filter Bank in MTD Radar," *Fire Control Radar Technology,* vol. 35, 2006, pp. 57–59.

[5] GuoRong, H., et al., "ASICs Design for an MTD Radar," *2nd Int. Conf. ASIC, 1996,* Shanghai, China, Oct. 21–24, 1996, pp. 69–72.

[6] Yanhang, L. I., and W. Xuegang, "Design of Radar MTD Based on ADSP-TS101," *Commun. Inf. Technol.,* no. 5, 2007, pp. 66–68.

[7] Yan-ping, L. I., "Radar Moving Target Detection System Based on Single Chip FPGA," *Shipboard Electron. Countermeasure,* vol. 31, no. 1, Feb. 2008, pp. 78–81.

[8] Gillespie, N. R., J. B. Higley, and N. MacKinnon, "The Evolution and Application of Coherent Radar Systems," *IRE Trans. Mil. Electron.,* vol. 5, no. 2, Apr. 1961, pp. 131–139.

[9] Barton, D. K., "A Half Century of Radar," *IEEE Trans. Microwave Theory and Techniques,* vol. 32, no. 9, Sept. 1984, pp. 1161–1170.

[10] Emslie, A. G., and R. A. McConnell, "Moving Target Indication," in *Radar Systems Engineering,* vol. 1 of MIT Radiation Lab. Series (L. N. Ridenour, ed.), New York: McGraw-Hill, 1947; Norwood, MA: Artech House (CD-ROM edition), 1999.

[11] Emslie, A. G., "MTI Using Coherent IF," MIT Radiation Lab. Series, Rep. No. 104, Aug. 22, 1945. Listed as reference in *Radar Systems Engineering,* vol. 1 of MIT Radiation Lab. Series (L. N. Ridenour, ed.), New York: McGraw-Hill, 1947, pp. 640–645; Norwood, MA: Artech House (CD-ROM edition), 1999.

[12] Emslie, A. G., "Moving Target Indication on MEW," MIT Radiation Lab. Series, Rep. No. 1080, Feb. 19, 1946. Listed as a reference in *Radar Systems Engineering,* vol. 1 of MIT Radiation Lab. Series (L. N. Ridenour, ed.), New York: McGraw-Hill, 1947, p. 645; Norwood, MA: Artech House (CD-ROM edition), 1999.

[13] Emslie, A. G., "Moving Object Radio Pulse System," U.S. Patent 2,543,448, Feb. 27, 1951.

[14] Emslie, A. G., "Moving Target Indication Radar System," U.S. Patent 2,555,121, May 29, 1951.

[15] Emslie, A. G., "Moving Target Detecting System," U.S. Patent 2,617,983, Nov. 11, 1952.

[16] Emslie, A. G., "Moving Object Radio Pulse-Echo System," U.S. Patent 2,659,076, Nov. 10, 1953.

[17] Emslie, A. G., "Moving Object Radio Pulse-Echo System," U.S. Patent 2,659,077, Nov. 10, 1953.

[18] Emslie, A. G., "Moving Target Indicating Radar System," U.S. Patent 2,710,398, Jun. 7, 1955.

[19] University of Maryland, "Innovation Hall of Fame," www.eng.umd.edu/html/ihof/inductees/smith.html.

[20] IEEE, "1984 Pioneer Award," *IEEE Trans. Aerosp. Electron. Syst.,* vol. 20, no. 31, May 1984, pp. 290–291.

[21] Perkins, L. C., H. B. Smith, and D. H. Mooney, "The Development of Airborne Pulse Doppler Radar," *IEEE Trans. Aerosp. Electron. Syst.,* vol 20, no. 3, May 1984, pp. 292–303.

[22] Fell, T. T., et al., "Pulse Doppler Radar System," U.S. Patent 3,011,166, Nov. 28, 1961.

[23] Smith, H. B., D. H. Mooney, Jr., and W. Ewanus, "Pulse Doppler Radar System," U.S. Patent 3,023,409, Feb. 27, 1962.

[24] IEEE, "2005 Pioneer Award," *IEEE Trans. Aerosp. Electron. Syst.,* vol. 42, no. 3, Jul. 2006, pp. 1171–1172.

[25] Stone, M., "2005 Pioneer Award to C.E. Muehe: Introductory Remarks," *IEEE Trans. Aerosp. Electron. Syst.,* vol. 42, no. 3, Jul. 2006, pp. 1173–1176.

[26] Muehe, C. E., "The Moving Target Detector," *IEEE Trans. Aerosp. Electron. Syst.,* vol. 42, no. 3, Jul. 2006, pp. 1177–1181.

[27] Bassford, R. S., W. Goodchild, and A. De La Marche, "Test and Evaluation of the Moving Target Detector (MTD) Radar," Lincoln Laboratories, Rep. No. FAA-RD-77-118, 1977. Available from DTIC as ADA047887.

[28] Kester, W., ed., *The Data Conversion Handbook*, New York: Newnes, 2005.

[29] Parker, M., *Digital Signal Processing 101: Everything You Need to Know to Get Started*, Cambridge, MA: Newnes, 2017.

[30] Morris, G. V. and L. Harkness, eds., *Airborne Pulsed Doppler Radar*, 2nd ed., Norwood, MA, 1996.

[31] Minkler, G. and J. Minkler, *CFAR: The Principles of Automatic Detection in Clutter*, Baltimore, MD: Magellan Book Company, 1990.

[32] Schleher, D. C., *MTI and Pulsed Doppler Radar with MATLAB*, 2nd Edition ed., Norwood: Artech House, 2010.

[33] Budge, M. C., Jr. and S. R. German, *Basic Radar Tracking*, Norwood, MA: Artech House, 2019.

Chapter 16

Signal Processor Analysis

16.1 INTRODUCTION

In this chapter, we derive equations for the signals on which the signal processor must operate. We trace the signal from its origin in the radar exciter to the output of the matched filter of the receiver. In the process, we discuss how the radar and environment affect signal characteristics. After we derive the signal model, we develop a general equation that we will later specialize to characterize the signal processors.

16.2 SIGNAL MODEL GENERATION

A simplified block diagram of a radar transmitter and receiver is shown in Figure 16.1. The block diagram contains only elements essential to our development. Specifically, it does not contain any IF amplifiers, filters, or the mixers needed to upconvert and downconvert the various signals, except for the STALO (STAble Local Oscillator), which we need to include because of phase noise. We have not lost any generality with this technique because we will use normalized, complex signal notation. This allows us to ignore IF processes.

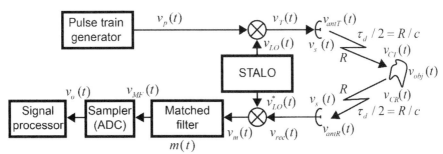

Figure 16.1 Transmitter, receiver, and signal processor.

Complex signal notation has an advantage of being easy to manipulate since the signals are represented by complex exponentials rather than sines and cosines. Operations such as filtering, sampling, and transforms are treated the same with complex signals and real signals. We must take care when using complex signals in nonlinear operations such as mixing. For example, in the transmit mixer of Figure 16.1, we used $v_{LO}(t)$, whereas on the receive mixer we used its conjugate, $v^*_{LO}(t)$. We knew we needed to do this based on real signal analyses. Specifically, we performed real signal analyses and used the results to determine what complex signal operations we needed to perform.

In Figure 16.1, $v_p(t)$ is the pulse train, a complex, baseband signal. This means that its energy, or power, is concentrated around 0 Hz, as opposed to some IF. As a note, signals that have a Doppler frequency are usually considered baseband signals, even though their energy is not truly concentrated around 0 Hz.

The typical $v_p(t)$ of interest is a sequence of rectangular pulses with a width of τ_p and a PRI of T. $v_p(t)$ could consist of a burst of pulses or a semi-infinite string of pulses, depending on the radar and the waveform. In an older, dish type of radar that tracks a single target, $v_p(t)$ would consist of a semi-infinite string of pulses. In phased array, multifunction radars, $v_p(t)$ would contain a burst of two pulses to hundreds of pulses. A graphical representation of $v_p(t)$ (actually $|v_p(t)|$) is shown in Figure 16.2.

In equation form, $v_p(t)$ is

$$v_p(t) = \sum_k \text{rect}\left[\frac{t-kT}{\tau_p}\right] = \text{rect}\left[\frac{t}{\tau_p}\right] * \sum_k \delta(t-kT) = \text{rect}\left[\frac{t}{\tau_p}\right] * i(t) \qquad (16.1)$$

where $\delta(t)$ is the Dirac delta and * denotes convolution. The summation notation denotes a summation over the number of pulses that make up $v_p(t)$—that is, the number of pulses in the burst.

The form of $v_p(t)$ implies that the pulses are unmodulated. A more general form would be

$$v_p(t) = p(t) * i(t) = \sum_k p(t-kT) \qquad (16.2)$$

where $i(t)$ is a train of unit impulses and $p(t)$ is a complex signal notation for a complicated waveform such as a phase-coded pulse or an LFM pulse (see Chapter 11).

The STALO signal, $v_{LO}(t)$, is of the form

$$v_{LO}(t) = e^{j\omega_c t} e^{j\phi(t)} \qquad (16.3)$$

In (16.3), $f_c = \omega_c/2\pi$ is the carrier frequency. $\phi(t)$ is termed the *phase noise* [1–5] on the STALO signal and represents the instability of the oscillator that generates the STALO signal. As implied by its name, $\phi(t)$ is a random process and is such that $\exp[j\phi(t)]$ is WSS. We will address phase noise later. Phase noise is included because it is often a limiting factor on clutter attenuation capabilities of the signal processor.

Figure 16.2 Depiction of $|v_P(t)|$.

In most radars, $v_{LO}(t)$ also includes an *amplitude noise* component such that

$$v_{LO}(t) = [1 + A(t)] e^{j\omega_c t} e^{j\phi(t)} \tag{16.4}$$

However, the amplitude noise, $A(t)$, is usually made very small by the radar designer and is normally considered to have a much smaller influence on signal processor performance than phase noise, $\phi(t)$. For this reason, amplitude noise is almost always ignored in signal processor analyses. Having said this, it should be noted that modern STALOs are becoming so stable that amplitude noise may soon overtake phase noise as the limiting factor in signal processor performance [6–12].[1]

$v_T(t)$ is the signal at the transmitter output and is given by

$$v_T(t) = v_p(t) v_{LO}(t) \tag{16.5}$$

To account for the fact that the antenna may be scanning (which is generally taken to mean the beam is rotating horizontally, as in a search radar), we include a scanning term, denoted $v_S(t)$. This scanning term captures the variation in directivity as the antenna beam scans past a target. If we are considering a tracking radar or a phased array radar that moves its beam in steps and transmits a burst of pulses at each beam position, then $v_S(t) = 1$. A standard form of $v_S(t)$ for the scanning case is [13, p. 134]

$$v_S(t) = e^{-t^2/2\sigma_{TS}^2} \tag{16.6}$$

where

$$\sigma_{TS}^2 = \frac{1}{2.77} \left(\frac{\alpha_B T_{SCAN}}{\pi} \right)^2 \tag{16.7}$$

T_{SCAN} is the scan period (in seconds) and α_B is the azimuth beamwidth (in radians). If the antenna is scanning in elevation instead of azimuth, we would replace α_B with ε_B, the elevation beamwidth. The form of (16.6) is based on the assumption that the main beam of the antenna can be adequately represented by a Gaussian type of function (see Chapter 17).

In practice, $v_S(t)$ is a periodic function with a period of T_{SCAN}. However, since the time period of interest in the signal processor is small relative to T_{SCAN}, it is assumed the radar beam scans by the target only one time. The time period of interest in the signal processor is the CPI.

---

[1] Another factor that affects signal processor performance is timing jitter [11, 12]. It is also normally ignored, but could become a limiting factor as the phase noise of STALOs continues to improve.

$v_{obj}(t)$ is the *object signal* and is our means of capturing the power spectrum properties of the clutter or target. $v_{obj}(t)$ is a random process and is assumed to be WSS.[2] For clutter we use $v_{obj}(t) = C(t)$, and for targets we use $v_{obj}(t) = T(t)$, where $T(t)$ represents the *target signal*.

The spectrum of $v_{obj}(t)$ is given by

$$V_{obj}(f) = \int_{-\infty}^{\infty} R_{obj}(\tau) e^{-j2\pi f \tau} d\tau \tag{16.8}$$

where

$$R_{obj}(\tau) = E\left\{v_{obj}(t+\tau) v_{obj}^*(t)\right\} \tag{16.9}$$

is the autocorrelation of $v_{obj}(t)$.

For clutter, we replace $V_{obj}(f)$ by $C(f)$, where $C(f)$ is the clutter spectrum, which is discussed in Chapter 17. For targets, we replace $V_{obj}(f)$ with

$$T(f) = \delta(f - f_d) \tag{16.10}$$

That is, we assume the target is represented by a single spectral line at the target Doppler frequency of f_d. As discussed in Chapter 3, we normally treat the target signal as a random process, since a deterministic signal can be considered a limiting case of a random signal. In this chapter, we further assume it is a WSS random process.

To complete our definitions, $v_{rec}(t)$ is the received signal after it goes through the antenna, $v_m(t)$ is the output of the receiver's mixer, and $v_{MF}(t)$ is the matched filter output. $v_o(t)$ is the sampled version of $v_{MF}(t)$ and is the signal that goes to the signal processor. The matched filter is usually assumed to be matched to a single pulse of the original pulse train, $v_p(t)$.

16.2.1 Signal Model: Time Domain Analysis

We now want to develop an equation for $v_o(t)$ and, ultimately, its power spectrum, $S_o(f)$. We start our analysis by noting that the mixer is a multiplication process. Thus, as indicated earlier, the signal sent to the antenna is

$$v_T(t) = v_p(t) v_{LO}(t) \tag{16.11}$$

If the antenna is scanning, its pattern modulates the amplitude of $v_T(t)$. We model this as a multiplication of $v_T(t)$ by $v_S(t)$. Thus, the signal radiated by the antenna is

$$v_{antT}(t) = v_p(t) v_{LO}(t) v_S(t) \tag{16.12}$$

Recall that we set $v_S(t) = 1$ if we consider the tracking problem or a phased array where the beam is fixed during the CPI.

---

[2] A note about stationarity: Realistically, none of the random processes we are dealing with are truly WSS. However, over the CPI, we can reasonably assume they are stationary (actually, cyclostationary). From random processes theory, we know that if a process is stationary in the wide sense over a CPI, then we can reasonably assume that it is WSS. This stems from the fact that we are interested only in the random process over the CPI.

After the signal leaves the antenna, it propagates a distance of R to the object (clutter or target). We represent this propagation by incorporating a delay into $v_{antT}(t)$. We denote this delay as $\tau_d/2$. We should also include an attenuation that depends on R. However, we will ignore range dependent attenuation for now. We will consider the effect of range attenuation at a later time.

With the above, the signal that arrives at the object is

$$v_{CT}(t) = v_{antT}(t - \tau_d/2)$$
$$= v_p(t - \tau_d/2) v_{LO}(t - \tau_d/2) v_S(t - \tau_d/2) \quad (16.13)$$

The object reradiates (see Chapter 2) the signal back to the radar and imposes its temporal, and thus spectral, characteristics on the reflected signal. We represent this reradiation operation by multiplying $v_{CT}(t)$ by $v_{obj}(t)$, the function that we use to represent the temporal properties of the object. We represent the reradiation operation by multiplication because the interaction of the signal with the clutter (or target) is essentially a modulation process. We learned this in Chapter 1 when we found the motion of a target caused a shift in the frequency of the signal (Doppler shift) and the amplitude of the return signal was a function of the target RCS [14].

The signal reflected by the object is

$$v_{CR}(t) = v_{CT}(t) v_{obj}(t)$$
$$= v_p(t - \tau_d/2) v_{LO}(t - \tau_d/2) v_S(t - \tau_d/2) v_{obj}(t) \quad (16.14)$$

and the signal at the receive antenna is

$$v_{antR}(t) = v_{CR}(t - \tau_d/2)$$
$$= v_p(t - \tau_d) v_{LO}(t - \tau_d) v_S(t - \tau_d) v_{obj}(t - \tau_d/2) \quad (16.15)$$

This signal next picks up the scan modulation and is then heterodyned by the receiver mixer to produce the matched filter input, $v_m(t)$. In equation form,

$$v_m(t) = v_{antR}(t) v_S(t) v_{LO}^*(t)$$
$$= v_p(t - \tau_d) v_{LO}(t - \tau_d) v_S(t - \tau_d) v_{obj}(t - \tau_d/2) v_S(t) v_{LO}^*(t) \quad (16.16)$$

We now want to study and manipulate this equation. We start by simplifying the equation and making some approximations. Since the antenna will not move much during the round-trip delay, τ_d, we can assume $v_S(t)$ does not change much over the time duration of τ_d. This means $v_S(t - \tau_d) \approx v_S(t)$. With this we get

$$v_m(t) = v_p(t - \tau_d) v_S^2(t) v_{obj}(t - \tau_d/2) v_{LO}(t - \tau_d) v_{LO}^*(t) \quad (16.17)$$

The output of the matched filter is

$$v_{MF}(t) = m(t) * v_m(t) \quad (16.18)$$

where $m(t)$ is the matched filter impulse response. Finally, the signal sent to the signal processor is

$$v_o(t) = v_{MF}(kT) \tag{16.19}$$

That is, $v_o(t)$ is a sampled and held version of $v_{MF}(t)$. As a note, in practice, many sets of $v_{MF}(kT)$ will be sent to the signal processor—one set for each range cell of interest.

The next step in the development is to manipulate (16.17) through (16.19) to eventually derive an equation for $S_o(f)$, the spectrum of $v_o(t)$. The development is very interesting, but also tedious and lengthy. As a result, we have moved it to Appendix 16A. We present the final results here and use them to begin our signal processor analyses.

16.2.2 Signal Model: Frequency Domain Analysis

From Appendix 16A we have that the spectrum input to the signal processor is

$$S_o(f) = \frac{1}{T} \sum_{l=-\infty}^{\infty} MF(f - l/T) S_r(f - l/T) \tag{16.20}$$

$MF(f)$ is the matched-range, Doppler cut of the cross ambiguity function of $p(t)$ and $q(t)$, where $q(t)$ is the signal to which the matched filter, $m(t)$, is matched (see Appendix 16A and Chapter 8, "Ambiguity Function"). Specifically,

$$MF(f) = \left| \int_{-\infty}^{\infty} p(t) q^*(t) e^{j2\pi ft} dt \right|^2 \tag{16.21}$$

Typically, $q(t) = p(t)$. For uncoded pulses, phase-coded pulses, and LFM pulses that do not incorporate weighting for range sidelobe reduction, $MF(f)$ is of the form

$$MF(f) = \left| \text{sinc}(f\tau_p) \right|^2 \tag{16.22}$$

where τ_p is the *uncompressed* pulsewidth.

From Appendix 16A we derived the spectrum

$$S_r(f) = \int_{-\infty}^{\infty} R_r(\tau) e^{-j2\pi f\tau} d\tau \tag{16.23}$$

where

$$R_r(\tau) = \overline{E\{r(t+\tau) r^*(t)\}} \tag{16.24}$$

is the averaged autocorrelation of $r(t)$. $r(t)$ is the part of (16.17) that doesn't contain $v_p(t-\tau_d)$ term. That is

$$r(t) \approx v_S^2(t) v_{obj}(t - \tau_d/2) v_{LO}(t - \tau_d) v_{LO}^*(t) = v_S^2(t) v_{obj}(t) \Phi(t) \tag{16.25}$$

where $\Phi(t)$ is the phase noise component of $v_{LO}(t)$ (see Appendix 16A).

The use of the averaged autocorrelation of $r(t)$ is necessary because the (generally) periodic nature of $v^2{}_S(t)$ makes $r(t)$ wide-sense cyclostationary (WSCS) instead of simply WSS (see Appendix 16A and Appendix 16B).

We can reasonably assume the random processes $v_{obj}(t)$ and $\Phi(t)$ are independent because the statistical properties of one has no influence on the statistical properties of the other. With this we can write

$$R_r(\tau) = R_{vs}(\tau) R_{obj}(\tau) R_\Phi(\tau) \tag{16.26}$$

where $R_{obj}(\tau)$ is given by (16.9),

$$R_\Phi(\tau) = E\{\Phi(t+\tau)\Phi^*(t)\} \tag{16.27}$$

and

$$R_{vs}(\tau) = \overline{v_S^2(t+\tau)v_S^2(t)} = \frac{1}{T_{scan}} \int_0^{T_{SCAN}} v_S^2(t+\tau)v_S^2(t)\,dt \tag{16.28}$$

is the averaged time autocorrelation of $v^2{}_S(t)$.

Since $R_r(\tau)$ is a product of autocorrelations, $S_r(f)$ is the convolution of their associated spectra. Thus,

$$S_r(f) = V_S(f) * V_{obj}(f) * \Phi(f) \tag{16.29}$$

The scanning function modulation spectrum, $V_S(f)$, has the form [13]

$$V_S(f) = \frac{1}{\sigma_S \sqrt{2\pi}} e^{-f^2/(2\sigma_S^2)} \tag{16.30}$$

where

$$\sigma_S = 0.265 \left(\frac{2\pi}{\alpha_B T_{scan}}\right) \tag{16.31}$$

and the variables are the same as in (16.6) and (16.7). If the radar antenna is not scanning, $V_S(f)$ reduces to

$$V_S(f) = \delta(f) \tag{16.32}$$

For clutter, we replace $V_{obj}(f)$ by the appropriate $C(f)$. The two forms of $C(f)$ we consider here are (see Chapter 17)

$$C_{Gauss}(f) = \frac{1}{\sigma_f \sqrt{2\pi}} e^{-f^2/2\sigma_f^2} \tag{16.33}$$

for a Gaussian spectrum model and

$$C_{Exp}(f) = \frac{r}{1+r}\delta(f) + \frac{1}{1+r}\frac{\beta\lambda}{4}e^{-\beta|\lambda f/2|} \qquad (16.34)$$

for an exponential spectrum model. For targets, we replace $V_{obj}(f)$ by the target spectrum $T(f)$, where

$$T(f) = \delta(f - f_d) \qquad (16.35)$$

and f_d is the target Doppler frequency.

$\Phi(f)$ represents the phase noise spectrum of the radar. As shown in Appendix 16A, we can write

$$v_{LO}(t - \tau_d)v_{LO}^*(t) = e^{j\omega_c \tau_d}e^{j\Delta\phi(t)} = v_{PH}(t) \qquad (16.36)$$

where $\Delta\phi(t)$ is the total transmit and receive phase noise of the STALO. We note that $\Delta\phi(t)$ is small relative to unity and write [1]

$$e^{j\Delta\phi(t)} \approx 1 + j\Delta\phi(t) = \Phi(t) \qquad (16.37)$$

With this we get the autocorrelation of $v_{PH}(t)$ as

$$\begin{aligned}R_{PH}(\tau) &= E\{v_{PH}(t+\tau)v_{PH}^*(t)\} = E\{\Phi(t+\tau)\Phi^*(t)\} \\ &= 1 + E\{\Delta\phi(t+\tau)\Delta\phi^*(t)\} = 1 + R_{\Delta\phi}(\tau)\end{aligned} \qquad (16.38)$$

where we made the tacit assumption that $\Delta\phi(t)$ is zero-mean and WSS.

From (16.38) we get the power spectrum of $v_{PH}(t)$ as

$$\Phi(f) = \int_{-\infty}^{\infty} R_{PH}(\tau)e^{-j2\pi f\tau}d\tau = \delta(f) + \Phi_{\Delta\phi}(f) \qquad (16.39)$$

If we assume $\Delta\phi(t)$ is white, we get

$$\Phi_{\Delta\phi}(f) = \Phi_0 \qquad (16.40)$$

where Φ_0 is termed the phase noise sideband amplitude of the phase noise spectrum, or simply *phase noise sideband level*. Later we will consider phase noise models where $\Delta\phi(t)$ is not white and investigate other forms of $\Phi_{\Delta\phi}(f)$. $\Phi_{\Delta\phi}(f)$ settles to a floor value of Φ_0 as f becomes large.

$\Phi_{\Delta\phi}(f)$ is caused by instability in the STALO circuitry. In dB terms, it has the units of dBc/Hz, which means dB relative to the power in the carrier of the radar, measured in a 1-Hz bandwidth, at some frequency relative to the carrier frequency. Rough estimates for Φ_0, the floor value, are −125 to −150 dBc/Hz for radars that use STALOs that employ very narrowband filters or phase-locked loops (such as klystron-based STALOs), −110 to −130 dBc/Hz for radars that use frequency multiplied crystal or digitally synthesized STALOs, and around −90 dBc/Hz for radars that use magnetron transmitters. To repeat, these values are rough estimates because the field of STALO technology is moving rapidly, especially crystal-based STALOs. Modern, well-designed radars that use good STALOs have phase noise values in the vicinity of -125 to -135 dBc/Hz. Some advanced radar designs appear to be pushing phase noise floor to -150 to -160 dBc/Hz.

If we ignore phase noise, $\Phi(f)$ reduces to

$$\Phi(f) = \delta(f) \tag{16.41}$$

$\delta(f)$ is the center spectral line, or carrier, and represents a pure sinusoid.

The power at the output of the sampler, which is also the power at the input to the signal processor, is given by (see Appendix 16A)

$$P_o = T \int_{-1/2T}^{1/2T} S_o(f) df = T \int_{-1/2T}^{1/2T} \frac{1}{T} \sum_{l=-\infty}^{\infty} MF(f - l/R) S_r(f - l/T) df \tag{16.42}$$

If we reverse the order of summation and integration, we get

$$P_o = \sum_{l=-\infty}^{\infty} \int_{-1/2T}^{1/2T} MF(f - l/T) S_r(f - l/T) df \tag{16.43}$$

In each of the integrals, we make the change of variables $\alpha = f - l/T$ to get

$$P_o = \sum_{l=-\infty}^{\infty} \int_{-l/T-1/2T}^{-l/T+1/2T} MF(\alpha) S_r(\alpha) d\alpha \tag{16.44}$$

We recognize (16.44) as an infinite sum of nonoverlapping integrals, which we can write as

$$P_o = \int_{-\infty}^{\infty} MF(f) S_r(f) df \tag{16.45}$$

or, using (16.29),

$$P_o = \int_{-\infty}^{\infty} MF(f) \left[V_S(f) * V_{obj}(f) * \Phi(f) \right] df \tag{16.46}$$

where we changed the variable of integration from α back to f. As a note, because power is preserved in the sampling process, P_o is also the power at the output of the matched filter.

P_o, as defined by (16.46), is a normalized power because each of the spectra in the brackets has an area of unity [if we ignore the phase noise part of $\Phi(f)$]. To properly scale this power for clutter and targets, we need to associate proper powers with their spectra. If we do this, we get clutter and target power at the matched filter output (which is the same as the power at sampler output and the power at the input to the signal processor) as

$$P_{Co} = P_C \int_{-\infty}^{\infty} MF(f)\left[V_S(f) * C(f) * \Phi(f)\right] df \qquad (16.47)$$

and

$$P_{So} = P_S \int_{-\infty}^{\infty} MF(f)\left[V_S(f) * T(f) * \Phi(f)\right] df \qquad (16.48)$$

where, for our purposes, P_C and P_S are scaling factors related to the clutter and target RCS and the various terms of the radar range equation. We assign values to these scaling factors based on SNR and CNR, which we can compute using the radar range equation.

16.2.3 Relation of P_C and P_S to the Radar Range Equation

We can write the SNR and CNR at the output of the matched filter as

$$SNR = P_{So}/P_{No} \quad \text{and} \quad CNR = P_{Co}/P_{No} \qquad (16.49)$$

Assuming we know P_{No}, we can solve (16.49) for P_{So} and P_{Co}. Further, for the ideal conditions where we assume $T(f) = C(f) = \delta(f)$, the antenna is not scanning, and we ignore phase noise (all of which are tacit assumptions we make when we use the radar range equation), the integrals of (16.47) and (16.48) are unity (see Exercise 2). From this, (16.47) and (16.48), we have

$$P_C = P_{Co} = (CNR)P_{No} \quad \text{and} \quad P_S = P_{So} = (SNR)P_{No} \qquad (16.50)$$

This gives us a means of scaling clutter and signal power relative to some arbitrary noise power through the radar range equation.

16.3 SIGNAL PROCESSOR ANALYSES

16.3.1 Background

Now that we have equations for the spectra at the input to the signal processor, we turn our attention to considering how to use them to perform signal processor analyses. We will consider sampled data signal processors. This could include signal processors using analog components or signal processors that are implemented with digital hardware. The characteristic that dictates we use sampled data techniques is the assumption that we are dealing with pulsed radars. Because of this, the signals into the signal processor are sampled once per PRI. The sampling can be performed by a sample-and-hold device for processors that includes analog devices, or an ADC for processors that uses digital signal processing. The process of performing the sampling and sending the samples to the signal processor is explained in the bulleted text following Figure 15.4. To summarize, each of the M_R range cells in a PRI interval is sampled every PRI (every T seconds) to gather N samples over the CPI (for a total of $M_R \times N$ samples). The N samples for each range cell are sent to, effectively, M_R signal processors to form the range Doppler matrix. We will be analyzing one of these signal processors (the signal processor for one range cell).

Therefore, as discussed in Appendix 16A, we will be concerned with a sampled-data system where the sample spacing is T.

We initially consider purely digital signal processors and assume we have a signal processor that has a z-transfer function of $H(z)$ and an equivalent frequency response of

$$H(f) = |H(z)|^2_{z=e^{j2\pi fT}} \tag{16.51}$$

We are using the form of frequency response function usually used in analyzing random processes because, by assumption, our clutter (and target) signal is a random process at the signal processor input.

The standard way of performing digital signal processor analyses in the frequency domain is to find $S_o(f)$ from (16.20) [with the appropriate $S_r(f)$], multiply it by $H(f)$, and integrate the result over $[-1/2T, 1/2T)$ to find the total signal power at the output of the signal processor. Recall that we use this approach because, for digital signals, the only valid frequency region is $[-1/2T, 1/2T)$.

As we did to find P_o, we propose a different approach. Rather than use $S_o(f)$ over $[-1/2T, 1/2T)$, we use $S_{MF}(f) = MF(f)S_r(f)$ and $H(f)$ over $(-\infty,\infty)$. As before, we multiply these and integrate to find the total signal power, except this time we integrate over $(-\infty,\infty)$. With this approach, we are "unfolding" (or "de-aliasing) $S_o(f)$ and $H(f)$ and then "refolding" them when we find the power. This approach has the advantage of avoiding the $S_o(f)$ summation of (16.20).

We digress to show that the approach we propose is valid in terms of computing the power out of the signal processor. We start by noting that the signal power at the output of the signal processor is (see Appendix 16A)

$$P_{out} = T \int_{-1/2T}^{1/2T} H(f) S_o(f) df \tag{16.52}$$

We substitute for $S_o(f)$ to get

$$P_{out} = T \int_{-1/2T}^{1/2T} H(f) \frac{1}{T} \sum_{l=-\infty}^{\infty} MF(f - l/T) S_r(f - l/T) df \tag{16.53}$$

and bring $H(f)$ inside of the sum to yield

$$P_{out} = T \int_{-1/2T}^{1/2T} \frac{1}{T} \sum_{l=-\infty}^{\infty} H(f) MF(f - l/T) S_r(f - l/T) df \tag{16.54}$$

We note $H(f)$ is periodic with a period of $1/T$. This allows us to replace $H(f)$ with $H(f - l/T)$ since $H(f - l/T) = H(f)$. Doing this, and reversing the order of summation and integration, results in

$$P_{out} = \sum_{l=-\infty}^{\infty} \int_{-1/2T}^{1/2T} H(f - l/T) MF(f - l/T) S_r(f - l/T) df \tag{16.55}$$

In each of the integrals of the sum, we make the change of variables $\alpha = f - l/T$ to get

$$P_{out} = \sum_{l=-\infty}^{\infty} \int_{-lT-1/2T}^{-lT+1/2T} H(\alpha) MF(\alpha) S_r(\alpha) d\alpha \qquad (16.56)$$

Finally, we recognize the above as an infinite sum of nonoverlapping integrals, which we can write as a single integral over $(-\infty,\infty)$. That is,

$$P_{out} = \int_{-\infty}^{\infty} H(f) MF(f) S_r(f) df \qquad (16.57)$$

which is the desired result. Note that we changed the variable of integration from α back to f.

The integral of (16.57) converges quickly for target and clutter spectra, and no phase noise. In that case $\Phi(f) = \delta(f)$ and $V_S(f)$ and $V_{obj}(f)$ are narrowband. This means $S_r(f)$ is also narrowband, and that $H(f)MF(f)S_r(f)$ is narrowband. This narrowband property means numerical evaluation of the integral, which is almost always required, can be performed over a small band of frequencies.

When we consider the phase noise component of $\Phi(f)$, we will need to use a different approach to finding P_{out} since, in that case, $S_r(f)$ will not be narrowband. A similar situation occurs when we consider receiver noise. We delay discussion of these situations until later, when we need them.

16.4 EXERCISES

1. Derive (16.22).

    2. Show that the integrals of (16.47) and (16.48) equal 1 for the case where $V_S(f) = C(f) = T(f) = \Phi(f) = \delta(f)$.

References

[1] Raven, R. S., "Requirements on Master Oscillators for Coherent Radar," *Proc. IEEE,* vol. 54, no. 2, Feb. 1966, pp. 237–243.

[2] Raven, R. S., "Correction to 'Requirements on Master Oscillators for Coherent Radar,'" *Proc. IEEE,* vol. 55, no. 8, Aug. 1967, p. 1425.

[3] Scheer, J. A., and J. L. Kurtz, eds., *Coherent Radar Performance Estimation,* Norwood, MA: Artech House, 1993.

[4] Rogers, R. G., *Low Phase Noise Microwave Oscillator Design* (Artech Microwave Library), Norwood, MA: Artech House, 1991.

[5] Lee, T. H., and A. Hajimiri, "Oscillator Phase Noise: A Tutorial," *IEEE J. Solid-State Circuits,* vol. 35, no. 3, Mar. 2000, pp. 326–336.

[6] Boroditsky, R., and J. Gomez, "Ultra Low Phase Noise 1 GHz OCXO," *IEEE Int. Frequency Control Symp.,* 2007 Joint with the 21st European Frequency and Time Forum, Geneva, Switzerland, May 29–June 1, 2007, pp. 250–253.

[7]   Poddar, A. K., and U. L. Rohde, "The Pursuit for Low Cost and Low Phase Noise Synthesized Signal Sources: Theory & Optimization," *2014 IEEE Int. Frequency Control Symp. (FCS)*, Taipei, Taiwan, May 19–22, 2014, pp. 1–9.

[8]   Hoover, L., H. Griffith, and K. DeVries, "Low Noise X-band Exciter Using a Sapphire Loaded Cavity Oscillator," *2008 IEEE Int. Frequency Control Symp. (FCS)*, Honolulu, HI, May 19–21, 2008, pp. 309–311.

[9]   Wenzel Associates, Inc., "Low Noise Crystal Oscillators > Sorcerer II," www.wenzel.com/wp-content/uploads/Sorcerer-II.pdf.

[10] Wenzel Associates, Inc., "Low Noise Crystal Oscillators > Golden MXO (PLO w/ Dividers)," www.wenzel.com/wp-content/uploads/GMXO-PLD.pdf.

[11] Budge, M.C., Jr., and S. M. Gilbert, "Timing Jitter Spectrum in Pulsed and Pulsed Doppler Radars," *Proc. IEEE Southeastcon '93*, vol. 4, Apr. 4–7, 1993.

[12] Budge, M.C., Jr., "Timing Jitter Characterization for Pulsed and Pulsed and Pulsed Doppler Radars," *Proc. IEEE Southeastcon '92*, vol.1, Apr. 12–15, 1992, pp. 199–201.

[13] Skolnik, M. I., *Introduction to Radar Systems*, 2nd ed., New York: McGraw-Hill, 1980.

[14] Budge, M. C., Jr., "EE 725: Advanced Radar Technique," www.ece.uah.edu/courses/material/EE725/index.htm.

[15] Tsui, J., *Digital Techniques for Wideband Receivers*, 2nd ed., Norwood, MA: Artech House, 2001.

[16] Kester, W., ed., *The Data Conversion Handbook*, New York: Newnes, 2005.

APPENDIX 16A: DERIVATION SIGNAL PROCESSOR INPUT SPECTRUM

We start the derivation at (16.17), which is

$$v_m(t) = v_p(t - \tau_d) v_S^2(t) v_{obj}(t - \tau_d/2) v_{LO}(t - \tau_d) v_{LO}^*(t)$$

For the first step, write the product of the last two terms as

$$v_{LO}(t - \tau_d) v_{LO}^*(t) = e^{j\omega_c(t - \tau_d)} e^{j\phi(t - \tau_d)} e^{-j\omega_c(t)} e^{-j\phi(t)} = e^{-j\omega_c \tau_d} e^{j\Delta\phi(t)} \quad (16A.1)$$

where $\Delta\phi(t) = \phi(t - \tau_d) - \phi(t)$. $\Delta\phi(t)$ represents the total (transmit and receive) phase noise in the radar. We note that $\Delta\phi(t)$ is small relative to unity so that [57, 58]

$$e^{j\Delta\phi(t)} \approx 1 + j\Delta\phi(t) = \Phi(t) \quad (16A.2)$$

With this $v_m(t)$ becomes

$$v_m(t) = v_p(t - \tau_d) v_S^2(t) v_{obj}(t - \tau_d/2) \Phi(t) \quad (16A.3)$$

Note that we dropped the phase term, $\exp(-j\omega_c \tau_d)$. We were able to do this because we can normalize it away in future calculations.

We further simplify (16A.3) by shifting the time origin by $-\tau_d$. This yields

$$v'_m(t) = v_m(t + \tau_d) = v_p(t) v_S^2(t + \tau_d) v_{obj}(t + \tau_d/2) \Phi(t + \tau_d) \quad (16A.4)$$

We argued earlier that $v_S(t)$ changes slowly relative to τ_d so that $v_S^2(t + \tau_d) \approx v_S^2(t)$. Also, $v_{obj}(t)$ and $\Phi(t)$ are WSS random processes. This means their means and autocorrelations do not depend on time origin. Thus, we can replace $v_{obj}(t + \tau_d/2)$ with $v_{obj}(t)$ and $\Phi(t + \tau_d)$ with $\Phi(t)$ and not change their means and autocorrelations [the autocorrelation is what we eventually use to find the power spectrum of $v_m(t)$]. With this we get

$$v_m(t) = v_p(t) r(t) \tag{16A.5}$$

where

$$r(t) = v_S^2(t) v_{obj}(t) \Phi(t) \tag{16A.6}$$

We dropped the prime and reverted to the notation $v_m(t)$ for convenience.

The next step in our derivation is to process $v_m(t)$ through the matched filter and then sample the matched filter output via the sampler/ADC (see Figure 16.1). Before we do this, we need to examine $v_m(t)$ more closely. If we substitute for $v_p(t)$ [see (16.2)] into (16A.5), we get

$$\begin{aligned} v_m(t) &= v_S^2(t) v_{obj}(t) \Phi(t) \sum_k p(t - kT) \\ &= \sum_k p(t - kT) v_S^2(t) v_{obj}(t) \Phi(t) \end{aligned} \tag{16A.7}$$

Since $v_{obj}(t)$ and $\Phi(t)$ are random processes, the product $r(t) = v_S^2(t) v_{obj}(t) \Phi(t)$ is also a random process. However, because $v_S(t)$ is periodic, $r(t)$ is not WSS. However, we show in Appendix 16B that $r(t)$ is wide sense cyclostationary (WSCS). As a result of this, we can use the averaged statistics of $r(t)$ and treat it as a WSS process in the following development. With this in mind, we write

$$v_m(t) = \sum_k p(t - kT) r(t) \tag{16A.8}$$

where we treat $r(t)$ as if it was a WSS random process. We note that $v_m(t)$ is not stationary because of the $p(t - kT)$ term. We address this in the following discussions.

If we represent the impulse response of the matched filter as $m(t)$, we can write the output of the matched filter as

$$v_{MF}(t) = m(t) * v_m(t) = \sum_k m(t) * \left[p(t - kT) r(t) \right] \tag{16A.9}$$

We normally derive $m(t)$ by saying the matched filter is matched to some signal $q(t)$. Recalling matched filter theory, this means we can write

$$m(t) = q^*(-t) \tag{16A.10}$$

As a reminder, the matched filter is termed here as a single-pulse matched filter. The matched filter is often matched to the transmit pulse, $p(t)$, in which case we would use

$$q(t) = p(t) \tag{16A.11}$$

When $p(t)$ is an LFM pulse, $m(t)$ could include an amplitude taper to reduce range sidelobes. In that case, $q(t)$ will not exactly equal $p(t)$. In the remainder of this derivation, we will use the more general form of (16A.10).

Substituting (16A.10) into (16A.9) yields

$$v_{MF}(t) = \sum_k q^*(-t) * \left[p(t-kT) r(t) \right] \quad (16A.12)$$

or

$$v_{MF}(t) = \sum_k \int_{-\infty}^{\infty} q^*(\tau - t) p(\tau - kT) r(\tau) d\tau \quad (16A.13)$$

where we replaced the convolution notation (∗) by the integral it represents.

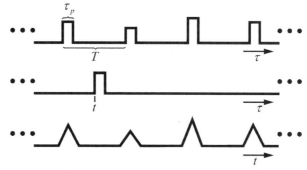

Figure 16A.1 Depictions of $|v_m(\tau)|$ (top plot), $|q^*(\tau - t)|$ (center plot), and $|v_{MF}(t)|$ (bottom plot).

Figure 16A.1 contains depictions of $|v_m(\tau)|$, $|q^*(\tau - t)|$, and $|v_{MF}(t)|$ for the case where $p(t)$ is an unmodulated pulse and $m(t)$ is matched to $p(t)$ [i.e., $q(t) = p(t)$]. As expected from matched filter theory, $v_{MF}(t)$ is a series of triangle-shaped pulses whose amplitudes depend on $r(t)$.

Since $v_m(t)$ is a nonstationary random process, so is $v_{MF}(t)$. This makes $v_{MF}(t)$ difficult to deal with since we do not have very sophisticated mathematical tools and procedures that allow us to efficiently analyze nonstationary random processes. Fortunately, because of the sampler/ADC, we do not need to deal directly with $v_{MF}(t)$. We will only work with samples of $v_{MF}(t)$.

$r(t)$ is a WSCS random process with an averaged autocorrelation of $R_r(\tau)$ and corresponding power spectral density of $S_r(f)$.

For now, we assume the sampler/ADC samples the output of the matched filter, $v_{MF}(t)$, once per PRI, T, at the peak of the matched filter response. We also, without loss of generality, assume the matched filter peaks occur at $t = lT$. With this we can write the output of the sampler/ADC as

$$v_o(l) = v_{MF}(t)\big|_{t=lT} = \sum_k \int_{-\infty}^{\infty} q^*(\tau - lT) p(\tau - kT) r(\tau) d\tau \quad (16A.14)$$

If we assume $p(t)$ and $q(t)$ are of the form

$$p(t) = p'(t)\text{rect}\left[\frac{t}{\tau_p}\right] \text{ and } q(t) = q'(t)\text{rect}\left[\frac{t}{\tau_p}\right] \tag{16A.15}$$

and $\tau_p < T/2$, then all of the terms of the summation of (16A.14) are zero except for the case where $k = l$.[3] With this (16A.14) reduces to

$$v_o(l) = \int_{-\infty}^{\infty} q^*(\tau - lT) p(\tau - lT) r(\tau) d\tau \tag{16A.16}$$

or, with the change of variables from l back to k,

$$v_o(k) = \int_{-\infty}^{\infty} q^*(\tau - kT) p(\tau - kT) r(\tau) d\tau \tag{16A.17}$$

To find the power spectrum of $v_o(k)$, we must first show $v_o(k)$ is WSS. To that end, we form

$$\begin{aligned}
R_o(k_1, k_2) &= E\{v_o(k_1) v_o^*(k_2)\} \\
&= E\left\{\left[\int_{-\infty}^{\infty} q^*(\tau - k_1 T) p(\tau - k_1 T) r(\tau) d\tau\right] \right. \\
&\quad \left. \left[\int_{-\infty}^{\infty} q^*(t - k_2 T) p(t - k_2 T) r(t) dt\right]^*\right\} \\
&= \int_{-\infty}^{\infty}\int_{-\infty}^{\infty} \left[q^*(\tau - k_1 T) p(\tau - k_1 T) q(t - k_2 T) p^*(t - k_2 T) \right. \\
&\quad \left. \overline{E\{r(\tau) r^*(t)\}}\right] d\tau dt
\end{aligned} \tag{16A.18}$$

From previous discussions, we note that

$$\overline{E\{r(\tau) r^*(t)\}} = R_r(\tau - t) \tag{16A.19}$$

where the overbar denotes the averaged expected value (see Appendix 16B). Using this for the expectation in (16A.18) gives

$$R_o(k_1, k_2) = \int_{-\infty}^{\infty}\int_{-\infty}^{\infty} q^*(\tau - k_1 T) p(\tau - k_1 T) \\
\times q(t - k_2 T) p^*(t - k_2 T) R_r(\tau - t) d\tau dt \tag{16A.20}$$

By making use of

---

[3] In (16A.15), $p'(t)$ and $q'(t)$ are the pulse modulations.

$$R_r(\tau) = \int_{-\infty}^{\infty} S_r(f) e^{j2\pi f\tau} df \tag{16A.21}$$

we can write

$$R_o(k_1,k_2) = \int_{-\infty}^{\infty}\int_{-\infty}^{\infty} q^*(\tau - k_1 T) p(\tau - k_1 T)$$
$$\times q(t - k_2 T) p^*(t - k_2 T) \left[\int_{-\infty}^{\infty} S_r(f) e^{j2\pi f(\tau - t)} df\right] d\tau dt \tag{16A.22}$$

We now make the change of variables, $\alpha = \tau - t$, $d\alpha = d\tau$, to write

$$R_o(k_1,k_2) = \int_{-\infty}^{\infty} q(t - k_2 T) p^*(t - k_2 T) \int_{-\infty}^{\infty} S_r(f)$$
$$\times \int_{-\infty}^{\infty} q^*(\alpha + t - k_1 T) p(\alpha + t - k_1 T) e^{j2\pi f\alpha} d\alpha df dt \tag{16A.23}$$

We next make the change of variables $\beta = \alpha + t - k_1 T$, $d\beta = d\alpha$ and get

$$R_o(k_1,k_2) = \int_{-\infty}^{\infty} q(t - k_2 T) p^*(t - k_2 T) \int_{-\infty}^{\infty} S_r(f)$$
$$\times \int_{-\infty}^{\infty} q^*(\beta) p(\beta) e^{j2\pi f(\beta - t + k_1 T)} d\beta df dt \tag{16A.24}$$

Rearranging yields

$$R_o(k_1,k_2) = \int_{-\infty}^{\infty} S_r(f) e^{j2\pi f k_1 T} \left(\int_{-\infty}^{\infty} q(t - k_2 T) p^*(t - k_2 T) e^{-j2\pi ft} dt\right)$$
$$\times \left(\int_{-\infty}^{\infty} q^*(\beta) p(\beta) e^{j2\pi f\beta} d\beta\right) df \tag{16A.25}$$

For the next change of variables, we let $\gamma = t - k_2 T$, $d\gamma = dt$, to yield

$$R_o(k_1,k_2) = \int_{-\infty}^{\infty} S_r(f) e^{j2\pi f(k_1 - k_2)T} \left(\int_{-\infty}^{\infty} q(\gamma) p^*(\gamma) e^{-j2\pi f\gamma} d\gamma\right)$$
$$\times \left(\int_{-\infty}^{\infty} q^*(\beta) p(\beta) e^{j2\pi f\beta} d\beta\right) df = R_o(k_1 - k_2) \tag{16A.26}$$

The first thing we note about (16A.26) is the right side is a function of $k_1 - k_2$. This constitutes the proof that $v_o(k)$ is WSS, since $R_o(k_1, k_2)$ depends on the separation in time, and not a time shift. The next thing we note is that the two integrals in the brackets are conjugates of each other. Finally, from ambiguity function theory, we recognize that we can write the product of the integrals as

$$\left| \int_{-\infty}^{\infty} p(\beta) q^*(\beta) e^{j2\pi f \beta} d\beta \right|^2 = \left| \chi_{pq}(0,f) \right|^2 \quad (16A.27)$$

where $|\chi_{pq}(0,f)|^2$ is the matched-range, Doppler cut of the cross ambiguity function of $p(t)$ and $q(t)$. In the remainder, we will use the notation $|\chi_{pq}(0,f)|^2 = MF(f)$. With all of the statements in this paragraph, we can write

$$R_o(k) = \int_{-\infty}^{\infty} MF(f) S_r(f) e^{j2\pi fkT} df \quad (16A.28)$$

where we let $k = k_1 - k_2$.

We next want to find the power spectrum of $v_o(k)$. We could do this by taking the discrete-time Fourier transform of $R_o(k)$. However, the math associated with this will probably be quite involved. We will take an indirect approach.

Let $v(t)$ be a WSS random process with an autocorrelation of $R(\tau)$ and a power spectrum of $S(f)$. Further, assume that we can sample $v(t)$ to get $v_o(k)$. That is,

$$v_o(k) = v(t)\big|_{t=kT} \quad (16A.29)$$

$v_o(k)$ is the same as the random process defined by (16A.14). From random processes theory, we can write

$$R_o(k) = R(\tau)\big|_{\tau=kT} \quad (16A.30)$$

Further, from the theory of discrete-time signals and their associated Fourier transforms, if $S(f)$ is the power spectrum of $v(t)$, the power spectrum of $v_o(k)$ is

$$S_o(f) = \frac{1}{T} \sum_{l=-\infty}^{\infty} S(f - l/T) \quad (16A.31)$$

From this same theory, we can write

$$R_o(k) = T \int_{-1/2T}^{1/2T} S_o(f) e^{j2\pi kfT} df \quad (16A.32)$$

If we substitute (16A.31) into (16A.32), we get

$$R_o(k) = T \int_{-1/2T}^{1/2T} \frac{1}{T} \sum_{l=-\infty}^{\infty} S(f - l/T) e^{j2\pi kfT} df \quad (16A.33)$$

or

$$R_o(k) = \sum_{l=-\infty}^{\infty} \int_{-1/2T}^{1/2T} S(f - l/T) e^{j2\pi k fT} df \qquad (16A.34)$$

We now make the change of variables $x = f - l/T$ to get

$$\begin{aligned} R_o(k) &= \sum_{l=-\infty}^{\infty} \int_{-1/2T-l/T}^{1/2T-l/T} S(x) e^{j2\pi k(x+l/T)T} dx \\ &= \sum_{l=-\infty}^{\infty} e^{j2\pi kl} \int_{-1/2T-l/T}^{1/2T-l/T} S(x) e^{j2\pi kxT} dx \qquad (16A.35) \\ &= \sum_{l=-\infty}^{\infty} \int_{-1/2T-l/T}^{1/2T-l/T} S(x) e^{j2\pi kxT} dx \end{aligned}$$

where we made use of $e^{j2\pi kl} = 1$.

We recognize that the last term is an infinite summation of integrals over nonoverlapping intervals and that the total of the nonoverlapping intervals covers the range of $x \in (-\infty, \infty)$. With this we can write

$$R_o(k) = \int_{-\infty}^{\infty} S(f) e^{j2\pi k fT} df \qquad (16A.36)$$

where we made the change of variables, $x = f$.

If we compare (16A.36) to (16A.28), we have

$$S(f) = MF(f) S_r(f) \qquad (16A.37)$$

With this and (16A.31), we arrive (16.20). That is,

$$S_o(f) = \frac{1}{T} \sum_{l=-\infty}^{\infty} MF(f - l/T) S_r(f - l/T)$$

Since we will need it later, we note that we can write the power in $v_o(k)$ as

$$P_o = R_o(0) = T \int_{-1/2T}^{1/2T} S_o(f) df \qquad (16A.38)$$

APPENDIX 16B: PROOF THAT *r(t)* IS WIDE-SENSE CYCLOSTATIONARY

In this appendix, we show that the process

$$r(t) = v_S^2(t) C(t) \Phi(t) \qquad (16B.1)$$

is wide-sense cyclostationary (WSCS). To show that $r(t)$ is WSCS, we must show

$$E\{r(t+kT_{SCAN}+\tau)r^*(t+kT_{SCAN})\} = E\{r(t+\tau)r^*(t)\} \quad (16B.2)$$

for some T_{SCAN}. That is, we must show that the autocorrelation of $r(t)$ is a periodic function of t.

We recall $C(t)$ and $\Phi(t)$ are WSS random processes. Thus, the product $C(t)\Phi(t)$ is also WSS. The function $v_S^2(t)$ is a deterministic function and is periodic with a period of T_{SCAN} where T_{SCAN} is the scan period of the antenna. If we form

$$R_r(t,\tau) = E\{r(t+\tau)r^*(t)\} \quad (16B.3)$$

we get

$$R_r(t,\tau) = v_S^2(t+\tau)v_S^2(t)E\{C(t+\tau)C^*(t)\}E\{\Phi(t+\tau)\Phi^*(t)\}$$
$$= v_S^2(t+\tau)v_S^2(t)R_C(\tau)R_\Phi(\tau) \quad (16B.4)$$

where we made use of the fact that $C(t)$ and $\Phi(t)$ are independent and WSS. In a similar fashion, we can write

$$R_r(t+kT_{SCAN},\tau) = v_S^2(t+kT_{SCAN}+\tau)v_S^2(t+kT_{SCAN})R_C(\tau)R_\Phi(\tau) \quad (16B.5)$$

However, since $v_S^2(t)$ is periodic with a period of T_{SCAN}, we have

$$v_S^2(t+kT_{SCAN}) = v_S^2(t) \quad (16B.6)$$

which leads to

$$R_r(t+kT_{SCAN},\tau) = v_S^2(t+\tau)v_S^2(t)R_C(\tau)R_\Phi(\tau) = R_r(t,\tau) \quad (16B.7)$$

which says $r(t)$ is WSCS.

From the theory of WSCS random processes, we can use the averaged autocorrelation of $r(t)$ to characterize the average behavior of $r(t)$. Specifically, in place of $R_r(t,\tau)$, we use

$$\bar{R}_r(\tau) = \frac{1}{T_{SCAN}} \int_{T_{SCAN}} R_r(t,\tau) dt \quad (16B.8)$$

where the integral notation means to perform the integration over one period of $R_r(t,\tau)$. As a note, (16B.8) shows a system will respond, on average, to $r(t)$ in the same manner as a WSS process that has the autocorrelation $\bar{R}_r(\tau)$. We will dispense with the overbar and use the notation $R_r(\tau)$.

Chapter 17

Clutter Model

17.1 INTRODUCTION

As indicated in Chapters 15 and 16, we will be concerned with characterizing the performance of signal processors against ground and rain clutter. To do so, we need to define suitable models for these clutter sources. We will be interested in both the amplitude and spectral characterization of the clutter. During these derivations, we will also discuss some characteristics of the radar that affect the clutter spectrum in order to capture their impact on signal processor performance. In this chapter, we confine ourselves to analyzing clutter for a stationary, ground-based radar.

17.2 GROUND CLUTTER MODEL

17.2.1 Ground Clutter RCS Model

A drawing we will use to develop the ground clutter model is shown in Figure 17.1 [1; 2, p. 63; 3; 4, p. 16–22; 5]. The top drawing represents a top view, and the bottom drawing represents a side view. For the initial development of the ground clutter model, we assume the earth is flat. Later, we will add a correction factor to account for the fact that the earth is not flat. The clutter model we develop is termed a *smooth earth* clutter model, which means we are not modeling specific terrain features such as trees, rocks, buildings, hills, or valleys.

The triangle and semicircle on the left of Figure 17.1 represents the radar, which is located at a height of h above the ground. When we discuss radar height, we are referring to the height of the *phase center* of the antenna. The phase center is usually taken to be the location of the feed for a reflector antenna or the center of the array for a phased array antenna [6–8].

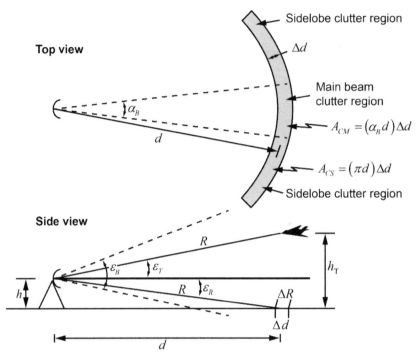

Figure 17.1 Geometry for a ground clutter model.

The dashed lines on the side and top views denote the 3-dB boundaries of the antenna main beam (one-way beamwidth)[1]. The angles ε_B and α_B denote the elevation and azimuth 3-dB beamwidths, respectively. The horizontal line through the antenna phase center is a reference line. It is not the elevation angle to which the main beam is steered.

The target is located at a range of R and an altitude of h_T. The elevation angle from the antenna phase center to the target is

$$\varepsilon_T = \sin^{-1}\left[(h_T - h)/R\right] \qquad (17.1)$$

In the geometry of Figure 17.1, the clutter patch of interest is also located at a range of R from the radar. In many applications, this is the region of clutter that is of interest because we are concerned with the clutter that competes with the target. However, for some cases, most notably MPRF and HPRF pulsed Doppler radars, the ground clutter that competes with the target will not be at the target range, but at a much shorter range. This is discussed later.

The width of the clutter patch along the R direction is ΔR. In most cases, ΔR is taken as the range resolution of the radar because almost all signal processors quantize the

---

[1] One could use two-way beam widths. However, these are usually more difficult to specify than are the one-way beam widths. Given the uncertainties in the clutter models relative to actual clutter, it is probably moot as to which should be used.

incoming signal into range cells that have a width of one range resolution cell. In some cases, a range resolution cell is large enough to cause problems with the accuracy of the ground clutter model. In that case, ΔR is taken to be smaller than a range resolution cell, and the signal processor calculations must include integration of clutter power across ΔR intervals.

The ground region that extends over ΔR at a range of R is an annulus centered on the radar. This is depicted in the top view where a portion of the annulus is illustrated by the shaded arc. For purposes of calculating the RCS of the ground in this annulus, it is divided into two regions. One of these, termed the *main beam clutter region*, represents the ground area illuminated by the main beam of the radar antenna. The other, termed the *sidelobe clutter region*, represents the ground area illuminated by the sidelobes of the radar antenna. We will assume the sidelobe clutter region extends from $-\pi/2$ to $\pi/2$. In other words, we assume there are no clutter returns from the back of the radar.[2] As implied by the statements above, the ground clutter model incorporates the transmit and receive antenna beam characteristics. In this development, we assume a monostatic radar that uses the same antenna for transmit and receive.

The magnitude of the clutter RCS depends on the size of the ground area illuminated by the radar (the annulus region discussed in the previous paragraph) and the *reflectivity* of the ground. We denote this reflectivity by the variable σ^0. Consistent with the previous discussions of target RCS, we can think of clutter reflectivity as the ability of the ground to absorb and reradiate energy (see Chapter 3). In general, clutter reflectivity depends on the type of ground (soil, water, asphalt, gravel, sand, grass, or trees) and its roughness. It also depends on moisture content and other related phenomena. Finally, clutter reflectivity also depends on the angle to the clutter patch (ε_R in Figure 17.1). Detailed discussions of σ^0 can be found in [4, 6, 9–19].

For the analyses we consider in this book, we use three values for σ^0: $\sigma^0 = -20$ dB, $\sigma^0 = -30$ dB, and $\sigma^0 = -40$ dB (Table 17.1). These are fairly standard values currently used for ground-based radars that operate in the 5- to 10-GHz range. The first case corresponds to moderate clutter and would be representative of trees, fields, and choppy water. The second value is for light clutter and would be representative of sand, asphalt, and concrete. The third value is for very light clutter and would be representative of smooth ice and smooth water [1–3, 5, 10, 14].

With the above, we can write the RCS of the main-beam ground area as

$$\sigma_{CM} = \sigma^0 A_{CM} = \sigma^0 \alpha_B d \Delta d \tag{17.2}$$

where the various parameters are shown in Figure 17.1. An assumption in this equation is that the azimuth beam width, α_B, is small so that the arc subtended by α_B can be assumed to be a straight line that is perpendicular to Δd (i.e., the mainlobe clutter patch is rectangular).

---

[2] This is approximately valid for ground-based radars. However, for airborne radars, there will be clutter returns from the entire annulus. Further, the Doppler frequency will vary around the annulus. For an example of this, see Example 4 of Chapter 24.

Table 17.1
Ground Clutter Backscatter Coefficients—Ground-Based Radar—Low Grazing Angles

Backscatter Coefficient, σ^0 (dB)	Comment
−20	Moderate clutter—trees, fields, choppy water
−30	Light clutter—sand, asphalt, concrete
−40	Very light clutter—smooth ice, smooth water

From the bottom part of Figure 17.1, we note that the clutter area is not located at beam center in the vertical direction. This means the clutter patch is not being fully illuminated, in elevation, by the antenna main beam. To account for this, we include a loss term (actually, a gain that is less than unity) that depends on the normalized directive gain pattern of the antenna. One approach is to use a normalized version of directive gain pattern of the actual antenna, $G(\alpha,\varepsilon)$, evaluated at $\alpha = 0$ [see (13.40) of Chapter 13]. An alternative is to use a generic pattern that provides a reasonable approximation of the actual (normalized) pattern, at least in the main beam [10, p. 150; 1; 2, pp. 147–148]. One of these generic patterns is the pattern of a linear array with uniform illumination. In the main beam region, this pattern is a $\text{sinc}^2(x)$ function and leads to the generic form,

$$G(\varepsilon) = \begin{cases} \text{sinc}^2\left[2.78\varepsilon/(\pi\varepsilon_B)\right] & |\varepsilon| < \frac{\pi}{2.78}\varepsilon_B \\ 0 & \text{elsewhere} \end{cases} \quad (17.3)$$

Another generic form that works reasonably well is the Gaussian approximation,

$$G(\varepsilon) = e^{-2.77(\varepsilon/\varepsilon_B)^2} \quad (17.4)$$

In (17.3) and (17.4), ε is the elevation angle off of beam center and ε_B is the elevation beam width of the antenna. Showing that the above models are reasonably good approximations in the main beam region is left as an exercise (Exercise 1). Of the two, the second is easier to use because it is not a piecewise function.

With this we can modify the equation for the main beam clutter (17.2) as

$$\sigma_{CM} = \sigma^0 \alpha_B d(\Delta d) G^2(\varepsilon_o + \varepsilon_R) \text{ m}^2 \quad (17.5)$$

where ε_o is the elevation pointing angle of the main beam and $\varepsilon_R = \sin^{-1}(h/R)$ is the depression angle to the ground patch. In some applications, we assume the main beam is pointed at the target so that $\varepsilon_o = \varepsilon_T$.

Equation (17.5) carries the assumption that the transmit and receive elevation antenna patterns are the same. If they are different, we would replace $G^2(\varepsilon)$ with $G_T(\varepsilon)G_R(\varepsilon)$, where $G_T(\varepsilon)$ and $G_R(\varepsilon)$ are the transmit and receive antenna patterns.

The basic approach for sidelobe clutter is the same as for the main beam clutter, but in this case, we need to account for the fact that the sidelobe clutter represents ground areas that are illuminated through the transmit antenna sidelobes and whose returns enter through the receive antenna sidelobes. The ground area of concern is the semicircular annulus (see Figure 17.1) excluding the main beam region. Relatively speaking, the ground area illuminated by the main beam is small compared to the ground area illuminated by the sidelobes. Because of this, it is reasonable to simplify calculations by including the main beam area in with the sidelobe area. With this the RCS of the clutter in the sidelobe region is

$$\sigma_{CS} = \sigma^0 (SL)^2 \pi d(\Delta d) \text{ m}^2 \tag{17.6}$$

where SL is the average antenna sidelobe level relative to the main beam peak. A typical value for SL is -30 dB or 0.001. However, it could be as low as -40 to -45 dB for "low sidelobe" antennas. The equation above includes $(SL)^2$ to account for the fact that the clutter is in the sidelobes of the transmit and receive antennas. If the sidelobes of the transmit and receive antennas are different, we would use $SL_T \times SL_R$ instead of $(SL)^2$.

To get the total clutter RCS from both the main lobe and the sidelobes, we assume the clutter signals are random processes that are uncorrelated from angle to angle. (We also assume the clutter signals are uncorrelated from range cell to range cell.) Since the clutter signals are uncorrelated random processes, and since RCS is indicative of energy or power (see Chapters 2 and 3), we can get the total clutter RCS by adding the main beam and sidelobe RCSs. Thus,

$$\sigma_C = \sigma_{CM} + \sigma_{CS} = \sigma^0 \left[G^2 (\varepsilon_o + \varepsilon_R) \alpha_B + \pi (SL)^2 \right] d(\Delta d) \text{ m}^2 \tag{17.7}$$

In this equation, the terms d and Δd are related to range, R, and range resolution, ΔR, by $d = R\cos\varepsilon_R$ and $\Delta d = \Delta R/\cos\varepsilon_R$ (see Figure 17.1). We use this to rewrite (17.7) as

$$\sigma_C = \sigma_{CM} + \sigma_{CS} = \sigma^0 \left[G^2 (\varepsilon_o + \varepsilon_R) \alpha_B + \pi (SL)^2 \right] R(\Delta R) \text{ m}^2 \tag{17.8}$$

For the final step, we need a term to account for the fact that the earth is round and not flat. We do this by including a *pattern propagation factor*. This pattern propagation factor allows the clutter RCS to gradually decrease as clutter cells move beyond the radar horizon. David Barton performed detailed analyses that led to sophisticated models for computing the pattern propagation effects [9, 12, 13, 20]. He also provided a simple approximation that works well. Specifically, he defined a loss factor as

$$L = 1 + (R/R_h)^4 \tag{17.9}$$

where R_h is the range to the radar horizon and is defined as

$$R_h = \sqrt{2(4R_E/3)h} \tag{17.10}$$

with $R_E = 6,371,000$ m being the mean radius of the earth. The 4/3 factor in the above equation invokes the "4/3 earth" model. This model states that, to properly account for clutter beyond the radar horizon, we need to increase the earth radius to effectively reduce its curvature. The 4/3 earth model is discussed in several references [9, 10, 21–24]. The derivation of (17.10) is left as an exercise (Exercise 2).

If we combine (17.9) and (17.8), we get

$$\sigma_C = \frac{\sigma^0 \left[G^2(\varepsilon_o + \varepsilon_R)\alpha_B + \pi(SL)^2 \right] R(\Delta R)}{1 + (R/R_h)^4} \text{ m}^2 \quad (17.11)$$

Figure 17.2 contains a plot of clutter RCS for a typical scenario. In particular, the radar uses a circular beam with a (one-way) beam width of 2°. Thus, $\alpha_B = \varepsilon_B = 2(\pi/180)$ rad. We used the Gaussian beam pattern of (17.4), assumed a sidelobe level of $SL = 0.001$ (−30 dB) and a used a range resolution of $\Delta R = 150$ m (a 1-μs pulse). The phase center of the antenna is at $h = 5$ m.

The three curves of Figure 17.2 correspond to beam-pointing angles (ε_o) of 0, 1/2, and 1 beamwidth above horizontal (i.e., 0°, 1°, and 2°). The clutter backscatter coefficient was assumed to be $\sigma^0 = 0.01$ (−20 dB).

The first observation from Figure 17.2 is that the ground clutter RCS is quite large for low beam elevation angles and short ranges. This means that for low altitude targets at short ranges (less than about 30 km), the clutter RCS is larger than that of typical aircraft targets, which have RCSs in the range of 6 to 10 dBsm [3, 10]. Thus, unless the radar includes signal processing to reduce clutter returns, they will dominate the target returns. At larger elevation angles, the problem is less severe because the ground is no longer being illuminated by the main beam, or it has moved beyond the clutter horizon.

The shape of the curves of Figure 17.2 requires some discussion. Examination of the equation for clutter RCS indicates that the numerator term increases with increasing range to the clutter. However, for ranges past the radar horizon, which is at a range of 9.2 km for this radar, the pattern propagation factor of (17.9) starts to predominate and reduces the clutter RCS. This is what causes the curves of Figure 17.2 to first increase and then decrease.

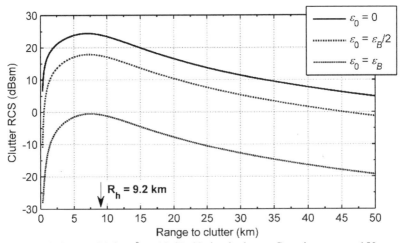

Figure 17.2 Ground clutter RCS for $\sigma^0 = -20$ dB, 2° circular beam, Gaussian pattern, 150 m range resolution, $h = 5$ m.

17.2.2 Ground Clutter Spectrum Model

The main signal characteristic we use to distinguish clutter from targets is Doppler frequency. Because of this, we need a model for the spectrum of signals returned from clutter.

The simplest Doppler spectrum model for ground clutter is to assume the Doppler frequency is zero. However, this is not strictly correct because, in most cases, the elements that make up ground clutter (leaves, grass, or waves, for example) are in motion and thus have a nonzero range rate. This will cause the Doppler frequency to have a small spread. The spread is important because, as we will learn, it is a significant factor in the ability of some signal processors, notably MTI and LPRF pulsed Doppler processors, to reject clutter.

Several models for the frequency spectrum of ground clutter have been proposed over the years [10, p. 152]. A standard model used in many texts is the Gaussian model, defined by

$$C_{Gauss}(f) = \frac{1}{\sigma_f \sqrt{2\pi}} e^{-f^2/2\sigma_f^2} \tag{17.12}$$

where f is frequency, in Hz, $\sigma_f = 2\sigma_v/\lambda$, in Hz, and σ_v is the velocity spread of the clutter, in m/s. Skolnik provides values of σ_v for several environment and wind conditions [25]. A sampling of these values is contained in Table 17.2.

The spectrum model currently believed to be the best for land clutter was developed by MIT Lincoln Laboratories as part of an extensive ground clutter characterization effort [14, 26]. The Lincoln Laboratories tests considered an environment that consisted of trees and "vegetation" and gathered data at low elevation angles and several frequencies. A form of the model presented in [27–29] is

$$C_{Exp}(f) = \frac{r}{1+r}\delta(f) + \frac{1}{1+r}\frac{\beta\lambda}{4}e^{-\beta|\lambda f/2|} \tag{17.13}$$

where β is a parameter that depends on wind speed and r is a parameter that apportions the spectrum between the spectral line at $f = 0$ and the spread part defined by the exponential. λ is the wavelength and $\delta(f)$ is the Dirac delta. Table 17.3 contains values of β provided in [14, 51–53] for different wind speeds. J. Barrie Billingsley points out that the entries for the first three wind conditions are based on measurements, but the entries for gale force winds are estimates. We term (17.13) the exponential spectrum model in this book.

Billingsley provides an equation for r, which is [14, p. 580]

$$r = 489.8 w^{-1.55} f_o^{-1.21} \tag{17.14}$$

where w is the wind speed in mph and f_o is the radar carrier frequency in GHz.

Table 17.2

Sample Values of σ_v

Environment	σ_v m/s
Sparse woods, calm winds	0.017
Wooded hills, 20-knot wind	0.22
Wooded hills, 40-knot wind	0.32

Source: [25].

Table 17.3

Clutter Spectrum Shape Parameters

| Wind Conditions | Wind Speed, w (mph) | Shape Parameter, β (s/m) | |
		Typical	Worst Case
Light air	1–7	12	—
Breezy	7–15	8	—
Windy	15–30	5.7	5.2
Gale force (est.)	30–60	4.3	3.8

Source: [14, p.578].

The β values given above have not, to the authors' knowledge, been extended to sea clutter. Skolnik provides a related parameter that we might use to infer a β value for sea clutter [25]. Specifically, he provides a standard deviation parameter, σ_v, which is used in the Gaussian clutter spectrum model. That parameter has the units of m/s, as opposed to the s/m (inverse velocity) units of β.

Skolnik provides a value of $\sigma_v = 0.22$ m/s for "wooded hills" in a 20-knot wind (23-mph wind). We note that $1/\sigma_v = 4.55$ s/m, which is somewhat close to the value of 5.2 to 5.8 listed in Table 17.3 for wind speeds between 15 and 30 mph. This, and a comparison of units, suggests an inverse relation between β and σ_v. For sea clutter, Skolnik has values of σ_v that range between 0.46 and 1.1 m/s. If we use the inverse relation between β and σ_v, we can speculate that reasonable values of β for sea clutter might be 0.91 s/m to 2.2 s/m. We might further speculate that the lower value corresponds to a high sea state and the larger value corresponds to a lower sea state.

In addition to spectrum spread, sea clutter can have a center value that depends on wind velocity and its direction relative to the radar. Nathanson provides a chart that shows a mean velocity of 3.4 m/s for sea state 4 and looking directly into the wind [3, p. 294][3]. He indicates this mean velocity could vary anywhere between –3.4 m/s and +3.4 m/s depending on the direction of the wind relative to the beam direction.

An implication of characterizing the clutter spectral properties by $C(f)$ is that the clutter is a WSS random process [30, 31]. Also, since

---

[3] Sea state 4 is termed a moderate sea state. It is associated with wave heights of 1.25 to 2.5 m.

$$\int_{-\infty}^{\infty} C(f)\,df = 1 \tag{17.15}$$

for the clutter spectrum models of (17.12) and (17.13) (see Exercise 4), the clutter spectrum is normalized to unity power. To get the actual clutter spectrum, we multiply $C(f)$ by the clutter power, which we compute from the clutter RCS and the radar range equation. $C(f)$ is the clutter spectrum model we used in Chapter 16.

17.3 RAIN CLUTTER MODEL

17.3.1 Rain Clutter RCS Model

Figure 17.3 contains a sketch of the geometry we use for the rain clutter model. We term the volume of the elliptical cone frustum the main beam clutter volume, V_{CM}, and use it to compute the rain clutter RCS in the main beam (the main beam is represented by the two slanted lines). If we assume the azimuth and elevation beam widths, α_B and ε_B, are small, we can treat the cone frustum as an elliptical cylinder and compute its volume as[4]

$$V_{CM} = \frac{\pi}{4}\alpha_B \varepsilon_B R^2 (\Delta R) \text{ m}^3 \tag{17.16}$$

Similar to the ground clutter case, the rain RCS (due to rain in the main beam region) is determined by multiplying V_{CM} by a rain reflectivity, or volumetric backscatter coefficient, η. That is

$$\sigma_C = \eta V_{CM} \tag{17.17}$$

Barton provides an equation for η as

$$\eta = 5.7 \times 10^{-14}\, r^{1.6}/\lambda^4 \quad (\text{m}^2/\text{m}^3) \tag{17.18}$$

where r is the rainfall rate in mm/hr and λ is the radar wavelength in m [2, p. 341]. Several examples of η based on this equation are given in Table 17.4.

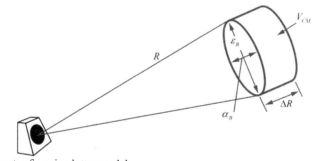

Figure 17.3 Geometry for rain clutter model.

---

[4] David Barton pointed out that the model of (17.16) assumes uniform illumination across the ellipse of Figure 17.3. To account for the fact that the illumination is not uniform, $\pi/4$ should be replaced by $1/1.77$. This ratio is based on a Gaussian beam shape.

Table 17.4
Example Rain Reflectivity Values [dB(m²/m³)]

Rain Type	Rainfall Rate (mm/hr)	Radar Carrier Frequency (GHz)			
		3	5	8	10
Drizzle	0.25	−102	−93	−85	−81
Light rain	1	−92	−84	−75	−72
Moderate rain	4	−83	−74	−66	−62
Heavy rain	16	−73	−64	−56	−52

Figure 17.4 contains plots of σ_C versus R for a rainfall rate of 4 mm/hr, the four frequencies of Table 17.4, $\alpha_B = \varepsilon_B = 2°$, and $\Delta R = 150$ m. Unlike ground clutter RCS, which increased then decreased with increasing range, rain clutter RCS continues to increase as range increases. This makes sense, as V_{CM} increases with increasing range.

Two assumptions not previously mentioned are that the rain RCS model assumes the rain occupies the entire main beam and that it is present at all ranges. The assumption that the rain occupies the entire main beam may be reasonable for pencil beam radars since their beam widths are, at most, a few degrees. For fan beam radars, this assumption becomes questionable. Also, David Barton noted that, in fact, the RCS would eventually decrease when the top of the beam moves above the top of the rain. To account for this, he states that, in (17.16), $\varepsilon_B R$ should be replaced by $h_m - h_0$, where $h_0 = R^2/2(4R_E/3)$ is the altitude at which the rain is at the horizon for the range R and h_m is the maximum rain altitude. He states that h_m is on the order of a few kilometers. To avoid a negative area, we suggest using the maximum of $h_m - h_0$ and zero.

The assumption that the rain is present at all ranges is also questionable. However, for preliminary investigations, it is probably a reasonable assumption since it would represent a worst-case scenario in terms of the clutter rejection the signal processor must provide.

Another assumption of the above formulation of σ_C is that it includes only main beam rain clutter. In general, this is a good assumption. To support this claim, we consider a specific example.

Suppose the antenna has a uniform sidelobe level of SL over a hemisphere of radius R centered on the radar. Further suppose rain is present over the volume encompassed by the hemisphere, and that the backscatter coefficient, η, is the same throughout the volume.

We are interested in the rain RCS in a hemispherical shell with a width of ΔR located at a range of R. A cross section of the shell is depicted in Figure 17.5.

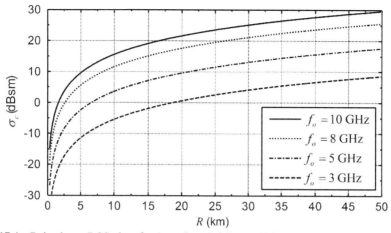

Figure 17.4 Rain clutter RCS plots for 4 mm/hr rainfall rate, 2° beamwidth, 150 m range resolution.

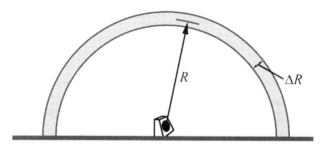

Figure 17.5 Side view of hemispherical shell used to compute sidelobe rain clutter RCS.

The approximate volume of the hemispherical shell is (see Exercise 6)

$$V_{shell} = \frac{2}{3}\pi R^2 (\Delta R) \qquad (17.19)$$

Thus, the RCS of the rain in the shell is

$$\sigma_{shell} = \eta V_{shell} (SL)^2 \qquad (17.20)$$

where we included *SL* to reflect the fact that the power (energy) returned from the rain in the shell is in the sidelobes of the transmit and receive antenna radiation patterns.

The ratio of sidelobe to main lobe RCS is

$$\frac{\sigma_{shell}}{\sigma_C} = \frac{\eta V_{shell} (SL)^2}{\eta V_{CM}} = \frac{\eta 2\pi R^2 (\Delta R)(SL)^2}{3\eta\pi\alpha_B\varepsilon_B R^2 (\Delta R)} = \frac{8(SL)^2}{3\alpha_B\varepsilon_B} \qquad (17.21)$$

As a specific example, we consider the case where the sidelobe levels of the transmit and receive antennas are uniformly 30 dB below the main beam peak ($SL = 10^{-3}$) and $\varepsilon_B = \alpha_B = 2°$. With this we get

$$\frac{\sigma_{shell}}{\sigma_C} = \frac{8(10^{-3})^2}{3(2\pi/180)^2} = 0.0022 \text{ or } -26.6 \text{ dB} \qquad (17.22)$$

In other words, the RCS of the rain clutter in the shell is well below the main beam RCS. We note that (17.21) has assumed the rain exists at all altitudes and ranges, which is clearly not realistic. Our only purpose for including it is to point out that, generally, the major contributor to rain RCS is the rain in the main beam. If the geometry is such that the rain is not in the main beam, the major contributor becomes rain in the sidelobes.

17.3.2 Rain Clutter Spectral Model

According to several sources [3, 9, 32], the spectral width of rain clutter depends on several factors including wind shear, turbulence, fall velocity of the rain, and variation in the fall velocity across the main beam. Of these, wind shear appears to be the main contributor to the width of the spectrum. Fred Nathanson provides an equation for wind shear as[5]

$$\sigma_{vshear} = 0.3 k R_{km} \varepsilon_B \text{ m/s} \qquad (17.23)$$

where R_{km} is range of the V_{CM} (see Figure 17.3) in km and $k = 4.0$ m/(sec-km) [3]. Nathanson states that this value of k is averaged over all azimuths. He goes on to say that σ_{vshear} is limited to 6 m/s for elevation beamwidths, ε_B, less than 2.5°. In (17.23), σ_{vshear} is the standard deviation of a Gaussian spectrum model. Nathanson indicates that the other three contributors (turbulence, fall velocity, and fall velocity variation), combined, are in the vicinity of 1.5 m/s.

Nathanson provides graphs of measured data that indicate the total rain velocity standard deviation, σ_{vrain}, is between 0.5 and 1.5 m/s for ranges below about 20 km and between 2 and 3 m/s for ranges of about 60 km. In another chart corresponding to a high shear case, he shows σ_{vrain} values that vary between 1 and 5 m/s, independently of range, over a 0- to 100-km range. Barton [9] shows measured data for a 2-D search radar that indicates a σ_{vrain} of 5 m/s. These values are summarized in Table 17.5.

σ_{vrain} is the standard deviation for a Gaussian spectrum model. Since there is no justification (i.e., no measurement data) for using an exponential spectrum model for rain, it should probably not be used.

Rain can have a mean velocity that depends on wind velocity and direction of the wind relative to the beam direction. Nathanson has an example that shows peak mean velocities of about 30 m/s at moderate altitudes [3, p. 294].

---

[5] Nathanson uses a factor of 0.42 instead of 0.3. However, in his model, ε_B is a two-way beamwidth rather than the one-way beamwidth used in (17.23). The value 0.3 is approximately $0.42/(2^{1/2})$.

Table 17.5
Sample Values of σ_{vrain} from Several Sources.

σ_{vrain} m/s	Source
<6	Nathanson [3]
0.5 to 1.5 at 20-km range	Nathanson Graph [3]
2 to 3 at 60-km range	Nathanson Graph [3]
1 to 5 for ranges between 0 and 100 km	Nathanson [3]
5	Barton [9]
1.8 to 4	Skolnik [25]

17.4 EXERCISES

1. Show that (17.3) and (17.4) provide reasonable approximations to the main beam of an antenna radiation pattern. Show this for a linear array with uniform weighting and a linear array with $\bar{n} = 6$, 30-dB Taylor weighting.

2. Derive (17.10).

3. Reproduce the plot of Figure 17.2.

4. Show that (17.13) satisfies (17.15).

5. Reproduce the plot of Figure 17.4.

6. Derive (17.19).

References

[1] Budge, M. C., Jr., "EE 725: Advanced Radar Technique," www.ece.uah.edu/courses/material/EE725/index.hm.

[2] Barton, D. K., *Radar Equations for Modern Radar*, Norwood, MA: Artech House, 2013.

[3] Nathanson, F. E., J. P. Reilly, and M. N. Cohen, *Radar Design Principles: Signal Processing and the Environment*, 2nd ed., New York: McGraw-Hill, 1991.

[4] Skolnik, M. I., ed., *Radar Handbook*, 3rd ed., New York: McGraw-Hill, 2008.

[5] Mahafza, B. R., *Radar Signal Analysis and Processing Using MATLAB*, New York: CRC Press, 2008.

[6] Blake, L. V., *Antennas*, Dedham, MA: Artech House, 1984.

[7] Schelkunoff, S. A., *Advanced Antenna Theory*, New York: Wiley & Sons, 1952.

[8] Stutzman, W. L., and G. A. Thiele, *Antenna Theory and Design*, New York: Wiley & Sons, 1981.

[9] Barton, D. K., *Radar System Analysis and Modeling*, Norwood, MA: Artech House, 2005.

[10] Skolnik, M. I., *Introduction to Radar Systems*, 3rd ed., New York: McGraw-Hill, 2001.

[11] Barton, D. K., ed., Radars, Vol. 5: *Radar Clutter* (Artech Radar Library), Dedham, MA: Artech House, 1975.

[12] Barton, D. K., "Ground Clutter Model," Raytheon Memo 7101-80-159, Feb. 8, 1980 (rev. Apr. 3, 1980).

[13] Barton, D. K., "Flat and Rolling Terrain Clutter Model," Raytheon Memo 7011-82-169, May 17, 1982.

[14] Billingsley, J. B., *Low-Angle Radar Land Clutter: Measurements and Empirical Models*, Norwich, NY: William Andrew, 2002.

[15] Feng, S., and J. Chen, "Low-Angle Reflectivity Modeling of Land Clutter," *IEEE Geoscience and Remote Sensing Lett.*, vol. 3, no. 2, Apr. 2006, pp. 254–258.

[16] Eaves, J. L., and E. K. Reedy, *Principles of Modern Radar*, New York: Van Nostrand Reinhold, 1987.

[17] Ulaby, F. T., and M. C. Dobson, *Handbook of Radar Scattering Statistics for Terrain* (Artech House Remote Sensing Library), Norwood, MA: Artech House, 1989.

[18] Barton, D. K., *Modern Radar System Analysis*, Norwood, MA: Artech House, 1988.

[19] Long, M. W., *Radar Reflectivity of Land and Sea*, 3rd ed., Norwood, MA: Artech House, 2001.

[20] Barton, D. K., "Land Clutter Models for Radar Design and Analysis," Proc. IEEE, vol. 73, no. 2, Feb. 1985, pp. 198–204.

[21] Kerr, D. E., ed., *Propagation of Short Radio Waves*, vol. 13 of MIT Radiation Lab. Series, New York: McGraw-Hill, 1951. Reprinted: Norwood, MA: Artech House (CD-ROM edition), 1999.

[22] Bean, B. R., and G. D. Thayer, "Models of the Atmospheric Radio Refractive Index," *Proc. IRE*, vol. 47, no. 5, May 1959, pp. 740–755.

[23] Bean, B. R., "The Radio Refractive Index of Air," *Proc. IRE*, vol. 50, no. 3, Mar. 1962, pp. 260–273.

[24] Doerry, A. W., "Earth Curvature and Atmospheric Refraction Effects on Radar Signal Propagation," Sandia Nat. Labs., Albuquerque, NM, Rep. No. SAND2012-10690, Jan. 2013.

[25] Skolnik, M. I., ed., *Radar Handbook*, 2nd ed., New York: McGraw-Hill, 1990.

[26] Bassford, R. S., W. Goodchild, and A. De La Marche, "Test and Evaluation of the Moving Target Detector (MTD) Radar," Lincoln Laboratories, Rep. No. FAA-RD-77-118, 1977. Available from DTIC as ADA047887.

[27] Billingsley, J. B., et al., "Impact of Experimentally Measured Doppler Spectrum of Ground Clutter on MTI and STAP," *Radar 97* (Conf. Publ. No. 449), Edinburgh, Scotland, Oct. 14–16, 1997, pp. 290–294.

[28] Greco, M., et al., "Analysis of Clutter Cancellation in the Presence of Measured L-band Radar Ground Clutter Data," *Rec. IEEE 2000 Int. Radar Conf.*, 2000, Alexandria, VA, May 7–12, 2000, pp. 422–427.

[29] Lombardo, P., et al., "Impact of Clutter Spectra on Radar Performance Prediction," IEEE Trans. Aerosp. Electron. Syst., vol. 37, no. 3, Jul. 2001, pp. 1022–1038.

[30] Papoulis, A., *Probability, Random Variables, and Stochastic Processes*, 3rd ed., New York: McGraw-Hill, 1991.

[31] Davenport, W. B., Jr., and W. L. Root, *An Introduction to the Theory of Random Signals and Noise*, New York: McGraw-Hill, 1958.

[32] IEEE Aerospace and Electronic Systems Society and IEEE New Hampshire Section, "Free Video Course in Radar Systems Engineering," (R. M. O'Donnell, lecturer), aess.cs.unh.edu.

Chapter 18

Moving Target Indicator (MTI)

18.1 INTRODUCTION

We are now ready to consider our first signal processor: a moving target indicator, or MTI. An MTI is a high-pass digital filter designed to reject clutter but not targets that are moving. A block diagram of a 2-pulse MTI is shown in Figure 18.1. It is termed a 2-pulse MTI because it operates on two pulses at a time. It successively subtracts the returns from two adjacent pulses. For signal processor buffs, it is a first-order, nonrecursive, high-pass, digital filter.

The block with z^{-1} represents a one PRI delay. In modern MTI processors, the delay is implemented using digital memory. In older, analog MTI processors, it was implemented using delay lines.

A time domain model of the filter is

$$v_{SP}(k) = v_o(k) - v_o(k-1) \tag{18.1}$$

Note that if $v_o(k) = K$, then $v_{SP}(k) = v_o(k) - v_o(k-1) = K - K = 0$. Thus, the MTI perfectly cancels DC, or zero-frequency, signals.

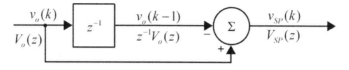

Figure 18.1 A 2-pulse MTI.

Taking z-transform of both sizes of (18.1), we get

$$V_{SP}(z) = V_o(z) - z^{-1}V_o(z) \tag{18.2}$$

which we solve to yield the filter transfer function

$$H_U(z) = \frac{V_{SP}(z)}{V_o(z)} = 1 - z^{-1} \tag{18.3}$$

where we use the subscript U to denote the fact that the filter transfer function is unnormalized. We will discuss normalization of the MTI shortly. From (18.3), we find the filter frequency response as

$$\begin{aligned} H_U(f) &= |H_U(z)|^2_{z=e^{j2\pi fT}} = |1 - z^{-1}|^2_{z=e^{j2\pi fT}} = |1 - e^{-j2\pi fT}|^2 \\ &= |e^{-j\pi fT}(e^{j\pi fT} - e^{-j\pi fT})|^2 = |e^{-j\pi fT}(2j\sin \pi fT)|^2 \\ &= 4\sin^2(\pi fT) \end{aligned} \tag{18.4}$$

A plot of $H_U(f)$ is shown in Figure 18.2 for the case where $T = 400$ μs.

18.1.1 MTI Response Normalization

Before we turn our attention to computing the clutter rejection capabilities of an MTI, we need to normalize the MTI response with respect to something. Without normalization, it is difficult to quantify the clutter rejection capabilities of the MTI because we have no reference. The instinct is to say the clutter rejection is a measure of the clutter power out of the MTI relative to the clutter power into the MTI. However, we can make the clutter power out of anything we want with the appropriate MTI gain. To avoid this problem, we normalize the MTI so that it has a noise gain of unity. In this way, we can compute the clutter rejection by comparing the CNR at the output of the MTI to the CNR at the input, since we have noise power as a common reference. In a similar fashion, we will be able to characterize the SNR improvement, or degradation, through the MTI.

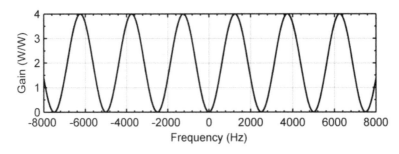

Figure 18.2 Frequency response of an unnormalized 2-pulse MTI, $T = 400$ μs.

We add a gain, K_{MTI}, to (18.1) and let $v_o(k) = n(k)$, where $n(k)$ is noise. We then write

$$v_{SPn}(k) = K_{MTI}\left[n(k) - n(k-1)\right] \tag{18.5}$$

We next assume the noise into the MTI is zero-mean, wide sense stationary (WSS), and white with a power of $P_{No} = E\{|n(k)|^2\}$. (As a reminder, P_{No} is the noise power at the sampler, and matched filter, output.) The noise power at the MTI output is

$$\begin{aligned}
P_{Nout} &= E\left\{|v_{SPn}(k)|^2\right\} = E\left\{K_{MTI}^2 |n(k) - n(k-1)|^2\right\} \\
&= E\left\{K_{MTI}^2 |n(k)|^2\right\} + E\left\{K_{MTI}^2 |n(k-1)|^2\right\} \\
&\quad - E\left\{K_{MTI}^2 n(k) n^*(k-1)\right\} - E\left\{K_{MTI}^2 n^*(k) n(k-1)\right\} \\
&= 2E\left\{K_{MTI}^2 |n(k)|^2\right\} = 2K_{MTI}^2 P_{No}
\end{aligned} \tag{18.6}$$

In (18.6), the cross expectations on the third line are zero because of the assumption that $n(k)$ is white and zero-mean. The relation $E\{K_{MTI}^2|n(k)|^2\} = E\{K_{MTI}^2|n(k-1)|^2\}$ comes from the assumption that $n(k)$ is WSS.

To get a noise gain of unity, that is $P_{Nout} = P_{No}$, we require $K_{MTI} = 1/(2)^{1/2}$. If we apply this to our previous derivation of $H(f)$, we get

$$H(f) = K_{MTI}^2 H_U(f) = \frac{1}{2}\left[4\sin^2(\pi f T)\right] = 2\sin^2(\pi f T) \tag{18.7}$$

A plot of the normalized frequency response, $H(f)$, is shown in Figure 18.3 for the case of $T = 400$ μs.

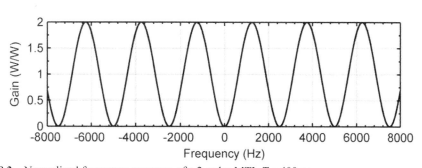

Figure 18.3 Normalized frequency response of a 2-pulse MTI, $T = 400$ μs.

18.2 MTI CLUTTER PERFORMANCE

Now that we have normalized the MTI response, we want to compute its *clutter attenuation*, *CA*, and *signal-to-clutter ratio* (SCR) *improvement*, I_{scr}. SCR is defined as the ratio of signal (target) power to clutter power. The IEEE Dictionary defines what we term SCR improvement as *MTI improvement factor* [1, p. 228, 2, p. 117, 118].

18.2.1 Clutter Attenuation

We start with clutter attenuation, which is defined as the ratio of the CNR at the input to the MTI to the CNR at the output of the MTI. The CNR at the input to the MTI is the CNR at the output of the sampler (and matched filter), and is given by the radar range equation (see Chapter 2). The CNR at the output of the MTI is the clutter power out of the (normalized) MTI divided by the noise power at the output of the MTI. However, because of the normalization, the noise power at the output of the MTI is equal to the noise power at the input. Thus, the clutter attenuation is the ratio of the clutter power at the input to the MTI and the clutter power at the output of the MTI. In equation form:

$$CA = \frac{CNR_{in}}{CNR_{out}} = \frac{P_{Co}/P_{No}}{P_{Cout}/P_{No}} = \frac{P_{Co}}{P_{Cout}} = \frac{P_C}{P_{Cout}} \tag{18.8}$$

where the numerator term of the last equality follows from the discussion that led to (16.50).

The clutter power at the output of the MTI is

$$P_{Cout} = \int_{-\infty}^{\infty} H(f) MF(f) S_{Cr}(f) df \tag{18.9}$$

where $S_{Cr}(f)$ is [see (16.29) and (16.47)]

$$S_{Cr}(f) = P_C V_S(f) * C(f) * \Phi(f) \tag{18.10}$$

With this

$$P_{Cout} = P_C \int_{-\infty}^{\infty} H(f) MF(f) \left[V_S(f) * C(f) * \Phi(f) \right] df = P_C G_C \tag{18.11}$$

where

$$G_C = \int_{-\infty}^{\infty} H(f) MF(f) \left[V_S(f) * C(f) * \Phi(f) \right] df \tag{18.12}$$

Using (18.11) in (18.8), we have the clutter attenuation, *CA*, as

$$CA = \frac{1}{G_C} \tag{18.13}$$

The form of (18.12) does not lend itself to a simple closed-form solution. However, we can obtain an approximate closed-form solution by making some assumptions about $V_S(f)$, $C(f)$, and $\Phi(f)$. For the first assumption, we temporarily ignore phase noise and let $\Phi(f) = \delta(f)$. We will include phase noise later.

We will derive approximate values for CA for the Gaussian and exponential clutter spectra models of Sections 18.2.2 and 18.3.2.

18.2.1.1 Gaussian Spectrum

The Gaussian model leads to the CA formulation that appears in most radar texts that discuss MTI [3, p. 41; 4, p. 319; 5, p. 2.11]. We will also use the $V_S(f)$ of (16.30). With this we have

$$V_S(f) * C(f) * \Phi(f) = \frac{1}{\sigma_S\sqrt{2\pi}}e^{-f^2/2\sigma_S^2} * \left(\frac{1}{\sigma_f\sqrt{2\pi}}e^{-f^2/2\sigma_f^2}\right) * \delta(f)$$
$$= \frac{1}{\sigma_T\sqrt{2\pi}}e^{-f^2/2\sigma_T^2} \qquad (18.14)$$

where $\sigma_T^2 = \sigma_f^2 + \sigma_S^2$, and we made use of the fact that the convolution of two Gaussian functions is another Gaussian function [5, p. 2.16] (as a reminder, σ_S is the spectrum spread due to scanning and σ_f is the spectrum width of the clutter). If the radar is not scanning, $V_S(f) = \delta(f)$ and we would use $\sigma_T = \sigma_f$.

Substituting (18.14) into (18.12) gives

$$G_{CGauss} = \int_{-\infty}^{\infty} H(f) MF(f) \frac{1}{\sigma_T\sqrt{2\pi}} e^{-f^2/2\sigma_T^2} df$$
$$= \int_{-\infty}^{\infty} 2\sin^2(\pi f T) MF(f) \frac{1}{\sigma_T\sqrt{2\pi}} e^{-f^2/2\sigma_T^2} df \qquad (18.15)$$

It can be shown (see Exercise 2) that

$$MF(f)e^{-f^2/2\sigma_T^2} \approx 1 \times e^{-f^2/2\sigma_T^2} \qquad (18.16)$$

for typical values of σ_T and τ_p. With this we can simplify (18.15) to

$$G_{CGauss} = \int_{-\infty}^{\infty} 2\sin^2(\pi f T) \frac{1}{\sigma_T\sqrt{2\pi}} e^{-f^2/2\sigma_T^2} df \qquad (18.17)$$

This integral does not have a simple, closed-form solution. However we can simplify the integral by observing that over the region of f where $\exp[-f^2/(2\sigma_T^2)]$ is large, $\pi f T$ is small and $\sin(\pi f T) \approx \pi f T$. Over the rest of f, $\exp[-f^2/(2\sigma_T^2)]$ is very small. Thus, the integrand is very small and adds little to the value of the integral. With this G_{CGauss} becomes

$$G_{Gauss} \approx \int_{-\infty}^{\infty} \frac{2(\pi f T)^2}{\sigma_T \sqrt{2\pi}} e^{-f^2/2\sigma_T^2} df = 2(\pi T)^2 \left(\frac{1}{\sigma_T \sqrt{2\pi}} \int_{-\infty}^{\infty} f^2 e^{-f^2/2\sigma_T^2} df \right) \quad (18.18)$$

From properties of Gaussian density functions [6], we recognize the term in parentheses as σ_T^2 and write

$$G_{CGauss} = 2\pi^2 T^2 \sigma_T^2 \quad (18.19)$$

From (18.13), with $PRF = 1/T$,

$$CA_{Gauss} = \frac{1}{G_{CGauss}} = \frac{1}{2\pi^2 T^2 \sigma_T^2} = 2\left(\frac{PRF}{2\pi\sigma_T}\right)^2 \quad (18.20)$$

which is the form found in many radar texts [3, 4].

18.2.1.2 Exponential Spectrum

For the exponential model, we use (17.13) and approximate $V_S(f)$ by a similar exponential model.[1] Specifically,

$$V_S(f) = \frac{\lambda \beta_S}{4} e^{-\lambda \beta_S |f|/2} \quad (18.21)$$

With this we have

$$V_{SExp}(f) * C_{Exp}(f) * \Phi(f) = \frac{1}{r+1} \left\{ \frac{\lambda \beta_S r}{4} e^{-\lambda \beta_S |f|/2} \right.$$

$$\left. + \left[\frac{\beta_S^2}{\beta_S^2 - \beta_C^2} \left(\frac{\lambda \beta_C}{4} e^{-\lambda \beta_C |f|/2} \right) - \frac{\beta_C^2}{\beta_S^2 - \beta_C^2} \left(\frac{\lambda \beta_S}{4} e^{-\lambda \beta_S |f|/2} \right) \right] \right\} \quad (18.22)$$

In (18.21) and (18.22), we used $\beta_S = 2/(\lambda \sigma_S)$, and β_C comes from Table 17.3. The derivation of (18.22) is left as an exercise. As a reminder, we are ignoring phase noise, so $\Phi(f) = \delta(f)$.

We next use (18.22) in (18.12) with $H(f) = 2\sin^2(\pi f T)$ to compute G_{Cexp}. While the resulting integral can be evaluated in closed form, the closed-form expression is somewhat complicated. A simpler form of the integral can be obtained by using $\sin(\pi f T) \approx \pi f T$. The result is (see Exercise 4)

$$G_{CExp} = \frac{(2\pi T)^2}{1+r} \left(\frac{2}{\lambda}\right)^2 \left\{ \frac{r}{\beta_S^2} + \frac{\beta_S^2 + \beta_C^2}{\beta_S^2 \beta_C^2} \right\} \quad (18.23)$$

For the nonscanning case, this reduces to

---

[1] The validity of this as the spectrum due to scanning is questionable. However, we use it anyway because Gaussian form of $V_S(f)$ convolved with the exponential clutter model does not lead to an easily computed closed form expression for G_C.

$$G_{CExp} = \frac{(2\pi T)^2}{1+r}\left(\frac{2}{\lambda \beta_C}\right)^2 \qquad (18.24)$$

The clutter attenuation is

$$CA_{Exp} = \frac{1}{G_{CExp}} \qquad (18.25)$$

18.2.2 SCR Improvement

We next examine SCR improvement, which is defined as the SCR out of the MTI divided by the SCR into the MTI, *averaged over all Doppler frequencies of interest*. Averaging is needed because the signal power out of the MTI will depend on the target Doppler frequency. As a result, the SCR improvement will be a function of Doppler frequency, which is cumbersome.

To compute the signal power at the MTI output, we use an equation of the form of (18.11), but replace $C(f)$ with $T(f)$ and P_C with P_S. The result is

$$P_{Sout} = P_S \int_{-\infty}^{\infty} H(f) MF(f) [V_S(f) * T(f) * \Phi(f)] df \qquad (18.26)$$

For target signals, we can ignore scanning and phase noise.[2] We do this by using $V_S(f) = \delta(f)$ and $\Phi(f) = \delta(f)$. With this and $T(f) = \delta(f - f_d)$, (18.26) becomes

$$P_{Sout} = P_S \int_{-\infty}^{\infty} H(f) MF(f) \delta(f - f_d) df = P_S H(f_d) MF(f_d) \qquad (18.27)$$

In most situations, we can assume $MF(f_d) = 1$ for Doppler frequencies of interest.[3] With this

$$P_{Sout} = P_S H(f_d) \qquad (18.28)$$

which tells us the signal power at the output of the MTI depends on Doppler frequency. This says the SNR improvement is a function of Doppler frequency and cannot be conveniently represented by a single number, as we would like. To circumvent this inconvenience, P_{Sout} is averaged over Doppler frequencies of interest.

From (18.7), the average of $H(f_d)$ is unity, as is the noise gain. Because of this, the *average* SNR gain through the MTI is unity. That is, $SNR_{out} = SNR_{in}$. With this and the clutter attenuation results from Section 18.2.1, we get

---

[2] We can ignore scanning because $V_S(f)$ will be narrow relative to the $H(f)$ and $M(f)$ for typical target Doppler frequencies. Also, the signal power due to phase noise is very small.
[3] This is a valid assumption for narrow pulses. However, for long, modulated pulses and large Doppler frequencies, it may be necessary to include the $MF(f_d)$ term.

$$I_{scr} = \frac{SCR_{out}}{SCR_{in}} = \frac{SNR_{out}/CNR_{out}}{SNR_{in}/CNR_{in}} = \frac{CNR_{in}}{CNR_{out}} = CA \qquad (18.29)$$

That is, the (average) SCR improvement is equal to the clutter attenuation. We note that the *peak* SCR improvement is $(CA)/(K_{MTI})^2$, or $2(CA)$ for the 2-pulse MTI, since the peak SNR gain through the MTI is $1/(K_{MTI})^2$ W/W.

18.3 GROUND CLUTTER EXAMPLE

For this example, we consider radar with a carrier frequency of 8 GHz and a PRI of $T = 400$ μs. We assume ground clutter that consists of wooded hills in a 20-knot wind and use $\sigma_v = 0.22$ m/s (Table 17.2) for the Gaussian spectrum model. From this, we compute the frequency spread as

$$\sigma_f = \frac{2\sigma_v}{\lambda} = \frac{2 \times 0.22}{0.0375} = 11.7 \text{ Hz} \qquad (18.30)$$

For the exponential spectrum model, we assume a 20-knot wind represents a windy condition and use $\beta_C = 5.5$ s/m from Table 17.3. With a change of units from knots to mph, we get a wind speed of $w = 23$ mph and use this in (17.14) to compute

$$r = 489.8 w^{-1.55} f_o^{-1.21} = 0.31 \qquad (18.31)$$

We first assume the case where the radar beam is stationary during the CPI, which we accommodate by setting $V_S(f) = \delta(f)$. With this we get $\sigma_T = \sigma_f = 11.7$ Hz.

The clutter attenuation and SCR improvement using the Gaussian spectrum model is

$$CA_{Gauss} = I_{scrGauss} = 2\left(\frac{PRF}{2\pi\sigma_T}\right)^2 = 2{,}313 \text{ W/W or } 33.6 \text{ dB} \qquad (18.32)$$

For the exponential spectrum model, we get

$$CA_{Exp} = I_{scrExp} = \frac{1+r}{(2\pi T)^2}\left(\frac{\lambda\beta_C}{2}\right)^2 = 2{,}206 \text{ W/W or } 33.4 \text{ dB} \qquad (18.33)$$

which is very close to the value we obtained with the Gaussian spectrum model.[4]

As an extension, we assume the same radar and clutter parameters, but use a scanning radar that has a 2-second scan period. We use the beam width associated with the example of Figure 17.2 (i.e., $\alpha_B = 2°$). From (16.31), we have

$$\sigma_S = 0.265\left(\frac{2\pi}{\alpha_B T_{SCAN}}\right) = 23.9 \text{ Hz} \qquad (18.34)$$

For the Gaussian spectrum model, we get a total spectrum width of

---

[4] David Barton indicated it would be worth noting that use of the exponential clutter model is important when values of $CA \gg 20$ dB are needed.

$$\sigma_T = \sqrt{\sigma_f^2 + \sigma_S^2} = 26.6 \text{ Hz} \tag{18.35}$$

It is interesting to note that scanning is the major contributor to the frequency spread. For the exponential spectrum model, we get $\beta_S = 2/(\lambda \sigma_S) = 2.24$ s/m. In this equation, we used the inverse relation between β and σ_v.

The resulting clutter attenuation for the Gaussian spectrum model, with scanning, is

$$CA_{Gauss} = I_{scrGauss} = 2\left(\frac{PRF}{2\pi\sigma_T}\right)^2 = 448.2 \text{ W/W or } 26.5 \text{ dB} \tag{18.36}$$

For the case of the exponential spectrum, we use (18.23) and (18.25) to obtain

$$CA_{Exp} = I_{scrExp} = 246.2 \text{ W/W or } 23.9 \text{ dB} \tag{18.37}$$

In this case, the clutter attenuation using the Gaussian spectrum model is slightly larger than with the exponential spectrum model.

Let us carry this example further and examine *SNR*, *CNR*, and *SIR*. SIR is the acronym for *signal-to-interference ratio* and is defined as

$$SIR = \frac{P_S}{P_C + P_N} = \frac{P_S/P_N}{P_C/P_N + 1} = \frac{SNR}{CNR + 1} \tag{18.38}$$

It is the ratio of the signal power to the total interference power. In a clutter environment, SIR is the parameter used to evaluate detection and tracking performance. For a target to be detected, the signal power must be greater than the total interference power by some margin (i.e., the detection threshold, see Chapter 6). Since SIR is a measure of signal power to total interference power, it is the quantity that should be used. The same argument applies to tracking performance.

In addition to the aforementioned parameters, for this example, we assume the additional radar parameters listed in Table 18.1.

Using the parameters of Table 18.1, the *SNR*, *CNR*, and *SIR* versus *R* at the sampler (and thus, matched filter) output is as shown in Figure 18.4. The *SNR* is reasonable, but the *SIR* is too low to support detection and track [a typical minimum SIR for detection is 13 dB (see Chapter 2) and a minimum SIR for track is in the same range [7]]. (The hook in the CNR plot is caused by the fact that we assumed the radar beam was pointed at the target, rather than at a fixed angle, as was the case for the plots of Figure 17.2.) Also, the Gaussian antenna pattern model [see (17.4)] was used in the clutter RCS generation routine.

Figure 18.5 contains plots similar to those of Figure 18.4 for the cases (nonscanning and scanning) where an MTI is used. Since the clutter attenuation was almost the same for the two spectrum models, only one set of plots is shown (for representative clutter attenuations of 33.5 and 25 dB). For the nonscanning case, the MTI provided enough clutter attenuation to give an SIR that remained above 13 dB (a value required to provide a P_d of 0.5 and P_{fa} of 10^{-6} on a Swerling 1 target—see Chapters 2 and 6) for ranges below 50 km. For the scanning case, the clutter attenuation was not quite adequate, and the SIR dipped to fairly low values at short ranges. This indicates that it might be necessary to consider a higher order MTI, with the hope that it will provide better clutter attenuation and, thus, SIR improvement.

Table 18.1

Radar Parameters

Peak power	50 kW
System noise temperature	1,000 K
Pulse width	4 µs (a 4-chip, Barker coded pulse)
Total losses for the target and clutter	10 dB
Height of the antenna phase center	5 m
Azimuth and elevation beamwidth	2°
Antenna directivity on transmit and receive	38 dB
rms antenna sidelobes	−30 dB
Clutter backscatter coefficient	−20 dB
Target RCS	6 dBsm
Ranges of interest	2 km to 50 km
PRF	2,500 Hz

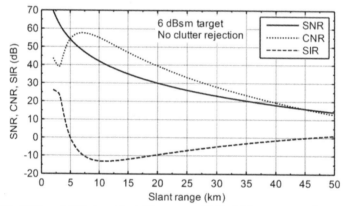

Figure 18.4 SNR, CNR, and SIR at matched filter output, Table 18.1 parameters, beam pointed at target.

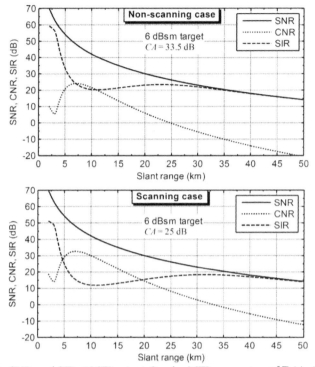

Figure 18.5 SNR, CNR, and SIR at MTI output, 2-pulse MTI, parameters of Table 18.1.

18.4 RAIN CLUTTER EXAMPLE

As an extension to the example of Section 18.3, we examine the behavior of the radar in rain clutter. We use the Gaussian spectrum model with $\sigma_v = 3$ m/s. This velocity spread is an intuitive average of the values given in Table 17.5. We compute the rain clutter RCS from (17.17) and (17.18) for a rainfall rate of 4 mm/hr and a carrier frequency of 8 GHz ($\eta = -66$ dBm^2/m^3). Although the mean velocity of rain is normally not zero, to this example we assume it is. We also ignore phase noise and assume a stationary beam.

To determine the clutter attenuation and SCR improvement, we use (18.20) with $\sigma_T = 2\sigma_v/\lambda = 160$ Hz. This gives

$$CA_{Gauss} = I_{scrGauss} = 2\left(\frac{PRF}{2\pi\sigma_T}\right)^2 = 12.4 \text{ W/W or } 10.9 \text{ dB} \qquad (18.39)$$

which is clearly not very large. Figure 18.6 illustrates the impact of the small value of CA. The top plot corresponds to the case where the MTI is not used, and the bottom plot corresponds to the case where the radar uses a 2-pulse MTI. As can be seen, the SIR is unacceptably low in both cases.

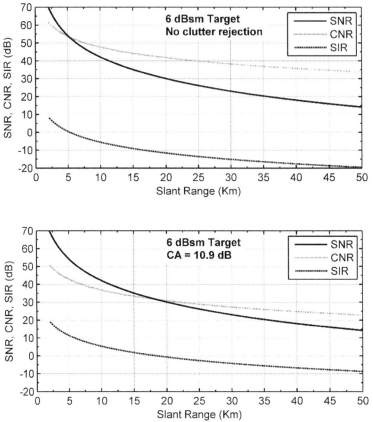

Figure 18.6 SNR, CNR, and SIR at MTI input (top) and output (bottom)—rain clutter, 2-pulse MTI.

18.5 PHASE NOISE

We next examine the impact of phase noise on the MTI clutter attenuation and SCR improvement. We assume the phase noise is white and use $\Phi_{\Delta\phi} = \Phi_0$ [see (16.39) and (16.40)]. With this we get $\Phi(f) = \delta(f) + \Phi_0$ for the phase noise (see Section 16.2.1), and

$$V_S(f) * C(f) * \Phi(f) = V_S(f) * C(f) * [\delta(f) + \Phi_0]$$
$$= V_S(f) * C(f) + \Phi_0 \qquad (18.40)$$

where we took advantage of

$$\int_{-\infty}^{\infty} [V_S(f) * C(f)] df = 1 \qquad (18.41)$$

to get Φ_0 for the last term.

This leads to

$$G_{C\Phi} = \int_{-\infty}^{\infty} H(f) MF(f) [V_S(f) * C(f) * \Phi(f)] df$$

$$= \int_{-\infty}^{\infty} H(f) MF(f) [V_S(f) * C(f) + \Phi_0] df \quad (18.42)$$

$$= G_C + \Phi_0 / \tau_p$$

where G_C is given by (18.19), (18.23), or (18.24), depending on the clutter spectrum model and whether or not the antenna is scanning. Derivation of the second term is left as an exercise. The resulting clutter attenuation is

$$CA = \frac{1}{G_{C\Phi}} = \frac{1}{G_C + \Phi_0 / \tau_p} \quad (18.43)$$

To get an idea of the impact of phase noise on the performance of MTI signal processors, we revisit the previous example and plot clutter attenuation versus phase noise level, Φ_0. This plot is shown in Figure 18.7 for scanning and nonscanning cases.

For the nonscanning case, the phase noise starts to degrade the clutter attenuation at a phase noise level of about −95 dBc/Hz. For the scanning case, the phase noise degradation is delayed until a phase noise level of about −85 dBc/Hz. The reason for this difference is due to the relative sizes in the denominator of (18.43). If G_C is small (meaning the clutter attenuation without phase noise is large), the Φ_0/τ_p term begins to predominate the overall clutter attenuation for relatively small values of Φ_0. However, if G_C is large (meaning the clutter attenuation without phase noise is small), Φ_0 must be fairly large before it begins to predominate the overall clutter attenuation.

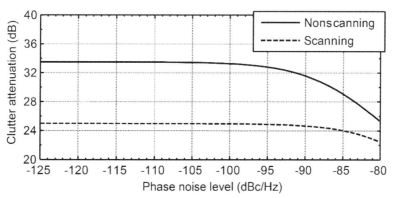

Figure 18.7 Phase noise effects on MTI clutter attenuation.

18.5.1 Higher Order MTI Processors

In the previous example, we found the 2-pulse MTI did not provide sufficient clutter attenuation to mitigate rain clutter. This leads to the question of how much clutter attenuation we could obtain if we use a 3-pulse, 4-pulse, or even higher order MTI. An alternate question is: What order MTI is needed to obtain a desired clutter attenuation?

To obtain an N_{MTI}-pulse MTI, we cascade $N_{MTI} - 1$-, 2-pulse MTIs. Specifically, if the transfer function of a 2-pulse MTI is $H(z)$, the transfer function of an N_{MTI}-pulse MTI is

$$H_{NMTI}(z) = K_{NMTI}\left[H_U(z)\right]^{N_{MTI}-1} \tag{18.44}$$

where the constant K_{NMTI} is used to normalize $H_{NMTI}(z)$ to provide unity noise gain.

The specific transfer functions for 2-, 3-, 4-, and 5-pulse MTIs are [6]

$$\begin{aligned}
H_2(z) &= K_{2MTI}\left(1 - z^{-1}\right) \\
H_3(z) &= K_{3MTI}\left(1 - z^{-1}\right)^2 = K_{3MTI}\left(1 - 2z^{-1} + z^{-2}\right) \\
H_4(z) &= K_{4MTI}\left(1 - z^{-1}\right)^3 = K_{4MTI}\left(1 - 3z^{-1} + 3z^{-2} - z^{-3}\right) \\
H_5(z) &= K_{5MTI}\left(1 - z^{-1}\right)^4 = K_{5MTI}\left(1 - 4z^{-1} + 6z^{-2} - 4z^{-3} + z^{-4}\right)
\end{aligned} \tag{18.45}$$

Note that the coefficients of the powers of z are binomial coefficients with alternating signs [3, p. 356; 5, p. 2.35].

Following the method we used for the 2-pulse MTI, we can compute the MTI gain as

$$K_{NMTI}^2 = \frac{1}{\sum_{m=0}^{N_{MTI}-1} b_m^2} \tag{18.46}$$

where the b_m are the binomial coefficients indicated above. Specific values of K^2_{NMTI} for the 2-, 3-, 4- and 5-pulse MTI are summarized in Table 18.2. K^2_{NMTI} for an N_{MTI}-pulse MTI with binomial coefficients is

$$K_{NMTI}^2 = \frac{\left[2(N_{MTI}-1)\right]!!}{2^{2(N_{MTI}-1)}\left[2(N_{MTI}-1)-1\right]!!} \tag{18.47}$$

where $(2m-1)!! = 1 \times 3 \times 5 \times \cdots \times (2m-1)$ and $(2m)!! = 2 \times 4 \times \cdots \times 2m$, $(0)!! = 1$.

If we extend the results of the 2-pulse analysis, we can write the normalized frequency response of an N_{MTI}-pulse MTI as

$$H_{NMTI}(f) = K_{NMTI}^2 \left[2\sin(\pi f T)\right]^{2(N_{MTI}-1)} \tag{18.48}$$

Figure 18.8 contains plots of the normalized frequency responses of 3- and 4-pulse MTIs, with $T = 400$ μs. Note that the peaks of the response become narrower, and the valleys become wider as the order of the MTI increases. This means we should expect higher clutter attenuation and SCR improvement as the MTI order increases.

We can compute the clutter attenuation for the general N_{MTI}-pulse MTI by extending the work we did for the 2-pulse MTI. For the Gaussian spectrum model, we again use the approximation that $\sin(\pi f T) \approx \pi f T$. With this we get

$$CA_{Gauss} = \frac{1}{G_{CGauss}} \tag{18.49}$$

where G_{CGauss} now becomes

$$G_{CGauss} = K_{NMTI}^2 \int_{-\infty}^{\infty} \frac{(2\pi f T)^{2(N_{MTI}-1)}}{\sigma_T \sqrt{2\pi}} e^{-f^2/2\sigma_T^2} df \tag{18.50}$$

Table 18.2

K_{NMTI}^2 for Various-Size MTIs

MTI Order—N_{MTI}	K_{NMTI}^2
2	1/2
3	1/6
4	1/20
5	1/70
N_{MTI}	$\dfrac{[2(N_{MTI}-1)]!!}{2^{2(N_{MTI}-1)}[2(N_{MTI}-1)-1]!!}$

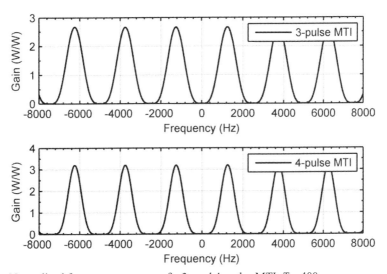

Figure 18.8 Normalized frequency response of a 3- and 4- pulse MTI, $T = 400$ μs.

Evaluation of this integral yields [5]

$$G_{CGauss} = K_{NMTI}^2 \left[2(N_{MTI} - 1) - 1 \right]!! \, (2\pi T \sigma_T)^{2(N_{MTI}-1)} \quad (18.51)$$

We can write the clutter attenuation as

$$\begin{aligned} CA_{Gauss} &= \frac{1}{G_{CGauss}} = \frac{1}{K_{NMTI}^2 \left[2(N_{MTI} - 1) - 1 \right]!!} \left(\frac{1}{(2\pi T \sigma_T)^{2(N_{MTI}-1)}} \right) \\ &= \frac{1}{K_{NMTI}^2 \left[2(N_{MTI} - 1) - 1 \right]!!} \left(\frac{PRF}{2\pi \sigma_T} \right)^{2(N_{MTI}-1)} \end{aligned} \quad (18.52)$$

As with the 2-pulse MTI case, we can show that the MTI gain, averaged across all expected target Doppler frequencies, is equal to unity. With this the SCR improvement, as before, is

$$I_{scrGauss} = CA_{Gauss} \quad (18.53)$$

Specific values of CA_{Gauss} and $I_{scrGauss}$ for a 3- and 4-pulse MTI are

$$I_{3scrGauss} = CA_{3Gauss} = 2 \left(\frac{PRF}{2\pi \sigma_T} \right)^4 \quad (18.54)$$

and

$$I_{4scrGauss} = CA_{4Gauss} = \frac{4}{3} \left(\frac{PRF}{2\pi \sigma_T} \right)^6 \quad (18.55)$$

Table 18.3 contains values of CA_{Gauss} for the nonscanning, scanning, and rain cases of Example 1. The clutter attenuation is large for the case of ground clutter. However, even with the 4-pulse MTI, clutter attenuation for rain clutter is not very good.

An equation for the clutter attenuation and SCR improvement for higher order MTIs and the exponential spectrum is

Table 18.3
Clutter Attenuation for Gaussian Spectrum Model—dB

MTI Order	Clutter Spectrum Conditions		
	Ground, Nonscanning	Ground, Scanning	Rain, Nonscanning
2	33.6	26.5	10.9
3	64.3	50.5	18.8
4	93.1	71.7	25

$$CA_{NMTIExp} = I_{NMTIscrExp} = \frac{1}{G_{CExp}} \quad (18.56)$$

where

$$G_{CExp} = \frac{K_{NMTI}^2 (2N_{MTI})!(2\pi T)^{2N_{MTI}}}{1+r}$$

$$\times \left\{ r\left(\frac{2}{\lambda \beta_S}\right)^{2N_{MTI}} + \frac{1}{\beta_S^2 - \beta_C^2}\left[\beta_S^2\left(\frac{2}{\lambda \beta_C}\right)^{2N_{MTI}} - \beta_C^2\left(\frac{2}{\lambda \beta_S}\right)^{2N_{MTI}}\right] \right\} \quad (18.57)$$

The derivation of (18.57) is tedious, but straightforward, and left as an exercise.

18.5.2 Staggered PRIs

Examination of the MTI frequency response plots of Figures 18.3 and 18.8 indicates that the SNR gain through the MTI can vary considerably with target Doppler frequency. This is quantified in Figure 18.9, which is a plot of the percent of time the MTI gain is above the value indicated on the horizontal axis. For example, the MTI gain is above −5 dB 73% of the time for the 2-pulse MTI, and 60% and 52% of the time for the 3- and 4-pulse MTIs. If we say, arbitrarily, that the MTI is blind when the gain drops below −5 dB, we can say the 2-pulse MTI is blind 27% of the time, and the 3- and 4-pulse MTIs are blind 40% and 48% of the time, respectively.

We can improve this situation by using staggered PRIs. That is, we use waveforms where the spacing between pulses changes on a pulse-to-pulse basis. With staggered PRIs, we "break up" the orderly structure of the MTI frequency response and "fill in" the nulls. We also reduce the peaks in the frequency response. The net effect is to provide an MTI frequency response that has fewer deep nulls and large peaks but rather a somewhat constant level. The response still has the null at zero frequency and still provides clutter rejection.

To analyze the frequency response of an MTI with a *staggered* PRI, we start by examining the output of the sampler for the staggered PRI waveform shown in Figure 18.10. We assume the sampler samples the matched filter output at a delay of τ_d, after each pulse. Thus, the sampler samples the output at τ_d, $\tau_d + T_1$, $\tau_d + T_1 + T_2$, $\tau_d + T_1 + T_2 + T_1$, and so forth. We further assume τ_d is such that we are sampling the matched filter output on the peak of its response to a target return.

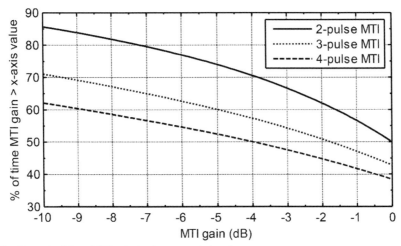

Figure 18.9 Percent of time MTI gain is above *x*-axis levels.

Figure 18.10 A 2-position stagger waveform.

If the target return consists of a (complex) sinusoid at a Doppler frequency of f_d, we can write the sampler output for the kth PRI as

$$v_o(k) = \begin{cases} e^{j2\pi f_d \tau_d} e^{j2\pi f_d T_k} & k \geq 0 \\ 0 & k < 0 \end{cases} \quad (18.58)$$

where

$$T_k = \sum_{l=0}^{k} PRI_l \quad \text{with} \quad PRI_0 = 0 \quad (18.59)$$

and PRI_l is the interpulse spacing of the lth PRI interval. For the waveform of Figure 18.10, $PRI_1 = T_1$, $PRI_2 = T_2$, $PRI_3 = T_1$, $PRI_4 = T_2$, and so forth.

For the 2-pulse MTI, we have, from (18.1), with the addition of K_{2MTI}

$$v_{SP}(k) = K_{2MTI}\left[v_o(k) - v_o(k-1)\right] \quad (18.60)$$

The output of the MTI after the first two pulses (i.e., at $k = 1$) is

$$\begin{aligned} v_{SP}(1) &= K_{2MTI}\left[v_o(1) - v_o(0)\right] \\ &= K_{2MTI}\left(e^{j2\pi f_d \tau_d} e^{j2\pi f_d T_1} - e^{j2\pi f_d \tau_d} e^{j2\pi f_d T_0}\right) \\ &= K_{2MTI} e^{j2\pi f_d \tau_d}\left(e^{j2\pi f_d T_1} - e^{j2\pi f_d T_0}\right) \end{aligned} \quad (18.61)$$

After the second two pulses, the output is

$$v_{SP}(2) = K_{2MTI}\left[v_o(2) - v_o(1)\right]$$
$$= K_{2MTI}e^{j2\pi f_d\tau_d}\left(e^{j2\pi f_d T_2} - e^{j2\pi f_d T_1}\right) \quad (18.62)$$

In general, after the kth pair of pulses, the output will be

$$v_{SP}(k) = K_{2MTI}\left[v_o(k) - v_o(k-1)\right]$$
$$= K_{2MTI}e^{j2\pi f_d\tau_d}\left(e^{j2\pi f_d T_k} - e^{j2\pi f_d T_{k-1}}\right) \quad (18.63)$$

We can extend this to an N_{MTI}-pulse MTI and write

$$v_{SP}(k) = K_{NMTI}\sum_{l=0}^{N_{MTI}-1}(-1)^l b_l v_o(k-l)$$
$$= K_{NMTI}e^{-j2\pi f_d\tau_d}\sum_{l=0}^{N_{MTI}-1}(-1)^l b_l e^{j2\pi f_d T_{k-l}} \quad (18.64)$$

where b_l are binomial coefficients defined by

$$b_l = \frac{(N_{MTI}-1)!}{l!(N_{MTI}-1-l)!} \quad (18.65)$$

We would start computing $v_{SP}(k)$ for $k = N_{MTI} - 1$ to allow time for the MTI transients to settle (more on this shortly).

If we were to plot $|v_{SP}(k)|^2$ versus k, we would find that it is not constant, as was the case for the unstaggered waveform. To smooth the variation with k, we average the $|v_{SP}(k)|^2$ over several k. Specifically, we form

$$V_{SP}(f_d) = \frac{1}{N_{pulse} - N_{MTI} + 1}\sum_{k=N_{MTI}-1}^{N_{pulse}-1}|v_{SP}(k)|^2 \quad (18.66)$$

N_{pulse} is determined by the waveform. We will discuss this shortly. The lower limit is $N_{MTI} - 1$ because we don't want to start using the MTI output until transients have settled. We added f_d as an argument of V_{SP} to recognize that it depends on Doppler frequency. As a note, the exponential on the outside of the sum of (18.64) goes away when we form $|v_{SP}(k)|^2$.

Because of the K_{NMTI} normalization, $V_{SP}(f_d)$ is power gain of the MTI at a frequency f_d. That is

$$H(f_d) = V_{SP}(f_d) = \frac{1}{N_{pulse} - N_{MTI} + 1}\sum_{k=N_{MTI}-1}^{N_{pulse}-1}|v_{SP}(k)|^2 \quad (18.67)$$

The averaging process just discussed is used in actual MTI implementations. This is illustrated in Figure 18.11.

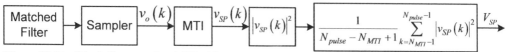

Figure 18.11 Block diagram of an MTI with output power averaging.

The form of (18.66) and (18.67) requires $N_{pulse} \geq N_{MTI}$, where N_{pulse} is the number of pulses in a burst, if the waveform, or burst, contains a finite number of pulses. We average the MTI output through the burst and record a single output at the end of the burst.

If the waveform consists of a semi-infinite string of pulses, a rule of thumb is to choose N_{pulse} as the maximum of N_{MTI} and the length of the PRI sequence (this applies to higher-order MTIs and staggered PRI waveforms). For example, suppose a sequence of PRIs was $T_1, T_2, T_3, T_2, T_1, T_2, T_3, T_2$, and so forth. This sequence repeats every four PRIs. That is, the PRI sequence is T_1, T_2, T_3, T_2, and has a length of 4. If we were using a 2-, 3-, or 4-pulse MTI, we would thus choose $N_{pulse} = 4$. If the order of the MTI was 5 or larger, we would choose N_{pulse} as the MTI order, N_{MTI}.

We can assemble the above into an algorithm for generating MTI frequency responses or, as they are commonly termed, *MTI velocity responses*. The latter name derives from the fact that we normally plot $H(f_d)$ versus v where $v = \lambda f_d/2$. The algorithm is as follows:

- Identify the number of pulses, N_{pulse}, in the burst, along with their PRIs. If the waveform is semi-infinite (semi-infinite burst of pulses), use $N_{pulse} = N_{PRI}$ where N_{PRI} is the number of PRIs in the PRI sequence.
- Compute the T_k using (18.59) for $k = 0$ to $N_{pulse} - 1$.
- Select the MTI order, N_{MTI}.
- Compute $v_{SP}(k)$ using (18.64) for $k = N_{MTI} - 1$ to $N_{pulse} - 1$ (without the exponential in front of the summation).
- Compute $H(f_d)$ using (18.67).
- Repeat the above steps for the f_d (or v) of interest.

As an example, we consider a burst of $N_{pulse} = 10$ pulses with repeating PRIs of $T_1 = 385$ μs and $T_2 = 415$ μs. We consider a 3-pulse MTI so $N_{MTI} = 3$. We need to compute $N_{pulse} - N_{MTI} + 1 = 10 - 3 + 1 = 8$ values of $v_{SP}(k)$ using[5]

$$v_{SP}(k) = K_{3MTI}\left[v_o(k) - 2v_o(k-1) + v_o(k-2)\right] \quad (18.68)$$

with

$$v_o(k) = e^{j2\pi f_d T_k} \quad (18.69)$$

for $k = N_{MTI} - 1$ to $N_{pulse} - 1$, or 2 to 9. We then average the eight values of $|v_{SP}(k)|^2$ to get $H(f_d)$ [see (18.67)]. The result of this is shown in Figure 18.12. In the figure, the horizontal axis is range rate and was computed using the conversion $\lambda f_d/2$ with $\lambda = 0.0375$ m (8-GHz RF). Actually, we started with range rates and computed the Doppler frequencies from them.

---

[5] Had the burst been semi-infinite, we would have used $K = 2$, the length of the PRI sequence.

The response with the staggered waveform still has a considerable variation in MTI gain as a function of range rate because we only used a 2-position stagger (a PRI sequence repeats after two PRIs). This can often be improved by using more than two values of T_k; that is, a higher position stagger with more interpulse periods. Skolnik discusses this in his *Radar Handbook* [5, Ch. 2].

To find the SNR gain through the MTI, we find the average signal gain from the MTI frequency response (e.g., Figure 18.12) and use this as the SNR gain. We can do this because the MTI is still normalized and provides unity noise gain. We often find the average MTI gain via the "eyeball" method; that is, we estimate it from the plot. A better method would be to numerically average the gain (in W/W) across the range rates of interest. The MTI gain indicated via the "eyeball" method for the response of Figure 18.12 is about 0 dB. The calculated gain is −0.08 dB.

To determine the clutter attenuation, we would, ideally, use the methods of Sections 18.2.1 and 18.2.2 with $H(f)$ replaced by the averaged frequency response of (18.67). An alternative to this would be an approximation where we use (18.51) or (18.57) with T replaced by the average of the PRIs in the burst or PRI sequence (for a semi-infinite burst of pulses).

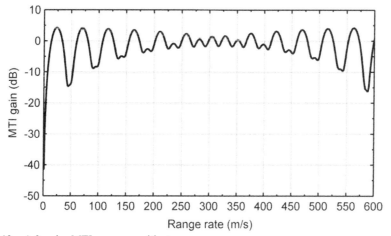

Figure 18.12 A 3-pulse MTI response with stagger.

18.5.3 MTI Transients

In the previous section, we noted that we would not use the MTI output until it had processed N_{MTI} pulses. That is, until $k = N_{MTI} - 1$ (recall, k starts at 0). We do this because the MTI is in its transient phase for the first $N_{MTI} - 1$ pulses. If the input contains clutter, the clutter rejection of the MTI is not realized until after the transient phase. As an illustration of this, consider a 3-pulse MTI where the input is a sequence of ones. That is, for $k \geq 0$, $v_o(k) = 1$, and for $k < 0$, $v_o(k) = 0$. By using these values in (18.68), we get

$$v_{SP}(0) = K_{3MTI}\left[v_o(0) - 2v_o(-1) + v_o(-2)\right] = K_{3MTI}\left[1 - 2\times 0 + 0\right] = K_{3MTI}$$
$$v_{SP}(1) = K_{3MTI}\left[v_o(1) - 2v_o(0) + v_o(-1)\right] = K_{3MTI}\left[1 - 2\times 1 + 0\right] = -K_{3MTI}$$
$$v_{SP}(2) = K_{3MTI}\left[v_o(2) - 2v_o(1) + v_o(0)\right] = K_{3MTI}\left[1 - 2\times 1 + 1\right] = 0 \qquad (18.70)$$
$$\vdots$$
$$v_{SP}(k) = K_{3MTI}\left[v_o(k) - 2v_o(k-1) + v_o(k-2)\right] = K_{3MTI}\left[1 - 2\times 1 + 1\right] = 0$$

That is, the output does not settle to zero until $k = 2 = N_{MTI} - 1$. To avoid having the transient affect detection and tracking functions that use the MTI output, the MTI output is usually gated off during the transient period.

18.6 EXERCISES

1. Show the approximation, $MF(f) = 1$, is valid for the parameters of Example 1.
2. Show the approximation, $\sin(\pi f T) \approx \pi f T$, is valid for clutter spectrum spread values considered in this chapter.
3. Derive (18.22).
4. Derive (18.23) and (18.24).
5. Show that

$$G_{C\Phi} = \Phi_0 \int_{-\infty}^{\infty} H(f) MF(f) df = \Phi_0/\tau_p$$

for the case where $H(f)$ is given by (18.7) and $MF(f)$ is given by (16.22). Assume $\tau_p \ll T$.

6. Repeat Example 1 and reproduce all of the plots.
7. Derive (18.47).
8. Derive (18.51).
9. Derive (18.57).
10. Derive (18.64).
11. Implement the algorithm of Section 18.5.2 and reproduce Figure 18.12.

12. Show that

$$\int_{-\infty}^{\infty} K_{NMTI}^2 \left[2\sin(\pi fT) \right]^{2(N_{MTI}-1)} df = 1$$

for all N_{MTI}.

References

[1] Barton, D. K., *Radar System Analysis and Modeling*, Norwood, MA: Artech House, 2005.

[2] Skolnik, M. I., *Introduction to Radar Systems*, 3rd ed., New York: McGraw-Hill, 2001.

[3] Schleher, D. C., *MTI and Pulsed Doppler Radar with MATLAB*, 2nd ed., Norwood, MA: Artech House, 2010.

[4] Barton, D. K., *Radar Equations for Modern Radar*, Norwood, MA: Artech House, 2013.

[5] Skolnik, M. I., ed., *Radar Handbook*, 3rd ed., New York: McGraw-Hill, 2008.

[6] Papoulis, A., *Probability, Random Variables, and Stochastic Processes*, 3rd ed., New York: McGraw-Hill, 1991.

[7] Budge, M. C. Jr. and S. R. German, *Basic Radar Tracking*, Norwood, MA: Artech House, 2018.

Chapter 19

Digital Pulsed Doppler Processors

19.1 INTRODUCTION

The exact origin of the phrase "pulsed Doppler" is not clear. It probably derives from early pulsed Doppler radars, which performed CW processing using pulsed waveforms. Classical CW radars work primarily in the frequency (and angle) domain, whereas pulsed radars work primarily in the time (and angle) domain. It is assumed that the phrase "pulsed Doppler" was coined when designers started using pulsed radars that worked primarily in the frequency, or Doppler, domain. Early pulsed Doppler radars used a 50% duty cycle and pulsed waveforms, and had virtually no range resolution capability, only Doppler resolution. The use of a pulsed waveform was motivated by the desire to use only one antenna and to avoid isolation problems caused by CW operation. Modern pulsed Doppler radars are actually low-, medium-, or high-PRF pulsed radars with typical duty cycles in the 5% to 10% range. They are used for both range and Doppler measurement.

Three classes of pulsed Doppler waveforms have evolved over the years:

1. The "classical" pulsed Doppler waveform has a high PRF and operates ambiguously in range, but is unambiguous in Doppler. High PRF (HPRF) waveforms have PRFs on the order of 50 to over 100 kHz, with pulse widths on the order of 0.5 to 2 µs.[1]
2. Medium PRF (MPRF) pulsed Doppler waveforms are ambiguous in both range and Doppler. These waveforms have PRFs in the approximate range of 10 to 50 kHz and pulse widths in the range of 2 to 10 µs. In some instances, the pulses are phase or frequency modulated to improve range resolution and to reduce clutter power entering the signal processor.
3. Low PRF (LPRF) waveforms are unambiguous in range and ambiguous in Doppler. LPRF waveforms have PRFs in the range of a few hundred Hz to 10 kHz and pulse widths in the range of 10 to 100 µs. LPRF waveforms almost always use phase- or frequency-modulated pulses to provide adequate range resolution and energy, and to reduce clutter power entering the signal processor.

---

[1] The PRF/pulsewidth ranges of the three classes of pulsed Doppler waveforms are approximate, not absolute, boundaries.

When we say the waveform is ambiguous in range, we mean the PRI is shorter than target range delays of interest. When we say the waveform is ambiguous in Doppler, we mean the PRF is smaller than the target Doppler frequencies of interest.

Some of the benefits of using pulsed Doppler waveforms in a radar are:

- The waveform can be used for detection of short- and long-range targets without the need to change pulse widths to maintain sufficient energy and counter blind range (recall that a radar is "blind" when target returns occur during the time of the transmit pulse).
- Pulsed Doppler processors can directly measure a target's range rate by measuring Doppler frequency. This can be helpful in tracking and mitigating ECMs such as range-gate pull-off (RGPO) [1–3].
- Pulsed Doppler processors are Doppler selective in that they can be designed to reject returns not at the target Doppler frequency. Because of this, pulsed Doppler processors are capable of mitigating clutter whose Doppler frequency is not zero, such as rain and chaff.
- Pulsed Doppler processors can provide both range and frequency information to the operator or computer. This can be used to detect and counter separating targets or various types of pull-off ECM such as RGPO, velocity deceptive jamming, or range and velocity deceptive jamming.

Some of the myths, or misconceptions, associated with radars that use pulsed Doppler waveforms are:

- *They are better at rejecting ground clutter*. From a radar system perspective, this is not totally correct. Pulsed Doppler processors (usually) provide higher SCR improvement than radars with MTI processors. However, with MPRF and HPRF waveforms, the SCR at the processor input is much lower than with waveforms used with MTI processors. This is because, in range ambiguous pulsed Doppler radars, the target must compete with clutter at much shorter ranges than the target's range. With LPRF waveforms, the target competes with clutter at the target range.
- *Pulsed Doppler radars are less susceptible to noise jamming*. This is not correct for broadband noise. Mitigation of broadband noise depends on the ratio of the target and jamming *energy* at the radar receiver input. This is not changed by the signal processor. Pulsed Doppler waveforms could help mitigate noise jamming if the jammer bandwidth is less than the radar PRF.

Some problems associated with pulsed Doppler waveforms and processors are:

- Pulsed Doppler signal processors are generally more complicated than MTI processors because of the added dimension of Doppler frequency. This extends to post processing such as detection logic and track algorithms.
- Local oscillators must have low phase noise since this is often a limiting factor in pulsed Doppler SCR improvement. Pulsed Doppler radars also have stringent timing jitter requirements since timing jitter translates to phase noise.
- MPRF and HPRF pulsed Doppler radar receivers must have large dynamic range to simultaneously accommodate large clutter returns from the first range

ambiguity and small signal returns from subsequent ambiguous regions. This extends to the ADC in radars where clutter rejection is performed entirely by the digital portion of the signal processor.

19.2 PULSED DOPPLER CLUTTER

The ground clutter environment in MPRF and HPRF pulsed Doppler radars is generally more severe than in pulsed radars that are unambiguous in range. This is because, in MPRF and HPRF pulsed Doppler radars, the signal returned from long-range targets must compete with clutter at short ranges.[2] This is illustrated in Figure 19.1. The solid triangle in the figure is a target return from the first (left-most) pulse in the burst of pulses and indicates that the target return does not arrive until several PRIs after the transmit pulse that caused the return. The dashed triangles are returns from the same target, but from different pulses. The solid, curved line through the solid triangle represents the clutter from the pulse immediately preceding the triangle, and from previous pulses. The dashed, curved lines are clutter returns related to previous pulses (and pulses before them). The significance of what signal comes from which pulse has to do with range attenuation. The target is at a range of R_{tgt} and will have a range attenuation proportional to $(R_{tgt})^4$. The clutter in the target range cell is at a range of R_{clut} and will undergo a range attenuation proportional to $(R_{clut})^3$ (recall that clutter attenuation varies as R^3 for ground clutter and R^2 for rain clutter). Since $R_{tgt} \gg R_{clut}$, the target return will undergo much more attenuation than the clutter return. The result of this is that the SCR at the input to the signal processor, in pulsed Doppler radars that use range ambiguous waveforms, is much lower than for the same scenario in radars that use range unambiguous waveforms.

We will explain this difference with the help of Figure 19.2. The top curve is a plot of SNR versus range and is applicable to both pulsed and pulsed Doppler waveforms that use the same pulse width. The middle curve is a plot of CNR for a radar that uses a range unambiguous waveform (e.g., LPRF waveform), at least over the 50-km range interval shown. In this case, the CNR continuously decreases with range. (The SNR, CNR, and SCR discussed here are the SNR, CNR, and SCR at the output of the matched filter.) The bottom curve is a plot of CNR for a radar that uses a range ambiguous waveform (e.g., a MPRF or HPRF waveform). In this case, the CNR decreases for a while and then resets to a large value. This reset occurs with every pulse of the waveform, which means the CNR stays large over the 50-km range of the plot. At the same time, the SNR is decreasing. Thus, the SCR will continually decrease as target range increases.

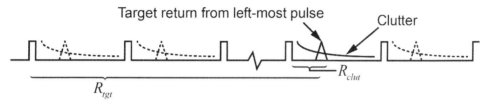

Figure 19.1 Target and clutter returns in an MPRF or HPRF pulsed Doppler radar.

---

[2] This is not the case for LPRF pulsed Doppler radars since LPRF waveforms are unambiguous in range.

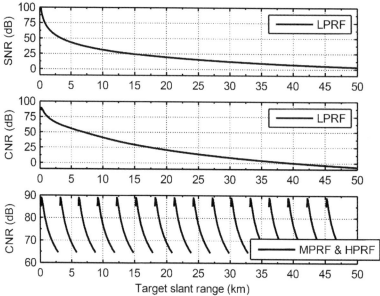

Figure 19.2 Plots of SNR and CNR for LPRF and MPRF, or HPRF, waveforms.

The net effect of the resetting of CNR and continual decrease in SNR is illustrated in Figure 19.3, which is a plot of SCR for an LPRF waveform and a MPRF or HPRF waveform. As can be seen, the SCR for the LPRF waveform initially decreases and then increases. However, the SCR for the MPRF/HPRF waveform continually decreases. Also, the SCR values for the MPRF/HPRF waveform are much lower than for the LPRF waveform. This means the pulsed Doppler signal processor must provide much larger SCR improvement for MPRF or HPRF waveform than it would for the LPRF waveform.

The aforementioned resetting phenomenon can be explained with the help of Figure 19.4, which shows notional clutter returns from three successive pulses, plus a composite return signal at the bottom.

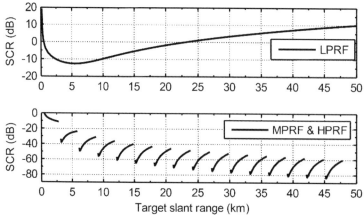

Figure 19.3 Plots of SCR for LPRF and MPRF, or HPRF, waveforms.

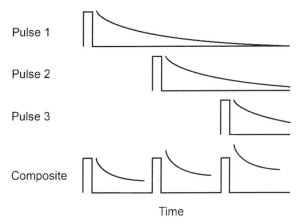

Figure 19.4 Illustration of clutter return resetting phenomenon.

The first pulse causes a clutter return that peaks after the pulse and decays as the range to the clutter increases. The same thing happens on the second and third pulses. As returns from the successive pulses are received, their power is added to the power from the previous pulses and causes the sum to increase after each pulse.

The bottom plot shows that not only does the composite return peak after each pulse, but each peak is a little larger than the previous peak because of residual clutter returns from previous pulses. In practice, this increase levels out with increasing pulse number because the contribution of earlier pulses decreases with range. The CNR resetting is sometimes termed *clutter folding*.

The discussion above indicates the clutter return in a particular range cell is the sum of the clutter returns from the current pulse and all previous pulses. Since the returns are from clutter at different ranges, and we assume the clutter returns from different ranges are uncorrelated, we sum the clutter powers. To derive the appropriate equations, we consider a clutter cell at a range R where R is greater than some start range, R_{start}, and less than some stop range, R_{stop}. R_{start} is usually chosen greater than a pulse width ($c\tau_p/2$) because the receiver is off during the transmit pulse and cannot fully process returns from clutter cells (or targets) at shorter ranges. R_{stop} is chosen to be less than $c\tau_{PRI}/2 - c\tau_p/2$ to allow time for the receiver to fully process pulse returns before it shuts off in preparation for transmit. $R_{PRI} = c\tau_{PRI}/2$ is the range associated with a PRI. When the radar receives a signal from clutter at a range R close to the most recent pulse, it also receives signals from clutter at $R + R_{PRI}$ due to the immediately prior pulse, $R + 2R_{PRI}$ from two pulses back, $R + 3R_{PRI}$ from three pulses back, and so forth. Since the powers from these returns add, the total power associated with the clutter return from the most recent pulse is

$$P_{Cpd}(R) = \sum_{k=0}^{N_{PRI}} P_C(R + kR_{PRI}) \qquad (19.1)$$

Where $P_C(R + kR_{PRI})$ is the clutter power associated with the clutter cell located at the range $R + kR_{PRI}$. As k increases, the associated $P_C(R + kR_{PRI})$ contributes less and less to the sum because of the R^3 roll off of clutter power. In many applications, the contribution becomes very small after only a few pulses.

A means of incorporating this clutter folding into the previous RCS model (Chapter 17) is as follows:

- Generate σ_C and CNR using the equations in Chapter 17 and the radar range equation. Extend the range to the point where the CNR is about 20 dB below its peak level. For HPRF waveforms, this will be about 10 PRIs. For LPRF waveforms, it will usually be one PRI, and for MPRF waveforms, it will be between 1 and 10 PRIs.
- Implement (19.1) for N_{PRI} equal to the number of PRIs determined in the previous step, and R between R_{start} and R_{stop} in steps of δR, where δR is the range resolution of the waveform.
- To generate a CNR plot like Figure 19.2, replicate $P_{Cpd}(R)$ for the number of PRIs needed to cover the range extent of interest.

To generate the associated SNR and SCR plots:

- Generate an array of SNR values over the range extent of interest.
- Blank the range cells in the regions $0 \leq R < R_{start}$ and $R_{stop} < R \leq R_{PRI}$ for each PRI.
- Generate the SCR by dividing the SNR array by the CNR array.

Figure 19.5 contains the result of implementing these algorithms for the parameters of Section 18.3 (MTI ground clutter example), with a waveform PRF of 50 kHz, summarized in Table 19.1.

The aforementioned procedures can also be used for rain clutter. The result of such an application for the rain clutter example of Section 18.4, and the 50-kHz waveform, is contained in Figure 19.6.

Table 19.1
Radar Parameters

Peak power	50 kW
System noise temperature	1,000 K
Pulsewidth	4 µs (a 4-chip, Barker coded pulse)
Total losses for the target and clutter	10 dB
Height of the antenna phase center	5 m
Azimuth and elevation beamwidth	2°
Antenna directivity on transmit and receive	38 dB
rms antenna sidelobes	−30 dB
Clutter backscatter coefficient	−20 dB
Target RCS	6 dBsm
Ranges of interest	2 km to 50 km
PRF	50 kHz

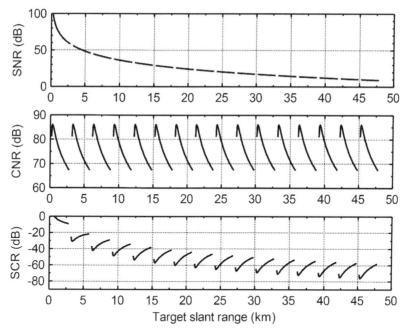

Figure 19.5 Plot of SNR, CNR, and SCR for the parameters of Section 18.3 and a 50-kHz PRF waveform—ground clutter.

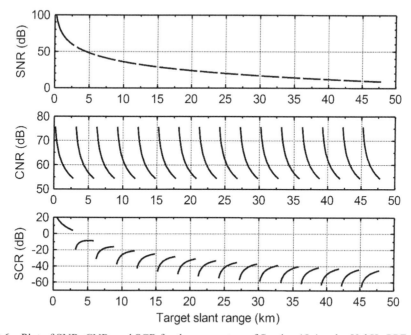

Figure 19.6 Plot of SNR, CNR, and SCR for the parameters of Section 18.4 and a 50-kHz PRF waveform—rain clutter.

19.3 SIGNAL PROCESSOR CONFIGURATION

The signal processor configuration we will use to evaluate clutter attenuation, SNR improvement, and, ultimately, SCR and SIR improvement, is illustrated in Figure 19.7. This is a generic digital, pulsed Doppler signal processor that is applicable to all processors considered in this chapter. It applies to processors for HPRF, MPRF, and LPRF pulsed Doppler waveforms. It is also similar to the configuration we used to analyze MTI processors, with the MTI replacing the high pass filter (HPF) and bandpass filter (BPF). The block diagram can be extended to hybrid processors (analog HPF and digital BPF) by moving the ADC to between the HPF and BPF and putting a sample-and-hold device in the current location of the ADC.

As before, the matched filter is matched to a single pulse of the transmit waveform. The ADC (analog to digital converter) samples the matched filter output, for a given range, once per PRI. For our analyses, we assume it samples on the peak of the matched filter response. As indicated in Chapters 15 and 18, the ADC actually generates several samples per PRI (one for each range gate) and stores them for processing after it has gathered samples for all pulses in the burst, or coherent processing interval (CPI). The samples within a PRI are usually spaced one range resolution cell apart, but are sometimes spaced a little closer to reduce range straddle loss.

In the previous paragraph, we indicated the sampler output from all pulses in a burst is stored and then sent to the processor. This would be the standard approach for radars that transmit waveforms in bursts, such as phased arrays. In older, dish-type radars, the waveform consists of a semi-infinite string of pulses, and the processor would process them continuously, mostly using analog hardware for the HPF and BPF. In those cases, the "burst," or CPI, for analysis purposes, is taken to be roughly the inverse of the BPF bandwidth.

Since we are using frequency-domain techniques in the analyses, they apply to both a burst of pulses and a semi-infinite string of pulses. One caveat regarding the burst of pulses is we assume processor transients have settled so that the frequency domain analyses apply (since they only apply to steady-state conditions). This is a consideration in the design and implementation of pulsed Doppler signal processors.[3]

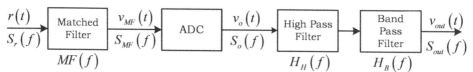

Figure 19.7 Pulsed Doppler signal processor.

---

[3] As will be shown later, there are ways to deal with transients if this causes problems.

The HPF following the ADC is used to reduce the clutter power located near zero Doppler. In addition to reducing clutter power, it also serves to reduce the dynamic range requirements on the BPF following the HPF. This HPF is usually included in processors for HPRF and MPRF waveforms because of their high-clutter attenuation requirements. The HPF can be omitted in LPRF pulsed Doppler processors since the clutter attenuation requirements of those waveforms are generally more modest. Having said this, in modern radars that use high dynamic range digital signal processors, it may be possible to eliminate the HPF and rely on the BPF to provide both clutter attenuation and SNR improvement.

In digital signal processors, the HPF is sometimes implemented before the ADC to limit the dynamic range of the signal into the ADC. In the past, it was thought that the dynamic range of the ADC needed to be greater than the SCR at the ADC input. However, recent analyses [4] indicate this is not the case. We will address the impact of ADC dynamic range in Section 19.4.2.

The final device in the signal processing chain is the BPF. In the diagram, we show it as a single BPF, which is all that is needed for these analyses. We assume the BPF is centered on the target Doppler frequency. We account for this not being the case in practice by including a Doppler mismatch loss in the radar range equation for the target signal (see Chapter 5). The main purpose of the BPF is to increase SNR (and thus SCR), although it also provides additional clutter rejection by reducing phase noise power.

In practice, the HPF output could feed several BPFs centered at different frequencies. The processor used during search would require enough BPFs to span the PRF [recall that, in sampled data systems, the sampler "folds" (or aliases, or wraps) the entire spectrum of the matched filter output into a frequency band between $-PRF/2$ and $+PRF/2$ (between $-1/2T$ and $+1/2T$)]. The processor used during track needs only a few BPFs since the target Doppler frequency is known reasonably well during track.

The implementation of pulsed Doppler signal processors has evolved over the years from all analog, to hybrid, to all, or almost all, digital. The evolution has generally been driven by the speed, availability, and cost of ADCs and digital signal processing components. Older radars (pre-1980s or so) used all-analog signal processors. Radars designed between about 1980 and 2000 used a hybrid mix of digital and analog components. Modern pulsed Doppler signal processors are almost exclusively digital. Some digitize the signal at the matched filter output, as in Figure 19.7. Others digitize the signal at the IF amplifier output and implement the matched filter in the digital domain (see Chapter 22).

In digital processors, the BPFs used in search are often implemented using FFTs or finite impulse response (FIR) filters with amplitude weighting to reduce Doppler sidelobes [5–7]. The FFT is attractive because, by default, its taps span the PRF. It is also computationally efficient. Since only a few BPFs are required in the signal processor used in the track channel, the FIR filters might be a better choice. It is not unusual that the HPF, when used, is implemented with an infinite impulse response (IIR) filter because the HPF generally requires sharp cutoff characteristics for good clutter rejection which would require high-order FIR filters.

19.4 DIGITAL SIGNAL PROCESSOR ANALYSIS TECHNIQUES

We analyze digital, pulsed Doppler signal processors using techniques very similar to those used for MTI processors. Specifically, we compute the clutter and target power at the output of the signal processor using equations similar to (18.11) for clutter and (18.27) for target signals with the MTI filter, $H(f)$, replaced by the HPF and BPF, $H_H(f)H_B(f)$. With this, we have

$$P_{Cout} = P_C \int_{-\infty}^{\infty} H_H(f) H_B(f) MF(f) \left[C(f) * \Phi(f) \right] df \qquad (19.2)$$

for the clutter signal and

$$P_{Sout} = P_S \int_{-\infty}^{\infty} H_H(f) H_B(f) MF(f) \delta(f - f_d) df \qquad (19.3)$$

for the target signal. We will discuss the noise shortly. The integral of (19.2) is usually computed numerically because a closed form solution is generally impossible to derive.

Because of the impulse function (Dirac delta), we can write P_{Sout} as

$$P_{Sout} = P_S H_H(f_d) H_B(f_d) MF(f_d) = P_S G_S \qquad (19.4)$$

where

$$G_S = H_H(f_d) H_B(f_d) MF(f_d) \qquad (19.5)$$

In most applications, the main lobe of $MF(f_d)$ (the matched-range Doppler cut of the ambiguity function) is much wider than the expected span of target Doppler frequencies so that $MF(f_d) \approx 1$. Also, the target Doppler frequency is normally assumed to be in the pass band of the HPF, and the BPF is assumed to be centered very close to the target Doppler frequency so that $H_H(f_d) \approx 1$ and $H_B(f_d) \approx 1$. Combining these observations leads to the conclusion that $G_S \approx 1$. We account for the fact that the various terms of (19.5) are not exactly unity by including a loss term in the radar range equation. However, the general form of G_S is useful for determining the limits the HPF might place on the ability of the radar to detect and track low Doppler targets, or targets whose Doppler frequency approaches a multiple of the PRF (ambiguous Doppler operation). In that case, the fact that $G_S \neq 1$ is not included in the losses.

Figure 19.8 contains a sketch of the various spectra discussed above. Note that because of sampling, the clutter spectrum, $C(f)$, and the target spectrum, $T(f)$, are repeated at intervals of $1/T$. Also, because the HPF and BPF are digital, their responses are periodic with a period of $1/T$. As indicated, $MF(f)$ is very wide relative to the other spectra.

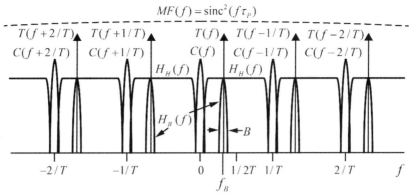

Figure 19.8 Spectra applicable to digital pulsed Doppler signal processor analyses.

For the clutter, $C(f)$ is one of the forms discussed in Sections 17.2.2 and 17.3.2. For $\Phi(f)$, we use the general form of (16.39). P_C is computed via CNR using the folded clutter discussed in Section 19.2. Using this, we have

$$P_{Cout} = P_C \int_{-\infty}^{\infty} H_H(f) H_B(f) MF(f) \{C(f) * [\delta(f) + \Phi_{\Delta\phi}(f)]\} df$$

$$= P_C \int_{-\infty}^{\infty} H_H(f) H_B(f) MF(f) C(f) df \qquad (19.6)$$

$$+ P_C \int_{-\infty}^{\infty} H_H(f) H_B(f) MF(f) [C(f) * \Phi_{\Delta\phi}(f)] df$$

$$= P_C [G_C + G_\phi]$$

where

$$G_C = \int_{-\infty}^{\infty} H_H(f) H_B(f) MF(f) [C(f) * \delta(f)] df$$

$$= \int_{-\infty}^{\infty} H_H(f) H_B(f) MF(f) C(f) df \qquad (19.7)$$

and

$$G_\phi = \int_{-\infty}^{\infty} H_H(f) H_B(f) MF(f) [C(f) * \Phi_{\Delta\phi}(f)] df \qquad (19.8)$$

We term G_C the *central line clutter gain* and G_ϕ the *phase noise clutter gain*. G_C is a measure of the ability of the processor to reject clutter if there was no phase noise. G_ϕ is a measure of the effect of phase noise on the ability of the signal processor to reject clutter. As we will show, for MPRF and HPRF pulsed Doppler radars, G_ϕ is usually much larger than G_C (however, both are typically much less than one). That is, phase noise is usually the limiting factor on the ability of the signal processor to reject clutter. The reverse is true for LPRF pulsed Doppler radars, that is G_C is usually larger than G_ϕ and phase noise is not usually the limiting factor.

We treat receiver noise (what we have called *noise*) differently from target and clutter signals because the target and clutter signal methodology does not apply to noise. For the former, we developed the appropriate equations by propagating a signal from the transmitter to the target or clutter, back to the radar, and through the receiver to the output of the sampler. Noise does not follow this paradigm.

Noise originates in all receiver stages (including the ADC), but the current practice is to reference it to the receiver input by specifying system noise temperature, T_s, or a system noise figure, F_n.[4] This is then used to compute the (white) noise power spectral density $N_0 = kT_s$ or $N_0 = kT_0 F_n$, depending on the noise model used (see Chapter 4). To be consistent with the terminology we have used thus far, we need noise power at the matched filter output, rather than the noise power spectral density at the receiver input. Since we can have $MF(f)$, we could compute the noise power at the matched filter output. However, there is an easier way to approach the problem. Specifically, we reference everything to the noise power at the matched filter output. That way, we do not need to specifically know P_N and we can, for analysis purposes, determine P_S and P_C from *SNR* and *CNR* at the matched filter output, which we can compute from the radar range equation.

Since the power at the output of a (theoretical) sampler (or ADC) is the same as the power at its input, we have $P_{No} = P_N$, the power at the matched filter output. Also, since the bandwidth of the noise out of the matched filter is much larger than the PRF, we can reasonably assume the noise at the output of the sampler is white (see Exercise 2). By definition, if the power associated with white sampled data (discrete time) noise has a value of P_{No}, its power spectral density also has a value of P_{No}; that is, $N_s = P_{No} = P_N$. With this, we can write the noise power at the signal processor output as

$$P_{Nout} = T \int_{-1/2T}^{1/2T} N_s H_H(f) H_B(f) df$$

$$= P_{No} T \int_{-1/2T}^{1/2T} H_H(f) H_B(f) df = P_{No} G_N \tag{19.9}$$

where

$$G_N = T \int_{-1/2T}^{1/2T} H_H(f) H_B(f) df \tag{19.10}$$

is the *noise gain* of the signal processor.

---

[4] Actually, the "receiver" noise also contains environment noise as discussed in Chapters 2 and 4.

We are now in a position to use (19.4), (19.6), and (19.9) to derive the SNR, CNR, and SCR gains through the signal processor. The SNR gain is

$$G_{SNR} = \frac{SNR_{out}}{SNR_o} = \frac{P_{Sout}/P_{Nout}}{P_{So}/P_{No}} = \frac{P_{Sout} P_{No}}{P_{So} P_{Nout}} = \frac{P_{So} P_{No}}{P_{So}(G_N P_{No})} = \frac{1}{G_N} \qquad (19.11)$$

The CNR gain is

$$\begin{aligned} G_{CNR} &= \frac{1}{CA} = \frac{CNR_{out}}{CNR_o} = \frac{P_{Cout}/P_{Nout}}{P_{Co}/P_{No}} = \frac{P_{Cout} P_{No}}{P_{Co} P_{Nout}} \\ &= \frac{\left[(G_C + G_\phi) P_{Co}\right] P_{No}}{P_{Co}(G_N P_{No})} = \frac{G_C + G_\phi}{G_N} \end{aligned} \qquad (19.12)$$

CA is the reciprocal of G_{CNR}. The SCR gain, or SCR improvement, is

$$\begin{aligned} G_{SCR} &= I_{scr} = \frac{SCR_{out}}{SCR_o} = \frac{P_{Sout}/P_{Cout}}{P_{So}/P_{Co}} = \frac{P_{Sout} P_{Co}}{P_{So} P_{Cout}} \\ &= \frac{P_{So} P_{Co}}{P_{So}\left[(G_C + G_\phi) P_{Co}\right]} = \frac{1}{(G_C + G_\phi)} \text{ or } \frac{G_{SNR}}{G_{CNR}} \end{aligned} \qquad (19.13)$$

19.4.1 Phase Noise and Range Correlation Effects

In high- and medium-PRF pulsed Doppler radars, phase noise is often the major factor that limits clutter attenuation and, as a result, SCR and SIR improvement. Because of this, we extend the phase noise model beyond the simple form of (16.40). In particular, we want to derive an expression for $\Phi_{\Delta\phi}(f)$. From (16.38) and (16.39), we can write

$$\Phi_{\Delta\phi}(f) = \int_{-\infty}^{\infty} R_{\Delta\phi}(\tau) e^{-j2\pi f\tau} d\tau \qquad (19.14)$$

where

$$R_{\Delta\phi}(\tau) = E\{\Delta\phi(t+\tau)\Delta\phi^*(t)\} \qquad (19.15)$$

is the phase noise autocorrelation function, and (see Appendix 16A)

$$\Delta\phi(t) = \phi(t - \tau_d) - \phi(t) \qquad (19.16)$$

In (19.16), $\tau_d = 2R_C/c$ is the time delay to the clutter (one of the point clutter sources that make up the clutter patch illuminated by the pulse, or prior pulses—see Section 19.2), and $\phi(t)$ is the local oscillator (LO) phase noise.

Using (19.16) in (19.15), we get

$$R_{\Delta\phi}(\tau) = E\left\{\left[\phi(t-\tau_d+\tau)-\phi(t+\tau)\right]\left[\phi(t-\tau_d)-\phi(t)\right]^*\right\}$$
$$= R_\phi(\tau) - R_\phi(\tau-\tau_d) - R_\phi(\tau+\tau_d) + R_\phi(\tau) \qquad (19.17)$$

where

$$R_\phi(\tau) = E\left\{\phi(t+\tau)\phi^*(t)\right\} \qquad (19.18)$$

is the autocorrelation of the LO phase noise. Substituting (19.17) into (19.14) results in

$$\Phi_{\Delta\phi}(f) = \int_{-\infty}^{\infty}\left[2R_\phi(\tau) - R_\phi(\tau-\tau_d) - R_\phi(\tau+\tau_d)\right]e^{-j2\pi f\tau}d\tau$$
$$= S_\phi(f)\left[4\sin^2(\pi f\tau_d)\right] \qquad (19.19)$$

where

$$S_\phi(f) = \int_{-\infty}^{\infty} R_\phi(\tau)e^{-j2\pi f\tau}d\tau \qquad (19.20)$$

is the LO phase noise spectrum.

Equation (19.19) is interesting because it indicates the phase noise component of the clutter return depends on the LO phase noise and the range delay to the (point) clutter source. This dependency is termed *range correlation* [8–13]. It indicates that returns from clutter at close range, due to phase noise, will be correlated and will cancel in the mixer where the LO signal is removed from the return signal. This assumes the same LO signal is used in the transmitter and receiver. If they use different LOs, there will be no correlation, and (19.19) would reduce to $\Phi_{\Delta\phi}(f) = 2S_\phi(f)$.

Equation (19.19) applies to a single, point source of clutter. Since clutter is distributed over a range extent, the spectrum of (19.19) must be integrated over the range region of interest. In this integration, we must also account for the variation of clutter power with range. Thus, to find the phase noise spectrum for a clutter region, we compute the integral

$$\Phi_{\Delta\phi}(f) = \int_\Re \frac{\Phi_{\Delta\phi}(f,R)}{R^3}dR = S_\phi(f)\left[\int_\Re \frac{4\sin^2(2\pi fR/c)}{R^3}dR\right]$$
$$= S_\phi(f)H_R(f,R_0) \qquad (19.21)$$

where $\Re$ is a region that contains the clutter ranges of interest. The R^3 factor accounts for the nominal cubic decrease in ground clutter power with range. For rain clutter, we would use R^2. R_0 is a reference range. It is the range to the front of the closest clutter patch. $\Re$ includes the ranges in the resolution cell of interest (the clutter cell containing the target) and all PRI multiples of that resolution cell.

The equation for $H_R(f,R_0)$, which is somewhat complicated, is included in Appendix 19A. This appendix also contains an approximation that works well. That approximation is

$$H_R(f, R_0) = \begin{cases} 8\sin^2(2\pi f R_0/c) & |f| < c/(12R_0) \\ 2 & |f| \geq c/(12R_0) \end{cases} \quad (19.22)$$

Figure 19.9 contains plots of $H_R(f,R_0)$ using the equation of Appendix 19A and the approximation of (19.22). As indicated, they match reasonably well. The plot was generated for a waveform with a 100-kHz PRF, a 1-μs unmodulated pulse, and $R_0 = 225$ m (R_0 is R_{start}). It is interesting to note that the curve levels out to 3 dB. This is because the clutter "voltage" [$v_{obj}(t)$ in Figure 19.1] is multiplied by $v_{LO}(t)$ on transmit and receive. Thus, the phase noise component of $v_{LO}(t)$ is added twice.

$S_\phi(f)$ is the phase noise spectrum of the LO. Figure 19.10 contains a sample phase noise spectrum for an 8.64-GHz LO. The LO signal was created by multiplying the frequency of a 320-MHz surface acoustic wave (SAW) oscillator by a factor of 27. The phase noise spectrum shown in Figure 19.10 represents mid-level technology in that the spectrum floor, Φ_0, is about −146 dBc/Hz.

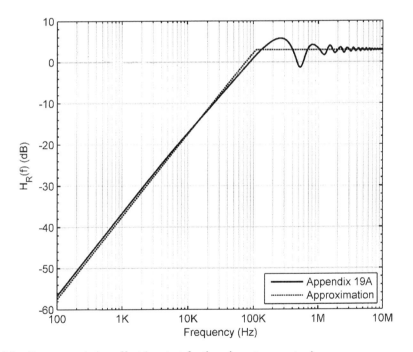

Figure 19.9 Range correlation effect (see text for the relevant parameters).

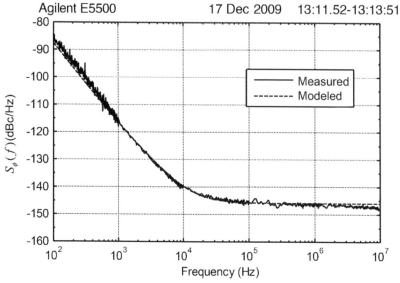

Figure 19.10 Measured and modeled phase noise plot. Model used (19.23) with $f_1 = 3$ kHz, $f_2 = 15$ kHz, $f_3 = 10$ Hz, and $\Phi_0 = -146$ dBc/Hz.. (*Source:* Bill Myles, Dynetics, Inc. Used with permission.)

The dashed line in Figure 19.10 was generated using a mathematical model developed by D. B. Leeson [14], with modifications suggested by Rick Poore in an Agilent Technologies report [15]. That model is

$$S_\phi(f) = \frac{\left(f^2 + f_2^2\right)\left(|f| + f_1\right)}{|f|^3 + f_3^3} \Phi_0 \qquad (19.23)$$

where $f_1, f_2,$ and f_3 are the corner frequencies indicated in Figure 19.11. As can be seen from Figure 19.10, the Leeson-Poore model fits the measured curve very well.

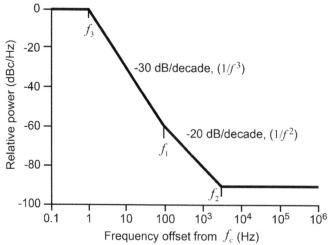

Figure 19.11 Bode plot of the modified Leeson phase noise spectrum model.

Figure 19.12 contains a plot of the $\Phi_{\Delta\phi}(f)$ that results from using the $H_R(f,R_0)$ of Figure 19.9 and the $S_\phi(f)$ model of (19.23). An interesting feature of Figure 19.12 is that the rise in $S_\phi(f)$ at low frequencies is partially canceled by $H_R(f,R_0)$ such that the net phase noise at low frequencies is very low. This is a general behavior in radars employing a common LO for transmit and receive, and is a function of clutter range.

The next step is to perform the convolution of $C(f)$ with $\Phi_{\Delta\phi}(f)$ that is indicated in (19.8). The result of using the Gaussian clutter spectrum, with $\sigma_v = 0.22$ m/s, is shown in Figure 19.13. As can be seen, convolving $\Phi_{\Delta\phi}(f)$ with $C(f)$ has almost no effect on the shape of the phase noise spectrum. This is expected because, relative to the variations in $\Phi_{\Delta\phi}(f)$, $C(f)$ is virtually an impulse function (Dirac delta). Thus, convolving $\Phi_{\Delta\phi}(f)$ with $C(f)$ produces almost the same result as convolving $\Phi_{\Delta\phi}(f)$ with a Dirac delta (recall that $C(f)$ is normalized to an area of unity).

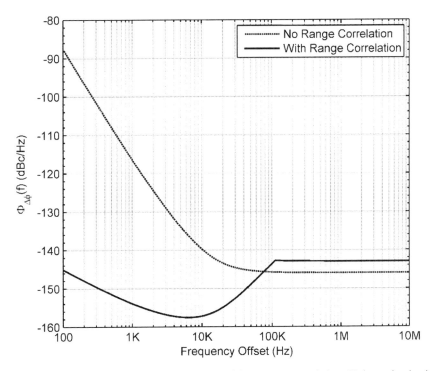

Figure 19.12 Total phase noise spectrum with and without range correlation. $\Re$ determined using a 100-kHz PRF, a 1-µs unmodulated pulse, and $R_0 = 225$ m.

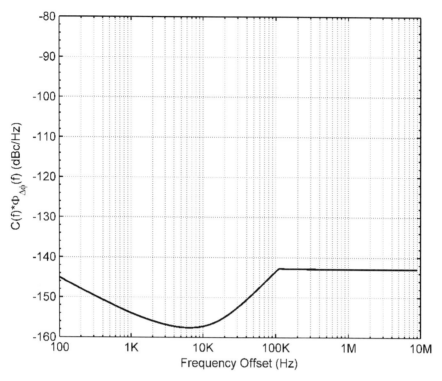

Figure 19.13 Plot of $C(f)^*\Phi_{\Delta\phi}(f)$ for a 100-kHz PRF, a 1-μs unmodulated pulse, and $R_0 = 225$ m.

19.4.2 ADC Considerations

A consideration in digital signal processors is the impact of the ADC on SCR improvement. The specific ADC properties of concern are the number of bits in the ADC, quantization noise, internal ADC circuit noise,[5] and ADC dynamic range. A common rule of thumb used to characterize the impact of the ADC on SCR improvement is to say the ADC imposes an absolute limit on performance of

$$I_{scr} = 20\log\left[\left(2^{N_{bit}} - 1\right)\frac{\sqrt{3}}{2}\right] \approx 6N_{bit} - 1 \text{ dB} \qquad (19.24)$$

where N_{bit} is the number of bits in the ADC [8, p. 233; 9, p. 139; 10, p. 2.74; 16, p. 93]. While I_{scr} is influenced by the number of bits in the ADC, the hard limit given by (19.24) is not valid when processing gain, phase noise, and other ADC properties are taken into consideration. A more representative equation for I_{scr} that includes the effects of G_{SNR}, G_{SCR}, phase noise, and the ADC, is [4]

---

[5] The level of quantization noise and noise generated within an ADC is very difficult to calculate theoretically. As a result of this difficulty, the amount of ADC and quantization noise is typically measured and stated by an ADC's manufacturer as the SNR of the ADC.

$$I_{scr} = \frac{G_{SCR}G_{SNR}P_{ADC}}{G_{SNR}P_{ADC} + G_{SCR}P_{ADC}\Phi_0 B + P_{NADC}B/F_s}$$
$$= \frac{G_{SCR}G_{SNR}P_{ADC}}{G_{SNR}P_{ADC} + G_{SCR}P_{ADC}\Phi_0/\tau_p + P_{NADC}/(\tau_p F_s)} \quad (19.25)$$

P_{ADC} is the level of the clutter at the ADC input relative to the ADC saturation level. We allow -6 dB to help avoid clutter fluctuations from occasionally causing ADC saturation.[6] The presence of P_{ADC} implies there is some type of gain control that monitors the clutter level into the ADC and adjusts the gain to keep the level 6 dB below ADC saturation (full-scale input).

The term P_{NADC} encompasses quantization noise, noise generated internally by ADC circuitry, and any additional dither noise added to assure linear operation of the ADC. It is referenced to full scale (see footnote 6). This raises an important issue concerning the ADC: For the ADC to preserve the relative sizes of signal, clutter, and noise after quantization, there must always be sufficient noise at the ADC input (see Section 22.8.1.5).

A reasonable value of P_{NADC} is [4]

$$P_{NADC} = 10^{-6[N_{bit}-1-\log_2(q)]/10} \quad (19.26)$$

where N_{bit} is the number of bits in the ADC and q is the number of quantization levels of the dither noise at the input to the ADC. Typical values of q are 1/2 to 1. [Note: if we say dither noise toggles the least significant bit (lsb) of the ADC, $q=1$; if it toggles the lower two bits, $q=3$ because the binary equivalent of 3 is the binary number 11.]

F_s is the ADC sample rate. It is normally taken to be the modulation bandwidth of the waveform if the ADC is operating on the baseband signal. For an unmodulated pulse, $F_s = 1/\tau_p$. If the radar uses IF sampling with digital down conversion, F_s can be much larger than the modulation bandwidth.

The I_{scr} equation of (19.25) is written in terms used for a pulsed Doppler signal processor. It is also applicable to the MTI processor with $G_{SNR} = 1$ and $G_C = 1/CA$.

19.5 SUMMARY AND RULES OF THUMB

Table 19.2 contains a summary of the results we obtained in the above discussions. It also contains some rules of thumb that were discussed, or will be discussed shortly.

To derive the rule of thumb for G_N, we assume the stopband of the HPF is much narrower than the PRF (narrower than $1/T$), and the BPF is ideal with a bandwidth of B and a gain of unity. We assume the BPF passband is centered at some frequency, f_B, in the passband of the HPF. Under these conditions,

---

[6] ADC full scale has been normalized to 0 dB in (19.25).

$$G_N = T \int_{-1/2T}^{1/2T} H_H(f) H_B(f) df = T \int_{-1/2T}^{1/2T} \text{rect}\left[\frac{f - f_B}{B}\right] df = BT \qquad (19.27)$$

We will later discuss the rule of thumb for the case where the bank of BPFs is replaced by an FFT (see Chapter 15).

Deriving the rule of thumb for G_ϕ is a little more involved. With $C(f) * \Phi_{\Delta\phi}(f) = \Phi_0$, we have

$$G_\phi = \Phi_0 \int_{-\infty}^{\infty} H_H(f) H_B(f) MF(f) df \qquad (19.28)$$

Table 19.2
Summary of Digital Pulsed Doppler Signal Processor Analysis Equations

Parameter	Equation	Rule of Thumb
CNR folding (based on (19.1) and related discussion	$CNR(R) = \sum_{k=0}^{\infty} CNR_1(R + kR_{PRI})$	The practical upper limit on the sum is usually 5 or 10
Target signal gain (19.8)	$G_S = H_H(f_d) H_B(f_d) MF(f_d)$	$G_S = 1$
Central line clutter gain (19.5)	$G_C = \int_{-\infty}^{\infty} H_H(f) H_B(f) MF(f) C(f) df$	Generally very small relative to G_ϕ and can be ignored for MPRF and HPRF waveforms, but is the predominant contributor for LPRF pulsed Doppler processors
Phase noise clutter gain (19.8)	$G_\phi = \int_{-\infty}^{\infty} H_H(f) H_B(f) MF(f)$ $\times \left[C(f) * \Phi_{\Delta\phi}(f) \right] df$	if $C(f) * \Phi_{\Delta\phi}(f) = \Phi_0$ $G_\phi = G_N \Phi_0 / \tau_p = \Phi_0 / (G_{SNR} \tau_p)$
Total clutter gain (19.6)	$G_{Ctot} = G_C + G_\phi$	$G_{Ctot} = G_\phi$ for MPRF and HPRF waveforms, but $G_{Ctot} = G_C$ for LPRF waveforms
Noise gain (19.10)	$G_N = T \int_{-1/2T}^{1/2T} H_H(f) H_B(f) df$	$G_N = BT$ B = BPF bandwidth
SNR gain (19.11)	$G_{SNR} = G_S / G_N$	$G_{SNR} = 1/G_N$ if use $G_S = 1$ G_{SNR} = FFT length for an FFT processor
CNR gain (19.12)	$G_{CNR} = (G_C + G_\phi)/G_N$	
Clutter attenuation (19.12)	$CA = 1/G_{CNR}$	
SCR gain/SCR improvement (19.13)	$I_{scr} = G_{SCR} = \dfrac{G_S}{G_C + G_\phi}$ or (19.25)	

We can approximate $MF(f)$ as an ideal LPF with a two-sided bandwidth of $1/\tau_p$. We assume T/τ_p is an integer, N, so that $H_H(f)$ and $H_B(f)$ will be repeated N times over the interval of $1/\tau_p$. We further use the G_N rule-of-thumb assumptions on $H_H(f)$ and $H_B(f)$ we used to arrive at (19.27). With this, and a little thought, G_ϕ becomes

$$G_\phi = \Phi_0 \int_{-\infty}^{\infty} H_H(f) H_B(f) \operatorname{rect}\left[f\tau_p\right] df = \Phi_0 \sum_N \int_{-1/2T}^{1/2T} \operatorname{rect}\left[\frac{f-f_B}{B}\right] df$$

$$= \Phi_0 NB = \Phi_0 \frac{T}{\tau_p} B = G_N \Phi_0 / \tau_p = \Phi_0 / G_{SNR} \tau_p$$

(19.29)

where we made use of (19.27). We leave it as an exercise to verify (19.29) via simulation. The remaining rules of thumb were discussed previously.

For the last term of (19.29), we used the fact that the HPF and BPF are normalized such that $G_S = 1$. Thus $G_{SNR} = G_S/G_N = 1/G_N$, or $G_N = 1/G_{SNR}$.

19.6 HPRF PULSED DOPPLER PROCESSOR EXAMPLE

To illustrate the above procedures, we consider three examples. The first is a pulsed Doppler radar that uses a PRF of 100 kHz and an unmodulated pulse with a width of 1 μs. The remaining radar, target, and clutter parameters are given in Table 19.3.

For this analysis, we assume the radar is searching and only consider the case where the radar beam is 1/2 beamwidth above 0°. We assume the target is flying radially toward the radar at the azimuth and elevation angle of the radar beam.[7] In the clutter RCS generation computer code, we use the $\operatorname{sinc}(x)$ antenna pattern of (17.3). Although we are interested in target ranges between 2 and 50 km, we must model clutter returns from much shorter ranges. We assume the receiver timing is such that the radar receives returns from clutter located at 225 m. This means the receiver is off during the transmit pulse (150 m) and for 1/2 pulsewidth after the transmit pulse. The receiver remains on until 1½ pulsewidths before the next transmit pulse. Thus, the receiver processes returns from clutter (and targets) over a range window that extends from 225 m to 1,275 m after the leading edge of the transmit pulse. Of course, it receives returns from multiples of this window repeated every PRI. During the time of 225 m before and after the leading edge of the transmit pulse, the receiver is off, which means the radar is blind during these times. In search, this does not generally pose a problem because targets will fly through the blind regions quickly. During track, it can pose problems. However, during track, pulsed Doppler radars adjust the PRF to assure the target is not in a range and/or Doppler blind region.

Figure 19.14 contains plots of SNR, CNR, SCR, and SIR at the matched filter output. At long ranges, the SNR is about −10 dB, which is too low to support detection and tracking. To raise the SNR to about 13 dB, the Doppler processor needs to provide about 23 dB of SNR gain (G_{SNR} = 23 dB).

---

[7] This target trajectory is obviously unrealistic. However, it is an assumption used in some search radar analyses. An alternate would be the more realistic assumption that the target is flying toward the radar at a constant altitude.

Table 19.3
Radar, Target, and Clutter Parameters for HPRF Pulsed Processor Example

Peak power	10 kW
Operating frequency	8 GHz
System noise temperature	1,500 K
PRF	100 kHz (PRI = 10 μs)
Burst length	7 ms (700 pulses per burst)
Pulsewidth	1 μs
Total losses for the target and clutter	6 dB
Height of the antenna phase center	3 m
Antenna gain	38 dB
Azimuth and elevation beamwidth	2°
Beam angle	Beam parked at ½ beamwidth above 0° elevation
RMS antenna sidelobes	30 dB below the peak gain
Clutter backscatter coefficient	−20 dB
Target RCS	−10 dBsm
Ranges of interest	2 km to 50 km

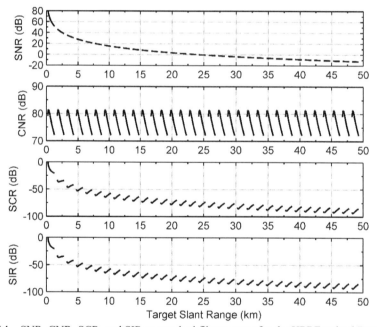

Figure 19.14 SNR, CNR, SCR, and SIR at matched filter output for the HPRF pulsed Doppler example.

The CNR has peaks of about 80 dB, which cause the SCR and SIR to be very low. It is estimated that the signal processor will need to provide 80 to 90 dB of clutter attenuation to raise the SCR and SIR to reasonable levels. The blank regions of the SNR and other plots are the regions where the receiver is gated off, as discussed above.

The signal processor consists of an HPF for clutter rejection followed by a bank of BPFs to provide SNR gain. For analysis purposes, we use only one BPF.

We want the radar to be able to detect and track targets with range rates down to about 40 m/s. This means we must choose the cutoff frequency of the HPF to be

$$f_{ch} < \frac{2\dot{R}_{min}}{\lambda} = \frac{2 \times 40}{0.0375} = 2,133 \text{ Hz} \tag{19.30}$$

We choose f_{ch} = 2,000 Hz, and use a fifth-order Butterworth HPF [17–20], to provide adequate clutter attenuation. An approximate $H_H(f)$ for this filter is

$$H_H(f) = 1 - \frac{1}{1 + (\beta/\beta_{ch})^{2 \times 5}} = \frac{(\beta/\beta_{ch})^{10}}{1 + (\beta/\beta_{ch})^{10}} \tag{19.31}$$

with

$$\beta = \tan(\pi f T) \tag{19.32}$$

and

$$\beta_{ch} = \tan(\pi f_{ch} T) \tag{19.33}$$

This response is derived from the $H(f)$ of an analog LPF by using the substitutions of (19.32) and (19.33) to make the response periodic with a period of $1/T$. Equations (19.32) and (19.33) are derived from the bilinear transform [18, 21, 22].

We typically want to choose the bandwidth of the BPF to be as small as possible since this sets the limit on SNR and SCR improvement. For the radar of this example, the burst length of 7 ms sets an absolute lower limit on bandwidth of about 140 Hz.[8] However, we must allow for transients in the HPF to give the output due to clutter time to decrease to the desired value. The duration of the transient is typically set by the HPF cutoff frequency. A rough rule of thumb is that the transients of a filter will settle in a time period equal to about five times the reciprocal of the HPF cutoff frequency. For the HPF of this example, this would be about 5/2,000 s or about 2.5 ms. We will allow 3.5 ms for HPF transients, which are often termed *clutter transients*, to settle. In other words, we gate the output of the HPF off for 3.5 ms and send the last 3.5 ms of pulses to the BPFs. The idea of gating the HPF output off to allow for clutter transients is sometimes termed *clutter gating*.[9]

Because of the clutter gating, the effective burst length available to the BPF is 3.5 ms. This sets an absolute lower limit on the BPF bandwidth of 1/(3.5 ms) or 287 Hz. We choose a bandwidth of f_{cb} = 350 Hz.

If we use the aforementioned rule of thumb that filter transients settle in $5/B$, where B is the filter bandwidth, we realize the BPF output will not reach steady state in 3.5 ms. Thus, the use of steady-state frequency analysis techniques is questionable. The alternative is to use time domain techniques, which is more difficult than using frequency

---

[8] This is somewhat of a "soft" lower limit. In some instances, we could use a BPF with a very small bandwidth and not be concerned with the filter response reaching steady state. Such a filter is sometimes termed a bandpass integrator.
[9] Typical clutter gate durations are ~ 1/3 to 1/2 of the dwell.

analysis techniques. However, for the case where we use a bandwidth that is larger than the reciprocal of the effective burst length (3.5 ms in this case), and the equivalent low pass filter order is fairly low (third order in this case since the bandpass filter is sixth order), frequency domain analysis techniques are usually accurate to within a dB or so. If the bandwidth is less than the reciprocal of the burst length, the BPF would act as a bandpass integrator (an integrator that is preceded by a frequency offset), and a different analysis technique would be needed. That analysis method is discussed later, when we discuss a LPRF pulsed Doppler signal processor.

We assume the BPF is a sixth-order Butterworth filter. For purposes of our analyses, we assume it is centered on the target Doppler frequency of $f_d = 8,000$ Hz (which corresponds to a target range rate of -150 m/s). An approximate $H_B(f)$ for the BPF is

$$H_B(f) = \frac{1}{1 + \left(\dfrac{\beta_{fT}}{\beta_{cb}/2}\right)^6} \tag{19.34}$$

where

$$\beta_{fT} = \tan\left[\pi(f - f_d)T\right] \tag{19.35}$$

and

$$\beta_{cb} = \tan(\pi f_{cb} T) \tag{19.36}$$

In these equations, $f_d = 8,000$ Hz and $f_{cb} = 350$ Hz. The BPF is derived from a third-order Butterworth LPF by using a frequency transformation derived from the bilinear transform, with a frequency shift to center the response at f_d.[10]

The clutter spectrum is the Gaussian model of Section 17.2.2 with $\sigma_v = 0.22$ m/s. Also, we use the $C(f) * \Phi_{\Delta\phi}(f)$ of Figure 19.13.

Since $MF(f_d) \approx 1$ and the target Doppler frequency is well within the passband of the HPF, we can use the rule of thumb that $G_S = 1$.

G_N is computed by numerically evaluating

$$G_N = T \int_{-1/2T}^{1/2T} H_H(f) H_B(f) df \tag{19.37}$$

with $H_H(f)$ and $H_B(f)$ from (19.31) and (19.34). This results in $G_N = 0.0037$ W/W. Alternately, we could have used the rule of thumb from Table 19.2 to arrive at a value of $G_N = 0.0035$ W/W. From this, the SNR gain is

$$\begin{aligned} G_{SNR} &= G_S / G_N \\ &= 270 \text{ W/W or } 24.3 \text{ dB} \end{aligned} \tag{19.38}$$

which is a little larger than the desired value of 23 dB.

---

[10] The filter defined by (19.34) is ideal in that it only has one passband at f_d rather than passbands at $\pm f_d$. Such a filter could be built with digital hardware that allows complex filter coefficients.

The center line clutter gain is

$$G_C = \int_{-\infty}^{\infty} H_H(f) H_B(f) MF(f) C(f) df \qquad (19.39)$$

$$= 4.35 \times 10^{-30} \text{ W/W or } -294 \text{ dB}$$

which, as predicted, is very small. The phase noise component of the total clutter gain is

$$G_\phi = \int_{-\infty}^{\infty} H_H(f) H_B(f) MF(f) \left[C(f) * \Phi_{\Delta\phi}(f) \right] df$$

$$= \Phi_0 \int_{-\infty}^{\infty} H_H(f) H_B(f) MF(f) df \qquad (19.40)$$

$$\approx 2 \times 10^{-11} \text{ W/W or } -107 \text{ dB}$$

The rule-of-thumb value (see Table 19.2) is also −107 dB. Equations (19.39) and (19.40) are computed using numerical integration. The second part of (19.40) implies we used Φ_0 instead of the $C(f)*\Phi_{\Delta\phi}(f)$ of Figure 19.13. This is justified in this case because the low region of the $C(f)*\Phi_{\Delta\phi}(f)$ curve extends from 0 to about 100 kHz, whereas the Φ_0 portion extends from 100 kHz to the matched filter cutoff frequency of about 500 kHz. Thus, the contribution of $C(f)*\Phi_{\Delta\phi}(f)$ to the integral is relatively small when compared to the contribution of the Φ_0 region. The proof of this is left as an exercise (Exercise 5)

With the above, the SCR improvement is

$$I_{scr} = G_{SCR}$$
$$= \frac{G_S}{G_C + G_\phi} = \frac{1}{G_\phi} \qquad (19.41)$$
$$= 107 \text{ dB}$$

which is quite large. The result of applying the gains to the plot of Figure 19.14 is shown in Figure 19.15. The plots of SCR and SNR are very similar. In fact, detailed examination indicates they are only different by 1 dB. Because of this, the SIR plot is about 3 dB smaller than either the SNR or SCR plots. More importantly, the SIR and SNR values are much improved over the values at the matched filter output, but still a little low at long ranges.

The above discussion did not include the effects of the ADC. If we were to include ADC effects, by using (19.25) instead of (19.41), I_{scr} would have been different. For example, if we use a 12-bit ADC with one bit of quantization noise (i.e., $q = 1$), I_{scr} would be limited to about 84 dB instead of the 107 dB given by (19.41). To get close to 107 dB, we would need a 16-bit ADC. With a 14-bit ADC, I_{scr} would be 96 dB. The effect of using a 12-bit or 14-bit ADC on the final plot of SIR is left as an exercise (Exercise 8).

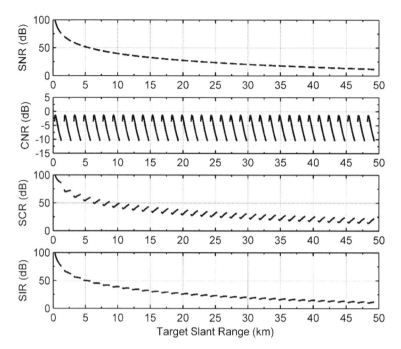

Figure 19.15 SNR, CNR, SCR, and SIR at the digital signal processor output for the HPRF pulsed Doppler example.

19.7 MPRF PULSED DOPPLER PROCESSOR EXAMPLE

For our second example, we consider a MPRF pulsed Doppler radar that uses a 25 kHz PRF waveform. This is a fourth of the PRF of the HPRF waveform of the previous example. Since we increased the PRI by a factor of four, from 10 μs to 40 μs, we also decided to increase the pulse width from 1μs to 4 μs. To maintain the same range resolution as the HPRF waveform, we use a 4-chip, Barker code (see Chapter 11) with a chip width of 1 μs. Finally, we maintain the 7-ms burst length of the HPRF waveform. This means there will be 175 pulses in the burst. The remainder of the radar, clutter, and target parameters are the same as in the HPRF example (Table 19.3). We still keep the receiver off for 1.25 pulse widths before and after the leading edge of the transmit pulse. This means the receiver processes returns from targets and clutter from 750 m after the pulse leading edges to 5250 m after the leading edges.

Figure 19.16 contains plots of CNR, SNR, SCR, and SIR at the matched filter output (and signal processor input). As with the HPRF example, the SNR is lower than desired and the SCR and SIR are quite low. This means the signal processor must provide both SNR improvement and a significant amount of clutter rejection, i.e., considerable SCR improvement. For this reason, we use the same signal processor configuration as in the HPRF example. However, to gain experience in how to accommodate an FFT in the analysis, we use an FFT in place of the bank of bandpass filters. We keep the same HPF as in the HPRF processor.

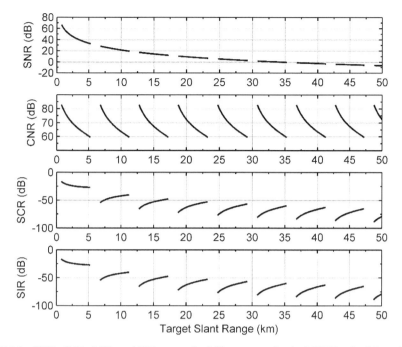

Figure 19.16 SNR, CNR, SCR, and SIR at matched filter output for the MPRF pulsed Doppler example.

As we mentioned in Chapter 15, the signal processor used during search will normally consist of a bank of BPFs that span the PRF, or, alternately, an FFT. Because of the way an FFT works, its output taps span $-PRF/2$ to $+PRF/2$. Further, the spacing between FFT taps is equal to PRF/N_{FFT}, where N_{FFT} is the number of taps in the FFT. This means the FFT is equivalent to a bank of BPFs that span the PRF and have center frequencies separated by PRF/N_{FFT}. The bandwidth of the equivalent BPF represented by each tap is also approximately PRF/N_{FFT}, assuming we don't use amplitude weighting to lower the Doppler sidelobes of the FFT. If amplitude weighting is used, the equivalent bandwidth would be a little wider than PRF/N_{FFT}. The increase in equivalent bandwidth would depend on the type of weighting used (see Appendix A).

If we compare Figures 19.16 and 19.14, we note that the SNR at the matched filter output is 6 dB higher in the MPRF case than in the HPRF case. This makes sense since the pulse width of the MPRF waveform is four times that of the HPRF waveform. With the HPRF processor, we specified a desired SNR gain of 23 dB. Given that the SNR at the matched filter output is 6 dB higher in the MPRF case than in the HPRF case, we can reduce the SNR gain requirement of the MPRF signal processor to 17 dB (23 dB minus 6 dB). As we will show, we can get close to that SNR improvement with a 64-point, or 64-tap, FFT.

Based on the above discussions, our MPRF signal processor consists of a fifth-order Butterworth HPF and a 64-point FFT. The HPF has a cutoff frequency of 2,000 Hz and we decided to use a 30-dB Chebyshev weighting on the FFT. Given the current detail of the system specifications we are using, we don't have a reason for specifying the FFT weighting. We included it for analysis purposes.

As with the HPRF case, we need to include a clutter gate between the HPF and the FFT to allow for HPF transients. In the HPRF example, we used a 3.5-ms gate. In this design, we only need to process returns from 64 pulses in the FFT. Given a burst length of 175 pulses, we could then allow 175−64 pulses, or 4.44 ms, for the HPF transient. In a "final design," we would want to investigate the clutter gate requirement in more detail. Two possibilities are to use the extra clutter gate time to shorten the burst, or to reduce the HPF cutoff frequency to improve Doppler frequency coverage.

In the HPRF signal processor design, we assumed a target Doppler frequency of 8,000 Hz and centered the BPF used in the analysis at that frequency. For the purposes of the FFT analysis, we want to select a target Doppler frequency that corresponds to the center frequency of one of the FFT taps and is close to 8,000 Hz. We will use a target Doppler frequency of 7,812.5 Hz, which would correspond to the 21st tap of the FFT. We computed the tap number from

$$n_{tap} = \left\lfloor \frac{N_{FFT}}{PRF} f_{cent} \right\rfloor + 1 \text{ for } 0 \leq f_{cent} < PRF/2 \qquad (19.42)$$

or

$$n_{tap} = \left\lfloor \frac{N_{FFT}}{PRF}(PRF/2 + f_{cent}) \right\rfloor \text{ for } -PRF/2 \leq f_{cent} < 0 \qquad (19.43)$$

In these equations, $\lfloor x \rfloor$ means round x to the nearest integer. Since we used a target Doppler frequency of 7,812.5 Hz, (19.42) would be the appropriate equation to use, and would yield $n_{tap} = 21$. As indicated in the HPRF signal processor example, we account for the fact that the target Doppler frequency may not be the same as that of an FFT tap center frequency by including a loss term in the radar range equation (see Chapter 5).

To compute the various parameters needed to find the SNR, CNR, and SCR at the signal processor output (at the 21st tap of the FFT), we need the frequency response of the 21st tap of the FFT. As discussed in Appendix 19B, the equation for the frequency response of the nth tap of a normalized FFT is

$$H_{FFT}(f) = \left| \frac{\sum_{n=0}^{N_{FFT}-1} w_n e^{-j2\pi n(f-f_{ntap})T}}{\sum_{n=0}^{N_{FFT}-1} w_n} \right|^2 \qquad (19.44)$$

where

$$f_{ntap} = \begin{cases} (n_{tap}-1)\dfrac{PRF}{N_{FFT}} & 1 \leq n_{tap} \leq \dfrac{N_{FFT}}{2} \\ (n_{tap}-N_{FFT})\dfrac{PRF}{N_{FFT}} & \dfrac{N_{FFT}}{2} < n_{tap} \leq N_{FFT} \end{cases} \qquad (19.45)$$

w_n is the amplitude weighting applied to the FFT (30-dB Chebyshev in this case), N_{FFT} is the FFT length (64 in this case), and T is the PRI (40 μs in this case). The denominator of (19.44) normalizes $H_{FFT}(f)$ so that $H_{FFT}(f_{ntap}) = 1$. This is consistent with the normalization used on the other filter responses and is needed by the equations used to find G_S, G_C, and so forth.

Since the pulse width is only 4 μs, we can again use the assumption that $MF(f_d) \approx 1$. Also, as in the HPRF example, the target Doppler frequency is well in the passband of the HPF. These, along with the fact that $H_{FFT}(f_{ntap}) = 1$ for $n_{tap} = 21$ and $f_d = f_{ntap}$, means we have $G_S = 1$. Thus, to find G_{SNR}, we only need to find

$$G_N = T \int_{-1/2T}^{1/2T} H_H(f) H_{FFT}(f) df \qquad (19.46)$$

When we (numerically) evaluate the integral, we get

$$G_N \approx \frac{1}{56} \text{ or } -17.5 \text{ dB} \qquad (19.47)$$

and

$$G_{SNR} = \frac{G_s}{G_N} = \frac{1}{1/56} = 56 \text{ or } 17.5 \text{ dB} \ ^{11} \qquad (19.48)$$

As an alternative to evaluating G_{SNR} using (19.48) and (19.46), we could have used a rule of thumb (see Table 19.2) that

$$G_{SNR} = G_{SNRFFTideal} = N_{FFT} \qquad (19.49)$$

That is, that the SNR gain of the combined HPF and FFT is equal to the SNR gain of an ideal FFT,[12] one where all $w_n = 1$. The basis of this rule of thumb is that, for MPRF and HPRF waveforms, where the cutoff frequency of the HPF is significantly less than the PRF, the SNR gain through the HPF is close to unity (0 dB).

To compute the SCR gain through the signal processor, we use [see (19.13)]

$$I_{scr} = G_{SCR} = \frac{G_S}{G_C + G_\phi} = \frac{1}{G_C + G_\phi} \qquad (19.50)$$

where G_C and G_ϕ are given by (19.7) and (19.8), and we use $H_{FFT}(f)$ in place of $H_B(f)$. If we want to include ADC effects, we would use (19.25).

To compute G_C we must evaluate the integral of (19.7). The result of evaluating that integral is

---

[11] As a note, the G_{SNR} of (19.48) includes the losses due to amplitude weighting. Because of this, FFT weighting loss (see Chapter 5) should not be included in the radar range equation.
[12] If we use $G_{SNR} = N_{FFT}$, we are not accounting for the FFT weighting. Because of this, the FFT weighting loss (see Chapter 5) should be included in the radar range equation.

$$G_C = \int_{-\infty}^{\infty} H_H(f) H_{FFT}(f) MF(f) C(f) df \approx 1 \times 10^{-23} \text{ w/w or } -230 \text{ dB} \quad (19.51)$$

which, as with the HPRF signal processor, is very small.

To compute G_ϕ, we use (19.8), with $H_B(f)$ replaced by $H_{FFT}(f)$, which is

$$G_\phi = \int_{-\infty}^{\infty} H_H(f) H_{FFT}(f) MF(f) \left[C(f) * \Phi_{\Delta\phi}(f) \right] df \quad (19.52)$$

In addition to replacing $H_B(f)$ by $H_{FFT}(f)$, the phase noise term, $\Phi_{\Delta\phi}(f)$, will be different because the R_0 needed in the range correlation term [see (19.21) and (19.22)] will be 750 m instead of the 225 m we used in the HPRF processor (recall that R_0 is equal to R_{start}, the range to the closest clutter cell). The derivation of the new $\Phi_{\Delta\phi}(f)$ is left as an exercise (see Exercise 13). As with the HPRF example, we could simplify the calculation of G_ϕ by assuming we can replace $C(f)*\Phi_{\Delta\phi}(f)$ with $\Phi_0 = -143$ dBc/Hz. The result of that approximation is

$$G_\phi = \Phi_0 \int_{-\infty}^{\infty} H_H(f) H_{FFT}(f) MF(f) df \approx 2 \times 10^{-11} \text{ W/W or } -107 \text{ dB} \quad (19.53)$$

which is the same value we got for the HPRF processor. As with the HPRF processor, we could have used the approximation

$$G_\phi = \Phi_0 / (G_{SNR} \tau_p) \quad (19.54)$$

to arrive at the same value. Using (19.50), we compute an SCR gain, or improvement of

$$I_{scr} = G_{SCR} = \frac{1}{G_C + G_\phi} = \frac{1}{G_\phi} = 107 \text{ dB} \quad (19.55)$$

If we apply the gains of $G_{SNR} = 17.5$ dB and $G_{SCR} = 107$ dB to the curves of Figure 19.16, we get the curves of Figure 19.17. As with the HPRF case, the SIR curve is virtually identical to the SNR curve, which means the signal processor has done an excellent job of attenuating the clutter. The SIR curve is still a little lower than desired at long ranges.

If we were to include ADC effects, we would find, using (19.25), that I_{scr} is limited to about 83 dB for a 12-bit ADC, 95 dB for a 14-bit ADC, and 104 dB for a 16-bit ADC. The latter two values would yield an SIR curve that is close to that of Figure 19.17. As a reference, the parameters used in (19.25) are $G_{SNR} = 56$, $P_{ADC} = -6$ dB, $G_C = -230$ dB, $\Phi_0 = -143$ dBc/Hz, $\tau_p = 4$ μs, and $F_s = 1$ MHz. P_{NADC} was computed from (19.26), using $q = 1$.

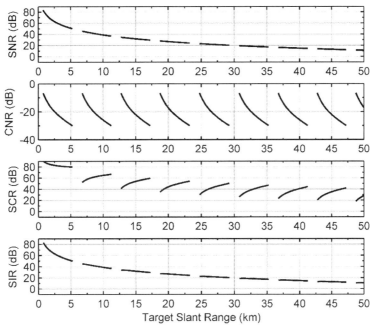

Figure 19.17 SNR, CNR, SCR, and SIR at digital signal processor output for the MPRF Pulsed Doppler Example.

19.8 LPRF PULSED DOPPLER SIGNAL PROCESSOR EXAMPLE

For this example, we consider a LPRF pulsed Doppler radar. We want to maintain approximately the same burst length, and we want the radar to operate range unambiguously at the maximum range of 50 km. To satisfy these constraints, we use a PRI of 400 µs, which will provide an unambiguous range of 60 km. We also use a burst of 18 pulses, which will result in a burst length of 7.2 ms. We will sacrifice some minimum range capability and choose a pulse width of 25-µs. To maintain the same range resolution as with the previous two examples, we use LFM modulation with a bandwidth of 1 MHz. The remainder of the radar, clutter, and target parameters are the same as in the HPRF and MPRF examples.

Figure 19.18 contains plots of CNR, SNR, SCR, and SIR at the matched filter output for this case. The curves do not exhibit the periodic behavior of Figures 19.14 to 19.17 because the waveform is unambiguous in range over the range interval of interest. Consistent with the 25-µs pulse width, the minimum range of the curves is 3.75 km. The SIR and SCR curves are virtually coincident because SCR is the main contributor to SIR. Recall that

$$SIR = \frac{P_S}{P_C + P_N} = \frac{1}{1/SNR + 1/SCR} \qquad (19.56)$$

From the curves of Figure 19.18, we note that *SCR* is much smaller than *SNR* for the ranges between 3.75 and 50 km. This means $1/SCR$ will be much larger than $1/SNR$, and that $1/SNR + 1/SCR \approx 1/SCR$. Thus, from (19.56), $SIR \approx SCR$.

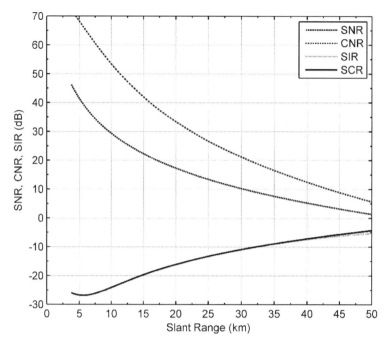

Figure 19.18 SNR, CNR, SIR, and SCR at the matched filter output for the LPRF pulsed Doppler processor example.

For the HPRF and MPRF waveforms of the previous two examples, the SCR ranged from zero to about −90 dB. With this waveform, the SCR ranges from about −5 to −30 dB, because there is no folding of the clutter in range [see (19.1)]. This is based on the assumption that clutter returns from ranges beyond 60 km will be much weaker than clutter returns between 3,750 m and the $c\tau_{PRI}/2 - 3.750$ m. The difference in SCR values is due to two factors: 1) the closest clutter cell is located at 3,750 m as opposed to 225 m for the HPRF and 750 m for the MPRF waveform, and 2) the target return only competes with clutter at its range, rather than with clutter at shorter ranges. Since the SCR is not extremely low, the signal processor does not need to provide the large clutter attenuation and SCR improvement of the HPRF and MPRF cases. Since high clutter attenuation is not needed, we decided to delete the HPF from the signal processor. Since the SNR is lower than desired, we still need to include a BPF, FFT, or some other type of coherent integrator to provide SNR improvement. The hope is the coherent integrator will also provide sufficient SCR improvement.

To gain experience in analyzing a different type of coherent integrator, we decided to use a digital bandpass integrator. As indicated earlier, in search we would use a bank of bandpass integrators that are designed such that their pass bands cover the PRF. For analysis purposes, we consider only one bandpass integrator whose pass band is centered on the target Doppler frequency.

If we assume a target Doppler frequency of $f_d = 8000$ Hz, centering the pass band of the bandpass filter on the target Doppler frequency causes a problem. Specifically, since the PRF of the waveform is 2.5 kHz ($PRF = 1/PRI = 1/400$ μs = 2.5 kHz), the target Doppler frequency is ambiguous, and the ambiguous Doppler frequency, f_{damb}, will be

$$f_{damb} = f_d - N_{amb} \times PRF \qquad (19.57)$$

where

$$N_{amb} = \left\lceil \frac{f_d}{PRF} \right\rceil \qquad (19.58)$$

and we use $\lceil x \rceil$ to denote truncation of x to the nearest integer toward zero (this is the MATLAB "fix" function). In this case, we have $N_{amb} = \lceil 8,000/2,500 \rceil = 3$ and $f_{damb} = f_d - N_{amb} \times PRF = 8,000 - 3 \times 2500 = 500$ Hz. Thus, we center the passband of the bandpass integrator at the ambiguous Doppler frequency of 500 Hz.

Figure 19.19 contains a block diagram of a digital bandpass integrator. It consists of a complex multiplier, a weight block, and a summer. One of the inputs to the complex multiplier, which is essentially a mixer, is a complex sinusoid at a frequency of $-f_{dcent}$. If the other input to the multiplier, $v_{HPF}(nT)$ in Figure 19.19, is also a (complex) sinusoid at a frequency of f_{dcent}, the output of the multiplier will be a constant. The weight block is a time-varying weight function that applies a different weight to each of the samples out of the multiplier. The weight function is used to control the sidelobes of the frequency response of the bandpass integrator. It is the same as the weights applied to the inputs of an FFT. The summer, which is a digital integrator, sums the outputs of the weight block to form the bandpass integrator output.

Another representation of the bandpass integrator is the FIR filter [18, 21, 22] of Figure 19.20 (also known as a transversal filter). In that block diagram, the blocks with z^{-1} in them represent digital delays of one PRI. The blocks with $w_n e^{j2\pi n (fdcent)T}$ in them are equivalent to the multiplier and weight function of Figure 19.19. They apply an amplitude weight of w_n and a frequency offset of $-f_{dcent}$ to each of the samples into the FIR filter. The output of the FIR filter is not used until all N_{tap} samples have been processed by the FIR filter.

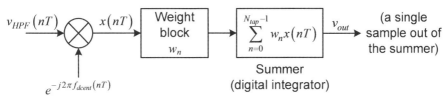

Figure 19.19 Digital bandpass integrator block diagram.

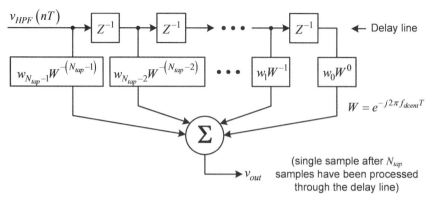

Figure 19.20 FIR filter representation of a digital bandpass integrator.

The normalized z-domain representation of the bandpass integrator can be written as

$$H_{BPI}(z) = \frac{\sum_{n=0}^{N_{tap}-1} w_n e^{j2\pi n f_{dcent} T} z^{-n}}{\sum_{n=0}^{N_{tap}-1} w_n} \qquad (19.59)$$

where N_{tap} is the number of samples (PRIs) processed by the bandpass integrator. The denominator term of (19.59) is used to normalize the frequency response of the bandpass integrator to unity at $f = f_{dcent}$. This is consistent with our approach with the BPF and FFT processors.

The frequency response of the bandpass integrator is

$$H_{BPI}(f) = \left| H_{BPI}(z) \right|^2_{z=e^{j2\pi fT}} = \left| \frac{\sum_{n=0}^{N_{tap}-1} w_n e^{-j2\pi n(f - f_{dcent})T}}{\sum_{n=0}^{N_{tap}-1} w_n} \right|^2 \qquad (19.60)$$

which is the same form as (19.44) for the FFT. Given the parallel we drew between the FFT and a BPF, and the fact that (19.60) is the same form as (19.44), we conclude that the bandpass integrator is also a band pass filter. The difference is that the BPF used in the HPRF signal processor is an IIR filter [18, 21, 22], while the FFT and band pass integrator are FIR filters.

Since the burst contains 18 pulses, we will use an $N_{tap} = 18$-tap bandpass integrator. From Figure 19.18, we note that about 40 dB of SCR improvement is needed to get the SCR, and hopefully SIR, above about 13 dB. Because of that, we choose to use a 40-dB Chebyshev weighting.

To compute G_C and G_ϕ, we use (19.7) and (19.8), with $H_H(f)H_B(f)$ replaced by $H_{BPI}(f)$. We use the same $C(f)$ as in the HPRF and MPRF examples. When evaluating $MF(f)$, we use $\tau_p = 25$ µs.

Since the range to the closest clutter cell is now 3.75 km, it is not clear whether range correlation will be much of a help in terms of reducing the effect of phase noise. To check this, we recreate the plot of Figure 19.13 with an R_0 value of 3,750 m instead of the 225 m used to generate Figure 19.13. The result of using the new R_0 is contained in Figure 19.21. As expected, because of the larger value of R_0, the phase noise spectrum is larger at low frequencies than the base value of $\Phi_0 = -143$ dBc/Hz. We still use the simplified expression to calculate G_ϕ [see (19.28)], but use a Φ_0 value of -130 dBc/Hz instead of -143 dBc/Hz. We chose -130 dBc/Hz as an "eyeball" value that would give the same result as evaluating the integral of (19.8). To obtain a more accurate value we would need to account for aliasing.

The results of performing the indicated calculations are $G_C = -40.4$ dB and $G_\phi = -95.4$ dB. Together these resulted in a $G_{SCR} = I_{scr}$ of

$$I_{scr} = G_{SCR} = \frac{1}{G_C + G_\phi} = 40.4 \text{ dB} \qquad (19.61)$$

As with the MTI example, and unlike the HPRF and MPRF examples, in this case G_C is the main driver on SCR improvement.

Because of the 40-dB Chebyshev weighting, we get a G_N of -11.4 dB, which translates to an SNR gain of 11.4 dB, which is less than desired. The result of applying G_{SNR} to the curves of Figure 19.18 is shown in Figure 19.22. The SIR is lower than desired at short and long ranges. Also, the SNR is lower than desired at long ranges. This means the next step in the design is to figure out how to improve SNR at the signal processor input. Since only a modest increase is needed, 1 or 2 dB, the best way to increase SNR would most likely be to reinvestigate the losses and system noise temperature (see Table 19.3) to see if they can be reduced.

Given the SCR improvement is only 40.4 dB for this signal processor, the ADC will not significantly degrade performance unless the number of bits in the ADC is less than about 8.

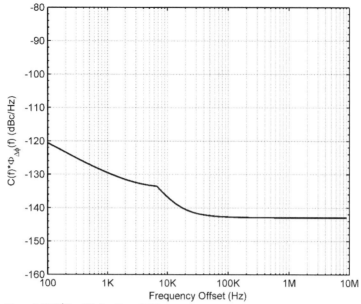

Figure 19.21 Plot of $C(f)^*\Phi_{\Delta\phi}(f)$ for $R_0 = 3750$ m.

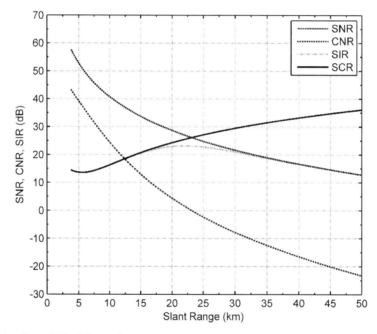

Figure 19.22 SNR, CNR, SCR, and SIR at digital signal processor output for the LPRF pulsed Doppler example.

19.9 EXERCISES

1. Reproduce the plots of Figure 19.2, 19.3, 19.5, and 19.6.

2. Show, by simulation, that the noise spectrum at the sampler output of Figure 19.7 is essentially constant. That is, that the noise is white. Use a matched filter matched to an unmodulated, 1-μs pulse and sample periods of 10 and 100 μs.

3. Derive (19.29) and verify it by simulation using the parameters of HPRF, MPRF, and LPRF Doppler signal processors.

4. Reproduce the plot of Figure 19.13.

5. Show that (19.40) is valid for the phase noise spectrum of Figure 19.13. Specifically, numerically evaluate the first integral of (19.40) using the spectrum of Figure 19.13.

6. Repeat HPRF Doppler signal processor example of Section 19.6 and produce the plots of Figures 19.14 and 19.15.

7. Repeat Exercise 6 using the rain clutter model from Chapter 18.

8. Repeat Exercise 6 for the case where the HPRF Doppler signal processor uses a 12-bit ADC. Repeat it for the case of a 14-bit ADC.

9. Verify the statements in the paragraph above Section 19.4.2 regarding the effect of the clutter spectrum on $\Phi_{\Delta\phi}(f)*C(f)$.

10. Repeat MPRF Doppler signal processor example of Section 19.7, and produce the plots of Figures 19.16 and 19.17.

11. Repeat Exercise 10 with the rain clutter model used in Chapter 18.

12. Prove that the SNR gain of the combined HPF and BPF, or the combined HPF and FFT, are practically the same for the HPRF and MPRF Doppler signal processor and waveform parameters of Sections 19.6 and 19.7.

13. Generate a plot like Figure 19.13 for the MPRF waveform and processor parameters.

14. Reproduce the plot of Figure 19.21.

15. Repeat the LPRF Doppler signal processor example of Section 19.8 and reproduce the plots of Figures 19.18 and 19.22.

16. Repeat Exercise 15 using the rain clutter model used in Chapter 18. What happens if the mean velocity of the rain is set to 10 m/s?

References

[1] E. J. Chrzanowski, *Active Radar Electronic Countermeasures*, Norwood, MA: Artech House, 1990.

[2] D. C. Schleher, *Introduction to Electronic Warfare*, Dedham, MA: Artech House, 1986.

[3] Naval Air Warfare Center Weapons Division (NAWCWD), *Electronic Warfare and Radar Systems Engineering Handbook*, 4th ed., NAWCWD Tech Commun Office, Point Mugu, CA, Rep No NAWCWD TP 8347, Oct. 2013.

[4]   M. C. Budge Jr. and S. R. German, "The Effects of an ADC on SCR Improvement," *IEEE Transactions, Aerospace and Electronic Systems,* Vols. 49, no. 4, October 2013, pp. 2463–2469.

[5]   F. J. Harris, "On the Use of Windows for Harmonic Analysis with the Discrete Fourier Transform," *Proceedings of the IEEE,* vol. 66, no. 1, January 1978, pp. 51-83.

[6]   A. D. Poularikas, *The Handbook of Formulas and Tables for Signal Processing*, Boca Raton, FL: CRC Press, 1999.

[7]   K. M. M. Prabhu, *Window Functions and Their Applications in Signal Processing*, Boca Raton, FL: CRC Press, 2014.

[8]   D. K. Barton, *Radar System Analysis and Modeling*, Boston, MA: Artech House, 2005.

[9]   M. I. Skolnik, *Introduction to Radar Systems*, third ed., New York: McGraw-Hill, 2001.

[10] M. I. Skolnik, Ed., *Radar Handbook*, 3rd ed., New York: McGraw-Hill, 2008.

[11] R. S. Raven, "Requirements on Master Oscillators for Coherent Radar," *Proceedings of the IEEE,* vol. 54, no. 2, February 1966, pp. 237-243.

[12] M. C. Budge Jr. and M. P. Burt, "Range correlation Effects on Phase and Amplitude Noise," *IEEE Proceedings Southeastcon '93,* 1993.

[13] M. C. Budge Jr. and M. P. Burt, "Range correlation effects on phase noise spectra," *Proceedings SSST 25th Southeastern Symposium on System Theory*, 1993, pp 492–496.

[14] D. B. Leeson, "A simple model of feedback oscillator noise spectrum," *Proceedings of the IEEE,* vol. 54, no. 2, Feb. 1966, pp. 329–330.

[15] R. Poore, "Overview on Phase Noise and Jitter," 17 May 2001. [Online]. Available: www.agilent.com. [Accessed 18 March 2015].

[16] P. A. Lynn, *Radar Systems*, New York: Van Nostrand Reinhold, 1988.

[17] S. Butterworth, "On the Theory of Filter Amplifiers," *Experimental Wireless & the Wireless Engineer,* vol. 7, October 1930, pp. 536–541.

[18] L. R. Rabiner and B. Gold, *Theory and Application of Digital Signal Processing*, Englewood: Prentice-Hall, 1975.

[19] J. D. Rhodes, *Theory of Electrical Filters*, John Wiley, 1976.

[20] T. W. Parks and C. S. Burrus, *Digital Filter Design*, John Wiley & Sons, 1987.

[21] A. V. Oppenheim and R. W. Schafer, *Discrete-Time Signal Processing*, Englewood Cliffs: Prentice-Hall, 1989.

[22] R. G. Lyons, *Understanding Digital Signal Processing*, 3rd ed., New York: Prentice Hall, 2011.

[23] I. S. Gradshteyn and I. M. Ryzhik, *Table of Integrals, Series, and Products*, Academic Press, 1980.

[24] W. D. Stanley, *Digital Signal Processing*, Reston, Virginia: Reston Publishing Company, Inc., 1975.

APPENDIX 19A: DERIVATION OF (19.21)

In this appendix, we present the equations necessary to compute (19.21). More specifically, we derive an equation for

$$H_R(f, R_0) = \int_{\Re} \frac{4\sin^2(2\pi fR/c)}{R^3} dR \qquad (19A.1)$$

We begin by examining the clutter spectrum at the input to the matched filter. The equation for that spectrum is

$$\begin{aligned}
S_C(f, R) &= P_C(R)C(f) * \Phi(f) \\
&= P_C(R)C(f) * \left[\delta(f) + \Phi_{\Delta\phi}(f, R)\right] \\
&= P_C(R)C(f) + P_C(R)C(f) * \Phi_{\Delta\phi}(f, R) \\
&= P_C(R)C(f) + C(f) * \left[P_C(R)\Phi_{\Delta\phi}(f, R)\right]
\end{aligned} \qquad (19A.2)$$

We included R in the argument of the clutter spectrum, $S_C(f)$, and the phase noise, $\Phi_{\Delta\phi}(f)$, to acknowledge that the spectrum is a function of range. $P_C(R)$ is the total clutter power at the input to the matched filter for a single, point clutter source at a range of R.

If we ignore the R^4 attenuation for clutter past the radar (clutter) horizon (which is reasonable because the predominant contributors to $P_C(R)$ for pulsed Doppler waveforms are at ranges close to the radar), $P_C(R)$ is of the form

$$P_C(R) = \frac{K}{R^3} \qquad (19A.3)$$

where K contains the rest of the terms of the radar range equation for clutter (see Chapter 17). With this we get

$$S_C(f, R) = \frac{K}{R^3}C(f) + KC(f) * \frac{\Phi_{\Delta\phi}(f, R)}{R^3} \qquad (19A.4)$$

To obtain the contribution of the clutter in the range region of interest, $\Re$, we integrate $S_C(f, R)$ over $\Re$. For pulsed Doppler waveforms, $\Re$ is defined by

$$\Re = \left[(R_0 + k\Delta R) \leq R \leq (R_0 + k\Delta R + \delta R) | k = 0, 1, \cdots\right] \qquad (19A.5)$$

where $\Delta R = cT/2$ is the range equivalent of the PRI and δR is the range resolution of the pulses of the burst. R_0 is the range to the front of the clutter cell closest to the radar. With this we get

$$S_C(f) = \sum_{k=0}^{N_{pulse}} \int_{R_{min}(k)}^{R_{max}(k)} S_C(f,R)\,dR$$

$$= \sum_{k=0}^{N_{pulse}} \int_{R_{min}(k)}^{R_{max}(k)} \left[\frac{K}{R^3}C(f) + KC(f) * \frac{\Phi_{\Delta\phi}(f,R)}{R^3}\right] dR \qquad (19A.6)$$

$$= KC(f) \sum_{k=0}^{N_{pulse}} \int_{R_{min}(k)}^{R_{max}(k)} \frac{dR}{R^3} + KC(f) * \sum_{k=0}^{N_{pulse}} \int_{R_{min}(k)}^{R_{max}(k)} \frac{\Phi_{\Delta\phi}(f,R)}{R^3} dR$$

$$= KC(f) \sum_{k=0}^{N_{pulse}} \int_{R_{min}(k)}^{R_{max}(k)} \frac{dR}{R^3} + KC(f) * \left[S_\phi(f) H_R(f,R_0)\right]$$

where

$$S_\phi(f) H_R(f,R_0) = \sum_{k=0}^{N_{pulse}} \int_{R_{min}(k)}^{R_{max}(k)} \frac{\Phi_{\Delta\phi}(f,R)}{R^3} dR \qquad (19A.7)$$

N_{pulse} is the number of PRIs needed for $S_C(f)$ to converge, $R_{min}(k) = R_0 + k\Delta R$ and $R_{max}(k) = R_0 + k\Delta R + \delta R$. Generally, N_{pulse} is much less than the number of pulses in the burst.

From (19.21), we have

$$\Phi_{\Delta\phi}(f,R) = S_\phi(f)\left[4\sin^2(2\pi f R/c)\right] \qquad (19A.8)$$

With this we get

$$S_\phi(f) H_R(f,R_0) = \sum_{k=0}^{N_{pulse}} \int_{R_{min}(k)}^{R_{max}(k)} \frac{S_\phi(f)\left[4\sin^2(2\pi f R/c)\right]}{R^3} dR$$

$$= S_\phi(f) \sum_{k=0}^{N_{pulse}} \int_{R_{min}(k)}^{R_{max}(k)} \frac{\left[4\sin^2(2\pi f R/c)\right]}{R^3} dR \qquad (19A.9)$$

which is of the form (19.21). From this we get

$$H_R(f,R_0) = \sum_{k=0}^{N_{pulse}} \int_{R_{min}(k)}^{R_{max}(k)} \frac{\left[4\sin^2(2\pi f R/c)\right]}{R^3} dR \qquad (19A.10)$$

To evaluate the integral of (19A.10), we use the trigonometric identity $\sin^2\theta = (1 - \cos 2\theta)/2$ and write

$$H_R(f, R_0) = 2 \sum_{k=0}^{N_{pulse}} \int_{R_{min}(k)}^{R_{max}(k)} \frac{[1 - \cos(4\pi f R/c)]}{R^3} dR$$

$$= 2 \sum_{k=0}^{N_{pulse}} \left[\int_{R_{min}(k)}^{R_{max}(k)} \frac{1}{R^3} dR - \int_{R_{min}(k)}^{R_{max}(k)} \frac{\cos(4\pi f R/c)}{R^3} dR \right] \quad (19A.11)$$

Evaluation of the first integral is simple. The basis for evaluating the second integral is (2.639.2) in [23, p. 187]

$$\int \frac{\cos x}{x^3} dx = \frac{\sin x}{2x} - \frac{\cos x}{2x^2} - \frac{\text{ci}(x)}{2} \quad (19A.12)$$

where

$$\text{ci}(x) = -\int_x^\infty \frac{\cos t}{t} dt \quad (19A.13)$$

is the cosine integral (8.230.2) in [23, p. 928].

After some manipulation,

$$H_R(f, R_0) = \frac{K}{P_{RT}} \sum_{k=0}^{N_{pulse}} G(R_0 + k\Delta R + \delta R) - G(R_0 + k\Delta R) \quad (19A.14)$$

with

$$G(R) = \left(\frac{4\pi f}{c}\right)^2 \text{ci}(4\pi f R/c) - \left(\frac{4\pi f}{c}\right) \frac{\sin(4\pi f R/c)}{R} - \frac{2\sin^2(2\pi f R/c)}{R^2} \quad (19A.15)$$

and

$$P_{RT} = \sum_{k=0}^{N_{pulse}} \int_{R_0 + k\Delta R}^{R_0 + k\Delta R + \delta R} \frac{K}{R^3} dR \quad (19A.16)$$

Figure 19A.1 contains a plot of $H_R(f, R_0)$ for $R_0 = 225$ m, $\Delta R = 1,500$ m, and $\delta R = 150$ m. It also contains the approximation

$$H_R(f, R_0) = \begin{cases} 8\sin^2(2\pi f R/c) & |f| < c/12R_0 \\ 2 & |f| \geq c/12R_0 \end{cases} \quad (19A.17)$$

that fits the exact curve well.

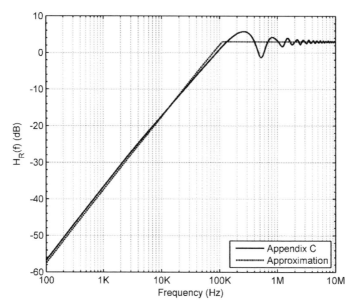

Figure 19A.1 Plots of (19A.14) and (19A.17) for $R_0 = 225$ m, $\Delta R = 1{,}500$ m, and $\delta R = 150$ m.

APPENDIX 19B: FFT FREQUENCY RESPONSE

We can write the signal at tap n_{tap} of an FFT as [24] as

$$X_{ntap} = \sum_{n=1}^{N_{FFT}} x(nT) w_n W^{n n_{tap}} \quad (19B.1)$$

where

$$W = e^{-j2\pi T(PRF/N_{FFT})} \quad (19B.2)$$

$x(nT)$ is the signal at the input to the nth tap, w_n is the weight applied to the signal at the nth tap, N_{FFT} is the FFT length, T is the spacing between input signals (the PRI), and $PRF = 1/T$ is the PRF of the signal at the FFT input. As a note, the collection of w_n $n \in [1, N_{FFT}]$ is the amplitude weighting applied to the FFT. For example, in the MPRF case, Section 19.7, a Chebyshev weighting was used.

If we substitute (19B.2) into (19B.1), we get

$$X_{ntap} = \sum_{n=1}^{N_{FFT}} x(nT) w_n e^{-j2\pi(nT)\left[(PRF/N_{FFT})n_{tap}\right]} \quad (19B.3)$$

In (19B.3), nT is a time variable and $f_{ntap} = (PRF \times (n_{tap}-1)/N_{FFT}$ is a frequency variable associated with tap n_{tap} of the FFT. We say tap n_{tap} is *tuned* to the frequency f_{ntap}.

We define the frequency response of tap n_{tap} as $|X_{ntap}|^2$ when the input is a complex sinusoid,

$$x(nT) = e^{j2\pi f(n-1)T} \quad (19B.4)$$

With this we get

$$\left|X_{ntap}(f)\right|^2 = \left|\sum_{m=0}^{N_{FFT}-1} e^{j2\pi fmT} w_{m+1} e^{-j2\pi mTf_{ntap}}\right|^2 = \left|\sum_{m=0}^{N_{FFT}-1} w_{m+1} e^{j2\pi m(f-f_{ntap})T}\right|^2 \quad (19B.5)$$

Where we made the change of variables, $m = n-1$. We also added f as an argument of X_{ntap} to acknowledge we are evaluating X_{ntap} as a function of f.

Figure 19B.1 contains a plot of $|X_{ntap}(f)|^2$ for the case where $w_n = 1$, $n \in [1, N_{FFT}]$ (uniform weighting), $T = 40$ μs, $n_{tap} = 5$, and $N_{FFT} = 16$. These values result in

$$f_{ntap} = \left(PRF \times (n_{tap}-1)\right)/N_{FFT} = \left(\tfrac{1}{T} \times (n_{tap}-1)\right)/N_{FFT} = 6.25 \text{ kHz} \quad (19B.6)$$

The plot of Figure 19B.1 exhibits the characteristics of a BPF. This is why we often think of an FFT as a bank of BPFs with center frequencies located at

$$f_{ntap} = \frac{(n_{tap}-1)}{N_{FFT}} PRF \quad n_{tap} \in [1, N_{FFT}] \quad (19B.7)$$

if we assume the taps span 0 to $PRF \times (1-1/N_{FFT})$ or

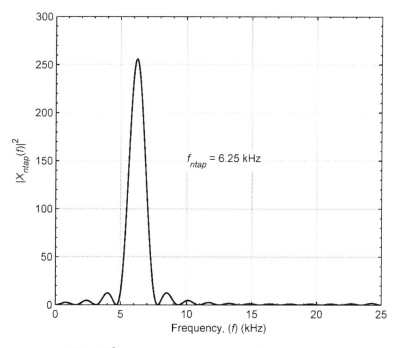

Figure 19B.1 Plot of $|X_{ntap}(f)|^2$ for $n_{tap} = 5$, $N_{FFT} = 16$, $T = 40$ μs, $w_n = 1$.

$$f_{ntap} = \begin{cases} (n_{tap} - 1)\dfrac{PRF}{N_{FFT}} & 1 \leq n_{tap} \leq \dfrac{N_{FFT}}{2} \\ (n_{tap} - N_{FFT})\dfrac{PRF}{N_{FFT}} & \dfrac{N_{FFT}}{2} < n_{tap} \leq N_{FFT} \end{cases} \quad (19\text{B}.8)$$

if we assume the taps span $-PRF/2$ to $(1-1/N_{FFT}) \times PRF/2$.

To be consistent with the conventions used in this chapter, we need to normalize $|X_{ntap}(f)|^2$ so that its peak value is unity. From (19B.5) and Figure 19B.1, we note the peak occurs at $f = f_{ntap}$, and is equal to

$$\left| X_{ntap}(f_{ntap}) \right|^2 = \left| \sum_{n=0}^{N_{FFT}-1} w_n e^{j2\pi n(f_{ntap} - f_{ntap})T} \right|^2 = \left| \sum_{n=0}^{N_{FFT}-1} w_n \right|^2 \quad (19\text{B}.9)$$

Thus, to arrive at the FFT frequency response we use in our analyses, we must divide $|X_{ntap}(f)|^2$ by $|X_{ntap}(f_{ntap})|^2$. This gives (19.44) or

$$H_{FFT}(f) = \left| \dfrac{\sum_{n=0}^{N_{FFT}-1} w_n e^{-j2\pi n(f - f_{ntap})T}}{\sum_{n=0}^{N_{FFT}-1} w_n} \right|^2 \quad (19\text{B}.10)$$

Chapter 20

Analog Pulsed Doppler Processors

20.1 INTRODUCTION

In analog processors, the HPF, $H_H(f)$, and the BPF, $H_B(f)$ (see Figure 19.7) are not periodic functions of frequency. As a result, we cannot use the same analysis techniques we used for digital pulsed Doppler processors and the MTI processor. Instead, we must compute the folded spectrum of the input signal, $S_o(f)$, and work with it. We must also replace the ADC of Figure 19.7 with a sampler and a hold device. Where we assume the hold device is a zero-order hold (ZOH). This configuration is illustrated in Figure 20.1.

The (power spectrum) frequency response of a ZOH is [1, 2]

$$H_Z(f) = T\text{sinc}^2(fT) \tag{20.1}$$

As before, we assume the receiver noise at the sampler output is white with a power of $P_{No} = P_N$, where P_N is the noise power at the matched filter output. Since the noise at the sampler output is a sampled signal, the amplitude of its power spectral density is also P_N. That is, $S_{oN}(f) = P_{No} = P_N$. The noise power spectrum at the output of the ZOH is thus

$$S_{ZN}(f) = H_Z(f)S_{oN}(f) = P_N H_Z(f) = P_N T\text{sinc}^2(fT) \tag{20.2}$$

and the noise spectrum at the processor output is

$$S_{Nout}(f) = P_N H_Z(f) H_H(f) H_B(f) \tag{20.3}$$

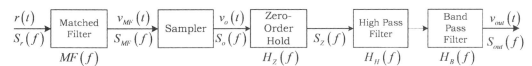

Figure 20.1 Analog Doppler processor block diagram.

The noise power at the processor output is

$$P_{Nout} = \int_{-\infty}^{\infty} S_{Nout}(f)\,df = P_N \int_{-\infty}^{\infty} H_Z(f) H_H(f) H_B(f)\,df = P_N G_N \quad (20.4)$$

where

$$G_N = \int_{-\infty}^{\infty} H_Z(f) H_H(f) H_B(f)\,df \quad (20.5)$$

is the noise gain through the processor. Figure 20.2 contains sketches of $S_{No}(f)$, $H_Z(f)$, $H_H(f)$, and $H_B(f)$. As shown, $H_Z(f)$ is a $\text{sinc}^2(x)$ function that has a first null at $f = \pm 1/T$.

An important point to note is that the BPF is centered below $1/2T$ (i.e., PRF/2). This is a requirement because of frequency folding, or aliasing. That is, all of the relevant frequency information in the signal folds into a region between $f = -1/2T$ and $f = +1/2T$.

We can develop a rule-of-thumb equation for G_N by making some simplifying assumptions about $H_H(f)$ and $H_B(f)$. We assume $H_B(f)$ is

$$H_B(f) = \text{rect}\left[\frac{f - f_B}{B}\right] \quad (20.6)$$

where f_B is in the passband of the HPF and B is small relative to $1/T$. We further assume the HPF has a passband gain of unity. With this we get

$$G_N = \int_{-\infty}^{\infty} T\text{sinc}^2(fT)\,\text{rect}\left[\frac{f - f_B}{B}\right] df \approx BT\text{sinc}^2(f_B T) \quad (20.7)$$

which is close to BT, the form we derived for the digital processor, see (19.27).

The target spectrum at the sampler output is (see Section 16.2.2)

$$S_{oS}(f) = \frac{P_S}{T} \sum_{l=-\infty}^{\infty} MF(f - l/T) S_{Sr}(f - f_d - l/T) \quad (20.8)$$

where $S_{rS}(f) = \delta(f - f_d)$ and P_S is the target signal power at the matched filter output. With this we have

$$S_{oS}(f) = \frac{P_S}{T} \sum_{l=-\infty}^{\infty} MF(f - l/T) \delta(f - f_d - l/T) \quad (20.9)$$

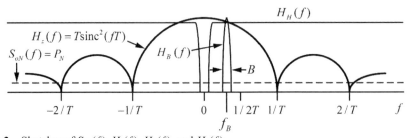

Figure 20.2 Sketches of $S_{No}(f)$, $H_Z(f)$, $H_H(f)$, and $H_B(f)$.

The target spectrum at the ZOH output is

$$S_{ZOHS}(f) = H_Z(f) S_{oS}(f)$$
$$= P_S \text{sinc}^2(fT) \sum_{l=-\infty}^{\infty} MF(f - l/T) \delta(f - f_d - l/T) \quad (20.10)$$

The spectrum at the signal processor output is

$$S_{Sout}(f) = S_{ZOHS}(f) H_H(f) H_B(f)$$
$$= P_S H_H(f) H_B(f) \text{sinc}^2(fT) \sum_{l=-\infty}^{\infty} MF(f - l/T) \delta(f - f_d - l/T) \quad (20.11)$$

Figure 20.3 contains depictions of the various signal-related spectra. In this figure, the BPF is centered on the target return. Note that even though there are many target spectral lines present in the ZOH output, only one is in the passband of the BPF. Again, note that the BPF and target spectral line of interest are in the range $f \in (-1/2T, +1/2T]$.

With the above, the target signal power at the processor output is

$$P_{Sout} = \int_{-\infty}^{\infty} S_{Sout}(f) df$$
$$= P_S H_H(f_d) H_B(f_d) \text{sinc}^2(f_d T) MF(f_d) = P_S G_S \quad (20.12)$$

where G_S is the signal gain, and is given by

$$G_S = H_H(f_d) H_B(f_d) \text{sinc}^2(f_d T) MF(f_d) \quad (20.13)$$

If we use the assumptions about $H_H(f)$ and $H_B(f)$ that we used for the noise rule of thumb of (20.7) and assume $f_B = f_d$, we obtain a rule-of-thumb equation for signal gain as

$$G_S = \text{sinc}^2(f_d T) MF(f_d) \quad (20.14)$$

If we further make the (reasonable) assumption $MF(f_d) \approx 1$, we get

$$G_S = \text{sinc}^2(f_d T) \quad (20.15)$$

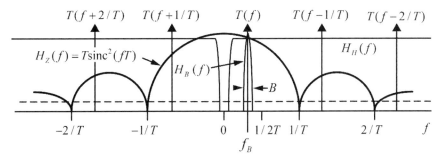

Figure 20.3 Sketches of $T(f)$, $H_Z(f)$, $H_H(f)$, and $H_B(f)$.

We can combine the results we obtained thus far to derive an equation for SNR gain through the processor as

$$G_{SNR} = \frac{SNR_{out}}{SNR_{MF}} = \frac{P_{Sout}/P_{Nout}}{P_S/P_N}$$
$$= \frac{P_{Sout}P_N}{P_S P_{Nout}}$$
$$= \frac{P_S G_S P_N}{P_S P_N G_N} \qquad (20.16)$$
$$= \frac{G_S}{G_N}$$

where G_S is given by (20.13) and G_N is given by (20.5). If we make use of the rules of thumb for G_S and G_N [see (20.15) and (20.7)], we get a rule-of-thumb equation for G_{SNR} as

$$G_{SNR} = \frac{\text{sinc}^2(f_B T)}{BT\,\text{sinc}^2(f_B T)} \qquad (20.17)$$
$$= \frac{1}{BT} = \frac{1}{G_N}$$

which is the same result we obtained for the digital signal processor (see Table 19.1).

From (16.20), with $S_r(f) = P_C S_{Cr}(f)$, the spectrum at the output of the sampler for the clutter signal is

$$S_{oC}(f) = \frac{P_C}{T}\sum_{l=-\infty}^{\infty} MF(f - l/T)S_{Cr}(f - l/T) \qquad (20.18)$$

where

$$S_{Cr}(f) = C(f) * \Phi(f) = C(f) * \left[\delta(f) + S_\phi(f) H_R(f, R_0)\right] \qquad (20.19)$$

The spectrum at the signal processor output is

$$S_{Cout}(f) = \frac{1}{T} H_H(f) H_B(f) \text{sinc}^2(fT) S_{oC}(f)$$
$$= P_C H_H(f) H_B(f) \text{sinc}^2(fT) \sum_{l=-\infty}^{\infty} MF(f - l/T) C(f - l/T)$$
$$+ P_C H_H(f) H_B(f) \text{sinc}^2(fT) \qquad (20.20)$$
$$\times \sum_{l=-\infty}^{\infty} MF(f - l/T) C(f - l/T) * \left[S_\phi(f - l/T) H_R(f - l/T, R_0)\right]$$
$$= P_C \left[S_{Coutc}(f) + S_{Cout\phi}(f)\right]$$

Figure 20.4 contains a sketch of the various spectrum components that make up $S_{Cout}(f)$.

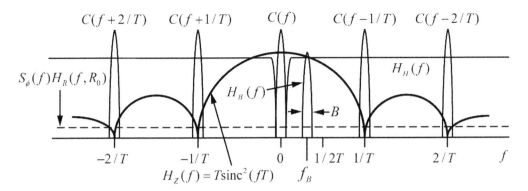

Figure 20.4 Sketches of $S_{\Delta\phi}(f)$, $C(f)$, $H_Z(f)$, $H_H(f)$, and $H_B(f)$.

The clutter power at the signal processor output is

$$P_{Cout} = \int_{-\infty}^{\infty} S_{Cout}(f)df = P_C\left[\int_{-\infty}^{\infty} S_{Coutc}(f)df + \int_{-\infty}^{\infty} S_{Cout\phi}(f)df\right] \quad (20.21)$$
$$= P_C(G_C + G_\phi)$$

with

$$G_C = \int_{-\infty}^{\infty} H_H(f)H_B(f)\text{sinc}^2(fT)\sum_{l=-\infty}^{\infty} MF(f-l/T)C(f-l/T)df \quad (20.22)$$

and

$$G_\phi = \int_{-\infty}^{\infty} \Big\{H_H(f)H_B(f)\text{sinc}^2(fT) \\
\times \sum_{l=-\infty}^{\infty} MF(f-l/T)C(f-l/T) \\
* \big[S_\phi(f-l/T)H_R(f-l/T,R_0)\big]\Big\}df \quad (20.23)$$

In general, these integrals must be evaluated numerically. However, for the case where the range correlation results in $S_\phi(f)H_R(f,R_0) = \Phi_0$, (20.23) reduces to

$$G_\phi = \Phi_0 \int_{-\infty}^{\infty} H_H(f)H_B(f)\text{sinc}^2(fT)\left[\sum_{l=-\infty}^{\infty} MF(f-l/T)\right]df \quad (20.24)$$

If we use the assumption we used to derive (20.7) and represent $MF(f)$ by an ideal LPF with a bandwidth of $1/\tau_p$, the summation of (20.24) reduces to N, where $N = T/\tau_p$ (see Chapter 19). If we carry this a step further and use the $H_H(f)$ and $H_B(f)$ assumptions we used to derive the rule-of-thumb equation for G_N, we can further reduce (20.24) to

570 Basic Radar Analysis

$$G_\phi = \frac{T}{\tau_p} \Phi_0 \int_{-\infty}^{\infty} \text{sinc}^2(fT) \text{rect}\left[\frac{f-f_B}{B}\right] df \approx \frac{\Phi_0}{\tau_p} BT\text{sinc}^2(f_B T)$$

$$= \frac{\Phi_0}{\tau_p} G_N$$

(20.25)

We can write the SCR gain of the signal processor as

$$G_{SCR} = \frac{SCR_{out}}{SCR} = \frac{G_S P_S / (G_C + G_\phi) P_C}{P_S / P_C} = \frac{G_S}{G_C + G_\phi}$$

(20.26)

The above equations and rules of thumb are summarized in Table 20.1.

Table 20.1
Summary of Analog Pulsed Doppler Signal Processor Analysis Equations

Parameter	Equation	Rule of Thumb
CNR folding based on Section 19.1 and related discussion	$CNR(R) = \sum_{k=0}^{\infty} CNR_1(R + kR_{PRI})$	The practical upper limit on the sum is usually 5 or 10 (see Chapter 19)
Target signal gain (20.13)	$G_S = H_H(f_d) H_B(f_d) \text{sinc}^2(f_d T) MF(f_d)$	$G_S = \text{sinc}^2(f_d T)$
Central line clutter gain (20.22)	$G_C = \int_{-\infty}^{\infty} \Big\{ H_H(f) H_B(f) \text{sinc}^2(fT)$ $\times \sum_{l=-\infty}^{\infty} MF(f-l/T) C(f-l/T) df \Big\}$	Generally very small and can be ignored for MPRF and HPRF waveforms, but is the predominant contributor for LPRF waveforms
Phase noise clutter gain (20.23)	$G_\phi = \int_{-\infty}^{\infty} \Big\{ H_H(f) H_B(f) \text{sinc}^2(fT)$ $\times \sum_{l=-\infty}^{\infty} MF(f-l/T) C(f-l/T) * S_{\Delta\phi}(f-l/T) \Big\} df$ $S_{\Delta\phi}(f) = \left[S_\phi(f) H_R(f, R_0)\right]$	if $C(f) * \Phi_{\Delta\phi}(f) = \Phi_0$ $G_\phi = G_N \Phi_0 / \tau_p$
Total clutter gain (Section 19.2)	$G_{Ctot} = G_C + G_\phi$	$G_{Ctot} = G_\phi$ for MPRF and HPRF waveforms, but $G_{Ctot} = G_C$ for LPRF waveforms
Noise gain (20.5)	$G_N = \int_{-\infty}^{\infty} T\text{sinc}^2(fT) H_H(f) H_B(f) df$	$G_N = BT\text{sinc}^2(f_B T)$ B = BPF bandwidth
SNR gain (20.16)	$G_{SNR} = G_S / G_N$	$G_{SNR} = 1/BT$
CNR gain	$G_{CNR} = (G_C + G_\phi)/G_N$	
Clutter attenuation	$CA = 1/G_{CNR}$	
SCR gain/SCR improvement	$G_{SCR} = \dfrac{G_S}{G_C + G_\phi}$	

20.2 ANALOG PULSED DOPPLER SIGNAL PROCESSOR EXAMPLE

To illustrate the analog processor analysis procedures, we consider the pulsed Doppler radar that uses a PRF of 100 kHz and an unmodulated pulse with a width of 1 μs (see Section 19.6). The remaining radar, target, and clutter parameters are given in Table 19.3 and discussed in Section 19.6. Also, when comparing Figure 20.1 to Figure 19.7, we note that the analog and digital signal processors are the same through the matched filter. Because of this, the plots of SNR, CNR, SCR, and SIR at the matched filter output are as shown in Figure 19.14. The plot of Figure 19.14 is repeated for convenience in Figure 20.5.

We will use the same types of filters as in Section 19.6; the only difference is that they are analog instead of digital. The frequency responses are

$$H_H(f) = 1 - \frac{1}{1+(f/f_h)^{10}} = \frac{(f/f_h)^{10}}{1+(f/f_h)^{10}} \qquad (20.27)$$

for the fifth-order Butterworth HPF and

$$H_B(f) = \frac{1}{1+\left(\dfrac{f-f_B}{B/2}\right)^6} \qquad (20.28)$$

for the sixth-order Butterworth BPF. In the above, $f_h = 2{,}000$ Hz, $f_B = 8{,}000$ Hz, and $B = 350$ Hz.

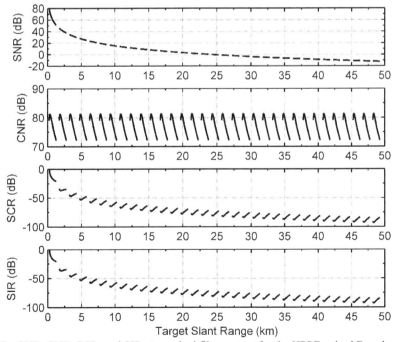

Figure 20.5 SNR, CNR, SCR, and SIR at matched filter output for the HPRF pulsed Doppler example.

G_N is computed by numerically evaluating

$$G_N = \int_{-\infty}^{\infty} T\text{sinc}^2(fT) H_H(f) H_B(f) df \tag{20.29}$$

This results in $G_N = 0.0036$ W/W. Alternately, we could use the rule of thumb from Table 20.1 to arrive at a value of $G_N = 0.0034$ W/W.

G_S is computed from

$$G_S = H_H(f_d) H_B(f_d) \text{sinc}^2(f_d T) MF(f_d) \tag{20.30}$$

which yields $G_S = 0.98$ W/W. The rule-of-thumb equation of Table 20.1 also results in $G_S = 0.98$ W/W.

Combining G_S and G_N results in a processor SNR gain of

$$G_{SNR} = \frac{G_S}{G_N} = 272.8 \text{ W/W or } 24.36 \text{ dB} \tag{20.31}$$

The central line and phase noise clutter gains are computed from

$$G_C = \int_{-\infty}^{\infty} \left\{ H_H(f) H_B(f) \text{sinc}^2(fT) \right. \\ \left. \times \sum_{l=-\infty}^{\infty} MF(f-l/T) C(f-l/T) df \right\} \tag{20.32}$$

and

$$G_\phi = \Phi_0 \int_{-\infty}^{\infty} H_H(f) H_B(f) \text{sinc}^2(fT) \left[\sum_{l=-\infty}^{\infty} MF(f-l/T) \right] df \tag{20.33}$$

We used the form of (20.33) to compute the phase noise clutter gain because we previously showed that, for all practical purposes, $S_\phi(f) H_R(f, R_0) = \Phi_0$, where $\Phi_0 = -143$ dBc/Hz for a clutter range of 225 m (see Section 19.6). The resulting values for these two gains are $G_C = -242$ dB and $G_\phi = -107$ dB.

With the above, the SCR improvement is

$$I_{scr} = G_{SCR} = \frac{G_S}{G_C + G_\phi} = \frac{1}{G_\phi} = 5 \times 10^{10} \text{ W/W or } 107 \text{ dB} \tag{20.34}$$

The results of applying these gains are shown in Figure 20.6. As expected, the performance results are very close when comparing them using the analog and digital processors.

For a hybrid signal processor, we would use the approach of this section to determine the noise, signal, and clutter spectra at the output of the analog portion of the processor. We would then use these in place of the matched filter frequency response, $MF(f)$, and the input signal spectrum, $S_r(f)$, in the digital processor analyses. We would compute the various powers out of the digital portion using (16.45) with the spectra out of the analog portion in place of $MF(f)S_r(f)$.

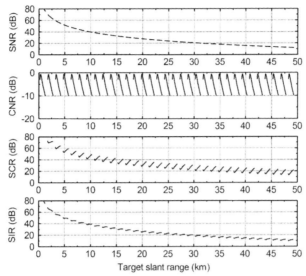

Figure 20.6 Plots of SNR, CNR, SCR, and SIR at analog processor output.

20.3 EXERCISES

1. Derive (20.7).

    2. Derive (20.20).

    3. Derive (20.25).

    4. Repeat the example of Section 20.2 and reproduce Figure 20.6.

References

[1] Rabiner, L. R., and B. Gold, *Theory and Application of Digital Signal Processing,* Englewood Cliffs, NJ: Prentice-Hall, 1975.

[2] Oppenheim, A. V., and R. W. Schafer, *Discrete-Time Signal Processing,* Englewood Cliffs, NJ: Prentice-Hall, 1989.

Chapter 21

Chaff Analysis

21.1 INTRODUCTION

Another type of clutter of concern in military radars is chaff. Chaff consists of short, tuned dipoles made of strips of aluminum foil or aluminum-coated pieces of glass, fiberglass, or Mylar which, when dispensed, bloom into a cloud that has a very large RCS [1–3].[1] Since the dipoles have large aerodynamic drag, they can remain aloft for long periods of time. The RCS, center velocity, and velocity spread of chaff can cause problems in tracking radars not designed to mitigate chaff. Specifically, the chaff has an initial velocity close to that of the dispensing aircraft. Since the chaff RCS is large, it can capture the radar tracking gates and cause the radar to track the chaff instead of the aircraft. Its ability to do this depends on how the chaff is dispensed, the characteristics of the radar track loops, and the type of signal processor. In general, pulsed Doppler signal processors are less susceptible to chaff than MTI processors and radars that do not use some type of clutter mitigation technique.

The time behavior of chaff RCS consists of four phases denoted as 1) the transient phase, 2) the bloom phase, 3) the mature phase, and 4) the decay phase. For self-screening chaff, the transient phase is the explosive birth phase of the chaff cloud and is the time a chaff cartridge is ejected from the aircraft and explodes to disperse the chaff dipoles.[2] During this time, the RCS is small since the dipoles are still tightly packed. This is also the time the initial velocity is highest and the velocity spread is highest. The latter property is most likely due to the fact that the explosion imparts widely differing velocities to the chaff particles.

The bloom phase begins immediately after the chaff cartridge has exploded and is characterized by rapid chaff cloud growth, which is important for masking the aircraft quickly and becomes more critical the faster the aircraft is traveling and the better

---

[1] Typically, chaff clouds are designed to cover the frequency bands associated with the intended victim radars. This is accomplished by mixing dipoles within the cloud that are cut to the appropriate lengths so that resonance occurs at each of the victim radar frequencies.
[2] Some chaff dispensers rely on air turbulence generated by the aircraft to disperse chaff packets. Others use small pyrotechnic charges shortly after the chaff leaves the aircraft. One explosive technique achieved an average time to bloom as short as 12 ms [11].

rejection the opposing radar has. This is also the time period in which the RCS rapidly increases because of the spreading of the chaff cloud. During this phase, the chaff center velocity and velocity spread decrease to a level that depends on factors such as wind speed, wind shear, turbulence, and fall velocity [3].

The mature phase is the time period when the full RCS of the chaff cloud is realized. During this period, the center frequency and velocity spread are likewise determined by factors such as wind shear, turbulence, and fall velocity [3].

In the decay phase, the chaff dipoles spread to the point where the chaff may no longer appear as a single cloud. As a result, the RCS decays and the velocity spread becomes narrower.

Several authors have presented discussions of chaff RCS, center frequency, and spectral spread (velocity spread). However, there is limited data on the transient behavior of chaff. We present models that might be useful in evaluating the transient behavior. The models use an exponential decay or increase since exponential functions are easy to use, and we have no justification for more complicated models. The model we propose for RCS dynamics is

$$\sigma_{chaff}(t) = (\sigma_{SS} - \sigma_I)(1 - e^{-t/\tau_\sigma}) + \sigma_I$$
$$= (\sigma_I - \sigma_{SS})e^{-t/\tau_\sigma} + \sigma_{SS} \quad \text{m}^2 \quad (21.1)$$

where σ_{SS} is the bloom RCS and σ_I is the RCS at the start of the bloom phase. τ_σ is the RCS time constant. Equation (21.1) is a variation of a model found in [4, 5] and allows an initial RCS other than 0 dBsm. Nathanson [1] provides an equation for σ_{SS} that he attributes to Schlesinger [6]. That equation is [1, p. 335]

$$\sigma_{SS} = 3,000 W_{chaff} / f_c \quad \text{m}^2 \quad (21.2)$$

where W_{chaff} is the weight of the chaff bundle, in pounds, and f_c is the radar operating frequency, in GHz. This equation applies to chaff made of aluminum foil that is 0.001 inch thick, $\lambda/2$ long, and 0.01 inch wide. He notes that one pound of this chaff at 3-GHz operating frequency would have an RCS of 1,000 m^2 [1, 7, 8]. Nathanson points out that one pound of chaff designed to cover a frequency range of 1 to 10 GHz might have a σ_{SS} of 60 m^2. It is not clear what value should be chosen for σ_I. A guess might be 0 to 5 dBsm since this is the beginning of the bloom phase.

Determining τ_σ is more difficult since there seems to be a dearth of published information in this area. Nathanson notes, "Chaff dipoles have high aerodynamic drag: their velocity drops to that of the local wind a few seconds after they are dispensed." Based on this, it seems as if a reasonable value of τ_σ might be 1 to 2 seconds.

The model we propose for chaff spectral width is

$$B(t) = (B_{SS} - B_I)(1 - e^{-t/\tau_B}) + B_I$$
$$= (B_I - B_{SS})e^{-t/\tau_B} + B_{SS} \quad \text{Hz} \quad (21.3)$$

In this equation, B_I is the 3-dB bandwidth at the beginning of the bloom phase, B_{SS} is the bandwidth at the end of the bloom phase, and τ_B is the time constant associated with the change in bandwidth. Nathanson notes that the components that affect the spectral spread of rain also affect the spectral spread of chaff. In particular, he notes that the shear component is the same as rain. According to a graph in his book [1], the spread due to shear can range from 0 to 5 m/s depending on altitude and beam elevation angle. He goes on to note that a reasonable value for the turbulence component is 0.7 to 1 m/s. He provides equations for the spread due to fall velocity, and spectrum broadening due to its elevation dispersion across the radar beam. Although he does not state it, we assumed that these values are for B_{SS}. If we were to summarize the various numbers into rule-of-thumb values, we would suggest a frequency range associated with a velocity range of 1 to 5 m/s (i.e., $B_{SS} = 2v/\lambda$ where v is between 1 and 5 m/s and λ is the wavelength of the radar signal.

It is more difficult to guess values for B_I. However, if we use the premise that B_I is larger than B_{SS}, a guess might be to choose B_I two to five times B_{SS}. As with the case of RCS, a reasonable value of τ_B might be 1 to 2 seconds.

Our proposed model for mean chaff radial velocity is

$$V(t) = (V_I - V_{wind}) e^{-t/\tau_V} + V_{wind} \quad \text{m/s} \tag{21.4}$$

where V_I is the radial velocity at the beginning of the bloom stage, V_{wind} is the radial velocity of the wind, and τ_V is a chaff deceleration time constant [5]. The chaff center frequency is therefore $f(t) = 2V(t)/\lambda$. As with rain, V_{wind} depends on the actual wind velocity and direction and the direction of the radar antenna beam. For chaff that is "dropped," V_I would equal the target range rate, or smaller if it is assumed the cartridge does not explode until it has separated from the aircraft by some distance. If the chaff cartridge is ejected forward of the aircraft, V_I could be considerably larger than the aircraft range rate. As with the other models, a reasonable value of τ_V might be 1 to 2 seconds. The separation between the chaff centroid and aircraft can be expressed as

$$\Delta R(t) = (V_I - V_{wind}) \left[-\tau_V \left(1 - e^{-t/\tau_V}\right) + t \right] \quad \text{m} \tag{21.5}$$

Given the nature of chaff, a suggested spectral model for chaff is the Gaussian[3] model given by (17.12) discussed in Section 17.2.2 [9]. This suggestion is based on the assertion that the velocities of the dipoles are most likely governed by a Gaussian distribution. Because of the direct relationship between velocity and frequency, this leads to a Gaussian frequency spectrum [10]. This spectral model for chaff can be expressed as

$$C(f) = \frac{P_{chaff}}{\sigma_B \sqrt{2\pi}} e^{-\frac{(f-f_m)^2}{2\sigma_B^2}} \tag{21.6}$$

---

[3] As indicated by our reviewer, the Gaussian model will not be exact. The Gaussian model is unbonded while the actual chaff spectrum width will be finite. However, we expect the Gaussian model will be sufficient.

where P_{chaff} is the total chaff power, f_m is the mean chaff frequency, and σ_B is the chaff frequency standard deviation. The chaff frequency standard deviation is related to the chaff 3-dB bandwidth by (see Exercise 2)

$$\sigma_B(t) = B(t)/2.35 \tag{21.7}$$

where $B(t)$ is the distance between 3 dB points of the chaff power spectrum given by (21.3).

21.2 CHAFF ANALYSIS EXAMPLE

To evaluate a radar's response to chaff, we would apply the chaff and target signals to the signal processor in question, using the dynamic characteristics of the chaff. For an initial analysis, we assume the track point stays centered on the aircraft. Utilizing models of the chaff and signal processing, we want to quantify I_{SCR}, which is a function of the mismatch between the track point and the chaff centroid. We will use numerical integration to determine I_{SCR}. Once the I_{SCR} is determined, we calculate the SCR at the output of the processor. The general sequence of computations used this preliminary analysis is:

- Select time.
- Compute chaff RCS from (21.1).
- Compute S/C input from $(S/C)_{IN} = \sigma_t/\sigma_C$.
- Compute chaff spectral width from (21.3) and $B_{SS} = 2v/\lambda$.
- Compute chaff center frequency from (21.4).
- Compute S/C improvement for the signal processor being analyzed.
- Compute output S/C from $(S/C)_{OUT} = I_{SCR}(\sigma_t/\sigma_C)$.
- Repeat for next selected time.

While more detailed analysis can be performed, the results of the procedure above generally provide a case useful for initial chaff performance assessments.

Figures 21.1 and 21.2 contain plots from a simulation of the use of chaff against a radar that uses MTI. For this example, the radar operates at 8 GHz and uses a staggered PRI waveform with PRIs of 520 and 572 μs. It uses a 3-pulse MTI, with wind compensation used to place the MTI notch at the final chaff velocity. The target RCS was set to 0 dBsm. The chaff parameters used in the examples are listed in Table 21.1.

For Figure 21.1, the aircraft dispensed a single chaff cartridge (sometimes referred to as a puff), and in Figure 21.2, the aircraft dispensed five chaff cartridges spaced 1 second apart. For the single chaff cartridge, the slowdown of the chaff cloud caused the target and chaff to separate and thus allowed the SCR to rise. We assumed the range gate of the radar remained with the target.

Chaff Analysis

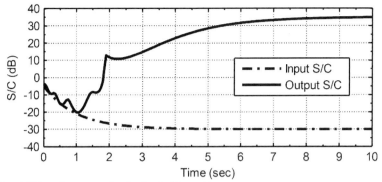

Figure 21.1 MTI performance for single chaff cartridge.

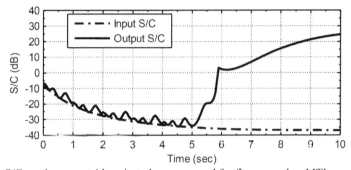

Figure 21.2 S/C vs. time—cartridge ejected every second for five seconds—MTI.

Table 21.1
Chaff Parameters Used in Chaff Simulation

Parameter	Value
Pulsewidth	30 μs
LFM bandwidth	1 MHz
Target range rate	150 m/s
Initial chaff RCS	5 dBsm
Steady-state chaff RCS	30 dBsm
Chaff RCS time constant	1 sec
Initial chaff bandwidth	1 kHz
Final chaff bandwidth	250 Hz
Chaff bandwidth time constant	2 sec
Location of MTI notch	5 m/s
Initial chaff velocity	150 m/s
Final chaff velocity	5 m/s
Chaff velocity time constant	1 sec

For the multiple cartridge case, the chaff continued to obscure the target since it was dispensing cartridges with a spacing equal to the various time constants associated with the chaff model. Again, we assumed the range gate remained with the target.

An interesting extension of this study would be to include a range track loop model to see what would be required for the chaff to cause the radar to break target lock and track the chaff.

Figures 21.3 and 21.4 contain plots like those of Figures 21.1 and 21.2 for a radar that uses a pulsed Doppler waveform and processor. For this example, the radar uses a 25-kHz, semi-infinite pulsed Doppler burst with unmodulated, 4-µs pulses. The signal processor consists of a fifth-order Butterworth HPF with a cutoff frequency of 1,000 Hz and a sixth-order Butterworth BPF with a bandwidth of 800 Hz. The range gate and BPF are centered on the target range and Doppler frequency, respectively. As illustrated in Figure 21.3 when one chaff cartridge was used, the SCR rose very quickly and stopped obscuring the aircraft in less than 0.5 seconds. This is because the chaff and target separate in Doppler frequency very quickly. Figure 21.4 shows the results of 10 chaff cartridges dispensed every 0.2 seconds. This extends the target masking due to chaff to about 2 seconds. This would seem to support an assertion that a pulsed Doppler processor might be effective in countering chaff.

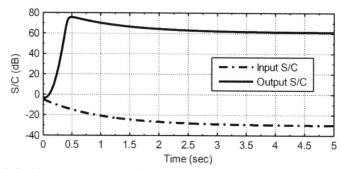

Figure 21.3 Pulsed Doppler performance for single chaff cartridge.

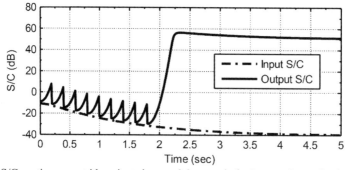

Figure 21.4 S/C vs. time—cartridge ejected every 0.2 seconds for 2 seconds—pulsed Doppler.

21.3 EXERCISES

1. Use (21.4) to derive the equation for the range to the chaff centroid [i.e., (21.5)].

    2. Derive the relationship between the standard deviation and the 3-dB bandwidth of the power spectrum relation of (21.7).

    3. Implement the chaff analysis algorithm discussed in this chapter for the case of an MTI processor and generate a plot like Figure 21.2. Use a target range rate of 300 m/s and 1.2 s between chaff puffs. Generate plots of chaff RCS, bandwidth, velocity, and center frequency as a function of time. How fast should chaff be dispensed to improve performance?

    4. Change the processor of Exercise 3 to a pulsed Doppler processor and generate a plot like Figure 21.4. Generate the other plots specified in Exercise 3. How fast should chaff be dispensed to mask the aircraft for 2 seconds? How many chaff bundles are necessary?

References

[1] Nathanson, F. E., J. P. Reilly, and M. N. Cohen, *Radar Design Principles: Signal Processing and the Environment,* 2nd ed., New York: McGraw-Hill, 1991.

[2] Schleher, D.C., *Introduction to Electronic Warfare,* Dedham, MA: Artech House, 1986.

[3] Rabiner, L. R., and B. Gold, *Theory and Application of Digital Signal Processing,* Englewood Cliffs, NJ: Prentice-Hall, 1975.

[4] Golden, A., *Radar Electronic Warfare,* New York: AIAA, 1987.

[5] Rohrs, R. J., "An Empirical Self-Protection Chaff Model Thesis," Air Force Institute of Technology, Wright Patterson Air Force Base, Ohio, Rep. No. 84D-54, 1984. Available from DTIC as AD-A151 928.

[6] Schlesinger, R. J., *Principles of Electronic Warfare,* Englewood Cliffs, NJ: Prentice-Hall, 1961.

[7] Cassedy, E. S., and J. Fainberg, "Back Scattering Cross Sections of Cylindrical Wires of Finite Conductivity," *IRE Trans. Antennas Propag.,* vol. 8, no. 1, Jan. 1960, pp. 1–7.

[8] Mack, C. L., Jr., and B. Reiffen, "RF Characteristics of Thin Dipoles," *Proc. IEEE,* vol. 52, no. 5, pp. 533–542, May 1964.

[9] Barlow, E. J., "Doppler Radar," *Proc. IRE,* vol. 37, no. 4, Apr. 1949, pp. 340–355.

[10] Papoulis, A., *Probability, Random Variables, and Stochastic Processes*, 3rd ed., New York: McGraw-Hill, 1991.

[11] Emslie, A. G., "Moving Object Radio Pulse-Echo System," U.S. Patent 2,659,077, Nov. 10, 1953.

Chapter 22

Radar Receiver Basics

22.1 INTRODUCTION

The general function of a radar receiver is to amplify, filter, shift frequency, and demodulate signal returns without distortion of the waveform modulation. The objective of a radar receiver is to facilitate discrimination between desired signals and unwanted noise and interference such as galactic noise, receiver noise, other radars, jamming, and clutter. In doing so, two principal design goals for a receiver are to ensure sufficient sensitivity to detect weak returns and to ensure adequate dynamic range to operate linearly for all expected return power levels.

There have been a number of receiver configurations used for radar over the decades (e.g., crystal, superregenerative, tuned RF, homodyne, heterodyne). However, the focus of this chapter is the superheterodyne receiver.[1] While the earliest receivers were entirely analog, receiver technology is rapidly trending more digital, less analog. In this chapter, we consider a superheterodyne receiver employing *direct IF sampling*.

Major Edwin Howard Armstrong invented the superheterodyne receiver in 1918 while serving with the U.S. Army Signal Corps during World War I [1–3]. Walter Hermann Schottky, working in Germany, conceived of the superheterodyne receiver independently [4]. Since then, the superheterodyne receiver has become predominant and is employed in virtually all radar receivers. For this and other contributions to the art of radio electronics, Armstrong was the first recipient of the Institute of Radio Engineers' (IRE) Medal of Honor in 1917 and was later awarded the American Institute of Electrical Engineers' (AIEE) Edison Medal in 1942 [5].

The superheterodyne receiver is based on the heterodyne principle invented by Reginald Aubrey Fessenden in 1902 [6]. Heterodyning is the process of combining or mixing signals with carrier frequencies of f_1 and f_2 to generate two other signals with frequencies of $f_1 + f_2$ (sum frequency) and $f_1 - f_2$ (difference frequency). Fessenden coined the term "heterodyne" from the Greek words for difference (hetero) and force (dyne) [7].

---

[1] Sometimes shortened to superhet receiver.

In early heterodyne receivers, the difference frequency was in the audible range so a telephone or telegraph signal could be heard in a headset [8].

Armstrong's superheterodyne receiver design was made possible by the first triode vacuum tube, called an audion, invented by Lee De Forest in 1906 [9–12]. The audion was the first vacuum tube device that could both detect and amplify signals [13, 14]. In contrast to the audio signal used by Fessenden's heterodyne receiver, Armstrong's superheterodyne uses a comparatively high (inaudible) frequency he termed the intermediate receiver frequency. Armstrong used the moniker "super" because of the supersonic IF.

22.2 SINGLE-CONVERSION SUPERHETERODYNE RECEIVER

A block diagram of a basic, single-conversion, superheterodyne receiver found in radars is given in Figure 22.1. Such a block diagram is sometimes referred to as a receiver chain. In current receiver vernacular, the components from the receiver input to the input to the detector are referred to as the RF front end. Earlier usage of the term stopped at the output of the first mixer. Single-conversion receivers are typically used in radars with bandwidth less than 20 MHz with limited tunable bandwidth.

Ideally, the receiver input consists of signals from targets of interest. One problem is there is a chance unwanted signals will also be present in the received signal. A BPF, called a preselector, is used to preserve the desired signal while eliminating unwanted signals (interference) such as those from other radars, jammers, or other sources of out-of-band energy [15].

The preselector is a low-loss device (typically < 1 dB), such as a cavity or waveguide filter, to minimize its impact on system noise figure. Parameters for various filter technologies over common operating regions (not necessary absolute limits) are summarized in Table 22.1. The preselector is usually low order, typically 2nd to 5th, to minimize unwanted ringing and overshoot.[2,3] Also, low-order filters generally have lower insertion loss, have less sensitivity to temperature, are less costly, and are smaller and lighter than higher-order filters.

The attenuators shown in Figure 22.1 are used to extend receiver dynamic range by reducing the power level of large returns in order to prevent saturating subsequent receiver components. The total attenuation necessary is typically distributed between RF and IF, as shown in Figure 22.1. The exact amount of attenuation required, and how it is distributed in a receiver chain, is determined via cascade analysis (see Section 22.7).

---

[2] Ringing is a decaying oscillating filter output caused by a transient filter input.
[3] When stating filter order, we are using the lowpass equivalent. The bandpass order is double.

Table 22.1
Bandpass Filter Implementation[,4]

Topology	Frequency (MHz)	3-dB Bandwidth[5] (%)	IL (dB)	Notes
Crystal	0.001 – 45	0.05 – 1.5	1 – 5	Crystal resonator, high Q achievable[6]
Lumped LC	0.03 – 10,00	1 – 200	0.8 – 5	Higher loss, wire wound inductors, parallel plate capacitors
Tubular	30 – 20,000	4 – 40	0.9	Very broad stopbands, very high rejection
Ceramic	300 – 6,000	0.3 – 30	0.5 – 5	Coaxial resonators, higher Q achievable with respect to lumped LC, extremely temperature stable
Cavity	30 – 40,000	0.2 – 30	0.3 – 3	Lower loss, narrower bandwidths, with respect to lumped LC, high power handling
Combline	500 – 40,000	3 – 18	0.3 – 1.7	High Q, extremely small packaging
Waveguide	2,000 – 94,000	0.1 – 20	0.3 – 2.5	Very low loss (typically < 0.5 dB), high power handling, half wavelength cavities, inductive iris coupling, extremely high Q achievable

Source: [16–21]

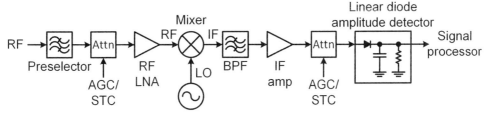

Figure 22.1 Superheterodyne receiver with linear amplitude detection.

Controlling the amount of attenuation is generally accomplished using automatic gain control AGC or sensitivity time control (STC) circuitry. AGC monitors the power level of a tracked target signal and adjusts receiver gain to establish a desired constant power level at the detector output.[7] STC uses an attenuation-versus-range (time) profile unrelated to target presence. When considering AGC, STC, and dynamic range, Barton notes that the following general rules apply:

- A target echo that saturates the receiver remains detectable in search radar, while angle tracking receivers (including both sum and difference channels in monopulse radar) must avoid saturation by the target to be tracked or measured.

---

[4] K&L microwave hosts a filter design wizard (www.klfilterwizard.com) that is useful for investigating filter options.
[5] 3-dB percentage bandwidth = (3-dB bandwidth / center frequency) × 100.
[6] Filter Q = center frequency / 3-dB bandwidth.
[7] Some radars use a default track gate location, such as the middle of the PRI, when not in track, to drive the AGC circuitry. In this case, a noise AGC (NAGC) is usually run in parallel and combined with the signal AGC. As a result, the noise level or signal level requiring the larger attenuation dominates control of the overall AGC.

- Clutter must not saturate the receiver or exceed the linear region in either search or tracking radars if high clutter attenuation is required. With modern digital signal processing, receiver gain control using a high-resolution clutter map offers an effective method of avoiding clutter nonlinearity on clutter peaks with minimal loss of target detections.
- Search radars cannot use AGC other than from a high-resolution clutter map because detection is required on small targets in the same beam as the strong signal on which conventional AGC is based.
- STC can be used in both search and tracking radars with LPRF waveforms (and possibly with MPRF if the STC action is limited to a small fraction of the PRI), never in HPRF radar. Receiver saturation on short-range clutter extends the region after the transmission in which targets are suppressed (eclipsed).
- A tracking radar can combine STC, AGC, and transmitter power variation to keep the target signal within the linear region, since echoes from ranges beyond the target need not be detectable.

When a low-PRF waveform is used, STC circuitry can be used to extend receiver dynamic range. STC reduces gain over the initial portion of the PRI according to an STC law, restoring full gain for the remainder of the PRI. The type of gain variation depends on the particular application. For example, when targets are expected to dominate the short-range returns, the law is $1/R^4$; if surface clutter is expected to dominate, the law is $1/R^3$ [22–24].

The portion of each PRI during the beam dwell over which STC is applied affects more than just a few μs (or meters of range), but typically tens of μs or km in range. The STC law must be established for a given application and applied without knowledge of target presence or power in any given beam dwell.

For illustration, Figure 22.2 shows a $1/R^4$ STC law designed to operate from 1.5 km to 10 km. The attenuation is held constant at 33 dB for the first 1.5 km and then decreases to 0 dB at 10 km. A $1/R^4$ STC law results in the receiver output power for targets being range independent (within the range STC is active). For a $1/R^3$ STC law, receiver output power for clutter becomes range independent.

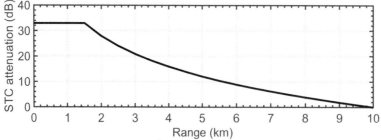

Figure 22.2 STC attenuation profile for $1/R^4$ STC law.

The RF amplifier of Figure 22.1 is a LNA used to establish the receiver noise figure. The receiver noise figure is also influenced by prior lossy components before and stages after the LNA. Lossy components prior to the LNA increase the noise figure by their insertion loss, dB for dB. This makes it important to minimize losses prior to the LNA.

The LNA also has fairly high gain so that, ideally, devices following the RF LNA do not significantly contribute to the overall receiver noise figure (see Section 4.6). For higher noise figure systems, an LNA gain of 20–25 dB is usually sufficient to ensure this. However, while the intent may be to allow negligible contributions from the stages following the LNA, this is not always realistic (especially when LNA technology achieves noise figures of 1 or 2 dB).

Early radars did not use RF amplifiers and, as a result, their noise figure was set by the mixer and components after the mixer. Skolnik notes that early receivers achieved noise figures of 12 to 15 dB, with 1960s vintage receivers having typical noise figures of 7 to 8 dB [22]. The noise figure of modern LNAs is typically in the range of 1 to 5 dB, and continues to fall as technology improves.

The RF LNA in older systems is usually tube-based technology, for example, TWT, backward wave amplifier (BWA), electrostatic amplifier (ESA), cyclotron wave electrostatic amplifier (CWESA), and electrostatic combined amplifier (ESCA). Development of devices based on electron beam cyclotron waves has been carried out largely in Russia by ISTOK and the Moscow State University. Early experiments were performed in the United States, but work was dropped in favor of solid-state amplifiers [25–27]. Tube amplifiers tend to be robust; some, such as the CWESA and ESCA, are essentially self-protected from overload [22, 25–28]. The self-protecting nature of cyclotron devices led to production by ISTOK[8] of the cyclotron protective device (CPD), functioning as receiver protection rather than LNA [29–32].

Radars are trending toward solid-state LNAs such as bipolar junction transistor (BJT), GaAs, and GaN. Solid-state LNAs require more care to protect from overload, but generally have a lower noise figure. The overall noise figures and gains of tube amplifiers versus solid state LNAs preceded by diode limiter overload protection circuitry are similar.

Solid-state LNAs must not only be protected from overload but from destruction by leakage from the transmitter. Whether this protector is considered part of the receiver or assigned to the duplexer, the use of a solid-state limiter is the current practice, and its loss must be included in calculating radar system noise figure.

The mixer is a nonlinear device used as a frequency heterodyne to translate the signal from the incoming RF to a desired IF using a reference frequency. In addition to the superheterodyne receiver, invention of the mixer is usually attributed to Armstrong as well [33]. The reference oscillator used for the RF to IF conversion in a receiver is referred to as the local oscillator (LO) [34]. This reference frequency is denoted as the LO frequency. This process is referred to as mixing.

---

[8] Former Soviet Union, Joint Stock Company, State Research & Production Corporation ISTOK, named after A. I. Shokin (www.istokmw.ru). ISTOK is the oldest Russian microwave organization.

The ideal mixer is a multiplier. Specifically, if the RF and LO signal are represented by

$$V_{RF}(t) = \cos(2\pi f_{RF}t)$$
$$V_{LO}(t) = \cos(2\pi f_{LO}t)$$
(22.1)

The mixing, or heterodyning, process is represented by

$$V_{IF}(t) = V_{RF}(t)V_{LO}(t) = \cos(2\pi f_{RF}t)\cos(2\pi f_{LO}t)$$
$$= \cos\left[2\pi(f_{RF}+f_{LO})t\right]/2 + \cos\left[2\pi(|f_{RF}-f_{LO}|)t\right]/2$$
(22.2)

Thus, an ideal mixer produces an output that contains the sum of the RF and LO frequencies and the difference of the RF and LO frequencies. The LO frequency can be either above or below the RF. The term high side mixer or high side injection is used when the $f_{LO} > f_{RF}$ and the term low side mixer or low side injection is used when the $f_{LO} < f_{RF}$ [35]. The absolute value in (22.2) underlines the fact that either $f_{LO} > f_{RF}$ or $f_{LO} < f_{RF}$ are valid RF to LO relationships.[9] For down conversion, the difference of the RF and LO frequencies is what we are after. The sum of the RF and LO frequencies is removed by the IF BPF in Figure 22.1.

The absolute value in (22.2) also means that there are two RF frequencies that result in the same IF. One is the desired RF used in the receiver; we term the other the image frequency. Any signal at the image frequency, when mixed with the target signal, results in another signal at the desired IF. Jammers can exploit this by placing interference at the image frequency. The image frequency is one of the unwanted signals the preselector must reject. It is not unusual for a preselector to provide more than 45 dB of image signal rejection.

Since there are two RF frequencies that will translate to the desired IF, an image reject filter (IRF) is sometimes necessary after the LNA to suppress unwanted receiver noise generated by the LNA in the image band. This has the benefit of preventing image noise from entering the passband, resulting in noise components from both the main and image responses adding, which would double the noise figure of the mixer. To minimize the effect of image noise, ~20 dB of image rejection is generally sufficient.

For frequency plans with a low IF, image filters can be difficult to implement because the image is too close to the passband. In these cases, an image reject (single sideband) mixer is often used instead of a filter. As the name implies, an image reject mixer uses phase canceling techniques to suppress the image in order to prevent the image sideband from converting to the IF passband. An image reject mixer can usually provide 25 to 35 dB of image suppression [36, p. 230]. The absence of an image reject filter after the LNA in the receiver in Figure 22.1 tacitly implies the use of an image reject mixer.

Figure 22.3 contains an illustration of an image frequency. In the figure, f_{RF} is 4,040 MHz and the desired f_{IF} of 40 MHz. The LO frequency we chose is $f_{LO} = f_{RF} - f_{IF} = 4,040 - 40 = 4,000$ MHz (low side). The image frequency is then $f_{IMAGE} = f_{RF} - 2f_{IF} = f_{LO} - f_{IF} = 3,960$ MHz. If we had used an LO where $f_{LO} > f_{RF}$ (high side), the LO frequency would be 4,080 MHz and the image would be 4,120 MHz.

---

[9] Strictly speaking, we do not need the absolute value since the cosine is an even function.

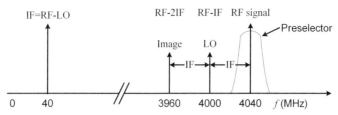

Figure 22.3 Low side mixer down conversion example.

Practical mixers are implemented by nonlinear devices such as diodes. As a result, in addition to signals at $|f_{RF} - f_{LO}|$ and $f_{RF} + f_{LO}$, signals at harmonics of f_{RF} and f_{LO} are generated. This can be expressed as

$$f = \pm m f_{RF} \pm n f_{LO} \qquad (22.3)$$

where m and n are integers [22, 33, 35]. These harmonics are unwanted mixer byproducts that are referred to as spurious signals, or simply spurs. The order of a spur is given by $|m| + |n|$. Except when an image reject mixer is necessary, double-balanced mixers are the most widely used in radar receivers. This is because a double-balanced mixer is designed to suppress the LO, the RF, and even ordered products at the output of the mixer. Double-balanced mixers also provide isolation between all mixer ports.[10]

The mixer spurs for the above example, up to $|m| + |n| = $ 10th order, are plotted in Figure 22.4. Only the spurs from 0 to 450 MHz and above −120 dBc are visible (dBc is referenced to the RF at the mixer input in Figure). The spurs shown in Figure 22.4 are multiples of the 40 MHz IF. The other spurs are either much farther away in frequency or are too low in power to be of concern. This is because as spur order increases, the level of the spur decreases. Predicting spur levels can be fairly complicated. Mixer spur levels are usually measured directly, or more often, manufacturer-supplied mixer spur tables provide the spurs levels for a particular mixer [37].

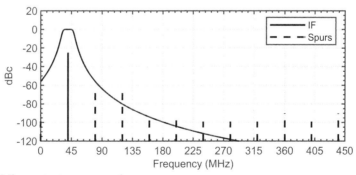

Figure 22.4 Mixer output spur example.

---

[10] Double-balanced mixers can be match-sensitive.

For this example, there are only a few spurs above −100 dBc within the 450 MHz shown. The BPF following the mixer in Figure 22.1 is used to reduce these spurs further. For illustration, the frequency response of a fourth-order Butterworth (BW) filter (named for British physicist Stephen Butterworth [38]) with a 16-MHz passband centered on an IF of 40 MHz is also shown in Figure 22.4. The spurs at 80 MHz and 120 MHz would be further suppressed by > 56 dB by the BPF shown. Since the closest spur to f_{IF} is 40 MHz away, the design of the post mixer BPF is easier than if the spur were, say, 5 MHz away, which would require a higher-order filter for the same amount of spurious rejection because the spur is closer to the filter passband.

Ultimately, spurs need to be suppressed to the point that no spurs would exceed the detection threshold, which is generally of greater concern in a radar that uses Doppler processing.

Considering the large number of spurs generated according to (22.3), there are sometimes spurs within the passband of the BPF following the mixer that are unavoidable. Beyond about ninth-order though, mixer spurs are usually low enough to be ignored, but not always. Selection of the f_{RF}, f_{LO}, and f_{IF} combination is an important consideration to insure that no low-order spurs fall within the IF passband.

Ideally, the IF BPF is a filter that is constant gain and linear phase (flat group delay) in the passband to minimize distortion. However, constant passband gain and linear phase are conflicting requirements in an analog filter (see Table 22.2). For example, a Butterworth filter is designed to have a maximally flat passband, but exhibits a nonlinear phase response. In contrast, a Bessel filter is characterized by a nearly linear phase response (maximally flat group delay), but is not constant amplitude. As such, the selected filter must necessarily be a compromise that partially meets both requirements. Alternatively, a filter with the desired gain response could be followed by an all-pass delay equalizer.

The order of the IF BPF is typically low (fourth or fifth) to avoid excessive ringing and overshoot. Lumped-element LC filter technology, which is suitable for lower frequencies, is typically used for implementation.

The typical IF for a single-conversion superheterodyne receiver, or the final IF for a multiconversion superheterodyne receiver is 30 to 100 MHz. In addition to the spurious considerations described earlier, a particular IF is selected because of the performance of available components. Generally speaking, as the IF is decreased, the cost of components goes down and performance improves. For example, a low IF simplifies the design of narrowband filters.

Table 22.2
General Filter Characteristics [17, 39, 40–42]

Approximation	Property	Pros	Cons	Notes
Butterworth	Maximally flat passband	Simple design, lowest loss	Slow roll-off rate, nonlinear phase	Monotonic attenuation, moderate phase distortion
Chebyshev Type I	Passband ripple	Steeper stopband slope with respect to BW and Cheby II[a]	Passband ripple, nonlinear phase	Poor phase response (group delay) with respect to BW
Chebyshev Type II	Stopband ripple	Flat passband, steeper stopband slope with respect to BW[a]	Stopband ripple, slower roll-off with respect to Cheby I, nonlinear phase	Poor phase response (group delay) with respect to BW, more linear than Cheby I
Elliptic	Passband/ stopband ripple	Steepest passband to stopband transition[a] (proportional to stopband ripple)	Passband and stopband ripple, nonlinear phase	Generally lowest order for given attenuation, poorest phase response
Bessel	Maximally flat group (time) delay[11]	Nearly linear phase, no signal distortion	Slowest roll-off rate (attenuation slope)	Monotonic attenuation, minimal overshoot/ringing, poor stopband attenuation

Source: [17, 39, 40–42]
[a] For the same filter order.

The IF amplifier is used to make up for the losses in the previous devices and to amplify the signal to desired levels for subsequent components. The detector is typically the last stage considered to be part of the receiver. When coherent processing is not required, linear- or square-law amplitude detectors are typically used (see Chapter 6) and the BPF is replaced by a pulse-matched filter. For coherent processing, a synchronous (or quadrature) detector[12] is used to preserve phase information in the signal, with matched filtering occurring in the signal processor. While the examples in this chapter presume matched filtering post ADC, some older radar systems implement the matched filter at IF and then I/Q detect for signal processing.

If we consider a frequency agile radar, where the RF can vary rapidly over a fairly large range, we need a wideband preselector and a higher IF. While increasing the IF simplifies image rejection by placing the image frequency further in the stopband of the preselector, the complexity of narrowband filter design is increased. Also, frequency agility complicates the issue of avoiding spurious mixer products within the IF passband greatly.

The approach typically used to alleviate these issues is to add more down conversion stages. The higher IF in the first frequency conversion stage has the benefit of good image rejection via the preselector [35]. The lower IF in the second stage of conversion enjoys the benefit of easier implementation of narrowband filters [35]. An additional key point to adopting multiple down conversion stages is that additional stages simplify the problem

---

[11] Group delay is a measure of the rate signals of different frequencies propagate through a filter.
[12] Synchronous detection, which could be considered somewhat of a misnomer, should not be confused with target detection, which does not actually occur in a quadrature baseband converter. Target detection generally occurs at the end of the signal processing chain.

of mixer spurs. Two down conversion stages are typically sufficient to ensure that passband spurs are at least eighth or ninth order (which have very low power levels).

22.3 DUAL-CONVERSION SUPERHETERODYNE RECEIVER

A block diagram of a dual-conversion superheterodyne receiver is shown in Figure 22.5. From the preselector to the IF attenuator output, the topology is identical to that of the single-conversion superheterodyne receiver of Figure 22.1. The principal differences are a wideband preselector, an agile first LO, and the inclusion of a second down conversion stage consisting of a mixer, filter, and amplifier cascade. Additionally, we have chosen to use an analog I/Q demodulator[13] to preserve phase information.[14]

The wideband preselector is used to limit frequencies to the agile range of interest. As with the single-conversion superheterodyne receiver, the preselector is a low-loss device to minimize its impact on the system noise figure. The preselector passband must be wide enough to accommodate the desired frequency agility band. A typical RF agility range is 100 to 500 MHz.

As with the single down conversion receiver, the attenuators are used to extend receiver dynamic range. Likewise, the LNA and prior lossy elements establish the overall noise figure of the radar, but must now be broadband. The IF amplifiers make up for losses in previous devices and amplify the signal to desired levels for subsequent components.

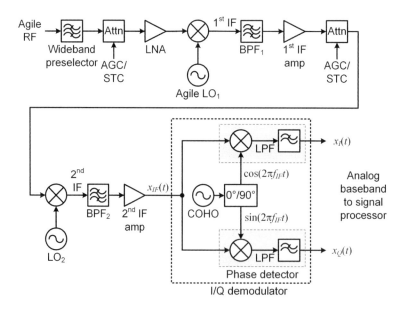

Figure 22.5 Dual-conversion superheterodyne receiver with synchronous detection.

---

[13] The I/Q converter doesn't demodulate the signal in the AM or FM radio sense. It merely shifts the signal spectrum to center it at zero frequency (DC).
[14] When phase coherency is required, the moniker STALO is sometimes used to describe the first LO.

To simplify subsequent IF filtering, we chose to use an agile first LO for the first down conversion stage. This agile LO_1 tracks with the RF, resulting in fixed first and second IFs. As a result, we only need one IF filter per down conversion stage. The first IF is selected to optimize suppression of the image frequency and other spurious signals generated by the first mixer. Because of the wide bandwidth, higher IFs generally simplify rejection of the image and spurs. Additionally, the first IF must be high enough to accommodate the RF agility bandwidth. A general rule of thumb is for the first IF to be at least 1.5 to 4 times the agility bandwidth to simplify component design.

The second IF is now analogous to the IF of a single down conversion superheterodyne receiver. The second IF is generally chosen to be low (<100 MHz) to simplify design of the narrowband IF filtering and other components. Spurious considerations are simplified because the bandwidth at the second IF is relatively narrow. Using a high IF followed by a low IF in our receiver design, we get the benefits of both.

Both IF BPFs in Figure 22.5 are used to reject mixer spurs. The bandwidth of the first IF filter is generally on the order of the RF channel spacing (e.g., 10 or 20 MHz) to reduce spurs while simplifying filter design. The bandwidth of the second BPF is usually somewhat narrower than the first, on the order of two to three times the modulation bandwidth, to ensure the modulation is undistorted (see Chapter 8). For example, if the radar uses an LFM waveform with a bandwidth of 2 MHz, the second IF BPF should have a bandwidth of about 6 MHz. Note that the choice of RFs, IFs, and LO frequencies is sometimes referred to as a frequency plan.

At this point, we should emphasize that we are generally not interested in the second IF in and of itself, but in the amplitude and phase information it carries via modulation. It is the modulation that contains the information we want, such as waveform, delay to target, and Doppler information.

This brings us to the I/Q demodulator[15] shown in Figure 22.5, which translates a real bandlimited signal, $x_{IF}(t)$, to baseband. Quadrature demodulation is usually performed after down conversion to a low IF, which for this example is the second IF.

The demodulator topology depicted in Figure 22.5 is the classical approach of splitting $x_{IF}(t)$ and then using two matched phase detectors using reference frequencies which are 90 degrees out of phase (in phase quadrature). Each detector is implemented as a mixer, which performs a down conversion of $x_{IF}(t)$ to baseband, followed by an LPF to remove unwanted harmonics. The difference here is that the desired IF is 0 Hz.

Recall that a real bandlimited IF signal can be represented as

$$\begin{aligned}
x_{IF}(t) &= A(t)\cos\left[2\pi f_{IF}t + \phi(t)\right] \\
&= A(t)\cos\left[\phi(t)\right]\cos(2\pi f_{IF}t) \\
&\quad -A(t)\sin\left[\phi(t)\right]\sin(2\pi f_{IF}t) \\
&= x_I(t)\cos(2\pi f_{IF}t) - x_Q(t)\sin(2\pi f_{IF}t)
\end{aligned} \quad (22.4)$$

---

[15] Terms such as I and Q detector, synchronous detector, quadrature detector, and coherent demodulator are used synonymously with quadrature demodulator.

where $A(t)$ and $\phi(t)$ represent the amplitude and phase modulation, respectively. The signals $x_I(t) = A(t)\cos[\phi(t)]$ and $x_Q(t) = A(t)\sin[\phi(t)]$ are the in-phase and quadrature baseband signals of interest. The output of the in-phase channel mixer is

$$x_I(t) = A(t)\cos\left[2\pi f_{IF}t + \phi(t)\right]\cos(2\pi f_{IF}t)$$
$$= \frac{1}{2}A(t)\cos\left[\phi(t)\right] + \frac{1}{2}A(t)\cos\left[4\pi f_{IF}t + \phi(t)\right] \tag{22.5}$$

After lowpass filtering, the I-channel phase detector output becomes

$$x_I(t) = A(t)\cos\left[\phi(t)\right] \tag{22.6}$$

Similarly, the Q-channel phase detector output becomes

$$x_Q(t) = A(t)\sin\left[\phi(t)\right] \tag{22.7}$$

Equations (22.6) and (22.7) contain all of the modulation information of (22.4) without the IF.

The quadrature detector uses an extremely stable reference oscillator for phase detection. This reference oscillator is usually of equal frequency to the IF and always phase coherent. The term coined for this oscillator is the coherent oscillator (COHO) [34], because it is related to the IF.

A problem with analog quadrature demodulation is amplitude and phase misalignments in the circuitry, which can cause imbalances between the I and Q channels. This imbalance can generate unwanted image and DC signals, which have a negative impact on subsequent signal processing. A means of avoiding such problems is to perform the quadrature detection with digital hardware [43]. The resulting receiver is termed a digital receiver and will be discussed in Section 22.8.

22.4 RECEIVER NOISE

As discussed in Chapter 2, the two main contributors to noise in radars are the environment (via the antenna) and thermal noise generated by the electronic components of the receiver [44]. As discussed in Chapter 4, the noise level present in a radar can be quantified in terms of equivalent/effective noise temperature or noise figure.

When considering radars with very low noise figures, where environmental noise is a major noise contributor, an effective noise temperature approach is favored [22]. For radars with larger noise figures (greater than about 7 dB), where receiver noise normally dominates environment noise, a noise figure approach is generally preferred [22]. Radars in the VHF band have such high environmental noise that the noise temperature characterization is appropriate even when receiver noise figures are not very low.

In this chapter, we take a measurement point of view, considering the receiver, and perhaps the signal processor in cascade, but not the entire radar. The importance of this is that analysis and measurement of the receiver or receiver and signal processor as a subsystem can be carried out under the assumption that the input is terminated in a resistor at 290 K. For this reason, we will use a noise figure approach as opposed to noise temperature approach.

Likewise, we will consider thermal noise generated in the receiver, not environment noise. When making receiver noise measurements, noise figure is used, and the reference temperature is $T_0 = 290$ K by definition [34]. Also, in a measurement setting, the receiver is usually disconnected from the antenna, with test equipment used to inject test signals into the receiver and to make measurements of various parameters such as gain, bandwidth, dynamic range, and noise figure.

While much of radar theory is concerned with ratios (e.g., SNR, CNR, SCR, SIR), when considering receivers, knowing absolute levels is a key consideration. One important level in a receiver is the noise level, often referred to as the noise floor, since receiver noise generally establishes the noise level competing with weak signals. The notable exception to this is for radars operating at frequencies below about 300 MHz. This is because of the steep increase in cosmic and other environmental noise below about 300 MHz (see Figure 2.5) resulting in receiver noise no longer dominating the system noise temperature and equivalently the system noise figure [45, 46].

Given a reference temperature of $T_0 = 290$ K for our measurements, the power spectral density, and thus noise power or noise floor in a receiver, is established at the output of the LNA. The noise floor at the output of each receiver stage can be determined using

$$P_n = kT_0 F B_n G \text{ W} \tag{22.8}$$

or in logarithmic form

$$P_n = -174 + G + B_n + F \text{ dBm} \tag{22.9}$$

where k is Boltzmann's constant, 1.38×10^{-23} W/(Hz K) and $T_0 = 290$ K. F, B_n, and G are the noise figure, bandwidth, and gain, respectively, up to the output of each receiver stage (determined by cascade analysis discussed in Section 22.7). In (22.8) and (22.9) and elsewhere, we use the same symbols to represent dB values and non-dB values, which should hopefully be clear looking at units and context (e.g., non-dB form uses multiplication and division, dB form uses addition and subtraction).

In using the forms of (22.8) and (22.9), we make the tacit assumption that bandwidths remain the same or decrease as one progresses through the various components of the receiver [47]. While not always true, this is a common assumption used for cascade analysis, where the overall RF to IF bandwidth is usually set by the last filter (or tuned amplifier) in the chain [48, p. 15]. We also make the implied assumption that there is sufficient gain ahead of any ADC to minimize the impact of an ADC's relatively high effective noise figure (e.g., 30 dB), which requires a lot of receiver gain prior to the ADC to make the ADC noise a small fraction of total noise (see Section 22.8.5.4).

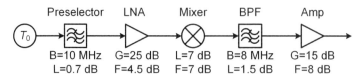

Figure 22.6 Noise floor example.

For example, consider the RF front end shown in Figure 22.6. The preselector we have chosen is a fourth-order Chebyshev Type I filter with 0.1-dB ripple and a 10-MHz bandwidth.[16] It has a loss of 0.7 dB, which means its gain is –0.7 dB. The second filter is a fifth-order Bessel filter with an 8-MHz bandwidth[17] and a gain of –1.5 dB. The amplifiers are assumed to have bandwidths equal to the preceding filter. Note also that we are driving the circuit under test with a calibrated noise source at standard temperature, T_0.

From Chapter 4, if we assume the filters are lossy passive devices, their noise figures equal their losses ($F_n = L$) [49–52]. We would like to calculate the noise power generated by the RF front end as well as the noise power after each component. Note that carrier frequency is irrelevant for this analysis.

At the cascade input, the noise is assumed white with a power spectral density expressed, in various units, as

$$kT_0 = 1.38 \cdot 10^{-23} \text{ W}/(\text{Hz} \times \text{K}) \times 290 \text{ °K} = 4 \cdot 10^{-21} \text{ W/Hz}$$
$$= -204 \text{ dBW/Hz} = -114 \text{ dBm/MHz} = -144 \text{ dBm/kHz} \quad (22.10)$$
$$= -174 \text{ dBm/Hz}$$

Until we impose a bandwidth, the noise power is theoretically infinite. Using the 10-MHz bandwidth from Figure 22.6 in (22.9), the noise power out of the preselector is

$$P_n = -174 + G + B_n + F$$
$$= -174 + (-0.7) + 10\log(10 \cdot 10^6) + 0.7 \quad (22.11)$$
$$= -104.0 \text{ dBm}$$

Looking at (22.11), we note that the noise power at the output of a lossy component is equal to the noise power at its input (assuming the input and output are both terminated in matched loads and everything is in thermal equilibrium). This is because for a lossy component the G and F in (22.9) and (22.11) have equal values but opposite signs and therefore cancel (see Section 4.9).

The noise power out of the RF amplifier is

---

[16] Based on Chebyshev polynomials, which are named for Russian mathematician Pafnuty Lvovich Chebyshev (Пафну́тий Льво́вич Чебышёв). Also transliterated as Chebychev, Chebysheff, Tchebychev.

[17] For Butterworth and Bessel filters above second order, the difference between 3-dB bandwidth and equivalent noise bandwidth is negligible.

$$P_n = -174 + G + B_n + F$$
$$= -174 + (25 - 0.7) + 10\log(10 \cdot 10^6) + (0.7 + 4.5) \quad (22.12)$$
$$= -74.5 \text{ dBm}$$

Considering (22.12), we can see that the preselector loss and noise figure doesn't affect the noise power at the LNA output (assuming the preselector is at the input temperature T_0), because for the lossy preselector, the input and output noise powers are the same.

This process is continued stage by stage. Note that after the amplifier, we need to use the Friis formula for cascade noise figure (see Section 22.7 and Chapter 4) [53]. The RF to IF bandwidth is 8 MHz. The gain, noise figure, and noise power out of the entire chain are 30.8 dB, 5.4 dB, and −68.7 dBm, respectively. The remaining details are left as an exercise.

We have now reflected on what is often considered the low end for signals in a receiver, namely, the noise floor. Signals below this level are said to be buried in the noise and cannot be discerned (without subsequent signal processing).[18] We now will consider the top end for signal level.

We have thus far assumed the amplifiers amplify signals in a linear fashion. In practice, this is not the case because above a certain input power level, the amplifier will saturate. This is because a finite DC voltage is used to power the amplifiers, which limits how large a signal can be linearly amplified. This leads to two important radar receiver concepts: the 1-dB compression point, and dynamic range. These terms apply equally to components and receivers, with some minor differences, as will be discussed.

22.5 THE 1-DB GAIN COMPRESSION POINT

Ideally, analog components amplify signals in a perfectly linear fashion. However, if overdriven by large input signals, the amplifier gain will become nonlinear. This, in turn, generates unwanted spurious signals. The compression point of a device, which is defined as the level of the *output* signal at which the gain of a device is reduced by a specific amount, is a useful index of the amount of distortion that can be accepted [34]. A specific index used in amplifier analyses is termed the 1-dB compression point. Consistent with the definition of compression point, it is the output signal level where the gain is reduced by 1 dB (from its nominal, constant value) [22].

The definition of 1-dB compression point of the previous paragraph is the formal definition used for components in general. For receivers, the standard definition is that it is the *input* level at which the gain decreases by 1 dB from its (nominally) constant value.

Figure 22.7 contains a plot of output power versus input power for a notional device. We will denote the 1-dB compression point as *P1*. To avoid ambiguity, we will use prefixes, with O designating output and I designating input, for example, *OP1* stands for

---

[18] Technically a signal needs to be about 5 to 8 dB below the noise floor for it not to make a discernable bump in an average detected output.

1-dB compression point at the output. The 1-dB compression point at the input is related to the 1-dB compression point at the output by [54, p. 541]

$$IP1 = OP1 - (G-1) \text{ dBm} \tag{22.13}$$

where G is the nominal device gain, in dB.

The general procedure used to measure the 1-dB compression point is to inject a signal and increase its amplitude until the gain is decreased by 1 dB [55]. This test is sometimes referred to as a transfer test because a transfer curve is usually generated. The transfer curve associated with a typical 1-dB compression point measurement is also illustrated in Figure 22.7.

For the example presented in Figure 22.7, we consider an IF amplifier with a gain, G, of 30 dB. We sweep the input test signal power from −40 dBm to 0 dBm, while measuring the output power. The 1-dB compression point at the output, $OP1$, occurred at 10 dBm. Using (22.13), the 1-dB compression point at the input is −19 dBm. Thus, we say that this amplifier has an output, 1-dB compression point of 10 dBm and an input, 1-dB compression point of −19 dBm.

The amplifier output saturated at an output power level, P_{SAT}, of 14 dBm. As a general rule of thumb, compression usually starts about 5 to 10 dB below the output 1-dB compression point. Similarly, saturation typically occurs around 3 to 6 dB higher than the output 1-dB compression point.

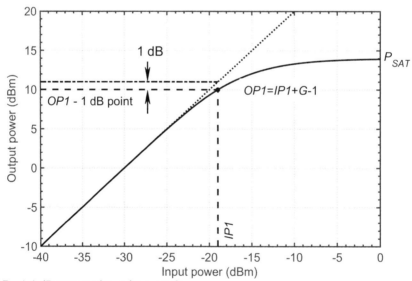

Figure 22.7 A 1-dB compression point example.

22.6 DYNAMIC RANGE

The dynamic range of a receiver, depicted by Figure 22.8, is commonly defined as the ratio of the maximum input signal that can be handled to the minimum signal input capable of being detected [22]. The maximum level is usually taken to be the 1-dB compression point because that is where we normally assume the device is departing from linear operation. The minimum level is often denoted MDS. There are, however, a number of variations of what is meant by MDS, as we shall see.

Dynamic range can be expressed as

$$DR = \text{maximum input power} - \text{minimum input power}$$
$$= IP1 - MDS \qquad (22.14)$$
$$= OP1 - (G-1) - MDS \text{ dBm}$$

There is no shortage of definitions concerning both dynamic range and the minimum levels used for determining the dynamic range of a receiver [56–59]. We also note that various definitions and terminology can potentially clash. To help explain and hopefully avoid some of the confusion about MDS, we will present MDS definitions applicable to radar receivers [36, 60–64]. We will, in short order, have three "standard" definitions that are commonly used in radar and radar receivers.

For this chapter, though, we define dynamic range in terms of the receiver sensitivity, which is taken as the minimum input signal required to produce a specified output signal having a specified signal-to-noise ratio [65]. For measurement purposes, we typically consider the output signal detectable when it is at or above the noise level, or SNR = 0 dB (S = N). Defining dynamic range in terms of receiver sensitivity allows us to measure receiver dynamic range without regard to noise sources from the environment. Thus, if we consider receiver noise without its preceding subsystems, the input noise will come from a resistive termination at 290 K by definition.

Likewise, receiver dynamic range can be characterized without considering the effects of signal processing that provides detectability of signals below noise level. However we often do measure the dynamic range of both the receiver and the receiver and signal processor combined.

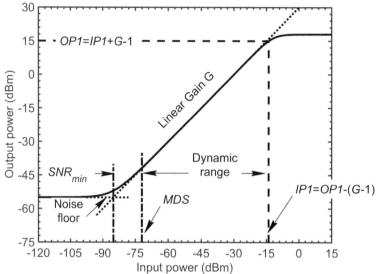

Figure 22.8 Dynamic range.

22.6.1 Sensitivity

As mentioned already, receiver sensitivity is defined as the minimum input signal required to produce an output signal with a specified SNR. Sensitivity is only concerned with internally generated receiver noise [60, p. 76]. This is because external noise from the antenna is not something we can control when designing a receiver. While a receiver's sensitivity can be expressed in terms of power or voltage, we usually use power in dBm.

A receiver's sensitivity is largely determined by the RF front-end components since the noise floor in a receiver is the limiting factor on receiver sensitivity. Also, this definition for sensitivity relates to power levels, not detection performance. It does not include specifications for probability of detection (P_d) or probability of false alarm (P_{fa}). In receiver vernacular, sensitivity is sometimes used synonymously with minimum detectable signal and minimum discernable signal, adding to the multiple MDS definition confusion [22, 35, 59].

The preselector input is a common input reference point used when defining or measuring sensitivity. When considering just the receiver, the output measurement point typically used is the IF amplifier output just prior to detection. When considering the receiver input to the signal processor output, the term "system sensitivity" is used. For this discussion, we confine ourselves to just the receiver and choose to use the input to the preselector and IF amplifier output prior to detection as our analysis, or measurement, points.

Receiver sensitivity can be expressed as[19]

---

[19] In (21.15), kT_0FB_n is an effective receiver input noise power rather than an actual (measurable) input noise power because it depends on the receiver noise figure, which can include RF losses. Nonetheless, (21.15) does provide the correct value for minimum usable signal power.

$$P_{S\min} = kT_0FB_n SNR_{\min} \text{ W} \tag{22.15}$$

or in dBm

$$P_{S\min} = -174 + 10\log(F) + 10\log(B_n) + 10\log(SNR_{\min}) \text{ dBm} \tag{22.16}$$

where P_{Smin} is the minimum signal level at the receiver input (preselector input) and SNR_{min} is a specified SNR (at the IF amplifier output). Using an SNR_{min} of either 0 dB or 3 dB is customary in the context of receiver sensitivity [22, 54, 59, 66]. Equations (22.15) and (22.16) are similar to (22.8) and (22.9) where gain is replaced by SNR, since the relevant quantity is SNR instead of absolute levels. Also, the terms in (22.15) and (22.16) are the overall bandwidth and noise figure of the receiver (or receiver/signal processor).

If we consider the RF front end shown in Figure 22.6, and stipulate a minimum acceptable SNR (SNR_{min}) of 3 dB, the sensitivity becomes

$$\begin{aligned}P_{S\min} &= -174 + 10\log(F) + 10\log(B_n) + 10\log(SNR_{\min}) \\ &= -174 + 5.4 + 10\log(8\cdot 10^6) + 3 = -96.6 \text{ dBm}\end{aligned} \tag{22.17}$$

If we compare P_{Smin} and P_n, we see that they differ. Specifically, they differ by $G-SNR_{min}$ rather than SNR_{min} because for (22.15) the SNR is specified at the output, but P_{Smin} is specified at the receiver input. For (22.8) P_n is specified at the output of the receiver, not the input.

Sensitivity, P_{Smin}, can be measured by injecting a calibrated target signal with a constant power level [i.e., Swerling case 0 (SW0)], at the receiver input and determining the SNR at the output of the receiver.[20] As discussed earlier, we typically use an $SNR_{min} = 1$, or 0 dB, for this measurement. This results in

$$P_{S\min} = kT_0FB_n \text{ W} \tag{22.18}$$

at the output of the radar's receiver (signal power = noise power) [59, 67]. This can be related to P_n using

$$P_n = GP_{S\min} = kT_0FB_nG \text{ W} \tag{22.19}$$

which is the same as (22.8). Skolnik refers to (22.19) as the *minimum signal of interest* [56].[21] Stephen Erst presents a test methodology based on the relationship between S/N and (S + N)/N rms voltage ratios the authors have put to good use in the field that can be used to determine when SNR = 0 dB at the output of the receiver (or the entire receiver signal processor chain) [60, pp. 76–78].

Measuring a receiver's sensitivity is important because it is closely tied to detection performance and can help determine or verify gain and loss terms in the radar range

---

[20] The SNR is often determined at the output of the signal processor instead.
[21] In the third edition of Skolnik's handbook, SNR = 0 is no longer stipulated with the term minimum signal of interest. Skolnik points out that digital signal processing techniques allow detection well below the receiver noise floor depending on the processing performed.

equation, overall noise figure, and overall bandwidth. Measuring receiver noise directly also results in a consistent quantity verifiable using only test equipment and a SW0 target signal injected directly into the RF front end.

22.6.1.1 Tangential Sensitivity

The criterion of SNR = 0 dB at the output of a receiver is sometimes referred to erroneously as tangential sensitivity (TSS). Tangential sensitivity, defined in [24, p. 456], corresponds (approximately) to SNR = 8 dB. TSS gets its name from the use of what is called a tangential signal to estimate sensitivity. Specifically, noise and signal plus noise are viewed on an oscilloscope, and the signal power is adjusted until the bottom of the signal-plus-noise trace aligns with, or is tangent to, the top of the noise-only trace (see Exercise 7).

TSS is generally accurate to within ±1 dB, and is influenced by the RF bandwidth, the video bandwidth, the noise figure, and the detector characteristic [58]. There is no theoretical value for tangential sensitivity, since it depends on a subjective matching of the peak level of noise with the minimum level of signal plus noise outputs on an A-scope display, neither level having a measureable value.

22.6.2 Minimum Detectable and Minimum Discernable Signal

Minimum detectable signal (MDS) for radar detection applications is the minimum signal power necessary to give reliable detection performance in the presence of white Gaussian noise; that is, the minimum signal level needed to give a specified P_d with a specified P_{fa} [34]. Specifying P_d and P_{fa} is necessary for this definition of MDS because minimum detectable signal is a statistical quantity.

Minimum detectable signal in the radar range equation can be expressed as

$$S_{min} = kT_0 F B_n L \cdot SNR_{min} \text{ W} \qquad (22.20)$$

where B_n is the equivalent noise bandwidth (see Chapter 2). Equation (22.20) is of the same form as (22.15), with the addition of a loss term L. In this context, MDS encompasses the entire receiver and signal processor and is related to the measureable quantity of (22.15). This definition of MDS is also concerned with both internally generated noise and externally generated noise.

Minimum discernable signal (MDS) is defined as "The minimum detectable signal for a system using an operator and display or aural device for detection" [34]. Including sensitivity, which is sometimes referred to as MDS, we now have three standard definitions for the acronym MDS.

22.6.3 Intermodulation Distortion

Dynamic range can also be defined in terms of spurious signals, which is called the spurious free dynamic range. While we use mixers as intentionally nonlinear devices to generate harmonics of the input signals, other active devices, such as amplifiers, are only approximately linear, and also act like mixers (just not very good ones), generating unwanted signals that are harmonics of the input. This phenomenon of generating spurious signals is termed intermodulation distortion (IMD). The term two tone is sometimes used when discussing intermodulation products because two tones are used to measure and characterize intermodulation distortion.

Like a mixer, if we let the desired frequencies at the input to an amplifier be f_1 and f_2, the spurious signals occur at frequencies of

$$\pm m f_1 \pm n f_2 \tag{22.21}$$

where m and n are integers. The harmonics that usually cause difficulties in receivers are the second- and third-order harmonics (recall the order of a harmonic $|m| + |n|$). This is because the second- and third-order harmonics tend to be the largest spurs and closest in frequency to f_1 and f_2, respectively.

Because they are apt to be the biggest spurs, intermodulation distortion performance is usually specified in terms of the second- and third-order intercept points. For superheterodyne receivers, the third-order intercept is most important because they are closest in frequency to the desired tones, making them problematic or impossible to filter out. The second-order intercept is more important than the third-order intercept point in homodyne receivers [55]. We will compute both in this chapter.

For example, let $f_1 = 30$ MHz and $f_2 = 31$ MHz. The third-order intermodulation frequencies are $2f_2 - f_1 = 32$ MHz and $2f_1 - f_2 = 29$ MHz. The second-order intermodulation frequencies occur farther away at 1 MHz and 61 MHz. This demonstrates a major problem. The second-order products, while potentially large, can possibly be filtered out. The third-order intermodulation products in this example cannot be filtered out because of their close proximity to the desired tones generating them.

Figure 22.9 depicts the concept of intercept points, which can be expressed in terms of both input (labels starting with *I*) and output levels (labels starting with *O*). In Figure 22.9, the first-, second-, and third-order input intercept points are denoted by *IP1*, *IIP2*, and *IIP3*, respectively. Similarly, the first-, second-, and third-order output intercept points are denoted by *OP1*, *OIP2*, and *OIP3*, respectively. The amplifier saturation level is labeled P_{sat}.

The second-order intercept point, *IIP2*, corresponds to a projected power level at which the second-order intermodulation product crosses a perfectly linear response. The second-order intermodulation product gain has a slope twice that of the linear gain of the desired input, when plotted in decibels.

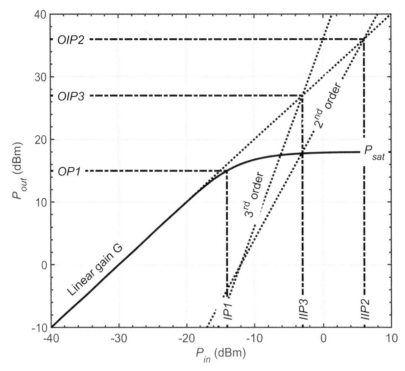

Figure 22.9 Diagram of second- and third-order intercept points.

Similarly, the third-order intercept point, *IIP3,* relates to a projected power level at which the third-order intermodulation product would intersect a projection of the linear gain. The third-order intermodulation product gain has a slope three times that of the desired input, when plotted in decibels. As a general rule of thumb, the third-order intercept point is 10 to 15 dB higher than the 1-dB compression point.

For example, increasing the desired (first-order) signal by 3 dB increases the second-order signal by 6 dB, and the 3rd-order signal by 9 dB. This results in the second- and third-order intermodulation distortion products rapidly becoming nearly the same amplitude as the desired input. Power exceeding the third-order intercept point ensures intermodulation distortion, though 3rd-order intermodulation can occur to some degree at any power level in a continuously differentiable nonlinearity.

It should be emphasized that at the saturation point, both the second- and third-order intercept points are much higher than the 1-dB compression point. Because of this, neither the second- or third-order intercept points can be measured directly. Instead, they are projected from measurements made using lower signal levels.

The spurious free dynamic range of a receiver is defined as the range over which a receiver does not compress an input signal and no spurious signal is above the receiver's noise floor [68]. The spurious free dynamic range is then defined in terms of the third-order input intercept point as [68, 69]

$$SFDR = \frac{2}{3}(IIP3 - \text{noise floor}) = \frac{2}{3}\left[IIP3 - (-174 + B_n + F)\right]$$
$$= \frac{2}{3}(IIP3 + 174 - B_n - F) \qquad (22.22)$$

Note that the SFDR of (22.22) is defined in terms of the minimum required signal being equal to the noise level of the receiver rather than minimum discernable or minimum detectable signals.

For example, consider a receiver with a third-order intercept of $IIP3 = -14$ dBm, an RF to IF bandwidth of 4 MHz, and a noise figure of 5 dB. Using (22.22), the spurious free dynamic range becomes

$$SFDR = \frac{2}{3}(IIP3 + 174 - B_n - F) = \frac{2}{3}(-14 + 174 - 66 - 5)$$
$$= 59.3 \text{ dB} \qquad (22.23)$$

22.6.4 Required Dynamic Range

Now that we can quantify dynamic range, it would be useful to estimate what the dynamic range needs to be. Radar echoes can cover a wide range (typically 90 to 100 dB) due to variations in target RCS, clutter RCS (see Chapter 17), and the $1/R^4$ range dependency. At a minimum we can predict the variation due to RCS and that due to $1/R^4$.

Recall from Chapter 2 that the signal power at the input to the receiver is

$$P_{Rx} = \frac{P_T G_T G_R \lambda^2 \sigma}{(4\pi)^3 R^4} \text{ W} \qquad (22.24)$$

where P_T is the peak transmit power, G_T denotes directivity of the transmit antenna, G_R denotes the directivity of the receive antenna, λ denotes the radar wavelength, R denotes the slant range from the radar to the target, and σ denotes the average target RCS or clutter RCS in a range resolution cell at range R.

Let us specify a minimum and maximum target or clutter RCS of σ_{min} and σ_{max}, respectively. Likewise, let us specify a minimum and maximum range of R_{min} and R_{max}, respectively. The maximum power present at the receiver is then

$$P_{Rx\,max} = \frac{P_T G_T G_R \lambda^2 \sigma_{max}}{(4\pi)^3 R_{min}^4} \text{ W} \qquad (22.25)$$

and the minimum power is

$$P_{Rx\,min} = \frac{P_T G_T G_R \lambda^2 \sigma_{min}}{(4\pi)^3 R_{max}^4} \text{ W} \qquad (22.26)$$

The minimum dynamic range necessary to accommodate for σ_{min} and σ_{max} and R_{min} and R_{max} is

$$DR_{min} = \frac{P_{Rx\,max}}{P_{Rx\,min}} = \frac{P_T G_T G_R \lambda^2 \sigma_{max}/(4\pi)^3 R_{min}^4}{P_T G_T G_R \lambda^2 \sigma_{min}/(4\pi)^3 R_{max}^4}$$
$$= \frac{\sigma_{max}/R_{min}^4}{\sigma_{min}/R_{max}^4} = \frac{\sigma_{max} \cdot R_{max}^4}{\sigma_{min} \cdot R_{min}^4} \qquad (22.27)$$

In logarithmic form

$$DR_{min} = \sigma_{max\,dB} - \sigma_{min\,dB} + 40\log(R_{max}) - 40\log(R_{min}) \text{ dB} \qquad (22.28)$$

Equation (22.28) gives a preliminary lower bound, since it does not take into account target fluctuation. Fluctuations in RCS can potentially span another 30 dB [22, p. 737].

For example, consider a radar transmitting a 10-μs pulse with a maximum instrumented range of 100 km. This radar is expected to accommodate RCSs between $\sigma_{min} = 0.001$ m² (−30 dBsm) and $\sigma_{max} = 1{,}000$ m² (30 dBsm). We determine R_{min} by means of the pulsewidth as c·τ/2 = 1.5 km. Using (22.28) results in

$$\begin{aligned} DR &= \sigma_{max} - \sigma_{min} + 40\log(R_{max}) - 40\log(R_{min}) \\ &= 30 - (-30) + 40\log(100 \cdot 10^3) - 40\log(1.5 \cdot 10^3) \\ &= 30 + 30 + 200 - 127 \\ &= 133 \text{ dB} \end{aligned} \qquad (22.29)$$

For this example, we need 133 dB of dynamic range. This does not mean our radar has 133 dB of dynamic range, but we would like it to. The required dynamic range of 133 dB is available only if the receiver gain is changed by STC, AGC, and/or change in transmitted power. This gets us back to the rules mentioned in Section 22.2, which depend on the radar objective and waveform.

It should be noted that the dynamic range of inputs in most surface-based radars is determined not by maximum *target* RCS and range, but maximum *clutter* RCS and range. Only for a tracking radar operating on targets more than one beamwidth above the surface is the maximum target RCS of concern. In search radar, the target remains detectable and reportable even if it is above the saturation level of the receiver. The calculations here should use maximum possible *clutter* RCS at the range where the lower edge of the radar beam first encounters the clutter source. Any number of land- and ship-based radars encounter clutter from structures that project into the beam at short range, which is why STC is used.

22.7 CASCADE ANALYSIS

Dynamic range and noise figure (sensitivity) considerations, along with how much gain is necessary at any particular receiver stage to get the final desired overall receiver gain and output signal level, leads to the necessity of considering component selection and gain distribution through the receiver.

For example, some receiver parameters, such as gain and dynamic range, are antithetical. As such, when selecting components for a receiver design, a certain amount of compromise is necessary. As another example, we may use a high-gain LNA as the first amplifier, but if we follow it with highgain high noise figure amplifiers, the LNA (and prior lossy components) will no longer dominate the total noise figure.

This makes it important to track noise and signal power levels through the stages of the receiver to ensure receiver operation remains linear (avoid exceeding the 1-dB compression point) over desired range of signal amplitude.

The tool used to evaluate the above considerations is called cascade analysis, which refers to the process of tracking parameters such as signal power, signal gain, noise gain, noise figure, 1-dB compression point, noise floor, dynamic range, and bandwidth through the stages of a receiver. Cascade analysis helps predict and evaluate the interaction between various receiver parameters and guides component selections.

As an example, to design a receiver's 1-dB compression point, we need to know in what order these devices reach the 1-dB compression point. Ideally, the last component in a receiver chain (e.g., the amplifier prior to detection) is the first and only device to saturate. Knowing which device saturates first, second, third, and so forth determines where to put gain control and how much is necessary.

22.7.1 Cascade Analysis Conventions

It should be noted that some parameters do not apply to some components. For instance, passive devices, such as a waveguide filter, do not compress or saturate. For these devices, we generally use a very high compression point, such as 150 dBm, to effectively remove these components from the cascade 1-dB compression point calculations.

It is conventionally assumed that the narrowest bandwidth component in the receiver chain sets the overall RF to IF bandwidth [35, p. 15]. This is frequently the last IF filter bandwidth. For components with comparatively broad bandwidths (e.g., several GHz), such as waveguide, isolators, and couplers, we generally use a large token bandwidth, for example, 900 MHz. With the cascade bandwidth taken to be the narrowest bandwidth up to the point being analyzed in the chain, using 900 MHz effectively removes these components from the cascade bandwidth calculations.

The noise figure of a passive device, such as an attenuator, is frequently considered equal to its loss (see Section 4.4.2). While this results in the noise level into and out of devices remaining unchanged, it should be stressed that the same cannot be said for signals. Similarly, for well-designed passive mixers, a common rule of thumb is that the noise figure of a mixer is equal to its conversion loss [70].[22] Conversion loss is the difference between the input RF power level and the output IF power [22, 54].

It is presumed all devices are impedance matched. While 1 Ω is often assumed when absolute power is not important, we will use a system impedance of $R_0 = 50$ Ω. A 50-Ω impedance is typical of RF and microwave devices used in radar receivers. Likewise, RF

---

[22] This convention is usually within 0.5 to 1 dB of a passive mixer's actual noise figure.

test equipment is usually matched to 50 Ω. When there is an impedance mismatch (e.g., an ADC with a 1-kΩ input) impedance matching must be used (or the loss accounted for).

22.7.2 Procedure

The general forms of cascade equations can quickly become unwieldy when applied to a receiver chain; for example, (4.43) and (4.44). As a result, cascade calculations are often performed iteratively, two stages at a time. This technique simplifies the analysis and is amenable to computer programming. The general procedure is as follows [22, 54]:

- Start at the first stage.
- Perform two-stage analysis on first two stages computing cascaded noise figure, gain, 1-dB compression point, and so forth.
- Replace these first two components by an equivalent single stage with the above cascaded parameter.
- Perform two-stage analysis on the equivalent component and the third stage.
- Repeat until all stages are included.

As we will see below, the various cascade equations are used in linear form, logarithmic form, or a combination thereof. The choice of form is generally a matter of programming simplicity or preference. For convenience, results are usually carried along in both linear and logarithmic forms. Likewise, since the various parameters can be referenced to either the device input or output, we generally compute the cascade for one, such as the output, and relate this to the input, carrying both results.

We will use the receiver chain shown in Figure 22.10 for illustration of the cascade of various receiver parameters. IL and CL denote insertion loss and conversion loss, respectively. *Conversion loss* is a term peculiar to passive mixers, and is the difference in output power at the output frequency and the input power at the input frequency, for a stated LO power level. LO power needs to be specified because a mixer's conversion loss is a function of LO power level. Mixers typically have a conversion loss of 4.5 to 9 dB [70]. The various component parameters are summarized in Table 22.3.

We note that for cascade analysis, the various RFs and IFs are not important beyond their impact on individual component parameters. Also, we need to keep in mind the compression point of an amplifier is usually defined referenced to the output, while for mixers the input is the typical reference point.

In Table 22.3, we set OP1, OIP3, and OIP2 to 150. An ADC has a hard saturation point that generates harmonics and intermodulation if exceeded. Setting the ADC OP1, OIP3, and OIP2 values to 150 dBm assumes that the magnitude of the input voltage (target+clutter+noise) will always be kept below the ADC saturation level.

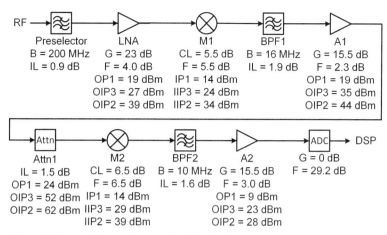

Figure 22.10 Example 1: superheterodyne receiver block diagram.

Table 22.3
Example 1: Device Specifications

Stage Description	G (dB)	F (dB)	OP1 (dBm)	OIP3 (dBm)	OIP2 (dBm)	BW (MHz)
PRE	−0.9	0.9	150	150	150	200
LNA	23.0	4.0	19.0	27.0	39.0	400
M1	−5.5	5.5	7.5	18.5	28.5	900
BPF1	−1.9	1.9	150	150	150	16
A1	15.5	2.3	16.0	32.0	44.0	400
Attn1	−1.5	1.5	24.0	52.0	62.0	900
M2	−6.5	6.5	6.5	22.5	32.5	900
BPF2	−1.6	1.6	150	150	150	10
A2	15.5	3.0	9.0	23.0	28.0	400
ADC	0.0	29.2	150	150	150	900

22.7.3 Power Gain

The cascaded gain, in dB, at the output of a particular device is the sum, in dB, of all of the prior stage gains to the point in the receiver chain of interest. It is sometimes more convenient to use linear gain and use the product of gains. The gain of a receiver, from the source through Device N is [48] [54]

$$G_T = G_1 G_2 \cdots G_N \quad \text{W/W} \tag{22.30}$$

$$G_T = G_1 + G_2 + \cdots + G_N \quad \text{dB} \tag{22.31}$$

Following the procedure outlined in Section 22.7.2, we first consider a two-stage cascade of the preselector and the LNA. Using (22.30) for $N = 2$, we get

$$G_C = G_1 G_2 = 10^{-0.9/10} \times 10^{23/10} = 0.813 \times 200$$
$$= 162.2 \text{ W/W} = 22.1 \text{ dB} \qquad (22.32)$$

Next we consider the mixer, using the results of (22.32), which results in

$$G_C = G_1 G_2 = 162.2 \times 10^{-5.5/10} = 162.2 \times 0.282$$
$$= 45.7 \text{ W/W} = 16.6 \text{ dB} \qquad (22.33)$$

This iterative, two-device cascade process is repeated as necessary. Note that as a result of considering two devices at a time, G_C, G_1, and G_2 in (22.33) have been redefined with regard to their meanings in (22.32).

22.7.4 Noise Figure and Noise Temperature

In a typical design, the receiver noise figure is established by the RF LNA and prior lossy elements. To ensure that subsequent components do not increase noise figure appreciably, we evaluate the noise figure through the entire receiver chain when selecting components. The Friis formula for the cascade of noise figure and noise temperature were covered in Section 4.5 of Chapter 4, specifically, (4.43) and (4.44) [53]. In keeping with our iterative procedure, we consider a two-stage cascade. The noise figure for a two-stage cascade is given in linear and logarithmic form, respectively, by [71–73]

$$F_C = F_1 + \frac{F_2 - 1}{G_1} = F_1\left(1 + \frac{F_2 - 1}{F_1 G_1}\right) \text{ W/W} \qquad (22.34)$$

$$F_C = F_1 + 10\log\left(1 + \frac{F_2 - 1}{F_1 G_1}\right) \text{ dB} \qquad (22.35)$$

Similarly, the equivalent/effective noise temperature for a two-stage cascade is given by [71–73]

$$T_C = T_1 + \frac{T_2}{G_1} = T_1\left(1 + \frac{T_2}{T_1 G_1}\right) \text{ K} \qquad (22.36)$$

where the relationship between noise figure and effective noise temperature is [71]

$$T_e = T_0(F - 1) \qquad (22.37)$$

We continue our example and find the noise figure of the first three devices. Applying (22.34), the combined noise figure of the preselector and LNA is

$$F_C = F_1 + \frac{F_2 - 1}{G_1} = 10^{0.9/10} + \frac{10^{4.0/10} - 1}{10^{-0.9/10}} = 1.23 + \frac{2.51 - 1}{0.813}$$
$$= 3.1 \text{ W/W} = 4.90 \text{ dB} \qquad (22.38)$$

The result is the sum of the LNA noise figure and the loss of the preceding device, which in this case is the preselector. This demonstrates the importance of minimizing losses prior to the LNA, which add dB for dB to overall noise figure [22, 71].

Using the results of (22.32) and (22.38), we proceed to the output of the mixer stage. Using (22.34) again we get

$$F_C = F_1 + \frac{F_2 - 1}{G_1} = 3.11 + \frac{10^{5.5/10} - 1}{162.2} = 3.11 + \frac{3.55 - 1}{162.2} \qquad (22.39)$$
$$= 3.1 \text{ W/W} = 4.92 \text{ dB}$$

Note that because of the high gain of the LNA, we see that there is very little change in noise figure due to the following mixer. As a result of considering two devices at a time, F_C, F_1, F_2, and G_1 in (22.39) have been redefined with regard to their meanings in (22.38).

22.7.5 1-dB Compression Point

The 1-dB compression point is explained in Section 22.5. Like gain and noise figure, we can determine the 1-dB compression point at each stage of the receiver. The 1-dB compression point at the output of a two-device cascade is [74, p. 58]

$$OP1_C = \left(\frac{1}{G_2 \cdot OP1_1} + \frac{1}{OP1_2} \right)^{-1} \text{ mW} \qquad (22.40)$$

which is usually used when specifying an amplifier. If we reference this at the input, typically used for receivers, we can use

$$IP1 = OP1 - (G - 1) \text{ dBm} \qquad (22.41)$$

where G is the net gain of the component chain. For our purposes, we first use (22.40) and then apply (22.41).

Considering the preselector and the LNA in cascade, we use (22.40) to get

$$OP1_C = \left(\frac{1}{G_2 \cdot OP1_1} + \frac{1}{OP1_2} \right)^{-1}$$
$$= \left(\frac{1}{10^{23/10} \cdot 10^{(150-30)/10}} + \frac{1}{79.4 \cdot 10^{-3}} \right)^{-1} \qquad (22.42)$$
$$= 79.4 \text{ mW} = 19 \text{ dBm}$$

where we use $OIP1_1$ of 150 dBm for the filter, since it does not compress. The 1-dB compression point is therefore equal to the compression point of the LNA for this case.

Next, we consider the mixer in cascade. We first note the 1-dB compression point is specified at the input. Relating this to the output, we get

$$OP1 = IP1 + (G - 1)$$
$$= 14 + (-5.5 - 1) = 7.5 \text{ dBm} \qquad (22.43)$$

Applying (22.40) again, we get

$$OPI_C = \left(\frac{1}{10^{(-5.5)/10} \cdot 79.4 \times 10^{-3}} + \frac{1}{10^{(7.5+30)/10}} \right)^{-1}$$

$$= \left(\frac{1}{0.282 \cdot 79.4 \times 10^{-3}} + \frac{1}{5.62} \right)^{-1} = \left(\frac{1}{22.4 \times 10^{-3}} + \frac{1}{5.62 \times 10^{-3}} \right)^{-1} \quad (22.44)$$

$$= 4.49 \text{ mW} = 6.53 \text{ dBm}$$

As before, we would continue the process iteratively.

22.7.6 Second-Order Intercept

As noted earlier, the second-order intercept is usually referenced at the output for amplifiers and at the input for mixers. The second-order intercept should be kept as high as possible. This is because power exceeding the second-order intercept point ensures intermodulation distortion, though second-order intermodulation can occur to some degree at any power level in a continuously differentiable nonlinearity. As a general rule of thumb, the second-order intercept point is 20 to 25 dB higher than the 1-dB compression point.

When performing a cascade analysis, we assign a suitably high intercept point to passive devices so they do not affect the overall system intercept point (>100 dBm is usually sufficient). This is because passive devices do not have intercept points.

The second-order intercept compression point of a two-device cascade, referenced to the device output, is given by [59, 61]

$$OIP2_C = \left(\sqrt{\frac{1}{OIP2_1}} + \sqrt{\frac{G_1}{OIP2_2}} \right)^{-2} = \left(\frac{1}{\sqrt{G_2}\sqrt{OIP2_1}} + \frac{1}{\sqrt{OIP2_2}} \right)^{-2} \text{ W} \quad (22.45)$$

or in logarithmic form [72]

$$OIP2_C = OIP2_2 - 20\log\left(1 + \sqrt{\frac{1}{G_2} \cdot \frac{OIP2_2}{OIP2_1}} \right) \text{ dBm} \quad (22.46)$$

We can reference the device input using

$$IIP2 = OIP2 - G \text{ dBm} \quad (22.47)$$

While we could use (22.45) or (22.46) to determine the cascade intercept point, we take a lesson from the 1-dB compression point example and note that since the second-order intercept for a passive filter is treated as essentially infinite, the second-order intercept for the cascade of the preselector and LNA is simply that of the LNA or 39 dBm. Adding the mixer to the cascade, we use (22.46) to get

$$OIP2_C = OIP2_2 - 20\log\left(1 + \sqrt{\frac{1}{G_2} \cdot \frac{OIP2_2}{OIP2_1}} \right)$$

$$= 28.5 - 20\log\left(1 + \sqrt{\frac{1}{0.282} \cdot \frac{0.707}{7.94}} \right) = 24.6 \text{ dBm} \quad (22.48)$$

The second-order intercept point is not always specified by manufacturers, or used in a cascade analysis because the spurious free dynamic range for a superheterodyne receiver is usually a function of the third-order intercept, which typically manifests before the second-order intercept. Also, second order intermodulation products typically fall outside the RF and/or IF passbands, but they can occur in-band for baseband signals.

22.7.7 Third-Order Intercept

The third-order intercept is an extrapolated value (see Section 22.6.3) that occurs when the output power of the desired input tones are equal to the third-order intermodulation power level (four tones total) [56]. Using the third-order intercept as a measure of linearity was first suggested by Avantek around 1964 [56]. The third-order intercept is usually specified at the output for amplifiers and at the input for mixers, but not always.

Power exceeding the 3rd-order intercept point ensures intermodulation distortion, though third-order intermodulation can occur to some degree at any power level in a continuously differentiable nonlinearity, so the higher the third-order intercept point the better. As a general rule of thumb, the third-order intercept point is typically 10 to 15 dB greater than the 1-dB compression point [55, p. 397].

Since the third-order intercept is lower (typically by about 10 dB) than the second-order intercept, third-order intermodulation products appear earlier than second-order products. This is why the third-order intercept is used to specify spurious free dynamic range. When receiver gains are significant, the third-order intercept of the last stage dominates the cascade.

The third-order compression point for two stages is given by [48, 75, 76]

$$OIP3_C = \left(\frac{1}{OIP3_1} + \frac{G_1}{OIP3_2}\right)^{-1} = \left(\frac{1}{G_2 \cdot OIP3_1} + \frac{1}{OIP3_2}\right)^{-1}$$
$$= \frac{OIP3_2}{1 + \frac{OIP3_2}{G_2 \cdot OIP3_1}} \quad \text{mW} \quad (22.49)$$

In logarithmic form, we get [60, 72]:

$$OIP3_C = OIP3_2 - 10\log\left(1 + \frac{1}{G_2} \cdot \frac{OIP3_2}{OIP3_1}\right) \text{dBm} \quad (22.50)$$

If we use the device input as a reference, we can relate to the device output using [55]

$$IIP3 = OIP3 - G \text{ dBm} \quad (22.51)$$

As before, by observing the very high intercept of the preselector, we can note that the cascaded third-order intercept point for the preselector and LNA is that of the LNA, or 27 dBm. Using (22.50), we now add the mixer to the cascade, resulting in

$$OIP3_C = OIP3_2 - 10\log\left(1 + \frac{1}{G_2} \cdot \frac{OIP3_2}{OIP3_1}\right)$$

$$= 18.5 - 10\log\left(1 + \frac{1}{0.828} \cdot \frac{708 \cdot 10^{-3}}{0.501}\right) \quad (22.52)$$

$$= 16.74 \text{ dBm}$$

Our results thus far are summarized in Table 22.4. The remaining cascade analysis is left as an exercise. For the last column in Table 22.4, we recall that the convention is for the narrowest bandwidth component to set the overall bandwidth [48, p. 15]. The cascade bandwidth is therefore set to the narrowest bandwidth of previous devices and the current device. A summary of cascade equations is provided in Table 22.5.

Figure 22.11 contains plots of the results of cascading gain and noise figure for the entire receiver chain of Figure 22.10. As we can see, the noise figure, and thus the sensitivity of the receiver, is dominated by the LNA. The only noticeable bump (~0.3 dB) in noise figure is because of the ADC, which we treated as a zero gain amplifier with a very large noise figure (see Section 22.8.5.4). The general up-down trend of gain tends to yield the largest dynamic range.

While the noise figure of Figure 22.11 is established by the RF LNA (and prior lossy elements), it should be noted that an LNA with 23-dB gain and no preceding RF attenuator is likely to be saturated by short-range clutter in a land-based radar. Either STC attenuation preceding the LNA or a lower LNA gain may be necessary to avoid saturation of the LNA. Either of these choices leads to a receiver noise figure higher than that of the LNA itself.

Figure 22.12 contains plots of the noise floor (see Section 22.4) and 1-dB compression point referenced to each device input. As indicated, the LNA and the preselector loss establishes the noise floor. Subsequent amplifiers add negligibly small amounts of noise. Passive components do not change the noise floor at all (assuming its input noise comes from a matched termination). A passive component can attenuate previously amplified noise, however, down to the thermal noise level.

The only lowering of noise floor is a result of narrowing bandwidth via the filters. The smooth decreasing trend of the 1-dB compression point indicates no components with a detrimentally low 1-dB compression point.

Table 22.4
Example 1: Cascade Example—First Three Stages

Device	Gain (dB)	F (dB)	OP1 (dBm)	IP1 (dBm)	OIP2 (dBm)	IIP2 (dBm)	OIP3 (dBm)	IIP3 (dBm)	BW (MHz)
Preselector	−0.9	0.9	150	150	150	150	150	150	200
LNA	22.1	4.90	19.0	−2.10	39.0	16.9	27.0	4.9	200
Mixer	16.6	4.93	6.53	−9.07	24.6	8.02	16.74	0.14	200

Table 22.5
Two-Stage Cascade Equations Used for Iterative Analysis

Parameter	Cascade Equation
Power gain	$G_C = G_1 G_2$ W/W $G_C = G_1 + G_2$ dB
Noise figure	$F_C = F_1 \left(1 + \dfrac{F_2 - 1}{F_1 G_1}\right)$ W/W $F_C = F_1 + 10\log\left(1 + \dfrac{F_2 - 1}{F_1 G_1}\right)$ dB
Noise temperature	$T_C = T_1\left(1 + \dfrac{T_2}{T_1 G_1}\right)$ K $T_C = T_1 + 10\log\left(1 + \dfrac{T_2}{T_1 G_1}\right)$ dBK
−1-dB compression point at output	$OP1_C = \left(\dfrac{1}{G_2 \times OP1_1} + \dfrac{1}{OP1_2}\right)^{-1}$ mW
−1-dB compression point at input	$IP1 = OP1 - (G - 1)$ dBm
second-order intercept point at output	$OIP2_C = OIP2_2 - 20\log\left(1 + \sqrt{\dfrac{1}{G_2} \cdot \dfrac{OIP2_2}{OIP2_1}}\right)$ dBm
second-order intercept point at input	$IIP2 = OIP2 - G$ dBm
third-order intercept point at output	$OIP3_C = OIP3_2 \Big/ \left(1 + \dfrac{OIP3_2}{G_2 \times OIP3_1}\right)$ mW $OIP3_C = OIP3_2 - 10\log\left(1 + \dfrac{1}{G_2} \cdot \dfrac{OIP3_2}{OIP3_1}\right)$ dBm
third-order intercept point at input	$IIP3 = OIP3 - G$ dBm
Bandwidth	$B_C = \min(B_1, B_2)$ Hz

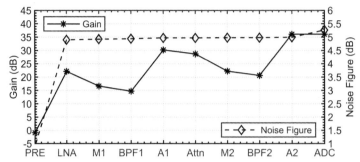

Figure 22.11 Example 1: gain and noise figure.

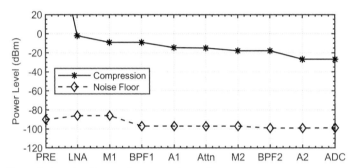

Figure 22.12 Example 1: noise floor and compression point.

The difference between the 1-dB compression point and the noise floor is one of the definitions for dynamic range (see Section 22.6), and is plotted in Figure 22.13. We note here that dynamic range is generally decreased by lossy components and active devices, and increased when the noise bandwidth is decreased. The mixer usually causes a dip in dynamic range because it is a lossy device early in the chain. For this example, the overall dynamic range is 72 dB and is constrained by the amplifier prior to the ADC, which is a limiting amplifier. We chose the limiting amplifier saturation point to be ~1 dB below the full-scale value of the ADC of 10 dBm to prevent potential damage to the ADC due to overloads.

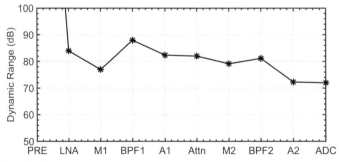

Figure 22.13 Example 1: dynamic range.

If we have an overall goal of 100 dB of dynamic range for our radar design, we would need about 30 dB of AGC and/or STC to extend the dynamic range of the receiver (see Section 22.2). Coherent integration also increases the overall dynamic range of the radar. The cascade analysis presented thus far can be extended past the receiver to the output of the signal processor by accounting for SNR improvements and losses after the receiver. For example, coherent integration of N pulses yields a $10 \cdot \log(N^2)$ dB increase in signal power and an increase of $10 \cdot \log(N)$-dB in noise power—likewise for various losses. Another way to capture this is to use the measured bandwidth of the signal processor. However, N^2/N is the most accurate way to calculate the SNR improvement due to coherent signal processing. Using the ratio of the preprocessing to post-processing bandwidths requires taking into account the effect of aliasing in the sampling process, including slow-time (PRI-to-PRI) sampling.

22.8 DIGITAL RECEIVER

As mentioned in Section 22.3, analog I/Q detectors suffer from I/Q channel imbalance issues. One method of avoiding imbalance problems is to use digital hardware to perform quadrature detection [43]. A block diagram of a wideband, frequency agile, digital, superheterodyne receiver or more simply, a digital receiver, is presented in Figure 22.14.

The use of direct IF sampling in the superheterodyne receiver of Figure 22.14 is what earns the moniker digital. Skolnik observes there is no unique definition for digital receiver [22, p. 742]. Yuanbin Wu and Jinwen Li suggest the use of direct IF sampling and direct digital synthesis (DDS) to generate the LO earns the moniker [77]. ADCs are normally used in digital receivers because digital signals are more reliable and more flexible than analog signals and offer reduced cost, size, weight, and power dissipation.

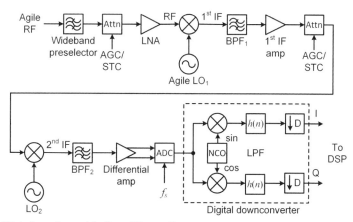

Figure 22.14 Digital receiver with direct IF sampling.

Up to the differential amplifier preceding the ADC, the configuration is the same as the double down conversion receiver of Section 22.3. However, we replaced the synchronous detector shown in Figure 22.5 with an ADC that is directly sampling the IF signal—thus the term, direct IF sampling. The amplitude and phase balance is much better (theoretically perfect) than that achieved by baseband sampling [43].[23] The quadrature references are now generated by a *numerically controlled oscillator* (NCO).

A single-ended input to differential output amplifier is shown driving the ADC in Figure 22.14 to make note of the fact that many high-performance (high sample rate, high dynamic range) ADCs are now being designed with differential inputs [78]. An RF transformer, such as the ADT4-1WT from Mini-Circuits, can also be used to couple into a differential input ADC [78]. Using differential inputs offers benefits such as better distortion performance, cancellation of even harmonics, and common-mode rejection of noise [79].

Since we are digitizing the IF signal, the analog signal needs to pass through an antialiasing filter (AAF) designed to pass expected modulation bandwidths prior to analog-to-digital conversion.[24] The AAF reduces noise bandwidth and assures that negligible amounts of signal and noise aliasing occur as a result of analog-to-digital conversion. For the receiver shown in Figure 22.14, the second IF BPF serves as an antialiasing filter in addition to eliminating spurious signals output of the second mixer.

22.8.1 Bandpass Sampling

We need to make a clarification when talking about the AAF. An AAF is classically lowpass in accordance with the Nyquist sampling theorem, which applies to lowpass signals (signals centered about 0 Hz). The Nyquist sampling theorem states that if a time-varying signal is sampled periodically, the sampling frequency should be at least twice the highest frequency component of the signal to prevent aliasing [78–81]. This theorem also bears the monikers of the Shannon sampling theorem [82, 83] and the Kotel'nikov sampling theorem [84] (as well as others). This theorem can be represented as

$$f_s \geq 2B \qquad (22.53)$$

where f_s is the sampling frequency, and B is the highest frequency contained in the signal. Equation (22.53) is referred to as the Nyquist criterion. The values $2B$ and $f_s/2$ are called the Nyquist rate and Nyquist frequency, respectively. Satisfying the Nyquist criterion allows the original signal to be perfectly recovered from the sampled values.

For example, let us consider sampling a 40-MHz IF and a 4-MHz bandwidth LFM waveform. According to the Nyquist criterion, we should use a sampling frequency of

$$f_s \geq 2B \geq 2(40 + 4/2) \geq 84 \text{ MHz} \qquad (22.54)$$

which corresponds to a Nyquist frequency of $f_s/2 = 42$ MHz. The driving factor is the 40-MHz IF, not the 4-MHz modulation, which contains the information in the signal.

---

[23] Direct IF sampling requires a higher sampling rate compared to baseband sampling the output of an I/Q demodulator.
[24] Antialiasing filters are also used with baseband digitization.

When using a signal centered about some IF, we often use a special case of the Nyquist sampling theorem, referred to as the bandpass sampling theorem[25], which only takes the signal bandwidth, B, that contains the information we want from the signal, into consideration [85–87]. Brigham explains that a bandpass signal can be reconstructed from samples if the sampling frequency, f_s, satisfies the relationships [85; 86, p. 322]

$$\frac{2f_H}{n} \leq f_s \leq \frac{2f_L}{n-1} \tag{22.55}$$

and

$$2 \leq n \leq \frac{f_H}{f_H - f_L} \tag{22.56}$$

where n is an integer (see Appendix 22A). The variables f_H and f_L are the highest and lowest frequency component of a signal, respectively. Expressing the minimum sample frequency in terms of signal bandwidth, $B = f_H - f_L$, we can use (22.55) and (22.56) to form [79]

$$f_s \geq \frac{2f_H}{f_H/(f_H - f_L)} = 2(f_H - f_L) = 2B \tag{22.57}$$

Equation (22.57) requires the sampling frequency used for direct IF sampling to be at least twice the modulation bandwidth B. The necessary AAF is now a bandpass filter rather than a lowpass filter.

When using direct IF sampling in a digital receiver, the signal is usually allowed to alias. In this way, the process of sampling acts as another down conversion stage [85–88].[26] For cases where the IF is undersampled, the criteria of (22.55) and (22.56) ensure that we avoid spectrum overlap corrupting the aliased signal bandwidth [86].

The concept of Nyquist zones, depicted in Figure 22.15, is often used to help visualize aliasing when using direct IF sampling [69]. Nyquist zones divide the frequency spectrum into an infinite number of frequency bands, each with a width equal to the Nyquist frequency, $f_s/2$ [79]. The signal spectrum of the signal being sampled can reside in any single Nyquist zone, but must not overlap any multipe of $f_s/2$ (which is the purpose of the antialiasing filter). There is either a frequency shifted duplicate of the signal spectrum or a frequency shifted mirror image (frequency reversal of the complex conjugate) of the signal spectrum within each Nyquist zone.

---

[25] Bandpass sampling is also referred to as undersampling, harmonic sampling, IF sampling, and direct IF-to-digital conversion [79].
[26] The term used for lower frequency radars where the RF is sampled is direct RF sampling.

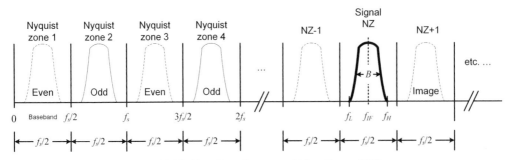

Figure 22.15 Analog spectrum divided into Nyquist zones. (*After:* Kester [79]).

Analog frequencies centered on f_{IF} in an odd Nyquist zone are downconverted via aliasing to a digital IF of

$$f_{IFadc} = f_s - \text{rem}(f_{IF}, f_s) \tag{22.58}$$

where rem(*a*,*b*) denotes the remainder after division of *a* divided by *b* [69, 89]. Similarly, analog frequencies in an even Nyquist zone alias to [69, 89]

$$f_{IFadc} = \text{rem}(f_{IF}, f_s) \tag{22.59}$$

When the signal spectrum occupies an even Nyquist zone, the result is a mirrored spectrum. A signal spectrum occupying an odd Nyquist zone aliases without mirroring the spectrum [79]. This is generally not a concern, as long as we know if the aliased spectrum is mirrored or not.

A design goal is to place the signal to be sampled in the center of a Nyquist zone. According to Walt Kester at Analog Devices, the sampling rate associated with Nyquist zone centers is given by [79, p. 81]

$$f_s = \frac{4 f_{IF}}{2NZ - 1} \tag{22.60}$$

where *NZ* is an integer corresponding to Nyquist zone. The largest Nyquist zone satisfying (22.55) is generally preferable, since it produces the lowest sample rate [79].

For example, let us again consider sampling a 4-MHz bandwidth LFM waveform at a 40-MHz IF. The minimum required sampling frequency according to (22.57) is 8 MHz. Increasing the sample rate increases the potential distance between images, allowing for more margin for the AAF. Let our ADC operate at 60 MSPS (megasamples per second). The resultant Nyquist zone is

$$NZ = 2\frac{f_{IF}}{f_s} + \frac{1}{2} = 2\frac{40}{60} + \frac{1}{2} = 1.8 \tag{22.61}$$

which is not an integer.

To place the 40-MHz IF in the second Nyquist zone, we would need f_s to be 53.33 MHz. If a *NZ* is not an integer, this indicates that our spectrum is not exactly centered on a Nyquist zone. Depending on the available clock frequencies, this may be unavoidable, or at least an acceptable compromise, with the exciter design (e.g., all clocks are a multiple of 10 MHz). So long as the AAF provides sufficient rejection of the aliased image, this is tolerable. For this reason, we will continue this example using a sampling rate of 60 MSPS.

22.8.2 Digital Down conversion

A *digital downconverter* (DDC) is used to translate the spectrum of interest from its digital IF to baseband I and Q samples. This use of digital down conversion eliminates an analog mixing stage and its attendant analog mixer and filter. One or more stages of down conversion usually follow or are incorporated into the DDC. The DDC shown in Figure 22.14 is typical of the type used for direct IF sampling.

A simplified block diagram showing the direct IF sampling and digital downconverter circuitry[27] of Figure 22.14 is provided in Figure 22.16. Digital down conversion is analogous to the I/Q demodulator of Figure 22.5 described in Section 22.3 followed by sampling using two ADCs, but provides better I and Q balance (theoretically perfect). The numerically controlled oscillator (typically provided by a DDS source) is set at the digital IF and the $h_{LPF}(n)$ are low pass filters used to reject the double frequency terms out of the NCO.

Following the digital downconversion is usually a decimation process whose decimation rate is selected according to waveform bandwidth. In this case, the LPFs shown in Figure 22.16, denoted by $h_{LPF}[n]$, not only reject the double frequency terms out of the NCOs but serve as antialiasing filters for the decimation. As such, the LPF cutoff frequency is determined by the desired decimation rate. The sampling rate after antialias filtering and decimation would be predicated on the modulation bandwidth of the sampled waveform, since a single DDC unit could operate on a number of waveforms.

The real IF signal entering the circuitry depicted in Figure 22.16 can be represented as

$$x_{IF}(t) = A(t)\cos\left[2\pi f_{IF} t + \phi(t)\right] \tag{22.62}$$

where $A(t)$ and $\phi(t)$ are amplitude and phase modulation, respectively. The bandpass filter, preceding the ADC, is a narrowband filter representing the RF to IF bandwidth (narrowest bandwidth in the receiver) up to the ADC, also serving as antialiasing filter. The bandwidth of the BPF is set to pass the waveform amplitude and phase modulation with minimal distortion. As such, the output of the bandpass filter becomes

$$x_{BPF}(t) = x_{IF}(t) = A(t)\cos\left[2\pi f_{IF} t + \phi(t)\right] \tag{22.63}$$

---

[27] Minus the differential amplifier, which has no bearing on DDC operation.

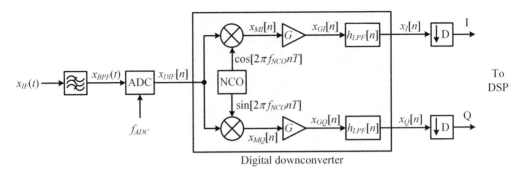

Figure 22.16 Direct IF sampling followed by digital down conversion.

The output of the ADC

$$x_{DIF}[n] = A[nT]\cos(2\pi f_{DIF}nT + \phi[nT]) \tag{22.64}$$

where $n = 0, 1, 2, 3, \ldots$, T is the sample period, and f_{DIF} is the, typically aliased, digital IF (see Section 22.8). Depending on the particular frequencies used, the signal may undergo aliasing (folding). We will stipulate that the criteria of the bandpass sampling theorem is satisfied (see Section 22.8.1) [85; 86, p. 322].

For illustration, we continue our example of a digital receiver with a second IF of 40 MHz sampled at a rate of 60 MSPS in a radar using a 4-MHz LFM waveform. The resultant spectrum out of the ADC is presented in Figure 22.17. This places our analog waveform in the second Nyquist zone (bands of frequency $f_s/2$ wide explained above) which is from 30 MHz to 60 MHz. Using (22.58), the aliased digital IF is rem(60, 40) = 20 MHz, as depicted in the plot in Figure 22.17.

Remember, since the sampled signal is real, the spectrum magnitude is an even function. Both images in Figure 22.17 both contain the same information, making one image redundant. Since only one image is necessary, we want to eliminate the other. To accomplish this, we first center either the upper of lower image on zero frequency, via mixing with quadrature outputs from a NCO. The quadrature NCO outputs can be represented as

$$\begin{aligned} x_{NCO_I}[n] &= \cos(2\pi f_{NCO}nT) \\ x_{NCO_Q}[n] &= \sin(2\pi f_{NCO}nT) \end{aligned} \tag{22.65}$$

Mixing with the NCO, results in

$$\begin{aligned} x_{M_I}[n] &= x_{DIF}[n]x_{NCO_I}[n] \\ &= A[nT]\cos(2\pi f_{DIF}nT + \phi[nT])\cos(2\pi f_{NCO}nT) \end{aligned} \tag{22.66}$$

and

$$x_{M_Q}[n] = x_{DIF}[n]x_{NCO_Q}[n]$$
$$= A[nT]\cos(2\pi f_{DIF}nT + \phi[nT])\sin(2\pi f_{NCO}nT) \quad (22.67)$$

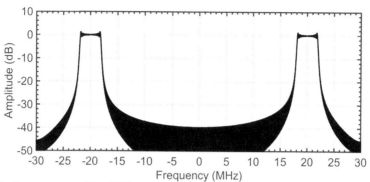

Figure 22.17 Spectrum out of the ADC.

Using appropriate trigonometric identities, we get

$$x_{M_I}[n] = \frac{A[nT]}{2}\cos(2\pi(f_{DIF} - f_{NCO})nT + \phi[nT])$$
$$+ \frac{A[nT]}{2}\cos(2\pi(f_{DIF} + f_{NCO})nT + \phi[nT]) \quad (22.70)$$

and

$$x_{M_Q}[n] = -\frac{A[nT]}{2}\sin(2\pi(f_{DIF} - f_{NCO})nT + \phi[nT])$$
$$+ \frac{A[nT]}{2}\sin(2\pi(f_{DIF} + f_{NCO})nT + \phi[nT]) \quad (22.71)$$

We now multiply the outputs of the mixers by gain G, which will be used to scale the mixer outputs later.

$$x_{G_I}[n] = G\frac{A[nT]}{2}\cos(2\pi(f_{DIF} - f_{NCO})nT + \phi[nT])$$
$$+ G\frac{A[nT]}{2}\cos(2\pi(f_{DIF} + f_{NCO})nT + \phi[nT]) \quad (22.72)$$

$$x_{G_Q}[n] = -G\frac{A[nT]}{2}\sin(2\pi(f_{DIF} - f_{NCO})nT + \phi[nT]) +$$
$$G\frac{A[nT]}{2}\sin(2\pi(f_{DIF} + f_{NCO})nT + \phi[nT]) \quad (22.73)$$

The NCO frequency can be set to $f_{NCO} = \pm f_{DIF}$, with the sign depending on which image is to be translated to zero. Setting the to $f_{NCO} = f_{DIF}$ results in

$$x_{G_I}[n] = G\frac{A[nT]}{2}\cos(\phi[nT]) + G\frac{A[nT]}{2}\cos(2\pi(2f_{DIF})nT + \phi[nT]) \quad (22.74)$$

and

$$x_{G_Q}[n] = -G\frac{A[nT]}{2}\sin(\phi[nT]) + G\frac{A[nT]}{2}\sin(2\pi(2f_{DIF})nT + \phi[nT]) \quad (22.75)$$

Setting the NCO frequency to $f_{NCO} = -f_{DIF}$ is left as an exercise.

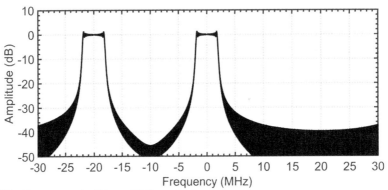

Figure 22.18 Spectrum out of the multiplier output.

Continuing our example, we can translate either the upper image (centered on 20 MHz) or the lower image (centered on –20 MHz) to baseband (centered on 0 Hz) using an NCO. The NCO generates two digital reference frequencies with quadrature phase. For this example, we choose 20 MHz as the NCO frequency. The lower image is translated up 20 MHz to baseband as depicted in the graph of Figure 22.18. The upper image is translated up to 40 MHz, but aliases (wraps in frequency) to –20 MHz. Recall that the spectrum generated by the Fourier transform is periodic.

The lowpass filters, $h_{LPF}[n]$, are used to remove the unwanted image at $2f_{DIF}$, which has been aliased to – 20 MHz, resulting in

$$x_I[n] = G\frac{1}{2}A[nT]\cos(\phi[nT]) \quad (22.76)$$

$$x_Q[n] = -G\frac{1}{2}A[nT]\sin(\phi[nT]) \quad (22.77)$$

This results in

$$\begin{aligned}x_{IQ}[n] &= G\frac{A[nT]}{2}\cos(\phi[nT]) - jG\frac{A[nT]}{2}\sin(\phi[nT]) \\ &= G\frac{1}{2}A[nT]e^{-j\phi[nT]}\end{aligned} \quad (22.78)$$

We recognize that, except for the factor of 1/2, (22.78) is the complex envelope of (22.64). While this factor of 1/2 can sometime cause confusion, we can choose to carry it along, or normalize it out, without loss of generality. So that we can take the real part of the complex representation (22.78) and have it equal to the original real signal we are representing (22.64), we choose $G = 2$.

The magnitude of the complex (baseband) signal is

$$\sqrt{V_I[nT]^2 + V_Q[nT]^2}$$
$$= \sqrt{\left\{\frac{1}{2}GA[nT]\cos(\phi[nT])\right\}^2 + \left\{-\frac{1}{2}GA[nT]\sin(\phi[nT])\right\}^2}$$
$$= \sqrt{\frac{1}{4}GA[nT]^2\cos^2(\phi[nT]) + \frac{1}{4}GA[nT]^2\sin^2(\phi[nT])} \qquad (22.79)$$
$$= \frac{1}{2}GA[nT]\sqrt{\cos^2(\phi[nT]) + \sin^2(\phi[nT])}$$
$$= \frac{1}{2}GA[nT]$$

Letting $G = 2$, we get

$$\sqrt{V_I[nT]^2 + V_Q[nT]^2} = A[nT] \qquad (22.80)$$

For our example, after translating the signal to baseband, we want to remove the image centered at -20 MHz. To do this, we use digital LPFs in the I and Q channels to reject the unwanted image, which are analogous to the LPFs used in the I/Q detector of Figure 22.5. Additionally, we need the LPFs to serve as antialiasing filters for a decimation by a factor of $D = 10$, discussed shortly. The LPFs used for this example are eighth-order Chebychev type I filters with 0.05 dB passband ripple and cutoff frequency of $0.8 \times (f_s/2)/D = 0.8 \times (60/2)/10 = 2.4$ MHz.[28] This equates to a filter cutoff of 1.2 times the single-sided LFM bandwidth. The digitally filtered output results in the I and Q terms we are after. The spectrum at the output of the LPFs and the LPF frequency response are depicted in Figure 22.19. At this point in our example, we have a complex baseband representation of our waveform, suitable for digital signal processing.

We now have a 4-MHz baseband signal that is sampled at 60 MSPS. Since the sample frequency is 15 times larger than the signal bandwidth, the signal is greatly oversampled. To reduce the amount of processing necessary, the signal is decimated or downsampled by a factor based on the modulation bandwidth.[29] It is important to note that since complex samples are produced after digital down conversion, a complex sample rate of $f_s \geq B$ (rather than $2B$ given by (22.53)) should be adequate. A slightly higher sampling rate (~1.5xB) is sometimes used to avoid affecting the sidelobe levels of LFM waveforms. Matching the LFM bandwidth exactly can potentially elevate the range sidelobe levels in the compressed waveform due to insufficient rejection of the aliased image.

---

[28] This is the default filter used by the MATLAB decimate function.
[29] The terms decimation and downsampling are sometimes used synonymously. The terms are sometimes differentiated. Decimation includes an antialias filter followed by downsampling. Down sampling is the process of keeping every N^{th} sample, discarding the rest.

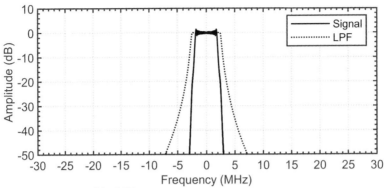

Figure 22.19 Spectrum out of the LPFs.

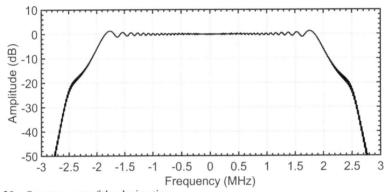

Figure 22.20 Spectrum out of the decimation.

Thus, relative to the sample rate we need for our example, we are oversampled by a factor of 60/(1.5×4) = 10. This means we can decimate the signal by a factor of 10 and still have a sample rate that is adequate for subsequent processing. For this reason, most DDCs include one or more stages of decimation. The output spectrum after decimation by a factor of 10 is provided in Figure 22.20.

22.8.3 Practical DDC

One problem in DDC design is how to design and implement the digital LPFs shown in Figure 22.16. Depicted in Figure 22.21 is a hardware DDC, patterned after the Analog Devices AD6620 67 MSPS decimating digital receiver chip [90]. DDC operation is the same as previously discussed. The topology of the LPFs and downsampling is different. The output of the multipliers feeds what are referred to as cascaded integrator comb (CIC) filters, which combine filtering with sampling rate decrease (decimation) or sampling rate increase (interpolation).[30] Following the second CIC filter is a programmable FIR (PFIR) filter that provides anti-aliasing filtering for a third stage of downsampling in this example.

---

[30] Comb filters get their name because their frequency response resembles a full-wave rectified sinewave and looks like a comb.

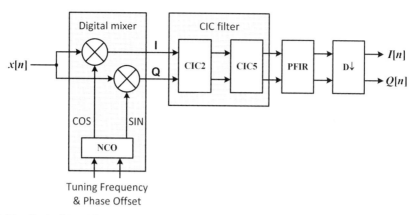

Figure 22.21 Typical DDC implementation.

CIC filters are a class of computationally efficient multirate linear phase narrowband low-pass FIR filters developed by Eugene Hogenauer as an efficient means of performing decimation and interpolation [91, 92]. CIC filters consists of only adders, subtractors and registers, requiring no multiplications, limited storage, and few data transfers [91–93]. Additionally, the CIC integrator and comb sections are independent of rate change.

CIC filters are typically used when the decimation (or interpolation) rate is large (above about 8 or 10) [56, 94]. For example, a 40 MHz IF sampled at 60 MSPS, given a signal bandwidth of 4 MHz. To achieve large rate changes, an FIR filter implementation would require fast multipliers and very long filters, which can potentially be a large DSP bottleneck [94]. As such, CIC filters are commonly used to mitigate the computational requirements of the first decimation stage in a DDC.

As we will see, CIC filters have a sinc-like frequency response, exhibiting a wide transition band, which results in passband droop. To correct this droop, a *compensation FIR* (CFIR) filter is usually used to flatten the passband [95]. For the example shown in Figure 22.21, the CFIR filter response can be incorporated into the PFIR filter. In some other DDC chips, the CFIR filter is a separate, distinct filter.

In this particular example, modeled after the AD6620, a fixed coefficient second-order CIC decimation filter is cascaded with a fixed coefficient fifth order CIC decimation filter, with programmable decimation rates of 2, 3 ... 16 and 1, 2 ... 32, respectively [90]. The PFIR is a programmable decimating RAM coefficient FIR filter, with a programmable decimation rate of 1, 2 ... 32, and processing up to 1,134 million taps per second [90]. The FPIR can accommodate up to 256 taps, storing 20-bit coefficients.

The DDC shown in Figure 22.21 is based on the AD6620 integrated circuit (IC). Additionally, DDCs are often implemented using application-specific integrated circuits (ASICs) and field-programmable gate arrays (FPGAs). FPGA vendors often provide ready-made DDC cores that can be used in their FPGAs [96, 97]. One benefit of using an FPGA implementation is a DDC becomes field upgradable should the design change.

22.8.4 CIC Filter Structure

The structure of a CIC filter, sometimes referred to as a Hogeneauer filter, consists of an equal number of integration-comb filter pairs separated by a sample rate changer, which adjusts the sample rate by a factor of R (either up or down) [91, 92]. The integrator section consists of N digital integrator stages. The comb section consists of N comb stages with a differential delay of M samples per stage [91]. The number of CIC stages has a direct bearing on the CIC frequency response. Examples of two-stage CIC filters, $N = 2$, used for decimation and interpolation are depicted in Figure 22.22 and Figure 22.23, respectively. The difference between the two topologies is the order if the integrator and comb stages. In a CIC decimation filter the integrators precede the comb filters and in a CIC interpolation (upsample) filter the combs precede the integrators.

For a CIC decimation filter, the sampling rate change downsamples the output of the last integrator stage, reducing the sample rate from f_s to f_s / R. For a CIC interpolation filter, the sampling rate change increases the sample rate by a factor of R. Increasing the sample rate is done by inserting R-1 zeros between samples [91]. We are primarily interested in CIC filters used for decimation.[31] In this case, the integrators operate at the high sample rate of f_s and the comb stages operate at the lower sample rate of f_s / R.

One of the important features of the CIC filter architecture for decimation is being able to tune the frequency response (for the same decimation factor) by selecting the appropriate number of filter pairs. In the CIC filter topology, the integrators (which consist only of additions, no multiplications) operate at the high input sampling rate (i.e., the ADC sample rate in DDC) and the comb filters operate the decimated sample rate (i.e., the ADC sample rate/R).

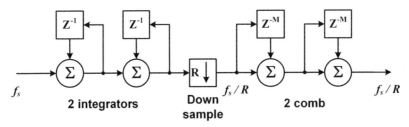

Figure 22.22 CIC decimation filter for $N = 2$.

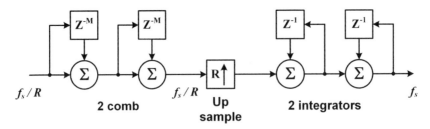

Figure 22.23 CIC interpolation filter for $N = 2$.

---

[31] To decimate by a rational factor, first interpolate (upsample), then decimate (downsample).

22.8.4.1 Basic Integrator

The frequency characteristic of the CIC filter can be derived from the transfer functions of its components, the integration filter, and the comb filter. In doing so, we will generally follow Hogenauer's explanation [91, 92].

First let us consider the integrator depicted in Figure 22.24. An integrator (accumulator) is a single-pole infinite impulse response (IIR) filter with unity feedback, which can be expressed as

$$y[n] = y[n-1] + x[n] \tag{22.81}$$

The corresponding z-domain transfer function is [91]

$$H_I(z) = \frac{1}{1-z^{-1}} \tag{22.82}$$

The cascade of N integrators is therefore

$$H_I^N(z) = \left(\frac{1}{1-z^{-1}}\right)^N \tag{22.83}$$

Hogeneauer notes that for a CIC decimator, unity feedback results in register overflow of all integrator stages [91]. He further stipulates that this is not a problem if the following conditions are met:

- The CIC filter is implemented with two's complement arithmetic or other number system which allows wrap-around between the most positive and most negative numbers.
- The range of the number system is equal to or exceeds the maximum magnitude expected at the output of the composite filter.

For CIC interpolators, the data is preconditioned by the comb section so that overflow will not occur in the integrator stages [91].

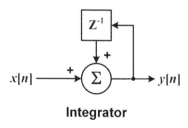

Integrator

Figure 22.24 CIC basic integrator.

22.8.4.2 Basic Comb Filter

Now let us consider the comb filter, depicted in Figure 22.25. A single comb filter is a differentiator, which can be expressed as

$$y[n] = x[n] - x[n-M] \tag{22.84}$$

For a CIC decimation filter, the comb filter operates at the lower sample rate of f_s / R. The corresponding z-domain transfer function for a comb filter is [95]

$$H_C(z) = 1 - z^{-M} \tag{22.85}$$

where M is the differential delay. The differential delay is a filter design parameter used to control the CIC filter's frequency response and null spacing [91]. M can be any positive integer, but in practice, is usually limited to 1 or 2 [91, 94, 95].

The single-stage comb filter system function referenced to the high sample rate, f_s, can be expressed as [91]

$$H_C = 1 - z^{-RM} \tag{22.86}$$

where we made use of the Noble identity (which allows us to move the comb filters across the sample rate change) [153, p. 25].

The transfer function of cascade of N comb filters is therefore

$$H_C^N = \left(1 - z^{-M}\right)^N \tag{22.87}$$

which can be expressed at the high sample rate as

$$H_C^N = \left(1 - z^{-RM}\right)^N \tag{22.88}$$

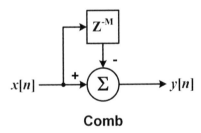

Comb

Figure 22.25 CIC basic comb filter.

22.8.4.3 CIC Filter Frequency Response

The general form of the z-domain system transfer function for a CIC filter consisting of N pairs of integrator and comb filters, for a sampling rate change of R, can be expressed as [91, 94, 95]

$$H(z) = H_I^N(z) H_C^N(z^R) = \left(\frac{1-z^{-RM}}{1-z^{-1}}\right)^N$$

$$= \left(\sum_{k=0}^{RM-1} z^{-k}\right)^N \tag{22.89}$$

where we made use of the Noble identity to determine an equivalent frequency response for an N-stage CIC filter [95; 93, p. 25]. It is left as an exercise to verify (22.89).

From (22.29), we see that the CIC filter response is the equivalent of N FIR filters cascaded, requiring RM storage registers per stage [92]. Because FIR filters are always stable, CIC filters are stable despite the integrators being IIR filters.

The power frequency response for a CIC filter can be expressed as [91]

$$|H(f)| = \left|\frac{\sin(\pi M f)}{\sin(\pi f / R)}\right|^N \tag{22.90}$$

22.8.4.4 Example: CIC Decimation Filter

Let us consider an example to illustrate the process of defining a CIC filter. Consider a 5-MHz bandwidth waveform sampled at 60 MSPS. We want to allow for a 6-MHz bandwidth (3 MHz LPF cutoff) to avoid distortion of the signal. The first step is defining the necessary decimation (rate change) factor. Observing the Nyquist sampling theorem, we determine a conservative candidate decimated sample rate (defined at the CIC filter output) of about 10 MHz as a good first estimate. This corresponds to a decimation factor of 6.

Applying the estimated decimation factor of 6 to samples occurring at 60 MSPS provides a basis for candidate CIC filter designs. Several CIC filter frequency responses, for various CIC filter lengths, a differential delay of one, and a decimation rate of 6, are plotted in Figure 22.26. The frequency characteristics exhibit a trade-off between roll-off and sidelobe level. As depicted in Figure 22.26, as roll-off increases, sidelobes decrease. Additionally, the frequency responses all have fixed nulls at the same frequency of f_s/R or 10 MHz, which is characteristic of CIC filters.

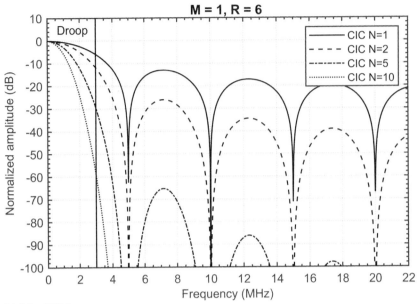

Figure 22.26 CIC frequency responses.

The process of selecting the number of stages that best suits our design begins by overlaying the 6 MHz (±3) bandwidth requirements of the waveform/signal processor with the roll off of the various CIC filters. The 6-MHz bandwidth requirement does not appear to be supported by any of the CIC filter responses plotted in Figure 22.26 due to the sinc(x) nature of the CIC frequency response drooping in the desired band (left of the solid vertical line). As discussed earlier, to address this disadvantage of CIC filters having a sinc-shaped passband, a compensation filter is commonly used in conjunction with a CIC filter to flatten the curvature of the CIC filter in the passband. The compensation filter follows the CIC filter as depicted in Figure 22.21. This particular type of compensation filter is referred to as an inverse sinc filter.

The design question now is not only what CIC filter to use, but what CIC filter, cascaded with a compensation filter, best suits our radar receiver design. One advantage of the CFIR filter following the CIC decimation filter is that the CFIR filter operates at the decimated sample rate, which for our example is 10 MSPS rather than the higher sample rate of 60 MSPS.

To achieve a flat passband over the band of interest using cascaded CIC and CFIR filters, the CFIR filter must have a magnitude response that is the inverse of the CIC filter frequency response. The inverse sinc filter response can be expressed as [95]

$$G(f) = \left| M \cdot R \frac{\sin(\pi f / R)}{\sin(\pi M \cdot f)} \right|^N \approx \left| \frac{\pi M \cdot f}{\sin(\pi M \cdot f)} \right|^N \quad (22.91)$$
$$= \left| \operatorname{sinc}^{-1}(M \cdot f) \right|^N$$

When the decimation factor is large, the compensation filter described by (22.91) is approximated by the inverse sinc function. The CFIR filter order can range from a few taps to over 100 taps. The longer the CFIR filter, the better the compensation. For example, the AD6620 FPIR can accommodate up to 256 taps [90]. A general rule of thumb is for the CFIR filter passband to be approximately one-fourth the frequency of the first null in the CIC frequency response to minimize added noise [98, p. 564].

Shown in Figure 22.27 and Figure 22.28 are two examples of CIC filter response, a corresponding CFIR filter design, and composite frequency response of the CIC-CFIR cascade.[32] The CIC filter response presented in each figure correspond to the CIC filter responses shown in Figure 22.26 for $N = 5$ and 10. The frequency extent of each plot shown in Figure 22.27 and Figure 22.28 is 10 MHz, which corresponds to the sample rate after decimation. Nyquist frequency is then, by definition, 5 MHz for the compensation filter design.

Looking first at Figure 22.27, showing the CIC filter response for $N = 5$ cascaded stages, we see the composite CIC-CFIR filter response in the passband looks flat over the specified 6 MHz (DC ± 3 MHz) band of interest. A potential problem can be seen in Figure 22.27, where the frequency response at 7 MHz is about -38 dB. As a generality, the frequency response should be no higher in the stopband than at the nearest close in first sidelobe. In the example shown in Figure 22.27, the first sidelobe is approximately -60 dB. As such, we surmise that there is probably a better CIC filter design option.

Shown in Figure 22.28 is our second CIC filter design option, with corresponding CFIR filter, where the number of CIC integration-comb filter pairs is increased from 5 to 10. As with the first example, the composite CIC-CFIR filter response plotted in Figure 22.28 exhibits a flat passband over the 6 MHz (DC ± 3 MHz) band of interest. The improvement of this second CIC filter design option compared to the first is lower sidelobes in the stopband. As depicted in Figure 22.28, the sidelobe at about 7 MHz is now -70 dB, which is well below the -60-dB first sidelobe level. As such, this configuration is selected as an acceptable design for our radar processor.

For our digital receiver design example, we specify that in both quadrature paths of the DDC we use a 10-stage CIC filter followed by a compensation filter. However, in practice, it is probable that there are multiple CIC filter stages in a DDC, which can equal a total of 10 CIC filter stages (as is the case for an AD6620, which uses a cascade of 2-stage and 5-stage CIC filters).

---

[32] For this example, the MATLAB Signal Processing Toolbox function fir2 was used to generate the coefficients for the CIC compensation filters. The fir2 function designs FIR filters with an arbitrary frequency response based on the frequency sampling method. The generated filter coefficients are real and symmetric.

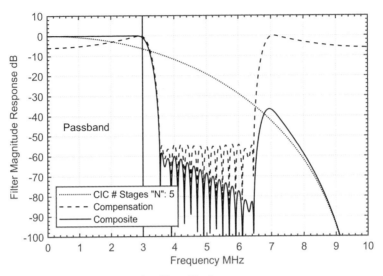

Figure 22.27 Cascaded CIC-compensation filter, $N = 5$.

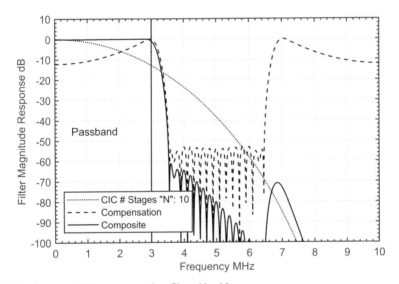

Figure 22.28 Cascaded CIC-compensation filter, $N = 10$.

22.8.5 Analog-to-Digital Converter

Incorporating an ADC into a radar receiver primarily affects dynamic range and sensitivity. Because of this, we will examine some key ADC parameters and how they factor into a receiver design. It is also important to understand the effects of noise present at the input to the ADC (usually called dither), quantization noise generated as a result of quantizing an input signal, and noise generated internally by the ADC due to circuit noise and timing instabilities.

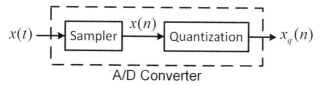

Figure 22.29 Block diagram of A/D converter (*After:* [99]).

An ADC, depicted functionally in Figure 22.29, performs the operations of sampling in time and quantizing in amplitude. Specifically, it samples and quantizes a continuous time signal, $x(t)$, to produce a digitized output, $x_q(n)$, that is a discrete time number sequence. This many-to-one mapping occurs because an ADC represents each signal sample using a finite number of binary digits (bits).[33]

ADCs are designed to operate on either unipolar or bipolar inputs. For our application, we will discuss bipolar converters, which are typical of ADCs used in digital down conversion. Representative input ranges for bipolar converters include ±1, ±2, ±2.5, ±5, and ±10 V,[34] with faster converters generally having smaller input ranges.

22.8.5.1 Quantization

The difference between the maximum, V_{max}, and the minimum, V_{min}, input values to an ADC is referred to as the full-scale range (FSR):

$$FSR = V_{max} - V_{min} \tag{22.92}$$

For example, the full-scale range of a bipolar ADC with a specified analog input range of ±1V is

$$FSR = V_{max} - V_{min} = 1 - (-1) = 2 \text{ V} \tag{22.93}$$

Given an output word length of b bits, we may represent $L = 2^b$ unique discrete levels, which are mapped to particular voltage levels depending on the full-scale voltage of the ADC. Each level at the ADC output is separated by[35]

$$\Delta = \frac{V_{max} - V_{min}}{2^b} = \frac{FSR}{L} \tag{22.94}$$

which is known synonymously as the quantization interval, a quanta, or the least significant bit (LSB) of the ADC [69]. As an example, for a word length of $b = 4$ bits, we can represent $L = 2^4 = 16$ discrete levels. For a full-scale range of 2V (±1V), the LSB size of the ADC becomes

---

[33] The number of bits in an ADC is also referred to as the resolution of the ADC.
[34] Unipolar converters can be used to convert bipolar signals by using a proper input driver to convert bipolar signals into unipolar signals.
[35] If the ADC encodes 2^b input voltage levels, including V_{max} and V_{min}, there are $2^b - 1$ steps between the encoded levels. Nonetheless, many authors use 2^b in the denominator to calculate the step size. This is a reasonable approximation when b is large.

$$\Delta = \frac{2}{16} = 125 \text{ mV} \tag{22.95}$$

The mapping from analog input to digital output (a nonlinear mapping) is typically performed via truncation or rounding. We will consider quantization via rounding, or [100, p. 11]

$$x_q = \text{round}(x/\Delta) \cdot \Delta = \text{round}\left(\frac{x \cdot 2^b}{FSR}\right) \cdot \frac{FSR}{2^b} \tag{22.96}$$

which is typical for digital signal-processing applications. Quantization via rounding also results in a quantization error that is symmetrical about zero, which is mathematically convenient.[36] For illustration, a full-scale sinusoidal input quantized by a 4-bit ADC with a full-scale range of 2V is depicted in Figure 22.30.

Note that using (22.96) results in 17 levels instead of 16. For quantizers with greater than about $L = 32$ levels (5 bits) or so, the effect of this extra level is negligible. In practical ADCs, the encoded range is usually $-L/2$ to $L/2 - 1$. We will choose to ignore this extra level, which simplifies simulation of ADC quantization.

22.8.5.2 Quantization Error

The difference in amplitude between the analog input to an ADC and the quantized digital output is referred to as the quantization error. The quantization error for the quantized sinusoid given in Figure 22.30 is shown in Figure 22.31.

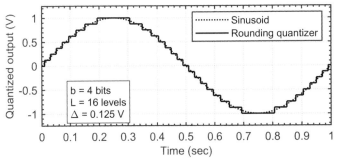

Figure 22.30 Quantized sinusoid, 4 bits, ±1V analog input range, 2V FSR.

---

[36] We have made the tacit assumption that inputs are confined to the linear range of the ADC. We will not consider input overload.

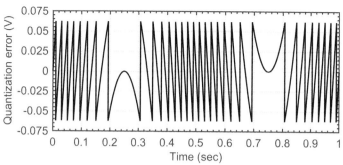

Figure 22.31 Quantization error for quantized sinusoid, 4 bits, ±1V analog input range, 2V FSR.

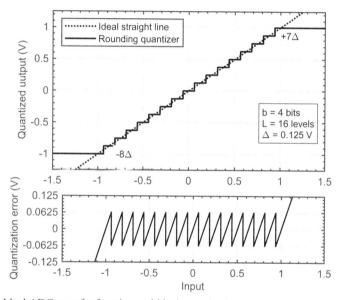

Figure 22.32 Ideal ADC transfer function and ideal quantization error.

The ideal transfer function for our example 4-bit, ±1V, bipolar ADC example is presented in Figure 22.32. For this example, the ideal quantization error shown in Figure 22.32 has an extent of $\pm \Delta/2$. Quantization errors in excess of $\pm \Delta/2$ indicate overload of the ADC.

Let us now quantify the mean-square value for the sawtooth quantization error shown in Figure 22.32. We will follow the derivation by Walt Kester presented in [78]. The equation for a sawtooth error can be expressed by

$$e(t) = st, \quad -\Delta/2s < t < \Delta/2s \text{ V} \tag{22.97}$$

where s is the slope. The mean-square value of the sawtooth error voltage may be derived as

$$E_q^2 = \overline{e^2(t)} = \frac{s}{\Delta}\int_{-\Delta/2s}^{\Delta/2s}(st)^2\, dt = \frac{\Delta^2}{12} \quad \text{V}^2 \tag{22.98}$$

and the rms value of the quantization error is then given by

$$E_q = \frac{\Delta}{\sqrt{12}} \quad \text{V} \tag{22.99}$$

where Δ is the LSB (volts).[37]

We could next quantify the mean-square value for the quantization error associated with a sinusoid, shown in Figure 22.31. However, for a sinusoidal signal that spans several quanta, (22.99) can serve as an approximation for the rms quantization error [79, p. 83].

Let us now compare the rms value of the quantization error to the rms value of a full-scale sinusoidal input signal. The full-scale rms voltage for a sinusoidal input is given by [78]

$$V_{rms} = \frac{FSR}{2\sqrt{2}} = \frac{\Delta 2^b}{2\sqrt{2}} \quad \text{V} \tag{22.100}$$

The ideal rms signal-to-noise ratio[38] in (W/W) with respect to quantization error for a full-scale sinusoid in an ideal ADC is then

$$SNR_q = \frac{V_{rms}^2}{E_q^2} = \frac{\left(FSR/2\sqrt{2}\right)^2}{\left(\Delta/\sqrt{12}\right)^2} = \left(\frac{\Delta 2^b/2\sqrt{2}}{\Delta/\sqrt{12}}\right)^2$$

$$SNR_q = \left(\frac{2^b \sqrt{12}}{\sqrt{8}}\right)^2 = \left(2^b \sqrt{3/2}\right)^2 \tag{22.101}$$

over the Nyquist bandwidth from dc to $f_s/2$ where f_s is the sampling frequency in Hz [101]. In decibel form, (22.101) becomes [102]

$$\begin{aligned}SNR_q &= 20\log\left[\left(2^b\sqrt{3/2}\right)^2\right] = 20\log(2^b) + 20\log(\sqrt{1.5}) \\ &= b\cdot 20\log(2) + 10\log(1.5) \\ &= 6.02b + 1.76 \quad \text{dB}\end{aligned} \tag{22.102}$$

Equation (22.102) represents the ideal signal-to-quantization error ratio of an ADC given sinusoidal inputs often quoted in literature [48, 49, 64, 79, 103, 104]. It should be emphasized that the achieved *SNR* for a practical ADC is always less than the theoretical SNR_q calculated from the number of bits. The theoretical performance of an ADC is however a useful gauge for comparison.

---

[37] Note that an impedance of $R_0 = 1\ \Omega$ is implied.
[38] We are calling this a signal to noise ratio although, strictly speaking it is not. More accurately, it is a signal-to-quantization error ratio.

In keeping with relating parameters in the receiver to absolute levels, let us put the full-scale input and rms quantization voltage in terms of dBm. The full-scale signal power into an ADC in dBm is given by [79, p. 133]

$$P_{FS} = 10 \cdot \log\left(\frac{V_{\text{Full scale rms}}^2}{R_0} \bigg/ 0.001 \right)$$
$$= 10 \cdot \log\left[\left(\frac{FSR}{2\sqrt{2}}\right)^2 \bigg/ R_0 \right] + 30 \quad \text{dBm} \tag{22.103}$$

where we recall the addition of 30 results in dBm instead of dBW. For ±1V input range to the ADC matched to a system impedance of 50 Ω, results in a full-scale power of 10 dBm. The quantization error power becomes

$$P_q = P_{FS} - SNR_q \quad \text{dBm} \tag{22.104}$$

Equation (22.102) is just one of several SNR_q equations associated with ADCs [50, 63]. Noise generated internally by ADCs is usually characterized as input referenced noise and expressed in terms of LSBs rms, corresponding to an rms voltage referenced to the ADC full-scale input range [79, p. 89; 105]. The level of quantization error can also be estimated via FFT [79, 102].

It should be noted that for these classical examples of quantization error, the error is completely deterministic. We have not yet approached quantization error analysis using a stochastic interpretation, which we will do shortly.

Even though we have not performed any stochastic analysis, e_q of (22.99) is frequently referred to synonymously as the rms quantization noise [69, 79]. Likewise, SNR_q of (22.102) is usually referred to synonymously as the signal-to-quantization noise ratio [69]. This is because, when approached from a stochastic point of view, the results are the same as (22.99) and (22.102) [104]. This is explored in Exercise 22.

22.8.5.3 Quantization Noise

While we can perform deterministic quantization error analysis, we usually treat quantization error as a random process and label it "quantization noise." We use this stochastic interpretation because we have the tools to handle random quantization error (i.e., quantization noise) but we do not have the tools needed to analyze deterministic quantization error in terms of its effect on signal processing.

The general assumptions used for quantization noise are [77, 100, 106]:

The quantization noise is additive and white.

The quantization noise is uncorrelated with the signal being quantized.

The quantization noise is uniformly distributed between ± Δ/2, resulting in zero mean and variance of $\Delta^2/12$ [see (22.98)].

The three assumptions listed above allow our treatment of quantization error as noise to be closer to being theoretically valid. The three assumptions are true if some amount of receiver noise is presented to the input of the ADC. The amount of noise required is set by making the expected standard deviation of the noise greater than the quantization level Δ.

Without noise present at the input of the ADC, the resulting quantization error can be deterministic and harmonic, which produces spurs. As ADC word lengths grow, however, internal instabilities in the ADC become more dominant, reducing the chance of spur-like quantization error. The additive noise present at the input to an ADC is often referred to as dither (Section 22.8.5.5).

Despite the third assumption listed above, Bennet notes that quantization noise is approximately Gaussian and essentially spread uniformly over the Nyquist bandwidth of dc to $f_s/2$ [101]. Interestingly, Widrow and Kollar refer to the general assumptions used for quantization noise listed above as more rumor than fact, but they do concede that these rumors are true under most circumstances, or are at least a very good approximation [100]. One important motivation for these assumptions is that they result in greatly simplified mathematical analysis since a nonlinear system now behaves like a linear system (the system has been linearized) [107–109].

Since we treat quantization noise at the output of the ADC as being uniformly distributed across the Nyquist bandwidth, we need to account for times when we filter the output of the ADC because the signal bandwidth in our receiver is less than the Nyquist bandwidth. This filtering eliminates quantization noise outside the signal bandwidth. To account for this, we modify (22.102) by including a processing gain, $(f_s/2)/B$, which results in [79]

$$SNR_q = 6.02b + 1.76 + 10\log\left(\frac{f_s/2}{B}\right) \text{ dB} \qquad (22.105)$$

where B is the signal bandwidth, or technically the filter bandwidth if it is wider than the modulation bandwidth.[39]

22.8.5.4 ADC Noise Figure

Having derived the ideal SNR performance for an ideal ADC, we would like to quantify how this compares to a practical ADC, which never achieves the ideal SNR. More importantly, we would like to incorporate the performance of a practical ADC into our receiver cascade analysis. The SNR of a practical ADC is difficult to predict analytically. The actual SNR of an ADC is generally provided by ADC manufacturers though.

ADC manufacturers usually measure SNR_{ADC} using a sinusoidal test signal at the ADC input. The test signal is usually full scale, or 0.5 to 1 dB below full scale (dBFS). Staying just under full scale at the input to the ADC is sometimes done because it results in better spurious behavior than a full-scale input.

Walt Kester at Analog Devices and James Karki at Texas Instruments both present a technique we can use to incorporate an ADC into a cascade analysis using an equivalent

---

[39] Sampling in excess of Nyquist can be used to take advantage of the processing gain resulting from a fixed amount of quantization noise being spread over a larger bandwidth is referred to, as might be expected, as oversampling.

noise figure for the ADC, which is derived using the measured ADC SNR$_{ADC}$ [79, 110, 111]. We will follow their example here. Specifically, an ADC can be thought of as a unity gain amplifier, with a given noise figure, and included in a cascade analysis (see Section 22.7). An equivalent ADC noise figure in excess of 30 dB is not unusual.

To derive the noise figure of an ADC, we first need to think about the power spectral density of the ADC noise, which includes quantization noise and noise internally generated by the ADC circuitry. Given SNR_{ADC} by the manufacturer, the ADC noise power can be expressed as

$$N_{ADC} = (P_{FS} - 1) - SNR_{ADC} \quad \text{dBm} \tag{22.106}$$

where P_{FS} is the full-scale power into the ADC given by (22.103). The factor of 1 subtracted from the full-scale input power in (22.106) is indicative of the manufacturer using a −1 dBFS test signal to measure SNR_{ADC}.

To express (22.106) in terms of power spectral density, we note SNR_{ADC} is specified for noise evenly distributed across the Nyquist bandwidth from dc to $f_s/2$. Adding a bandwidth term to relate (22.106) to a 1-Hz bandwidth, we get the ADC noise power spectral density given by James Karki in his derivation, which is [111]

$$N_{ADC} = (P_{FS} - 1) - SNR_{ADC} - 10\log(f_s/2) \quad \text{dBm/Hz} \tag{22.107}$$

Equation (22.107) represents the power spectral density of quantization noise and noise internally generated by the ADC combined.

In formulating noise figure, we also need the noise into the ADC. The reference power spectral density into the ADC from thermal noise, is given by

$$N_I = kT_0 = -174 \quad \text{dBm/Hz} \tag{22.108}$$

Recall that noise figure can be expressed as

$$F = \frac{S_I/N_I}{S_O/N_O} \tag{22.109}$$

Noting that the ideal gain through the ADC is 1, for example, $S_I = S_O$, and substituting (22.107) and (22.108) into (22.109), we get

$$\begin{aligned} F_{ADC} &= \frac{S_I/N_I}{S_O/N_O} = \frac{S_I/N_I}{S_O/(N_I + N_{ADC})} = \frac{1/N_I}{1/(N_I + N_{ADC})} \\ &= 1 + \frac{N_{ADC}}{N_I} \end{aligned} \tag{22.110}$$

We can approximate (22.110) as given by Walt Kester (see Exercise 12) as [79, p. 102; 112]

$$\begin{aligned} F_{ADC} &\cong N_{ADC} - N_I \\ &= (P_{FS} - 1) + 174 \text{ dBm} - SNR_{ADC} - 10\log(f_s/2) \quad \text{(dB)} \end{aligned} \tag{22.111}$$

As an example, let the full-scale power into the ADC be +10 dBm, corresponding to an analog input range of ±1V and let the input resistance of the ADC be 50 Ω [see (22.103)]. We assume a 14-bit converter, operating at f_s = 60 MSPS, with a specified SNR of 74.8 dB. Substitution into (22.111) yields

$$F_{ADC} = (10-1) + 174 - 74.8 - 10\log(60 \cdot 10^6/2) \qquad (22.112)$$
$$= 33.4 \text{ dB}$$

We have made the tacit assumption of matched impedances, which are typically 50 Ω. The input impedance of an ADC is not always 50 Ω. ADC input impedances of 200 Ω and 1 KΩ are not unusual. To avoid impedance mismatch, one practice is to use an impedance matching transformer to match the system impedance to the ADC impedance [79, 111].

James Karki uses the example of an ADC with an input impedance of 200 Ω. Matching the 50 Ω system output impedance to a 200 Ω ADC input impedance requires a 1:4 impedance ratio (1:2 turns ratio) transformer [111]. Compared to (22.111), the ADC noise figure is reduced by the 4:1 impedance ratio of the system and the ADC impedance, which can be expressed as

$$\Delta F = 10\log(R_{ADC}/R_0) \qquad (22.113)$$

For this example, the ADC noise figure is reduced by

$$\Delta F = 10\log(R_{ADC}/R_0) = 10\log(200/50) = 6 \text{ dB} \qquad (22.114)$$

Gain Prior to ADC

An important design consideration is to establish the right amount of amplified receiver noise to act as dither at the input to the ADC (see Section 22.8.5.5) [63]. We can use the ADC noise figure of (22.111) to determine how much gain (and analog noise figure) is necessary prior to the ADC in order to minimize its impact on overall system noise figure while still dithering the input. The receiver gain in combination with the full-scale level of the ADC, determines the maximum signal input to the receiver. As such, there is a trade-off between system noise figure and maximum input signal or dynamic range.

For illustration, we will consider a two-stage cascade of the RF front end followed by an ADC, depicted in Figure 22.33. The gain and noise figure of the analog portion of a digital receiver are encompassed in G_{RF} and F_{RF}, respectively. For the ADC stage, G_{ADC} = 1, and F_{ADC} is the NF of the ADC.

Let ΔNF represent the amount of acceptable degradation in the overall receiver noise figure due to the ADC noise figure. By comparing the noise figure of the receiver front end and ADC cascade to the noise figure of the receiver front end, we can write (see Exercise 22) [113]

$$G_{RF} = 10\log\left(\frac{10^{F_{ADC}/10} - 1}{10^{\Delta NF/10} - 1}\right) - F_{RF} \qquad (22.115)$$

Figure 22.33 ADC cascade.

A general rule of thumb for ΔNF is to allow a few tenths of dB increase in noise figure due to the ADC. This offers a reasonable compromise between gain and sensitivity. For example, let the noise figure of the RF front end be 5 dB with an ADC noise figure of 30 dB. In our design, let the acceptable amount of degradation be $\Delta NF = 0.4$ dB [112]. The necessary amount of RF front end gain up to the ADC becomes

$$G_{RF} = 10\log\left(\frac{10^{30/10}-1}{10^{0.4/10}-1}\right) - 5 = 10\log\left(\frac{1{,}000-1}{1.1-1}\right) - 5 = 35.2 \text{ dB} \quad (22.116)$$

Determining how much dither noise is applied to the ADC input by using this technique is left as an exercise. We note that 35.2 dB is fairly high compared to the 20- to 25-dB RF LNA gain generally required by an analog design to establish noise figure. This is due to the very high noise figure of the ADC.

In addition to performing a cascade analysis using an equivalent noise figure for the ADC, there are a number of other approaches and guidelines used for establishing the correct amount of dither into the ADC. One general rule is to use 1 to 1.5 bits of dither. Lyons suggests an rms level of 1/3 to 1 LSB voltage level for wideband dither and 4 to 6 LSB voltage levels for out-of-band dither [98, p. 708].

Barton characterizes the *quantizing noise voltage* E_q added by the ADC as [73, p. 220]

$$E_q = \frac{\Delta}{\sqrt{12}} = \frac{V_{max}}{2^{b-1}\sqrt{12}} \quad (22.117)$$

where Δ is the voltage corresponding to the LSB of the A/D converter, V_{max} is the peak voltage that corresponds to the full ADC output, and b is the number of ADC bits that express the peak voltage that varies over $\pm V_{max}$. Equation (22.117) is a combination of (22.94) and (22.99).

Similarly to the ADC noise figure approach discussed earlier, Barton recommends adjusting the gain prior to the ADC such that the rms noise voltage at the output of the ADC, resulting from thermal noise and quantization, is [73, p. 220]

$$E_{nout} = q\Delta = q\sqrt{12}E_q \quad (22.118)$$

where Barton suggests $q \approx 1.5$, which is a constant chosen to provide a practical compromise between the conflicting needs of dynamic range and small quantizing noise (e.g., sensitivity). This results in a thermal noise power at the input to the ADC that is $12q^2$ times the *quantizing noise power*, E_q^2 (R = 1 Ω). It is left as an exercise to see how this approach compares to the ADC noise figure approach [110, 111].

22.8.5.5 Dither

Dithering is the deliberate use of a small amount of noise at the input to an ADC that is uncorrelated with the signal to be digitized. This noise is usually referred to as dither noise" or simply dither. One purpose of dither is to counter the effects of quantization noise by controlling the statistical properties of quantization error. More specifically, dither eliminates the periodicity of the quantization error due to a sinusoidal input signal, whitening the spectrum of the quantization error. Another is to linearize the characteristics of the ADC, thus improving the effective resolution of the ADC.

Dithering is imperative in radar receivers for a number of reasons. For instance, digitizing sinusoidal signals can result in quantization noise that is highly correlated, resulting in spurious signals at harmonics of the input. Dithering randomizes the quantization error, reducing spurious levels.

Dithering is also very important when considering weak or subquanta signals [114]. A weak signal that exercises only a single quanta results in clipping, causing numerous spectral harmonics. Subquanta signals, which would not exercise even a single quanta, are irrevocably lost due to the ADC. Dithering preserves the information from weak or subquanta inputs (including their power ratios) by whitening the signal and clutter components of the ADC input. Dithering causes these signals to exercise at least a few quanta, allowing signal and clutter components to be recovered via coherent integration.[40]

Some of the earliest work on the ability of dither to extend ADC dynamic range via coherent integration was published in 1963 by G. G. Furman in two RAND Corp reports [107, 108]. Furman considered sinusoidal and sawtooth dither signals, asserting that dither improves quantizer performance by enabling coarse quantizers to emulate ultrafine ones [109]. Vanderkooy and Lipshitz, showed that by the use of dither, the resolution of an ADC can be improved to well below the least significant bit [115]. Oppenheim emphasized that to preserve dynamic range, at least the lowest level of the ADC must be dithered by noise [106, p. 309].

Dither noise can be generated in a number of ways. One common method employed in radar is to use amplified thermal noise from the receiver front end (see Section 22.8.5.4), where the receiver gain is designed to establish the desired level of noise into the ADC (see Section 22.8.5.5). This type of dither is bandlimited according to the RF-IF bandwidth, which is typically less than the Nyquist frequency. Similarly, we can inject random noise from a calibrated external noise source [98]. The downside of these approaches is that the dither falls within the passband of the receiver, resulting in a loss in sensitivity. In some radars clutter can provide adequate ADC dithering, particularly if the clutter bandwidth spans multiple Doppler resolution cells (i.e., the clutter signal is not a pure sinusoid). This is true in SAR, for example.

There are techniques aimed at avoiding this loss of sensitivity by removing the dither after it has served its purpose. One approach is referred to as subtractive dither, which uses digitally generated pseudorandom noise. The pseudorandom noise is converted to

---

[40] Integration time (and its attendant bandwidth) predicates the level to which very small signals may be recovered.

analog and added to the signal into the ADC. It is then removed via subtraction after conversion.[41] An analog variation of this, referred to as out-of-band dither, is to use band-limited dither, usually low-frequency noise, which is designed to be rejected by subsequent digital filtering [98, 116].

22.9 RECEIVER CONFIGURATIONS

We close this chapter with a brief discussion of some receiver configurations and discuss some of their balance, alignment, and calibration requirements. Figure 22.34 contains a simplified block diagram of a three-channel, monopulse receiver (MF = matched filter, SP = signal processor) [117, p. 168]. A monopulse receiver is used on radars where there is a requirement to provide a three-dimensional measurement of target position. These radars typically measure a range-related quantity, Δr, and two orthogonal, angle-related quantities, Δu and Δv. Δr is usually measured relative to an expected target range, such as the output of a range tracker. Δu and Δv are angle quantities relative to boresight, which is the direction the radar beam is pointing. Δr, Δu, and Δv can be combined with other range and angle parameters to determine the target location relative to some coordinate system such as a Cartesian coordinate system centered at the radar.

The term "monopulse" derives from the fact that the radar, ideally, measures Δr, Δu, and Δv based on the return from a single (mono) pulse, or a burst of pulses if the radar is using coherent processing. The modifier "three-channel" derives from the use of separate receivers, or channels, for each parameter: Δr, Δu, and Δv.

The three signals processed by the receiver channels are formed in the feed/array, indicated notionally, on the left of the diagram. In a radar that uses a reflector antenna or a space-fed phased array (see Chapter 13), the device is the antenna feed and in a constrained-feed phased array, the device is the array itself. In one of the simplest forms, the feed consists of four horn antennas spaced close together. In practice, the feed can consist of several horn antennas where some of the horn antennas are multimode [48; 117, p. 165; 118–123]. Multihorn, multimode feeds are used when there is a desire or requirement to simultaneously provide sidelobe control of both sum and difference antenna patterns. In a constrained-feed phased array, again in the simplest form, the array is divided into four quadrants to provide the necessary signals.

---

[41] This type of dither is incorporated into some ADCs and controlled by an enable bit.

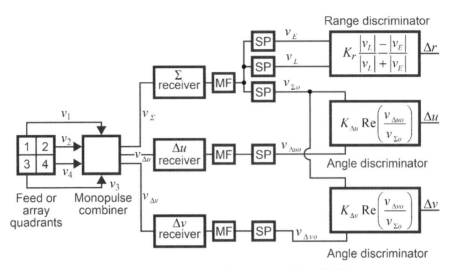

Figure 22.34 Three-channel monopulse receiver and processor. (*After*: [117].)

The outputs of the four ports of the feed, or the four quadrants of the array, are combined in the monopulse combiner [117, pp. 141, 160] to create the three signals used by the monopulse receiver and the subsequent monopulse processor. In one case, the four outputs are summed to form the *sum*, or Σ, signal. One of the orthogonal angle channel signals (e.g., Δv) is formed by summing the signals from ports 1 and 2, summing the signals from ports 3 and 4, and subtracting the two sums. This is termed the Δv *difference channel signal*. The other orthogonal angle channel signal, Δu, is formed by summing the signals from ports 1 and 3, summing the signals from ports 2 and 4, and subtracting the two sums. In equation form [117, p. 160]

$$v_\Sigma(t) = v_1(t) + v_2(t) + v_3(t) + v_4(t)$$
$$v_{\Delta v}(t) = [v_1(t) + v_2(t)] - [v_3(t) + v_4(t)] \quad (22.119)$$
$$v_{\Delta u}(t) = [v_1(t) + v_3(t)] - [v_2(t) + v_4(t)]$$

We note that we are forming the sums of voltages with the sums being performed at the RF. The implication of this is that the relative phases and amplitudes of $v_1(t)$ through $v_4(t)$ must be preserved in the monopulse combiner for all RFs of interest. This places restrictions on the combiner. Also, as discussed in Section 22.7.4, the combiner must be a low loss device since its loss contributes directly to receiver noise figure.

If we plot normalized versions of $v_\Sigma(t_1)$, $v_{\Delta u}(t_1)$, and $v_{\Delta v}(t_1)$ as we vary the target location relative to boresight, we have a normalized *sum* and two normalized *difference* voltage patterns. In these expressions, t_1 is time the target return is present. Examples of a sum and one of the difference patterns are contained in Figure 22.35 [normalized to the peak of $v_\Sigma(t_1)$] [117, pp. 153, 159]. The difference voltage plot has two main beams and, more importantly, is zero when the target is at boresight. Also, the sign and magnitude of the difference voltage is directly related to the location of the target relative to boresight. This is the information we use to determine Δu and Δv, the target angles relative to boresight.

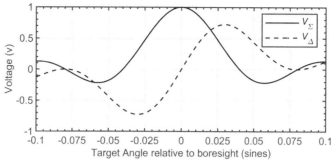
Figure 22.35 Sum and difference patterns.

After the monopulse combiner, the $v_\Sigma(t)$, $v_{\Delta u}(t)$, and $v_{\Delta v}(t)$ signals are sent to three identical receiver channels. A key term here is "identical." The receivers must have identical gain and phase characteristics over their entire operating frequency range. Also, the gain and phase characteristics should be independent of signal amplitudes. This is important because the $v_\Sigma(t)$ and $v_\Delta(t)$ signals can have much different amplitudes. If the target is at, or close to, boresight, $v_\Sigma(t)$ will be large and $v_\Delta(t)$ will be small. If the receivers do not provide the same gain and phase shifts to $v_\Sigma(t)$ and $v_\Delta(t)$, the subsequent processing used to determine Δu and Δv will not give the expected result.

Since the three channels are not generally identical, the receivers and the Δu and Δv formation circuits/algorithms must be calibrated, which is usually accomplished by creating *discriminator curves*. This can be done by radiating a test signal from a test tower in the far field of the antenna and moving the antenna boresight while measuring Δu and Δv. The plots of Δu and Δv versus the angle between the test signal and boresight are the discriminator curves. The Δu discriminator curve is generated for $\Delta v = 0$ and vice versa. Note that calibration is especially important in digital receivers where the ADC can introduce significant nonlinearities at small signal levels (see Section 22.8).

The calibration and alignment must be performed at several frequencies within the operating band of the radar since phase errors can be caused by path length differences in the combiner and other plumbing between the feed outputs (outputs of the four horns or four array quadrants) and the first mixer (see Figure 5.11). Also, gain and phase characteristics of the three RF amplifiers will most likely be different over the RF operating range. Since the characteristics of the various receiver components can change over relatively short periods of time, it is often necessary that calibration be performed regularly. This is most often accomplished by injecting a test signal, termed a *pilot pulse*, into the receiver front end and determining the amount of phase and gain imbalance between channels. Channel balance is then maintained by controlling attenuators and phase shifters in each receiver channel accordingly (or in the calculating of the monopulse output if implemented via computer) [124, p. 69].

As indicated earlier, the Δu and Δv signals are formed in the angle discriminators. The angle discriminators of Figure 22.22 would apply to a reflector antenna or a space-fed phased array antenna because the form of Δ (Δu or Δv) is

$$\Delta = K_\Delta \text{real}\left(\frac{v_{\Delta o}}{v_{\Sigma o}}\right) \qquad (22.120)$$

For a constrained-feed phased array, the real operator would be replaced by the imaginary operator since the angle information in this type of an array is contained in the imaginary part of $v_{\Delta o}/v_{\Sigma o}$ [124].

In (22.120), $v_{\Sigma o} = G_\Sigma v_\Sigma(t_1)$ and $v_{\Delta o} = G_\Delta v_\Delta(t_1)$ where Δ could be Δu or Δv and t_1 is the time at which the matched filter output is sampled (hopefully at the target range delay). G_Δ and G_Σ are the total, complex voltage gains of the sum and difference receivers, from the feed output to the inputs of the discriminators. K_Δ is a scale factor that converts the ratio to an angle. Figure 22.36 contains a plot of Δ versus u for the sum and difference pattern plots of Figure 22.35. The angle, u, is the angle between the antenna boresight and the LOS to the target, and has the units of sines (see Chapter 13). K_Δ was chosen so that the slope of the curve is unity.

The solid curve corresponds to the balanced case where $G_\Delta = G_\Sigma$ and the dashed curve corresponds to the case where $G_\Delta = (2^{1/2})G_\Sigma$, a 3-dB gain imbalance. As can be seen, the slope of the discriminator curve is no longer unity for the imbalance case. This can have an impact on track loop performance in that the slope of the discriminator curve directly affects the closed-loop bandwidth of the track loop, and thus the track accuracy.

Figure 22.37 contains a plot for the case where $|G_\Delta| = |G_\Sigma|$ but where the phases differ by 30°. As with the gain imbalance, the phase imbalance caused a change in the slope of the discriminator curve.

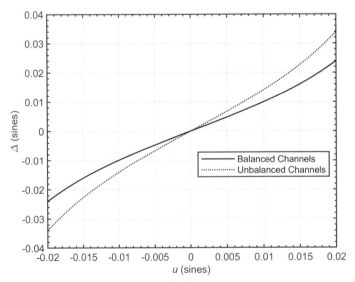

Figure 22.36 Angle discriminator—amplitude imbalance.

Receiver calibration can also affect the output of the range discriminator, but not as much as for angle. This is because of the way the Δr signal is usually formed. Specifically, the Δr signal is formed by some variation of the equation [117, p. 95]

$$\Delta r = K_r \frac{|V_L| - |V_E|}{|V_L| + |V_E|} \tag{22.121}$$

where $|V_L|$ and $|V_E|$ are termed the late and early gate signals. The form of (22.121) assumes digital signal processing. In analog processing, this ratio is sometimes formed differently and in a fashion where channel balance can affect the output of the range discriminator [117]. The early and late gate signals and are defined by

$$\begin{aligned} |V_E| &= |G_\Sigma v_{\Sigma o} (\tau_{trk} - \Delta\tau)| \\ |V_L| &= |G_\Sigma v_{\Sigma o} (\tau_{trk} + \Delta\tau)| \end{aligned} \tag{22.122}$$

where τ_{trk} is the expected target range delay from the range tracker and $\Delta\tau$ is an offset about τ_{trk}. Typically, $\Delta\tau$ is one-half of the compressed pulsewidth [117]. This is illustrated in Figure 22.38 for the case of an ideal, unmodulated pulse. K_r is chosen so that Δr has the desired units (e.g., m).

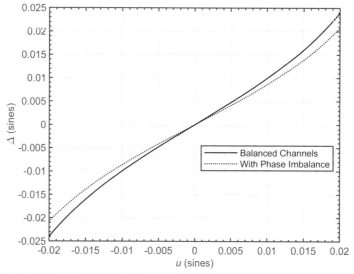

Figure 22.37 Angle discriminator—phase imbalance.

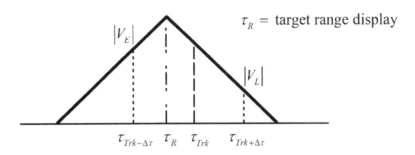

Figure 22.38 Illustration of range samples. (*After*: [117].)

The forms of (22.121) and (22.122) mean that receiver calibration and channel balance are not an issue in range tracking because G_Σ cancels in the numerator and denominator of (22.121). A calibration factor that must be considered is pulse shape at the output of the matched filter or imbalances in the signal processing between the matched filter output and the input to the range discriminator. As with the angle discriminator, these can be accounted for by calibration.

Early radars tried to conserve hardware by using two receiver channels instead of three. An example of one such implementation is shown in Figure 22.39. In this diagram, the Δ signal is switched between the Δu and Δv signals from the combiner and the difference signal is added to and subtracted from the sum signal to form $v_\Sigma(t) + v_\Delta(t)$ and $v_\Sigma(t) - v_\Delta(t)$. The combiner effectively un-combines the sum and difference signals from the monopulse combiner to recover the original signals from the left and right (or upper and lower) halves of the aperture.[42] While this implementation saves hardware, it also doubles the amount of time required to determine all three of the Δr, Δu, and Δv parameters.

The use of $v_\Sigma(t) + v_\Delta(t)$ and $v_\Sigma(t) - v_\Delta(t)$ relieves some of the gain and phase linearity issues in that both $v_\Sigma(t) + v_\Delta(t)$ and $v_\Sigma(t) - v_\Delta(t)$ are about the same size when the radar is tracking the target. However, gain and phase imbalance becomes more of an issue. In the example of Figure 22.39, the Δ signal is formed as

$$\Delta = K_\Delta \left\{ \ln \left| G_{\Sigma-\Delta} \left[v_\Sigma(t_1) - v_\Delta(t_1) \right] \right| - \ln \left| G_{\Sigma+\Delta} \left[v_\Sigma(t_1) + v_\Delta(t_1) \right] \right| \right\}$$
$$= K_\Delta \ln \frac{\left| G_{\Sigma-\Delta} \left[v_\Sigma(t_1) - v_\Delta(t_1) \right] \right|}{\left| G_{\Sigma+\Delta} \left[v_\Sigma(t_1) + v_\Delta(t_1) \right] \right|} \quad (22.123)$$

If $|G_{\Delta+\Sigma}| = |G_{\Delta-\Sigma}|$ and $|v_\Delta(t_1)| \ll |v_\Sigma(t_1)|$, (22.123) reduces to [124, p. 167]

---

[42] This results in more equal signal levels through the ADCs but imposes more stringent requirements on receiver channel matching and calibration after the combiner. For this reason the combiner is often located at the end of the IF chain just prior to baseband conversion and/or A/D conversion.

$$\Delta \approx K_\Delta \frac{|v_\Delta(t_1)|}{|v_\Sigma(t_1)|} \text{sgn}\left[v_\Delta(t_1)\right] \qquad (22.124)$$

If $|G_{\Delta+\Sigma}| \neq |G_{\Delta-\Sigma}|$, Δ can vary significantly from this ideal value and can even lead to bias errors in the angle tracker.

To mitigate the problems caused by channel imbalance in two-channel receivers, some early radars also reversed the signals into the receiver on alternate dwells (pulses or coherent processing intervals). One example of this is illustrated in Figure 22.40. In this case, the receiver channels alternately carry $v_\Sigma(t)$ $-/+$ $v_\Delta(t)$ and $v_\Sigma(t)$ $+/-$ $v_\Delta(t)$. This will cause errors due to channel imbalance to average out over time. However, now the update rate has been decreased by a factor of four relative to full monopulse.

Instead of reducing the number of receiver channels, we can increase them to four and eliminate the monopulse combiner. This is illustrated in Figure 22.41. With this configuration, we process $v_1(t)$ through $v_4(t)$ in separate receivers and form the sum and difference signals at the output of the signal processors. If we use a digital receiver or a digital signal processor, we would have an implementation of *digital beamforming*. This technique would enhance flexibility in that we could form (tightly spaced) *multiple simultaneous beams*, or simultaneously implement amplitude and phase comparison monopulse (for a constrained-feed phased array or an active array) or perform some other angle functions such as sidelobe cancellation. The price paid for this flexibility is that receiver balance, calibration, and alignment, over the operational RFs of the radar, become much more important. If the receivers are not properly balanced, we would likely introduce significant angle bias errors and could significantly degrade the monopulse discriminator. Depending on how the signal processor outputs are combined to determine range error, this could also be seriously degraded by channel imbalance.

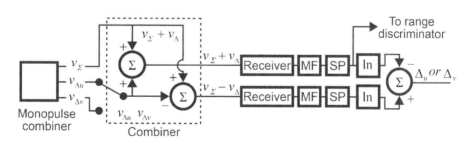

Figure 22.39 Two-channel monopulse receiver (*After*: [117].)

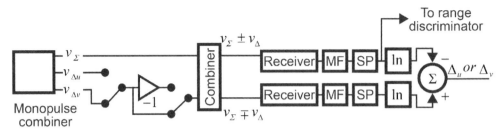

Figure 22.40 Two-channel receiver with Δ sign switching (*After*: [117]).

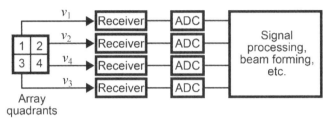

Figure 22.41 Four-channel receiver.

An extension to the four-channel receiver that is being implemented in some modern radars [125, 126] is to divide the array (usually an active array) into a large number of subarrays with a separate receiver, and possibly, signal processor, for each subarray. With this the concept of digital beamforming can be expanded to include multiple simultaneous beams, adaptive nulling, difference pattern sidelobe control, and/or some form of super-resolution technique (e.g., multiple signal classification (MUSIC) [127]), that has been only theoretically considered in the past. However, as implied by the above discussions, this added capability comes at the cost of a more stringent balance requirements. A counter to this would be to move the ADCs closer to the LNA output of the T/R modules and have a true digital radar, wherein calibration and alignment can be performed digitally.

22.10 EXERCISES

5. Given an RF of 10 GHz, a first IF of 60 MHz, and assuming a low side LO, what does the LO frequency need to be? What is the image frequency? If we use a forth-order Butterworth filter with an 8-MHz passband as preselector, how much image rejection is provided?

6. Repeat Exercise 1 using a first IF of 540 MHz and assuming a high side LO.

7. Calculate the noise bandwidth of a Butterworth filter for orders 1 through 5. How does the noise bandwidth compare to the 3-dB bandwidth? Noise bandwidth is given by [73, p. 198]

$$B_n \equiv \frac{1}{|H(f_0)|^2} \int_{-\infty}^{\infty} |H(f)|^2 \, df \quad \text{Hz} \tag{22.125}$$

where $H(f)$ is the frequency response and f_0 is the center of the frequency response.

8. Show that the units in (22.10) are correct.

9. Simulate Gaussian noise passed through a 4th order Butterworth filter with a 3-dB bandwidth of 10 MHz at a sample rate of 100 MHz. Use a noise figure of 5 dB and a gain of 0 dB. Look at the ensemble average of the output for 100 runs in the time domain. Does the result correlate to (22.8)? Hint: To generate the input noise power, use $B = 100$ MHz.

10. For the RF chain shown in Figure 22.6, calculate the gain, noise figure, and noise power at each device output.

11. Simulate noise and signal plus noise for a 10-μs pulse in additive white Gaussian noise passed through a 4th order Butterworth filter with a 3-dB bandwidth of 8 MHz. Run the simulation as a sample rate of 100 MHz. Use a noise figure of 5 dB and a gain of 0 dB to generate the input noise. Look at the time domain output and adjust the SNR until the TSS requirement is met. Is the SNR as expected?

12. Complete the cascade analysis for the receiver chain shown in Figure 22.10. Do your results match those indicated in Figures 22.11, 22.12, and 22.13?

13. We want to choose a sample rate to center our signal in a Nyquist zone so that we maximize the amount of spacing between images, simplifying filter design. For a 5-MHz bandwidth LFM waveform on a 30 MHz IF, what is the required ADC rate to center the signal in the second Nyquist zone? What is the aliased digital IF? What is the spacing between images?

14. Using the parameters from Exercise 9, generate a figure like Figure 22.17, 18, 19 and 20.

15. Generate a baseband 3-MHz bandwidth LFM waveform with a 100-μs pulsewidth. Using a sample rate of 3 MHz, digitally match filter (pulse compress) the waveform. Repeat for sample rates of 4.5 MHz, 6 MHz, and 7.5 MHz. Compare the sidelobe levels of the compressed pulse. Is there a benefit to oversampling slightly?

16. Why is the noise figure expression of an (22.111) approximation? Hint: derive in terms of the linear definition for noise figure.

17. Given an RF of 8 GHz and a first IF of 30 MHz, what is the LO for highside injection? What is the image frequency? If we use a 2nd order Butterworth as a preselector, how much image rejection do we get? If we need 60 dB of image rejection, what is the minimum filter order?

18. Using the parameters of Exercise 13, calculate all of the spurious frequencies generated by mixing up to the eighth order. What is the nearest frequency to the passband of the preselector? Are there any spurious responses in the passband of the preselector?

19. Consider two tones at $f_1 = 60$ MHz and $f_2 = 63$ MHz. What are the frequencies of the 2nd- and 3rd-order intermodulation products? Which tones are nearest the

desired signals in frequency? Which tones are farthest away? Is it practical to reject some or all of the intermodulation products by filtering?

20. A receiver has a 3rd order input intercept point of –10 dBm and a noise figure of 7 dB and a bandwidth of 4 MHz. What is the spurious-free dynamic range?

21. Considering a radar transmitting a 1-MHz bandwidth LFM with a 40 μs pulsewidth with a maximum instrumented range of 120 km. This radar is to accommodate target RCSs between σ_{min} = 0.01 m^2 (–20 dBsm) and σ_{max} = 1,000 m^2 (30 dBsm). What is the minimum dynamic range we need to design the radar for?

22. Consider a 16-bit ADC operating at 40 MSPS with a bipolar input of ±10 V, what is the full scale power in dBm given a 50 Ω system impedance? What is the LSB? If the manufacturers specified SNR is 78 dB, what is the effective noise figure of the ADC?

23. The ADC described in Exercise 18 has a 1 kΩ impedance instead of the 50 Ω impedance of the rest of the receiver chain. If we use an inductive impedance transformer, what is the effect on the noise figure of the ADC?

24. Consider the cascade shown in Figure 22.33. If the receiver and ADC have noise figures of 7 dB and 30 dB, respectively, if we allow ΔNF = 0.3 dB of noise figure degradation, how much gain is needed in the receiver front end to use amplified thermal noise from the RF front end as dither? Determine the thermal noise power into the ADC and the total noise power at the output of the ADC.

25. Treat the quantization error voltage as a random variable ε with a uniform probability density function spanning ±q/2 with an amplitude of 1/q. Calculate the mean-square and root mean-square values. How do they compare to the results of (22.99) and (22.102).

26. Derive (22.115). Hint: start with the cascade noise figure of the receiver front end and the ADC and compare to the noise figure of the receiver front end.

27. For the example associated with Figure 22.33, determine how much dither noise is applied to the ADC input.

28. Using the parameters from Exercise 20, determine the gain necessary prior to the ADC, the thermal noise into the ADC, and the total noise at the output of the ADC using Barton's approach given by (22.117) and (22.118). How does this compare to the results of Exercise 20? If the results differ, what ΔNF would be needed to match the results?

29. Prove (22.89).

References

[1]   Armstrong, E. H., "Some Recent Developments in the Audion Receiver," *Proceedings of the Institute of Radio Engineers*, vol. 3, no. 3, pp. 215-238, 1915.

[2]   Armstrong, E. H., "Method of receivering High Frequency Oscillations". Patent 1,342,885, 8 June 1920.

[3]   Armstrong, E. H., "The Super-Heterodyne-Its Origin, Developement, and Some Recent Improvements," *Proceedings of the Institute of Radio Engineers*, vol. 12, no. 5, pp. 539-552, October 1924.

[4]   Schottky, W. H., "On the Origin of the Super-Heterodyne Method," *Proceedings of the Institute of Radio Engineers*, vol. 14, no. 5, pp. 695-698, October 1926.

[5]   Brittain, J. E., "Electrical engineering Hall of Fame-Edwin H. Armstrong,," *Proceedings of the IEEE*, vol. 92, no. 3, pp. 575-578, March 2004.

[6]   Fessenden, R. A., "Wireless Signaling". Patent 706740, 12 August 1902.

[7]   Belrose, J. S., "Reginald Aubrey Fessenden and the birth of wireless telephony," *Antennas and Propagation Magazine, IEEE*, vol. 44, no. 2, pp. 38-47, April 2002.

[8]   Brittain, J. E., "Electrical engineering hall of fame: Reginald A. Fessenden," *Proceedings of the IEEE*, vol. 92, no. 11, pp. 1866-1869, November 2004.

[9]   De Forest, L., "Space Telegraphy". Patent 879,532, 18 February 1908.

[10]   De Forest, L., "The Audion: A new receiver for wireless telegraphy," Proceedings of the American Institute of Electrical Engineers, vol. 25, no. 10, pp. 719-747, October 1906.

[11]   De Forest, L., "The audion-detector and amplifier," *Proceedings of the Institute of Radio Engineers*, vol. 2, no. 1, pp. 15-29, March 1914.

[12]   Brittain, J. E., "Electrical Engineering Hall of Fame: Lee de Forest," *Proceedings of the IEEE*, vol. 93, no. 1, pp. 198-202, January 2005.

[13]   Hong, S., *Wireless, From Marconi's Black-Box to the Audion*, Cambridge, Massachusetts: The MIT Press, 2001.

[14]   Godfrey, D. G. and F. A. Leigh, Eds., *Historical Dictionary of American Radio*, Westport, CT: Greenwood Publishing Group, Inc., 1998.

[15]   Pound, R. V., *Microwave Mixers*, vol. 16 of MIT Radiation Lab. Series, New York: McGraw-Hill, 1948.

[16]   Saad, T. S., R. C. Hansen and G. J. Wheeler, Eds., *Microwave Engineers' Handbook*, vol. 1, Dedham, MA: Artech House, Inc., 1971.

[17]   Matthaei, G. L., L. Young and E. M. T. Jones, *Microwave Filters, Impedance-Matching Networks, and Coupling Structures*, Norwood, MA: Artech House, 1980.

[18]   K&L Microwave, "K&L Product Catalog," [Online]. Available: www.klmicrowave.com. [Accessed 2019].

[19]   Cobham, "Cobham Microwave RF & Microwave Filters," [Online]. Available: www.cobham.com. [Accessed 2019].

[20]   Pond, C. W., "Crystal Filter Transient Behavior," *Proceedings of the 6th Quartz Devices Conference*, 1984.

[21]   Belov, L. A., S. M. Smolskiy, and V. N. Kochemasov, *Handbook of RF, Microwave, and Millimeter-Wave Components*, Boston: Artech house, 2012.

[22]   Skolnik, M. I., *Introduction to Radar Systems*, Third ed., New York: McGraw-Hill, 2001.

[23]   Hall, J. S., Ed., *Radar Aids to Navigation*, vol. 2 of MIT Radiation Lab. Series, New York: McGraw-Hill, 1948.

[24]   Van Voorhis, S. N., *Microwave Receivers*, vol. 23 of MIT Radiation Lab. Series, New York: McGraw-Hill, 1948.

[25]   Manheimer, W. M. and G. Ewell, "Cyclotron Wave Electrostatic and Parametric Amplifiers," Naval Research Laboratory, Washington, 1997.

[26]   Manheimer, W. M. and G. W. Ewell, "Electrostatic and parametric cyclotron wave amplifiers," Plasma Science, IEEE Transactions on, vol. 26, no. 4, pp. 1282-1296, Aug. 1998.

[27]   Vanke, V. A. and N. S. Hiroshi Matsumoto, "Cyclotron Wave Electrostatic Amplifier," Journal of Radioelectronics, 1999.

[28]   Budzinsky, Y. A. and S. Kantyuk, "A new class of self-protecting low-noise microwave amplifiers," Microwave Symposium Digest, 1993., *IEEE MTT-S International*, vol. 2, pp. 1123-1125, 14-18 June 1993.

[29]   Budzinskiy, Y. A., S. V. Bykovskiy, V. E. Kotov and O. A. Savrukhin, "On the increase of the frequency band of cyclotron protective device," *15th International Crimean Conference Microwave & Telecommunication Technology*, Sevastopol, Crimea, 2005.

[30]   Budzinskiy, Y. and S. Bykovskiy, "Cyclotron protective device with increased frequency band," Vacuum Electronics Conference, 2009. IVEC '09. *IEEE International*, pp. 60-61, 28-30 April 2009.

[31]   Budzinskiy, Y. A., S. V. Bykovskiy, I. I. Golenitskiy and N. G. Dukhina, "Electron optical system of Cyclotron Protective Device," *24th International Crimean Conference Microwave & Telecommunication Technology (CriMiCo)*, Sevastopol, 2014.

[32]   Budzinskiy, Y. A., S. V. Bykovskiy and V. G. Kalina, "Engineering calculation of cyclotron protective devices," *24th International Crimean Conference Microwave & Telecommunication Technology*, Sevastopol, 2014.

[33]   Maas, S. A., *Microwave Mixers*, 2nd ed., Norwood, MA: Artech House, 1993.

[34]   Jay, F., Ed., *IEEE Standard 100, IEEE Standard Dictionary of Electrical and Electronics Terms*, Fourth ed., New York: The Institute of Electrical and Electronics Engineers, Inc., 1988.

[35]   Carson, R. S., *Radio Communications Concepts: Analog*, New York: John Wiley & Sons, 1990.

[36]   Poisel, R. A., *Electronic Warfare Receivers and Receiving Systems*, Norwood, MA: Artech House, 2014.

[37]   Faria, D., L. Dunleavy and T. Svensen, "The Use of Intermodulation Tables (IMT) for Mixer Simulation," April 2008. [Online]. Available: www.agilent.com. [Accessed 22 April 2015].

[38]   Butterworth, S., "On the Theory of Filter Amplifiers," *Experimental Wireless & the Wireless Engineer*, vol. 7, pp. 536-541, October 1930.

[39]   Blinchikoff, H. J. and A. I. Zverev, *Filtering in the Time and Frequency Domains*, New York: John Wiley & Sons, Inc., 1976.

[40]   Rhodes, J. D., *Theory of Electrical Filters*, John Wiley, 1976.

[41]   Van Valkenburg, M. E., *Analog Filter Design*, New York: Oxford University Press, 1982.

[42]   Thede, L., *Practical Analog and Digital Filter Design*, Norwood, MA: Artech house, 2004.

[43]   Waters, W. M. and B. R. Jarrett, "Bandpass Signal Sampling and Coherent Detection," *Aerospace and Electronic Systems, IEEE Transactions*, Vols. AES-18, no. 6, pp. 731-736, Nov. 1982.

[44]   Johnson, J. B., "Thermal Agitation of Electricity in Conductors," *Physical Review*, vol. 32, pp. 97-109, July 1928.

[45]   Blake, L. V., "A Guide to Basic Pulse-Radar Maximum-Range Calculation, Part 1 - Equations, Definitions, and Aids to Calculation," Washington, 1969.

[46]   Blake, L. V., "A Guide to Basic Pulse-Radar Maximum-Range Calculation, Part 2 - Derivations of Equations, Bases of Graphs,and Additional Explanations," Washington, 1969.

[47] Goldberg, H., "Some Notes on Noise Figures," *Proceedings of the IRE*, vol. 36, no. 10, pp. 1205-1214, Oct. 1948.

[48] Tsui, J., *Digital Techniques for Wideband Receivers*, Second Edition ed., Boston: Artech House, 2001.

[49] Pozar, D. M., *Microwave Engineering*, 4th ed., Hoboken, NJ: John Wiley & Sons, Inc., 2012.

[50] Rezavi, B., *RF Microelectronics, Upper Saddle River*, NJ: Prentice Hall, 1998.

[51] Egan, W. F., *Practical RF System Design*, Hoboken, New YJrsey: John Wiley & Sons, Inc., 2003.

[52] Pozar, D. M., *Microwave and RF Design of Wireless Systems*, New York: John Wiley & Sons, 2001.

[53] Friis, H. T., "Noise Figures of Radio Receivers," *Proceedings of the IRE*, vol. 32, no. 7, pp. 419-422, July 1944.

[54] Ludwig, R. and G. Bogdanov, *RF Circuit Design: Theory & Applications*, 2nd ed., Prentice Hall, 2008.

[55] Lee, T. H., *The Design of CMOS Radio-Frequency Integrated Circuits*, 2nd ed., Cambridge Univ. Press, 2003.

[56] Skolnik, M. I., Ed., *Radar Handbook*, 3rd ed., New York: McGraw-Hill, 2008.

[57] Barton, D. K., *Radar System Analysis and Modeling*, Boston, MA: Artech House, 2005.

[58] Avionics Department, *Electronic Warfare and Radar Systems Engineering Handbook*, 4th ed., Point Mugu, CA: Naval Air Warfare Center Weapons Division, 2013.

[59] Rohde, U. and J. Whitaker, *Communications Receivers*, 3rd ed., New York: McGraw-Hill, 2001.

[60] Erst, S. J., *Receiving Systems Design*, Dedham: Artech House, Inc., 1984.

[61] Tsui, J. B., *Microwave Receivers and Related Components*, Los Altos, CA: Peninsula Publishing, 1985.

[62] Tsui, J., *Digital Microwave Receivers*, Norwood, MA: Artech House, 1989.

[63] Tsui, J., *Special Design Topics in Digital Wideband Receivers*, Norwood, MA: Artech House, 2010.

[64] Pace, P. E., *Advanced Techniques for Digital Receivers*, Norwood, MA: Artech House, 2000.

[65] IEEE 100, *The Authoritative Dictionary of IEEE Standards Terms*, 7th ed., New York: IEEE, 2000.

[66] Carr, J. J., *Practical Radio Frequency Test & Measurement*: A Technician's Handbook, New York: Newnes, 2002.

[67] Poberezhskiy, Y. S., *On Dynamic Range of Digital Receivers*, IEEE Aerospace Conference ed., 2007, pp. 1-17.

[68] Dixon, R. C., *Radio Receiver Design*, New York: Marcel Dekker, Inc., 1998.

[69] Reed, J. H., *Software Radio: A Modern Approach to Radar Engineering*, Upper Saddle River, NJ: Prentice Hall, 2002.

[70] Marki, F. and C. Marki, "Mixer Basics Primer," 2010. [Online]. Available: www.markimicrowave.com. [Accessed 13 April 2015].

[71] Vizmuller, P., *RF Design Guide : Systems, Circuts and Equations*, Boston, MA: Artech House, 1995.

[72] Norton, D. E., "The Cascading of High Dynamic Range Amplifiers," *Microwave Journal*, vol. 16, no. 6, pp. 57-71, June 1973.

[73] Barton, D. K., *Radar Equations for Modern Radar*, Boston: Artech House, 2013.

[74] East, P. W., *Microwave System Design Tools and EW Applications*, 2nd ed., MA: Artech House, 2008.

[75] P. B. Kenington, High Linearity RF Amplifier Design, Norwood, MA: Artech House, 1999.

[76] Aguilar, H. Jardon, R. Acevo Herrera and G. Monserrat Galvan Tejada, "Intermodulation interception points of nonlinear circuits connected in cascade," *Electrical and Electronics Engineering*, 2004. (ICEEE). 1st International Conference on, pp. 11-16, 8-10 Sept. 2004.

[77] Wu, Y. and J. Li, "The Design of Digital Radar Receivers," *IEEE National Radar Conference*, 1997.

[78] Kester, W., Ed., *Analog-Digital Conversion*, Analog Devices, Inc., 2004.

[79] Kester, W., Ed., *The Data Conversion Handbook*, New York: Newness, 2005.

[80] Nyquist, H., "Certain Factors Affecting Telegraph Speed," *Bell System Technical Journal*, vol. 3, pp. 324-346, April 1924.

[81] Nyquist, H., "Certain Topics in Telegraph Transmission Theory," *Transactions of the American Institute of Electrical Engineers*, vol. 47, no. 2, pp. 617-644, April 1928.

[82] Shannon, C. E., "A mathematical theory of communication," *The Bell System Technical Journal*, vol. 27, no. 3, pp. 379-423, July 1948.

[83] Shannon, C. E., "Communication in the Presence of Noise," *Proceedings of the IRE*, vol. 37, no. 1, pp. 10-21, January 1949.

[84] Kotelnikov, V. A., "On the Capacity of the 'Ether' and Cables in Electrical," *Proc. 1st All-Union Conf. Technological Reconstruction of the Commun. Sector and Low-Current Eng.*, Moscow, 1933.

[85] Vaughan, R. G., N. L. Scott and D. R. White, "The theory of bandpass sampling," *Signal Processing, IEEE Transactions*, vol. 39, no. 9, pp. 1973-1984, Sep. 1991.

[86] Brigham, E. O., *The Fast Fourier Transform and Its Applications*, Englewood Cliffs, NJ: Prentice-Hall, Inc., 1988.

[87] Hill, G., "The benefits of undersampling," *Electron Design*, pp. 69-79, July 1994.

[88] Coulson, A. J., R. G. Vaughan and M. A. Poletti, "Frequency Shifting Using Bandpass Sampling," *IEEE Transactions on Signal Processing*, vol. 42, no. 6, pp. 1556-1559, June 1994.

[89] Akos, D. M., M. Stockmaster, J. Tsui and J. Caschera, "Direct bandpass sampling of multiple distinct RF signals," *Communications, IEEE Transactions*, vol. 47, no. 7, pp. 983-988, 1999.

[90] Analog Devices, "AD6620 Data Sheet: 67 MSPS Digital Receive Signal Processor," Analog Devices, Inc., 2001.

[91] Hogenauer, E., "A class of digital filters for decimation and interpolation," *IEEE International Conference on Acoustics, Speech, and Signal Processing*, Denver, CO, 1980.

[92] Hogenauer, E. B., "An economical Class of Digital Filters for Decimation and Interpretation," *IEEE Transactions on Acoustics, Speech, and Signal Processing*, Vols. ASSP-29, no. 2, pp. 155-162, April 1981.

[93] Harris, F. J., *Multirate Signal Processing for Communication Systems*, Upper Saddle River: Prentice Hall, 2004.

[94] Donadio, M. P., "CIC Filter Introduction," 18 July 2000. [Online]. Available: https://dspguru.com/files/cic.pdf. [Accessed 17 Monember 2019].

[95] Altera, "Application Note 455: Understandinc CIC Filters," Altera Corporation, 2007.

[96] Altera, "CIC MegaCore Function User Guide," Altera Corporation, 2010.

[97] XILINK, "Application Note: Virtex-5 Family - Designing Efficient Digital Up and Down Converters for Narrowband Systems," XILINX, Inc., 2008.

[98] Lyons, R. G., *Understanding Digital Signal Processing*, 3rd ed., New York: Prentice Hall, 2011.

[99]   Rabiner, L. R. and B. Gold, *Theory and Application of Digital Signal Processing*, Englewood: Prentice-Hall, 1975.

[100] Widrow, B. and I. Kollar, *Quantization Noise*, New York: Cambridge University Press, 2008.

[101] Bennett, W. R., "Spectra of Quantized Signals," *Bell System Technical Journal*, vol. 27, pp. 446-472, 1948.

[102] Kester, W., "MT-001 Tutorial: Taking the Mystery out of the Infamous Formula, "SNR = 6.02N + 1.76dB," and Why You Should Care," 2009. [Online]. Available: www.analog.com/static/imported-files/tutorials/MT-001.pdf.

[103] Williston, K., *Digital Signal Processing: World Class Designs*, New York: Newness, 2009.

[104] Orfanidis, S. J., *Introduction to Signal Processing*, Englewood Cliffs, New Jersey: Prentice Hall, 1996.

[105] Kester, W., "MT-004 Tutorial: The Good, the Bad, and the Ugly Aspects of ADC Input Noise—Is No Noise Good Noise?," 2009. [Online]. Available: www.analog.com/static/imported-files/tutorials/MT-004.pdf.

[106] Oppenheim, A. V., Ed., *Applications of digital Signal processing*, Englewood Cliffs: Prentice-Hall, 1978.

[107] Furman, G. G., "Removing the Noise from the Quantization Process by Dithering: Linearization," 1 February 1963. [Online]. Available: www.dtic.mil/cgi-bin/GetTRDoc?AD=AD296598&Location=U2&doc=GetTRDoc.pdf.

[108] Furman, G. G., "Improving the Quantization of Random Signals by Dithering," United States Air Force Project RAND, 1963.

[109] Gray, R. M. and T. G. Stockham, "Dithered Quantizers," *IEEE Transactions on Information Theory*, vol. 39, no. 3, pp. 805-812, May 1993.

[110] Kester, W., "MT-006 Tutorial: ADC Noise Figure—An Often Misunderstood and Misinterpreted Specification," 2009. [Online]. Available: www.analog.com/static/imported-files/tutorials/MT-006.pdf.

[111] Karki, J., "Calculating noise figure and third-order intercept in ADCs," Texas Instruments Incorporated, 2003. [Online]. Available: www.ti.com/sc/analogapps. [Accessed 23 March 2015].

[112] Li, Z., L. Ligthart, P. Huang, W. Lu and W. v. d. Zwan, *Trade-off between Sensitivity and Dynamic Range in Designing Digital Radar Receivers*, International Conference on Microwave and Millimeter Wave Technology, ICMMT ed., 2008.

[113] Li, Z., L. Ligthart, P. Huang, W. Lu, W. v. d. Zwan, E. Lys and O. Krasnov, "Design Considerations of the RF Front-end for high Dynamic Range Digital RADAR Receivers," pp. 1-4, 2008.

[114] Budge, M. C. Jr. and S. R. German, "The Effects of an ADC on SCR Improvement," *Aerospace and Electronic Systems, IEEE Transactions*, Vols. 49, no. 4, pp. 2463-2469, October 2013.

[115] Vanderkooy, J. and S. P. Lipshitz, "Resolution Below the Least Significant Bit in Digital Systems with Dither," Vols. 32, No. 3, 1984.

[116] Wu, Y., *The Application of All-digital Array Receiver to OTH Radar*, International Conference on Radar, CIE '06 ed., 2006, pp. 1-4.

[117] Budge, M. C. Jr. and S. R. German, *Basic Radar Tracking*, Norwood, MA: Artech House, 2019.

[118] Elliott, R. S., *Antenna Theory and Design*, Revised Edition, New York: Wiley & Sons, 2003.

[119] Barton, D. K., *Modern Radar System Analysis*, Norwood, MA: Artech House, 1988.

[120] Barton, D. K., *Radars, Volume 1, Monopulse Radar*, Dedham, MA: Artech House, 1975.

[121] Hannan, P. W., "Optimum feeds for all three modes of a monopulse antenna I: Theory," *IRE Transactions on Antennas and Propagation*, vol. 9, no. 5, pp. 444-454, September 1961.

[122] Hannan, P. W., "Optimum feeds for all three modes of a monopulse antenna II: Practice," *IRE Transactions on Antennas and Propagation*, vol. 9, no. 5, pp. 454-461, September 1961.

[123] Ricardi, L. and L. Niro, "Design of a Twelve-Horn Monopulse Feed," *in 1958 IRE International Convention Record*, New York, NY, 1961.

[124] Sherman, S. M. and D. K. Barton, *Monopulse Principles and Techniques*, Second ed., Norwood, MA: Artech House, 2011.

[125] Wirth, W.D., *Radar Techniques Using Array Antennas*, London, UK: The Institution of Electrical Engineers, 2001.

[126] Brookner, E., "Phased-Array and Radar Astounding Breakthroughs—An Update," in *IEEE 2008 Radar Conf. (RADAR '08)*, 2008.

[127] Schmidt, R. O., "Multiple Emitter Location and Signal Parameter Estimation," *IEEE Trans. Antennas Propag.*, vol. 34, no. 3, pp. 276,280, March 1986.

APPENDIX 22A: DIGITAL DOWN CONVERSION USING BAND-PASS SAMPLING

PRELIMINARIES

In this Appendix we want to discuss the use of band-pass sampling to accomplish down conversion of a modulated signal. In particular, suppose we have the real signal

$$v_{IF}(t) = a(t)\cos 2\pi f_{IF} t + b(t)\sin 2\pi f_{IF} t \qquad (22A.1)$$

where f_{IF} is some intermediate frequency (IF) and $a(t)$ and $b(t)$ are real, baseband, or low-pass, signals. We assume that $A(f) = \Im[a(t)]$ and $B(f) = \Im[b(t)]$ are confined to the region $\pm \Delta f_I$. We further assume that $f_{IF} \Box \Delta f_I$. These assumptions are tantamount to stating that $v_{IF}(t)$ is a band-pass signal and that $V_{IF}(f) = \Im[v_{IF}(t)]$ has energy only in the bands $f \in [f_{IF} - \Delta f_L, f_{IF} + \Delta f_H]$ and $f \in [-f_{IF} - \Delta f_H, -f_{IF} + \Delta f_L]$. We note that, in general, even though $A(f)$ and $B(f)$ are conjugate symmetric about $f = 0$ (i.e., $A(-f) = A^*(f)$) the two spectral components of $V_{IF}(f)$ are not necessarily conjugate symmetric about $\pm f_{IF}$. That is, $v_{IF}(t)$ is, in general, a single side-band signal.

PROBLEM STATEMENT

In his text [74], Brigham states that if the sample frequency, f_s, satisfies the relationships [73, 74, p. 322]

$$\frac{2f_H}{n} \le f_s \le \frac{2f_L}{n-1} \qquad (22A.2)$$

and

$$2 \le n \le \frac{f_H}{f_H - f_L} \qquad (22A.3)$$

where n is an integer, then the sampled version of $v_{IF}(t)$, $v_{dc}(kT)$, will be a down-converted form of $v_{IF}(t)$ (with possible spectral inversion). That is,

$$v_{dc}(kT) = a(kT)\cos 2\pi f_{dc} kT \pm b(kT) \sin 2\pi f_{dc} kT \qquad (22A.4)$$

where $f_{dc} < f_{IF}$, $T = 1/f_s$ and the $\pm$ sign on the second term depends on whether ($-$ sign) or not ($+$ sign) spectral inversion occurs. In the above $f_H = f_{IF} + \Delta f_H$, $f_L = f_{IF} - \Delta f_L$ and n is an integer.

DERIVATION OF BRIGHAM'S EQUATIONS

We now want to derive Brigham's equations and present some observations related to their use. To present the derivation we need to make some observations about sampling and the assumptions associated with Brigham's equations.

From sampling theory we can write the spectrum of $v_{dc}(kT)$ as [74, p. 21, 25]

$$V_{dc}(f) = f_s \sum_{k=-\infty}^{\infty} V_{IF}(f + kf_s). \qquad (22A.5)$$

This equation tells us that we form $V_{dc}(f)$ by shifting and adding the spectrum of $v_{IF}(t)$. It also tells us that $V_{dc}(f)$ is periodic (with a period of f_s). We will make use of these two observations to derive Brigham's equations.

The main thing we want to avoid when we sample $v_{IF}(t)$ is the destruction of $a(t)$ and/or $b(t)$. If we were to choose $f_s \ge 2f_H$ the Shannon sampling theorem assures us that this will not happen. However, we want to sample at $f_s < 2f_H$, and still preserve the integrity of $a(t)$ and $b(t)$.

Let us define $V_{IF}^+(f)$ as the part of $V_{IF}(f)$ between f_L and f_H and $V_{IF}^-(f)$ as the part of $V_{IF}(f)$ between $-f_H$ and $-f_L$. These two parts of $V_{IF}(f)$, along with f_{IF}, contain all of the information needed to recover $a(t)$ and $b(t)$.

If we were to do anything to destroy the nature of $V_{IF}^+(f)$ or $V_{IF}^-(f)$ we would not be able to recover $a(t)$ or $b(t)$. The only way in which the nature of $V_{IF}^-(f)$ will be destroyed, after sampling, is if any shifted versions of $V_{IF}^+(f)$ occupy the band between $-f_H$ and $-f_L$. Likewise, the only way in which the nature of $V_{IF}^+(f)$ will be destroyed, after sampling, is if any shifted versions of $V_{IF}^-(f)$ occupy the band between f_L and f_H

. Because of the symmetry of $V_{IF}(f)$ and the periodic nature of $V_{dc}(f)$ the two previous statements imply each other. Thus, we only need to address one of them.

Suppose in Equation (22A.5) we consider a $k = n-1$ where $V_{IF}^{-}(f + (n-1)f_s)$ lies just to the left of $V_{IF}^{+}(f)$. That is, $V_{IF}^{-}(f + (n-1)f_s)$ lies in a band of frequencies just below f_L. This implies that

$$-f_L + (n-1)f_s \leq f_L \tag{22A.6}$$

since $-f_L + (n-1)f_s$ is the upper edge of $V_{IF}^{-}(f + (n-1)f_s)$.

Now let's look at the location of $V_{IF}^{-}(f + nf_s)$. That is, shift $V_{IF}^{-}(f)$ up by one more f_s. Since $V_{IF}^{-}(f + (n-1)f_s)$ was just below $V_{IF}^{+}(f)$, $V_{IF}^{-}(f + nf_s)$ will either overlap $V_{IF}^{+}(f)$ or lie in a band just above the band of $V_{IF}^{+}(f)$. If the former condition occurs we will have spectrum overlap and the nature of $a(t)$ and/or $b(t)$ will be destroyed. Thus, the desired condition is that $V_{IF}^{-}(f + nf_s)$ lie above $V_{IF}^{+}(f)$. This means that the lower edge of the band of $V_{IF}^{-}(f + nf_s)$ must be greater than the upper edge of the band of $V_{IF}^{+}(f)$. Since the lower edge of the band of $V_{IF}^{-}(f + nf_s)$ is at $-f_H + nf_s$ and the upper edge of the band of $V_{IF}^{+}(f)$ is at f_H we have the requirement that

$$-f_H + nf_s \geq f_H . \tag{22A.7}$$

If Equations (22A.6) and (22A.7) are satisfied for some f_s and (positive) n then there will be no overlap of the shifted versions of $V_{IF}^{-}(f)$ and $V_{IF}^{+}(f)$. Because of the periodic nature of $V_{dc}(f)$ there will also be no overlap of any of the shifted versions of $V_{IF}^{-}(f)$ with any of the shifted versions of $V_{IF}^{+}(f)$. Thus, $V_{dc}(f)$ will contain only shifted replicas of $V_{IF}^{-}(f)$ and $V_{IF}^{+}(f)$. As we will show shortly, there is one, and only one, replica of $V_{IF}^{-}(f)$ and $V_{IF}^{+}(f)$ in the desired region of $f \in [-f_s/2, f_s/2]$. This means that sampling has preserved the nature of $a(t)$ and $b(t)$ and has moved the signal from a band of $f \in [-f_h, f_h]$ to the band $f \in [-f_s/2, f_s/2]$. Thus, the sampling has accomplished down-conversion.

We now want to manipulate Equations (22A.6) and (22A.7) to get Brigham's equations. We can manipulate Equations (22A.6) and (22A.7) to get

$$f_s \leq 2f_L/(n-1) \tag{22A.8}$$

and

$$f_s \geq 2f_H/n . \tag{22A.9}$$

We can then combine Equations (22A.8) and (22A.9) to get the first of Brigham's equations. That is

$$2f_H/n \leq f_s \leq 2f_L/(n-1). \tag{22A.10}$$

Equation (22A.10) leads to the further requirement that

$$2f_H/n \leq 2f_L/(n-1) \tag{22A.11}$$

or

$$n \leq f_H/(f_H - f_L). \tag{22A.12}$$

We further examine Equation (22A.10) for specific values of n. First, if $n = 0$ we have

$$\infty \leq f_s \leq -2f_L \tag{22A.13}$$

which is nonsensical. Thus we require $n > 0$.

If $n = 1$ we have

$$2f_H \leq f_s \leq \infty \tag{22A.14}$$

which is a restatement of the normal sample rate requirement imposed by the Shannon sampling theorem. Since we are really interested in the under-sampling case, i.e., $f_s < 2f_H$, we can eliminate this n and require

$$n \geq 2. \tag{22A.15}$$

If we combine Equations (22A.15) and (22A.12) we get the second of Brigham's equations or

$$2 \leq n \leq f_H/(f_H - f_L). \tag{22A.16}$$

DISCUSSION AND THOUGHTS

Spectral Inversion

We consider an n and f_s that satisfy Brigham's equations. We further consider $V_{IF}^-(f + (n-1)f_s)$ and $V_{IF}^+(f)$. We recall that these two spectra are adjacent and that $V_{IF}^-(f + (n-1)f_s)$ is to the left of (in a lower frequency band than) $V_{IF}^+(f)$. The left (low) edge of $V_{IF}^-(f + (n-1)f_s)$ is at $-f_H + (n-1)f_s$ and the right (high) edge of $V_{IF}^+(f)$ is at f_H. The center between these two bounds, and thus between $V_{IF}^-(f + (n-1)f_s)$ and $V_{IF}^+(f)$, is

$$f_c = \frac{(f_H + (-f_H + (n-1)f_s))}{2} = \frac{n-1}{2} f_s. \tag{22A.17}$$

We note that $V_{IF}^-(f+(n-1)f_s)$ and $V_{IF}^+(f)$ are conjugate symmetric about f_c.

If n is odd, $n-1$ will be even and $f_c = ((n-1)/2)f_s$ will be an integer multiple of f_s. This means that a replica of $V_{IF}^-(f+(n-1)f_s)$ and $V_{IF}^+(f)$ will be centered at $f = 0$. Further, the positive and negative frequency components will be conjugate symmetric and in the same relative location as $V_{IF}^-(f)$ and $V_{IF}^+(f)$. That is, there will be no spectral inversion.

We now consider $V_{IF}^+(f)$ and $V_{IF}^-(f+nf_s)$. The left edge of $V_{IF}^+(f)$ is at f_L and the right edge of $V_{IF}^-(f+nf_s)$ is at $-f_L + nf_s$. Thus, the center of these two spectra will be at

$$f_c = \frac{(f_L + (-f_L + nf_s))}{2} = \frac{n}{2}f_s. \qquad (22A.18)$$

If n is even, $f_c = (n/2)f_s$ will be an integer multiple of f_s. This means that a replica of $V_{IF}^+(f)$ and $V_{IF}^-(f+nf_s)$ will be centered at $f = 0$ and the two translated spectra will be conjugate symmetric about $f = 0$. However, the locations will be inverted relative to $V_{IF}^-(f)$ and $V_{IF}^+(f)$. That is, the down-converted signal will undergo spectral inversion.

In summary:

- If n is odd, the spectrum of the sampled signal will be centered at zero with no spectral inversion relative to the original IF signal.
- If n is even, the spectrum of the sampled signal will be centered at zero but will undergo a spectral inversion relative to the original IF signal.

There is one, and only one, replica of $V_{IF}^-(f)$ and $V_{IF}^+(f)$ in the interval $[-f_s/2, f_s/2]$

We first need to show that there is one replica of $V_{IF}^-(f)$ and $V_{IF}^+(f)$ in an interval that if f_s wide. To do so, we consider the adjacent $V_{IF}^-(f+(n-1)f_s)$ and $V_{IF}^+(f)$ we discussed earlier. Recall that we considered an n and f_s so that $V_{IF}^-(f+(n-1)f_s)$ was to the left of (lower in frequency than) $V_{IF}^+(f)$. The left edge (lower frequency bound) of $V_{IF}^-(f+(n-1)f_s)$ is at $-f_H + (n-1)f_s$ and the right edge of $V_{IF}^+(f)$ is at f_H. Thus, the total span of $V_{IF}^-(f+(n-1)f_s)$ and $V_{IF}^+(f)$ is

$$\Delta f = f_H - (-f_H + (n-1)f_s) = 2f_H - (n-1)f_s. \qquad (22A.19)$$

But, from Brigham's equations we have

$$2f_H \le nf_s. \qquad (22A.20)$$

If we combine Equations (22A.19) and (22A.20) we get

$$\Delta f = 2f_H - (n-1)f_s \leq nf_s - (n-1)f_s = f_s \qquad (22A.21)$$

which established the fact that one replica of $V_{IF}^-(f)$ and $V_{IF}^+(f)$ lie in an interval f_s wide. Further, since the replicas of $V_{IF}^-(f)$ and $V_{IF}^+(f)$ are symmetrically located about $f = 0$, the two replicas lie in the interval $[-f_s/2, f_s/2]$

We now want to show that no more than one (full) replica of $V_{IF}^-(f)$ and $V_{IF}^+(f)$ lies in an interval f_s wide. Recall that $V_{IF}^-(f+(n-1)f_s)$, $V_{IF}^+(f)$ and $V_{IF}^-(f+nf_s)$ lie in adjacent frequency bands from low to high frequencies. From above, the left edge of $V_{IF}^-(f+(n-1)f_s)$ is at $-f_H + (n-1)f_s$. The left edge of $V_{IF}^-(f+nf_s)$ is at $-f_H + nf_s$. Thus, these two left edges are separated by

$$\delta f = (-f_H + nf_s) - (-f_H + (n-1)f_s) = f_s. \qquad (22A.22)$$

Thus, if all of $V_{IF}^-(f+(n-1)f_s)$ is in an interval f_s wide, $V_{IF}^-(f+nf_s)$ cannot be in the same interval, and vice versa. With this we have established that only one replica of $V_{IF}^-(f)$ and $V_{IF}^+(f)$ lie in the interval $[-f_s/2, f_s/2]$.

"Carrier" frequency of the sampled signal

We want to relate f_{dc} of Equation (22A.4) to f_{IF}, the sample frequency, f_s, and n. We have established that if n is odd that the center of $V_{IF}^-(f+(n-1)f_s)$ and $V_{IF}^+(f)$ would be at $f_c = ((n-1)/2)f_s$. To shift this center to $f = 0$ we need to subtract f_c from all of the frequency components. This means that the translated IF will become

$$f_{dc} = f_{IF} - f_c = f_{IF} - ((n-1)/2)f_s. \qquad (22A.23)$$

Similarly, if n is even, the center of $V_{IF}^+(f)$ and $V_{IF}^-(f+nf_s)$ would be at $f_c = (n/2)f_s$. To shift this center to $f = 0$ we need to subtract f_c from all of the frequency components. This means that the translated IF will become

$$f_{dc} = f_{IF} - f_c = f_{IF} - (n/2)f_s. \qquad (22A.24)$$

Recovery of $a(kT)$ and $b(kT)$

Based on the discussions above, the output of the sampler will be the modulated signal of Equation (22A.4). This mean that $a(kT)$ and $b(kT)$ can be recovered by digital, quadrature demodulation of $v_{dc}(kT)$.

The only issue not addressed is whether f_s satisfies the Shannon sampling theorem relative to the bandwidths of $a(t)$ and $b(t)$. The answer is yes. It is left as an exercise for the reader to prove this assertion.

Chapter 23

Introduction to Synthetic Aperture Radar Signal Processing

23.1 INTRODUCTION

The term synthetic aperture radar (SAR) derives from the fact that the motion of an aircraft (airplane, satellite, or UAV, for example) is used to artificially create, or *synthesize,* a very long, linear array, or aperture. The reason for creating a long array is to provide the ability to resolve targets that are closely spaced in angle, or cross range (usually azimuth). This, in turn, is driven by one of the main uses of SAR: to image the ground or targets. In both cases, the radar needs to be able to resolve very closely spaced scatterers. Specifically, resolutions of less than a meter to a few meters are needed. To realize such resolutions in the range coordinate, the radar uses wide bandwidth waveforms. To realize such resolutions in cross range, very long antennas are required.

To get an idea of what we mean by "long" antenna, we consider an example. Suppose we are trying to image a ground patch at a range of 20 km. To do so, we want a cross-range resolution of 1 m. We can approximately relate cross-range resolution, δy, to antenna beam width, θ_B, and range to the ground patch, R, by

$$\delta y \approx R\theta_B \tag{23.1}$$

as shown in Figure 23.1. For desired $\delta y = 1$ m and $R = 20$ km, we get required $\theta_B = 5 \times 10^{-5}$ rad or about 0.003°!

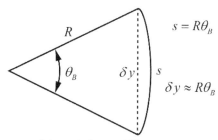

Figure 23.1 Relation of cross-range distance to beamwidth.

The beamwidth of a linear array with uniform illumination can be approximately related to antenna length by [1, 2]

$$\theta_B = \lambda/L \tag{23.2}$$

If we assume the radar of the above example operates at X-band and $\lambda = 0.03$ m, we get

$$L = \lambda/\theta_B = 600 \text{ m} \tag{23.3}$$

Clearly, it would not be practical to use a real antenna that is as long as six football fields. Instead, a SAR synthesizes such antenna by using aircraft motion and signal processing. An interesting property illustrated by (23.1) and (23.3) is that the resolution and SAR antenna length depends on wavelength. This means that if a certain resolution is desired and there are limits on how long the synthetic array can be made, we are driven to shorter wavelength, or higher frequency, radars. This will also affect down-range resolution since it is related to waveform bandwidth, and large waveform bandwidths are easier to obtain at higher operating frequencies.

23.2 BACKGROUND

According to a paper by H. D. Griffiths [3], the concept of aperture synthesis was introduced by Ryle and Hawkins in the 1940s or 1950s in relation to their work in radio astronomy. However, the recognized father of SAR, as it is known today, is Carl A. Wiley, who conceived of the concept in 1951 and termed it *Doppler beam sharpening* [4–6]. Shortly after that, in 1952, scientists at the University of Illinois experimentally demonstrated the concept [2]. Since that time, SAR has found wide use in both commercial and military applications [7–12].

23.2.1 Linear Array Theory

Before we discuss SAR processing, we consider some properties of SAR. We start with a review of linear arrays since a SAR synthesizes a linear array. Suppose we have a $2N + 1$ element linear array[1] as shown in Figure 23.2. We have a target located at some x_i, y_i that emits an E-field $E_o e^{j2\pi f_o t}$ that eventually reaches each antenna element. We can write the E-field at the nth element as

$$E_n(t) = E_o(r_{i,n}) e^{j2\pi f_o(t - r_{i,n}/c)} = E_o(r_{i,n}) e^{-j2\pi r_{i,n}/\lambda} e^{j2\pi f_o t} \tag{23.4}$$

where we included the function of $r_{i,n}$ to indicate that the magnitude of the E-field intensity at the nth element depends on the range from the target to that element. However, in general, $E_o(r_{i,n})$ will, for all practical purposes, be the same at each element.

---

[1] We chose an odd number of elements to simplify some of the notation to follow.

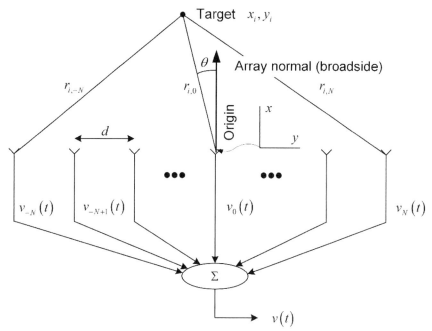

Figure 23.2 2N + 1 element linear array.

The resulting voltage at the output of the nth element is

$$v_n(t) = V_o(r_{i,n}) e^{-j2\pi r_{i,n}/\lambda} e^{j2\pi f_o t} \tag{23.5}$$

As with $E_o(r_{i,n})$, we can assume $V_o(r_{i,n})$ is the same at all elements and replace $V_o(r_{i,n})$ with V_o. V_o is the magnitude of the voltage out of each element.

We can write $r_{i,n}$ as

$$r_{i,n} = \sqrt{x_i^2 + (y_i - nd)^2} = \sqrt{x_i^2 + y_i^2 + n^2 d^2 - 2ndy_i} \\ = \sqrt{r_0^2 + n^2 d^2 - 2nd r_0 \sin\theta} \tag{23.6}$$

where r_0 is the range from the target to the center element, and we used $y_i = r_0 \sin\theta$ (see Figure 23.2).

Consistent with the linear array theory of Chapter 13, we claim $r_0 \gg nd$ and approximate $r_{i,n}$ as[2]

$$r_{i,n} = \sqrt{r_0^2 - 2nd r_0 \sin\theta} = r_0 \sqrt{1 - \frac{2nd}{r_0}\sin\theta} \approx r_0 - nd\sin\theta \tag{23.7}$$

We next substitute (23.7) into the exponent of $v_n(t)$ (23.5) to get

---

[2] As we develop the detail of SAR processing, we will find we need to abandon this approximation. For now, we take note of this and proceed.

$$v_n(t) = V_o e^{-j2\pi r_0/\lambda} e^{j2\pi f_o t} e^{j\frac{2\pi nd}{\lambda}\sin\theta} \tag{23.8}$$

To form the total output of the array, we sum the $v_n(t)$ to get

$$v(t) = \sum_{n=-N}^{N} v_n(t) = V_o e^{-j2\pi r_0/\lambda} e^{j2\pi f_o t} \sum_{n=-N}^{N} e^{j\frac{2\pi nd}{\lambda}\sin\theta} \tag{23.9}$$

We next form a scaled antenna "power" pattern (see Chapter 13) as[3]

$$P(\theta) = \frac{|v(t)|^2}{(2N+1)} = \frac{P_S}{(2N+1)^2} \left\{ \frac{\sin\left[\frac{(2N+1)\pi d}{\lambda}\sin\theta\right]}{\sin\left(\frac{\pi d}{\lambda}\sin\theta\right)} \right\}^2 \tag{23.10}$$

where P_S is the normalized power returned from the target.

When we formulated the antenna power pattern as above, we were interested in how $P(\theta)$ varied with *target angle*, θ. As given in (23.10), the peak of $P(\theta)$ occurs at a target angle of $\theta = 0$ as shown in Figure 23.3, which is a plot of $P(\theta)$ for $P_S = 1$ W.

As an extension to the above, we steer the beam to an angle of θ_S by including a linear phase shift across the array elements as shown in Figure 23.4. In Chapter 13, we found we can do this by multiplying the $v_n(t)$ by a_n and letting $a_n = \exp(-j2\pi(d/\lambda)\sin\theta_S)$. With this we get

$$v(t) = \sum_{n=-N}^{N} a_n v_n(t) = V_o e^{-j2\pi r_0/\lambda} e^{j2\pi f_o t} \sum_{n=-N}^{N} e^{-j\frac{2\pi nd}{\lambda}\sin\theta_S} e^{j\frac{2\pi nd}{\lambda}\sin\theta} \tag{23.11}$$

which leads to a more general $P(\theta)$ of

$$P(\theta) = \frac{P_S}{(2N+1)^2} \left\{ \frac{\sin\left[\frac{(2N+1)\pi d}{\lambda}(\sin\theta - \sin\theta_S)\right]}{\sin\left[\frac{\pi d}{\lambda}(\sin\theta - \sin\theta_S)\right]} \right\}^2 \tag{23.12}$$

In standard array theory, we are interested in how $P(\theta)$ varies with θ for a *fixed* θ_S. In this case the peak of $P(\theta)$ would occur at $\theta = \theta_S$, as shown in the example of Figure 23.5.

---

[3] We multiplied the $R(\theta)$ of Chapter 13 by P_S and $(2N+1)$ to convert it to the power pattern.

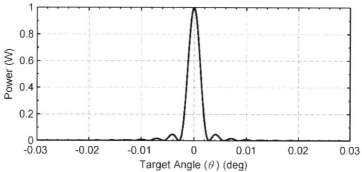

Figure 23.3 Power pattern vs. target angle, $P_S = 1$ W.

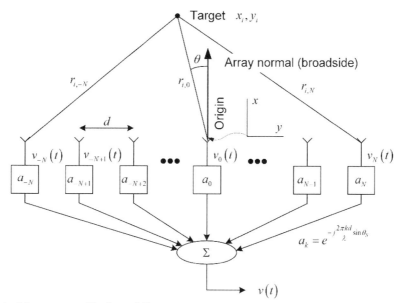

Figure 23.4 Linear array with phase shifters.

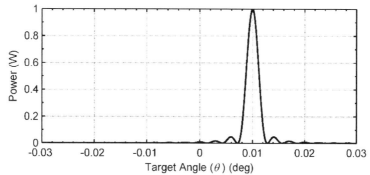

Figure 23.5 Power pattern vs. target angle—beam steered to 0.01°, $P_S = 1$ W.

23.2.2 Transition to SAR Theory

In SAR theory, we need to reorient ourselves by thinking of the target angle, θ, as being fixed and examining how $P(\theta)$ *varies with* θ_S. In other words, we consider a fixed, θ, and plot $P(\theta_S)$. An example plot of $P(\theta_S)$ for $\theta = 0.01°$ (i.e., the target location is fixed at 0.01°) is shown in Figure 23.6. In this plot, $P(\theta_S)$ peaks *when the beam is steered to an angle of* $\theta_S = 0.01°$.

Figure 23.7 contains a plot of $P(\theta_S)$ for the case where there is a target at 0.01° and a second target at −0.02°. Further, the second target has twice the RCS (radar cross section), and thus twice the power, of the first target. Here we note that the plot of $P(\theta_S)$ tells us the location of the two targets and their relative powers. This is the type of information we want when we form SAR images.

$P(\theta_S)$ gives us information in one dimension. To form an image, the other information we use is $P(r)$, the power out of the matched filter for a target range of r. We compute $P(\theta_S)$ and $P(r)$ for various values of θ_S and r and then plot $|P(\theta_S)|\|^{1/2}$, with $P_S = P(r)$ as intensities on a rectangular grid. The discrete values of θ_S and r will be separated by the angle resolution of the SAR array and the range resolution of the waveform. The resulting image is a preliminary version of a SAR image. We say it is a preliminary version of a SAR image because it lacks the focusing step needed to form an actual SAR image. It is also a preliminary image because it is based on a one-way antenna pattern. We will develop more specific SAR generation topics in the ensuing pages.

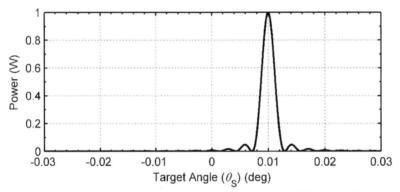

Figure 23.6 Power pattern vs. beamsteering angle—target located at 0.01°, $P_S = 1$ W.

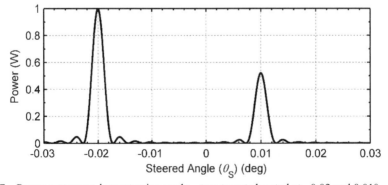

Figure 23.7 Power pattern vs. beamsteering angle—two targets located at −0.02 and 0.01°.

23.3 DEVELOPMENT OF SAR-SPECIFIC EQUATIONS

With the above background, we now address issues associated with forming $P(\theta_S)$ in practical SAR situations. We begin by modifying the above array theory so that it more directly applies to the SAR problem.

In standard array theory, we generate a *one-way* antenna pattern because we consider an antenna radiating toward a target (the transmit antenna case) or a target radiating toward an array (the receive antenna case). In SAR theory, we consider a *two-way* problem since we transmit and receive from each element of the synthetic array. If we refer to Figure 23.2, we can think of each element as the position of the SAR aircraft as it transmits and receives successive pulses. When the aircraft is located at $y = nd$, the normalized transmit "voltage" is[4]

$$v(t) = e^{j2\pi f_o t} \qquad (23.13)$$

The resultant received signal (voltage) from a scatterer[5] at x_i, y_i is

$$v_{i,n}(t) = \sqrt{P_{Si}}\, v(t - 2r_{i,n}/c) = \sqrt{P_{Si}}\, e^{j2\pi f_o (t - 2r_{i,n}/c)}$$
$$= \sqrt{P_{Si}}\, e^{-j4\pi r_{i,n}/\lambda}\, e^{j2\pi f_o t} \qquad (23.14)$$

where P_{Si} is the return signal power and is determined from the radar range equation. $r_{i,n}$ is the range to the ith scatterer when the aircraft is at $y = nd$.

We note that the difference between (23.5) and (23.14) is that the latter has twice the phase shift as the former because of the two-way range delay, $2r_{i,n}/c$.

Modifying (23.7) as

$$r_{i,n} = \sqrt{r_0^2 - 2ndr_0 \sin\theta} = r_0\sqrt{1 - \frac{2nd}{r_0}\sin\theta} \approx r_0 - nd\sin\theta \qquad (23.15)$$

and repeating the math of Section 23.2.1, we get the equation for the power pattern of a SAR antenna as

$$P(\theta_S) = \frac{P_{Si}}{(2N+1)^2} \left\{ \frac{\sin\left[\frac{2(2N+1)\pi d}{\lambda}(\sin\theta_S - \sin\theta)\right]}{\sin\left[\frac{2\pi d}{\lambda}(\sin\theta_S - \sin\theta)\right]} \right\}^2 \qquad (23.16)$$

Figure 23.8 contains plots of $P(\theta_S)$ for the standard linear array [(23.12)] and the SAR array [(23.16)]. In both cases, we used $P_{Si} = 1$ W. The notable difference between the two plots is that the width of the main beam of the SAR array is half that of the standard linear array. This leads to one of the standard statements in SAR books that a SAR has twice the resolution capability of a standard linear array [13]. In fact, this is not quite true. If we were to consider the *two-way* antenna pattern of a standard linear array, we would find

---

[4] In our initial discussions, we will be concerned only with cross-range imaging and can thus use a CW signal. We will consider a pulsed signal when we add the second dimension.
[5] We are changing terminology from "target" to "scatterer" since the latter is common in SAR theory.

that its beamwidth lies between the one-way beamwidth of a standard linear array and the beamwidth of a SAR array. The reason that the two-way beamwidth of a standard linear array is not equal to the beamwidth of a SAR array has to do with the interaction between "elements" in the two arrays. In a standard linear array, each receive element receives returns from all of the elements of the transmit array. However, in the SAR array, each receive "element" receives returns only from itself.

Adapting (23.2), we have, for the SAR array,

$$\theta_B = \lambda/2L \qquad (23.17)$$

If we combine this with the equation for cross-range distance (23.1), we get, again for the SAR array,

$$\delta y \approx R\theta_B = R\lambda/2L \qquad (23.18)$$

which is termed the cross-range resolution of the SAR. This equation indicates that the cross-range resolution of a SAR can be made arbitrarily small (fine) by increasing the length of the SAR array. In theory, this is true for a spotlight SAR [14, 15]. In the case of strip map SAR, the size of the actual antenna on the SAR aircraft (the element of the SAR array) is the theoretical limiting factor on resolution. In either case, there are several other factors related to phase coherency that place further limits on the cross-range resolution.

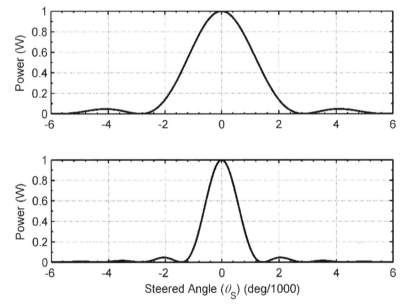

Figure 23.8 Power patterns for a standard linear array (top plot) and a SAR array (bottom plot) – $P_S = 1$ W.

23.4 TYPES OF SAR

There are several types of SARs and SAR modes of operation including the strip map SAR, spotlight SAR, scan SAR, step scan SAR, circular SAR, monopulse SAR, interferometric SAR, forward-looking squint mode, backward-looking squint mode, multilook mapping SAR, Doppler beam sharpening, ground mapping, and moving target display [1–3, 5, 7, 13–14], to name a few. Scan SAR and the others are basically extensions of strip map SAR and spotlight SAR. Thus, we will concentrate on the latter two types of SAR.

Figures 23.9 and 23.10 contain illustrations of the geometry associated with strip map and spotlight SAR, respectively. With strip map SAR, the actual antenna remains pointed at a fixed angle as the aircraft flies past the area being imaged. This angle is shown as 90° in Figure 23.9 but can, in theory, be almost any angle. For spotlight SAR, the actual antenna is steered to constantly point toward the area being imaged. The term "strip map" derives from the fact that this type of SAR can continually map strips of the ground as the aircraft flies by. The term "spotlight" derives from the fact that the actual antenna constantly illuminates, or spotlights, the region being imaged.

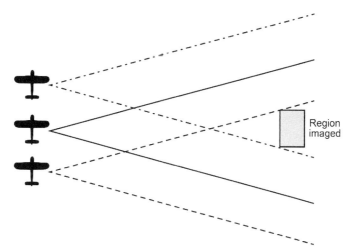

Figure 23.9 Strip map SAR geometry.

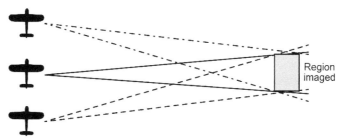

Figure 23.10 Spotlight SAR geometry.

As might be deduced from Figure 23.9, a limitation of the strip map SAR geometry is that the region imaged during any one processing interval must remain in the actual antenna beam during that processing interval. This does not limit the total area imaged since a strip map SAR can perform imaging continuously, dropping off data as it picks up new data. It only limits the size of the area imaged and the cross-range resolution in any one processing interval.

For the case of spotlight SAR, the antenna is always pointed at the region being imaged so that the length of the synthetic array can, in theory, be as large as desired. In practice, the length of the synthetic array for the spotlight SAR is limited by other factors such as coherency and signal processing limitations. Since the cross-range resolution of a SAR is related to the length of the synthetic array, spotlight SARs can usually attain finer cross-range resolution than strip map SARs.

23.4.1 Theoretical Limits for Strip Map SAR

The theoretical limit on cross-range resolution for a strip map SAR can be deduced with the help of Figure 23.11. As illustrated in this figure and discussed previously, the point to be imaged must be in the actual antenna beam over the processing interval (the CPI). We discuss how this limits resolution in the following.

The cross-range span of the main beam of the actual antenna is

$$L = r_i \theta_{ANT} \tag{23.19}$$

where r_i is the perpendicular range from the aircraft flight path to the point being imaged. In the geometry of Figure 23.11, note that the point being imaged will remain in the actual antenna beam as the aircraft traverses a distance of L. Thus, the length of the synthetic array applicable to any CPI is L.

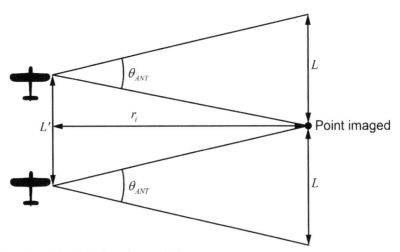

Figure 23.11 Resolution limit for strip map SAR.

Using (23.2), we can write the beam width of the actual antenna as

$$\theta_{ANT} = \lambda/L_{ANT} \tag{23.20}$$

where L_{ANT} is the horizontal width of the actual antenna. If we substitute (23.20) into (23.19), we get

$$L = r_i \lambda/L_{ANT} \tag{23.21}$$

which we can combine with (23.18) to get

$$\delta y \approx \frac{r_i \lambda}{2 r_i \lambda/L_{ANT}} = L_{ANT}/2 \tag{23.22}$$

Thus, the finest cross-range resolution a strip map SAR can achieve is half of the horizontal width of the actual antenna. This cross-range resolution applies only to the case where a point is being imaged. The resolution for a finite-sized area will be slightly worse, as shown in the next subsection.

23.4.2 Effects of Imaged Area Width on Strip Map SAR Resolution

Figure 23.12 illustrates a case where the width of the region to be imaged is w. It can be observed from this figure that $L' = L - w$. From this we conclude that the modified cross-range resolution is

$$\delta y' \approx \frac{r_i \lambda}{2L'} = \frac{\delta y}{1 - w/L} > \delta y \tag{23.23}$$

In practice, the term w/L will be small so that $\delta y' \approx \delta y$. As an example of this, we consider the earlier example where the SAR processed returns over $L = 600$ m. From (23.18), the resulting resolution for a point target is, in theory, $\delta y = 0.5$ m. Suppose we wanted to image an area with a width of 50 m. For this case, we would need to shorten the distance over which we process returns to $L' = L - w = 550$ m. As a result, from (23.23), the resolution would be 0.546 m instead of 0.5 m.

These discussions of the relation between distance over which we process returns (the length of the synthetic array during the CPI) and resolution are based on the assumption that the antenna directivity is constant over θ_{ANT} and zero elsewhere. This will clearly not be the case in an actual SAR. Therefore, the relation between resolution and L_{ANT} should be considered as an approximate limitation rather than a hard constraint.

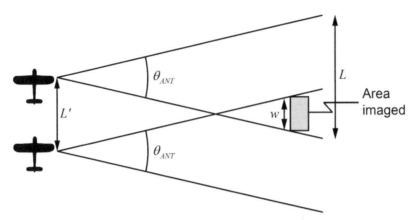

Figure 23.12 Effect of finite area width on strip map SAR resolution.

23.5 SAR SIGNAL CHARACTERIZATION

To formulate a SAR processor, we need to characterize the signal that the SAR processor will operate on. Although our previous discussions treated SAR cross-range imaging as an antenna problem, for the rest of the development we will cast the problem in the Doppler domain. We make this change because experience indicates that the Doppler formulation is easier to understand than the antenna formulation. Also, the Doppler formulation is consistent with other texts and articles that discuss SAR processing [16–24].

23.5.1 Derivation of the SAR Signal

As we have done thus far, we will initially consider only the cross-range problem. We will later extend the discussions to the down-range and cross-range problem. Since we are considering the cross-range problem, we start by considering a normalized CW transmit signal of the form

$$v_T(t) = e^{j2\pi f_o t} \qquad (23.24)$$

From the geometry of Figure 23.13, the appropriately normalized signal returned from the "i^{th}" scatterer located at x_i, y_i is

$$v_{iRF}(t) = \sqrt{P_{Si}}\, v_T(t - 2r_i(t)/c) = \sqrt{P_{Si}}\, e^{j2\pi f_o(t - 2r_i(t)/c)} \qquad (23.25)$$

where

$$r_i(t) = \sqrt{x_i^2 + \left[y_i - d(t)\right]^2} \qquad (23.26)$$

and $d(t)$ is the y position of the aircraft at some time t. For now, we are assuming the aircraft (platform) is at an altitude of zero. The extension to nonzero altitude is straightforward.

If we assume the aircraft is flying at a constant velocity of V, and $t = 0$ occurs when the aircraft is at $y = 0$ of Figure 23.13, we get

$$d(t) = Vt \tag{23.27}$$

We assume the total time for the aircraft to travel a distance of L is T_L and that the aircraft starts at $-L/2$ when $t = -T_L/2$. With this we get

$$L = T_L V \tag{23.28}$$

The area to be imaged has a cross-range width of w and a down-range length of l. The region is centered in cross range at $y = 0$ and in down range at $x = r_0$.

In (23.25), P_{Si} is the normalized signal power associated with the ith scatterer. It is related to scatterer RCS through the radar range equation. Thus, P_{Si} characterizes the relative sizes of the scatterers in the imaged area. $(P_{Si})^{1/2}$ is analogous to brightness or contrast in a photographic image.

We can rewrite (23.25) as

$$v_{iRF}(t) = \sqrt{P_{Si}} e^{j2\pi f_o t} e^{-j4\pi r_i(t)/\lambda} \tag{23.29}$$

Since the information needed to form the image is in the second exponential term, we eliminate the first exponential term by heterodyning (which is done in the actual radar) to yield the baseband signal

$$v_i(t) = e^{-j2\pi f_o t} v_{iRF}(t) = \sqrt{P_{Si}} e^{-j4\pi r_i(t)/\lambda} \tag{23.30}$$

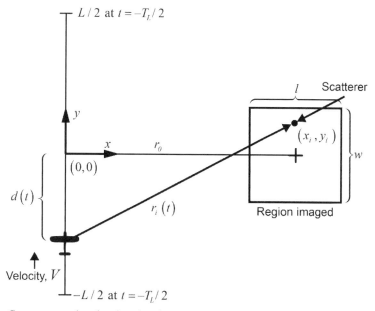

Figure 23.13 Geometry used to develop signal representation—cross-range imaging.

If we have N_s scatterers in the image region, the resulting composite baseband signal would be

$$v(t) = \sum_{i=1}^{N_s} v_i(t) = \sum_{i=1}^{N_s} \sqrt{P_{Si}} e^{-j4\pi r_i(t)/\lambda} \qquad (23.31)$$

23.5.2 Examination of the Phase of the SAR Signal

Since the information we seek is in the phase of $v_i(t)$, we examine it. To proceed, we examine $r_i(t)$, which we can write as

$$\begin{aligned} r_i(t) &= \sqrt{x_i^2 + [y_i - d(t)]^2} \\ &= \sqrt{x_i^2 + y_i^2 - 2y_i V t + V^2 t^2} = \sqrt{r_i^2 - 2y_i V t + V^2 t^2} \\ &= r_i \sqrt{1 - 2y_i V t / r_i^2 + (Vt/r_i)^2} \end{aligned} \qquad (23.32)$$

We note that $r_i \approx r_0$, $y_i \ll r_0$ and $Vt \ll r_0 \; \forall \; Vt \in [-L/2, L/2]$. This means the second and third terms of the last square root are small relative to 1. This, in turn, allows us to write

$$r_i(t) \approx r_i \left(1 - \frac{y_i V t}{r_i^2} + \frac{1}{2}(Vt/r_i)^2 \right) = r_i - y_i V t / r_i + \frac{1}{2} V^2 t^2 / r_i \qquad (23.33)$$

Substituting this into (23.30) yields

$$v_i(t) = \sqrt{P_{Si}} e^{-j4\pi r_i/\lambda} e^{j4\pi (y_i V/\lambda r_i) t} e^{-j2\pi (V^2/\lambda r_i) t^2} \qquad (23.34)$$

23.5.2.1 Linear Phase, or Constant Frequency, Term

The first exponential is a phase caused by range delay to the scatterer. For a single scatterer, the exponential is of no concern because it will disappear when we form the magnitude of the processed version of $v_i(t)$. For multiple scatterers, however, the exponential can cause constructive and destructive interference, which leads to speckle in SAR images [13, 18]. Speckle is usually mitigated by image processing techniques, including multi-look processing [13, 25].

The second term of (23.34) is a linear phase term or a term that we associate with frequency, which is

$$f_{yi} = 2 y_i V / \lambda r_i \qquad (23.35)$$

This tells us $v_i(t)$ has a frequency component that depends on scatterer cross-range position, y_i. f_{yi} also depends on the aircraft velocity, V, and the radar wavelength, λ. However, both of these are known (and fixed). Finally, f_{yi} also depends on r_i. If we assume all of the scatterers are at the same $x_i = r_0$ (which we can do here because we are concerned only with the cross-range problem), and we note that $y_i \ll r_0$, we get the previous assertion that

$$r_i = \sqrt{x_i^2 + y_i^2} \approx r_0 \quad (23.36)$$

From this discussion, we conclude that we can determine y_i if we can measure f_{yi}. Specifically,

$$y_i = \frac{\lambda f_{yi} r_0}{2V} \quad (23.37)$$

23.5.2.2 Quadratic Phase, or LFM, Term

The third exponential of (23.34) is a quadratic phase, or linear frequency modulation, term that causes problems in image formation. We can write the quadratic phase as

$$\phi_Q(t) = -2\pi \left(\frac{V^2}{\lambda r_i}\right) t^2 \quad (23.38)$$

With the previous assumption of $r_i \approx r_0$, $\phi_Q(t)$ is approximately the same for all scatterers. This means we can remove it by a mixing or heterodyning process during the digital signal processing.[6] If we do this, we will be left with only the magnitude, constant phase term, and the y_i-dependent frequency term. This is what we want. The process of removing the quadratic phase is known as focusing.

23.5.3 Extracting the Cross-Range Information

Once we remove the quadratic phase, we have

$$v_{Ii}(t) = e^{j2\pi(V^2/\lambda r_0)t^2} v_i(t) = \sqrt{P_{Si}} e^{-j4\pi r_i/\lambda} e^{j2\pi f_{yi} t} \quad (23.39)$$

for a single scatterer. For the more general case of N_s scatterers, we have

$$v_I(t) = e^{j2\pi(V^2/\lambda r_0)t^2} v(t) = \sum_{i=1}^{N_s} \sqrt{P_{Si}} e^{-j4\pi r_i/\lambda} e^{j2\pi f_{yi} t} \quad (23.40)$$

The forms of (23.39) and (23.40) tell us we can extract the information we want by taking the Fourier transform of $v_{Ii}(t)$ or, more generally, $v_I(t)$. From our experience with Fourier transforms, this will give us a response that has peaks at the frequencies f_{yi}. The heights of the peaks will be proportional to $(P_{Si})^{1/2}$. If we use (23.37) to plot this as amplitude $[(P_{Si})^{1/2}]$ versus y_i, we have a one-dimensional image.[7] We compute the Fourier transform of $v_{Ii}(t)$ using

$$V_{Ii}(f) = \int_{-\infty}^{\infty} v_{Ii}(t) e^{-j2\pi f t} dt \quad (23.41)$$

We recognize that $v_i(t)$, and thus $v_{Ii}(t)$, is measured only over $t \in [-T_L/2, T_L/2]$. Thus, we assume $v_{Ii}(t)$ is zero outside of these limits and write

---

[6] Note that this is similar to stretch processing, wherein we remove the quadratic phase in the mixer.
[7] Again, note the similarity to stretch processing.

$$V_{Ii}(f) = \int_{-T_L/2}^{T_L/2} v_{Ii}(t) e^{-j2\pi ft} dt = \int_{-T_L/2}^{T_L/2} \sqrt{P_{Si}} e^{-j4\pi r_i/\lambda} e^{j2\pi(f_{yi}-f)t} dt \qquad (23.42)$$

$$= \left(\sqrt{P_{Si}}/T_L\right) e^{-j4\pi r_i/\lambda} \text{sinc}\left[\left(f - f_{yi}\right)T_L\right]$$

where, as a reminder,

$$\text{sinc}(x) = \sin(\pi x)/(\pi x) \qquad (23.43)$$

Figure 23.14 contains a plot $|V_{Ii}(f)|$ versus $(f - f_{yi})T_L$. Note that the response has a peak at $f - f_{yi} = 0$, or at $f = f_{yi}$, and that the peak has a height of $|V_{Ii}(f)| = (P_{Si})^{1/2}$. The width of the peak is $1/T_L$, which means that the SAR image will have a resolution of

$$\delta f = 1/T_L \qquad (23.44)$$

If we change the horizontal axis to y using the relation [see (23.37)]

$$y = \frac{\lambda f r_0}{2V} \qquad (23.45)$$

we get the plot of $|V_{Ii}(y)|$ versus $(y - y_i)$ of Figure 23.15. This plot has a peak at $y = y_i$ with a height of $(P_{Si})^{1/2}$.

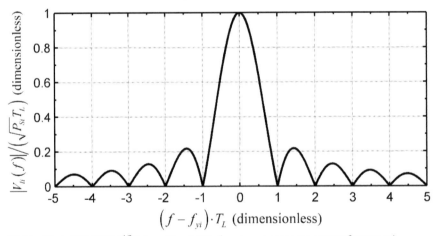

Figure 23.14 Plot of $|V_{Ii}|/[(P_{Si})^{1/2}T_L]$ vs. $(f - f_{yi})T_L$ (i.e., target response versus frequency).

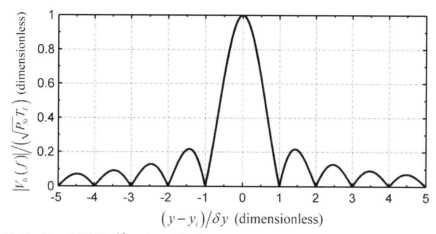

Figure 23.15 Plot of $|V_{Ii}|/(P_{Si})^{1/2}$ vs. $(y - y_i)/\delta y$ (i.e., target response versus cross-range position).

From (23.44), if the resolution of Figure 23.14 is $\delta f = 1/T_L$, the resolution of Figure 23.15 is

$$\delta y = \frac{\lambda r_0}{2V} \delta f = \frac{\lambda r_0}{2} \frac{1}{T_L V} = \frac{\lambda r_0}{2L} \tag{23.46}$$

since $L = T_L V$. This is the same as the resolution we obtained from the linear array approach [see (23.18)].

23.6 PRACTICAL IMPLEMENTATION

In the previous section, we established that the processing methodology we must use to form an image is to form a Fourier transform of $v_I(t)$ [or $v_{Ii}(t)$]. However, this approach makes the tacit assumption that $v_I(t)$ is a continuous function of time. Thinking ahead to when we will consider both cross-range and down-range imaging, we realize the SAR will transmit a pulsed signal rather than a CW signal. Because of this, we recognize that $v_I(t)$ will not be a continuous-time signal but a discrete-time signal with samples spaced by the radar PRI. In recognition of this, we replace $v_I(t)$ with $v_I(kT)$ or $v_I(k)$ where we are using k to represent the kth PRI, or pulse when we consider pulsed signals.

23.6.1 A Discrete-Time Model

For a single scatterer [i.e., $v_{Ii}(t)$], we get [from (23.34)]

$$v_{Ii}(k) = \sqrt{P_{Si}} e^{-j4\pi r_i/\lambda} e^{j4\pi(y_i V/\lambda r_i)(kT)} e^{-j2\pi(V^2/\lambda r_i)(kT)^2} \tag{23.47}$$

After we (digitally) remove the quadratic phase term, the signal we process to form the image is

$$v_{Ii}(k) = \sqrt{P_{Si}} e^{-j4\pi r_i/\lambda} e^{j4\pi(y_i V/\lambda r_i)Tk} \tag{23.48}$$

Since $v_{Ii}(k)$ (and $v_I(k)$ for multiple scatterers) is a discrete-time signal, we use the discrete-time Fourier transform (DTFT). Specifically, we find

$$V_{Ii}(f) = \sum_{k=-\infty}^{\infty} v_{Ii}(k) e^{-j2\pi f kT} \qquad (23.49)$$

As with the *continuous*-time Fourier transform, we limit the sum by considering that we gather data only from $-L/2$ to $L/2$ or for $|t| \leq T_L/2$. If we use $t = kT$, the limits on k become

$$|k| \leq T_L/2T = K_L \qquad (23.50)$$

where it is understood that we round, or truncate, $T_L/2T$ to the nearest integer.

Combining (23.50) with (23.49), (23.48), and (23.35), we get

$$\begin{aligned} V_{Ii}(f) &= \sqrt{P_{Si}} e^{-j4\pi r_i/\lambda} \sum_{k=-K_L}^{K_L} e^{j2\pi(f_{yi}-f)Tk} \\ &= \sqrt{P_{Si}} e^{-j4\pi r_i/\lambda} \frac{\sin\left[\pi(2K_L+1)(f-f_{yi})T\right]}{\sin\left[\pi(f-f_{yi})T\right]} \end{aligned} \qquad (23.51)$$

We note that this is similar to (23.42).

Figure 23.16 contains a normalized plot of $|V_{Ii}(f)|$ versus $(f - f_{yi})T_L$. As can be seen, it has a peak at $f - f_{yi} = 0$, as did Figure 23.14. However, it also has peaks at $f - f_{yi} = \pm 1/T$. In fact, if we recall the theory associated with discrete-time signals and the DTFT, we recognize that $|v_{Ii}(f)|$ will have peaks at $f - f_{yi} = \pm n_{peak}/T$, where n_{peak} is an integer. All peaks except the one corresponding to $n_{peak} = 0$ are ambiguities and are undesirable. In terms of SAR, they result in what are termed *ghost images*. Basically, if a there are scatterers outside of the image area that $f - f_{yi} = \pm n_{peak}/T$ for those scatterers, they will appear as ghosts inside of the image ares. T and the characteristics of the SAR antenna are usually chosen to avoid these ghosts since they can result in misleading SAR images. The SAR antenna was mentioned because it acts as a spatial antialiasing filter [13, 17, 18].

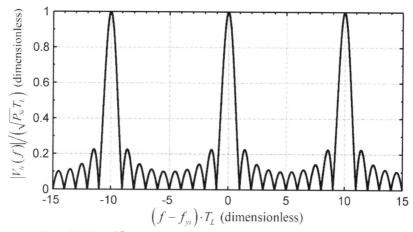

Figure 23.16 Plot of $|V_{Ii}|/(P_{Si})^{1/2}$ vs. $(f - f_{yi})T_L$ using a discrete-time signal with $T = 0.1T_L$.

23.6.2 Other Considerations

As we did before, we want to change the horizontal axis of Figure 23.16 to cross-range distance rather than frequency. To do so, we use (23.45). This results in the plot of $|v_{Ii}(y)|$ shown in Figure 23.17. The ambiguities (that produce ghosts) are shown in this figure and are located at

$$y_{ambig} - y_i = \pm \frac{\lambda r_0}{2V} \frac{1}{T} \tag{23.52}$$

Equation (23.52) tells us that we want to choose the PRI such that all scatterers lie within $\pm 1/2$ ambiguity. That is, we want to choose the PRI such that all y_i satisfy

$$y_i \leq \frac{1}{2} \frac{\lambda r_0}{2V} \frac{1}{T} \tag{23.53}$$

All scatterers of interest lie within the imaged area; therefore, we want to choose the PRI such that

$$w \leq \frac{\lambda r_0}{2V} \frac{1}{T} \tag{23.54}$$

In fact, the PRI is usually chosen such that

$$w \ll \frac{\lambda r_0}{2V} \frac{1}{T} \tag{23.55}$$

to be sure the SAR antenna beam (of the physical antenna on the SAR platform) adequately attenuates targets outside of the imaged region. Because of the constraint of (23.53), the SAR processor will form an image of an area wider than w. The desired image is determined by truncating the generated image to the desired width.

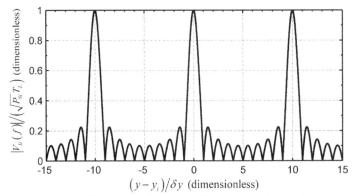

Figure 23.17 Plot of $|V_{Ti}|/(P_{Si})^{1/2}$ vs. $(y-y_i)/\delta y$ using a discrete-time signal with $T = 0.1T_L$.

We can turn (23.54) around and use it to find an upper bound on PRI. Specifically, we solve (23.54) for T to yield

$$T \le \frac{\lambda r_0}{2V} \frac{1}{w} \qquad (23.56)$$

or from (23.55)

$$T \ll \frac{\lambda r_0}{2V} \frac{1}{w} \qquad (23.57)$$

If we consider an earlier example where $r_0 = 20$ km, $\lambda = 0.03$ m, and $w = 50$ m, and consider an aircraft velocity of $V = 50$ m/s, we get

$$T \ll 120 \text{ ms} \qquad (23.58)$$

which is an easy constraint to satisfy. When we consider down-range imaging, we impose a lower limit on T to satisfy unambiguous range operation. However, that lower limit is generally well below the upper limit of (23.58).

23.7 AN ALGORITHM FOR CREATING A CROSS-RANGE IMAGE

To summarize the above as an algorithm we can implement to form a cross-range image.

- Assume a baseband, CW signal [see (23.30) and (23.31)]
- Sample this signal at intervals of T and generate $2K_L+1$ samples where
    - $T \ll \dfrac{\lambda r_0}{2V} \dfrac{1}{w}$ (23.57)
    - $K_L = T_L/2T$ (23.50)
    - $T_L = L/V$ (23.28)
    - $L = r_0 \lambda / 2\delta y$ (23.46)

In these equations, λ, r_0, V, w, and δy are desired, known parameters. The samples are taken for kT between $-T_L/2$ and $T_L/2$ or for k between $-K_L$ and K_L.

- Remove the quadratic phase by multiplying the sampled signal by

$$v_h(k) = e^{j2\pi\left(\frac{V^2}{\lambda r_0}\right)(kT)^2} \tag{23.59}$$

This gives $v_h(k)$ for a single scatterer and $v_t(k)$ for several scatterers.

- Compute the DTFT of $v_h(k)$ or $v_t(k)$, as appropriate. This is most easily done using an FFT. The minimum FFT length is $2K_L + 1$, although we usually choose the FFT length to be a power of 2 greater than $2K_L + 1$. In "real" applications, we often adjust various SAR parameters so that $2K_L + 1$ is close to a power of 2. For purposes of problems discussed herein, we choose an FFT length much greater (4 to 16 times) than $2K_L + 1$ so that the resulting frequency plot is smooth.
- If L_{FFT} is the length of the FFT, the frequency spacing between output FFT taps is

$$\Delta f = \frac{1}{TL_{FFT}} \tag{23.60}$$

After the front and rear halves of the FFT outputs are swapped (to place the zero-frequency tap in the center of the FFT output), the frequencies of the taps are

$$f = m_f \Delta f \quad m_f \in [-L_{FFT}/2, L_{FFT}/2 - 1] \tag{23.61}$$

Transform the frequency scale to cross range using (23.45) and plot the magnitude of the FFT output versus y. This does not produce an image, but instead produces a linear plot as shown in Figure 23.16.

- To generate a pseudo image, create an array of zeros where the number of columns, N_{col}, is equal to the number of samples needed to cover the width, w, of the image area. Set the number of rows, N_{row}, equal to N_{col}. (This is a somewhat arbitrary choice and can be changed.) Finally, replace row $N_{row}/2$ with the FFT outputs that cover w. The resulting array is then used to create the pseudo image.

23.8 EXAMPLE 1 - GENERATION OF A CROSS-RANGE SAR IMAGE

To illustrate the above, we consider a specific example. The parameters of this example are given in Table 23.1.

Table 23.1
Parameters Used in SAR Example 1

Parameter	Value
Width of image area, w	50 m
Depth of image area, l	50 m
Range to image area center, r_0	20 km
SAR wavelength, λ	0.03 m
Aircraft velocity, V	50 m/s
Synthetic array length, L	600 m
Number of scatterers, N_S	3
Scatterer locations, (x_i, y_i)	$(r_0, 0), (r_0, 20), (r_0, -15)$ m
Scatterer powers, P_{Si}	1, 0.25, 0.09 W

Given these, we can compute some of the SAR parameters indicated in the algorithm description. Specifically:

$$T_L = L/V = 600/50 = 12 \text{ s} \tag{23.62}$$

and

$$T \ll \frac{\lambda r_o}{2Vw} = \frac{(0.03)(20,000)}{(2)(50)(50)} = 120 \text{ ms} \tag{23.63}$$

We will choose a PRI of 50 ms. That is, we choose

$$T = 50 \text{ ms} \tag{23.64}$$

This gives

$$K_L = \frac{T_L}{2T} = \frac{12}{(2)(0.05)} = 120 \tag{23.65}$$

With this the SAR starts sampling at $t = -6$s and samples until $t = 6$ s. The samples are taken every $T = 50$ ms and a total of $2K_L + 1 = 241$ samples are used. This means that we need, as a minimum, a 256-point FFT. However, to produce a smooth plot, we use a 2,048-point FFT.

We note that, since we chose $T = 50$ ms, the actual width of the area included in the image is

$$w_{actual} = \frac{\lambda r_0}{2VT} = \frac{(0.03)(20,000)}{(2)(50)(0.05)} = 120 \text{ m} \tag{23.66}$$

To form the image, we discard the FFT outputs outside of the range of ±25 m (after the conversion from frequency to y position).

The cross-range resolution of the SAR image is

$$\delta y = \frac{r_0 \lambda}{2L} = \frac{(20,000)(0.03)}{(2)(600)} = 0.5 \text{ m} \tag{23.67}$$

This means we should be able to distinguish scatterers separated by about 1 m or greater, and maybe down to 0.5-m separation if their relative powers and phases allow this.

Before processing the SAR signal using the previously discussed algorithm, we need to generate the SAR signal. To do so, we use (23.31) with $N_S = 3$. We generate 241 samples of $v(t)$ starting at $t = -6$ s and ending at $t = 6$ s in steps of $T = 0.05$ s. Specifically, we generate $r_i(t)$, $i = 1, 2, 3$, using (23.26) and (23.27). We then combine these with the P_{Si} values in (23.30) to compute the three $v_i(t)$. Finally, we sum the three $v_i(t)$ to form $v(t)$.

Figure 23.18 is a linear plot of the $|V_I(y)|$ for -25 m $\leq y \leq 25$ m, and Figure 23.19 is a pseudo image. The pseudo image was created by starting with an array of zeros that had 101 rows (which is $l/\delta y + 1$) and a number of columns equal to the number of y values in the linear plot. The $|V_I(y)|$ values from the linear plot were loaded into the 51st row of the array, and the pseudo image was generated using image plotting software. The image of

Figure 23.19 is a negative image. That is, large amplitudes are black and zero is white. This was done to make the experimental images look better, while also conserving printer ink.

In examining Figure 23.18, we note that $|V_I(y)|$ has three peaks at the y positions of the scatterers. Further, the heights of the peaks are $(P_{Si})^{1/2}$. The image (Figure 23.19) shows three dots at the given scatterer positions, and the dots are different shades of gray, indicating different amplitudes.

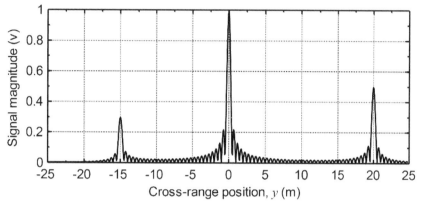

Figure 23.18 Linear plot of $|V_I(y)|$—three scatterers at −15 m, 0 m, 20 m.

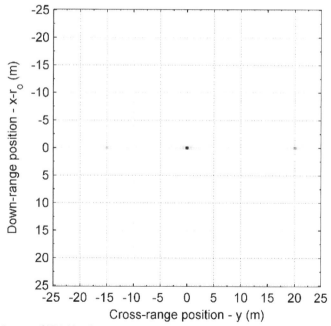

Figure 23.19 Image of $|V_I(y)|$—three scatterers at −15 m, 0 m, 20 m.

To check the aforementioned resolution statement, the simulation was rerun with scatterer y positions of -1, 0, and 1 m. The results are shown in Figures 23.20 and 23.21. The linear plot clearly shows three peaks, but the relative amplitudes are somewhat different than those of Figure 23.18. This is due to the sidelobes of the Fourier transform response function and the way the responses to the three scatterers constructively and destructively combine. The presence of the three scatterers can also be seen in the image of Figure 23.21.

As another interesting experiment, the quadratic phase removal step of the SAR processing algorithm was eliminated. The results are shown in Figures 23.22 and 23.23. (The original scatterer locations of -15, 0, and 20 m were used.) As can be seen, the peaks are spread and the image is blurred in the y direction. In SAR terminology, we say the image is not focused. In fact, as indicated earlier, the process of removing the quadratic phase is sometimes termed *focusing* of the SAR image.

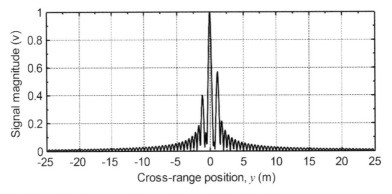

Figure 23.20 Linear plot of $|V_I(y)|$—three scatterers at -1 m, 0 m, 1 m.

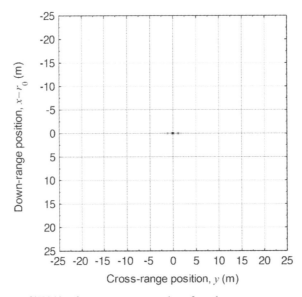

Figure 23.21 Image of $|V_I(y)|$—three scatterers at -1 m, 0 m, 1 m.

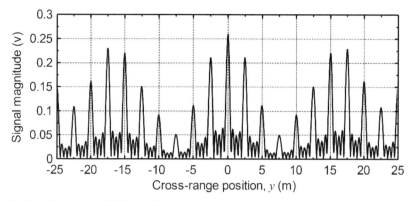

Figure 23.22 Linear plot of $|V_I(y)|$—three scatterers at -15 m, 0 m, 20 m, without quadratic phase removal.

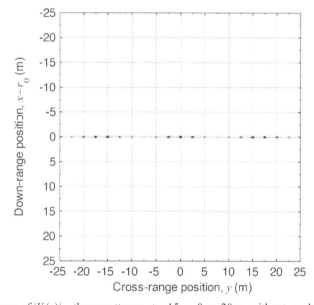

Figure 23.23 Image of $|V_I(y)|$—three scatterers at -15 m, 0 m, 20 m, without quadratic phase removal.

23.9 DOWN-RANGE AND CROSS-RANGE IMAGING

We now extend the previous work to both down-range and cross-range imaging. We will also extend the problem to include a more general case of *squinted SAR* [2, 13, 15, 20]. Squinted SAR is normally associated with strip map SAR, but the development here also applies to spotlight SAR. As before, we will start by defining the signal that the SAR processor must work with, since this will give insight into how to process the signal.

23.9.1 Signal Definition

The geometry of interest is a modification of the geometry of Figure 23.13 and is contained in Figure 23.24. The main difference between Figure 23.13 and Figure 23.24

is that in Figure 23.13, the center of the imaged area lies on the x-axis of the coordinate system, while in Figure 23.24 it does not. This offset of the imaged area center will result in additional Doppler considerations, plus a phenomenon termed *range cell migration* (RCM) [13], both of which complicate SAR processing. Another minor difference is that the coordinates of the scatterer are relative to the center of the imaged area. We did this as a convenience.

Since we are considering both down-range and cross-range imaging, the transmit waveform will be pulsed instead of CW. In practical SAR, the pulses are phase-coded, usually with LFM, to achieve the dual requirements of large bandwidth for fine range resolution and long duration to provide sufficient energy. In this development, we will use narrow, uncoded (unmodulated) pulses to avoid complicating the development with pulse coding and the associated matched filter or stretch processing. The extension to coded pulses is relatively straightforward. The use of narrow, uncoded pulses also helps clarify the concept of RCM correction (RCMC).

Given the above, we write the transmit signal as

$$v_T(t) = e^{j2\pi f_o t} \sum_k \text{rect}\left[\frac{t-kT}{\tau_p}\right] \quad (23.68)$$

where, as a reminder,

$$\text{rect}[x] = \begin{cases} 1 & |x| \leq 1/2 \\ 0 & |x| > 1/2 \end{cases} \quad (23.69)$$

and τ_p is the pulsewidth. The sum notation means a sum over all k and is used to indicate that the waveform is, in theory, an infinite duration pulse train. We will later make it finite duration.

The signal from a single scatterer at (x_i, y_i) (see Figure 23.24) is

$$v_{iRF}(t) = \sqrt{P_{Si}} e^{j2\pi f_o(t - 2r_i(t)/c)} \sum_k \text{rect}\left[\frac{t - kT - 2r_i(t)/c}{\tau_p}\right] \quad (23.70)$$

where

$$r_i(t) = \sqrt{(x_0 + x_i)^2 + (y_0 + y_i - Vt)^2} \quad (23.71)$$

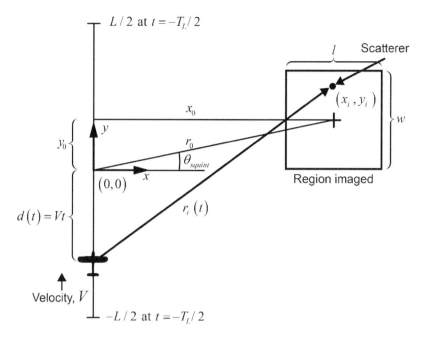

Figure 23.24 Geometry for down- and cross-range imaging.

23.9.1.1 Removal of the Carrier and Gross Doppler

As before, the first operation we perform is removal of the carrier. However, in addition, we will also remove what we term *gross Doppler* [13]. Removal of gross Doppler is necessary in some applications in that this Doppler is large relative to the PRF and has the potential of causing problems with aliasing and Doppler ambiguities (ghosts). In practice, the gross Doppler is either predicted from SAR platform data or is estimated from a spectral analysis of the return signal [13].

To determine the gross Doppler, we examine the phase of the returned (RF) signal. From (23.70), this phase is

$$\phi_{RF}(t) = 2\pi f_o \left(t - 2r_i(t)/c\right) = 2\pi f_o t - 4\pi r_i(t)/\lambda \qquad (23.72)$$

We can find the frequency as

$$f_{RF} = \frac{1}{2\pi}\frac{d\phi_{RF}(t)}{dt} = f_o - \frac{2\dot{r}_i(t)}{\lambda} = f_o + f_d(t) \qquad (23.73)$$

The first term is the carrier frequency and the second is the Doppler frequency. We define the gross Doppler, f_{dg}, as the Doppler frequency at $x_i = 0$, $y_i = 0$ and $t = 0$. That is,

$$f_{dg} = -\frac{2\dot{r}_i(t)}{\lambda}\bigg|_{t=0, x_i=0, y_i=0} = \frac{2V}{\lambda}\frac{y_0}{r_o} = \frac{2V}{\lambda}\sin\theta_{squint} \qquad (23.74)$$

In (23.74), θ_{squint} is the *squint angle* [2, 13, 15, 20]. For the unsquinted SAR we considered in the CW development, θ_{squint} was zero because y_0 was zero. This also made $f_{dg} = 0$.

Given the above, we remove f_o and f_{dg} from the received signal by multiplying $v_{iRF}(t)$ by the heterodyne signal,

$$v_h(t) = e^{-j2\pi(f_o + f_{dg})t} \tag{23.75}$$

With this we get the baseband signal,

$$v_i'(t) = v_h(t) v_{iRF}(t) = e^{-j2\pi(f_o + f_{dg})t} v_{iRF}(t)$$
$$= \sqrt{P_{Si}} e^{-j2\pi f_{dg} t} e^{-j4\pi r_i(t)/\lambda} \sum_k \text{rect}\left[\frac{t - kT - 2r_i(t)/c}{\tau_p}\right]^8 \tag{23.76}$$

23.9.1.2 Single-Pulse Matched Filter

The next step in processing is to send $v_i'(t)$ through a matched filter matched to the transmit pulse. The (normalized) output of the matched filter is

$$v_i(t) = \sqrt{P_{Si}} e^{-j2\pi f_{dg} t} e^{-j4\pi r_i(t)/\lambda} \sum_k \text{tri}\left[\frac{t - kT - 2r_i(t)/c}{\tau_p}\right] \tag{23.77}$$

where

$$\text{tri}(x) = \begin{cases} 1 - |x| & |x| \leq 1 \\ 0 & |x| > 1 \end{cases} \tag{23.78}$$

23.9.1.3 Generation of the Sampled Signal

Recall that for the CW case, we sampled $v_i(t)$ at intervals of T. We will do the same for the pulsed case. However, for each pulse (each T) we will also subsample $v_i(t)$ at intervals of τ_p, the pulsewidth. We start sampling, relative to each transmit pulse, at some

$$\tau_{min} = \frac{2(x_0 - l/2)}{c} \tag{23.79}$$

This is the minimum range delay between the front edge of the imaged region and the SAR platform. We continue sampling to

$$\tau_{max} = \frac{2 r_{max}}{c} \tag{23.80}$$

where

$$r_{max} = \sqrt{(x_0 + l/2)^2 + (|y_0| + w/2 + L/2)^2} \tag{23.81}$$

---

[8] It appears as if f_{dg} is still in $v_i'(t)$ but it is not. There is an f_{dg} in the second exponential term, which is removed by the f_{dg} in the first exponential term.

r_{max} is the maximum range between the SAR platform and the back of the imaged region.[9] Between τ_{min} and τ_{max}, we obtain approximately

$$M = \frac{\tau_{max} - \tau_{min}}{\tau_p} \tag{23.82}$$

range samples. We do this $2K_L + 1$ times to form $M \times (2K_L + 1)$ samples, which we will collect into an M by $2K_L + 1$ element array for further processing.

Mathematically, we sample $v_i(t)$ at

$$t = kT + \tau_{min} + m\tau_p \tag{23.83}$$

where m is the range cell number. This gives

$$v_i(m,k) = \sqrt{P_{Si}} e^{-j2\pi f_{dg}(kT+\tau_{min}+m\tau_p)} e^{-j4\pi r_i(kT+\tau_{min}+m\tau_p)/\lambda}$$
$$\text{tri}\left[\frac{\tau_{min} + m\tau_p - 2r_i(kT + \tau_{min} + m\tau_p)/c}{\tau_p}\right] \tag{23.84}$$

Equation (23.84) is the equation that generates the samples we use in the SAR processor simulations discussed in the upcoming sections and in the exercises. A separate $v_i(m,k)$ array is generated for each simulated scatterer, and the composite return is created by summing the $v_i(m,k)$ across i. m varies from 0 to $M - 1$ and k varies from $-K_L$ to K_L.

We previously defined an upper bound on T [see (23.56) and (23.57)] based on the width of the imaged area. A lower bound on T is that it must be such that $T > \tau_{max} - \tau_{min}$. Since $\tau_{max} - \tau_{min}$ is usually on the order of μs and the upper bound on T is on the order of ms, both of these bounds on T are easy to satisfy unless the imaged area becomes very wide and deep.

Figure 23.25 contains a simplified block diagram of the operations that have been discussed thus far. The diagram is a functional representation of the operations that would be performed in an actual SAR receiver. The voltage symbols above the line would apply when one thinks of processing returns from a single scatterer. The symbols below the line, without the i subscript, would apply when one thinks of processing returns from more than one scatterer. The actual SAR receiver will need to perform many other functions such as mixing and amplification to convert the RF signal to the digital, baseband signal sent to the SAR processor. As a note, the mixing operation does not always remove all of the gross Doppler. This sometimes needs to be removed as part of the subsequent processing [13].

---

[9] The limits on τ are applicable for small squint angles. For larger squint angles we would need to use a slightly different approach. This is discussed in the appendix.

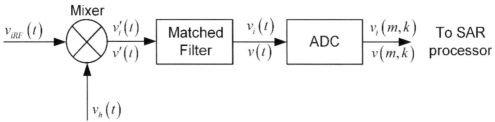

Figure 23.25 Preliminary SAR processing block diagram.

23.9.2 Preliminary Processing Considerations

If we were to directly extend our CW processing methodology, we would, for each m, remove a quadratic phase term and then perform a DFT across k. Unfortunately, the situation is complicated by the range sampling so that this straightforward approach is not directly applicable. We will need to first perform an interim step of RCMC.

23.9.2.1 Range Cell Migration Correction

Figure 23.26 contains a plot of $|v_i(m,k)|$. The axes are range cell number, m, and cross-range sample number, k. White corresponds to a level of zero and black corresponds to the maximum value of $|v_i(m,k)|$. The plot is not an image of the form of Figures 23.19, 23.20, or 23.23. It is a means of representing the magnitude of the analog to digital converter (ADC) output as a function of two variables, m and k. The plot was generated for a single scatterer at $(x_i, y_i) = (0,0)$ m, the center of the imaged region. $|v_i(m,k)|$ was generated using the parameters of example 1 (Section 23.8, Table 23.1) with the added parameters of $(x_0, y_0) = (20000, 200)$ m, which defines the center of the imaged area. With this specification, we are considering a squinted SAR with a squint angle of $\theta_{squint} = \sin^{-1}(200/20000) = 0.0573°$ [see (23.74)]. We assume a pulse with a width of $\tau_p = 3.33$ ns, which translates to a range resolution of 0.5 m. As indicated earlier, this is an unrealistic pulse for an actual SAR. We use it only to avoid having to complicate the development by considering long, modulated pulses.

For each pulse, we start sampling in range at $\tau_{min} = 133.1667$ μs and stop at $\tau_{max} = 133.5459$ μs. We used (23.79), (23.80), and (23.81) to compute τ_{min} and τ_{max}. With this τ_{min} and τ_{max}, and the pulsewidth of 3.33 ns, we will have $M = 114$ range samples [see (23.82)]. Since we want sample 0 to correspond to r_0, we let m vary from -50 to 63. From example 1, k varies from -120 to 120 for a total of 241 cross-range samples.

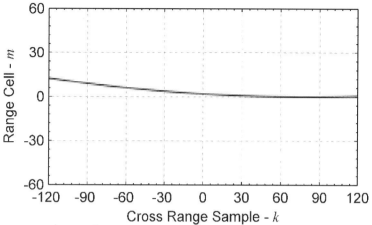

Figure 23.26 Plot of $|v_i(m,k)|$ for a single scatterer at $(x_i, y_i)=(0,0)$ m.

Since we are considering a single scatterer in the center of the imaged area, our initial expectation is that the return should be located at range cell zero ($m = 0$) on all pulses. However, this is not correct because range cell location of the return depends on the *range* to the scatterer, *not its x location*. The range to the scatterer on the kth pulse is

$$r_i(kT) = \sqrt{(x_0 + x_i)^2 + (y_0 + y_i - VkT)^2} \qquad (23.85)$$

which means it will vary with k. This is why the line in Figure 23.26 is curved. The curving of the line is the aforementioned RCM. This name derives from the fact that the return from a single scatterer *migrates* across several range cells.

Range cell migration becomes a problem when we apply the quadratic phase correction and then take the Fourier transform to form the image. To form the image, we want to adapt the procedure we developed for the CW case and perform these operations for each range cell (for each row of the M by $2K_L + 1$ array of $|v_i(m,k)|$). However, because of RCM we cannot do this. Instead we should apply the quadratic phase removal and Fourier transform procedures to the range and cross-range samples along the curved line. Since this is difficult, we take the approach of "straightening" the curved line of Figure 23.26 [13]. Said another way, we will remove the effect of RCM through the process of range cell migration correction, or RCMC.

There are several methods of applying RCMC [13]. All of them involve some type of interpolation, and some are more effective than others. In this book, we discuss a technique based on the Fourier transform. It derives from a property of Fourier transforms that a linear phase gradient applied in the frequency domain will result is a time shift in the time domain.

A characteristic of the Fourier transform technique is that it moves all range cells the same amount for a particular k. This is adequate for small squint angles. However, for a squint angle more than a few degrees, it is a questionable approach because, in that case, different range cells must be moved different amounts. This is discussed further in [13].

23.9.2.2 RCMC Algorithm

As indicated above, the Fourier transform RCMC algorithm takes advantage of the time shift property of the Fourier transform. We consider a time function $v(t)$ with a Fourier transform

$$V(f) = \int_{-\infty}^{\infty} v(t) e^{-j2\pi ft} \quad (23.86)$$

We next consider a shifted version of $v(t)$, $v_{Shift}(t) = v(t - \tau)$. The Fourier transform of $v_{Shift}(t)$ is

$$V_{Shift}(f) = \int_{-\infty}^{\infty} v_{Shift}(t) e^{-j2\pi ft} dt = \int_{-\infty}^{\infty} v(t-\tau) e^{-j2\pi ft} dt$$
$$= \int_{-\infty}^{\infty} v(\gamma) e^{-j2\pi f(\gamma+\tau)} d\gamma = e^{-j2\pi f\tau} V(f) \quad (23.87)$$

Equation (23.87) says that if we want to shift some $v(t)$ by some τ we

- Find the Fourier transform of $v(t)$, $V(f)$.
- Multiply $V(f)$ by $e^{-j2\pi f\tau}$ [apply a linear phase gradient to $V(f)$].
- Find the inverse Fourier transform of the result.

This is the essence of the RCMC algorithm. We develop the algorithm for a single scatterer at $(x_i, y_i) = (0,0)$ m and apply it to all scatterer locations.

A suggested algorithm is as follows: from (23.85), the minimum value of $r_i(kT)$, for $x_i = y_i = 0$, occurs when $y_0 - kVT = 0$ and is equal to x_0. We decide that we want this range to correspond to a down-range delay of $\tau = 0$. For each k we compute

$$\Delta\tau(k) = 2\left(\sqrt{x_0^2 + (y_0 - VkT)^2} - x_0\right)/c \quad (23.88)$$

This $\Delta\tau(k)$ then becomes the range correction based on the assumption that $\tau = 0$ when $y_0 - kVT = 0$. We use this with the Fourier transform method to move the samples in range. The specific algorithm is

- For each k compute $\Delta\tau(k)$ from (23.88).
- Compute the Fourier transform (discrete-time Fourier transform) of $v_i(m,k)$.
- Multiply the Fourier transform by $e^{-j2\pi f\Delta\tau(k)}$.
- Compute the inverse Fourier transform.

The FFT can be used to approximate the Fourier transform and inverse Fourier transform. The length of the FFT should be the next power of 2 that is equal or greater than M. In the example used here, a 128-point FFT was used since $M = 114$. When the inverse FFT is computed, the last 14 samples (of the 128) are discarded. The frequency values would be computed from $f = s/(N_{FFT}\tau_p)$, $s \in [-N_{FFT}/2, N_{FFT}/2 - 1]$. Be sure that the FFT algorithm you use places the zero-frequency tap in the center of the FFT output.

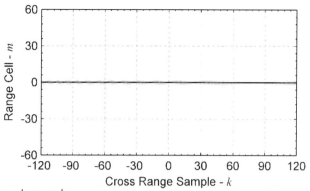

Figure 23.27 Plot of $|v_i(m,k)|$ for a single scatterer at $(x_i, y_i) = (0,0)$ m with RCMC.

The result of applying the above methodology to the plot of Figure 23.26 is shown in Figure 23.27. As can be seen, the curved line of Figure 23.26 is now a straight line, but somewhat blurred. The blurring is caused by the fact that the output of our matched filter is not matched to the type of interpolation the Fourier transform performs. The Fourier transform uses a sinc(x) interpolation but our matched filter output is a triangle function. If we had modeled our matched filter output as a sinc(x) function (which we would get for an LFM pulse), the line of Figure 23.27 would be a straight line with no blurring.

The RCMC methodology discussed above was derived for a scatterer at the center of the imaged region. There is a question of whether it will perform RCMC for all other scatterers in the imaged region. To address this question, we consider two examples. In the first, we place three scatterers at $y_i = 0$ m and $x_i = -23$ m, 0 m, and 23 m (range sample, or cell, numbers of −46, 0, and 46). The resulting uncorrected plot of $v(m,k)$ is shown in Figure 23.28 and the RCM-corrected image is shown in Figure 23.29. It will be noted that there are three straight lines located at $m = -46$, 0, and 46 in the RCMCed image.

As another example, we place the three scatterers at $(x_i, y_i) = (-23,23), (0,0), (23,-23)$ m. That is, at diagonal corners and the center of the imaged area. The resulting uncorrected and corrected plots of $v(m,k)$ are shown in Figures 23.30 and 23.31. Careful examination of Figure 23.30 shows that the three curved lines are not exactly the same. Also, the top and bottom straight lines of Figure 23.31 are not exactly horizontal. In some applications, this can cause problems and an interim processing step must be used to eliminate the problem.

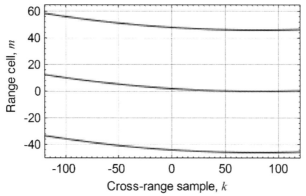

Figure 23.28 Plot of $|v(m,k)|$ for a three scatterers at $(x_i, y_i) = (-23,0), (0,0), (23,0)$ m.

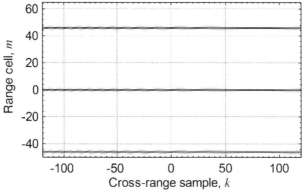

Figure 23.29 Plot of $|v(m,k)|$ for a three scatterers at $(x_i, y_i) = (-23,0), (0,0), (23,0)$ m after RCMC.

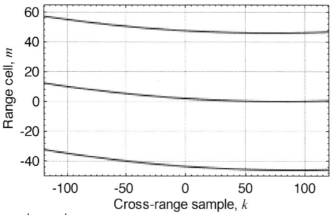

Figure 23.30 Plot of $|v(m,k)|$ for a three scatterers at $(x_i, y_i)=(-23,23), (0,0), (23,-23)$ m.

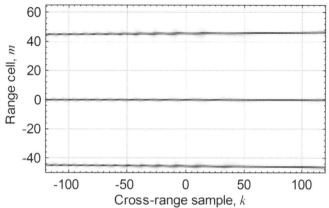

Figure 23.31 Plot of $|v(m,k)|$ for a three scatterers at $(x_i, y_i) = (-23, 23), (0,0), (23, -23)$ m after RCMC.

23.9.3 Quadratic Phase Removal and Image Formation

Now that we have an algorithm that performs RCMC, we need to develop an algorithm for removing the quadratic phase. We will want to remove the quadratic phase from the RCMCed signal. The information we need is in the phase of $v_i(m,k)$ (for a single scatterer, $v(m,k)$ for multiple scatterers) at the peak of the tri(x) function (i.e., along the curved ridge before RCMC).

If we refer to $v_i(t)$ of (23.77), we find we want to examine the information in the phase of $v_i(t)$ at

$$t = kT + 2r_i(t)/c \tag{23.89}$$

A problem with this equation is that t appears on both sides and is embedded in $r_i(t)$ on the right side. As a result, solving for t will involve the solution of a rather complicated equation. To avoid this, we seek a simpler approach. Specifically, we ask the question: Does the phase of $v_i(t)$ vary slowly enough to allow the use of an approximate value of t?

We write the phase of $v_i(t)$, from (23.77), as

$$\phi(t) = -2\pi f_{dg} t - 4\pi r_i(t)/\lambda \tag{23.90}$$

From calculus, we know that we can relate variations of $\phi(t)$ to variations of t by

$$\Delta\phi(t_0) = \left.\frac{\partial \phi(t)}{\partial t}\right|_{t_0} \Delta t \tag{23.91}$$

Computing the partial derivative, we get

$$\Delta\phi(t_0) = -2\pi \left[f_{dg} - \frac{2V(y_0 - Vt_0)}{\lambda r_i(t_0)} \right] \Delta t \tag{23.92}$$

We are interested in the variation of $\phi(t)$ over the times we are taking measurements. Specifically, from $t = kT + \tau_{min}$ to $t = kT + \tau_{max}$. We use $t_0 = kT + \tau_{min}$ and let $\Delta t = \tau_{max} - \tau_{min} = \Delta\tau$. With this we have

$$\Delta\phi(kT + \tau_{min}) = -2\pi\left\{f_{dg} - \frac{2V[y_0 - V(kT + \tau_{min})]}{\lambda\sqrt{x_0^2 + [y_0 - V(kT + \tau_{min})]^2}}\right\}\Delta t \quad (23.93)$$

Figure 23.32 contains a plot of $\Delta\phi(kT + \tau_{min})$ versus k as the top plot. For reference, the bottom curve is a plot of pulse-to-pulse phase change versus k. Note that the pulse-to-pulse phase change ranges between about $-1{,}000°$ and $+1{,}000°$ while the phase variation, or phase error, over $\Delta\tau$ is between $-0.006°$ and $+0.006°$. This indicates that $\phi(t)$ varies slowly over $\Delta\tau$, and thus, it will be reasonable to compute $\phi(t)$ at $kT + \tau_{min}$, or even at kT, rather than via the more accurate form of (23.90).

Given this, we now examine $\phi(kT)$ to formulate a quadratic phase correction scheme. We write

$$\begin{aligned}\phi(kT) &= -2\pi f_{dg}kT - 4\pi r_i(kT)/\lambda \\ &= -2\pi\left(f_{dg}kT + \frac{2\sqrt{x_0^2 + (y_0 - VkT)^2}}{\lambda}\right) \\ &\approx -2\pi\left\{f_{dg}kT + \frac{2}{\lambda}\left[r_0 - \frac{2y_0VkT}{r_0} + \frac{(VkT)^2}{r_0}\right]\right\} \\ &= \frac{-4\pi r_0}{\lambda} - 2\pi\left(f_{dg} - \frac{2y_0V}{\lambda r_0}\right)kT - 2\pi\frac{(VkT)^2}{\lambda r_0}\end{aligned} \quad (23.94)$$

where we made use of (23.33) to approximate the square root.

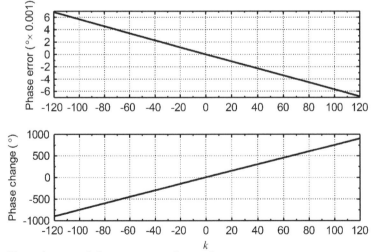

Figure 23.32 Phase change and phase error vs. pulse number.

The first term of the last equality of (23.94) is a constant phase that we do nothing about. The second term is zero, since, by (23.74), $f_{dg} = 2Vy_0/(\lambda r_0)$. Finally, the third term is the quadratic phase that we want to eliminate. It will be noted that this quadratic phase term is exactly the same as the quadratic phase term in the CW problem. Thus, to perform the quadratic phase correction, we multiply each row of the RCMCed signal array by

$$v_q(k) = e^{j2\pi \left(\frac{V^2}{\lambda r_0}\right)(kT)^2} \qquad (23.95)$$

We note that the quadratic phase correction was computed for the scatterer at the center of the image area. However, this correction is usually adequate for all other scatterers.

We are now in a position to formulate an algorithm for creating a cross-/down-range image.

23.10 ALGORITHM FOR CREATING A CROSS- AND DOWN-RANGE IMAGE

An algorithm for creating a cross- and down-range image is:

- Assume a sampled baseband signal of the form given by (23.84) (for a single scatterer—for multiple scatterers, we would sum across i). Note: this signal has had the gross Doppler, f_{dg}, removed, even though the term appears in the equation.
- Perform RCMC. The RCMC is applied to all range cells for each k.
- Perform the quadratic phase correction by multiplying the returns for each range cell by the $v_q(k)$ of (23.95).
- Take the FFT across pulses, for each range cell. As before, be sure the FFT algorithm you use places the zero-frequency tap in the center of the FFT output.
- Transform the frequency and range delay axes of the output of the FFTs to cross-range and down-range, and plot the image.

Figure 23.33 contains an update to the block diagram of Figure 23.25 that includes the image generation algorithm discussed above.

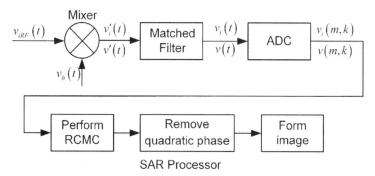

Figure 23.33 SAR processor block diagram.

23.11 EXAMPLE 2: CROSS- AND DOWN-RANGE SAR IMAGE

We extend Example 1 of Section 23.8 to include cross- and down-range imaging. Table 23.2 is a repeat of Table 23.1 with additions and modifications consistent with the cross- and down-range image generation methodology.

As with Example 1, we have $K_L = 120$ so that k goes from -120 to 120, and we transmit 241 pulses over a time period of -6 to 6 seconds. We use the 3.33-ns: unmodulated pulse we considered in the RCMC discussions. Recall that since our T is smaller than the minimum dictated by the width of the imaged area, our SAR image will need to be trimmed in cross range before we plot the image.

We start the range sampling at τ_{min} and let m vary from -50 to 63 as we did when we performed RCMC. As a result, the down-range extent of the image will be -25 m to 32 m relative to scene center. Since we are interested only in a down-range extent of -25 m to 25 m, we will also trim the down-range coordinate of the image.

In Example 1, we used an FFT length that was longer than the number of samples because we wanted a smooth linear plot. Since we are forming only an image for this example, we can limit the FFT length to the nearest power of two greater than $2K_L + 1$. Since $2K_L + 1$ is 241, a 256-point FFT will suffice.

Figure 23.34 contains the image for this example. Note that the three dots are approximately where they should be. The center dot is at (0,0) m and is fairly sharp. This is expected since the RCMC and quadratic phase correction is based on a scatterer at the center of the imaged area. The other two dots are somewhat smeared and are offset slightly in the cross-range direction. The offset is due to a residual Doppler, and the smearing is due to a residual quadratic phase.

Table 23.2
Parameters Used in SAR Example 2

Parameter	Value
Width of image area, w	50 m
Depth of image area, l	50 m
SAR wavelength, λ	0.03 m
Aircraft velocity, V	50 m/s
Synthetic array length, L	600 m
Number of scatterers, N_s	3
Waveform PRI, T	50 ms
Down-range resolution, δ_x	0.5 m
Center of imaged area (x_0, y_0)	(20000, 200) m
Scatterer locations, (x_i, y_i)	$(-23, 0), (0, 0), (23, 0)$ m
Scatterer powers, P_{Si}	1, 1, 1 W

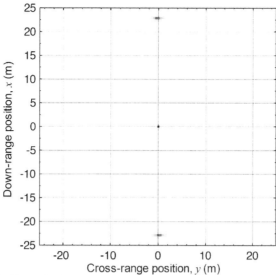

Figure 23.34 Image for Example 2.

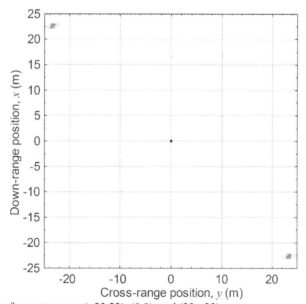

Figure 23.35 Image for scatterers at (−23,23), (0,0), and (23,−23) m.

Figure 23.35 contains an image that resulted when the three scatterers were placed at (−23,23), (0,0), and (23,−23) m. Again, the center dot is reasonably sharp, but the other two dots are offset in the cross-range dimension and smeared in both the cross- and down-range dimensions. The cross-range offset is due to the aforementioned residual Doppler, and the smearing is due to the residual quadratic phase. The down-range smearing is due to the imperfect RCMC discussed in association with Figures 23.30 and 23.31.

23.12 AN IMAGE-SHARPENING REFINEMENT

We noted in the generation of Figure 23.34 that there was a slight skewing of the upper and lower dots. Given that the skewing was in opposite directions at the top and bottom, we surmise that it is due to a frequency shift, and possibly FM slope variation (cross range residual quadratic phase) that is dependent on the down-range location of the scatterer, x_i. We now examine this further.

For a scatterer at $(x_0 + x_i, y_0)$ we have

$$r_i(t) = \sqrt{(x_0 + x_i)^2 + (y_0 - Vt)^2} \tag{23.96}$$

where we are temporarily using $kT = t$ for convenience. We manipulate this as

$$r_i(t) = \sqrt{(x_0 + x_i)^2 + y_0^2 - 2y_0 Vt + V^2 t^2} = \sqrt{r_i^2 - 2y_0 Vt + V^2 t^2}$$

$$\approx r_i - \frac{y_0 V}{r_i} t + \frac{V^2}{2r_i} t^2 \tag{23.97}$$

where $r_i = r_i(0) = \sqrt{(x_0 + x_i)^2 + y_0^2}$ is the range to the ith scatterer at $t = 0$.

With this the phase of $v_i(t)$ is [see (23.90)]

$$\phi(t) = -2\pi \left(f_{dg} t + 2r_i(t)/\lambda \right) = -2\pi \left(\frac{2r_i}{\lambda} + \left(f_{dg} - \frac{2y_0 V}{\lambda r_i} \right) t + \frac{V^2}{\lambda r_i} t^2 \right) \tag{23.98}$$

During the quadratic phase removal step, we essentially add

$$\phi_q(t) = 2\pi \frac{V^2}{\lambda r_0} t^2 \tag{23.99}$$

to the above phase to get a corrected phase of

$$\phi_c(t) = -2\pi \left(\frac{2r_i}{\lambda} + \left(f_{dg} - \frac{2y_0 V}{\lambda r_i} \right) t + \left(\frac{V^2}{\lambda r_i} - \frac{V^2}{\lambda r_0} \right) t^2 \right) \tag{23.100}$$

We first examine the linear phase, or frequency, term. We write it as

$$\phi_f(t) = -2\pi \left(f_{dg} - \frac{2y_0 V}{\lambda r_i} \right) t \tag{23.101}$$

Recalling that $f_{dg} = 2y_0 V / \lambda r_0$, we have (as a note, y_0/r_0 is the sine of the squint angle)

$$\phi_f(t) = -2\pi \frac{2y_0 V}{\lambda} \left(\frac{1}{r_0} - \frac{1}{r_i} \right) t \tag{23.102}$$

The range to the ith scatterer at $t = 0$ is

$$r_i = \sqrt{(x_0 + x_i)^2 + y_0^2} = \sqrt{x_0^2 + 2x_i x_0 + x_i^2 + y_0^2}$$
$$= \sqrt{r_0^2 + 2x_i x_0 + x_i^2} \approx r_0 + \frac{x_i x_0}{r_0} \approx r_0 + x_i \qquad (23.103)$$

where we made use of $r_0^2 \gg 2x_i x_0 + x_i^2$, $2x_i x_0 \gg x_i^2$, and $r_0 \approx x_0$, since $x_0^2 \gg y_0^2$. With this we have

$$\phi_f(t) = -2\pi \frac{2 y_0 V}{\lambda} \left(\frac{x_i}{r_0^2} \right) t \qquad (23.104)$$

where we used $r_0 r_i \approx r_0^2$.

Taking the derivative of (23.104), we see that we have a residual frequency of

$$\Delta f = -\frac{2 y_0 V}{\lambda r_0} \left(\frac{x_i}{r_0} \right) \qquad (23.105)$$

When the scatterer is at scene center, $x_i = 0$ and thus $\Delta f = 0$. That is, there is no frequency offset. When $x_i \neq 0$, there will be a residual frequency offset, which will lead to a cross-range offset.

To see if the frequency offset could be the cause of the skewing in Figure 23.34, we recall that cross-range position is related to frequency by [see (23.45)]

$$y = \frac{\lambda r_0}{2V} f \qquad (23.106)$$

With this and (23.105) we can write

$$\Delta y = \frac{\lambda r_0}{2V} \Delta f = -\frac{x_i y_0}{r_0} \qquad (23.107)$$

In our case, $r_0 = 20$ km, $y_0 = 200$ m, $x_i = 25$ m, and

$$\Delta y = -\frac{25 \times 200}{20,000} = -0.25 \text{ m} \qquad (23.108)$$

or half of a cross-range resolution cell, which is about the shift noted in Figure 23.34. This leads us to conclude that it might be a good idea to include a range-cell-dependent frequency correction when we apply the quadratic phase correction. When such a correction was included, the image of Figure 23.36 was obtained. As the figure shows, the skewing is no longer present.

Very careful examination of Figure 23.36 reveals a slight cross-range smearing of the upper and lower dots relative to the center dot. From our experience with stretch processing, we postulate that this could be due to the residual quadratic phase term of (23.100).

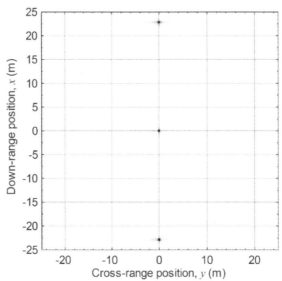

Figure 23.36 Case of Figure 23.34 with additional Doppler correction.

From (23.100) we can write the residual quadratic phase term as

$$\phi_q(t) = -2\pi\left(\frac{V^2}{\lambda r_i} - \frac{V^2}{\lambda r_0}\right)t^2 = \pi(\Delta\alpha)t^2 \qquad (23.109)$$

With approximations similar to the previous development, we obtain

$$\Delta\alpha = -\frac{2V^2}{\lambda}\left(\frac{1}{r_i} - \frac{1}{r_0}\right) \approx \frac{2V^2}{\lambda r_0}\frac{x_i}{r_0} = \alpha_0 \frac{x_i}{r_0} \qquad (23.110)$$

which is a residual quadratic phase that depends on the x location of the scatterer. This indicates that we should apply a residual quadratic phase correction that is range dependent. The result of applying this correction is contained in Figure 23.37. Very careful examination of this figure reveals that all three dots are equally sharp in the cross-range direction.

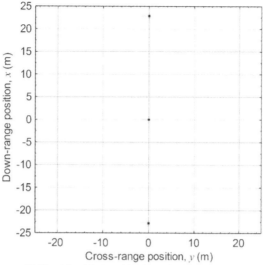

Figure 23.37 Case of Figure 23.32 with added residual quadratic phase correction.

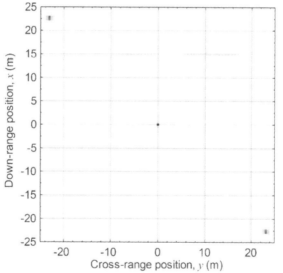

Figure 23.38 Case of Figure 23.35 with additional Doppler and quadratic phase correction.

Figure 23.38 contains an image equivalent to Figure 23.35 with the aforementioned residual frequency and quadratic phase corrections included. As can be seen, the dots of Figure 23.38 seem to be slightly more focused in the cross-range direction than those of Figure 23.35. However, the upper and lower dots are still smeared in the down-range direction. As discussed earlier, this down-range smearing is caused by the fact that the RCM is due to THE cross-range position of the scatterer, whereas the RCMC is based on a scatterer at zero cross range. Cummings and Wong [13] allude to an alternate RCMC algorithm that corrects this problem. We will not discuss it here. The reader is referred to [13, pg. 235].

Note that the images in Figures 23.34 through 23.38 exhibit some smearing in cross range. This is due to sidelobes of the Fourier transform operation used to create the cross-range dimension of the image. The smearing can be reduced by applying sidelobe reduction weighting to the input to the FFT. However, such weighting will slightly degrade cross-range resolution.

23.13 CLOSING REMARKS

The discussions presented in this chapter are very preliminary when compared to the body of literature on SAR processing. The technique presented is a bare, basic image formation method, with the exception of the image refinement technique of Section 23.12. There are several texts that discuss other image formation and sharpening techniques [13, 17, 18]. Many of these provide sharper images but are also more difficult to implement and run slowly when compared to the technique discussed herein.

The technique discussed herein is applicable to both strip map and spotlight SAR for the case where the SAR platform is moving in a straight line. There is another class of spotlight SAR termed *circular SAR*. In this type of SAR, the SAR platform follows a circular path relative to some point in the imaged area. The techniques developed in this chapter are not applicable to this type of SAR because the RCMC technique developed herein cannot be directly extended to the circular SAR case. The most common techniques applicable to circular SAR appear to be a matched filter technique and a technique termed *back projection* [23, 24], both of which require a large amount of computation and computer time. These techniques are also applicable to the type of SAR considered in this chapter.

In the derivations of this chapter, it was (somewhat unrealistically) assumed that the SAR platform was flying in the x-y plane (i.e., at an altitude of zero). The extension to a nonzero, but constant, altitude is straightforward. In essence, when the nonzero altitude case is considered, the image that results is in slanted plane. The points in this slanted plane can be mapped to the ground by a coordinate transformation.

The assumption that the SAR platform was flying at a *constant* altitude, cross-range position, and velocity is reasonable for satellite-based SAR because satellite trajectories are very stable and, over L, reasonably straight relative to the imaged area, which is also reasonably flat over w. For aircraft-based SAR, this is not the case. In this type of SAR, an interim step of "straightening" the aircraft trajectory must be performed [2, 16].

The discussions herein make the assumption that synthetic antenna length (distance the SAR platform travels) and the dimensions of the imaged area are small compared to the slant range to the imaged area. If this is not the case, a somewhat more complicated method of accounting for SAR platform motion must be used [7, 15, 16]. Also, RCM and RCMC become more of an issue.

The developments of this chapter were based on the assumption that the transmit signal was a narrow, unmodulated pulse. As was indicated, such a pulse is unrealistic in practical SARs because it would dictate high peak power to get a reasonable SNR at the matched filter output. Most practical SARs use LFM pulses of reasonable length. The only impact of this as it relates to the processing presented herein is that the matched filter of Figures 23.25 and 23.33 must be matched to an LFM pulse rather than an unmodulated pulse. In some instances where extremely high bandwidth pulses are used (to get fine down-range resolution), stretch processing may be necessary.

Finally, one of the assumptions is that the phase and frequency of the transmit signal are fixed over the processing interval. In other words, the signal remains coherent over the processing interval. This could become questionable for long processing intervals. In any event, it is something that must be considered when designing the SAR sensor and determining the size of the image area and the attainable cross-range resolution.

23.14 EXERCISES

1. Derive (23.10) starting from (23.9).

2. Re-create the plot of Figure 23.7 using an appropriately modified version of (23.11).

3. Generate the plots of Figure 23.8 but add a third plot that is the two-way, normalized, radiation pattern for a linear array. Discuss the relation between the beamwidths of the three plots.

4. Re-create Figures 23.14 and 23.15.

5. Implement SAR signal generation and processing routines using the methodology of Section 23.7. Test your routines by duplicating Figures 23.18 through 23.21. Use the parameters in Table 23.1. When you set up your signal generation routine, make it general enough to accommodate any number of scatterers located at any position and with any powers. Make it general enough to accommodate any sample period and any SAR array length.

6. In the signal generation code from Exercise 5, decrease the number of scatterers to one centered at $y_1 = 0$ with an amplitude of unity. In your SAR processing code, do not perform the quadratic phase correction. Finally, decrease the sample period to $T = 10$ ms and form the pseudo image. Is this what you expected? Explain.

   As an interesting experiment, try a few different values of T to see what happens to the pseudo image. Discuss your results.

7. In the signal generation code from Exercise 5, place scatterers at $y = 20$ and $y = 30$ and give them amplitudes of unity. Process the signal from the two scatterers through your SAR processor and produce the pseudo image. Is the pseudo image what you expected? Explain.

8. Change T to its maximum value of 120 ms and repeat Exercise 7.

9. Implement a SAR signal generation algorithm as described in Section 23.9 and generate the plot of Figure 23.26.

10. Implement a RCMC algorithm and reproduce the image of Figure 23.27. Generate the images of Figures 23.28 through 23.31.

11. In the discussion of Figure 23.27, it was indicated that the blurring was caused by the fact that an unmodulated pulse was used in the signal generation routine. This type of pulse is not ideally compatible with the use of the Fourier transform to perform

interpolation. If the signal generation routine had used an LFM pulse, the resulting matched filter output would have been more compatible with Fourier transform interpolation, and the blurring to the line would not be present. The output of a matched filter for an LFM pulse can, for the purposes of this exercise, be approximated by

$$v_i(m,k) = \sqrt{P_{Si}} e^{-j2\pi f_{dg}(kT+\tau_{min}+m\tau_p)} e^{-j4\pi r_i(kT+\tau_{min}+m\tau_p)/\lambda}$$
$$\operatorname{sinc}\left[\frac{\tau_{min}+m\tau_p - 2r_i(kT+\tau_{min}+m\tau_p)/c}{\tau_p}\right] \quad (23.111)$$
$$\times \operatorname{rect}\left[\frac{\tau_{min}+m\tau_p - 2r_i(kT+\tau_{min}+m\tau_p)/c}{\tau_u}\right]$$

where τ_p is the compressed pulsewidth and τ_u is the width of the uncompressed LFM pulse. For this exercise, use the compressed pulsewidth of Exercise 9 and use an uncompressed pulsewidth of 50 μs. Use this equation in the signal generation code you developed for Exercise 9 and generate plots like Figures 23.26 through 23.31. You should note that the blurring in Figure 23.37 is now significantly reduced.

12. Extend the RCMC algorithm of Exercise 10 to include the image formation algorithm of Sections 23.6.6 and 23.6.7. Reproduce the figures of Example 2 (Figures 23.34 and 23.35). Place five scatterers in the imaged area and generate the resulting image. Use the same amplitude for the five scatterers. Are the scatterers where you expected them to be? Explain.

    Try this exercise with the unmodulated pulse discussed in the text and with the LFM pulsed introduced in Exercise 11.

13. Implement the image sharpening algorithms discussed in Section 23.12 and reproduce images like those of Figures 23.36 through 23.38. Repeat this with the five scatterers you used in Exercise 12.

14. Use the SAR processor you developed in the previous exercises to create an image from the data in either the file named Trinity.txt or the file named Trinity.mat. The file named Trinity.mat is a MATLAB mat file which you can read with the command "load Trinity." This will cause $v(m,k)$ [see (23.84) and (23.85)] to be loaded into the 114- by 241-element complex array with the name RD. The Trinity.txt file is a text file that contain 241 columns of data with 228 entries in each column. The first 114 rows of the file are the real part of $v(m,k)$, and the last 114 rows are the imaginary part of $v(m,k)$. The image generated by the SAR processor will be a photo since $v(m,k)$ was generated from a photo using (23.84) and (23.85). The photo can be found in the file Trinity.jpg. The various parameters that were used to generate the signal are those of Table 23.2. Thus, your processing algorithm should use the same parameters. When you form the image, do not use a negative as discussed in the text (unless you want to see a negative of the photo). Also, when you form the image, turn the axis labels off so the image will look like a photo. Try the processor with and without the image refinement algorithms of Section 23.12.

15. For this exercise, you will use some actual SAR data to form an image. The data was obtained from the RADARSAT1 spaced-based SAR platform. The data is a subset of the SAR data found on a compact disc that accompanies [13]. The files were preprocessed to put them in a form that is compatible with the signals discussed in this chapter. Specifically, the signals were preprocessed to create $v(m,k)$. The preprocessed data is contained in the text file labeled SARdata.txt. The file contains two columns of data. The first column is the real part of $v(m,k)$ and the second column is the imaginary part of $v(m,k)$ [see (23.84) and (23.85)]. The file has $1{,}024 \times 1{,}536 = 1{,}572{,}864$ rows. After you load the data file, reshape it into a 1,024-by-1,536 array of complex numbers. Specifically, the array should contain $v(m,k)$ for $m = 1$ to 1,024 and $k = 1$ to 1,536.

A photo of the imaged region of the RADAR SAT 1 data [13] is in the lower left part Figure 23.39. The dark area is water and the gray area is land. The two projections into the water are docks. The data supplied is for an image of the larger dock and the edge of the smaller dock. The image you create will also show a ship or two that is not in the photo.

The geometry for this case is somewhat different than the one indicated in Figure 23.24. Specifically, the squint angle, θ, is negative for this data. Also, the imaged area is well behind the satellite. Because of these factors, the $\Delta\tau(k)$ (23.86) used for RCMC must be changed to

$$\Delta\tau(k) = 2\left(\sqrt{x_0^2 + (y_0 - VkT)^2} - r_{min}\right)\Big/c \quad (23.112)$$

where

$$r_{min} = \sqrt{x_0^2 + (|y_0| - L/2 - w/2)^2} \quad (23.113)$$

The reason for this change is that the minimum range, for RCMC purposes, is the distance between the position of the satellite at $y = L/2$ and the upper left corner of the imaged area. x_0 was also redefined, as shown in Figure 23.40.

The various SAR parameters you need are contained in Table 23.3. You should be able to compute the other parameters from those given in the table.

As with Exercise 14, create a positive image. It may be necessary to adjust the contrast of the final image. If you use MATLAB, this can be done through the clim parameter of the imagesc image generation routine. A value of clim that seems to work is clim = [3,000 15,000].

Table 23.3
SAR Parameters for Exercise 15

Parameter	Value
L	8,624 m
x_0	993.4627 km
y_0	−27.466 km
f_{dg}	−6,750 Hz
V	7,062 m/s
T	$1/PRF$
PRF	1,256.98 Hz
λ	0.05657 m
τ_p	$(1/32.317)$ μs

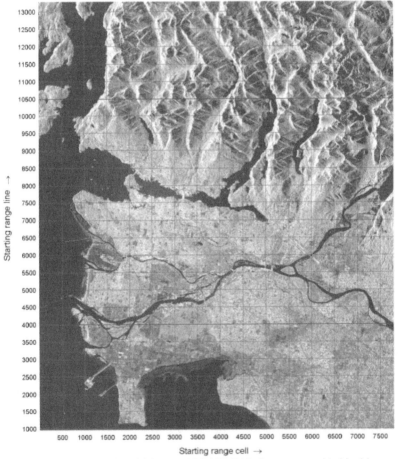

Figure 23.39 Photo of the region for which RADARSAT1 SAR data was provided in this exercise. (RADARSAT Data © Canadian Space Agency/Agence Spatiela Canadienne 2002—All Rights Reserved.)

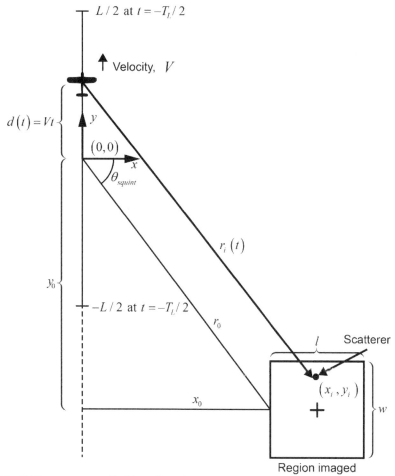

Figure 23.40 SAR geometry applicable to Exercise 15.

References

[1] Skolnik, M. I., ed., *Radar Handbook,* 3rd ed., New York: McGraw-Hill, 2008.

[2] Hovanessian, S. A., *Introduction to Synthetic Array and Imaging Radars,* Dedham, MA: Artech House, 1980.

[3] Griffiths, H. D., "Developments in Modern Synthetic Aperture Radar," *Proc. 2007 IEEE Radar Conf.,* Boston, MA, Apr. 17–20, 2007, pp. 734–739.

[4] Wiley, C. A., "Pulsed Doppler Radar Methods and Apparatus," U.S. Patent 3,196,436, Jul. 20, 1954.

[5] Wiley, C. A., "Synthetic Aperture Radars: A Paradigm for Technology Evolution," *IEEE Trans. Aerosp. Electron. Syst.,* vol. 21, no. 3, May 1985, pp. 440–443.

[6] Love, A. W., "In Memory of Carl A. Wiley," *IEEE Antennas Propag. Soc. Newsletter,* vol. 27, no. 3, Jun. 1985, pp. 17–18.

[7] Morris, G. V., *Airborne Pulsed Doppler Radar,* Norwood, MA: Artech House, 1988.

[8] Cutrona, L. J., and G. O. Hall, "A Comparison of Techniques for Achieving Fine Azimuth Resolution," *IRE Trans. Military Electron.,* vol. 6, no. 2, Apr. 1962, pp. 119–121.

[9] Lee, H., "Extension of Synthetic Aperture Radar (SAR) Technique to Undersea Applications," *IEEE J. Oceanic Eng.,* vol. 4, no. 2, Apr. 1979, pp. 60–63.

[10] Reigber, A., et al., "Very-High-Resolution Airborne Synthetic Aperture Radar Imaging: Signal Processing and Applications," *Proc. IEEE,* vol. 101, no. 3, Mar. 2013, pp. 759–783.

[11] Elachi, C. et al., "Spaceborne Synthetic-Aperture Imaging Radars: Applications, Techniques, and Technology," *Proc. IEEE,* vol. 70, no. 10, Oct. 1982, pp. 1174–1209.

[12] Ottl, H., and F. Valdoni, "X-Band Synthetic Aperture Radar (X-SAR) and Its Shuttle-Borne Application for Experiments," *Microwave 17th European Conf. 1987,* Rome, Italy, Sept. 7–11, 1987, pp. 569–574.

[13] Cumming, I. G., and F. H. Wong, *Digital Processing of Synthetic Aperture Radar Data: Algorithms and Implementation,* Norwood, MA: Artech House, 2005.

[14] Carrara, W. G., et al., *Spotlight Synthetic Aperture Radar: Signal Processing Algorithms,* Norwood, MA: Artech House, 1995.

[15] Brookner, E., *Radar Technology,* Dedham, MA: Artech House, 1977.

[16] Kovaly, J. J., *Synthetic Aperture Radar,* Dedham, MA: Artech House, 1976.

[17] Mensa, D. L., *High Resolution Radar Imaging,* Dedham, MA: Artech House, 1981.

[18] Oliver, C., and S. Quegan, *Understanding Synthetic Aperture Radar Images,* Norwood, MA: Artech House, 1998.

[19] Rihaczek, A. W., *Principles of High-Resolution Radar,* New York: McGraw-Hill, 1969. Reprinted: Norwood, MA: Artech House, 1995.

[20] Wehner, D. R., *High Resolution Radar,* Norwood, MA: Artech House, 1987.

[21] Brown, W. M., and L. J. Porcello, "An Introduction to Synthetic-Aperture Radar," *IEEE Spectrum,* vol. 6, no. 9, 1969, pp. 52–62.

[22] Currie, A., "Synthetic Aperture Radar," *Electron. & Comm. Eng. J.,* vol. 3, no. 4, Aug. 1991, pp. 159–170.

[23] Fitch, J. P., *Synthetic Aperture Radar,* New York: Springer-Verlag, 1988.

[24] Soumekh, M., Synthetic *Aperture Signal Processing with MATLAB Algorithms*, New York: Wiley & Sons, 1999.

[25] Argenti, F. et al., "A Tutorial on Speckle Reduction in Synthetic Aperture Radar Images," *IEEE Geoscience Remote Sensing Mag.,* vol. 1, no. 3, Sept. 2013, pp. 6–35.

Chapter 24

Introduction to Space-Time Adaptive Processing

24.1 INTRODUCTION

In this chapter, we provide an introduction to space-time adaptive processing, or STAP. When we discuss radars, we normally consider the processes of beamforming, matched filtering, and Doppler processing separately. By doing this, we are forcing the radar to separately operate in the three domains: space for beamforming, *fast time* for matched filtering, or *slow time* for Doppler processing. This separation of functions sacrifices capabilities because the radar does not make use of all available information or *degrees of freedom*.

Suppose we have a linear phased array that has N elements. In terms of beamforming, to maximize the target return and minimize returns from interference (e.g., clutter, jammers, and noise), we say that we have $2N$ degrees of freedom (N amplitude dimensions and N phase dimensions). If we also process K pulses in a Doppler processor, we say we have an additional $2K$ degrees of freedom. With normal processing methods, whereby we separate beamforming and Doppler processing, we have a total of $2K + 2N$ degrees of freedom. If we were to consider that we could simultaneously perform beam forming and Doppler processing, we would have $2KN$ degrees of freedom. This is the premise of the "ST" part of STAP.

Figure 24.1 might provide further help in visualizing this. It contains a depiction of angle-Doppler space. Each of the squares corresponds to a particular angle and Doppler. There are N beam positions and K Doppler cells. The dark square indicates a beam position and Doppler cell that contains interference. With standard processing techniques, we would suppress the interference by independently placing a null at the beam position and Doppler cell containing the interference. The beam null is denoted by the crosshatched squares, and the Doppler null is denoted by the dotted squares. With this approach, the process of suppressing the interference will also cause any signals in the cross-hatched and dotted regions to be suppressed, including target signals. This happens because we separately process in angle and Doppler space.

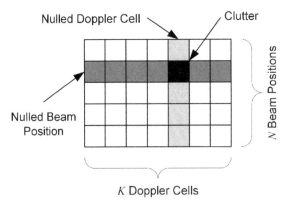

Figure 24.1 Clutter nulling using conventional methods.

With STAP, we would, ideally, simultaneously process in angle and Doppler space. With this simultaneous processing, the processor can be made to place a null at only the angle and Doppler of the interference (at the location of the dark square in Figure 24.1). Thus, it is possible to suppress only interference, and not suppress other signals that might be located at the same angle or Doppler of the interference.

According to [1], it appears that the concept of STAP was first introduced in a 1973 paper by Brennan and Reed [2]. STAP has been, and still is, extensively studied in applications such as SAR, GMTI, MIMO radar, array antennas, tracking radar, SONAR, early warning, and jamming suppression [3–7]. Despite the relatively high processing burden, there are many implemented and fielded STAP platforms [8–10].

We begin the discussion of STAP by first discussing spatial processing (the "S") and then temporal processing (the "T"). We next discuss how these are combined to perform space-time processing. Following that, we briefly discuss some topics related to the "A," or adaptive, part of STAP.

The general approach used in STAP is to design the processor to maximize signal-to-interference-plus-noise ratio (SINR) [11–13]. This is the same as the approach used in the matched filter development of Chapter 8. In fact, for the case where the interference is white in the space-time domain, the space-time processor is equal to the space-time representation of the signal. That is, the space-time processor is matched to the signal. As a further illustration of the relation between the matched filter and STAP, we note that one of the Cauchy-Schwarz inequalities is used to design the matched filter and the space-time processor [14, 15].

24.2 SPATIAL PROCESSING

As indicated, we begin the STAP development by first considering spatial processing, or *beamforming*. We start by considering the signal and receiver noise and then address a combination of signal, receiver noise, and interference, such as jamming or clutter.

24.2.1 Signal Plus Noise

We start with the N element linear array shown in Figure 24.2.[1] In that figure, it is assumed that the target is located at an angle of θ_s relative to array normal. From linear array theory (see Chapter 13) we can write the output of the array as[2]

$$V(\theta_s) = \sum_{n=0}^{N-1} a_n \sqrt{P_S} e^{-j2\pi nd \sin\theta_s/\lambda} \tag{24.1}$$

where P_S is the signal power from the target at each of the array elements. It is the signal power term of the radar range equation, without the receive directivity term (see Chapter 2).

We define

$$W^H = \begin{bmatrix} a_0 & a_1 & \cdots & a_{N-1} \end{bmatrix} \tag{24.2}$$

and

$$S(\theta_s) = \begin{bmatrix} s_0 & s_1 & \cdots & s_{N-1} \end{bmatrix}^T = \begin{bmatrix} 1 & e^{-j2\pi d\sin\theta_s/\lambda} & \cdots & e^{-j2\pi(N-1)d\sin\theta_s/\lambda} \end{bmatrix}^T \tag{24.3}$$

W^H is the complex weight vector from Chapter 13[3] and $S(\theta_s)$ is the target, or signal, *steering vector*. The superscript H denotes the Hermitian, or conjugate-transpose operation [16] (this notation will come into play shortly). W^H is also sometimes thought of as weights in a spatial filter. Using (24.2) and (24.3), we can write $V(\theta_s)$ as

$$V(\theta_s) = \sqrt{P_S} W^H S(\theta_s) \tag{24.4}$$

We assume there is a separate receiver connected to each element. This makes the noise at each of the antenna elements of Figure 24.2 uncorrelated. This is depicted in Figure 24.3 by the separate n_n in each block. The n_n are complex random variables that we assume are zero-mean and uncorrelated. That is

$$E\{n_n n_l^*\} = 0 \quad n \neq l \tag{24.5}$$

---

[1] We will restrict the development to linear arrays as a convenience. The extension to a planar array is reasonably straightforward.
[2] Consistent with other developments in this book, we are using complex signal notation as a convenient means of representing the amplitude and phase of RF or IF signals.
[3] In Chapter 13, we used W instead of W^H. We made the switch here to be more consistent with the notation used in STAP.

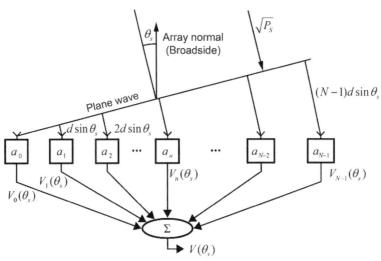

Figure 24.2 Linear phased array.

We further stipulate

$$P_N = E\{|n_n|^2\} \tag{24.6}$$

where P_N is the noise power at the input to each of the a_n of Figures 24.2 and 24.3. Equation (24.6) implies the noise power is the same at the output of each receiver. Strictly speaking, this is not necessary. We included it here as a convenience. The noise voltage at the output of the summer of Figure 24.3 can be written as

$$V_{No} = \sum_{n=0}^{N-1} a_n n_n = W^H N \tag{24.7}$$

where

$$N = \begin{bmatrix} n_0 & n_1 & \cdots & n_{N-1} \end{bmatrix}^T \tag{24.8}$$

As a point of clarification, the signal and noise in the above equations are at the output of the matched filter of each receiver. That is, the weights a_n are applied to the signal and noise at the outputs of the matched filters. More specifically, the signal-plus-noise (plus interference) is sampled at the output of the matched filters and then sent to the processor. Ideally, the samples are taken at a time corresponding to the range delay to the target to be sure that the signal is present in the matched filter output. If the target range delay is not known, several range (time) samples and processors will be needed.

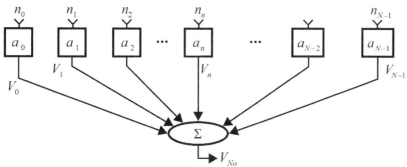

Figure 24.3 Array with only noise.

Receivers are needed at each element to implement STAP in its pure form. If we are willing to give up spatial degrees of freedom, receivers could be applied to groups of elements, or *subarrays* (see Chapter 14). However, with this approach, we limit where the STAP can place nulls. Further discussion of subarraying and STAP can be found in STAP literature [3, 7, 12].

The STAP design criterion is maximization of SINR (or SNR for the noise-only case) at the processor output. Therefore, we need to develop equations for the signal and noise power at the processor output. From (24.4), the signal power at the processor output is

$$P_{So} = |V(\theta_s)|^2 = P_S |W^H S(\theta_s)|^2 \tag{24.9}$$

Since the noise is a random process, we write the noise power at the output of the summer as

$$P_{No} = E\{|V_{No}|^2\} = E\{|W^H N|^2\} = W^H E\{NN^H\} W = W^H R_N W \tag{24.10}$$

In (24.10)

$$R_N = E\{NN^H\} = E\left\{\begin{bmatrix} n_0 \\ n_1 \\ \vdots \\ n_{N-1} \end{bmatrix} \begin{bmatrix} n_0^* & n_1^* & \cdots & n_{N-1}^* \end{bmatrix}\right\} = P_N I \tag{24.11}$$

and is termed the *receiver noise covariance matrix*. In (24.11), I is the identity matrix. With (24.11), the output noise power becomes

$$P_{No} = P_N W^H W = P_N \|W\|^2 \tag{24.12}$$

The SNR at the output of the summer is

$$SNR = \frac{P_{So}}{P_{No}} = \frac{P_S |W^H S(\theta_s)|^2}{P_N \|W\|^2} \tag{24.13}$$

At this point we invoke one of the Cauchy-Schwarz inequalities [14, 15]. In particular, we use

$$\left|W^H S(\theta_s)\right|^2 \leq \|W\|^2 \|S(\theta_s)\|^2 \tag{24.14}$$

with equality when

$$W = \kappa S(\theta_s) \tag{24.15}$$

where κ is an arbitrary, complex constant, which we will set to unity. With this we get

$$SNR = \frac{P_S \left|W^H S(\theta_s)\right|^2}{P_N \|W\|^2} \leq \frac{P_S \|W\|^2 \|S(\theta_s)\|^2}{P_N \|W\|^2} = \frac{P_S \|S(\theta_s)\|^2}{P_N} = \frac{P_S}{P_N} N \tag{24.16}$$

where we made use of [see (24.3)]

$$\|S(\theta_s)\|^2 = N \tag{24.17}$$

Equation (24.16) tells us that the SNR at the array output has an upper bound equal to the sum of the SNRs at (the outputs of the matched filters of the receivers attached to) each element. Further, the actual SNR at the array output will equal the upper bound if W is chosen according to (24.15), that is W is *matched* to $S(\theta_s)$.

24.2.2 Signal Plus Noise and Interference

We now consider a case where we have interference that is correlated across the array. This interference could be clutter and/or jammers. The appropriate model for this situation is given in Figure 24.4. In this figure, n_{Ii} represents the interference "voltage" and is a zero-mean, complex, random variable. The subscript i is used to represent the ith interference source (which we will need shortly when we consider multiple interference sources). The fact that the same random variable is applied to each of the antenna elements makes the random variables at the outputs of the array correlated. We write $V_{Ii}(\phi_i)$ as

$$V_{Ii}(\phi_i) = \sum_{n=0}^{N-1} a_n n_{Ii} e^{-j2\pi knd\sin\phi_i/\lambda} = W^H N_{Ii} \tag{24.18}$$

where

$$N_{Ii} = n_{Ii} D(\phi_i) \tag{24.19}$$

and

$$D(\phi_i) = \begin{bmatrix} 1 & e^{-j2\pi d\sin\phi_i/\lambda} & \cdots & e^{-j2\pi(N-1)d\sin\phi_i/\lambda} \end{bmatrix}^T \tag{24.20}$$

is the steering vector for the ith interference source.

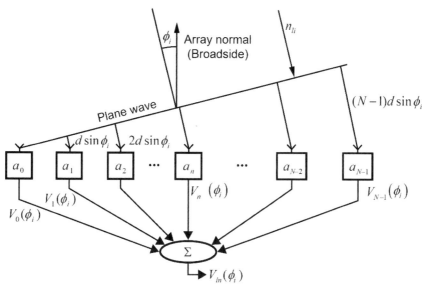

Figure 24.4 Array with interference.

We accommodate multiple interference sources by simply summing the voltages for the multiple sources. Specifically,

$$\mathbf{N}_I = \sum_{i=1}^{N_I} \mathbf{N}_{Ii} = \sum_{i=1}^{N_i} n_{Ii} D(\phi_i) \qquad (24.21)$$

We further assume the N_i interference sources are independent so that

$$E\{n_{Ii} n_{Ik}^*\} = \begin{cases} P_{Ii} & i=k \\ 0 & i \neq k \end{cases} \qquad (24.22)$$

The interference power (from the N_i interference sources) is

$$P_{Io} = E\{|W^H \mathbf{N}_I|^2\} = W^H E\{\mathbf{N}_I \mathbf{N}_I^H\} W = W^H R_I W \qquad (24.23)$$

In (24.23)

$$R_I = E\{\mathbf{N}_I \mathbf{N}_I^H\} = \sum_{i=1}^{N_i} \sum_{k=1}^{N_i} E\{n_{Ii} n_{Ik}^*\} D(\phi_i) D^H(\phi_k)$$
$$= \sum_{i=1}^{N_i} P_{Ii} D(\phi_i) D^H(\phi_i) \qquad (24.24)$$

where we made use of (24.22).

Combining (24.10) with (24.23), we get the total noise plus interference power as

$$P_{N+I} = P_{No} + P_{Io} = W^H (R_N + R_I) W = W^H R W \qquad (24.25)$$

and write the signal-to-interference-plus-noise ratio (SINR) at the output of the summer as

$$SINR = \frac{P_{So}}{P_{N+I}} = \frac{P_S |W^H S(\theta_s)|^2}{W^H R W} \quad (24.26)$$

As before, we want to choose the spatial filter that maximizes SINR. To do this using the Cauchy-Schwarz inequality, we need to manipulate (24.26). We start by noting that because of the receiver noise, R will be positive definite [14]. This means we can define a matrix, $R^{1/2}$, such that $R = R^{1/2} R^{1/2}$. Further, $R^{1/2}$ is Hermitian and its inverse, $R^{-1/2}$, exists and is Hermitian [11, 14]. We use this to write

$$SINR = \frac{P_S |W^H R^{1/2} R^{-1/2} S(\theta_s)|^2}{W^H R^{1/2} R^{1/2} W} = \frac{P_S |W_R^H S_R(\theta_s)|^2}{\|W_R\|^2} \quad (24.27)$$

where $W_R = R^{1/2} W$ and $S_R(\theta_s) = R^{-1/2} S(\theta_s)$.

Equation (24.27) has the same form as (24.13). Thus, we conclude that the SINR is maximized when

$$W_R = \kappa S_R(\theta_s) \quad (24.28)$$

If we let $\kappa = 1$ and substitute for W_R and $S_R(\theta_s)$ we get the solution

$$W = R^{-1} S(\theta_s) \quad (24.29)$$

The net effect of the above equation is that the weight vector, W, is ideally selected to place the main beam on the target and simultaneously place nulls at the angular locations of the interference sources. We used the qualifier "ideally" because it is possible that the algorithm will not place the main beam at the target angle or a null at the interference angle. This might happen if the target and interference angles were close to each other (see Exercise 7).

A critical part of this development is that the total interference consists of both receiver noise and other interference sources. The inclusion of receiver noise is what makes the R matrix positive definite and thus nonsingular. If R was singular, R^{-1} would not exist, and we would need to use another approach for finding W. On occasion, R will become ill-conditioned because the jammer-to-noise ratio (JNR) is large. If this happens, alternate methods of finding W may be needed. One of these is to use a mean-square criterion such as least-mean-square estimation or pseudo inverse [17–21]. Another is termed *diagonal loading,* which is discussed later.

24.2.3 Example 1: Spatial Processing

As an example, we consider a 16-element linear array with ½ wavelength element spacing ($d/\lambda = \frac{1}{2}$). We assume that we have a per-element SNR of 0 dB (at the output of the matched filters of the receivers associated with each of the elements); that is, $P_S/P_N = 1$ W/W. We have two noise jammers with per-element JNRs of 40 dB (again, at the outputs of the matched filters). The target is located at an angle of zero, and the jammers are located at angles of +18° and −34°. The selected jammer angles place the jammers on the second and fourth sidelobes of the antenna pattern that results from using uniform illumination (see Figure 24.5). If we normalize powers to a noise power of one watt, the above specifications lead to the following parameters $P_S = 1$, $P_N = 1$, $P_{J1} = 10^4$, $P_{J2} = 10^4$, $\theta_s = 0$, $\phi_1 = 18°$, and $\phi_2 = -34°$.

For the first case, we consider only receiver noise (no jammers). From (24.15) with $\kappa = 1$, we have

$$W = S(\theta_s) = S(0) = \begin{bmatrix} 1 & 1 & \cdots & 1 \end{bmatrix}^T \quad (24.30)$$

and $SNR_{max} = 16 P_S / P_S = 16$ W/W or 12.04 dB. The weight vector, W, results in an array with uniform weighting, or uniform illumination (see Chapter 13). A plot of the normalized radiation pattern for this case is shown as the dotted curve in Figure 24.5, which is mostly obscured by the solid curve. As a note, the patterns of Figure 24.5 were generated using

$$R(\theta) = \frac{|V(\theta)|^2}{\max(|V(\theta)|^2)} = \frac{|W^H S(\theta)|^2}{\max(|W^H S(\theta)|^2)} \quad (24.31)$$

where θ was varied from −90° to 90°.[4]

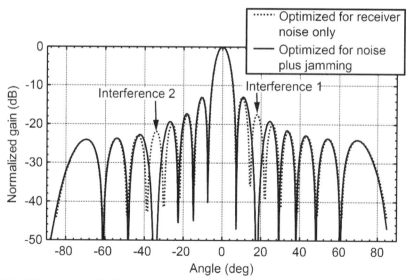

Figure 24.5 Normalized radiation pattern with and without optimization—16-element linear array.

---

[4] As a caution $R(\theta)$ is the radiation pattern, which is not to be confused with the covariance matrix, R [without (θ)].

If we use the W given by (24.30) and include the two interference sources, the SINR at the processor output is about -24 dB. If we include the interference properties in the calculation of W by using (24.29), the SINR increases to 12 dB, which is close to the noise-only case of about 12.04 dB ($10\log 16$). To accomplish this, the algorithm chose the weights to place nulls in the antenna pattern at the locations of the interference sources. This is illustrated by the solid curve of Figure 24.5, which is a plot of the radiation pattern when the new weights are used.

24.3 TEMPORAL PROCESSING

The temporal processing part of STAP is most often thought of as Doppler processing. In particular, we consider the returns (signal and interference) from several pulses and, similar to spatial processing, weight and sum them. As with spatial processing, we choose the weights to maximize SINR at the output of the processor. The input to the Doppler processor is the output of the matched filter. Thus, we need to characterize the signal, noise, and interference at the matched filter output.

24.3.1 Signal

We consider a transmit waveform that consists of a string of K pulses and write it as

$$v_T(t) = e^{j2\pi f_o t} \sum_{k=0}^{K-1} p(t-kT) \qquad (24.32)$$

where $p(t)$ is a general representation of a pulse and T is the spacing between pulses, or PRI (see Chapter 1). As examples, for an unmodulated pulse

$$p(t) = \operatorname{rect}\left[\frac{t}{\tau_p}\right] \qquad (24.33)$$

and for an LFM pulse

$$p(t) = e^{j\pi\alpha t^2} \operatorname{rect}\left[\frac{t}{\tau_p}\right] \qquad (24.34)$$

where τ_p is the (uncompressed) pulsewidth and α is the LFM slope (see Chapter 8). The exponential term in (24.32) represents the carrier part of the transmit signal (see Chapter 1).

The normalized return signal, from a point target, is a delayed and scaled version of $v_T(t)$. We define it as

$$v_R(t) = \sqrt{P_S}\, v_T(t - 2r(t)/c) = \sqrt{P_S}\, e^{j2\pi f_o(t-2r(t)/c)} \sum_{k=0}^{K-1} p(t - 2r(t)/c - kT) \qquad (24.35)$$

where P_S is the signal power at the matched filter output and $r(t)$ is the range to the target.

If the target is moving at a constant range rate, we can write $r(t)$ as

$$r(t) = r_0 + \dot{r}t \tag{24.36}$$

where r_0 is the target range at $t = 0$ and $\dot{r}$ is the range rate (see Chapter 1). We usually set $t = 0$ at the center of the first pulse of the train of K pulses.

With (24.36) $v_R(t)$ becomes

$$v_R(t) = \sqrt{P_S}\, e^{-j4\pi r_0/\lambda} e^{j2\pi f_o t} e^{j2\pi f_d t} \sum_{k=0}^{K-1} p(t - \tau_r - kT) \tag{24.37}$$

In (24.37), $f_d = -2\dot{r}/\lambda$, $\tau_r = 2r_0/c$ and $\lambda = c/f_o$ is the wavelength of the transmit signal. If we assume that the phase across the pulse is constant, we can write

$$v_R(t) = \sqrt{P_S}\, e^{-j4\pi r_0/\lambda} e^{j2\pi f_o t} \sum_{k=0}^{K-1} e^{j2\pi f_d kT} p(t - \tau_r - kT) \tag{24.38}$$

In the receiver, we heterodyne to remove the carrier, normalize away the first exponential, and process the signal through the matched filter to obtain

$$v_M(t) = \sqrt{P_S} \sum_{k=0}^{K-1} e^{j2\pi f_d kT} m(t - \tau_r - kT) \tag{24.39}$$

where $m(t)$ is the response of the matched filter to $p(t)$. We assume $m(t)$ is normalized to a peak value of $m(0) = 1$.

For the next step, we sample $v_M(t)$ at times $t = \tau_{RC} + kT$. That is, we sample the output of the matched filter once per PRI at a time τ_{RC} relative to the center of each transmit pulse.[5] The result is a sequence of samples we denote as

$$v_k(f_d) = \sqrt{P_S}\, e^{j2\pi f_d kT} m_{RC} \quad k \in [0, K-1] \tag{24.40}$$

where m_{RC} is the (generally complex) value of $m(t - \tau_r - kT)$ evaluated at $t = \tau_{RC} + kT$. If we sample the matched filter output at its peak, we will have $\tau_{RC} = \tau_r$ and $m_{RC} = 1$.

24.3.2 Noise

The noise at the matched filter output is also sampled at $t = \tau_{RC} + kT$. This produces a sequence of K, uncorrelated, zero-mean, random variables with equal variances (and mean-square values, or powers) of P_N. We denote these as

$$n_k \quad k \in [0, \ K-1] \tag{24.41}$$

As a note, it is not necessary that the noise samples be uncorrelated and have equal variances; however, this is the standard assumption when discussing STAP [11].

Assuming we sample the matched filter output at its peak when only signal and noise are present, the signal power in each sample is P_S and the noise power for each sample is P_N. Thus, the SNR at the sampler output, and the input to the processor, is $SNR = P_S/P_N$.

---

[5] This carries the assumption that the radar is operating unambiguously in range.

24.3.3 Interference

We assume the interference bandwidth is small relative to the transmit waveform PRF (PRF = 1/T). More specifically, we assume the interference signal at the sampler output is a wide-sense stationary, zero-mean, bandlimited random process with an autocorrelation given by

$$R_I(k) = E\{v_I(k+l)v_I^*(l)\} \tag{24.42}$$

where $v_I(k)$ is the interference voltage at the sampler output. It is equal to the output of the matched filter, sampled at $t = \tau_{RC} + kT$, when the input is the signal returned from the interference.

In general, $R_I(k)$ is a complicated function of k. For the special case where the interference is a tone with a Doppler frequency of f_I and a random amplitude with a mean-square value (power) of P_I, $R_I(k)$, becomes

$$R_I(k) = P_I e^{j2\pi f_I kT} \quad k \in [0 \quad K-1] \tag{24.43}$$

24.3.4 Doppler Processor

We assume the Doppler processor is a K-length FIR filter with coefficients of ω_k. If the input to the processor is $v_{in}(k)$, the output, after K samples have been processed, is

$$V_{Io} = \sum_{k=0}^{K-1} \omega_k v_{in}(k) = \Omega^H V_{in} \tag{24.44}$$

where

$$\Omega^H = [\omega_0 \quad \omega_1 \quad \cdots \quad \omega_{K-1}] \tag{24.45}$$

and

$$V_{in} = [v_{in}(0) \quad v_{in}(1) \quad \cdots \quad v_{in}(K-1)]^T \tag{24.46}$$

When the input is the signal, we have, from (24.40)

$$v_{in}(k) = v_k(f_d) = \sqrt{P_S} e^{j2\pi f_d kT} m_{RC} \tag{24.47}$$

If we further assume the sampler samples the matched filter output at $t = \tau_R + kT$, we have $m_{RC} = 1$. Using (24.46) we have

$$V_{in}^{signal} = \sqrt{P_S} S(f_d) = \sqrt{P_S} \left[1 \quad e^{j2\pi f_d T} \quad \cdots \quad e^{j2\pi f_d (K-1)T} \right]^T \tag{24.48}$$

The signal voltage at the Doppler processor output is

$$V_{So}(f_d) = \sqrt{P_S} \Omega^H S(f_d) \tag{24.49}$$

For the noise, we write

$$V_{in}^{noise} = \mathrm{N} = \begin{bmatrix} n_0 & n_1 & \cdots & n_{K-1} \end{bmatrix}^T \qquad (24.50)$$

and the output of the Doppler processor is

$$V_{No} = \Omega^H \mathrm{N} \qquad (24.51)$$

We write the interference input to the Doppler processor as

$$V_{in}^{interference} = \mathrm{N}_I = \begin{bmatrix} v_I(0) & v_I(1) & \cdots & v_I(K-1) \end{bmatrix}^T \qquad (24.52)$$

and the processor output as

$$V_{Io} = \Omega^H \mathrm{N}_I \qquad (24.53)$$

As with the spatial processing case, we choose the Ω that maximizes SINR at the Doppler processor output. Thus, we need an equation for the peak signal power, P_{So}, and the total, average interference power, $P_{No} + P_{Io}$, at the processor output. By using the sum of the noise and interference powers, we are assuming the receiver noise and the interference are uncorrelated. This is a standard assumption.

The peak signal power is

$$P_{So} = |V_{So}(f_d)|^2 = P_S |\Omega^H S(f_d)|^2 \qquad (24.54)$$

and the average noise power is

$$P_{No} = E\{|V_{No}|^2\} = \Omega^H E\{\mathrm{NN}^H\} \Omega \qquad (24.55)$$

Since we assumed the noise samples were uncorrelated and had equal power,

$$E\{\mathrm{NN}^H\} = P_N I \qquad (24.56)$$

and

$$P_{No} = P_N \|\Omega\|^2 \qquad (24.57)$$

As indicated earlier, I is the identity matrix.

The interference power at the processor output is

$$P_{Io} = E\{|V_{Io}|^2\} = \Omega^H E\{\mathrm{N}_I \mathrm{N}_I^H\} \Omega = \Omega^H R_I \Omega \qquad (24.58)$$

where

$$R_I = E\{N_I N_I^H\} = E\left\{\begin{bmatrix} v_I(0) \\ v_I(1) \\ \vdots \\ v_I(K-1) \end{bmatrix} \begin{bmatrix} v_I^*(0) & v_I^*(1) & \cdots & v_I^*(K-1) \end{bmatrix}\right\}$$

$$= \begin{bmatrix} R_I(0) & R_I(-1) & \cdots & R_I(1-K) \\ R_I(1) & R_I(0) & \cdots & R_I(0) \\ \vdots & \vdots & \ddots & \vdots \\ R_I(K-1) & R_I(K-2) & \cdots & R_I(0) \end{bmatrix} \tag{24.59}$$

and $R_I(k)$ is defined in (24.42).

For the case where the interference is a tone,

$$R_I = E\{N_I N_I^H\} = P_I D(f_I) D^H(f_I) \tag{24.60}$$

where

$$D(f_I) = \begin{bmatrix} 1 & e^{j2\pi f_I T} & \cdots & e^{j2\pi f_I (K-1)T} \end{bmatrix}^T \tag{24.61}$$

For multiple interference sources

$$R_{IT} = \sum_i R_{Ii} \tag{24.62}$$

where the sum is taken over the total number of interference sources.

The SINR at the Doppler processor output is

$$SINR = \frac{P_{So}}{P_{No} + P_{Io}} = \frac{P_S |\Omega^H S(f_d)|^2}{\Omega^H R \Omega} = \frac{P_S |\Omega^H S(f_d)|^2}{\Omega^H (R_{IT} + P_N I)\Omega} \tag{24.63}$$

This is the same form as in the spatial processing case. Applying those results here gives

$$\Omega = R^{-1} S(f_d) \tag{24.64}$$

24.3.5 Example 2: Temporal Processing

As an example, we consider a Doppler processor with $K = 16$. We assume an input SNR of 0 dB (at the output of the matched filter). That is, $P_S/P_N = 1$ W/W. We also assume we sample the matched filter output at $t = \tau_R + kT$. We have two tone interferences with JNRs of 40 dB (again, at the output of the matched filter). We assume a PRF of 1,000 Hz, which gives $T = 0.001$ s. The target is located at a Doppler frequency of zero, and the interferences are located at Doppler frequencies of 217 Hz and –280 Hz. These Doppler frequencies place the interferences on the second and fourth sidelobes of the Doppler processor frequency response that results from using uniform weighting. The above specifications lead to the following parameters: $P_S = 1$, $P_N = 1$, $P_{I1} = 10^4$, $P_{I2} = 10^4$, $f_d = 0$, $f_{I1} = 217$ Hz, and $f_{I2} = -280$ Hz (we again normalize the noise power to 1 W).

For the first case, we consider only receiver noise. From (24.64) with $R = P_N I$ we have

$$\Omega = S(f_d) = S(0) = \begin{bmatrix} 1 & 1 & \cdots & 1 \end{bmatrix}^T \quad (24.65)$$

And $SNR_{max} = 16\, P_S/P_N = 16$ W/W or 12.04 dB. The weight vector, Ω, results in a Doppler processor with uniform weighting. A plot of the normalized frequency response of the Doppler processor is shown as the dotted curve in Figure 24.6, which is mostly obscured by the solid curve. As a note, the frequency responses of Figure 24.6 were generated using

$$F(f) = \frac{\left|\Omega^H S(f)\right|^2}{\max\left(\left|\Omega^H S(f)\right|^2\right)} \quad (24.66)$$

where f was varied from $-PRF/2$ to $PRF/2$, or -500 Hz to 500 Hz.

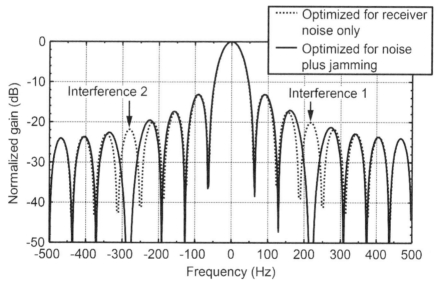

Figure 24.6 Normalized frequency response with and without optimization—16-tap Doppler processor.

If we use the Ω given by (24.65) and include the two interferences, the SINR at the processor output is about –22 dB. If we include the interference in the calculation of Ω by using (24.64), the SINR increases to 12 dB, which is close to the noise-only case of about 12.04 dB (10log16). To accomplish this, the algorithm chose the weights to place nulls in the frequency response of the Doppler processor at the Doppler frequencies of the interferences. This is illustrated by the solid curve of Figure 24.6, which is a plot of the frequency response when the new weights are used.

24.4 ADAPTIVITY ISSUES

We have discussed both the space and time parts of STAP. However, we have not addressed the adaptive part. Since the target and interference angles and Dopplers could change every dwell (sequence of K pulses), the target steering vector and the R matrices must be recomputed on each dwell. This means that new weights would be computed on each dwell to *adapt* to the target and interference environment—thus the adaptive part. In Section 25.6, we discuss another aspect of adaptivity that involves measuring the environment to estimate the R matrix.

24.5 SPACE-TIME PROCESSING

We now address the issue of combined space and time processing. In space-time processing, rather than form a function of angle or a function of Doppler, we combine spatial and temporal equations for the signal [(24.1) and (24.49)] to form a combined function of angle and Doppler at the output of the space-time processor. In equation form, we write

$$V(\theta_s, f_d) = \sqrt{P_s} \left(\sum_{n=0}^{N-1} a_n e^{-j2\pi nd \sin\theta_s/\lambda} \right) \left(\sum_{k=0}^{K-1} \omega_k e^{j2\pi k f_d T} \right)$$
$$= \sqrt{P_s} \sum_{n=0}^{N-1} \sum_{k=0}^{K-1} a_n \omega_k e^{-j2\pi nd \sin\theta_s/\lambda} e^{j2\pi k f_d T} \quad (24.67)$$

We recognize the above as a sum of KN terms. Generalizing the product of the weights to KN distinct weights we get

$$V(\theta_s, f_d) = \sqrt{P_s} \sum_{n=0}^{N-1} \sum_{k=0}^{K-1} w_{n,k} e^{-j2\pi nd \sin\theta_s/\lambda} e^{j2\pi k f_d T} \quad (24.68)$$

We next organize the weights into a general weight vector, w, and the $e^{-j2\pi nd \sin\theta_s/\lambda}$ $e^{j2\pi k f_d T}$ terms into a generalized steering vector, S, and write $V(\theta_s, f_d)$ in matrix form as

$$V(\theta_s, f_d) = \sqrt{P_s} w^H S(\theta_s, f_d) \quad (24.69)$$

Extending the interference representation of Sections 24.2 and 24.3, we can write the interference at the space-time processor output as

$$V_{N+I} = w^H \mathbf{N}_{N+I} \tag{24.70}$$

where

$$\mathbf{N}_{N+I} = \mathbf{N} + \mathbf{N}_I = \mathbf{N} + \sum_{i=1}^{N_i} n_{Ii} D(\phi_i, f_i) \tag{24.71}$$

In (24.71), **N** is the receiver noise and $D(\phi_i, f_i)$ is the steering vector to the interference in angle-Doppler space. With this representation of interference, we are limiting ourselves to tone interferences.

We use the techniques discussed in Sections 24.2 and 24.3 to place the "main beam" in angle-Doppler space on the target and to place nulls at the angle-Doppler locations of the interferences. Specifically, we find that the optimum weight vector is given by

$$w = \kappa R^{-1} S(\theta_s, f_d) \tag{24.72}$$

where

$$R = E\{\mathbf{N}_{N+I} \mathbf{N}_{N+I}^H\} \tag{24.73}$$

and κ is an arbitrary, complex constant that we normally set to unity.

At this point, we need to further discuss the signal and interference steering vectors, $S(\theta_s, f_d)$ and $D(\theta_i, f_i)$, and how to compute R. We note that the exponential terms of (24.67) and (24.68) contain all possible KN combinations of $e^{-j2\pi nd \sin \theta_s / \lambda}$ and $e^{j2klf_d T}$. We organize the N exponentials containing θ_s into a vector

$$S(\theta_s) = \begin{bmatrix} 1 & e^{-j2\pi d \sin \theta_s / \lambda} & \cdots & e^{-j2\pi(N-1)d \sin \theta_s / \lambda} \end{bmatrix}^T \tag{24.74}$$

and the K exponentials containing f_d in to a vector

$$S(f_d) = \begin{bmatrix} 1 & e^{j2\pi f_d T} & \cdots & e^{j2\pi(K-1)f_d T} \end{bmatrix}^T \tag{24.75}$$

We next use these vectors to form a matrix

$$\mathbf{S}(\theta_s, f_d) = S(\theta_s) S^T(f_d) \tag{24.76}$$

that contains all KN combinations of the elements of $S(\theta_s)$ and $S(f_d)$. To form the KN element vector, $S(\theta_s, f_d)$, we concatenate the columns of $\mathbf{S}(\theta_s, f_d)$. The $D(\theta_i, f_i)$ vector for each interference source is formed in a similar fashion.

From (24.71) and (24.73), we can form R as

$$\begin{aligned} R &= E\{\mathbf{N}_{N+I} \mathbf{N}_{N+I}^H\} = E\{(\mathbf{N} + \mathbf{N}_I)(\mathbf{N} + \mathbf{N}_I)^H\} \\ &= E\{\mathbf{NN}^H\} + E\{\mathbf{N}_I \mathbf{N}_I^H\} = R_N + R_I \end{aligned} \tag{24.77}$$

where we made use of the standard assumption that the receiver noise, **N**, and interference, n_I, are independent.

There are N receivers and matched filters, and each receiver processes K pulses though the matched filter and sampler. Thus, we will have KN receiver noise samples. We assume they are all zero-mean, uncorrelated, and have equal powers of P_N. Thus,

$$R_N = P_N I \qquad (24.78)$$

where I is an KN by KN identity matrix.

For each interference we have

$$\begin{aligned} R_{Ii} &= E\left\{ \left[n_{Ii} D(\phi_i, f_i) \right] \left[n_{Ii} D(\phi_i, f_i) \right]^H \right\} = E\left\{ |n_{Ii}|^2 \right\} D(\phi_i, f_i) D^H(\phi_i, f_i) \\ &= P_{Ii} D(\phi_i, f_i) D^H(\phi_i, f_i) \end{aligned} \qquad (24.79)$$

where P_{Ii} is the power associated with the ith interference. With this we get

$$R = P_N I + \sum_i P_{Ii} D(\phi_i, f_i) D^H(\phi_i, f_i) \qquad (24.80)$$

where the sum is taken over the total number of interference sources.

With some thought, it should be clear that the dimensionality of the STAP problem has increased substantially when compared to only spatial or temporal processing. If we perform STAP separately in angle and Doppler, we would need to compute $K + N$ complex weights. If we simultaneously perform STAP in angle and Doppler space, we must compute KN complex weights. To complicate the problem further, remember that we need to compute a separate set of weights for each range cell that is processed. This represents a considerable computational burden. To minimize the burden, much of today's research in STAP is concerned with avoiding the computation of KN weights while still trying to maintain acceptable performance [11].

24.5.1 Example 3: Space-Time Processing

As an illustration of the space-time processing, we extend Examples 1 and 2 to a full space-time processor. We again assume a 16-element array and a Doppler processor that uses 16 pulses. We use the classical STAP approach and process all $16 \times 16 = 256$ signal-plus-noise-plus-interference samples in one processor with 256 weights. (Recall that we do this for each range cell of interest.) We assume the target is located at an angle of zero and a Doppler frequency of zero. The element spacing is ½ wavelength and the PRF is 1,000 Hz. The single-pulse, per-element SNR is 0 dB (at the outputs of the matched filters). We consider two tone interference sources. They are located at angles of $+18°$ and $-34°$. Their Doppler locations, corresponding to the above angles, are 217 Hz and -280 Hz, respectively. The JNRs of the two interference sources are 50 dB. With these specifications, we get the following parameters: $P_S = 1$, $\theta_s = 0$, $f_d = 0$, $P_N = 1$, $P_{I1} = 10^5$, $P_{I2} = 10^5$, $\phi_1 = 18°$, $\phi_2 = -34°$, $f_1 = 217$ Hz and $f_2 = -280$ Hz.

We compute R using (24.78) through (24.80). Since $\theta_s = 0$ and $f_d = 0$

$$S(\theta_s, f_d) = S(0,0) = \begin{bmatrix} 1 & 1 & \cdots & 1 \end{bmatrix}^T \qquad (24.81)$$

or a vector of 256 ones. Finally, we compute w using (24.72) with $\kappa = 1$.

In an actual STAP implementation, we would compute the output of the STAP processor using

$$V_o = w^H V_{in} \qquad (24.82)$$

where V_{in} is a vector that contains the KN outputs from the samplers in each receiver. The first N elements of V_{in} are the outputs from the N receivers on the first pulse. The next N elements are the outputs from the N receivers on the second pulse, and so forth.

For this example, we want to generate a three-dimensional plot of the processor output as a function of angle and frequency. We can do this in several ways. One would be to use (24.69) and compute

$$G(\theta, f) = \frac{\left|w^H S(\theta, f)\right|^2}{\max\left(\left|w^H S(\theta, f)\right|^2\right)} \qquad (24.83)$$

for θ and f of interest. An alternate method would be to use the FFT to implement [see (24.68)]

$$V(\theta, f) = \sum_{n=0}^{N-1} \sum_{k=0}^{K-1} w_{n,k} e^{-j2\pi nd \sin\theta/\lambda} e^{j2\pi fT} \qquad (24.84)$$

and use

$$G(\theta, f) = \frac{\left|V(\theta, f)\right|^2}{\max\left(\left|V(\theta, f)\right|^2\right)} \qquad (24.85)$$

The weight vector is formed into a two-dimensional weight matrix, W, by reversing the algorithm used to form $S(\theta_s, f_d)$ and $D(\phi_i, f_i)$. That is, we let the first column of W be the first N elements of w, the second column be the second N elements, and so forth. We next compute $V(\theta, f)$ by computing the Fourier transform of W using a two-dimensional (2-D) FFT. Finally, $G(\theta, f)$ is computed using (24.85).

The results of this process are shown in Figures 24.7 and 24.8. The figures are contour plots where shading is used to indicate power in dB. The bar to the right provides the relation between power level and shading. The y-axis is $\sin(\theta)$ and has the units of sines (see Chapter 13). This vertical axis scaling was chosen because it was compatible with the routine used to generate the plots. A 512- by 512-, 2-D FFT (rather than a 16- by 16-, 2-D FFT) was used to generate the plots. This was done to provide a plot that showed the gradations in power level.

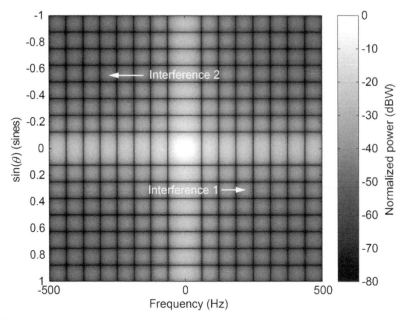

Figure 24.7 Angle-Doppler map—weights based on only receiver noise.

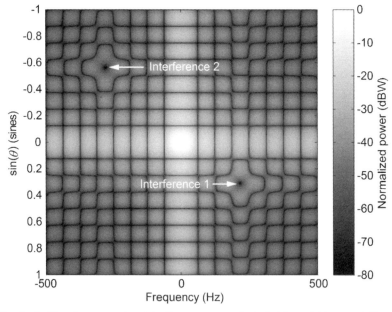

Figure 24.8 Angle-Doppler map—two interference sources included in weight computation.

Figure 24.7 is a plot of $G(\theta,f)$ for the case where the interference consisted of only receiver noise. Since the target was located at $(\theta_s,f_d) = (0,0)$, the resulting weight, w, was a vector of 256 ones. As expected, the peak of $G(\theta,f)$ occurs at $(0,0)$. Note that the two interference sources are located on the peaks of two angle-Doppler sidelobes $G(\theta,f)$.

Because of this, the only rejection of these sources offered by the processor is due to the amplitudes of the sidelobes relative to the response at (0,0). The SINR for this case was −13.4 dB. When the interference sources were omitted, the SNR was the expected, noise-limited value of 10log(256) = 24.1 dB.

Figure 24.8 is a plot of $G(\theta,f)$ for the case where the interference sources were included in the weight computation. The two nulls at the locations of the interferences are clearly visible, as is the main beam at (0,0). With this set of weights, the SINR was 24.1 dB, which is the noise-limited value.

As an experiment, the optimization was extended to include two desired targets: one at (0,0) and the other at (θ_{s2},f_{d2}) = (39°, 217 Hz).[6] Both targets had the same normalized power of $P_{S1} = P_{S2} = 1$. The second target was placed so that its Doppler frequency was the same as one of the interference sources. However, it was separated in angle from the interference source. The interference sources were left at the locations shown in Figures 24.7 and 24.8.

Figure 24.9 contains $G(\theta,f)$ for the case where the weight computation was based on only receiver noise. As can be seen, the calculated weights are such that there are two main lobes at the locations of the two targets. The distortion in the angle-Doppler map is due to the interaction of the two targets. Specifically, the targets were placed so that one was on the peak of a sidelobe of the other. When the interference sources were omitted, the SNR was about 21.1 dB for each of the targets. However, when the interference sources were included, the SINR for each of the targets was −31.1 dB. The noise-only SNR of 21.1 dB is 3 dB less than the single target case because of the presence of two targets rather than one.

Figure 24.10 corresponds to the case where the two interference sources were included in the computation of *w*. As would be expected, the peaks at the locations of the targets are still present. However, the weights have altered the angle-Doppler sidelobe structure to place a null at the angle location of the interference sources that was at the same Doppler frequency as one of the targets. For this case, the combined SINR at the output of the processor was about 21.2 dB for target 1 [the target at (0,0)] and 21 dB for the other target, which is about the same as the noise only case. This indicates that the weight calculation algorithm chose the weights so that both interference sources were greatly attenuated, and the two "main beams" were about the same.

As noted in footnote 6, the examples of this section are "academic." In practice, it is unlikely that interference would be at only two specific angle-Doppler locations (or that we would want to place beams on two targets at the same time). More likely, the interference would be a line through angle-Doppler space. This might be the situation encountered in an airborne radar application where STAP was used to mitigate ground clutter. We consider this in Example 4.

---

[6] As a note, this example is not realistic because the two targets are widely separated in angle. However, it is intended as an illustrative example. To illuminate the targets, the radar would need to have a very wide transmit beam, which would have a negative impact on SNR.

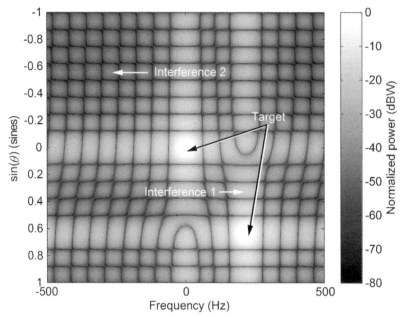

Figure 24.9 Angle-Doppler map—weights based on only receiver noise—two targets.

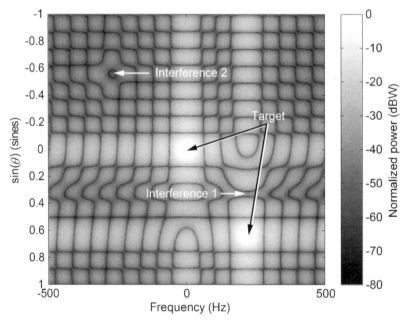

Figure 24.10 Angle-Doppler map—two interference sources included in weight computation—two targets.

24.5.2 Example 4: Airborne Radar Clutter Example

As another example of STAP, we consider the simplified airborne radar problem shown in Figure 24.11. The aircraft in the center of the concentric circles contains a search radar (e.g., airborne warning and control system (AWACS)) that is flying at an altitude of 3 km in the direction of the arrow at a velocity of 100 m/s. The target is also at an altitude of 3 km and is flying in the direction shown at a velocity of 50 m/s. At the time of interest, the angle to the target is $\alpha_T = -30°$. The range to the target, $r_T = 10$ km.

To simplify the example, we (unrealistically) assume the antenna consists of 16 omnidirectional (isotropic) radiators that are located on the bottom of the aircraft. The array is oriented along the length of the aircraft and the element spacing is ½ wavelength. The radar transmits 16 pulses. Thus, the antenna and waveform are consistent with Example 3. We will use STAP to form a beam and nulls in azimuth-Doppler space. We assume the radar is using an operating frequency of 3 GHz and a PRI of $T = 200$ µs. Since we do not need it, we will leave the pulse width unspecified.

The ring of Figure 24.11 represents the ground region illuminated by the radar at the range to the target (10 km). The radar will also illuminate clutter at ranges of 10 km plus ranges corresponding to multiples of the PRI. That is, at ranges of $r_T + mcT/2$, where m is an integer and c is the speed of light. For this example, we ignore the clutter returns at multiples of the PRI.

We assume the per-pulse and per-element SNR and SCR are 0 dB and −50 dB, respectively. Assuming a normalized noise power of $P_N = 1$ W, the normalized signal power is $P_S = 1$ W, and a normalized interference (clutter) power is $P_I = 10^5$ W. The powers are defined at the output of the single-pulse matched filter.

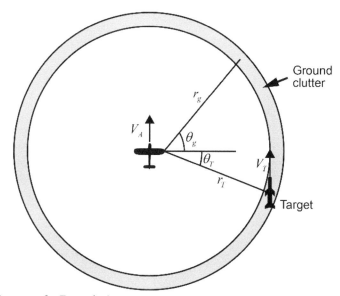

Figure 24.11 Geometry for Example 4.

Given that the aircraft altitude is $h_A = 3$ km and the range to the ground clutter is $r_g = 10$ km, the ground range to the clutter ring is

$$d_g = \sqrt{r_g^2 - h_g^2} = 9.54 \text{ km} \tag{24.86}$$

We can use this, along with V_A, to compute the Doppler frequency of the ground clutter as

$$f_{dg} = \frac{2V_A d_g}{\lambda r_g} \sin\theta_g = \frac{2V_A d_g}{\lambda r_g} u_g = 1.91 u_g \text{ kHz} \tag{24.87}$$

where we note that u_g varies from -1 to 1 as θ_g varies from $-\pi/2$ to $+\pi/2$.

Since the aircraft and the target are at the same altitude, we can write the equation for the target Doppler frequency at the radar as

$$f_{dT} = \frac{2(V_A - V_T)}{\lambda} \sin(\theta_T) = -0.5 \text{ kHz} \tag{24.88}$$

The target is located at $(\theta_T, f_{dT}) = (-\pi/6, -0.5 \text{ kHz})$ in angle-Doppler space.

Rather than being concentrated at a point in angle-Doppler space, the clutter is distributed along a line defined by (24.87). This is illustrated in Figure 24.12, which is a plot like Figure 24.7 with the "beam" in angle-Doppler space steered to (θ_T, f_{dT}), the target location. The white line is a plot of (24.87) and the black circle indicates the target location. The brightest square is the main beam and the other squares are sidelobes. The vertical axis is $u = \sin(\theta)$ and the horizontal axis is frequency, f, in kilohertz (kHz). For this example, we assumed the clutter spectrum width was zero. In practice, the width will be not be zero because of internal clutter spectral spread, aircraft motion, and finite azimuth beamwidth. As can be seen, the clutter "line" skirts the main beam and passes close to the target. We did this intentionally to stress the STAP algorithm.

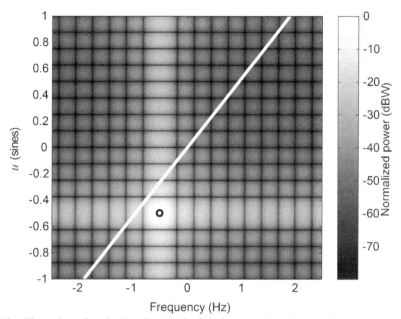

Figure 24.12 Illustration of angle-Doppler plot for interference (white line) and target (black circle), overlaid on the unoptimized angle-Doppler contour plot.

In its basic form, the STAP algorithm developed in this chapter is designed to accommodate only point sources of interference in angle and Doppler. However, we can approximate the continuous line of Figure 24.12 by a series of closely spaced point sources. We choose the point sources so that the spacing between them is much less than the angle and Doppler resolution of the waveform and linear array.

As a reminder, the Doppler resolution of the waveform is equal to the reciprocal of the CPI duration, or $1/(16T)$ in this case. The angle resolution of the linear array is approximately equal to the reciprocal of its length, in wavelengths, which is $16(\lambda/2)$ in this case. To satisfy the point source spacing requirement, we represented the line by 40 point sources. We set the angle spacing, Δu, between the point sources to the length of the line (2 sines) divided by 40. We computed the corresponding f_{dg} from (24.87).

To compute the R matrix, we need to form 40 interference, angle-Doppler steering vectors. The ith angle-Doppler steering vector, $D(i\Delta u, i\Delta f)$, is a $16 \times 16 = 256$ element vector whose elements are given by

$$\exp\left[j2\pi\left(k(i\Delta f)T - n(i\Delta u)d\right)\right] \; n \in [0,15], \; k \in [0,15] \quad (24.89)$$

with $\Delta u = 2/40$ and $\Delta f = \Delta f_{dg} = 1{,}900\Delta u$ [see (24.87)].

With this we use (24.80) to form the R matrix as

$$R = P_N I + P_I \sum_{i=1}^{40} D(i\Delta u, i\Delta f) D^H(i\Delta u, i\Delta f) \quad (24.90)$$

Finally, we use (25.72), with $\kappa = 1$, to find the optimum weight. We use (25.74), (25.75), and (25.76), with $\theta_s = \theta_T$ and $f_s = f_{dT}$, to find $S(\theta_T, f_{dT})$. The result of computing the weights and applying them in the STAP processor is shown in Figure 24.13. Note that there is now a deep notch where the white line of Figure 24.12 was located. The SINR before optimization was −31 dB. After optimization, it was 23.5 dB, which is close to the noise limited case of 24.1 dB. This means the STAP processor has effectively attenuated the clutter. As with Figure 24.12, the black circle is the target location and the white square is the main beam.

In this example, we knew the location of the target in range, angle, and Doppler and we knew the angle-Doppler distribution of the ground clutter. We also knew the SNR and SCR at the matched filter output for each antenna element and pulse. If the radar was conducting a search, we would effectively know the range, angle, and Doppler of interest for each search interrogation, but not the target angle and Doppler. We would compute $S(\theta_T, f_{dT})$ at the angle and Doppler of interest. We could predict the angle-Doppler distribution of the clutter using aircraft navigation data.

As a note, since we assumed a linear array of omnidirectional elements, when the STAP algorithm formed an angle-Doppler beam at $(\theta_T, f_{dT}) = (-\pi/6, -0.5 \text{ kHz})$, it formed another one at $(\theta_T, f_{dT}) = (\pi + \pi/6, -0.5 \text{ kHz})$. We ignored this second beam. This second beam would most likely be suppressed by the individual element patterns in an actual array.

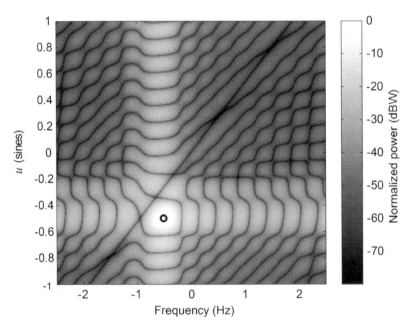

Figure 24.13 Angle-Doppler contour plot with the optimum weights. The black circle is the target location.

24.6 ADAPTIVITY AGAIN

In our work so far, we assumed we knew the various parameters needed to compute the optimum weights. In particular, we assumed we had enough information to compute the R matrix. In most applications, this is not the case, and we must estimate R through measurements. This is part of the adaptive part of STAP: the environment is probed and the results are used to experimentally formulate the R matrix. A potential procedure for doing this follows.

For each antenna element (T/R module) and pulse, we sample the combined noise and interference in range cells we believe contain the interference but not the target.[7] We then use the samples to estimate R for each range cell. Specifically, if we write the combined noise and interference voltage on a particular sample as V^l_{N+I}, we can form an estimate of R as

$$\hat{R} = \frac{1}{L}\sum_{l=1}^{L} V^l_{N+I} \left(V^l_{N+I}\right)^H \tag{24.91}$$

where L is the number of samples taken. As a point of clarification, it should be noted that V^l_{N+I} is a KN element vector.

A question that arises is: How large does L need to be? If $L = 1$, we will be multiplying a KN element vector by its Hermitian to produce an KN by KN matrix. This matrix will have a rank of 1 since it was formed as the outer product of two vectors and thus has only one independent column. This means that $\hat{R}$ has only one nonzero eigenvalue, is thus singular, and $\hat{R}^{-1}$ does not exist. Because of this, solving for w by the previous method will not work.

Given V^l_{N+I} consists of random variables, there is a chance that $\hat{R}$ will have a rank equal to L (for $L \leq KN$). Thus, to have any chance of obtaining a $\hat{R}$ that is nonsingular, at least KN samples of V^l_{N+I} must be taken. As L becomes larger, $\hat{R}$ will converge to reasonable approximation of R, and will (hopefully) be nonsingular. A relation that gives an idea of how large L must be is [11]

$$\rho = \frac{L - KN + 2}{L + 1} \quad L \geq KN \tag{24.92}$$

In this equation, ρ is the ratio of achievable SINR with $\hat{R}$ and the SINR improvement when the actual R is used. For $L = KN$

$$\rho = \frac{KN - KN + 2}{KN + 1} = \frac{2}{KN + 1} \tag{24.93}$$

which says that the SINR improvement actually achieved will be significantly less than the theoretical SINR improvement possible with the actual R. As a specific example, in Examples 3 and 4, $KN = 256$. Therefore, the expected SINR based on 256 samples of V^l_{N+I}

---

[7] In practice, we can allow the range cells to contain the target return if the overall SNR and SIR is very small for each antenna element and pulse.

will be 2/257 or about 21 dB below the optimum SINR improvement. If we increase L to $2KN$ or 512 samples we would get

$$\rho = \frac{2KN - KN + 2}{2KN + 1} = \frac{1}{2} \quad (24.94)$$

Thus, the expected SINR improvement obtained by using $\hat{R}$ would be about 3 dB below the optimum SINR improvement. However, we note that this represents a large number of samples, which will require extensive time and radar resources. Also, for the aircraft case of Example 4, the environment could change before the STAP algorithm could gather enough samples to form the $\hat{R}$ matrix. We will briefly address this in the next section.

24.7 PRACTICAL CONSIDERATIONS

In practice, it may be possible to use fewer samples of V^t_{N+1} if we have a reasonable estimate of the receiver noise power. We would use the aforementioned approximation to form an estimate of R_I, the interference covariance matrix. If we term this estimate $\hat{R}_I$, we would form $\hat{R}$ from

$$\hat{R} = \hat{R}_I + P_N I \quad (24.95)$$

where P_N is the receiver noise power estimate (per antenna element and pulse). This approach is termed *diagonal loading* [11, 22, 23]. Adding the term $P_N I$ ensures that $\hat{R}$ will be positive definite and that $\hat{R}^{-1}$ will exist.

With this method, the number of samples, L, can theoretically be as small as the anticipated number of interference sources [11]. Note that this will generally be much smaller than KN.

This method can have problems in that sometimes $\hat{R}$ can become ill-conditioned [14], which can cause the optimization to put nulls in the wrong locations. To circumvent this problem, it may be necessary to use more samples in the computation of $\hat{R}_I$ and/or artificially increase P_N. Taking more samples is problematic because this requires an extra expenditure of time and radar resources. However, increasing P_N will cause the SINR improvement to degrade, potentially to unacceptable levels.

For the aircraft clutter problem, it may be possible to use aircraft information such as altitude or velocity to form somewhat of an analytical estimate of the clutter distribution over angle-Doppler space. Still another approach suggested in [24] is somewhat of an extension of the method used in sidelobe cancellation. Specifically, a portion of the array would be used to gather data and another portion would be used in the actual STAP algorithm. This would reduce the degrees of freedom available to the STAP algorithm, but it may make it possible to obtain sufficient clutter rejection

More information about these and other practical aspects of STAP can be found in [3, 11–13, 24].

24.8 EXERCISES

1. Show that (24.1) follows from (24.4).
2. Derive the form of (24.11). Specifically, show that R_n is a diagonal matrix.
3. Derive (24.17).
4. Derive (24.24). Specifically, explain why the double sum reduces to a single sum.
5. Derive (24.27) starting with (24.26).
6. Implement a spatial optimization algorithm and generate the plot of Example 1.
7. Repeat Exercise 6 with interference 1 located at 4° instead of 18°. This places the interference slightly more than ½ beam width from the target. You will note that the algorithm places a null in the main beam and moves the peak of the main beam slightly off of the target.
8. Derive (24.37) using (24.36) and (24.35).
9. Implement a temporal optimization algorithm and generate the plot of Example 2.
10. Implement a space-time optimization algorithm and generate the four plots of Example 3.
11. Repeat Exercise 10 with the second target located at $(\theta_{s2}, f_{d2}) = (34°, -217 \text{ Hz})$. Note the difference in the angle-Doppler maps when compared to Figures 24.9 and 24.10.

References

[1] Melvin, W. L., "A STAP Overview," *IEEE Aerosp. Electron. Syst. Mag.,* vol. 19, no. 1, Jan. 2004, pp. 19–35.

[2] Brennan, L. E., and I. S. Reed, "Theory of Adaptive Radar," *IEEE Trans. Aerosp. Electron. Syst.,* vol. 9, no. 2, Mar. 1973, pp. 237–252.

[3] Klemm, R., ed., *Applications of Space-Time Adaptive Processing,* London: Institute of Engineering and Technology, 2004.

[4] Sjogren, T. K. et al., "Suppression of Clutter in Multichannel SAR GMTI," *IEEE Trans. Geosci. Remote Sens.,* vol. 52, no. 7, Jul. 2014, pp. 4005–4013.

[5] Wei, L. et al., "Application of Improved Space-Time Adaptive Processing in Sonar," *2012 Int. Conf. Comput. Sci. Electron. Eng. (ICCSEE),* vol. 3, Hangzhou, China, Mar. 23–25, 2012, pp. 344–348.

[6] Ahmadi, M., and K. Mohamed-pour, "Space-Time Adaptive Processing for Phased-Multiple-Input–Multiple-Output Radar in the Non-Homogeneous Clutter Environment," *IET Radar, Sonar & Navigation,* vol. 8, no. 6, Jul. 2014, pp. 585–596.

[7] Dan, W., H. Hang, and Q. Xin, "Space-Time Adaptive Processing Method at Subarray Level for Broadband Jammer Suppression," *2011 IEEE Int. Conf. Microwave Tech. & Comput. Electromagnetics (ICMTCE),* Beijing, China, May 22–25, 2011, pp. 281–284.

[8] Zei, D., et al., "Real Time MTI STAP First Results from SOSTAR-X Flight Trials," *2008 IEEE Radar Conf.,* Rome, Italy, May 26–30, 2008, pp. 1–6.

[9]   Paine, A. S. et al., "Real-Time STAP Hardware Demonstrator for Airborne Radar Applications," *2008 IEEE Radar Conf.,* Rome, Italy, May 26–30, 2008, pp. 1–5.

[10] Nohara, T. J. et al., "SAR-GMTI Processing with Canada's Radarsat 2 Satellite," *IEEE 2000 Adaptive Syst. Signal Proc., Commun., Control Symp.,* Lake Louise, Alberta, Canada, Oct. 1–4, 2000, pp. 379–384.

[11] Guerci, J. R., *Space-Time Adaptive Processing for Radar,* 2nd ed., Norwood, MA: Artech House, 2014.

[12] Klemm, R., *Principles of Space-Time Adaptive Processing,* 3rd ed., London: Institute of Engineering and Technology, 2006.

[13] Klemm, R., *Space-Time Adaptive Processing: Principles and Applications,* London: Institute of Engineering and Technology, 1998.

[14] Franklin, J. N., *Matrix Theory,* Englewood Cliffs, NJ: Prentice-Hall, 1968.

[15] Urkowitz, H., *Signal Theory and Random Processes,* Dedham, MA: Artech House, 1983.

[16] Shores, T. S., Applied Linear Algebra and Matrix Analysis, New York: Springer, 2007.

[17] Widrow, B., et al., "Adaptive Antenna Systems," *Proc. IEEE,* vol. 55, no. 12, Dec. 1967, pp. 2143–2159.

[18] Nitzberg, R., *Adaptive Signal Processing for Radar,* Norwood, MA: Artech House, 1992.

[19] Nitzberg, R., *Radar Signal Processing and Adaptive Systems,* Norwood, MA: Artech House, 1999.

[20] Manolakis, D. G., et al., *Statistical and Adaptive Signal Processing: Spectral Estimation, Signal Modeling, Adaptive Filtering, and Array Processing*, Norwood, MA: Artech House, 2005.

[21] Golan, J. S., *The Linear Algebra a Beginning Graduate Student Ought to Know*, 2nd ed., New York: Springer, 2012.

[22] Gabriel, W. R., "Using Spectral Estimation Techniques in Adaptive Processing Antenna Systems," *IEEE Trans. Antennas Propag.,* vol. 34, no. 3, Mar. 1986, pp. 291–300.

[23] Carlson, B. D., "Covariance Matrix Estimation Errors and Diagonal Loading in Adaptive Arrays," *IEEE Trans. Aerosp. Electron. Syst.,* vol. 24, no. 4, Jul. 1988, pp. 397–401.

[24] Wirth, W. D., *Radar Techniques Using Array Antennas,* London: IEE Press, 2001.

Chapter 25

Sidelobe Cancellation

25.1 INTRODUCTION

A sidelobe canceller (SLC) attempts to mitigate interference by using signals from an auxiliary channel to remove the interference from the main channel signal [1]. This is in contrast to STAP, or more accurately SAP (spatial adaptive processing), which mitigates interference by modifying the main (sum) channel antenna pattern (see Chapter 24). An SLC does not modify the main channel antenna pattern; instead it modifies the main channel signal by subtracting a weighed form of the auxiliary channel signal from the main channel signal. Because of this, SLC can be used with all types of antennas. This is in contrast to SAP, which, practically speaking, can only be used with array antennas that provide an output from each element, or from a large number of subarrays (see Chapter 14).

The SLC design criterion is based on minimizing the interference at in the main channel of the radar. The basic SLC design methodology is an application of Wiener filtering, the theory used in communication systems for mitigation of multipath and other types of interference signals [2–5].

An SLC is designed to operate against active electronic attack (EA) devices (jammers), and not against clutter or passive interference such as chaff. It is usually assumed the EA signal is noise-like with a bandwidth that exceeds the IF bandwidth of the radar receiver. However, this is not a requirement, and a SLC can cancel narrowband interference. We note, however, that the time required for a SLC to gather sufficient interference data is inversely proportional to the interference bandwidth at the point where the interference data is obtained. This creates the possibility that the SLC may not have time to gather the data needed to cancel the interference (e.g., radar cannot schedule enough time for interference sensing due to higher resource priorities). Also, it is usually assumed the EA signal is entering the radar antenna through one of the sidelobes of the main antenna radiation pattern. The fact that SLC *cancels* interference entering the radar through the main antenna *sidelobes* is believed to be the origin of the term *sidelobe cancellation* (also abbreviated SLC).

Paul W. Howells invented the sidelobe canceller in the 1960s and was awarded a patent for it on August 24, 1965 [6]. Shortly thereafter, Sydney P. Applebaum published a classified report on his analysis of Howells' SLC [7]. Since that time, the SLC implementation invented by Howells is usually referred to as the Howells-Applebaum SLC.

The original Howells-Applebaum SLC was an analog, closed-loop, servomechanism device, which was later implemented as digital loops. With the advent of high-speed analog-to-digital converters and high-speed digital signal processors, SLCs have evolved into open-loop, digital implementations. Both types of SLC involve calculation of a weight function. With a closed-loop SLC, the main channel output signal is used to continually update the weight. That is, there is continuous feedback between the SLC output and the weight calculation. With the open-loop SLC, the weight is (usually) calculated once per dwell (or PRI) and applied during the dwell. There is no feedback during the time the SLC is canceling the interference.

Both types of SLC are discussed in this chapter. We begin by discussing interference cancellation, the theory on which SLC is based. We then describe the open-loop implementation of the SLC. We next discuss SLC weight computation via the gradient technique and use it to derive the Howells-Applebaum form of the SLC. We close the chapter with a brief discussion of *sidelobe blanking* (SLB), which also operates on interference entering through the main antenna sidelobes. While SLC attempts to cancel interference, the sidelobe blanker simply causes the radar to ignore range and/or Doppler cells in which the interference is obscuring the desired return.

25.2 INTERFERENCE CANCELLER

Figure 25.1 contains a functional block diagram of the interference canceller we will consider. The top antenna represents the main antenna of the radar, and the bottom antenna is an auxiliary antenna used to gather information on the interference signal. The block with $w^*(t)$ is a gain, or weight. The arrow through the box indicates that the weight is adjusted based on the error signal, $v_e(t)$. The error signal is formed by subtracting a weighted version of the auxiliary channel signal, $v_a(t)$, from the main channel signal, $v_m(t)$. In equation form[1]

$$v_e(t) = v_m(t) - w^*(t)v_a(t) \qquad (25.1)$$

where * denotes the complex conjugate. The error signal is sent to the rest of the radar receiver and signal processor.

---

[1] Throughout this chapter, we are using complex signal notation. We also assume the various signals are scaled to properly account for their relative power levels.

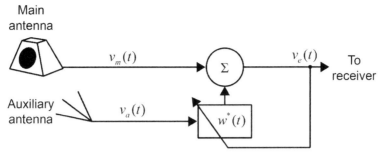

Figure 25.1 Interference canceller block diagram.

If the interference canceller is working correctly, $v_e(t)$ will contain only returns received through the main beam of the main antenna, which we will consider to be the desired, or target, returns.[2] Indeed, suppose $v_m(t)$ consists of a desired signal, $v_s(t)$, and an interference signal, $v_I(t)$. That is,

$$v_m(t) = K_1 v_s(t) + K_2 v_I(t) \qquad (25.2)$$

The auxiliary channel signal, $v_a(t)$, also consists of $v_s(t)$ and $v_I(t)$, but in different proportions. That is

$$v_a(t) = K_3 v_s(t) + K_4 v_I(t) \qquad (25.3)$$

Suppose we are able to choose the weight, $w^*(t)$, as

$$w^*(t) = K_2/K_4 \qquad (25.4)$$

With this we get

$$\begin{aligned}
v_e(t) &= v_m(t) - w^*(t) v_a(t) \\
&= K_1 v_s(t) + K_2 v_I(t) - \frac{K_2}{K_4}\left[K_3 v_s(t) + K_4 v_I(t)\right] \qquad (25.5) \\
&= K_1 v_s(t) + K_2 v_I(t) - \frac{K_2 K_3}{K_4} v_s(t) - K_2 v_I(t) = \left(K_1 - \frac{K_2 K_3}{K_4}\right) v_s(t)
\end{aligned}$$

Thus, as hoped, the error signal consists of a scaled version of the desired signal and no interference signal. This tells us the configuration of Figure 25.1 has the potential to accomplish the desired objective and gives us incentive to develop a more practical algorithm.

---

[2] The signal entering through the main beam could also contain returns from clutter and/or jammers. From the SLC perspective, they are also "desired" signals that the SLC is not designed to cancel. That job falls to the signal and data processors.

25.3 INTERFERENCE CANCELLATION ALGORITHM

We first derive an algorithm for cancellation of a single interference signal and then extend it to the case where there are multiple interference signals.

25.3.1 Single Interference Signal

We assume both the desired and the interference signals, $v_s(t)$ and $v_I(t)$, are complex random processes. We need this assumption because the interference signals of interest are noise-like, and we have established (in earlier chapters) that we should treat target return signals as random processes. We consider them as complex because they have random amplitudes and phases. We further assume $v_I(t)$ is zero-mean and wide-sense stationary (WSS). We assume $v_s(t)$ is zero-mean, but we cannot assume it is WSS because, in general, it is a pulsed signal. Since $v_s(t)$ and $v_I(t)$ are complex, zero-mean random processes, so are $v_m(t)$ and $v_a(t)$. We assume $v_s(t)$ and $v_I(t)$ are independent, since they are presumably from different sources. As a note, $v_m(t)$ and $v_a(t)$ will also contain components due to receiver noise. We will ignore the receiver noise for now, but will consider it in the more general development of Section 25.3.3.

Since $v_s(t)$, $v_I(t)$, $v_m(t)$, and $v_a(t)$ are zero-mean, complex random processes, for some $t = t_1$, $v_s(t_1)$, $v_I(t_1)$, $v_m(t_1)$, and $v_a(t_1)$ are zero-mean, complex random variables. We will denote these as v_s, v_I, v_m, and v_a. Also, because of the WSS assumption, the mean-square value, or power, of $v_I(t_1)$ is independent of t_1.

We define the error signal at $t = t_1$ as

$$v_e = v_m - w^* v_a \qquad (25.6)$$

We note that v_e is zero-mean.

We now define a criterion for determining w. In STAP (Chapter 24), we used maximization of SINR as the design criterion. For the SLC, we use minimization of the mean-square value of v_e as the criterion. In equation form, we choose w according to

$$w_{opt} = \min_w E\{|v_e|^2\} = \min_w E\{|v_m - w^* v_a|^2\} \qquad (25.7)$$

We use the magnitude of v_e because it is complex; we use the expected value, $E\{x\}$, because v_e is a random variable; and we use the square because it is reasonably easy to work with. Equation (25.7) is termed a *least mean-square* (LMS) criterion, which appears in Wiener filter theory [2, 4, 8].

From Wiener filter theory, a necessary and sufficient condition for w to minimize the mean-square error is to choose it so that

$$\nabla J\big|_{w=w_{opt}} = \frac{\partial J}{\partial w}\bigg|_{w=w_{opt}} = \frac{\partial E\{|v_e|^2\}}{\partial w}\bigg|_{w=w_{opt}} = 0 \qquad (25.8)$$

The symbol ∇ denotes the gradient operator.

It can be shown that (25.8) (see Exercise 1) reduces to

$$E\{v_a v_e^*\} = E\{v_a(v_m^* - w_{opt} v_a^*)\} = 0 \qquad (25.9)$$

Equations (25.8) and (25.9) are also known as the *orthogonality condition*. The orthogonality condition, (25.8) and (25.9), tells us the optimum weight, w_{opt}, is chosen so that the error signal, v_e, is orthogonal, in a statistical sense, to the auxiliary channel signal, v_a.

Solving (25.9) for w_{opt} gives

$$w_{opt} = \frac{E\{v_a v_m^*\}}{E\{|v_a|^2\}} \qquad (25.10)$$

which is one form of the Wiener-Hopf equation [2–4, 9].

25.3.2 Simple Canceler Example

To illustrate the procedure of Section 25.3.1, we consider an example using (25.1) through (25.3). Using (25.2) and (25.3), we get

$$\begin{aligned}
E\{v_a v_m^*\} &= E\{(K_3 v_s(t_1) + K_4 v_I)(K_1^* v_s^*(t_1) + K_2^* v_I^*)\} \\
&= E\left\{\begin{array}{l} K_1^* K_3 v_s(t_1) v_s^*(t_1) + K_2^* K_3 v_s(t_1) v_I^* \\ + K_1^* K_4 v_s^*(t_1) v_I + K_2^* K_4 v_I v_I^* \end{array}\right\} \\
&= K_1^* K_3 E\{|v_s(t_1)|^2\} + K_2^* K_4 E\{|v_I|^2\} \\
&= K_1^* K_3 P_s(t_1) + K_2^* K_4 P_I
\end{aligned} \qquad (25.11)$$

where we made use of the assumption that $v_s(t)$ and $v_I(t)$ are independent and zero mean. In (25.11), $P_s(t_1)$ is the power (or energy) of the desired signal at $t = t_1$, and P_I is the power (or energy) of the interference signal.[3] The powers are measured at the point where the SLC is implemented. This could be before or after the matched filter. We left the time parameter on v_s because we will discuss this time dependency later in this example.

The denominator of (25.10) is

---

[3] The parenthetical term "(or energy)" is included to indicate that these quantities could be interpreted as either power or energy. In the future, we will use the term "power," but the reader should keep in mind that they could refer to either power or energy.

$$E\{|v_a|^2\} = E\{(K_3v_s(t_1) + K_4v_I)(K_3^*v_s^*(t_1) + K_4^*v_I^*)\}$$

$$= E\begin{Bmatrix} K_3K_3^*v_s(t_1)v_s^*(t_1) + K_3K_4^*v_s(t_1)v_I^* \\ +K_3^*K_4v_s^*(t_1)v_I + K_4K_4^*v_Iv_I^* \end{Bmatrix} \quad (25.12)$$

$$= |K_3|^2 E\{|v_s(t_1)|^2\} + |K_4|^2 E\{|v_I|^2\}$$

$$= |K_3|^2 P_s(t_1) + |K_4|^2 P_I$$

Combining (25.10) through (25.12) gives

$$w_{opt}(t_1) = \frac{E\{v_a v_m^*\}}{E\{|v_a|^2\}} = \frac{K_1^* K_3 P_s(t_1) + K_2^* K_4 P_I}{|K_3|^2 P_s(t_1) + |K_4|^2 P_I} \quad (25.13)$$

If we assume $P_I \gg P_s(t_1)$ (the interference signal is much larger than the desired signal, at the faces of the main and auxiliary antennas), (25.13) reduces to

$$w_{opt}(t_1) = \frac{K_2^* K_4}{|K_4|^2} = \left(\frac{K_2}{K_4}\right)^* \quad (25.14)$$

which is the solution we postulated in Section 25.2.

If $P_s(t_1) \gg P_I$, (25.13) reduces to

$$w_{opt}(t_1) = \left(\frac{K_1}{K_3}\right)^* \quad (25.15)$$

Substituting this into (25.1) gives the disturbing result,

$$v_e(t) = v_m(t) - w_{opt}^*(t_1)v_a(t)$$

$$= [K_1v_s(t) + K_2v_I(t)] - \left(\frac{K_1}{K_3}\right)[K_3v_s(t) + K_4v_I(t)]$$

$$= K_1v_s(t) - K_1v_s(t) + K_2v_I(t) - \frac{K_1K_4}{K_3}v_I(t) \quad (25.16)$$

$$= \left(K_2 - \frac{K_1K_4}{K_3}\right)v_I(t)$$

Equation (25.16) says that, unlike (25.5), if we compute the weight based on data at the time the desired signal is present, and if the desired signal is much larger than the interference signal, the SLC will cancel the desired signal and pass the interference.

This result leads to the observation that, if possible, the weight, w_{opt}, should be based on measurements obtained when the input to the main and auxiliary antennas contains only the interference signal. This may be possible if the samples are obtained shortly before the transmit pulse (or before a CPI), since it is unlikely that any desired echo signals will be present at that time.

As another extension of this example, we examine the case where there are two independent interference sources, $v_{I1}(t)$ and $v_{I2}(t)$, with powers P_{I1} and P_{I2}. We assume $P_{I1} \gg P_s(t_1)$ and $P_{I2} \gg P_s(t_1)$. For this example, we have

$$v_m = K_1 v_s(t_1) + K_2 v_{I1} + K_5 v_{I2} \quad (25.17)$$

and

$$v_a = K_3 v_s(t_1) + K_4 v_{I1} + K_6 v_{I2} \quad (25.18)$$

With this we get

$$\begin{aligned} E\{v_a v_m^*\} &= E\{(K_3 v_s(t_1) + K_4 v_{I1} + K_6 v_{I2})(K_1^* v_s^*(t_1) + K_2^* v_I^* + K_5^* v_{I2}^*)\} \\ &= K_1^* K_3 E\{|v_s(t_1)|^2\} + K_2^* K_4 E\{|v_{I1}|^2\} + K_5^* K_6 E\{|v_{I2}|^2\} \\ &= K_1^* K_3 P_s(t_1) + K_2^* K_4 P_{I1} + K_5^* K_6 P_{I2} \end{aligned} \quad (25.19)$$

and

$$\begin{aligned} E\{|v_a|^2\} &= E\{(K_3 v_s(t_1) + K_4 v_{I1} + K_6 v_{I2})(K_3^* v_s^*(t_1) + K_4^* v_{I1}^* + K_6^* v_2^*)\} \\ &= |K_3|^2 P_s(t_1) + |K_4|^2 P_{I1} + |K_6|^2 P_{I2} \end{aligned} \quad (25.20)$$

The optimum weight is

$$w_{opt}(t_1) = \frac{K_1^* K_3 P_s(t_1) + K_2^* K_4 P_{I1} + K_5^* K_6 P_{I2}}{|K_3|^2 P_s(t_1) + |K_4|^2 P_{I1} + |K_6|^2 P_{I2}} \quad (25.21)$$

or with the assumptions $P_{I1} \gg P_s$ and $P_{I2} \gg P_s$:

$$w_{opt}(t_1) = \frac{K_2^* K_4 P_{I1} + K_5^* K_6 P_{I2}}{|K_4|^2 P_{I1} + |K_6|^2 P_{I2}} \quad (25.22)$$

We note that $w_{opt}(t_1)$ is a function of not just K_2 and K_4 as in (25.14), but also K_5, K_6, P_{I1}, and P_{I2}. If we substitute this into (25.6), we get

$$\begin{aligned} v_e(t) &= v_m(t) - w_{opt}^*(t_1) v_a(t) \\ &= K_1 v_s(t) + K_2 v_{I1}(t) + K_5 v_{I2}(t) \\ &\quad - \left(\frac{K_2^* K_4 P_{I1} + K_5^* K_6 P_{I2}}{|K_4|^2 P_{I1} + |K_6|^2 P_{I2}}\right)^* [K_3 v_s(t) + K_4 v_{I1}(t) + K_6 v_{I2}(t)] \end{aligned} \quad (25.23)$$

It is not clear if either $v_{I1}(t)$ or $v_{I2}(t)$ will be canceled or even reduced. Also, the impact of the SLC on both the signal and interferences will depend on the interference powers.

This leads to the observation made by Applebaum in his original SLC analysis [7] that the SLC may not be able to cancel all interference signals if the number of interferences exceeds the number of auxiliary channels.

Before we discuss SLC performance further, we will extend the development to the case of multiple interferences and multiple auxiliary channels. We will also add noise to $v_a(t)$ and $v_m(t)$.

25.3.3 Multiple Interference Sources

Figure 25.2 contains a functional block diagram configuration for multiple interferences and multiple auxiliary channels. We assume one main channel, N auxiliary channels, and K interference sources. The output of each auxiliary channel, $v_{an}(t)$, is multiplied by a weight, w_n^*. The results are then summed and subtracted from the main channel signal, $v_m(t)$, to form the error signal, $v_e(t)$. The equation for $v_e(t)$ is[4]

$$v_e(t) = v_m(t) - \mathbf{w}^H \mathbf{v}_a(t) \tag{25.24}$$

where

$$\mathbf{w}^H = \begin{bmatrix} w_1^* & w_2^* & \cdots & w_N^* \end{bmatrix} \tag{25.25}$$

and

$$\mathbf{v}_a(t) = \begin{bmatrix} v_{a1}(t) & v_{a2}(t) & \cdots & v_{aN}(t) \end{bmatrix}^T \tag{25.26}$$

The superscripts H and T denote the conjugate-transpose (Hermitian) and transpose, respectively.

As before, the design criterion is minimization of the mean-square error. That is,

$$\mathbf{w}_{opt} = \min_{\mathbf{w}} E\{|v_e|^2\} = \min_{\mathbf{w}} E\{|v_m - \mathbf{w}^H \mathbf{v}_a|^2\} \tag{25.27}$$

where $v_e = v_e(t_1)$, $v_m = v_m(t_1)$, and so forth. Extending (25.8) to the vector case results in

$$\nabla J\big|_{\mathbf{w}=\mathbf{w}_{opt}} = \frac{\partial J}{\partial \mathbf{w}}\bigg|_{\mathbf{w}=\mathbf{w}_{opt}} = \frac{\partial E\{|v_e|^2\}}{\partial \mathbf{w}}\bigg|_{\mathbf{w}=\mathbf{w}_{opt}} = 0 \tag{25.28}$$

The gradient, ∇J, is

$$\nabla J = \begin{bmatrix} \dfrac{\partial J}{\partial w_1} & \dfrac{\partial J}{\partial w_2} & \cdots & \dfrac{\partial J}{\partial w_N} \end{bmatrix}^T \tag{25.29}$$

---

[4] Matrices are denoted in bold.

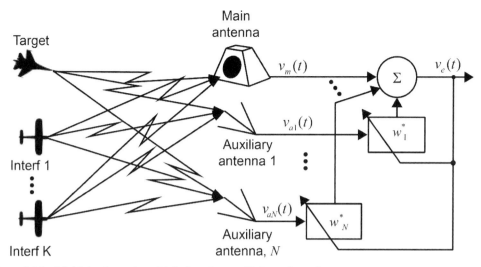

Figure 25.2 Multiple-channel, multiple-interference SLC configuration.

Using (25.29) in (25.28), with (25.27), results in

$$\nabla J\big|_{\mathbf{w}=\mathbf{w}_{opt}} = -E\left\{\left[\mathbf{v}_a\left(v_m^* - \mathbf{v}_a^H \mathbf{w}_{opt}\right)\right]^*\right\} = 0 \qquad (25.30)$$

which we can solve to give

$$\mathbf{w}_{opt} = R^{-1}\boldsymbol{\eta} \qquad (25.31)$$

where

$$R = E\left\{\mathbf{v}_a \mathbf{v}_a^H\right\} \qquad (25.32)$$

and

$$\boldsymbol{\eta} = E\left\{\mathbf{v}_a v_m^*\right\} \qquad (25.33)$$

R is a covariance matrix much like the one discussed in Chapter 24.

25.4 SLC IMPLEMENTATION CONSIDERATIONS

Now that we have a general formulation of the SLC equations, we consider topics we need for analysis and modeling a SLC.

25.4.1 Form of $v_m(t)$ and $v_a(t)$

From Figure 25.2, we note that the signal in the main channel, $v_m(t)$, and each of the auxiliary channels, $v_{an}(t)$, are functions of the desired signal, $v_s(t)$, and all of the interference signals, $v_{I\,k}(t)$. As indicated earlier, they will also include a noise component, $n_m(t)$ and $n_{an}(t)$. With this we can write

$$v_m(t) = A_m(u_{tgt}, v_{tgt})v_s(t) + \mathbf{A}_m \mathbf{v}_I(t) + n_m(t) \qquad (25.34)$$

where $A_m(u_{tgt}, v_{tgt})$ is the (complex) "voltage directivity" of the main antenna in the direction of the desired signal and

$$\begin{aligned}\mathbf{A}_m &= \begin{bmatrix} A_m(u_1, v_1) & A_m(u_2, v_2) & \cdots & A_m(u_K, v_K) \end{bmatrix} \\ &= \begin{bmatrix} A_{m1} & A_{m2} & \cdots & A_{mK} \end{bmatrix}\end{aligned} \qquad (25.35)$$

is a vector (a row vector) of voltage directivities of the main antenna in the directions of the K interference sources. The angles, u and v, are sine space angles (see Chapter 13). The $A(u,v)$ in (25.35) are given by

$$A(u,v) = \sum_{m=0}^{M_{elt}-1} \sum_{n=0}^{N_{elt}-1} a_{mn} e^{j2\pi m(u-u_0)d_x/\lambda} e^{j2\pi n(v-v_0)d_y/\lambda} \qquad (25.36)$$

for an array with rectangular packing (see Chapter 13). (u,v) is the location of the desired signal source or interference, as appropriate, and (u_0, v_0) is the direction to which the beam is steered. These were the functions used to compute the radiation pattern [i.e., $R(u,v) = |A(u,v)|^2$] in Chapter 13.

$v_I(t)$ is a vector (a column vector) of interference signals represented by

$$\mathbf{v}_I(t) = \begin{bmatrix} v_{I1}(t) & v_{I2}(t) & \cdots & v_{IK}(t) \end{bmatrix}^T \qquad (25.37)$$

and $n_m(t)$ is the noise in the main channel.

$v_a(t)$ is a vector of auxiliary channel signals and is given by (25.19). Each of the auxiliary channel signals (each of the elements in the vector) is of the form

$$v_{an}(t) = A_{an}(u_{tgt}, v_{tgt})v_s(t) + \mathbf{A}_{an}\mathbf{v}_I(t) + n_{an}(t) \qquad (25.38)$$

In (25.38), $n_{an}(t)$ is the noise in the nth auxiliary channel, and $A_{an}(u_{tgt}, v_{tgt})$ is the voltage directivity of the nth auxiliary channel antenna in the direction of the desired signal source. A_{an} is a complex vector of voltage directivities in the directions of the interference sources. It has the same form as A_m. That is,

$$\begin{aligned}\mathbf{A}_{an} &= \begin{bmatrix} A_{an}(u_1, v_1) & A_{an}(u_2, v_2) & \cdots & A_{an}(u_K, v_K) \end{bmatrix} \\ &= \begin{bmatrix} A_{an1} & A_{an2} & \cdots & A_{anK} \end{bmatrix}\end{aligned} \qquad (25.39)$$

The complex directivity, $A_{an}(u,v)$, contains a phase that depends on the pointing angle to the kth interference source, (u_k, v_k), and the location of the nth auxiliary antenna relative to the main antenna. If the location of the center (phase center) of the nth auxiliary antenna relative to the center (phase center) of the main antenna is (x_n, y_n, z_n), this phase is (see Appendix 25A)

$$\phi_{nk} = 2\pi \left[(x_n/\lambda)u_k + (y_n/\lambda)v_k + (z_n/\lambda)\sqrt{1 - u_k^2 - v_k^2} \right] \qquad (25.40)$$

The $A_{an}(u,v)$ also contain a phase that accounts for any inherent phase shifts of the main and auxiliary channels. This carries the tacit assumption that the main and auxiliary channels are calibrated so that the phase can be determined. If the SLC determines R and η from measurements, all of the amplitudes and phases of the $A_{an}(u,v)$, and the $A_m(u,v)$ will be accounted for by the measurement process and thus do not need to be known.

The $v_a(t)$ vector is given by

$$\mathbf{v}_a(t) = \mathbf{A}_a(u_{tgt}, v_{tgt}) v_s(t) + A_a \mathbf{v}_I(t) + \mathbf{n}_a(t) \tag{25.41}$$

where

$$\mathbf{v}_a(t) = \begin{bmatrix} v_{a1}(t) & v_{a2}(t) & \cdots & v_{aN}(t) \end{bmatrix}^T \tag{25.42}$$

$$\mathbf{A}_a(u_{tgt}, v_{tgt}) = \begin{bmatrix} A_{a1}(u_{tgt}, v_{tgt}) e^{j\phi_{1,tgt}} & \cdots & A_{aN}(u_{tgt}, v_{tgt}) e^{j\phi_{N,tgt}} \end{bmatrix}^T \tag{25.43}$$

$$A_a = \begin{bmatrix} A_{a11}e^{j\phi_{11}} & A_{a12}e^{j\phi_{12}} & \cdots & A_{a1K}e^{j\phi_{1K}} \\ A_{a21}e^{j\phi_{21}} & A_{a22}e^{j\phi_{22}} & \cdots & A_{a2K}e^{j\phi_{2K}} \\ \vdots & \vdots & \ddots & \vdots \\ A_{aN1}e^{j\phi_{N1}} & A_{aN2}e^{j\phi_{N2}} & \cdots & A_{aNK}e^{j\phi_{NK}} \end{bmatrix} \tag{25.44}$$

$$\mathbf{n}_a(t) = \begin{bmatrix} n_{a1}(t) & n_{a2}(t) & \cdots & n_{aN}(t) \end{bmatrix}^T \tag{25.45}$$

and $v_I(t)$ is given by (25.37).

25.4.2 Properties of $v_s(t)$, $v_I(t)$, $n_m(t)$, and $n_{an}(t)$

The standard assumption is that the interference sources are independent and generate noise-like signals. Thus, we assume the elements of $v_I(t)$ are independent, zero-mean, WSS, random processes. We also assume the noises are zero-mean, WSS, random processes, and the elements of $v_I(t)$ and the noises, $n_m(t)$ and $n_a(t)$, are mutually independent. With this the covariance matrix of $v_I(t)$ is

$$R_I = E\{\mathbf{v}_I(t)\mathbf{v}_I^H(t)\} = \begin{bmatrix} P_{I1} & 0 & \cdots & 0 \\ 0 & P_{I2} & \cdots & 0 \\ \vdots & \vdots & \ddots & \vdots \\ 0 & 0 & \cdots & P_{IK} \end{bmatrix} \tag{25.46}$$

That is, R_I is a diagonal matrix of the interference powers.

The main and auxiliary noise powers are $P_{nm} = E\{|n_m(t)|^2\}$ and $P_{nan} = E\{|n_{an}(t)|^2\}$, respectively. The fact that these powers are represented by constant values is due to the WSS assumption. We collect the noise powers of the auxiliary channels into an auxiliary channel noise covariance matrix that we write as

$$R_{an} = \begin{bmatrix} P_{na1} & 0 & \cdots & 0 \\ 0 & P_{na2} & \cdots & 0 \\ \vdots & \vdots & \ddots & \vdots \\ 0 & 0 & \cdots & P_{naN} \end{bmatrix} \quad (25.47)$$

The power in the desired signal voltage is $P_s(t) = E\{|v_s(t)|^2\}$. This power is not constant because $v_s(t)$ may or may not be present at the time of interest (the time when the weights are computed). $v_s(t)$ is independent of $v_I(t)$, $n_m(t)$, and $n_{an}(t)$.

25.4.3 Scaling of Powers

To be able to simulate and analyze a SLC, the various powers indicated in Section 25.4.2 must be specified. To avoid the difficulty of directly specifying the various powers, we will work with SNR and interference-to-noise ratio, JNR.[5] We suggest the procedure outlined below. This procedure is based on the assumption that the main and auxiliary channel receivers use a matched filter and that the SLC is implemented after the matched filter.

- Compute the SNR from the radar range equation.
- Compute the JNR of each interference source at the radar main antenna from

$$JNR_k = \frac{E_{Jk}}{E_N} = \frac{(P_{Ik}/B_{Ik})G_R\lambda^2}{(4\pi)^2 R_k^2 L_I kT_o F_n} \quad (25.48)$$

where (P_{Ik}/B_{Ik}) is the *effective radiated energy* of the kth interference source, and R_k is the range to the kth interference source. G_R is the receive directivity of the main antenna in the direction of the interference, L_I captures the receive losses in the main channel associated with the interference, and F_n is the noise figure of the main channel receiver.[6]

- Normalize A_m (main array voltage directivity) so its magnitude at (u_0,v_0) is unity.
- Scale the A_{an} (auxiliary array voltage directivities) so their magnitudes at some (u_{0an},v_{0an}) are at some level relative to unity, and account for any difference in receive losses between the main and auxiliary channel. This is a somewhat standard way of specifying the directivities of the auxiliary antennas. That is, their directivities are often specified as being a certain number of dB below the main antenna directivity. The directivity of the auxiliary antennas should be above the sidelobe levels of the main antenna to prevent the SLC from significantly raising the noise floor of the main channel.
- Set $P_{nm} = 1$ W. This, along with the use of SNR and JNR, means that all of the powers are normalized relative to a main channel noise power of 1 W at the matched filter output.

---

[5] We will use the acronym JNR for interference-to-noise ratio. JNR is the abbreviation for jammer-to-noise ratio.
[6] In (25.48) we are using kT_oF_n as the noise energy. An alternate would be to use kT_s where T_s is the system noise temperature (see Chapter 4).

- Compute the auxiliary channel noise powers from $P_{nan} = (F_{an}/F_n)$, where the F_{an} are the system noise figures of the auxiliary channels. This allows for different noise powers in the various receivers.
- Compute the signal and interference powers using $P_s = SNR$ and $P_{Ik} = JNR_k$.

25.4.4 Two Auxiliary Channel Open-Loop SLC Example

To illustrate the previously discussed procedure, we consider an example where we have two interference sources and two auxiliary channels. For the example, we consider a 16-element linear array with uniform weighting. The element spacing is $d = \lambda/2$, making the total effective length of the array $16\lambda/2$. We assume the beam is steered to $u = 0$. We assume the center of the array is located at $x = 0$.

The two auxiliary antennas are located at $x_1 = -10\lambda$ and $x_2 = 12\lambda$. We assume both auxiliary antennas are isotropic radiators with a normalized directivity of -15 dB relative to the peak directivity of the main channel antenna. The noise figures of the two auxiliary channel receivers are the same as the noise figure of the main channel receiver. The main and auxiliary channels use matched filters, and the SLC is implemented after the matched filter.

The two interference sources are located at $u_1 = \sin(18°)$ and $u_2 = \sin(-34°)$, and their JNRs are $JNR_1 = 40$ dB and $JNR_2 = 50$ dB. The desired signal source (the target) is located at $u_{tgt} = 0$ and the SNR is 20 dB.

Figure 25.3 contains a depiction of the antenna geometry, and the various parameters are listed in Table 25.1.

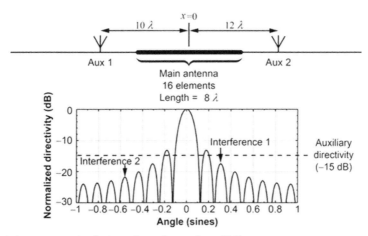

Figure 25.3 Antenna geometry for two-channel open-loop SLC.

Table 25.1
Parameters for Two-Channel Open-Loop SLC

$u_1 = \sin(18°)$, $JNR_1 = 40$ dB
$u_2 = \sin(-34°)$, $JNR_2 = 50$ dB
$u_{tgt} = 0$, $SNR = 20$ dB
$P_{nm} = P_{na1} = P_{na2} = 1$ W[7]
$P_s = 10^2$ W, $P_{I1} = 10^4$ W, $P_{I2} = 10^5$ W
$\phi_{1,tgt} = 4\pi(12\lambda/\lambda)u_{tgt}$ rad, $\phi_{21} = 4\pi(12\lambda/\lambda)u_1$ rad, $\phi_{22} = 4\pi(12\lambda/\lambda)u_2$ rad
$\phi_{2,tgt} = 4\pi(-10\lambda/\lambda)u_{tgt}$ rad, $\phi_{11} = 4\pi(-10\lambda/\lambda)u_1$ rad, $\phi_{12} = 4\pi(-10\lambda/\lambda)u_2$ rad
$A_m(u_{tgt}, v_{tgt}) = 1$, $\mathbf{A}_m = [0.1334\ \ 0.0809]$
$A_{a1}(u_{tgt}, v_{tgt}) = A_{a11} = A_{a12} = 10^{-15/20}$, $A_{a2}(u_{tgt}, v_{tgt}) = A_{a21} = A_{a22} = 10^{-15/20}$
(constant gain auxiliary antennas)

We assume the weight, $\mathbf{w}_{opt}$, is calculated before the transmit pulse and that there is no desired signal present. With this we get (see Exercise 2)

$$R = E\{\mathbf{v}_a \mathbf{v}_a^H\} = A_a R_I A_a^H + R_{an} \tag{25.49}$$

where, from (25.44),

$$A_a = \begin{bmatrix} A_{a11}e^{j\phi_{11}} & A_{a12}e^{j\phi_{12}} \\ A_{a21}e^{j\phi_{21}} & A_{a22}e^{j\phi_{22}} \end{bmatrix}$$

$$= \begin{bmatrix} 10^{-15/20}e^{j4\pi(-10\lambda/\lambda)u_1} & 10^{-15/20}e^{j4\pi(-10\lambda/\lambda)u_2} \\ 10^{-15/20}e^{j4\pi(12\lambda/\lambda)u_1} & 10^{-15/20}e^{j4\pi(12\lambda/\lambda)u_2} \end{bmatrix} \tag{25.50}$$

from (25.46), (25.47), and (25.35)

$$R_I = \begin{bmatrix} P_{I1} & 0 \\ 0 & P_{I2} \end{bmatrix} = \begin{bmatrix} 10^4 & 0 \\ 0 & 10^5 \end{bmatrix} \tag{25.51}$$

$$R_{an} = E\{\mathbf{n}_a(t)\mathbf{n}_a^H(t)\} = \begin{bmatrix} P_{an1} & 0 \\ 0 & P_{an2} \end{bmatrix} = \begin{bmatrix} 1 & 0 \\ 0 & 1 \end{bmatrix} \tag{25.52}$$

$$\boldsymbol{\eta} = E\{\mathbf{v}_a v_m^H\} = A_a R_I \mathbf{A}_m^H \tag{25.53}$$

and A_m is given in Table 25.1.

The resulting weight vector, using (25.31), is

$$\mathbf{w} = R^{-1}\boldsymbol{\eta} = \begin{bmatrix} -0.4943 - j0.2215 \\ -0.9263 - j0.0911 \end{bmatrix} \tag{25.54}$$

---

[7] Recall that we are referencing everything to a normalized, main channel power of 1 W.

25.4.5 Performance Measures

A standard measure of the performance of a SLC is the *cancellation ratio* (CR), which is defined as the total interference power in v_e if the SLC was not present ($w = 0$) divided by the total interference power when the SLC is active [7–10]. By total interference power, we mean the combined power of the interference sources and total noise. In equation form

$$CR = \frac{E\left\{|v_e(t)|^2_{w=0,\,no\,signal}\right\}}{E\left\{|v_e(t)|^2_{no\,signal}\right\}} = \frac{E\left\{|v_m(t)|^2_{no\,signal}\right\}}{E\left\{|v_m(t) - \mathbf{w}^H_{opt}\mathbf{v}_a(t)|^2_{no\,signal}\right\}} \quad (25.55)$$

where we made use of (25.24). With the terms delineated earlier, (25.55) reduces to

$$CR = \frac{\mathbf{A}_m R_I \mathbf{A}^H_m + P_{nm}}{\left(\mathbf{A}_m - \mathbf{w}^H_{opt} A_a\right) R_I \left(\mathbf{A}_m - \mathbf{w}^H_{opt} A_a\right)^H + P_{nm} + \mathbf{w}^H_{opt} R_{an} \mathbf{w}_{opt}} \quad (25.56)$$

which is 26 dB for the example of Section 25.4.4.

Another measure of performance is a comparison of the SINR without the SLC (i.e., $w = 0$) and the SINR with the SLC. We can compute the SINR without the SLC as (see Exercise 2)

$$SINR_{before} = \frac{E\left\{|v_m(t)|^2_{signal\,only}\right\}}{E\left\{|v_m(t)|^2_{no\,signal}\right\}} = \frac{|A_m(u_{tgt}, v_{tgt})|^2 P_s}{\mathbf{A}_m R_I \mathbf{A}^H_m + P_{nm}} \quad (25.57)$$

or −9.2 dB for the example of Section 25.4.4. This low value of SINR is due mainly to the interference source at −34°. That source has a JNR of 50 dB that is attenuated by the −22 dB sidelobe (see Figure 25.3). This alone would result in an SINR of −8 dB (SNR − JNR − SLL = 20 − 50 + 22). The remaining 1.2-dB degradation is due to the other interference source and main channel noise.

With the SLC on, the SINR is

$$SINR_{after} = \frac{E\left\{|v_e(t)|^2_{signal\,only}\right\}}{E\left\{|v_e(t)|^2_{no\,signal}\right\}}$$

$$= \frac{|A_m(u_{tgt}, v_{tgt}) - \mathbf{w}^H_{opt} A_a(u_{tgt}, v_{tgt})|^2 P_s}{\left(\mathbf{A}_m - \mathbf{w}^H_{opt} A_a\right) R_I \left(\mathbf{A}_m - \mathbf{w}^H_{opt} A_a\right)^H + P_{nm} + \mathbf{w}^H_{opt} R_{an} \mathbf{w}_{opt}} \quad (25.58)$$

or 18.6 dB for the example of Section 25.4.4.

The SINR at the output of the SLC is close to the SNR of 20 dB specified in the problem definition (of Section 25.4.4). The SLC resulted in a signal power increase of about 2 dB and a noise power increase of about 3.3 dB. This interesting coincidence meant the SNR (exclusive of the interference sources) went down by about 1.3 dB, which means the SLC was quite effective at removing almost all of the interference due to the interference sources.

The increase in noise power is due to the last two terms in the denominator of (25.58). These terms tell us the overall noise level will be equal to the main channel receiver noise plus some portion due to the noise in the auxiliary channels. In this particular example, the magnitude of the weight vector was such that the auxiliary channels added a noise power slightly different from the noise power of the main channel [about 1.2 W versus the (normalized) main channel noise of 1 W].

The reason the auxiliary channel noises did not add much to the overall receiver noise was because the directivity of the auxiliary channel receivers was greater than the directivity of the main antenna sidelobes containing the interferences (see Figure 25.3). Had the directivities of the auxiliary antennas been below the sidelobe levels, the SLC weights would have had a magnitude greater than unity. This would have amplified the noises in the auxiliary channels and caused the overall receiver noise to increase substantially. This is considered further in Exercise 7.

25.4.6 Practical Implementation Considerations

While the methods discussed in Sections 25.4.1 through 25.4.4 are suitable for analyzing sidelobe cancellers, they cannot be directly used in an actual SLC implementation because the various parameters (e.g., Table 25.1) are not known a priori. As a result, the various expected values must be estimated based on measurements of $v_m(t)$ and $v_a(t)$. Strictly speaking, the expected values are ensemble averages and cannot be evaluated from a single of $v_m(t)$ and $v_a(t)$ measurement. To obtain a valid ensemble average, we would need to average across many radars, desired signals, environments, and interference sources (all of the same type and in the same location). Clearly, this is not possible since we have only one radar, and so on. To get around this problem, we invoke the concept of ergodicity [11, 12]. This concept states that if a random process is ergodic, ensemble averages can be replaced by time averages. Proving that a process is ergodic is very difficult, if not impossible. However, it is a standard assumption that appears to be valid as long as one is confident that the processes are at least WSS.

We will assume the measurements are made right before the transmitted pulse. This is necessary to ensure the interference and noises satisfy the WSS restriction. For phased array antennas, we impose the additional constraint that the measurements are made after the main and auxiliary beams have been steered to their new location and after any local oscillators and such have been retuned.

Figure 25.4 contains a possible timing diagram illustrating how the SLC power estimation and weight computation would fit into an overall radar timeline. As shown, time is allotted at the end of a PRI for (1) frequency retuning, (2) beam steering, (3) SLC power estimation, (4) SLC weight computation, and (5) receiver noise measurement (for AGC or detection threshold determination, not SLC). For a high-PRF burst waveform, there is not sufficient time before every pulse to compute SLC weights, so they are computed before the burst and held throughout the burst.

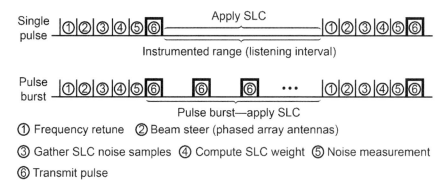

Figure 25.4 SLC timing diagram.

Actually, if the radar performs coherent processing (MTI, pulsed Doppler, coherent integration), the weights are usually computed and held constant for the CPI. If the weights were computed before each pulse, they could affect the pulse-to-pulse phase characteristics of the main channel target and clutter signals, which could degrade the clutter rejection and/or SNR improvement of the signal processor.

The frequency retuning and beam steering operations could be reversed; however, both must be performed before the SLC weight computation. The frequency retuning must precede the SLC weight computation because it affects the phase shifts in the main and auxiliary channels. The beamsteering must precede the SLC weight computation because it establishes the main antenna sidelobe levels and the directivities of the auxiliary antennas (both amplitude and phase). The noise measurement is made after the SLC weight calculation because the SLC will affect the noise floor in the main receiver.

We further assume there is no desired return signal when the SLC samples are taken, only interference and receiver noise. We assume the interference and receiver noise samples are ergodic in the mean and autocorrelation [11, 12]. If we assume a digital implementation of the SLC and use samples of $v_a(t)$ and $v_m(t)$, ergodicity tells us we can estimate R and η using

$$\hat{R} = \frac{1}{L_{samp}} \sum_{l=1}^{L_{samp}} \mathbf{v}_a(l) \mathbf{v}_a^H(l) \tag{25.59}$$

and

$$\hat{\eta} = \frac{1}{L_{samp}} \sum_{l=1}^{L_{samp}} \mathbf{v}_a(l) v_m^*(l) \tag{25.60}$$

where $v_a(l)$ and $v_m(l)$ are samples of the auxiliary and main channel signals. We then use these to form the weight estimate as

$$\hat{\mathbf{w}} = \hat{R}^{-1} \hat{\eta} \tag{25.61}$$

We use this weight estimate throughout the PRI, or CPI if the radar performs coherent processing of the desired return signals. That is, for all range cells in the PRI or CPI, we use

$$v_e(m) = v_m(m) - \hat{\mathbf{w}}^H \mathbf{v}_a(m) \tag{25.62}$$

where m is the range cell index.

The idea that we can use the same weight throughout the PRI or CPI is a consequence of the WSS assumption.[8] This means that $\hat{R}$ and $\hat{\eta}$, and thus $\hat{w}$, are constant.

The method of determining the SLC weight based on estimates of R and η is termed the *sample matrix inversion* (SMI) technique [10, 13–16]. Its name derives from the fact that the weights are found using (25.61), which involves the inversion of a matrix based on samples of $v_a(l)$.

As with STAP, there is a question of how many samples are needed to obtain a reliable estimate of R and η. If the SLC has N auxiliary channels, $v_a(l)$ will have N elements and $\hat{R}$ will be an N-by-N matrix. Thus, the minimum number of samples needed is N. Otherwise, $\hat{R}$ will be singular. As indicated in Chapter 24, Nitzburg [5], Reed, Mallet, and Brennan [14] developed an efficiency parameter for STAP. That parameter indicated how the SINR improvement using a SMI approach would deviate from some theoretical SINR based on complete knowledge of the system, desired signal, interference, and noise. If we adapt that parameter to the SLC case, we would have

$$\rho = \frac{L_{samp} - N + 2}{L_{samp} + 1} \quad L_{samp} \geq N \tag{25.63}$$

As an example, if we had $N = 3$ auxiliary channels and used $L_{samp} = 3$, we would get

$$\rho = \frac{3 - 3 + 2}{3 + 1} = \frac{1}{2} \tag{25.64}$$

and would expect an improvement that is about 3 dB less than theoretical. Doubling the number of samples would increase this to about 1.5 dB less than theoretical. This assumes there will be inaccuracies only in $\hat{R}$. There will also be inaccuracies in $\hat{\eta}$ because of the limited number of samples. Because of this, additional samples will be needed to account for the measurement of $v_m(l)$.

The spacing between samples should be equal to or greater than the inverse of the bandwidth of the interference signal(s) at the point where the interference power is computed. For interference whose bandwidth is greater than the bandwidth of the receiver components up to where the interference power is measured, the sample spacing should be the inverse of that bandwidth. If the interference (and noise) powers are measured after the matched filter, the spacing between the samples should be at least the inverse of the waveform modulation bandwidth. This will ensure the samples are uncorrelated and thus, that the estimate will not have a bias.

---

[8] As a note, the assumption of ergodicity implies that the interferences and noises are WSS.

This bandwidth requirement can have an impact on how much of the radar timeline is allocated to SLC. For an SLC to have the ability to counter narrowband interference, a significant amount of time must be allotted to the power estimation phase of the SLC weight computation. With modern hardware, the weight computation should be fairly quick, possibly in the order of microseconds.

25.4.7 Two Auxiliary Channel Open-Loop SLC Example with SMI

To investigate the relation between L_{samp} and expected SLC performance, the SMI technique of Section 25.4.6 was implemented for the system of Section 25.4.4. L_{samp} was varied from 2 to 40. L_{samp} samples of $v_a(l)$ and $v_m(l)$ were used to compute $\hat{R}$ and $\hat{\eta}$ using (25.59) and (25.60). These were then used in (25.61) to compute $\hat{w}$. Next, one more sample of $v_a(l)$ and $v_m(l)$ was chosen to compute $v_e(l)$ from (25.62). These were used to compute the powers indicated in the numerator and denominator of (25.55) In all cases $v_s(l)$ was set to zero since we wanted to compute the cancellation ratio. Finally, these powers were averaged over 10,000 Monte Carlo runs and used to compute CR using (25.55).

Figure 25.5 contains a plot of CR versus L_{samp}. According to (25.63), it was expected that ρ would be $(2 - 2 + 2)/(2 + 1) = 2/3$ for $L_{samp} = 2$. With this the CR should have been about 1.8 dB below the theoretical value of 26 dB [see (25.56)]. Clearly, this did not happen. However, as L_{samp} increased, the SMI method did give an ultimate cancellation ratio very close to the theoretical value. The resulting weight magnitudes converged to about 0.53 and 0.92, which is also consistent with the theoretical values of (25.54).

Based on this one example, it would seem that (25.63) should be considered only as a guide to how large L_{samp} should be. To reiterate a previous statement, (25.63) was derived for a STAP application and not for a SLC application.

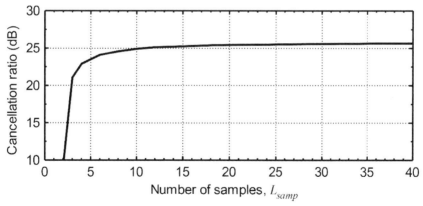

Figure 25.5 Cancellation ratio versus L_{samp}.

Figure 25.5 indicates the *CR* is within 1 dB of its theoretical value with $L_{samp} = 10$. If the SLC was designed to handle a broadband jammer, and had the samples been taken at the output of a matched filter matched to a 1-μs pulse, 10 μs would be needed to gather the interference samples needed to compute $\hat{R}$ and $\hat{\eta}$. However, if the SLC were to have the requirement that it cancel interference with a bandwidth of 100 kHz (and the samples were taken at the matched filter output), a 100-μs data gathering period would be needed since the samples would need to be spaced 1/(100 kHz) = 10 μs apart.

25.5 HOWELLS-APPLEBAUM SIDELOBE CANCELLER

The SMI (open-loop SLC) methodology discussed in Section 25.4 requires the use of high-speed digital processors to compute $\hat{R}$ and $\hat{\eta}$ and solve for $\hat{w}$. Such processors are available to modern radar designers but were not available to radar designers in the earlier years of SLC. Designers of those radars had to use an analog SLC. Most used the Howells-Applebaum, closed-loop, SLC, or a modification thereof.

25.5.1 Howells-Applebaum Implementation

The weight calculation technique on which the Howells-Applebaum SLC is based is termed a gradient search technique [2, 4, 8, 17]. The gradient search technique is also sometimes termed the LMS technique. It is used extensively for interference mitigation in communications equipment and in other applications such as noise-canceling headphones. The gradient search technique iteratively computes weights to eventually minimize the mean-square error

$$e = E\left\{\left|v_e(t_1)\right|^2\right\} \quad (25.65)$$

In the implementation of the technique, the expected value is approximated by the simple square error, or

$$e(k) = \left|v_e(t_1)\right|^2 \quad (25.66)$$

The gradient algorithm is given by the equation

$$w_{k+1} = w_k - \mu \nabla_e \quad (25.67)$$

In (25.67), ∇_e is the gradient of the error evaluated at w_k and is given by

$$\nabla_e = \left.\frac{\partial e(k)}{\partial w}\right|_{w=w_k} = \left.\frac{\partial \left|v_m(t_1) - w^* v_a(t_1)\right|^2}{\partial w}\right|_{w=w_k}$$

$$= -\left[v_m^*(t_1) - w_k v_a^*(t_1)\right] v_a(t_1) \quad (25.68)$$

$$= -v_e^*(t_1)_k v_a(t_1)$$

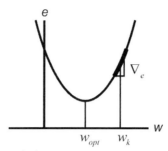

Figure 25.6 Illustration of gradient technique.

Basically, the gradient is used to update the latest estimate by adding a correction that is proportional to the negative of the slope, or gradient, of the error evaluated at the latest estimate. This is illustrated in Figure 25.6. In this figure, $w_k > w_{opt}$ and we note the slope is positive. We also note we want w_{k+1} to be less than w_k if we are to move toward w_{opt}. Thus, we see that we want to move in a direction that is opposite to the sign of the slope. With some thought, we also note that if w_k is far away from w_{opt}, we would like to change w_k by a large amount, whereas if w_k is close to w_{opt}, we want to change w_k by a small amount. Thus, the amount of change is related to the magnitude of the slope. This is what the algorithm of (25.67) does.

The parameter μ controls the rate at which the estimate approaches w_{opt}. If μ is small, w_k will approach w_{opt} in small steps; if μ is large, w_k will approach w_{opt} in large steps. If μ is too small, convergence will be very slow. However, if μ is too large, the solution could diverge. Thus, choosing μ is one of the important aspects of implementing a Howells-Applebaum SLC.

Equation (25.67) is a difference equation. However, early Howells-Applebaum SLCs were implemented in the continuous time domain. We can convert (25.67) to a differential equation of the form

$$\frac{dw(t)}{dt} = -\mu_c \nabla_e = \mu_c v_e^*(t) v_a(t) \tag{25.69}$$

where we have made use of (25.68). We changed the parameter μ to μ_c to denote it is different for discrete-time and continuous-time implementations.

Equation (25.69) also contains another subtle change relative to (25.67). Specifically, in (25.69), we allow the error signal and auxiliary channel signal to change with time as the weight is being updated. In (25.67), we used one sample of the error signal and auxiliary channel signal to iterate on the weight. Allowing the error signal and auxiliary signals to change over time incorporates averaging into the analog SLC loop. It is also what makes the SLC a closed loop implementation.

If we represent (25.69) as a block diagram, we have the functional block diagram of the Howells-Applebaum SLC shown in Figure 25.7.

25.5.2 IF Implementation

The Howells-Applebaum loop is sometimes implemented at some IF. As such, the lower multiply of Figure 25.7 is generally performed by a mixer, whereas the upper multiply is a variable gain amplifier. The $v_m(t)$, $v_a(t)$, and $v_e(t)$ are IF signals, while $w^*(t)$ is a baseband signal. The conjugation on the right side (the block with * in it) is implemented as a 90° phase shift. The block diagram of Figure 25.7 uses complex signal notation. In an actual IF implementation, quadrature signals are used to capture the operations implied by the complex signal notation.

An example block diagram for an IF implementation is contained in Figure 25.8. In this figure, the circles with crosses are mixers and the squares with crosses are variable gain amplifiers. The gain is bipolar. That is, the weight can vary the amplifier gain and the sign of the product depending on the signs of the weight components.

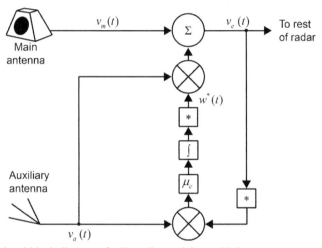

Figure 25.7 Functional block diagram of a Howells-Applebaum SLC.

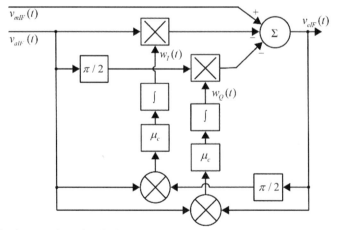

Figure 25.8 IF implementation of an SLC.

The blocks with integral signs in them are typically implemented using lowpass filters, where the bandwidth of the lowpass filter is set somewhat lower than the reciprocal of the integration time of the SLC.

The block diagrams of Figures 25.7 and 25.8 leave the impression that the SLC is continually updating the weights. This is not necessary, or even desirable. The loop could be allowed to update the weights during some time period before the transmit pulse and then hold the weights for the rest of the PRI, or CPI for the case where the radar performs coherent processing (see Figure 25.4).

25.5.3 Single-Loop Howells-Applebaum SLC Example

As an example, a digital version of the Howells-Applebaum SLC of Figure 25.7 was implemented. We chose to use a digital implementation because it was easier to program. In the digital version, the integrator of Figure 25.7 is replaced by a summer. Another variation that was needed was to normalize the value of $v_a(k)$ by dividing by $|v_a(k)|$. We found this necessary because without it, the convergence time and stability of the SLC were very dependent on the interference power. In an actual SLC, this normalization would be performed by some type of instantaneous AGC, such as an IF limiter [5; 18, p. 119].

Figure 25.9 contains a block diagram of the Howells-Applebaum SLC that was implemented. The $v_m(k)$ and $v_a(k)$ signals were created using the parameters in Table 25.1 of Section 25.4.4. Specifically, we used the parameters corresponding to interference source 1 and auxiliary channel 1. As before, $v_s(k)$ was set to zero since we were concerned with only the calculation of the weights. The particular area of interest in this example was the variation of the weight and the cancellation ratio as a function of sample number, k (time), for different values of interference power. The output of the simulation for JNRs (interference power levels; see Section 25.4.4) of 40, 50, and 60 dB are contained in Figure 25.10. The left graph contains plots of $|w(k)|$ versus k and the right graph contains plots of $CR(k)$ versus k. As with the example of Section 25.4.7, the curves are based on 10,000 Monte Carlo runs. The value of μ used in the simulation was 0.005 and was chosen, by experiment, to provide a reasonable settling time for the three JNRs.

As expected, the time it takes the SLC to reach steady state increases as the jammer power decreases. However, in all three cases, the SLC reached steady state by about 70 samples. If we were to assume that the radar uses a waveform with a compressed pulse width of 1 µs, and if we assume the samples are spaced 1 µs apart to satisfy the independence requirement, the SLC would settle in about 70 µs. This means that about 70 µs would be needed for the SLC to stabilize (see Figure 25.4) before the weight was held and used. This example demonstrates that the JNR affects both convergence time and CR. This is because the effective loop gain of the SLC is proportional to the JNR. This proportionality is sometimes described as a potential disadvantage of the Howells-Applebaum SLC [18, pp. 119–120].

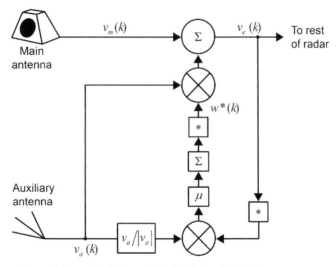

Figure 25.9 Howells-Applebaum single-loop SLC simulation block diagram.

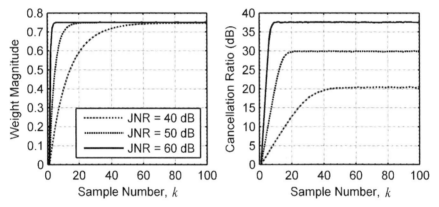

Figure 25.10 Howells-Applebaum single-loop SLC simulation results.

25.5.4 Two-Loop Howells-Applebaum SLC Example

As an extension of the single-loop Howells-Applebaum example of Section 25.5.4, we now consider a two-loop Howells-Applebaum SLC, a block diagram, which is shown in Figure 25.11. The block diagram of Figure 25.11 is similar to Figure 25.9 except there are two auxiliary channels and two weight calculation paths. Each auxiliary channel signal drives a separate weight calculation path. However, consistent with Figure 25.2, the single error signal, $v_e(t)$, is common to both weight calculation paths. Thus, the two loops will attempt to work in context to cancel the two interference signals.

We implemented the two-loop Howells-Applebaum SLC of Figure 25.11 using the parameters in Table 25.1 of Section 25.4.4. As with the single-loop example, we set $v_s(t)$ to zero since we were only interested in computing the SLC weights and the cancellation ratio. We also again used a μ of 0.005 and averaged the weights and cancellation ratio over 10,000 Monte Carlo runs.

The results of this experiment are shown in Figure 25.12. The top plot contains the magnitudes of the two weights, and the bottom plot contains the cancellation ratio, both versus number of samples, k. The weight magnitudes converge to about 0.54 and 0.93, which is consistent with the theoretical values of (25.54) and the values of 0.53 and 0.92 for the SMI technique of Section 25.4.7. However, the cancellation ratio is only about 22.5 dB, which is considerably lower than the theoretical value of 26 dB [see (25.56)] and the value of almost 26 dB provided by the SMI technique of Section 25.4.7.

Another interesting comparison is that the SMI technique needed about 10 samples of interference and noise, whereas the Howells-Applebaum implementation took about 200 samples to reach a steady-state cancellation ratio. Experimentation indicated that the convergence time could be decreased by increasing μ, but at the expense of cancellation ratio. Conversely, the cancellation ratio could be increased by decreasing μ, but at the expense of convergence time.

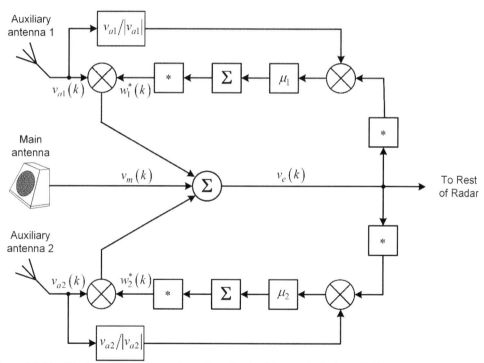

Figure 25.11 Block diagram of a two-loop Howells-Applebaum single-loop SLC.

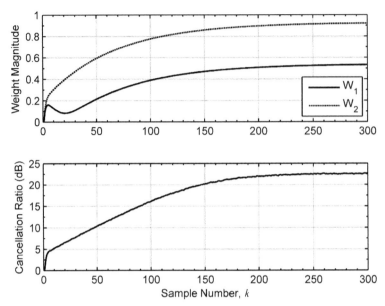

Figure 25.12 Howells-Applebaum two-loop SLC simulation results.

25.6 SIDELOBE BLANKER

Rather than trying to cancel interference, the SLB causes the radar to ignore (blanks) those range (or range/Doppler for pulsed Doppler processors) cells where the signal in range (or range/Doppler) cell of the auxiliary channel is larger than the signal in the range (or range/Doppler) cell of the main channel by some specified amount. A simplified functional block diagram of the circuity that accomplishes this is shown in Figure 25.13. As with the SLC, the SLB operates on signals from the main channel and an auxiliary channel. In fact, the auxiliary channel receiver used for the SLC could also be used for the SLB. One arrangement would be to process the signals through the SLC to try to cancel interference, and then use the SLB to blank range (or range/Doppler) cells that contain interferences that were not rejected by the SLC. An example of such an interference would be random pulses from some source (another radar, for example). Such random pulses would not be rejected by the SLC. However, the SLB would detect their presence and blank the appropriate range (or range/Doppler) cells [10; 19, p. 368; 21]. Another example suggested by our reviewer is that it "is commonly used in airborne MPRF radars for rejecting sidelobe clutter as well as jammers. In those applications the auxiliary channel is referred to as a 'guard channel'.

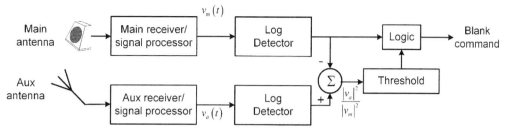

Figure 25.13 Functional diagram of a sidelobe blanker.

The main and auxiliary channels contain the same receiver and signal processors. The log detectors, which operate on the outputs of the signal processors, contain square-law detector and logarithm circuits. The subtraction of the two log detector outputs effectively forms the logarithm of the ratio of the auxiliary and main channel powers in each range (or range/Doppler) cell. If this ratio exceeds some threshold, T, the logic block outputs a blank command for that range (or range/Doppler) cell. If the ratio is below the threshold, the logic block outputs a no-blank command. The set of blank/no-blank signals is sent to the detection logic or computer, along with the range (or range/Doppler) output of the main receiver/signal processor. The process of actually blanking the cells is performed in the detection logic or the computer, depending on the specific implementation.

The normal design criterion for the SLB is for the directivity of the auxiliary antenna to be larger than the sidelobe levels of the main antenna, but well below the directivity of the main antenna. Thus, if the interference is entering through the sidelobes of the main antenna, and is large enough, $\log(|v_a|^2/|v_m|^2)$ would be greater than T and the logic block would issue a blank command for the range (or range/Doppler) cell being tested.

For a signal entering through the main lobe of the main antenna, $|v_m|^2$ would be much larger than $|v_a|^2$. This means $\log(|v_a|^2/|v_m|^2)$ would be less than T and the logic block would not issue a blank command. If the *interference* was entering through the main lobe of the main antenna, $\log(|v_a|^2/|v_m|^2)$ would also be less than T and the logic would not issue a blank command. Thus, the SLB would not be helpful in mitigating main lobe interference.

In a 1968 paper, Louis Maisel showed that, as might be expected, an SLB can affect both false alarm and detection probability [20]. In that paper, Maisel discussed how these probabilities were affected by the interaction between the threshold T and the relation between the auxiliary channel directivity and the sidelobe levels of the main antenna. An interesting observation from his paper is that the auxiliary channel directivity should be well above the sidelobe levels of the main antenna, but well below the maximum directivity of the main antenna. Maisel's analysis, where the detection was limited to the case of a single radar pulse with a Marcum, or Swerling 0, target fluctuation, was later expanded to account for arbitrary numbers of pulses integrated and additional target fluctuation models based on the gamma distribution [22, 23].

25.7 EXERCISES

1. Derive (25.9) from (25.8). As a hint, if $w = a + jb$, and $J = |e|^2$ [24],

$$\nabla J = \frac{1}{2}\left[\frac{\partial e}{\partial a}e^* + \frac{\partial e^*}{\partial a}e - j\left(\frac{\partial e}{\partial b}e^* + \frac{\partial e^*}{\partial b}e\right)\right] \qquad (25.70)$$

2. Derive (25.49), (25.53), (25.56), and (25.57).

3. Rewrite (25.57) and (25.58) in terms of P_s, R_I, P_{nm}, R_{an}, A_a, A_m, and A_a, without w_{opt}.

4. Repeat Example 2.

5. Repeat Example 3.

6. Repeat Example 4.

7. Extend Exercise 4 to generate a plot of SINR versus auxiliary antenna directivity relative to the main antenna directivity. Let the relative auxiliary antenna directivity vary from −10 to −30 dB relative to the peak directivity of the main antenna. Plot the ratio of the noise powers with and without the SLC [i.e., $(P_{nm} + w^H_{opt}R_{an}w_{opt})/P_{nm}$]. The results of this exercise will demonstrate why the rule of thumb is that the directivity of the auxiliary antenna should be well above the sidelobes of the main antenna.

8. Move the first interference source of Example 2 from the second sidelobe to the third sidelobe of the main antenna directivity pattern and repeat the example. As a note, the sign of $A_m(u,v)$ is negative on the third sidelobe.

9. Decrease the number of interference sources to one and repeat Example 2. Does the SLC still work?

10. Increase the number of interferences sources of Example 2 to three by adding a third interference source at $u_3 = \sin(-18°)$. Assign it a JNR of 40 dB. Does the SLC still work?

11. Extend Example 2 to accommodate three auxiliary channels and see if it rejects the three interferences of Exercise 10. Place the third auxiliary antenna at 6λ. Assume its directivity is the same as the other two auxiliary antennas.

References

[1] Applebaum, S. P., "Adaptive Arrays," *IEEE Transactions on Antennas and Propagation*, vol. 24, no. 5, September, 1976, pp. 585−598.

[2] Manolakis, D. G. et al., *Statistical and Adaptive Signal Processing: Spectral Estimation, Signal Modeling, Adaptive Filtering, and Array Processing*, Norwood, MA: Artech House, 2005.

[3] Barkat, M., *Signal Detection and Estimation*, 2nd ed., Norwood, MA: Artech House, 2005.

[4] Haykin, S., *Adaptive Filter Theory*, 3rd ed., Upper Saddle River, NJ: Prentice-Hall, 1996.

[5] Nitzberg, R., *Adaptive Signal Processing for Radar*, Norwood, MA: Artech House, 1992.

[6] Howells, P. W., "Intermediate Frequency Side-Lobe Canceller," U.S. Patent 3,202,990, Aug. 24, 1965.

[7]   Applebaum, S. P., "Steady-State and Transient Performance of the Sidelobe Canceller," Special Projects Laboratory, Syracuse Univ. Res. Corp., Syracuse, NY, Apr. 8, 1966. Available from DTIC as AD 373326.

[8]   Nitzberg, R., *Radar Signal Processing and Adaptive Systems*, Norwood, MA: Artech House, 1999.

[9]   Wiener, N., *Extrapolation, Interpolation, and Smoothing of Stationary Time Series*, New York: Wiley, 1949. Reprinted: Cambridge, MA: The M.I.T. Press, 1964.

[10] Farina, A., *Antenna-Based Signal Processing Techniques for Radar Systems*, Norwood, MA: Artech House, 1992.

[11] Papoulis, A., *Probability, Random Variables, and Stochastic Processes*, 3rd ed., New York: McGraw-Hill, 1991.

[12] Urkowitz, H., *Signal Theory and Random Processes*, Dedham, MA: Artech House, 1983.

[13] Guerci, J. R., *Cognitive Radar: The Knowledge-Aided Fully Adaptive Approach*, Norwood, MA: Artech House, 2010.

[14] Reed, I. S., et al., "Rapid Convergence Rate in Adaptive Arrays," *IEEE Trans. Aerosp. Electron. Syst.*, vol. 10, no. 6, Nov. 1974, pp. 853–863.

[15] Gerlach, K., "Adaptive Array Transient Sidelobe Levels and Remedies," *IEEE Trans. Aerosp. Electron. Syst.*, vol. 26, no. 3, May 1990, pp. 560–568.

[16] Fenn, A. J., *Adaptive Antennas and Phased Arrays for Radar and Communications*, Norwood, MA: Artech House, 2008.

[17] Widrow, B., et al., "Adaptive Antenna Systems," *Proc. IEEE*, vol. 55, no. 12, Dec. 1967, pp. 2143–2159.

[18] Lewis, B. L., F. F. Kretschmer, and W. W. Shelton, *Aspects of Radar Signal Processing*, Norwood, MA: Artech House, 1986.

[19] Barton, D. K., *Radar System Analysis and Modeling*, Norwood, MA: Artech House, 2005.

[20] Maisel, L. J., "Performance of Sidelobe Blanking Systems," *IEEE Trans. Aerosp. Electron. Syst.*, vol. 4, no. 2, Mar. 1968, pp. 174–180.

[21] Maisel, L. J., "Noise Cancellation Using Ratio Detection," IEEE Trans. Inf. Theory, vol. 14, no. 4, Jul. 1968, pp. 556–562.

[22] Shnidman, D. A., and S. S. Toumodge, "Sidelobe Blanking with Integration and Target Fluctuation," IEEE Trans. Aerosp. Electron. Syst., vol. 38, no. 3, Jul. 2002, pp. 1023–1037.

[23] Shnidman, D. A., and N. R. Shnidman, "Sidelobe Blanking with Expanded Models," *IEEE Trans. Aerosp. Electron. Syst.*, vol. 47, no. 2, Apr. 2011, pp. 790–805.

[24] Fischer, R. F. H., *Precoding and Signal Shaping for Digital Transmission*, New York: Wiley & Sons, 2002.

APPENDIX 25A: DERIVATION OF ϕ (25.40)

Figure 25A.1 contains a depiction of the geometry used to derive ϕ, the phase difference caused by the path length difference between the interference and the main and auxiliary antennas. r is the range from the center of the main antenna to the interference, and r_a is the range from the center of the auxiliary antenna to the interference. The center of the main antenna is located at the origin of a coordinate system, and the center of the auxiliary antenna is located at (x_a, y_a, z_a). We want to find $r_a - r$ as r approaches infinity (the far field condition). We can write

$$r = \sqrt{x_I^2 + y_I^2 + z_I^2} \tag{25A.1}$$

and

$$r_a = \sqrt{(x_I - x_a)^2 + (y_I - y_a)^2 + (z_I - z_a)^2} \tag{25A.2}$$

Manipulating (25A.2) gives

$$r_a = \sqrt{r^2 + x_a^2 + y_a^2 + z_a^2 - 2x_a x_I - 2y_a y_I - 2z_a z_I} \tag{25A.3}$$

recognizing that r^2 is much larger than the rest of the terms and using

$$\sqrt{1+x} \approx 1 + x/2 \quad \text{for } x \ll 1 \tag{25A.4}$$

results in

$$r_a = r + \frac{x_a^2 + y_a^2 + z_a^2}{2r} - \frac{x_a x_I}{r} - \frac{y_a y_I}{r} - \frac{z_a z_I}{r} \tag{25A.5}$$

or, since r is large relative to the numerator of the term after the plus sign, we can drop that term and write

$$r_a = r - \frac{x_a x_I}{r} - \frac{y_a y_I}{r} - \frac{z_a z_I}{r} \tag{25A.6}$$

With this we get

$$\Delta r = r_a - r = -\frac{x_a x_I}{r} - \frac{y_a y_I}{r} - \frac{z_a z_I}{r} \tag{25A.7}$$

We recognize that $u_I = x_I/r$ and $v_I = y_I/r$. Also

$$\frac{z_I}{r} = \frac{\sqrt{r^2 - x_I^2 - y_I^2}}{r} = \sqrt{1 - \frac{x_I^2}{r^2} - \frac{y_I^2}{r^2}} = \sqrt{1 - u_I^2 - v_I^2} \tag{25A.8}$$

Thus,

$$\Delta r = -x_a u_I - y_a v_I - z_a \sqrt{1 - u_I^2 - v_I^2} \tag{25A.9}$$

The phases due to range delay from the interference to the centers of the main and auxiliary antennas are

$$\phi_r = -2\pi f_c \tau_r = -2\pi r/\lambda \tag{25A.10}$$

and

$$\phi_{r_a} = -2\pi f_c \tau_{r_a} = -2\pi r_a/\lambda \tag{25A.11}$$

Thus, the difference of the two phases is

$$\phi = \phi_{r_a} - \phi_r = -2\pi\Delta r/\lambda = 2\pi\left(x_a u_I + y_a v_I + z_a\sqrt{1-u_I^2-v_I^2}\right)\Big/\lambda \qquad (25A.12)$$

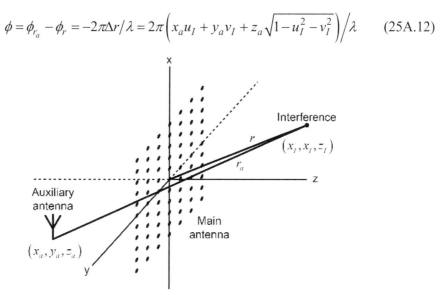

Figure 25A.1 Geometry for calculating ϕ.

Chapter 26

Advances in Radar

26.1 INTRODUCTION

As with many areas of science and engineering, the field of radar is advancing at an ever-increasing rate. This includes advances in both theory and hardware. Two of the newer theory areas are multiple-input, multiple-output (MIMO) radar [1, 2] and cognitive radar [1, 3–5]. However, other areas of theory being studied include advanced phased array system techniques [6, 7], advanced waveforms [4, 8], and advanced tracking algorithms [9–11]. On the hardware side, there have been significant advances in virtually all subsystems. Examples of these include direct digital synthesizers (DDSs), extremely quiet local oscillators, advances in transmit/receive (T/R) module technology, very low noise RF amplifiers, high-speed ADCs with large dynamic range, high-speed digital signal processors, and incredibly fast computers with massive memory.

26.2 MIMO RADAR

The concept of MIMO has been used in the wireless and cell phone industry for the past several years [12]. It provides a means of having several cell phones operate on the same frequency and uses orthogonal, or almost orthogonal, waveforms to separate the signals. Radar engineers are currently analyzing the application of this methodology to radars [1, 2, 13, 14].

Probably the simplest example of where MIMO might be applied to radars is in multistatic radar. As one example, suppose we have N widely spaced radars operating on the same frequency. We want to use the radars to perform tracking by the method of trilateration. This is illustrated in Figure 26.1 for the case of $N = 3$. Each radar can transmit a single beam and can form multiple receive beams over an angular sector. If Radar 2 illuminates a target and the return signal is in one of the multiple beams of the other two radars, all three radars will receive the signal from the target. If the control center knows the locations of the three radars, and knows which receive beam contains the target return, it can use this information to refine the estimate of target position via trilateration.

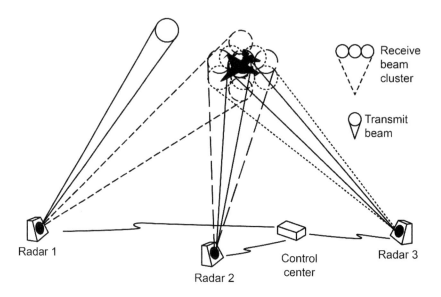

Figure 26.1 Tracking by trilateration.

If only one radar is transmitting, this is a fairly straightforward problem. However, if more than one radar illuminates the target, the receive functions of the radars need a means of knowing which radar transmitted the signal they received. This is where the concept of MIMO enters. If the three radars transmit orthogonal, or nearly orthogonal, waveforms, such as different PRN-coded pulses (see Chapter 11), the sources of the received signals would be known. With that knowledge, the control center can more precisely determine the target location through trilateration. This is but one example of a MIMO radar application. Others can be found in [2].

26.3 COGNITIVE RADAR

The general idea of a cognitive radar is that it can sense and adapt to its environment to improve its operation. In the limit, researchers talk about giving the radar "human" qualities, a laudable but questionable end objective.

Different authors view cognitive radar from different perspectives. For example, Haykin [5] approaches it from an adaptive tracking perspective, Guerci [3, 15] approaches it from a STAP perspective, and Pillai and his colleagues [4] approach it from a waveform selection perspective.

"Cognition" in radar has been around for a long time. A simple example is the sidelobe canceller. It senses the radar environment and adapts to it by adjusting the weights of the canceller algorithm to minimize the interference (see Chapter 25).

Another example is a frequency agile radar that performs clear channel search. Specifically, it interrogates its environment seeking frequency bands that contain the minimum interference. It then chooses its operating frequency to use one of the clear channels. An example of this would be in a sky wave, over-the-horizon radar that depends on clear channel search to find operating frequencies free of interference from other sources.

Other examples include

- Clutter maps that the radar uses to avoid clutter or to change processor characteristics to better mitigate the clutter based on intensity, Doppler characteristics, and/or spatial characteristics.
- Maneuver detectors that adjust track loop characteristics to accommodate target maneuvers.
- Interacting multiple model (IMM) trackers that change track filters depending on target types and/or kinematic characteristics.
- Constant false alarm rate (CFAR) algorithms that change detection thresholds based on the sensed noise or jamming environment.

These are but a few examples of "cognitive" radar techniques that have been in use for many years.

26.4 OTHER ADVANCEMENTS IN RADAR THEORY

Although MIMO and cognitive radar are the current "hot" radar research topics, there are other radar theory advancements that are being studied. Three examples of these are

- *Advanced phased array techniques:* There has been a good deal of research aimed at taking advantage of the ability to do digital beamforming in modern, active phased array radars. One example of this includes the adaptive nulling discussed Chapter 24. Other examples include advanced angle super-resolution techniques such as MUSIC or the examples mentioned in [7], such as adaptive-adaptive array processing or principal components processing.
- *Advanced waveforms:* Analysts are constantly developing new waveforms or adapting waveforms from other fields such as communications [16]. Examples of the former are recent finds in minimum peak sidelobe codes [17–20] and polyphase Barker codes [21, 22]. Examples of adaptations of communications waveforms are Costas coding, Huffman codes, and codes based on pseudorandom noise sequences [8]. Actually, the theories behind advanced waveforms have been studied for a long time [23–26]. However, it is because of the invention of the DDS and high-speed digital signal processors that they can now be more widely and easily used in radars. In fact, "advanced" waveforms have been used in radars in the past, although not to the extent now possible with DDSs and high-speed digital signal processing.
- *Advanced tracking algorithms:* Up until fairly recently, digital track algorithms consisted of g-h (α-β), g-h-k (α-β-γ) and low-order, or partitioned [27, 28], Kalman filters. The main reason for this was speed and memory limitations of digital computers. With the development of high-speed computers and advanced programming languages, researchers are investigating more advanced filters such as high-order Kalman filters, unscented Kalman filters, IMM filters, and particle filters [9–11, 29, 30]. The latter offer the potential of improved tracking

performance and also improved target detection through the use of a technique termed "track before detect" [31].

26.5 HARDWARE ADVANCEMENTS

Some of the more interesting advancements have taken place, and are still taking place, in the area of hardware. One of these is the DDS, or Direct Digital Synthesizer. This device is used to digitally generate waveforms and has almost unlimited waveform generation capability. This means that waveforms we could previously consider only in theoretical and simulation studies can now be easily generated in radar hardware. This includes waveforms such as nonlinear FM, PRN-coded pulses, minimum peak sidelobe waveforms, and the other types of waveforms discussed in Chapter 11 and elsewhere [4, 8]. The DDS, along with modern digital signal processing, will allow changing waveforms "on the fly" to contend with changes in the environment and/or to counter jamming. The DDS will also be an enabler of MIMO and cognitive radars.

DDSs are advancing to the point where they are suitable for agile LO frequency synthesis and can generate high-bandwidth, high-BT product waveforms. An example would be the AD9858, which is a 1-gigasample-per-second (GSPS) DDS [32]. This DDS is capable of generating a frequency-agile analog output sine wave at up to 400 MHz. The AD9858 also includes an automatic frequency sweeping feature, simplifying LFM generation for chirped radar. Another notable example is the AD9914, which is a 3.5 GSPS DDS with 12-bit digital-to-analog converter (DAC). It can generate frequencies up to 1.4 GHz and includes 12-bit amplitude scaling for fast amplitude hopping [33].

Another significant hardware advancement is the solid-state T/R module [34, 35]. From an operational perspective, T/R modules will improve radar reliability because they allow elimination of the single point of failure that exists with transmitters that use a single, high-power transmit tube. Also, since they are solid-state, they do not use high-voltage power supplies, which would enhance reliability and safety.

The newer generation of T/R modules that are based on gallium nitride (GaN) transistors have five times the power density compared to gallium arsenide (GaAs) transistors [36]. This increased power density requires fewer GaN transistors in parallel for a given output power, which lends itself to wideband operation. This is because the matching networks tend to be the limiting factor for bandwidth. Needing fewer GaNs requires simpler matching networks, which exhibit wider frequency response in general [36]. GaN transistors also offer the promise of allowing variable output power. This will allow transmit amplitude weighting to reduce antenna sidelobes. It may also lead to the ability to use amplitude weighting (in time) across the transmit waveform. Neither of these capabilities can be easily obtained with tube transmitters or T/R modules that use older technology such as GaAs or silicon. To operate at maximum efficiency, these types of transmitters must be operated at full power output (class C).

The ability to time weight could lead to the development of a new class of waveforms that take advantage of the ability to amplitude weight on transmit [25, 26]. It will be interesting to see where research in this area leads.

T/R modules will be a key element of MIMO and cognitive radar designs that are currently being analyzed. They will also be key elements in radars that use advanced beamforming or STAP algorithms. The reason for this is that radars that use STAP, MIMO, or cognitive ideas will need access to, and control of, individual antenna elements or, at a minimum, a fairly large number of subarrays.

Another transmitter (and receiver) component that has advanced significantly over the past few years is the local oscillator of the exciter. Only a few years ago, oscillators with phase noise sidebands of about −150 dBc/Hz, at frequencies far removed from the carrier, were difficult to find. Today, it is not uncommon to see oscillators with phase noise sideband levels of −170 dBc/Hz [37–41].

A key hardware element on the receive side is the ADC. Recent technology has pushed ADCs to sample rates in the hundreds of MHz to low GHz and dynamic ranges in the 16- to 20-bit range [42–46]. For example, the AD9680 from Analog Devices is a 14-bit ADC operating at 1 GSPS [47]. As another example, Texas Instruments offers ADCs for direct RF sampling that are 12 bit and operate at rates up to 3.6 GSPS (ADC12Dxx00RF family) [48–50]. Pushing sampling rates further (at the expense of ADC bits), ApisSys produces a 10-bit, 10-GSPS ADC on a 3U VPX board [51]. Because of this, digital receivers (see Chapter 22) are being studied and implemented in radars, taking advantage of direct IF or RF sampling. Further, the ADC is being moved closer to the RF front end of the radar. It would not be unreasonable to expect that in the near future, radars could use direct RF sampling and a completely digital receiver. However, such a capability will most likely introduce a new set of problems, such as maintaining the selectivity while retaining the capability of rapid frequency agility, the amount of data that must be moved through the receiver and processor, processor speed, and others.

A major limitation of moving the ADC closer to the antenna will be the effective noise figure of ADCs, which currently are on the order of 20 dB to 40 dB. If such an ADC were placed directly at the antenna, the receiver noise figure will be in this 20- to 40-dB range because it will set the noise figure of the ADC (see Chapter 4). Even if the ADC were moved to immediately after the LNA, the large noise figure of the ADC would have a significant impact on the overall receiver noise figure. For example, suppose we consider an LNA with a noise figure of 2 dB and a gain of 25 dB. If the noise figure of the ADC is 30 dB, the overall noise figure of the combined LNA and ADC will be

$$F_{LNA+ADC} = F_{LNA} + \frac{F_{ADC}-1}{G_{LNA}} = 6.8 \text{ dB} \qquad (26.1)$$

which is significantly larger than the 2-dB noise figure of the LNA. Until ADC noise figure values are reduced to more reasonable values, the ADC will need to reside further down the receiver chain so as to minimize its impact on overall receiver noise figure.

This ability to digitize the received signal early in the receiver chain, coupled with high-speed and small footprint digital signal processors, will be instrumental in MIMO and cognitive radars currently being considered. The reason for this is that such radars will need to process data from a very large number of T/R modules (or subarrays) [36]. This will require a very large number of receivers, something that will be difficult to achieve without small, digital devices that, individually, consume small amounts of power. Also, the advantages of digitizing at the module level rather than at the subarray level may not, at least in the near term, be sufficient to justify the cost. It will be interesting to see how the technology and techniques progress in this area.

Two other key hardware advancements are small, high-speed digital signal processors and computers. These are currently key elements in advanced radar signal processors and will become more important enablers in future radar concepts.

26.6 CONCLUSION

Even though this book focuses on basic radar analysis, when we wrote it, we had in mind the future theory and hardware advances mentioned in this chapter. This was part of our motivation for including the detailed mathematics and some of the advanced topics discussed in the various chapters. It is our belief that a thorough understanding of the basic radar theory presented in this book is critical to implementing advanced theories and making use of state-of-the art hardware.

References

[1] Melvin, W. L., and J. A. Scheer, eds., *Principles of Modern Radar, Vol. II: Advanced Techniques*, Edison, NJ: Scitech, 2013.

[2] Li, J., and P. Stoica, eds., *MIMO Radar Signal Processing*, New York: Wiley & Sons, 2009.

[3] Guerci, J. R., *Cognitive Radar: The Knowledge–Aided Fully Adaptive Approach*, Norwood, MA: Artech House, 2010.

[4] Pillai, U., et al., *Waveform Diversity, Theory and Applications*, New York: McGraw-Hill, 2011.

[5] Haykin, S., "Cognitive Radar: A Way of the Future," *IEEE Signal Proc. Mag.*, vol. 23, no. 1, Jan. 2006, pp. 30–40.

[6] Wirth, W. D., *Radar Techniques Using Array Antennas*, London: IEE Press, 2001.

[7] Brookner, E., "Active-Phased-Arrays and Digital Beamforming: Amazing Breakthroughs and Future Trends," *2008 IEEE Radar Conf.*, Rome, Italy, May 26–30, 2008, Session TU-S2.1.

[8] Levanon, N., and E. Mozeson, *Radar Signals*, New York: Wiley-Interscience, 2004.

[9] Julier, S. J., and J. K. Uhlmann, "Unscented Filtering and Nonlinear Estimation," *Proc. IEEE*, vol. 92, no. 3, Mar. 2004, pp. 401–422.

[10] Herman, S., and P. Moulin, "A Particle Filtering Approach to FM-Band Passive Radar Tracking and Automatic Target Recognition," *IEEE Aerosp. Conf. Proc.*, vol. 4, 2002, pp. 1789–1808.

[11] Mazor, E., et al., "Interacting Multiple Model Methods in Target Tracking: A Survey," *IEEE Trans. Aerosp. Electron. Syst.*, vol. 34, no. 1, Jan. 1998, pp. 103–123.

[12] Jankiraman, M., *Space-Time Codes and MIMO Systems*, Norwood, MA: Artech House, 2004.

[13] Sen, S., and A. Nehorai, "OFDM MIMO Radar with Mutual-Information Waveform Design for Low-Grazing Angle Tracking," *IEEE Trans. Signal Processing*, vol. 58, no. 6, Jun. 2010, pp. 3152–3162.

[14] Sen, S., and A. Nehorai, "OFDM MIMO Radar for Low-Grazing Angle Tracking," *2009 Conf. Rec. 43rd Asilomar Conf. Signals, Syst., and Comput.*, Nov. 1–4, 2009, pp. 125–129.

[15] Guerci, J. R., *Space-Time Adaptive Processing for Radar*, 2nd ed., Norwood, MA: Artech House, 2014.

[16] Mow, W. H., K.-L. Du and W. H. Wu, "New Evolutionary Search for Long Low Autocorrelation Binary Sequences," *IEEE Trans Aerosp. Electron. Syst.*, vol. 51, no. 1, Jan. 2015, pp. 290–303.

[17] Coxson, G., and J. Russo, "Efficient Exhaustive Search for Optimal-Peak-Sidelobe Binary Codes," *IEEE Trans. Aerosp. Electron. Syst.*, vol. 41, no. 1, Jan. 2005, pp. 302–308.

[18] Nunn, C. J., and G. E. Coxson, "Best-Known Autocorrelation Peak Sidelobe Levels for Binary Codes of Length 71 to 105," *IEEE Trans. Aerosp. Electron. Syst.*, vol. 44, no. 1, Jan. 2008, pp. 392–395.

[19] Leukhin, A. N., and E. N. Potekhin, "Binary Sequences with Minimum Peak Sidelobe Level up to Length 68," *Int. Workshop on Coding and Cryptography,* Bergen, Norway, Apr. 15–19, 2013.

[20] Leukhin, A. N., and E. N. Potekhin, "Optimal Peak Sidelobe Level Sequences Up to Length 74," *2013 European Radar Conf. (EuRAD),* Nuremberg, Germany, Oct. 9–11, 2013, pp. 495–498.

[21] Gabbay, S. "Properties of Even-Length Barker Codes and Specific Polyphase Codes with Barker Type Autocorrelation Functions," Naval Research Laboratory, Washington, DC, Rep. 8586, Jul. 12, 1982. Available from DTIC as AD-A117415.

[22] Nunn, C. J., and G. E. Coxson, "Polyphase Pulse Compression Codes with Optimal Peak and Integrated Sidelobes," *IEEE Trans. Aerosp. Electron. Syst.*, vol. 45, no. 2, Apr. 2009, pp. 775–781.

[23] Rihaczek, A. W., *Principles of High-Resolution Radar,* New York: McGraw-Hill, 1969. Reprinted: Norwood, MA: Artech House, 1995.

[24] DeLong, D., and E. M. Hofstetter, "On the Design of Optimum Radar Waveforms for Clutter Rejection," *IEEE Trans. Inf. Theory,* vol. 13, no. 3, Jul. 1967, pp. 454–463.

[25] Rummler, W. D., "Clutter Suppression by Complex Weighting of Coherent Pulse Trains," *IEEE Trans. Aerosp. Electron. Syst.*, vol. 2, no. 6, Nov. 1966, pp. 689–699.

[26] Rummler, W. D., "A Technique for Improving the Clutter Performance of Coherent Pulse Train Signals," *IEEE Trans. Aerosp. Electron. Syst.*, vol. 3, no. 6, Nov. 1967, pp. 898–906.

[27] Johnson, G., "Choice of Coordinates and Computational Difficulty," *IEEE Trans. Automatic Control,* vol. 19, no. 1, Feb. 1974, pp. 77–78.

[28] Mehra, R. K., "A Comparison of Several Nonlinear Filters for Reentry Vehicle Tracking," *IEEE Trans. Automatic Control,* vol. 16, no. 4, Aug. 1971, pp. 307–319.

[29] Ye, B., and Y. Zhang, "Improved FPGA Implementation of Particle Filter for Radar Tracking Applications," *2nd Asian-Pacific Conf. Synthetic Aperture Radar (APSAR 2009),* Xian, Shanxi, Oct. 26–30, 2009, pp. 943–946.

[30] Budge, M. C., Jr., "A Variable Order IMM Algorithm," *Proc. Huntsville Simulation Conf.,* Huntsville, AL, 2005.

[31] Boers, Y., et al., "A Track Before Detect Algorithm for Tracking Extended Targets," *IEE Proc. Radar, Sonar and Navig.,* vol. 153, no. 4, Aug. 2006, pp. 345–351.

[32] Analog Devices, "AD9858 1 GSPS Direct Digital Synthesizer," 2003. www.analog.com.

[33] Analog Devices, "AD9914, 3.5 GSPS Direct Digital Synthesizer with 12-Bit DAC," 2014. http://www.analog.com.

[34] Manqing, W., "Digital Array Radar: Technology and Trends," *2011 IEEE CIE Int. Conf. Radar,* vol. 1, Chengdu, China, Oct. 24–27, 2011, pp. 1–4.

[35] Li, T., and X.-G. Wang, "Development of Wideband Digital Array Radar," *2013 10th Int. Comput. Conf. Wavelet Active Media Tech. Inf. Process. (ICCWAMTIP),* Chengdu, China, Dec. 17–19, 2013, pp. 286–289.

[36] Farina, A., et al., "AESA Radar — Pan-Domain Multi-Function Capabilities for Future Systems," *2013 IEEE Int. Symp. Phased Array Syst. & Tech.,* Waltham, MA, Oct. 15–18, 2013, pp. 4–11.

[37] Wenzel Associates, Inc., "Low Noise Crystal Oscillators > Golden MXO (PLO w/ Dividers)," www.wenzel.com.

[38] Wenzel Associates, Inc., "Low Noise Crystal Oscillators > Sorcerer II," www.wenzel.com.

[39] Boroditsky, R., and J. Gomez, "Ultra Low Phase Noise 1 GHz OCXO," *IEEE Int. Frequency Control Symp., 2007 Joint with the 21st European Frequency and Time Forum,* Geneva, Switzerland, May 29–Jun. 1, 2007, pp. 250–253.

[40] Hoover, L., H. Griffith, and K. DeVries, "Low Noise X-band Exciter Using a Sapphire Loaded Cavity Oscillator," *2008 IEEE Int. Frequency Control Symp. (FCS),* Honolulu, HI, May 19–21, 2008, pp. 309–311.

[41] Poddar, A. K., and U. L. Rohde, "The Pursuit for Low Cost and Low Phase Noise Synthesized Signal Sources: Theory & Optimization," *2014 IEEE Int. Frequency Control Symp. (FCS),* Taipei, Taiwan, May 19–22, 2014, pp. 1–9.

[42] Jha, A. R., "High Performance Analog-to-Digital Converters (ADCs) for Signal Processing," *2nd Int. Conf. Microwave and Millimeter Wave Technology Proc.*, Beijing, China, Sept. 14–16, 2000, pp. 32–35.

[43] Smith, A., "High Sensitivity Receiver Applications Benefit from Unique Features in 16-Bit 130Msps ADC," October 2005. cds.linear.com/docs/en/design-note/DSOL44.pdf.

[44] Linear Technology Corporation, "LTC2207/LTC2206 16-Bit, 105Msps/80Msps ADCs," 2006. cds.linear.com/docs/en/datasheet/22076fc.pdf.

[45] Linear Technology Corporation, "LTC2338-18, 18-Bit, 1Msps, ±10.24V True Bipolar SAR ADC," 2013. cds.linear.com/docs/en/datasheet/233818f.pdf.

[46] Linear Technology Corporation, "LTC2378-20, 20-Bit, 1Msps, Low Power SAR ADC with 0.5ppm INL," 2013. cds.linear.com/docs/en/datasheet/237820fa.pdf.

[47] Analog Devices, "AD9680 Data Sheet: 14-Bit, 1 GSPS JESD204B, Dual Analog-to-Digital Converter," 2014. www.analog.com.

[48] Texas Instruments, "RF-Sampling and GSPS ADCs, Breakthrough ADCs Revolutionize Radio Architectures," 2012. www.ti.com/gigadc.

[49] Texas Instruments, "ADC12D1800, 12-Bit, Single 3.6 GSPS Ultra High-Speed, ADC," Jan. 2014. http://www.ti.com/lit/ds/symlink/adc12d1800.pdf.

[50] Texas Instruments, "ADC12J4000 12-Bit 4 GSPS ADC with Integrated DDC," Sept. 2014. http://www.ti.com/lit/ds/symlink/adc12j4000.pdf.

[51] ApisSys, "AV101, 10-bit 10 GSPS ADC and Signal Processing 3U VPX Board," Sept. 2013. www.apissys.com/pdf/AV101.pdf.

Appendix A

Data Windowing Functions

Table A.1 contains continuous and discrete time forms of some common window functions whose uses include range, Doppler, and antenna sidelobe reduction [1–19]. The discrete time window functions are in causal symmetric form (identical endpoints), which is generally used for FIR filter design. Periodic forms, characterized by a missing (implied) endpoint to accommodate periodic extension, are generally used for spectral estimation (divide by N versus $N-1$).

Table A.1
Window Functions[1]

Window	Continuous, $w(t)$ $0 \leq t \leq T$	Discrete, $w[n]$ $0 \leq n \leq N-1$
Rectangular (Dirichlet)	1	1
Gaussian[a] (Weierstrass)	$\exp\left\{-\dfrac{1}{2}\left[\alpha\left(\dfrac{2t}{T}-1\right)\right]^2\right\}$	$\exp\left\{-\dfrac{1}{2}\left[\alpha\left(\dfrac{2n}{N-1}-1\right)\right]^2\right\}$
$\cos^\kappa(x)$	$\cos^\kappa\left(\dfrac{\pi t}{T}-\dfrac{\pi}{2}\right)=\sin^\kappa\left(\dfrac{\pi t}{T}\right)$	$\cos^\kappa\left(\dfrac{\pi n}{N-1}-\dfrac{\pi}{2}\right)=\sin^\kappa\left(\dfrac{\pi n}{N-1}\right)$
$\cos^\kappa(x)$ on a pedestal ($0\leq\alpha\leq1.0$)	$\alpha+(1-\alpha)\cos^\kappa\left(\dfrac{\pi t}{T}-\dfrac{\pi}{2}\right)$	$\alpha+(1-\alpha)\cos^\kappa\left(\dfrac{\pi n}{N-1}-\dfrac{\pi}{2}\right)$
Generalized Hamming ($0.5\leq\alpha\leq1.0$)	$\alpha-(1-\alpha)\cos\left(\dfrac{2\pi t}{T}\right)$	$\alpha-(1-\alpha)\cos\left(\dfrac{2\pi n}{N-1}\right)$
Hamming	$0.54-0.46\cos\left(\dfrac{2\pi t}{T}\right)$	$0.54-0.46\cos\left(\dfrac{2\pi n}{N-1}\right)$

---

[1] The terms windowing and weighting, are sometimes associated with time domain and frequency domain application, respectively [18, p. 192].

Table A.1 (continued)

Window	Continuous, $w(t)$ $0 \leq t \leq T$	Discrete, $w[n]$ $0 \leq n \leq N-1$
Hann [$\cos^2(x)$, raised cosine, Hanning, von Hann]	$0.5 - 0.5\cos\left(\dfrac{2\pi t}{T}\right)$ $= \sin^2\left(\dfrac{\pi t}{T}\right)$	$0.5 - 0.5\cos\left(\dfrac{2\pi n}{N-1}\right)$ $= \sin^2\left(\dfrac{\pi n}{N-1}\right)$
Kaiser-Bessel[b]	$\dfrac{I_0\left[\pi\alpha\sqrt{1-\left(\dfrac{2t}{T}-1\right)^2}\right]}{I_0(\pi\alpha)}$	$\dfrac{I_0\left[\pi\alpha\sqrt{1-\left(\dfrac{2n}{N-1}-1\right)^2}\right]}{I_0(\pi\alpha)}$
Blackman	$0.42 - 0.50\cos\left(\dfrac{2\pi t}{T}\right)$ $+0.08\cos\left(\dfrac{4\pi t}{T}\right)$	$0.42 - 0.50\cos\left(\dfrac{2\pi n}{N-1}\right)$ $+0.08\cos\left(\dfrac{4\pi n}{N-1}\right)$
Exact Blackman	$7938/18608$ $-\dfrac{9240}{18608}\cos(2\pi t/T)$ $+\dfrac{1430}{18608}\cos(4\pi t/T)$	$7938/18608$ $-\dfrac{9240}{18608}\cos\left[2\pi n/(N-1)\right]$ $+\dfrac{1430}{18608}\cos\left[4\pi n/(N-1)\right]$
Triangular (Bartlett, Fejer)	$1 - \left|\dfrac{2t}{T} - 1\right|$	$1 - \left|\dfrac{2n}{N-1} - 1\right|$
Minimum Blackman-Harris (3-term)	0.42323 $-0.49755\cos(2\pi t/T)$ $+0.07922\cos(4\pi t/T)$	0.42323 $-0.49755\cos\left[2\pi n/(N-1)\right]$ $+0.07922\cos\left[4\pi n/(N-1)\right]$
Minimum Blackman-Harris (4-term)	0.35875 $-0.48829\cos(2\pi t/T)$ $+0.14128\cos(4\pi t/T)$ $-0.01168\cos(6\pi t/T)$	0.35875 $-0.48829\cos\left[2\pi n/(N-1)\right]$ $+0.14128\cos\left[4\pi n/(N-1)\right]$ $-0.01168\cos\left[6\pi n/(N-1)\right]$
Minimum Blackman-Nuttall	0.3635819 $-0.4891775\cos(2\pi t/T)$ $+0.1365995\cos(4\pi t/T)$ $-0.0106411\cos(6\pi t/T)$	0.3635819 $-0.4891775\cos\left[2\pi n/(N-1)\right]$ $+0.1365995\cos\left[4\pi n/(N-1)\right]$ $-0.0106411\cos\left[6\pi n/(N-1)\right]$

Table A.1 (continued)

Window	Continuous, $w(t)$ $0 \leq t \leq T$	Discrete, $w[n]$ $0 \leq n \leq N-1$
Flat-top	0.21557895 $-0.41663158\cos(2\pi t/T)$ $+0.277263158\cos(4\pi t/T)$ $-0.083578947\cos(6\pi t/T)$ $+0.006947368\cos(8\pi t/T)$	0.21557895 $-0.41663158\cos[2\pi n/(N-1)]$ $+0.277263158\cos[4\pi n/(N-1)]$ $-0.083578947\cos[6\pi n/(N-1)]$ $+0.006947368\cos[8\pi n/(N-1)]$
Bartlett-Hann	$0.62 - 0.48\left\|\dfrac{t}{T} - \dfrac{1}{2}\right\|$ $-0.38\cos\left(\dfrac{2\pi t}{T}\right)$	$0.62 - 0.48\left\|\dfrac{n}{N-1} - \dfrac{1}{2}\right\|$ $-0.38\cos\left(\dfrac{2\pi n}{N-1}\right)$
Nuttall	0.355768 $-0.487396\cos(2\pi t/T)$ $+0.144232\cos(4\pi t/T)$ $-0.012604\cos(6\pi t/T)$	0.355768 $-0.487396\cos[2\pi n/(N-1)]$ $+0.144232\cos[4\pi n/(N-1)]$ $-0.012604\cos[6\pi n/(N-1)]$
Lanczos	$\text{sinc}\left(\dfrac{2t}{T} - 1\right)$	$\text{sinc}\left(\dfrac{2n}{N-1} - 1\right)$
Truncated Taylor ($0 \leq \alpha \leq 1.0$)	$\dfrac{1+\alpha}{2} - \dfrac{1-\alpha}{2}\cos\left(\dfrac{2\pi t}{T}\right)$	$\dfrac{1+\alpha}{2} - \dfrac{1-\alpha}{2}\cos\left(\dfrac{2\pi n}{N-1}\right)$
Parabolic (Riesz, Bochner)	$1 - \left(\dfrac{2t}{T} - 1\right)^2$	$1 - \left(\dfrac{2n}{N-1} - 1\right)^2$
Bohman	$\left(1 - \left\|\dfrac{2t}{T} - 1\right\|\right)\cos\left(\pi\left\|\dfrac{2t}{T} - 1\right\|\right)$ $+ \dfrac{1}{\pi}\sin\left(\pi\left\|\dfrac{2t}{T} - 1\right\|\right)$	$\left(1 - \left\|\dfrac{2n}{N-1} - 1\right\|\right)\cos\left(\pi\left\|\dfrac{2n}{N-1} - 1\right\|\right)$ $+ \dfrac{1}{\pi}\sin\left(\pi\left\|\dfrac{2n}{N-1} - 1\right\|\right)$

[a] The parameter α is inversely proportional to sidelobe level. Values for α of 2.5 to 3.5 are typical.

[b] I_0 is the zero-order modified Bessel function of the first kind. Sometimes $\beta = \pi\alpha$ is used in the expression. The parameter β is inversely proportional to sidelobe level. Values for β of 2.0π to 3.5π are representative.

References

[1] Kunt, M., *Digital Signal Processing,* Norwood, MA: Artech House, 1986.

[2] Tsui, J., *Digital Microwave Receivers: Theory and Concepts*, Norwood, MA: Artech House, 1989.

[3] Tsui, J., *Digital Techniques for Wideband Receivers,* 2nd ed., Norwood,, MA: Artech House, 2001.

[4] Harris, F. J., "On the Use of Windows for Harmonic Analysis with the Discrete Fourier Transform," *Proc. IEEE,* vol. 66, no. 1, Jan. 1978, pp. 51–83.

[5] Nuttall, A. H., "Some Windows with Very Good Sidelobe Behavior," *IEEE Trans. Acoust., Speech, Signal Process.,* vol. 29, no. 1, Feb. 1981, pp. 84–91.

[6] Cook, C. E., and M. Bernfeld, *Radar Signals: An Introduction to Theory and Application,* New York: Academic Press, 1967. Reprinted: Norwood, MA: Artech House, 1993.

[7] John, D. G. M., and G. Proakis, *Digital Signal Processing, Principles, Algorithms, and Applications*, 2nd ed., New York: Macmillan, 1992.

[8] Lyons, R. G., *Understanding Digital Signal Processing,* 3rd ed., Englewood Cliffs, NJ: Prentice Hall, 2011.

[9] Oppenheim, A. V., ed., *Applications of Digital Signal Processing,* Englewood Cliffs, NJ: Prentice Hall, 1978.

[10] Rabiner, L. R., and B. Gold, *Theory and Application of Digital Signal Processing,* Englewood Cliffs, NJ: Prentice Hall, 1975.

[11] Schaefer, R. T., R. W. Schaefer, and R. M. Mersereau, "Digital Signal Processing for Doppler Radar Signals," *Proc. 1979 IEEE Int. Conf. on Acoust., Speech, Signal Process.,* Washington, DC: 1979, pp. 170–173.

[12] Blackman, R. B., and J. W. Tukey, *The Measurement of Power Spectra from the Point of View of Communications Engineering*, New York: Dover, 1958.

[13] Harris, F. J., "Windows, Harmonic Analysis, and the Discrete Fourier Transform," Department of the Navy, Undersea Surveillance Department, Naval Undersea Center, San Diego, CA, Rep. No. NUC TP 532, Sept. 1976.

[14] Poularikas, A. D., *Handbook of Formulas and Tables for Signal Processing*, Boca Raton, FL: CRC Press, 1999.

[15] Rapuano, S., and F. J. Harris, "An Introduction to FFT and Time Domain Windows," *IEEE Instrum. Meas. Mag.,* vol. 10, no. 6, Dec. 2007, pp. 32–44.

[16] Hamming, R. W., *Digital Filters*, 2nd ed., Englewood Cliffs, NJ: Prentice-Hall, 1983.

[17] Kuo, F. F. and J. F. Kaiser, (Eds.), *System Analysis by Digital Computer*, New York: John Wiley & Sons, 1966.

[18] Elliot, D. F. and K. R. Rao, *Fast Transforms: Algorithms, Analysis, Applications*, New York: Academic Press, 1982.

[19] Parzen, E., "Mathematical Considerations in the Estimation of Spectra," *Technometrics,* vol. 3, 1961, pp. 167–190.

Acronyms and Abbreviations

°C	Degrees Celsius
°F	Degrees Fahrenheit
3-D	Three-Dimensional
AAF	Anti-Aliasing Filter
AC	Alternating Current
ADC	Analog-to-Digital Converter
AESS	Aerospace and Electronic Systems Society
AGC	Automatic Gain Control
AIEE	American Institute of Electrical Engineers
ARSR	Air Route Surveillance Radar
AWGN	Additive White Gaussian Noise
BJT	Bipolar Junction Transistor
BPF	Bandpass Filter
BT	Time-Bandwidth
BWA	Backward Wave Amplifier
CA	Cell Averaging
CFAR	Constant False Alarm Rate
CL	Conversion Loss
CNR	Clutter-to-Noise Ratio
COHO	COHerent Oscillator
CPD	Cyclotron Protective Device
CPI	Coherent Processing Interval
CR	Cancellation Ratio
CRPL	Central Radio Propagation Laboratory
CW	Continuous Wave
CWESA	Cyclotron Wave Electrostatic Amplifier
dB	Decibel
dB(m)	Decibel Relative to 1 Meter
dB(W-s)	Decibel Relative to 1 Watt-Second

dB/km	Decibels Per Kilometer
dB/m	Decibels Per Meter
dBFS	Decibel Below Full Scale
dBi	Antenna Directivity (Gain) Relative to the Directivity of an Isotropic Antenna
dBm	Power Level Relative to 1 Milliwatt
dBsm	Area in Square Meters Relative to 1 m^2
dBV	Voltage Level Relative to 1 Volt Root Mean Square
dBW	Power Level Relative to 1 Watt
DC	Direct Current
DDC	Digital Downconverter
DDS	Direct Digital Synthesis (or Synthesizer)
deg	Degree
DFT	Discrete-time Fourier Transform
DOF	Degrees of Freedom
EA	Electronic Attack
ECM	Electronic Countermeasure
E-field	Electric Field
EIA	Electronic Industries Association
ERP	Effective Radiated Power
ESA	Electrostatic Amplifier
ESCA	Electrostatics Combined Amplifier
FFT	Fast Fourier Transformer
FIR	Finite Impulse Response
FM	Frequency Modulation
FMCW	Frequency Modulated Continuous Wave
FSR	Full-Scale Range
ft	Foot
GaAs	Gallium Arsenide
GaN	Gallium Nitride
GHz	Gigahertz
GMTI	Ground Moving Target Indication
GO	Greatest Of
GPS	Global Positioning System
GSPS	Giga Sample Per Second
HF	High Frequency
HPF	Highpass Filter
Hz	Hertz
IEEE	Institute of Electrical and Electronics Engineers

IF	Intermediate Frequency
IFFT	Inverse FFT
IIR	Infinite Impulse Response
IL	Insertion Loss
IMD	Intermodulation Distortion
IMM	Interacting Multiple Model
IRE	Institute of Radio Engineers
IRF	Image Reject Filter
J/J	Joules per Joule
JAN	Joint Army Navy
JNR	Jammer-to-Noise Ratio
K	Kelvin
kft	Kilofoot
kHz	Kilohertz
km	Kilometer
kW	Kilowatt
LFM	Linear Frequency Modulation
LMS	Least Mean-Square
LNA	Low Noise Amplifier
LO	Local Oscillator
LOS	Line of Sight
LPF	Lowpass Filter
lsb	Least Significant Bit
LSR	Linear Shift Register
LTI	Linear Time Invariant
m	Meter
m/s	Meter Per Second
m/μs	Meter Per Microsecond
m^2	Square Meter
MDS	Minimum Detectable Signal
MDS	Minimum Discernable Signal
MF	Matched Filter
MHz	Megahertz
mi	Mile
MIMO	Multiple-Input, Multiple-Output
MIT	Massachusetts Institute of Technology
MKS	Meter, Kilogram, Second
mm	Millimeter Wave
mm	Millimeter

mm/hr	Millimeter Per Hour
mph	Mile Per Hour
MSPS	Mega Samples Per Second
MTD	Moving Target Detector
MTI	Moving Target Indicator
MUSIC	Multiple Signal Classification
NAGC	Noise Automatic Gain Control
$\bar{n}$	Number of Constant Level Sidelobes
NCO	Numerically Controlled Oscillator
NLFM	Nonlinear FM
nmi	Nautical Mile
Np/m	Neper per meter
ns	Nanosecond
PPI	Plan Position Indicator
PRF	Pulse Repetition Frequency
PRI	Pulse Repetition Interval
PRN	Pseudo Random Noise
rad^2	Steradian
Radar	RAdio Detection And Ranging
RC	Resistor Capacitor
RCM	Range Cell Migration
RCMC	RCM Correction
RCS	Radar Cross Section
RF	Radio Frequency
RGPO	Range-Gate Pull Off
RLC	Resistor Inductor Capacitor
rms	Root Mean Square
rpm	Revolutions Per Minute
RRE	Radar Range Equation
Rx	Receive
s	Second
SALT	Strategic Arms Limitation Talk
SAP	Spatial Adaptive Processing
SAR	Synthetic Aperture Radar
SAW	Surface Acoustic Wave
SCR	Signal-to-Clutter Ratio
SINR	Signal-to-Interference-Plus-Noise Ratio
SIR	Signal-to-Interference Power Ratio
SLB	Sidelobe Blanking

SLC	Sidelobe Cancellation
SMI	Sample Matrix Inversion
SNR	Signal-to-Noise Ratio
SO	Smallest Of
SONAR	SOund Navigation And Ranging
STALO	STAble Local Oscillator
STAP	Space-Time Adaptive Processing
STC	Sensitivity Time Control
T/R and TR	Transmit/Receive
TNR	Threshold-to Noise Ratio
TSS	Tangential Sensitivity
TWT	Traveling Wave Tube
Tx	Transmit
UAV	Unmanned Aerial Vehicle
UHF	Ultra High Frequency
USAAF	United States Army Air Forces
VHF	Very High Frequency
W	Watt
w.r.t.	With Respect To
W/Hz	Watt Per Hertz
W/m^2	Watt Per Square Meter
W/W	Watt Per Watt
WG	Waveguide
W-m^2	Watt-Meter-Square
W-s	Watt-Second
WSCS	Wide-Sense Cyclostationary
WSS	Wide-Sense Stationary
ZOH	Zero-Order Hold
μs	Microsecond
℧/m	Mho Per Meter
Ω	ohm

About the Authors

Dr. Mervin C. Budge

Mervin Budge, Ph.D., is chief scientist of Dynetics Inc., where he is responsible for overall technical quality. He also conducts research in the area of radars and air defense systems, with emphasis on foreign radars and air defense systems. In this role, he conducts analysis and design of advanced signal processing and antenna methodologies and intelligence assessments of foreign radars and integrated air defense systems. This includes performance analyses and specific radar designs.

Dr. Budge has served as an adjunct professor at the University of Alabama in Huntsville (UAH) since 1973 and previously served as a part-time professor at the Southeastern Institute of Technology, an instructor at Texas A&M University, and both a graduate and undergraduate teaching assistant at the University of Louisiana in Lafayette. He currently teaches courses in radar, radar tracking, and Kalman filters at UAH.

He holds a Ph.D. in electrical engineering from Texas A&M University and master's and bachelor's degrees in electrical engineering from the University of Louisiana in Lafayette. Dr. Budge's graduate research included "Prediction of Clear Air Turbulence," sponsored by NASA Marshall Space Flight Center; "Space Shuttle Aeroelastic Stability and Control Properties," sponsored by NASA Johnson Space Center; "Parameter Identification for Linear Systems," a Themis Grant; "Electric Power System Control," sponsored by Philco-Ford; and "Analytical Techniques for Predicting Flow Fields in Oil Wells." His graduate thesis and dissertational were in the areas of the effect of interaction in multivariable control systems and the design of recursive and nonrecursive digital filters. He has authored more than three dozen publications.

Dr. Budge's honors include Outstanding Educator Award from the Huntsville Section of IEEE; the Dynetics R. Duane Hays Award for Technical Excellence; the Outstanding Engineer Award from the Huntsville Section IEEE; Outstanding Ph.D. Student, Electrical Engineering Department, Texas A&M University; and Outstanding Ph.D. Student, Texas A&M University. He holds memberships in Eta Kappa Nu, Tau Beta Pi, Sigma Xi, Phi Kappa Phi, the Institute of Electrical and Electronic Engineers, Phi Mu Alpha Sinfonia, and Kappa Delta Pi.

Shawn R. German

Shawn R. German is a senior principal engineer at Dynetics Inc., where he has been employed since 1995. He is responsible for performing radar system design and analysis. His current areas of research include radar receivers, advanced signal processing methodologies, CFAR processing, detection theory and AESAs. He is responsible for generating assessments of ground based and airborne radars, seekers, and integrated air defense systems, including performance analyses and specific radar designs. Mr. German has served as a subject matter expert in a number of radar disciplines including: system test and measurement, RF and IF system testing, and performance analysis. He has served as a consultant on various radar related technologies and theories.

Mr. German's expertise includes the following areas: radar system repair and alignment, radar system and subsystem performance analysis, environment and target modeling, waveform analysis and design, signal processor analysis, radar system design, radar subsystem design, signal processor hardware prototyping, development of mathematical models from measured data, radar system and subsystem simulation and emulation, radar ECM and ECCM, and phased array and AESA theory.

Mr. German is a senior member of the Institute of Electrical and Electronic Engineers and holds memberships in Eta Kappa Nu, Tau Beta Pi, Phi Kappa Phi, and the Association of Old Crows. He holds bachelor's and master's degrees in electrical engineering from Mississippi State University and an associate's degree in electronics technology from the Mississippi Gulf Coast Community College. He is completing a Ph.D. in electrical engineering from the University of Alabama in Huntsville (UAH). His undergraduate research in "Error Control Coding in Satellite Communication" was sponsored by NASA.

Mr. German has served as a graduate teaching assistant in the Electrical Engineering Department at Mississippi State University. He currently teaches internal classes at Dynetics covering topics such as general radar theory, detection theory, digital receivers, filter theory, and analog/digital signal processing. He has coauthored several publications.

Index

1 dB gain compression point. *See* compression point, 1 dB
1-D scanning, 93
2-D scanning, 93
3-dB points, 26
absorption coefficient for oxygen, 120
absorption coefficient for water vapor, 122
active electronically steered arrays, 411
AD6620, 626
adaptive detection threshold, 168
adaptive nulling
   example, 724
adaptivity, 732, 743
ADC. *See* analog-to-digital converter
ADC noise figure, 783
advanced phased array techniques, 779, 781
advanced tracking algorithms, 781
advanced waveforms, 779
AESA, 14, 411
   noise figure, 440
AGC. *See* automatic gain control
airborne STAP
   example, 739
Alexander Graham Bell, 11
Alfred Clebsch, 45
Alfred de la Marche, 456
Alfred G. Emslie, 456
ambiguity
   range, 5
ambiguity function, 271, 286
   Barker code, 298
   cross, 271
   derivation of, 272
   Frank polyphase, 296
   LFM, 297
   LFM pulse, 279

   mismatched PRN, 307
   numerical calculation of, 281
   PRN, 303
   unmodulated pulse, 276
   Woodward, 271
ambiguous Doppler, 522
ambiguous range, 4, 522
amplitude
   detection, 228, 235
   detector, 228, 236
amplitude noise, 465
amplitude taper, 371
analog Doppler processor, 565
analog-to-digital converter
   noise figure, 640
analog-to-digital converter, 634
   full-scale power, 639
   full-scale range, 635
   full-scale rms, 638
   ideal transfer function, 637
   least significant bit, 635
   quanta, 635
   quantization, 635
   quantization error, 636
   quantization interval, 635
   quantization noise, 639
analog-to-digital converter
   SNR, 641
analog-to-digital converter
   gain prior to, 642
angle-Doppler map
   airborne STAP, 742
angle-Doppler response
   optimum, 736
angle-Doppler space, 717
antenna
   aperture, 27
   beamwidth, 25

directivity, 23, 24, 25, 27, 39
effective area, 27
gain, 24
loss, 22, 24
noise temperature, 29, 30, 39
pattern, 24
phase center, 483
phased array, 27
sidelobes, 24, 26
antenna broadside, 362
antenna directivity, 27
antenna pattern
Gaussian model, 486
$sinc^2(x)$ model, 486
antenna pattern models, 108
anti-aliasing filter, 618
aperture, 25
effective, 27
aperture plane, 400
Armstrong, Edwin Howard, 583
array factor, 365
Arthur C. Omberg, 19
AT-11 Kansan, 50
atmospheric loss, 117
attenuator, 74
audion, 584
automatic gain control, 585
Avantek, 613
average power, 21, 40
average power-aperture product, 34
backward wave amplifier, 587
bandpass sampling theorem, 619
Barker code
table of, 300
Barker coded pulses, 297
baseband, 14
baseband representation, 129
baseband signal notation, 13
beam area, 35
beam boadening, 438
beam dwell, 6
beam forming, 717, 718
linear phased array, 720
signal plus noise, 719
signal plus noise and interference, 722

TDU, 423
beam packing, 94
dense, 94
sparse, 94
beam spoiling, 438
beam steering, 372
beamshape loss, 235
beamwidth, 25, 26, 361, 371
binary integration, 228
binary integrator, 235
bistatic, 2
blind range, 7, 522
blind velocity, 102
Boltzmann's constant, 20, 69, 595
Brennan and Reed
STAP, 718
broadside, 362
BT product, 221
Butterworth, Stephen, 590
CA-CFAR, 105
calibration
pilot pulse, 647
cancellation ratio, 761
carrier frequency, 8
cascade, 76
effective noise temperature, 78
equivalent noise temperature, 78
noise figure, 76, 78
cascade analysis, 606
1 dB compression point, 611
2nd order intercept, 612
3rd order intercept, 613
conventions, 607
equations, 615
gain, 609
noise figure, 610
noise temperature, 610
procedure, 608
cascaded attenuators, 81
Cauchy-Schwarz in STAP, 722
Cauchy-Schwarz inequality, 212
cell under test, 168
cell-averaging CFAR, 168
scale factor, 169
threshold, 169
cells, 168

central line clutter gain, 531
CFAR, 100
CFAR (*see* constant false alarm rate), 163
chaff, 575
   radial velocity, 577
   RCS, 576
   self screening, 575
   spectral width, 576
   time behavior, 575
Charles E. Muehe, 456
Chebyshev, Pafnuty Lvovich, 596
chirp, 219
chi-square, 238
Christian Hülsmeyer, 1
CIC filter, 628
   comb filter, 630
   compensation filter, 627
   frequency response, 631
   integrator, 629
   Noble identity, 630
   programmable FIR filter, 626
circular polarization, 398
circulator, 87
clutter, 163, 164
   folding, 525
   ground, 455, 457
   rain, 455, 457
   rejection, 455
   rejection, HPRF, 458
   rejection, LPRF, 458
   rejection, MPRF, 458
   transient, 525
clutter attenuation, 500, 504, 512
clutter folding, 525, 526
clutter gating, 543
clutter model, 164, 483
   Gamma, 164
   geometry, 483
   ground, 483
   K-distribution, 164
   log-normal, 164
   log-Weibull, 164
   rain, 491
   Rayleigh, 164
   smooth earth, 483
   Weibull, 164
clutter nulling, 718
clutter transient, 525
clutter transients, 543
cognitive radar, 779, 780
coherent demodulator. *See* I/Q demodulator
coherent integration, 227, 228, 236
coherent oscillator, 594
coherent processing interval, 528
coherent processing interval (CPI), 460
COHO. *See* coherent oscillator
coincidence detection, 228
combined noise temperature, 73
combiner/splitter, 417
complementary error function, 239
complex signal notation, 12
compressed pulsewidth, 221
compression filter, 286
compression point
   1 dB, 597
conjugate-transpose, 719
constant false alarm rate, 100, 163
   constant, 168
   detection, 163
   detector, 167
   guard cells, 168
   lagging cells, 168
   leading cells, 168
   multiplier, 168
   problems, 198
   processing, 163
   reference cells, 168
   reference window, 168
   scale factors, 199
   threshold, 168
   window, 168
conversion loss, 607
corner turn memory, 460
cosmic noise, 595
Costas array, 314
Costas pulse, 316
Costas sequence, 314, 315
Costas waveform, 316
Costas waveforms, 313
coupling ratio, 88

covariance matrix estimate
   STAP, 744
CPI, 460, 528, 676
CR. *See* cancellation ratio
cross ambiguity function, 271
cross range, 667
CRPL Exponential Reference
   Atmosphere, 123
cumulative
   detection, 248
cumulative detection probability, 248, 267
cumulative probability, 227
CW radar, 1
cyclotron protective device, 587
cyclotron wave electrostatic amplifier, 587
D. B. Leeson, 536
D. O. North, 209, 227
Darlington, Sidney
   waveforms, 286
data windowing functions, 787
David K. Barton, 19, 227
David Middleton, 209
dB, 11
De Forest, Lee, 584
decibel
   arithmetic, 11
   dBi, 11, 370
   dBm, 11
   dBsm, 11
   dBV, 11
   dBW, 11
decibel relative to an isotropic radiator, 370
decision rule, 170
degrees of freedom, 717
density function, 128
   chi-squared, 52
   exponential, 52
desired signal, 749
detection
   probability, 34
   range, 34
   threshold, 34
detection analysis, 232

detection probability, 127, 128, 144
detection theory, 127
diagonal loading, 744
Dickey, Robert H.
   waveforms, 286
difference channel, 646
difference pattern, 646
digital bandpass integrator, 553, 554
digital beam forming, 651
digital downconverter, 621
digital receiver, 617
digital signal processor, 530
diode limiter, 88, 587
direct IF sampling, 617, 618, 621
Direct IF sampling, 583
direct RF sampling, 619
directional coupler, 88
directive gain, 20, 23, 369
directive gain pattern, 367
directivity, 20, 23, 25
discrete time Fourier transform, 377
discriminator
   angle, 648
   range, 649
discriminator curve, 647
distortion
   pattern, 422
dither, 644
   noise, 644
Doppler
   frequency, 10
   shift, 10
Doppler beam sharpening, 668
Doppler frequency, 7, 10
Doppler processing
   STAP, 726
Doppler processor
   STAP, 728
Doppler shift, 10
double balanced mixer, 589
down chirp, 219
down conversion, 588
down range, 679
dual threshold detection, 228
dual threshold detector, 235
duplexer, 87

duty cycle, 35
dynamic range, 522, 599
   required, 605
edge taper, 386
effective aperture, 27, 40
effective area, 27
effective noise temperature, 73, 78
effective radiated power, 24, 39
EIA designation, 112
Electronic Industries Association, 112
electrostatic amplifier, 587
electrostatic combined amplifier, 587
element packing, 383
element pattern, 376
elliptical, 398
environmental noise, 595
equivalent noise temperature, 73, 78
Ernest J. Wilkinson, 87
ERP, 24
error function, complementary, 239
Euler identity, 13
false alarm probability, 127, 146, 149, 154
false alarm time, 154
fast time, 717
feed, 384
   constrained, 384
   parallel, 90
   series, 90
   space, 384
   stripline, 90
feedback shift register, 301, 302
feedback tap configurations, 301
ferrite limiter, 88
Fessenden, Reginald Aubrey, 583
FFT, 228, 229, 234, 282, 378, 459
FFT frequency response, 562, 564
field point, 364
filter order, 584
Finn and Johnson, 167, 174
fixed threshold detector, 163
FM Waveforms, 287
four channel receiver, 651
Fourier transform, 281
Fowle
   nonlinear FM synthesis, 293

Fowle, nonlinear FM synthesis, 289
Frank
   code, 296
   matrix, 295
Frank polyphase
   ambiguity function, 296
Frank Polyphase Code, 295
frequency agile, 591
frequency agility, 63
frequency bands, 2
frequency coding, 286
frequency hop waveforms, 286, 307, *See* step frequency waveforms
frequency response
   optimized, 731
Fresnel integral, 330
   cosine, 330
   sine, 330
Friis formula, 597
   cascade noise figure, 610
   cascade noise temperature, 610
gallium arsenide, 415
gallium nitride, 415
gamma function
   incomplete, 237
GaN, 415
GaS, 415
Gaussian, 230, 232
generalized Barker codes, 299
GOCA CFAR (*see* greatest of CFAR), 181
Golomb, S. W.
   PRN codes, 301
grating lobes, 361, 430
   with non-uniform subarrays, 437
greatest of CFAR
   multiplier, 182
   noise power estimate, 182
   P_{fa}, 182
   threshold, 182
gross Doppler, 693
ground clutter, 523
   airborne STAP, 740
ground clutter spectrum, 489
   Gaussian model, 489
ground range, 4

guard cells, 168
Guglielmo Marconi, 358
Gustav Mie, 49
Hansen and Sawyers, 181
Harald Trap Friis, 73
harmonics, 589
Harry B. Smith, 456
Henry Wallman, 209
Hermann Rohling, 188
Hermitian, 719, 724, *See* conjugate-transpose
heterodyne, 583
heterodyning, 583, 588
high PRF, 521
high side injection, 588
high side mixer, 588
Hogeneauer filter, 628
horizontally polarized, 398
HPRF, 521
I and Q detector. *See* I/Q demodulator
I/Q demodulator, 592, 593
IF representation, 129
image
   frequency, 588
   noise, 588
   reject filter, 588
incomplete gamma function, 237
independent, 248, 257
instrumented range, 7
integration
   binary, 228
   coherent, 227, 228, 229, 231, 234, 235
   noncoherent, 227, 235, 236, 242, 243
interference, 717
   multiple, 723
   plus noise, 723
   steering vector, 723
interference canceller, 748
interference signal, 749
intermodulation distortion, 602
interpolation, 697
inverse sinc function, 633
isolator, 87
isotropic, 358
   radiator, 358

isotropic radiator, 24
ISTOK, 587
J. H. Van Vleck, 209
J. I. Marcum, 127
John William Strutt, 45
Johnson noise, 28, 69
Johnson, John Bertrand, 69
joint density, 130, 131, 142
Karki, James, 640
Kenneth A. Norton, 19
Kester, Walt, 640
Kotel'nikov sampling theorem, 618
least mean-square, 750
least significant bit, 635
left-circular, 398
LFM, 329
   and the sinc function, 326
   bandwidth, 219, 278, 330
   Fourier transform, 330
   frequency response high BT product, 327
   frequency response, low BT produce, 327
   matched filter, 330
   pulse, 329
   slope, 219, 278, 330
   spectrum, 331
LFM pulse, 219, 278, 285
LFM with Amplitude Weighting, 287
linear array, 363
linear frequency modulation, 219, 329
linearly polarized, 398
LNA, 88, 411, 415, *See* low noise amplifier
local oscillator, 587
Lord Rayleigh, 45
loss
   TDU, 444
losses, 20, 28
   atmospheric, 95, 117
   beamshape, 93
   CA-CFAR, 105
   CFAR, 100, 101, 104, 105
   circulator, 86
   directional coupler, 86
   Doppler straddle, 101, 104

duplexer, 86
example, 90, 95, 98, 99, 106
feed, 86
GO-CFAR, 104
isolator, 86
matched filter, 100
mismatch, 92
mode adapter, 86
phase shifter, 91
power divider, 86
preselector, 86
propagation, 95
radar, 85
radome, 91
rain attenuation, 117
range straddle, 101, 104
receiver protection, 86
rectangular waveguide, 111
RF, 85
rotary joint, 86
scalloping, 102
scan, 92
SO-CFAR, 105
straddle, 104
T/R switch, 86
transmit, 85
waveguide, 86, 88, 89
waveguide attenuator, 86, 99
waveguide switch, 86
weighting, 102
Louis Maisel, 773
low noise amplifier, 88, 411
low PRF, 521
low side injection, 588
low side mixer, 588
LPRF, 521
magic T, 111
main beam, 361
main beam clutter region, 485
Marcum Q function, 149, 233
marginal density, 131, 140
matched
   Doppler, 273
   range, 273
matched filter, 29, 209, 214, 227, 261
   FFT based, 288

LFM, 330
matched filter response
   LFM, 287
   NLFM, 293
maximal length sequences, 301
maximize SINR, 718, 721
medium PRF, 521
MIMO, 779, *See* multiple-input, multiple-output
MIMO Radar, 779
minimum detectable signal, 600, 602
minimum discernable signal, 600, 602
minimum peak sidelobe codes, 299
   table of, 300
minimum selected cell averaging
   CFAR, 192
   algorithm, 193
minimum signal of interest, 601
mismatched
   Doppler, 273
   range, 273
mismatched filter, 286
Mismatched PRN Processing, 303
mixer
   conversion loss, 607
   double balanced, 589
   harmonics, 589
   ideal, 588
   image reject, 588
   practical, 589
   spur, 589
mode adapter, 88
modulation
   amplitude, 285
   frequency, 285
   linear frequency, 285
   non-linear frequency, 285
   phase, 285
   quadratic phase, 285
m-of-n
   detection, 227, 228, 250
monopulse, 645
   calibration, 647
   combiner, 646
   gain imbalance, 648
   phase imbalance, 648

three-channel, 645
two-channel, 650
    sign switching, 651
monostatic, 2
moving target indicator, 497
MPRF, 521
MSCA CFAR (*see* minimum selected cell averaging CFAR), 192
MTI, 455, 497
    clutter performance, 500
    Improvement Factor, 500
    response normalization, 498
    Transients, 518
    velocity response, 516
multiple PRIs, 6
multiple signal classification, 652
multiple simultaneous beams, 651
multiple target environment, 178
multiple-input, multiple-output, 779
MUSIC. *See* multiple signal classification
NCO, 618, 621
nepers, 115
NLFM, 286
NLFM frequency and phase plots, 293
noise
    cosmic, 595
    dither, 644
    energy, 20, 28
    environmental, 595
    factor, 20
    figure, 20, 29, 74
    floor, 595
    power, 31
    power spectral density, 28
    receiver, 594
    temperature, 20, 28
    thermal, 28, 595
noise covariance matrix, 721
noise figure, 73, 74, 78, 79
    AESA, 440
        example, 444
    analog-to-digital converter, 640
    array, 443
    attenuator, 74
    calculation AESA, 440
    subarray, 441
    T/R module, 415, 440
noise gain, 532
noncoherent integration, 227, 235, 236, 242, 243
nonlinear FM, 285, 289
nonlinear FM synthesis, 289
Norbert Wiener, 209
number of samples, 743
numerically controlled oscillator, 618
Nyquist
    criterion, 618
    frequency, 618
    rate, 618
    zones, 619
Nyquist criterion
    for FFT matched filter, 289
Nyquist sampling theorem, 618
Nyquist, Harry Theodor, 69
optimum weight
    airborne STAP, 742
    temporal, 730
optimum weight vector
    space-time, 733
ordered statistic CFAR, 188
orthogonal waveforms, 779
orthogonality condition, 751
OS CFAR (*see* ordered statistic CFAR), 188
overlapped subarrays, 433
oxygen resonance frequencies, 120
P. A. Bakut, 238
packing
    rectangular, 384
    triangular, 384
packing factor, 35
parabolic reflector, 399
Parseval's theorem, 214, 290
passive component, 80
pattern
    difference, 646
    sum, 646
pattern propagation factor, 487
Paul Dirac, 337
Paul W. Howells, 748
peak transmit power, 20, 21

permeability of free space, 114
permittivity of free space, 114
PESA, 411
Peter Swerling, 127
phase code
   Barker, 297
   Frank polyphase, 295
   generalized Barker, 299
   minimum peak sidelobe, 299
   polyphase Barker, 299
   PRN, 300
   pseudo mandom noise, 300
phase coded pulse
   general equation for, 294
phase detectors, 593
phase modulation, 285
phase noise, 464, 508, 533
   Leeson-Poore model, 536
phase noise clutter gain, 531
phase noise spectrum, 535
phase steering, 372, 375
phased array
   element packing, 384
   linear, 363
   N-element, 363
   radiation pattern, 382
   shape, 383
   sidelobes, 387
   two element, 358
   weighting, 387
Philip M. Woodward, 271
pilot pulse, 647
planar arrays, 380
Planck's law, 70
   Planck constant, 70
polarization, 397
   circular, 398
   elliptical, 398
   linear, 398
   slant, 398
polyphase Barker codes, 299
post-detection integration, 235
power divider, 87
   coupling loss, 87
   insertion loss, 87
   Wilkinson, 87

power gain, 24
practical considerations
   STAP, 744
preselector, 88, 584, 592
PRF, 5
PRI, 5
primary target, 171
PRN
   ambiguity function, 303
   ambiguity function for mismatched
      waveform, 307
   code generator, 302
   coded waveforms, 300
   feedback tap configurations, 301
   maximal length sequences, 301
   mismatched processing, 303
   optimum phase shift, 306
   sequence, 301
probability
   cumulative, 227, 258
   cumulative detection, 267
   detection, 34, 234, 236, 248
   false alarm, 248, 249
   m-of-n, 251
probability density function
   exponential, 171
   Rayleigh, 171
   Rayleigh power, 171
processing interval, 676
processing window, 7
processor
   analog, 456
   HPRF, 455
   LPRF, 455
   MPRF, 455
   MTI, 455
   pulsed Doppler, 455
pulse
   constant amplitude, 285
   LFM, 285
   phase coded, 285
pulse compression, 39, 286
pulse repetition frequency, 5
pulse repetition interval, 5
pulsed Doppler clutter, 523

pulsed Doppler processor, 455, 521, 565
pulsed Doppler signal processor, 528
pulsed radar, 1, 4
pulsewidth, 8
Q function, 233
quadratic phase, 681
quadratic phase coding, 285
quadrature demodulator. *See* I/Q demodulator
quadrature detector. *See* I/Q demodulator
quanta, 635
quantization, 635
   rounding, 636
quantization interval, 635
quantization noise, 639
   asumptions, 639
radar
   block diagram, 14
   cognitive, 779
   frequency bands, 2, 3
   multiple-input, multiple-output, 779
   origin of term, 1
radar cross section, 20, 26, 45
radar frequency bands, 3
radar losses, 85
radar range equation, 19, 39
   basic, 19
   search, 34
   summary, 39
radar receiver, 70
radar types, 1
   continuous wave, 1
   pulsed, 1
RADARSAT, 714
radiated power
   effective, 24
   ERP, 24
radiation pattern, 360, 363, 365
   optimized, 725
radiator
   isotropic, 358
radius of the earth, 119, 487
rain attenuation, 97, 117
rain clutter
   spectral model, 494
rain clutter RCS, 491
random variables, 130
range
   detection, 34
   slant, 20
range cell migration, 692
range cell migration correction, 696
range correlation, 534
range correlation effect, 535
range cut
   ambiguity function, of, 285
   matched Doppler, 285
range delay, 4
range discriminator, 649
range gate walk, 234
range measurement, 2, 3, 4
range rate, 7
range resolve, 6
range sidelobes, 286
range-Doppler matrix, 460
range-rate measurement, 3
rank, 743
Rayleigh density, 137
Rayleigh's energy theorem, 214
RCM. *See* range cell migration
RCMC. *See* range cell migration correction
RCS, 20, 26, 45
   AT-11 Kansan, 51
   flat plate, 49
   perfectly conducting sphere, 46, 48
   simple shapes, 46
real signal notation, 12
receive elements, 361
receiver, 79, 583
   chain, 584
   configurations, 645
   digital, 617
   four channel, 651
   function, 583
   noise floor, 595
   sensitivity, 600
   three-channel, 645
   two-channel, 650
receiver noise, 532, 594

reciprocity, 361
rect, 9
rect(x), 9
rectangle function, 9
rectangular envelope, 285
rectangular packing, 384, 408
rectangular pulse, 216
rectangular waveguide
   attenuation, 111
   dominant mode, 114
   pressurized, 113
   standard specifications, 113
reference temperature, 595
reference window, 168
reflection coefficient, 92
reflectivity, 485
RF amplifier, 79
RF front end, 584
Rice model, 51, 127
right-circular, 398
Ronald S. Bassford, 456
room temperature, 29
rotary joint, 88
SALT I, 37
sample matrix inversion, 764
sampling theorem
   bandpass, 619
   Kotel'nikov, 618
   Nyquist, 618
   Shannon, 618
SAR, 329
   image, 672, 690
   image focusing, 690
   platform, 678
   processing interval, 676
   processor, 678, 695
   signal characterization, 678
   spotlight, 675
   strip map, 675
scan angle
   and subarray size, 418
scan loss, 246
scan period, 465
scanning
   1-D, 93
   2-D, 93

scatterer, 679
Schottky, Walter Hermann, 583
SCR improvement, 504, 533, 538
search
   radar, 34
   sector, 34, 35
   solid angle, 40
search radar range equation, 40
search solid angle, 43
self masking, 178
sensitivity, 600
   measuring, 601
   tangential, 602
sensitivity time control, 585
   law, 586
Shannon sampling theorem, 618
sidelobe blanker, 772, 773
sidelobe cancellation, 744, *See* sidelobe canceller
sidelobe canceller, 748
   auxiliary antenna, 748
   closed loop, 748
   coherent processing, 763
   Howells-Applebaum, 766
   main antenna, 748
   open loop, 748
   sample matrix inversion, 764
   SMI, 764
   timing, 763
   weight, 748
sidelobe clutter region, 485
sidelobes, 24, 26
signal energy, 20, 28
signal processor, 271, 272
signal-to-clutter ratio improvement, 500
signal-to-noise ratio, 19
   energy, 31
   power, 31
signal-to-noise ratio, 31
simultaneous multiple beams, 427
Singer, James
   PRN codes, 301
single pulse SNR, 227
SINR
   maximimized, 724
slant range, 4, 20

SLB. *See* sidelobe blanker
slow time, 717
smallest of CFAR, 105, 184
   noise power estimate, 185
SMI. *See* sample matrix inversion
smooth earth clutter model, 483
SNR, 19, 31, 227, 228, 229, 231, 238
   improvement, 455
   single pulse, 227, 238
SO-CFAR, 105, 184, *See* smallest of CFAR
solid-state LNA, 587
Space-Time Adaptive Processing. *See* STAP
space-time processing, 718, 732
   example, 734
spatial filter, 719, 724
spatial processing, 718
spectrum analyzer, 332
spherical correction, 386
spotlight SAR, 675
spurious free dynamic range, 604
spurious signal, 589
   order, 589
   passband, 590
square law
   detection, 235
   detector, 235, 240
staggered PRIs, 513
standard atmosphere, 95
STAP, 717
   practical considerations, 744
STC. *See* sensitivity time control
steering, 372
   phase, 372
   time delay, 372
steering vector, 719, 722
   airborne clutter, 741
   interference, 722
   interference, space-time, 733
   interference, temoral, 730
   signal, space-time, 732
   space-time, 733
step frequency waveforms
   Doppler effects, 311
   range ambiguities, 309

range resolution, 309
Stephen Oswald Rice, 127
stretch processing, 329
stretch processor, 332
   block diagram, 332, 343
   configuration, 332
   expanded model, 340
   heterodyne duration, 334
   implementation, 338
   mixer output, 334
   modification, 342
      heterodyne LFM slope, 343
      mixers, 342
   operation, 334
   range delay, 333
   range resolution, 336
   range-rate effects, 340
      range bias, 348
   slope mismatch effects, 344
   SNR, 336
   SNR loss, 338
   spectrum analyzer, 335
      FFT length, 339
   timing, 333
strip map SAR, 675, 676
subarray, 416, 721
   non-uniform, 437
   overlapped, 433
subarray combiner, 417
subarray size
   scan angle, 418
   waveform bandwidth, 418
subarraying
   one level, 417
   two level, 417
sum pattern, 646
superheterodyne receiver, 584
   dual conversion, 592
   single conversion, 584
SW0/SW5, 128, 135, 140, 160, 237, 241
SW1, 51, 237, 242
SW1/SW2, 128, 136, 139, 160
SW2, 51, 237, 243
SW3, 52, 237, 244
SW3/SW4, 128, 138, 142, 160

SW4, 52, 238, 244
Swerling, 228, 238
Swerling 0, 51
Swerling 1, 51
Swerling 2, 51
Swerling 3, 51
Swerling 4, 51
Swerling 5, 51
Swerling fluctuation models, 53
Swerling models, 127
Swerling RCS models, 51
Sydney P. Applebaum, 748
synchronous detector. *See* I/Q demodulator
synthetic aperture radar, 329, *See* SAR
system noise figure, 76, 78
system noise temperature, 29, 39, 73
T/R module, 89, 413
T/R modules
  MIMO, 783
tangential sensitivity, 602
target masking, 178
Taylor weights, 406
telemobiloscope, 1
temperature
  reference, 595
temporal processing, 726
  example, 730
thermal noise, 28, 69, 595
three-channel monopulse, 645
threshold multiplier, 168
threshold-to-noise ratio, 163
time delay, 3
time delay steering, 416
time delay units, 416
time-bandwidth product, 221
time-delay steering, 372, 373
TNR, 232, 238
TNR (*see* threshold-to-noise ratio), 163
transmit
  beam spoiling, 438
transmit loss, 22
transmit power
  average, 21
  peak, 21
transmit/receive module, 413

transversal filter, 553
triangle function, 308
triangular packing, 384, 409
trilateration, 779
Trunk, 184
T-type attenuator, 75
two-channel monopulse, 650
  sign switching, 651
unambiguous range, 5
uncompressed pulsewidth, 221, 468
unmodulated pulse, 216, 274
up chirp, 219
usable range
  maximum, 7
  minimum, 7
Van Vleck-Weisskopf formula, 121
vertically polarized, 398
video, 235
visible space, 361, 383
Vladimir Aleksandrovich Kotel'nikov, 282
VSWR, 92
wave number, 365
waveform coding, 285
waveguide, 85
  magic T, 111
  surface resistivity, 116
wavelength, 10, 20
weight, 749
weight vector, 719
weighting
  Bartlett, 103
  Blackman, 103
  Blackman-Harris, 103
  Chebychev, 103, 371
  $\cos^n$, 371
  $\cos^n(x)$, 103
  elliptically symmetric, 387
  Gaussian, 103
  Hamming, 103
  Hann, 103
  multiplicative, 387
  Nuttall, 103
  rectangular, 103
  Taylor, 103, 371
  uniform, 359

wide sense stationary, 130, 134, 466
wideband waveforms, 416, 427
Wiener filter, 747, 750
William Goodchild, 456
William J. Caputi, 329
window
   figures of merit, 107
   periodic, 107
   symmetric, 107
Woodward
   ambiguity function, 271
WR number, 112
WSS, 466, *See* wide sense stationary
Yang and Kim, 192, 193

Recent Titles in the Artech House Radar Series

Dr. Joseph R. Guerci, Series Editor

Adaptive Antennas and Phased Arrays for Radar and Communications, Alan J. Fenn

Advanced Techniques for Digital Receivers, Phillip E. Pace

Advances in Direction-of-Arrival Estimation Sathish Chandran, editor

Airborne Pulsed Doppler Radar, Second Edition, Guy V. Morris and Linda Harkness, editors

Basic Radar Analysis, Second Edition, Mervin C. Budge, Jr. and Shawn R. German

Basic Radar Tracking, Mervin C. Budge, Jr. and Shawn R. German

Bayesian Multiple Target Tracking, Second Edition , Lawrence D. Stone, Roy L. Streit, Thomas L. Corwin, and Kristine L Bell

Beyond the Kalman Filter: Particle Filters for Tracking Applications, Branko Ristic, Sanjeev Arulampalam, and Neil Gordon

Cognitive Radar: The Knowledge-Aided Fully Adaptive Approach, Joseph R. Guerci

Computer Simulation of Aerial Target Radar Scattering, Recognition, Detection, and Tracking, Yakov D. Shirman, editor

Control Engineering in Development Projects, Olis Rubin

Design and Analysis of Modern Tracking Systems, Samuel Blackman and Robert Popoli

Detecting and Classifying Low Probability of Intercept Radar, Second Edition, Phillip E. Pace

Digital Techniques for Wideband Receivers, Second Edition, James Tsui

Electronic Intelligence: The Analysis of Radar Signals, Second Edition, Richard G. Wiley

Electronic Warfare in the Information Age, D. Curtis Schleher

Electronic Warfare Target Location Methods, Second Edition, Richard A. Poisel

ELINT: The Interception and Analysis of Radar Signals, Richard G. Wiley

EW 101: A First Course in Electronic Warfare, David Adamy

EW 102: A Second Course in Electronic Warfare, David Adamy

EW 103: Tactical Battlefield Communications Electronic Warfare, David Adamy

FMCW Radar Design, M. Jankiraman

Fourier Transforms in Radar and Signal Processing, Second Edition, David Brandwood

Fundamentals of Electronic Warfare, Sergei A. Vakin, Lev N. Shustov, and Robert H. Dunwell

Fundamentals of Short-Range FM Radar, Igor V. Komarov and Sergey M. Smolskiy

Handbook of Computer Simulation in Radio Engineering, Communications, and Radar, Sergey A. Leonov and Alexander I. Leonov

High-Resolution Radar, Second Edition, Donald R. Wehner

Highly Integrated Low-Power Radars, Sergio Saponara, Maria Greco, Egidio Ragonese, Giuseppe Palmisano, and Bruno Neri

Introduction to Electronic Defense Systems, Second Edition, Filippo Neri

Introduction to Electronic Warfare, D. Curtis Schleher

Introduction to Electronic Warfare Modeling and Simulation, David L. Adamy

Introduction to RF Equipment and System Design, Pekka Eskelinen

Introduction to Modern EW Systems, Andrea De Martino

An Introduction to Passive Radar, Hugh D. Griffiths and Christopher J. Baker

Introduction to Radar using Python and MATLAB®, Lee Andrew Harrison

Linear Systems and Signals: A Primer, JC Olivier

Meter-Wave Synthetic Aperture Radar for Concealed Object Detection, Hans Hellsten

The Micro-Doppler Effect in Radar, Second Edition, Victor C. Chen

Microwave Radar: Imaging and Advanced Concepts, Roger J. Sullivan

Millimeter-Wave Radar Targets and Clutter, Gennadiy P. Kulemin

MIMO Radar: Theory and Application, Jamie Bergin and Joseph R. Guerci

Modern Radar Systems, Second Edition, Hamish Meikle

Modern Radar System Analysis, David K. Barton

Modern Radar System Analysis Software and User's Manual, Version 3.0, David K. Barton

Monopulse Principles and Techniques, Second Edition, Samuel M. Sherman and David K. Barton

MTI and Pulsed Doppler Radar with MATLAB®, Second Edition,
   D. Curtis Schleher

Multitarget-Multisensor Tracking: Applications and Advances Volume III, Yaakov
   Bar-Shalom and William Dale Blair, editors

Non-Line-of-Sight Radar, Brian C. Watson and Joseph R. Guerci

Precision FMCW Short-Range Radar for Industrial Applications,
   Boris A. Atayants, Viacheslav M. Davydochkin, Victor V. Ezerskiy, Valery S.
   Parshin, and Sergey M. Smolskiy

Principles of High-Resolution Radar, August W. Rihaczek

Principles of Radar and Sonar Signal Processing, François Le Chevalier

Radar Cross Section, Second Edition, Eugene F. Knott, et al.

Radar Equations for Modern Radar, David K. Barton

Radar Evaluation Handbook, David K. Barton, et al.

Radar Meteorology, Henri Sauvageot

Radar Reflectivity of Land and Sea, Third Edition, Maurice W. Long

Radar Resolution and Complex-Image Analysis, August W. Rihaczek and
   Stephen J. Hershkowitz

Radar RF Circuit Design, Nickolas Kingsley and J. R. Guerci

Radar Signal Processing and Adaptive Systems, Ramon Nitzberg

Radar System Analysis, Design, and Simulation, Eyung W. Kang

Radar System Analysis and Modeling, David K. Barton

Radar System Performance Modeling, Second Edition, G. Richard Curry

Radar Technology Encyclopedia, David K. Barton and Sergey A. Leonov, editors

Radio Wave Propagation Fundamentals, Artem Saakian

Range-Doppler Radar Imaging and Motion Compensation, Jae Sok Son, et al.

Robotic Navigation and Mapping with Radar, Martin Adams,
   John Mullane, Ebi Jose, and Ba-Ngu Vo

Signal Detection and Estimation, Second Edition, Mourad Barkat

Signal Processing in Noise Waveform Radar, Krzysztof Kulpa

Signal Processing for Passive Bistatic Radar, Mateusz Malanowski

Space-Time Adaptive Processing for Radar, Second Edition, Joseph R. Guerci

Special Design Topics in Digital Wideband Receivers, James Tsui

Systems Engineering of Phased Arrays, Rick Sturdivant, Clifton Quan, and Enson Chang

Theory and Practice of Radar Target Identification, August W. Rihaczek and Stephen J. Hershkowitz

Time-Frequency Signal Analysis with Applications, Ljubiša Stanković, Miloš Daković, and Thayananthan Thayaparan

Time-Frequency Transforms for Radar Imaging and Signal Analysis, Victor C. Chen and Hao Ling

Transmit Receive Modules for Radar and Communication Systems, Rick Sturdivant and Mike Harris

For further information on these and other Artech House titles, including previously considered out-of-print books now available through our In-Print-Forever® (IPF®) program, contact:

Artech House	Artech House
685 Canton Street	16 Sussex Street
Norwood, MA 02062	London SW1V HRW UK
Phone: 781-769-9750	Phone: +44 (0)20 7596-8750
Fax: 781-769-6334	Fax: +44 (0)20 7630-0166
e-mail: artech@artechhouse.com	e-mail: artech-uk@artechhouse.com

Find us on the World Wide Web at: www.artechhouse.com